Advanced UAV Aerodynamics, Flight Stability and Control

Aerospace Series

Introduction to Nonlinear Aeroelasticity	Grigorios Dimitriadis	April 2017
Introduction to Aerospace Engineering with a Flight Test Perspective	Stephen Corda	March 2017
Adaptive Aeroservoelastic Control	Tewari	March 2016
Theory and Practice of Aircraft Performance	Kundu, Price and Riordan	November 2015
The Global Airline Industry, Second Edition	Belobaba, Odoni and Barnhart	July 2015
Modeling the Effect of Damage in Composite Structures: Simplified Approaches	Kassapoglou	March 2015
Introduction to Aircraft Aeroelasticity and Loads, 2nd Edition	Wright and Cooper	December 2014
Aircraft Aerodynamic Design: Geometry and Optimization	Sóbester and Forrester	October 2014
Theoretical and Computational Aerodynamics	Sengupta	September 2014
Aerospace Propulsion	Lee	October 2013
Aircraft Flight Dynamics and Control	Durham	August 2013
Civil Avionics Systems, 2nd Edition	Moir, Seabridge and Jukes	August 2013
Modelling and Managing Airport Performance	Zografos, Andreatta and Odoni	July 2013
Advanced Aircraft Design: Conceptual Design, Analysis and Optimization of Subsonic Civil Airplanes	Torenbeek	June 2013
Design and Analysis of Composite Structures: With Applications to Aerospace Structures, 2nd Edition	Kassapoglou	April 2013
Aircraft Systems Integration of Air-Launched Weapons	Rigby	April 2013
Design and Development of Aircraft Systems, 2nd Edition	Moir and Seabridge	November 2012
Understanding Aerodynamics: Arguing from the Real Physics	McLean	November 2012
Aircraft Design: A Systems Engineering Approach	Sadraey	October 2012
Introduction to UAV Systems 4e	Fahlstrom and Gleason	August 2012
Theory of Lift: Introductory Computational Aerodynamics with MATLAB and Octave	McBain	August 2012
Sense and Avoid in UAS: Research and Applications	Angelov	April 2012
Morphing Aerospace Vehicles and Structures	Valasek	April 2012
Gas Turbine Propulsion Systems	MacIsaac and Langton	July 2011
Basic Helicopter Aerodynamics, 3rd Edition	Seddon and Newman	July 2011
Advanced Control of Aircraft, Spacecraft and Rockets	Tewari	July 2011
Cooperative Path Planning of Unmanned Aerial Vehicles	Tsourdos et al	November 2010
Principles of Flight for Pilots	Swatton	October 2010
Air Travel and Health: A Systems Perspective	Seabridge et al	September 2010
Design and Analysis of Composite Structures: With applications to aerospace Structures	Kassapoglou	September 2010
Unmanned Aircraft Systems: UAVS Design, Development and Deployment Austin		April 2010
Introduction to Antenna Placement & Installations	Macnamara	April 2010
Principles of Flight Simulation	Allerton	October 2009
Aircraft Fuel Systems	Langton et al	May 2009
The Global Airline Industry	Belobaba	April 2009
Computational Modelling and Simulation of Aircraft and the Environment: Volume 1 - Platform Kinematics and Synthetic Environment	Diston	April 2009
Handbook of Space Technology	Ley, Wittmann Hallmann	April 2009
Aircraft Performance Theory and Practice for Pilots	Swatton	August 2008
Aircraft Systems, 3rd Edition	Moir & Seabridge	March 2008
Introduction to Aircraft Aeroelasticity and Loads	Wright & Cooper	December 2007
Stability and Control of Aircraft Systems	Langton	September 2006
Military Avionics Systems	Moir & Seabridge	February 2006
Design and Development of Aircraft Systems	Moir & Seabridge	June 2004
Aircraft Loading and Structural Layout	Howe	May 2004
Aircraft Display Systems	Jukes	December 2003
Civil Avionics Systems	Moir & Seabridge	December 2002

Advanced UAV Aerodynamics, Flight Stability and Control

Novel Concepts, Theory and Applications

Edited by

Dr Pascual Marqués
President at Marques Aviation Ltd, United Kingdom

Dr Andrea Da Ronch
New Frontiers Fellow and Lecturer of Aircraft Structural Design
University of Southampton, United Kingdom

WILEY

This edition first published 2017

Registered Offices
John Wiley & Sons, Inc., 111 River Street, Hoboken, NJ 07030, USA
John Wiley & Sons Ltd, The Atrium, Southern Gate, Chichester, West Sussex, PO19 8SQ, UK

Editorial Office
The Atrium, Southern Gate, Chichester, West Sussex, PO19 8SQ, UK

For details of our global editorial offices, customer services, and more information about Wiley products visit us at www.wiley.com.

Wiley also publishes its books in a variety of electronic formats and by print-on-demand. Some content that appears in standard print versions of this book may not be available in other formats.

Library of Congress Cataloging-in-Publication data applied for

Hardback ISBN: 9781118928684

Cover Design: Wiley
Cover Image: Courtesy of Sönke Dierks - Director of Aircraft Design (Marques Aviation).

Set in 10/12pt Warnock by SPi Global, Pondicherry, India

10 9 8 7 6 5 4 3 2 1

Dedication

To our colleague Prof. Ramesh K. Agarwal, Honorary Fellow of the Royal Aeronautical Society of London. The Fellowship represents the world's highest distinction for aerospace achievement and was awarded to Prof. Agarwal in 2015 in recognition of his outstanding leadership in aerospace education and research, and for advancing the development of a wide range of flying vehicles through ingenious application of computational fluid dynamics. Prof. Agarwal is also William Palm Professor of Engineering at Washington University in St Louis, USA.

Contents

List of Contributors

Emaid A. Abdul Retha
Unmanned Vehicle University, Phoenix, AZ, USA

Ramesh K. Agarwal
Washington University in St. Louis, MO, USA

Ubaidullah Akram
Lockheed Martin Commercial Flight Training, Sassenheim, the Netherlands

George N. Barakos
School of Engineering, University of Liverpool, UK
School of Engineering, University of Glasgow, UK

R. Bardera Mora
Low-speed Experimental Aerodynamics Laboratory, National Institute for Aerospace Technology, Madrid, Spain

Ron Barrett
The University of Kansas, Lawrence, KS, USA

Wen-Hua Chen
Department of Aeronautical and Automotive Engineering, Loughborough University, UK

Kelly Cohen
Department of Aerospace Engineering and Engineering Mechanics, University of Cincinnati, OH, USA

Antonio J. Conesa Torres
Experimental Aerodynamics Branch, INTA, Madrid, Spain

Marco Cristofaro
AVL List GmbH, Graz, Austria

Andrea Da Ronch
University of Southampton, UK

Ranjan Ganguli
Department of Aerospace Engineering, Indian Institute of Science, Bengaluru, India

Qinglei Hu
Department of Control Science and Engineering, Harbin Institute of Technology, Harbin, China

Daniel J. Inman
Department of Aerospace Engineering, University of Michigan, Ann Arbor, MI, USA

Catherine S. Johnson
School of Engineering, University of Liverpool, UK
School of Engineering, University of Glasgow, UK

Jalal Karimi
Space Research Institute, MUT, Tehran, Iran

Maryam Kiani
Center for Research and Development in Space Science and Technology, Sharif University of Technology, Tehran, Iran

Bo Li
Department of Control Science and Engineering, Harbin Institute of Technology, Harbin, China

Cunjia Liu
Department of Aeronautical and Automotive Engineering, Loughborough University, UK

Pascual Marqués
Marques Aviation Ltd, Southport, UK

Giuseppe Pezzella
Italian Aerospace Research Centre, Capua, Italy

Seid H. Pourtakdoust
Center for Research and Development in Space Science and Technology, Sharif University of Technology, Tehran, Iran

Qiulin Qu
Beijing University of Aeronautics and Astronautics, Beijing, China

Nicholas Schwartz
Department of Aerospace Engineering and Engineering Mechanics, University of Cincinnati, OH, USA

Oksana Stalnov
Faculty of Aerospace Engineering, Technion—Israel Institute of Technology, Haifa, Israel

Nikolaos D. Tantaroudas
European Dynamics, SA, Athens, Greece

Mark B. Tischler
US Army Research, Development and Engineering Command, Moffett Field, CA, USA

Antonio Viviani
University of Campania 'Luigi Vanvitelli', Aversa, Italy

Ya Wang
Department of Mechanical Engineering, SUNY Stony Brook, NY, USA

Wei Wei
Department of Electrical Engineering, University of Cincinnati, OH, USA

Mark Woodgate
School of Engineering, University of Liverpool, UK
School of Engineering, University of Glasgow, UK

Bing Xiao
Department of Control Science and Engineering, Harbin Institute of Technology, Harbin, China

Youmin Zhang
Department of Mechanical and Industrial Engineering, Concordia University, Montreal, Quebec, Canada

Series Preface

From their early, cautious entry into the field of aviation many years ago, unmanned vehicles are now almost ubiquitous in many applications from domestic, industrial, government/official and military. The range of configurations now includes fixed wing, multi-rotorcraft, adaptive wing, and space re-entry vehicles, in both remotely piloted and autonomous modes of operation. As a result there are many classes of unmanned vehicle in existence, and many types within each class, developed by many manufacturers. They are all capable of carrying some form of payload, including sensors, and of relaying sensor information to the ground. Despite these advances there is still much to be learned; like conventional manned aircraft the range of applications and the demand for improved performance will continue to grow.

This book addresses aspects of aerodynamic design, stability and control, adaptive control, modelling and wind tunnel testing of vehicles for the range of unmanned aerial types available and introduces many novel concepts. The techniques introduced are valuable, not only for the design of vehicle itself, but also as the basis of control law design for remote and autonomous flight control systems in order to achieve high integrity and safe operation. The application of the techniques described will provide an important basis for the gathering of evidence to support certification of such vehicles.

The *Aerospace Series* has continued to provide practical, topical and relevant information for people working in the aerospace industry, including engineering professionals and operators, allied professions such as commercial and legal executives, and also engineers in academia. The range of topics is intended to be wide ranging, covering design and development, manufacture, operation and support of aircraft, as well as topics such as infrastructure operations and developments in research and technology.

This book continues that tradition. It is a compendium of subjects written by a wide range of authors, each experienced in their field. Each subject is described with comprehensive illustrations and mathematical proofs. The book introduces novel concepts and techniques. It forms a sound repository of knowledge for students, academics and practitioners in the field of unmanned vehicle design.

Peter Belobaba, Jonathan Cooper and Allan Seabridge

Preface

Advanced UAV Aerodynamics, Flight Stability and Control: Novel Concepts, Theory and Applications has been written for university students and lecturers, aerodynamics researchers, aerospace engineers, and aircraft designers and manufacturers. Readers will need to have foundational knowledge of aerodynamics and principles of flight stability and control. Nonetheless, Chapter 1 provides an introduction to Aerodynamics and Flight Stability for the benefit of readers who do not have a conventional engineering background. The technical level of the book is 'advanced', and includes many equations and data. However, descriptive narratives are also used to elucidate concepts in language that is not excessively technical. Each Chapter is illustrated using diagrams and/or photos in colour.

The book is divided into two parts. Part I addresses novel concepts in unmanned aircraft aerodynamics and contains chapters on fixed-wing (airplanes), rotary wing (helicopter), and hybrid aircraft. Part II presents novel concepts in unmanned aircraft flight stability and control and includes chapters on fixed-wing (airplanes) and quad-rotor aircraft. The reader will find in this textbook very sophisticated design concepts for unmanned aerospace vehicles and analyses of the aerodynamic environment, including aerodynamics of re-entry vehicles, aeroservoelastic analysis of very flexible aircraft, wind tunnel testing, ground-effect aerodynamics, dynamics modelling and system identification of helicopters, radial basis function neural networks, optimization of rotor blades with approximate BERP tips, helicopter trailing-edge flaps, blade twist using smart actuation, aerodynamics of hover-to-dash convertible UAVs, closed-loop flow control, autonomous gust alleviation, virtual flight simulation using Cfd, flow-structure modification using plasma actuation, constrained motion planning and trajectory optimization, autonomous space navigation, adaptive fault-tolerant attitude control of spacecraft, multi-rotor Vtol Uav dynamics and stability, and system identification and flight control of quadrotors.

Advanced UAV Aerodynamics, Flight Stability and Control: Novel Concepts, Theory and Applications is the first book dedicated to unveiling the secrets of the rapidly developing field of unmanned aircraft aeronautics. Aeronautics is a wide-ranging and multidisciplinary specialism covering several disciplines and domains of engineering that together enable the aerospace industry to produce unique, technologically advanced unmanned aircraft. This book makes the wealth of knowledge and experience of world-class researchers and industry experts available to aerospace practitioners and university students.

Aerodynamics is the underlying science that underpins aircraft flight. Much of the efficiency of modern unmanned aircraft is due to the ability of engineers to model aerodynamic flows using sophisticated computational fluid dynamics. This book promotes an understanding of the underlying physical laws and the mathematical treatment of the topic, as well as more advanced theory and innovations in flight stability and control engineering.

Technological innovation and increasingly diverse applications are two key drivers of the rapid expansion of unmanned aircraft technology. The global defence budget for unmanned aircraft procurement is expanding, and the market for civilian unmanned aircraft is expected to surpass it. The book presents the state of the art of unmanned aircraft aerodynamic models of different configurations, real-world flight stability, and control strategy validation in experimental platforms. This advanced level treatment of the aerodynamics of both fixed-wing and rotary-wing flight will appeal to aircraft design engineers and undergraduate and graduate students of aeronautics, as well as practising engineers looking for a refresher textbook on the subject.

Key Features

- Modern treatment and real-world research on aerodynamics and flight stability.
- Presentation of important contemporary trends in flight control systems.
- Detailed and rigorous exposition and examples, with illustrations.
- Ancillary PowerPoint presentations for lecturers available on an accompanying website.

We are enormously grateful to the authors, who are world class experts in the field, for dedicating part of their precious time to contribute to the book. In chapter order, contributions are provided by Dr. Giuseppe Pezzella, Prof. Antonio Viviani, Nikolaos N. Tantaroudas, Dr. Andrea Da Ronch, Dr. Rafael Bardera Mora, Dr. Qiulin Qu, Prof. Ramesh K. Agarwal, Prof. Wen-Hua Chen, Dr. Cunjia Liu, Prof. Ranjan Ganguli, Dr. Catherine S. Johnson, Dr. Mark Woodgate, Prof. George N. Barakos, Dr. Pascual Marqués, Dr. Ron Barrett-Gonzalez, Dr. Oksana Stalnov, Prof. Daniel J. Inman, Dr Ya Wang, Ubaidullah Akram, Marco Cristofaro, Dr. Antonio Jesús Conesa Torres, Prof. Seid H Pourtakdoust, Dr. Jalal Karimi, Dr. Maryam Kiani, Prof. Qinglei Hu, Dr. Bing Xiao, Dr. Bo Li, Dr Youmin Zhang, Dr. Emaid A. Abdul Retha, Wei Wei, Dr. Mark B. Tischler, Nicholas J. Schwartz, and Prof. Kelly Cohen.

Pascual Marqués and
Andrea Da Ronch

1

Advanced UAV Aerodynamics, Flight Stability and Control: An Introduction

Pascual Marqués

Marques Aviation Ltd, Southport, UK

'For some years I have been afflicted with the belief that flight is possible to man.'
Wilbur Wright, 13 May 1900.

This introductory chapter is divided into two main sections: Section 1.1 on unmanned aircraft aerodynamics and Section 1.2 on flight stability, and control. The chapter addresses fundamental principles of aerodynamics, flight stability and control and forms a knowledge base for the student of aerospace engineering before proceeding to more advanced chapters in this book. The chapter includes classroom problems.

1.1 Unmanned Aircraft Aerodynamics

1.1.1 Introduction: UAV Categories and Configurations

Unmanned aerial vehicle (UAV) size categories range from nano air vehicles (NAV) with a wing span of only 4 cm to high-altitude long-endurance (HALE) aircraft with a wing span of 35 m or more. In between, UAV categories include micro (MAV), mini, close-range, medium-range or tactical and medium-altitude long-endurance (Figure 1.1). The fluid medium in which an NAV operates is highly viscous, whereas the fluid flow around large (normally manned) aircraft is dominated by inertial effects. Consequently, aerodynamic characteristics vary considerably according to the size of the vehicle.

The aeronautical configuration of a UAV is closely related to its operational mission requirements and dictated by airspeed, endurance and operational range. Whether the vehicle is fixed-wing or rotary-wing is determined by the speed requirements. HALE surveillance aircraft necessitate a high aspect ratio (AR) wing for flight at high altitude. In contrast, Unmanned Combat Air Vehicles (UCAVs) operate at high airspeed and perform rapid manoeuvers and therefore have low AR wings. Civilian or military applications that involve operation from off-board a ship benefit from vertical take-off and

Advanced UAV Aerodynamics, Flight Stability and Control: Novel Concepts, Theory and Applications,
First Edition. Edited by Pascual Marqués and Andrea Da Ronch.

Companion website: http://www.wiley.com/go/marques/advanced_UAV_aerodynamics

Figure 1.1 General Atomics RQ-1A Predator. (Photo: USAF Museum). (*See insert for color representation of this figure.*)

landing capability of the aircraft. Fixed-wing configurations include *tailplane aft*, *tailplane forward* or *canard*, and *tailless* types.

The conventional tailplane aft designs usually have the horizontal stabiliser positioned aft on the fuselage or connected to the wings by booms. The aircraft's centre of gravity (CG) is often forward of the aerodynamic centre (AC), which creates a nose-down pitching moment. To this negative moment must be added the nose-down moment characteristic of cambered wings. The net nose-down moment is balanced by a down-load on the tailplane.

In a canard configuration, both the foreplane and the CG are located forward of the wing. Equilibrium in pitch is achieved by the positive lift generated by the foreplane. The canard design is aerodynamically more efficient than the tailplane aft design, as both the horizontal stabiliser and the wing produce positive lift.

Tailless types include the flying wing and delta configurations. Tailless types have sweptback wings and an effective tail. The airfoils at the wingtips are set at a lower incidence than the airfoils of the inner wing, in a washout configuration to provide stability in pitch. The absence of a horizontal stabiliser reduces profile drag. However, wing sweepback in the flying wing leads to poor lift distribution, high induced drag and negative lift at high airspeeds. Delta wings have a low AR, poor lift distribution and higher induced drag caused by high span loading.

In the majority of UAVs, the powerplant is mounted at the rear of the fuselage. This arrangement makes the front of the aircraft available for the installation of a payload and allows an unobstructed view forward. There are also aerodynamic advantages when using a rear-mounted propeller. The friction drag induced by the slipstream ahead of the pusher propeller is less than the drag generated by a front-mounted tractor propeller. Ubiquitous types of rotary-wing UAVs are the *single-main rotor* and *quad rotor* (Figure 1.2). Other alternatives in design are the *co-axial rotor* and *hybrid* configurations such as the *tilt-rotor* and *tilt-wing*.

Figure 1.2 Aeryon Scout VTOL MAV with gyro-stabilized camera payload. Photo: Dkroetsch CC-BY-3.0.

1.1.2 Theoretical Aerodynamics

Aerodynamic analysis of novel concepts in UAV design can be conducted using a number of methodologies that range from traditional aerodynamic theories to modern computational fluid dynamics (CFD). *Thin-airfoil theory* is an analytical method that predicts lift as a function of angle of attack and assumes idealised incompressible inviscid flow. The theory can be applied to airfoils of thickness no greater than 12% of the chord (c) at low angles of attack (below the stall). Thin-airfoil theory was developed by German–American mathematician Max Munk and further refined by British aerodynamicist Hermann Glauert in the 1920s. The theory provides a sound theoretical foundation for modern aerodynamic theories. Prandtl's *lifting-line theory* is a mathematical model for the prediction of the lift distribution along the span of a three-dimensional wing (Figure 1.3). In the model, the vortex strength varies elliptically along the wingspan and the loss in vortex strength is shed as a vortex-sheet from the trailing edge. Lift distribution is obtained from the wing geometry (constituent airfoil, taper, twist) and freestream conditions; that is, air density and flow velocity. Modified versions of the classical lifting-line theory are used to compute the lift distribution in geometrically- or aerodynamically-twisted wings.

The *vortex panel method* is a numerical method that permits the computation of airfoil lift assuming ideal flow in which the effects of compressibility and viscosity are negligible. In this method, the shape of the airfoil surface is reconstructed using a series of vortex panels or line segments arranged to form a closed polygon. Vortex sheets mimic the boundary layer around the airfoil. The vortex sheets represent miniature vortices that give rise to circulation, hence lift.

1.1.3 Flight Regimes and Reynolds Number

The Reynolds number (Re) is a dimensionless number that indicates the ratio of inertial forces to viscous forces for given flow conditions. The concept is named after Osborne Reynolds who introduced its use in 1883. The Re characterizes different flow regimes.

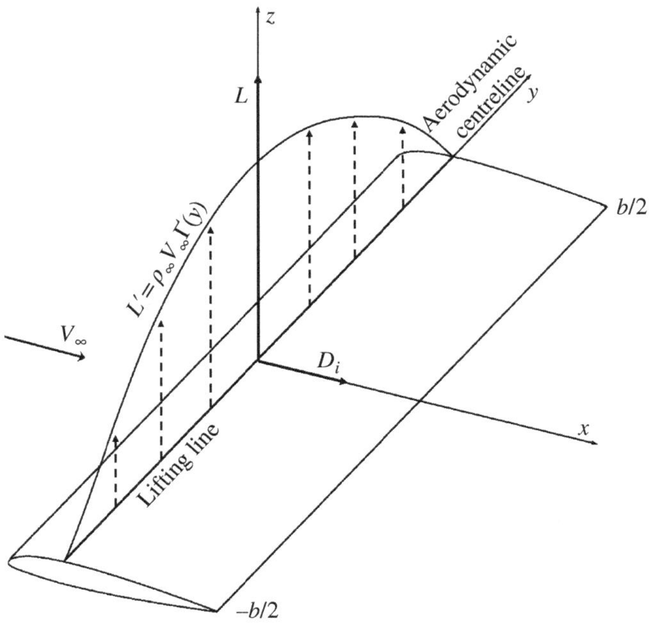

Figure 1.3 Prandtl's classical lifting-line theory. L', lift per unit span; ρ_∞, air density; V_∞, freestream velocity; $\Gamma(y)$, circulation; D_i, induced drag.

Laminar flow occurs at low Res, where viscous forces dominate. Turbulent flow occurs at high Res where inertial forces predominate. Small UAVs operate in the Re regime between 3×10^4 and 5×10^5. For operation at low Re, the design of efficient airfoils is critical. At $3 \times 10^4 \leq \text{Re} \leq 7 \times 10^4$, relatively thick airfoils (≥6% thickness) can experience considerable hysteresis as a result of the lift and drag forces caused by laminar separation and transition to turbulent flow. Below Re values of 5×10^4, laminar separation occurs and the flow does not transition and it does not reattach to the airfoil surface. At $7 \times 10^4 \leq \text{Re} \leq 2 \times 10^5$, extensive laminar flow over the surfaces of the airfoil can be attained, which reduces airfoil drag. However, in some airfoils a laminar separation bubble forms in this flight regime. At Re values above 2×10^5, airfoils become more efficient. The bubble is shorter and parasite drag decreases. The flight regime of large UAVs is in the region of $\text{Re} \geq 3 \times 10^6$. At high Re values, the laminar boundary layer transitions to turbulent a short distance downstream of the wing's leading edge. Laminar separation and separation bubbles do not occur.

1.1.4 Airfoils for UAVs

The cost of operation of a UAV can be reduced with airfoil optimization and improvements in the vehicle's aerodynamic efficiency. Considerations when selecting an airfoil for a UAV include a high maximum lift coefficient (c_{lmax}), high lift-to-drag ratio (c_l/c_d), high endurance factor ($c_l^{3/2}/c_d$), effectiveness at low Re values, low pitching moment coefficient to minimize the load on the tail, mild stall characteristics, insensitivity to surface roughness caused by rain or dust, good flap performance, and minimal airfoil complexity for ease of manufacture. Airfoils originally designed for operation at high Re for manned aircraft ($3 \times 10^6 \leq \text{Re} \leq 6 \times 10^6$) are often adapted for UAVs that operate in the low-Re flight regime (e.g., $5 \times 10^5 \leq \text{Re} \leq 1.5 \times 10^6$). The aerodynamic performance of airfoils operating in off-design conditions of low Re deteriorates proportionally to the reduction in Re, manifested as a lower c_{lmax}, higher drag, worsened c_l/c_{d}, and possible lift coefficient (c_l) hysteresis behaviour as the airspeed and Re fluctuate during flight. Hysteresis is caused by the bursting of the laminar separation bubble at low Re values. Therefore, when adapting an airfoil for a small UAV it is important to examine the aerodynamic performance and hysteresis of the airfoil in the off-design low-Re flight regimes in which the vehicle will operate.

An example of an airfoil adapted for small UAVs is the Selig Donnovan SD7032 (Figure 1.4, top). The relatively simple geometry of this airfoil makes wing manufacture uncomplicated. The SD7032 has a c_{lmax} of 1.5, and lower drag than the early-design NACA 23012, of similar geometry. The SD7032 performs well at low Re values through the full range of values of c_l.

Designed for extensive laminar flow over both the upper and lower surfaces, the Wortmann FX61-147 (Figure 1.4, centre) has a thickness ratio of 14.7%, camber of 3.18% and c_{lmax} of 1.5 (Reneaux *et al.*, 1997). Its considerable thickness allows the low drag bucket to be maintained for a large range of values of c_l. However, the large airfoil thickness incurs a higher minimum drag coefficient. At relatively low flight speeds (say,

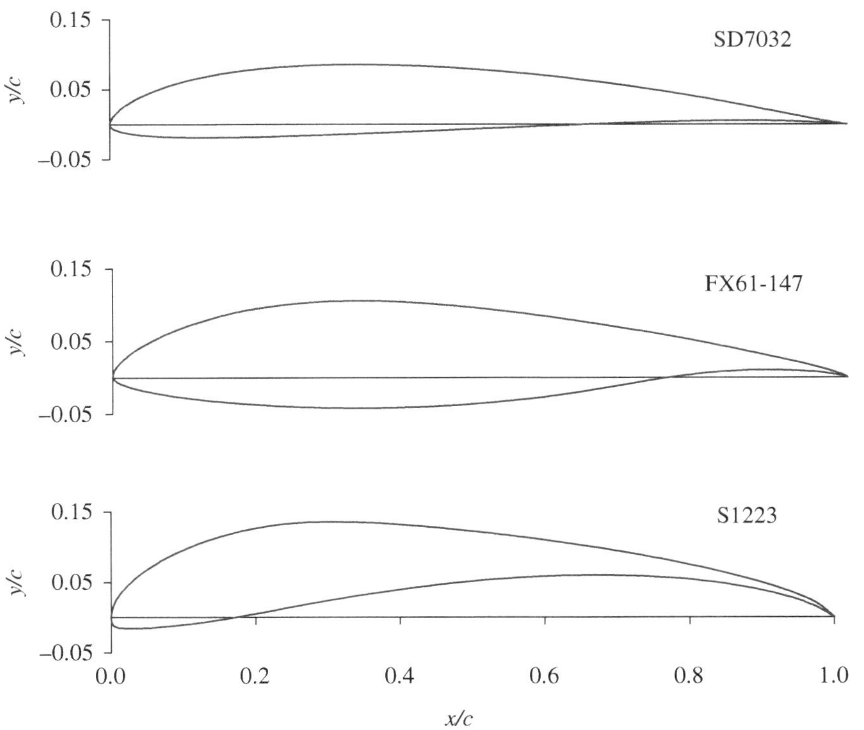

Figure 1.4 Airfoil coordinates.

a low Re of 1.5×10^6), this airfoil generates low drag due to its extensive laminar flow characteristics. Transition of the boundary layer from laminar to turbulent is caused by laminar separation bubbles. At high values of c_l, the finite transition region moves upstream on the upper surface and at low values of c_l a loss of laminar flow occurs in the lower surface. The FX61-147 performs fairly well in small UAVs. Nonetheless, attention need to be paid to the movement of the finite transition region towards the leading edge and the extent of the laminar separation bubble with fluctuations in flight speed and the associated hysteresis.

The S1223 airfoil (Figure 1.4, bottom) of 12.1% thickness ratio and 8.9% camber is a popular choice for small UAVs (Ma *et al.*, 2010). This airfoil emerged from the high-lift design philosophy, which is characterised by concave pressure recovery with aft loading. At the low Re of 1.5×10^5, the S1223 exhibits a c_{lmax} of 2.2, high c_l/c_{d} and acceptable stall characteristics. Higher values of c_l are achieved with the installation of vortex generators on the upper surface located at 17% c and also through the use of a Gurney flap. However, the high camber and complex shape of the S1223 makes wing manufacture intricate and more expensive. High lift is essential for UAVs that operate at low speeds with the wing near the c_{lmax} for loitering, cruising and landing. The S1223 has a high c_{lmax}, which is approximately 25% higher than for an FX63-137 airfoil. The S1223 also shows good flight performance compared with the NACA 0012 and CLARK-Y, thus implying that a high cambered airfoil is preferable. The higher lift of the S1223 incurs a drag penalty, but a better c_l/c_{d} ratio can be obtained using airfoil optimization. Because of their inherent laminar boundary layer, low-Re airfoils have low skin-friction drag. Optimization of the low-Re S1223 airfoil involves preserving low-friction drag by modifying both the upper and lower surfaces and delaying transition (Ma *et al.*, 2010). Optimization using CFD has resulted in the S1223 OPT 2 airfoil, which shows increased lift and efficiency, and therefore good aerodynamic characteristics in off-design conditions. At low values of Re ($7.5 \times 10^4 \leq \mathrm{Re} \leq 2.0 \times 10^6$), increased camber and reduced thickness provide more favourable lift, drag and endurance factor characteristics. A flexible airfoil, such as the UF airfoil of 6% thickness and 8.9% camber, yields better overall performance than a similar rigid profile in the unsteady oscillating environment characteristic of turbulent flight conditions (Shyy *et al.*, 1999).

1.1.5 Wing Geometry

Induced drag is caused by the wingtip vortex and it is a by-product of the lift force generated by the wing. Wings with higher AR yield less induced drag and enhance UAV flight efficiency. The lift force generated by a wing is a function of several parameters: the constituent airfoil, wing geometry, air density, airspeed and angle of attack. The UAV designer must consider the different planform shapes and evaluate the spanwise lift distribution of each variant. The wing planform of a UAV can be rectangular, elliptical, straight tapered, sweptback, sweptback tapered, semi tapered, forward swept tapered, delta, double delta, crescent, ogival, blended wing-body, joined, or some other shape (Figure 1.5). A wing of elliptical planform is the most efficient. It produces a constant lift distribution along the span and minimum induced drag. However, a purely elliptical wing is difficult to build. A rectangular planform is uncomplicated to construct, but the wingtips are 'idle' - they generate little lift. The rectangular planform has the advantage that the wing root stalls before the wing tips. Thus, aileron function is

Figure 1.5 Sweptback tapered planform in a Northrop Grumman RQ-4 Global Hawk. Photo: Jim Gordon.

maintained, as the ailerons are positioned outboard on the wing. A practical compromise is to use a straight wing that is tapered towards the tip; this improves the spanwise lift distribution. In a sweptback tapered planform, the c_l increases gradually towards the wingtip. Such increasing lift distribution is even more prominent in the delta and double-delta planforms. The tips tend to stall at a lower angle of attack than the wing root, which is undesirable in aircraft. Aileron function deteriorates, and the UAV tends to drop one wing and possibly enter a spin from which the vehicle is difficult to recover.

Tip stall in sweptback-tapered, delta and double-delta wings is prevented by twisting the wing in a washout configuration, so that the wing tips are set at a lower angle of incidence than the wing root. Geometric and/or aerodynamic twists is employed in UAVs to reduce the strength of the tip vortex and the associated induced drag, control the stall so that the wing root stalls first, and maintain aileron function. Improved wing performance is attained using an optimised amount of total twist together with an optimised elliptical distribution of the twist along the span. The work of Boschetti *et al.* (2012) has shown that application of local twist near the wingtip region is also effective in reducing induced drag. Phillips (2005) has developed an adaptive control surface, the twisteron, which modifies the shape of the tailing edge of the wing in response to instantaneous flight conditions to maximize lift and diminish induced drag.

1.1.6 Wingtip Devices

The strength of the wingtip vortices, downwash and induced drag are reduced using winglets (Azlin *et al.*, 2011). Winglets increase the effective AR of the wing. The use of winglets is often dictated by aircraft aesthetics. While the winglets are intended to restrict the tip vortex, it is difficult to prove their effectiveness in practice. Induced drag is inversely proportional to the square of the freestream velocity. Therefore, reducing vortex drag is important during slow flight, and when the angle of attack and values of c_l are high. There is little vortex drag to be saved at high speed. Thus, winglets create additional parasite drag at high flight speeds. The Whitcomb winglet is a carefully designed tip device that extracts some of the energy from the wingtip vortex. The device weakens the tip vortex and turns energy into forward propulsion. However, the Whitcomb winglet has several limitations. The precise angle of attack of the winglet changes as the strength of the wingtip vortex varies with flight speed and wing angle of attack. The winglet is efficient for only a narrow range of flight speeds. Nonetheless, UAVs operate mostly at one speed and optimisation of the winglet is possible, although winglet optimisation requires extensive research. Most of the winglet lift acts laterally, which tends to bend the winglet and add load to the wing structure. The lift generation produces localised vortices and therefore vortex drag at the tips of the winglets. The use of winglets for applications in UAVs is an area that requires further research. Wingtip fences, raked wingtips, non-planar wingtips, upswept tips, Hoerner tips, and tip sails offer appealing alternatives in UAV design.

1.1.7 High-lift Devices

Multi-element wings increase the maximum lift coefficient of the wing during takeoff and landing. Multi-element systems consist of movable leading- and trailing-edge devices such as the Krueger flap and the Fowler flap, respectively. At high wing angles of attack, slotted Krueger flaps allow the higher pressure on the underside of the slat to travel through the slot, gaining velocity and energising the boundary layer over the upper wing surface. There are different categories of trailing-edge flap systems, which vary in complexity; these include plain, split, slotted and Fowler flaps. The plain flap is the most common in UAVs due to its simplicity. A split flap yields greater increases in lift, but it creates more drag and is costlier to construct. In a slotted flap, airflow from under the wing passes through the gap between the main wing element and the flap and energises the turbulent boundary layer over the upper surface of the flap. This leads to better boundary-layer adherence, and allows a larger angle of attack before the stall, a higher c_l and lower drag than the plain flap. Deflection of a Fowler flap entails rearward extension of the flap, which increases both the wing area and camber. The Fowler flap yields generous gains in lift, but it is mechanically complex and difficult to construct. Flaperons incorporate a flap for high lift and an aileron for roll control in the same control surface. The flaperon is not integral with the wing but it is attached to the underwing and acts like an external wing flying in close formation with the main wing. A flaperon can increase c_l by 1.0 but adds a drag penalty. Elevons contain the elevators and ailerons on the same mechanical control to provide combined pitch and roll control.

With reduced UAV flight speed and increasing angles of attack during the approach to landing, the boundary layer becomes unstable and tends to separate. The boundary layer can be energised using a vortex generator. These small vanes are of rectangular or triangular shape and are positioned obliquely at an angle of attack with respect to the local airflow. Each vane creates a tip vortex. Vortex generators have a height of 80% of the boundary-layer height and draw energetic rapidly-moving air from outside the slow-moving boundary layer into contact with the aircraft skin. They delay flow separation and stall, and improve the effectiveness of the control surfaces.

Gurney flaps are mechanically-simple but highly-effective devices that increase the maximum lift of the wing, vertical fin or horizontal stabiliser. The Gurney flap is a small flat tab of a height of 1–2% of c. The flap projects from the trailing edge, at right angles to the pressure side, and runs along the full span of the lifting surface. The underlying principle behind the Gurney flap is that it improves the aerodynamic performance of the constituent airfoil. The flap increases the pressure on the pressure side and reduces the pressure on the suction side. The boundary layer remains attached at the trailing edge on the suction side.

1.1.8 Aeroelasticity

Aircraft performance is influenced by aeroelastic phenomena, arising from the interaction of elastic, inertial and aerodynamic forces. There is a strong link between aeroelasticity and loads. UAV structures are not completely rigid. Aeroelastic phenomena occur when structural deformations cause changes in aerodynamic forces. The additional aerodynamic forces further increase the structural deformations, which lead to greater aerodynamic forces in a self-feeding fashion. These interactions may attenuate until a condition of equilibrium is reached, or may diverge catastrophically as a result of resonance.

Aeroelasticity can be categorised as *steady* (*static*) or *dynamic*. Steady aeroelasticity involves the interaction between aerodynamic and elastic forces on an elastic structure. Phenomena studied under the category of steady aeroelasticity include *divergence* and *control surface reversal*. Divergence occurs when a lifting surface deflects under an aerodynamic load. The applied load increases and the twisting effect on the structure intensifies. The increased load deflects the structure further, eventually bringing it to its load limit and to failure. Control surface reversal is the loss (or reversal) of the expected response of a control surface due to structural deformation of the main lifting surface. Examples of dynamic aeroelastic phenomena include flutter, dynamic response and buffeting. Flutter is a self-feeding and potentially destructive form of vibration where aerodynamic forces on an object couple with a structure's natural mode of vibration to produce rapid periodic motion. When the energy input by the aerodynamic excitation is larger than that dissipated by damping in the system, the amplitude of vibration will increase, resulting in a self-exciting oscillation. Dynamic response is the response of the UAV to gusts and atmospheric disturbances. Buffeting is defined as high-frequency instability caused by airflow separation or shock wave oscillations from one object striking another. It is a random forced vibration. Generally, it affects the tail unit of the aircraft structure due to the wake downstream of the wing.

Trailing edge micro-actuators can facilitate the aeroelastic control of a UAV by providing flutter suppression in a flexible wing (Kroo *et al.*, 2003). Aeroelastic problems can be solved by stiffening the structure or altering mass distribution. However, such methods incur a performance penalty. For UAVs, active control of the aeroelastic behaviour is the best solution. Actuator technology makes use of small devices that are capable of generating forces sufficient for aeroelastic control. Small simple trailing-edge devices do not require accurate servo control. For example, the Miniature Trailing Edge Effector (MiTE) consists of a large number of micro devices of a height of 1–5% of c that deploy along the wing span and are deflected vertically into the flow. Their operation is similar to that of a Gurney flap.

1.1.9 Helicopter and Quad Rotor UAVs

Current trends in rotorcraft UAV design include vibration and noise reduction, rotor performance improvement, active blade tracking, and stability augmentation. Helicopter UAVs can be affected by a number of problems, including high vibratory loads, noise generation, poor performance, instabilities, difficulties in maintaining rotor-blade track, and limited load capacity and forward flight speeds (Wilbur and Wilkie, 2005). These problems originate from the highly variable aerodynamic environment that the rotor system encounters as it rotates in forward flight. The normal approach to eliminating these problems has been to modify the periodic aerodynamic loads. The simplest method is to modify the rotor-blade pitch at harmonic frequencies above the rotational frequency of the rotor.

Higher harmonic control reduces vibration and noise through use of active control of the rotor swashplate to change the pitch at the root of the blades. Another active control method, known as individual blade control (IBC) uses hydraulically actuated pitch links to achieve control of each of the blades independently. A concern when using IBC is obtaining the necessary hydraulic power in the rotating system. Current active rotor-control concepts include on-blade controls, such as trailing-edge flaps or strain-induced blade twist actuated by piezoelectric devices. On-blade active control is a feasible and efficient approach to solving rotor-system problems. Nonetheless, active rotor-control systems can be complex, expensive to develop, and bring inherent risks when implemented in a vehicle that will carry human beings. A recommended approach for the successful development of such technologies is experimentation using UAVs. Active-twist helicopter rotor systems have been explored recently. This type of rotor uses piezoelectric fiber composite actuators embedded within the composite blade structure. When excited by an electrical voltage, the actuators produce strain-induced twisting motions of the blade. Test results indicate that significant vibration and noise reduction can be achieved.

Small quad rotors have an important role in civil and military operations. A quad rotor is a vertical takeoff and landing (VTOL) air vehicle that can travel omni-directionally and also hover. The vehicle is powered by four motors attached to a rigid cross frame (Sanchez *et al.*, 2011). Quad rotor control is achieved by varying the rotational speed of the rotors. Each rotor produces combined thrust and torque that generate the main pitch, roll and yaw movements of the rotorcraft. The front and rear rotors rotate counter-clockwise, while the left and right rotors rotate clockwise, thus cancelling gyroscopic effects and aerodynamic torques in stationary trimmed flight. Vertical motion is controlled using collective throttle input. Forward and backward motion is achieved by adjusting the differential speed of the front and rear motors, which causes the quad-rotor to tilt at a pitch angle. Left and right flight is accomplished by controlling the differential speed of the right and left motors, tilting the craft at a roll angle. The two sets of rotors turning in opposite directions permit yawing of the vehicle. Yaw rotation is attained by increasing or decreasing the speed of the front and rear motors while also adjusting the speed of the lateral motors (Figure 1.6). The total thrust is kept constant to maintain altitude. Most designs use fixed-pitch propellers that are controlled by individually varying the rotational speed of each of the four rotors. Future variable-pitch designs could increase the operational envelope of the quad rotor. In fact, blade-element numerical codes are able to capture the effects of the blade geometry and forward flight to explore variable-pitch capability.

1.1.10 Adaptive Wing Technology

Aircraft wings represent a compromise that will allow the aircraft to fly in different conditions. However, the aircraft's aerodynamic performance in each condition is sub-optimal (Barbarino *et al.*, 2011). Adaptive wings change their geometry during flight to

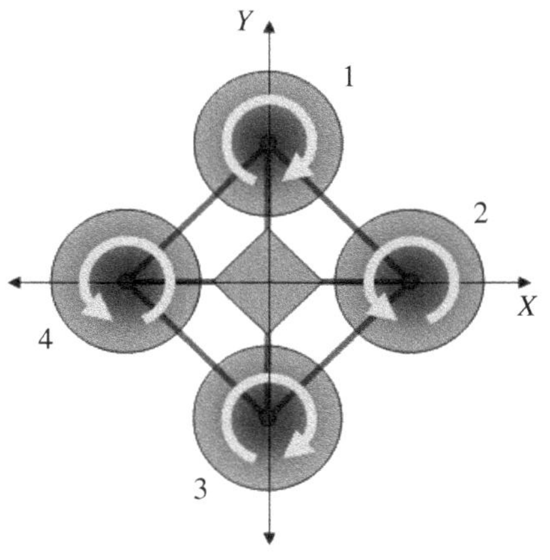

Figure 1.6 Yaw torque and control in a quad rotor aircraft. Rotors 1 and 3 spin in one direction and rotors 2 and 4 spin in the opposite direction, yielding opposing reaction torques. Image: Gabriel Hoffmann.

optimise the wing configuration. Geometrical parameters susceptible to morphing include: airfoil (camber and thickness), planform (chord, span, sweep), and out-of-plane (twist, dihedral/gull, span-wise bending). Modern smart materials allow the blending of morphing and smart structures for the design of the next generation of UAVs.

The morphing wing is a biomimetic technology that has been applied to UAVs by means of changes in wing cross-section, planform, and spar position. Yang *et al.* (2006) investigated a morphing wing with variable camber using shape memory alloy (SMA) actuators. SMAs have the unique effects of shape memory and pseudoelasticity, which result from phase transformations between martensite (rapid cooling) and austenite (heating) phases induced by changes in temperature or applied stress. SMAs allow high recovery forces and large deformations. Lift force increases without a parallel increase in drag. Monitoring of aeroelastic unstable phenomena at various angles of attack reveals limit-cycle oscillations with constant frequency at flutter speed. Activation of the SMA actuators considerably reduces the amplitude of the limit-cycle oscillations. Such findings suggest that a smart composite wing with integrated smart sensors and actuators is feasible for applications in UAVs and MAVs with relatively simple and less stiff wing structures. Gamboa *et al.* (2007) present a morphing approach for wing-drag reduction in small UAVs. A tool is used that searches for an optimum airfoil geometry and wing planform shape, given geometric wing constraints. The aerodynamic analysis consists of two stages. First, the two-dimensional aerodynamic coefficients based on angle of attack and Re are obtained using the airfoil flow field solver XFOIL (an interactive program for the design and analysis of subsonic isolated airfoils). Instead of using control points to define the airfoil surface, control points are used to represent the airfoil camber line and thickness distribution. Second, a non-linear lifting-line-method algorithm gives the lift distribution and induced drag. Aerodynamic shape optimization is carried out using the sequential quadratic programming constrained optimization algorithm DONLP2. Computational analysis shows that 14–30% drag reduction is attainable at different stages of flight. A novel gust-load alleviation technique based on active flow control was developed by Xiaoping *et al.* (2011). The dynamic response of the quasi ‘Global Hawk’ airfoil was investigated using arrays of jets during gust simulation based on the unsteady Navier–Stokes equations. Five kinds of flow-control technique were introduced, including steady blowing, steady suction, unsteady blowing, unsteady suction and synthetic jets. Physical analysis of the effect on gust-load alleviation was then used to produce new guidelines for practice. The numerical results showed that a new technique called 'active flow-control' could effectively suppress gust-induced fluid disturbances.

Planform morphing aims to emulate nature in aerial vehicles, enabling highly adaptive and fluidic motions with freedom of configuration (Barbarino *et al.*, 2011). The concept of morphing aircraft dates back to the Wright brothers' first powered aircraft. To control the vehicle, the pilot shifted his hips, which in turn actuated tendons that twisted the wings. The concept of 'wing-warping' was soon abandoned and today most aircraft use conventional hinged control surfaces. Planform morphing is used for mission adaptation, and control morphing allows vehicle manoeuvring. Planform morphing involves varying wingspans, sweeps and tail configurations such that the AR, planform area and static margin are modified. Control morphing involves smaller changes that alter the flowfield slightly, enabling the aircraft to manoeuvre, but without significantly affecting its dynamics. Combinations of planform and control morphing permit advanced manoeuvring capabilities; for example, using wing twist and asymmetric wing extension for effective roll control.

1.1.11 Hybrid Aircraft

Fixed-wing UAVs are able to fly for long periods at high speeds and their mechanical design is relatively simple. However, they require runways or additional launch and recovery equipment. Rotary-wing UAVs do not need forward airspeed for flight and manoeuvring, and do not require any infrastructure for takeoff and landing. However, rotary wing UAVs are mechanically complex and have low flight speeds and endurance. Hybrid UAVs integrate both horizontal and vertical flight capabilities. They are, however, more mechanically complex and more difficult to control. Tilt-rotor UAVs possess adequate stability, energy efficiency and controllability. Examples of UAV's with dual tilt-rotors include the Bell Eagle Eye, the Smart UAV of KARI and the BIROTAN, while UAVs with dual tilt-wings include the HARVee and the UAV of the Universita di Bologna. The dual-rotor UAVs require cyclic controls on the propellers for stabilization and manoeuvring, and are thus more complex. Electric-powered quad tilt-wing UAVs are capable of VTOL in addition to horizontal flight (Cetinsoy *et al.*, 2012). Greater stability is attained with a quad tilt-wing UAV, such as the QUX-02 of the Japan Aerospace Exploration Agency.

The aerodynamic design of the quad tilt-wing UAV developed by Cetinsoy *et al.* (2012) aims to generate sufficient lift for flight while minimizing drag. The fuselage shape is a rectangular prism with rounded nose and back sections, and gradually decreasing thickness for streamlining. When viewed from the top, the fuselage resembles a symmetrical wing. This water-drop shape gives a low drag coefficient. The wing planform is adjusted by considering the speed range, wing-span constraints, wing efficiency and interactions with the fuselage. Airfoil chord length, maximum camber ratio and thickness were determined using simulations in the ANSYS® environment. The slipstream velocity from the rotors needs to be taken into account, since the rotors have a marked effect on the behaviour of the wings. The use of greater chord lengths and large winglets, instead of greater wing thickness, increases efficiency. In fact, at high angles of attack, the large chord provides a large inclined surface against the air flow. At high speeds, a thicker wing causes greater drag. It is known that thin wings with low camber experience leading-edge separation at lower angles of attack, which leads to stalls. However, the wings of the quad tilt-wing UAV are almost fully submerged in the slipstream of the propellers and the high-speed slipstream prevents air separation at high angles of attack and gives additional lift. The low AR of 4 and the rectangular planform cause severe spanwise airflow at high angles of attack. The spanwise airflow reduces wing aerodynamic efficiency due to strong wingtip vortices and pressure equalization in the wingtip region. The large winglets restrict the tip vortex. The lift of the rear wings is adversely affected by the downwash of the front wings. This causes imbalances in the transition and horizontal flight modes as the centre of mass of the vehicle is located between the aerodynamic centres of the front and rear wings. The downwash from the two front wings reduces the effective angle of attack of the rear wings. To account for the downwash configuration, the rear wings are set at a higher angle of incidence.

1.1.12 Transonic and Supersonic Flight Regimes

The behaviour of supersonic airflow is very different to that of subsonic flow. In front of a UAV, the fluid builds up a stagnation pressure as its impact with the vehicle brings the moving fluid to a rest. At subsonic speeds, the pressure disturbance propagates upstream, changing the flow pattern ahead of the aircraft. At supersonic speeds, the pressure disturbance does not propagate upstream and when the airflow strikes the front of the UAV, it is forced to change its properties (temperature, density, pressure, and Mach number) in a violent and irreversible fashion, producing a shock wave. At flight speeds near the speed of sound, supersonic flow and wave drag may occur over parts of the wing. Wave drag is considerably higher than induced drag, and must be avoided to improve UAV performance. Wave drag is prevented using a very thin wing that has a slowly changing airfoil profile over a wide chord. Limitations of very thin wings include insufficient space for fuel storage and the landing gear, and less torsional stiffness. Wing sweep prevents the development of the shock wave at the leading edge (Figure 1.7).

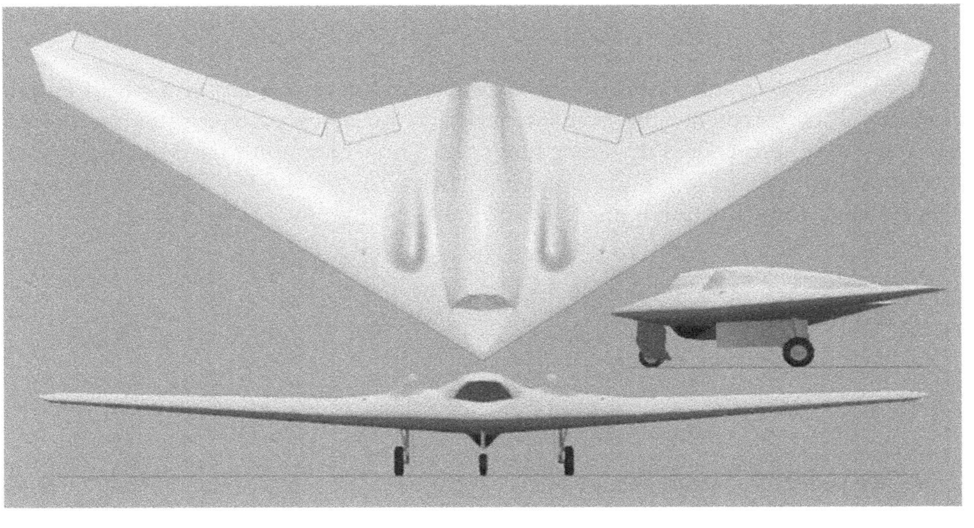

Figure 1.7 Pronounced wing sweepback in a Lockheed Martin RQ-170 Sentinel. Image: Truthdowser.

1.1.13 Wind Tunnel Testing and Computational Fluid Dynamics

There is a scarcity of airfoil experimental data at low values of Re. The use of wind tunnel testing to design and optimize new airfoils is time consuming and costly. With the emergence of high-speed electronic computers, numerical simulation is frequently used in airfoil design. CFD is an effective method of studying the complex flow field around the airfoil, allowing rapid evaluation of preliminary airfoil designs, and reducing cost, time and the need for multiple experiments. CFD consists of a combination of aerodynamic analysis and optimization of the airfoil geometry for a given set of constraints. To ensure good aerodynamic performance for a range of fight speeds and conditions, multi-objective optimization is preferable to single-objective optimization. Therefore, the aircraft is optimised for a selected airspeed while preserving good performance in other flight regimes. For example, the design of an efficient high-lift low-Re airfoil for UAV applications can be achieved using hierarchical multi-objective optimization that combines the direct search optimization algorithm EXTREM with XFOIL to determine aerodynamic performance (Park *et al.*, 2008). Airfoil shape optimization comprises four steps.

- Describe or parameterize the airfoil shape.
- Select an appropriate aerodynamic analysis method.
- Implementing an effective CFD optimization algorithm.
- Identify a suitable multi-objective optimization method.

Design optimization commences by selecting a baseline airfoil. The new airfoil may be represented in the computer programme by Eqs. (1.1) and (1.2).

$$y_{\text{upper}}(x) = y_{\text{ou}} + \sum_{7}^{k=1} c_k f_k(x) \tag{1.1}$$

$$y_{\text{lower}}(x) = y_{\text{ol}} + \sum_{7}^{k=1} c_{k+7} f_k(x) \tag{1.2}$$

where, y_{ou} and y_{ol} are the ordinates of the upper and lower surface of the base airfoil, respectively, k represents the number of design variables, c_k represents the design variables and f_k symbolises the shape functions. Three commonly used shape parameterization methods are the polynomial, the Hicks–Henne and the Wagner functions. The Hicks–Henne function is given by Eq. (1.3).

$$f_k(x) = \begin{cases} x^{0.25}(1-x)e^{-20x}, & k=1 \\ sin^3\left(\pi x^{e(k)}\right), & k>1 \end{cases} \tag{1.3}$$

where,

$$e(k) = \frac{log0.5}{logx_k}, \quad 0 \le x_k \le 1. \tag{1.4}$$

During shape parameterization, a linear superposition method can be used and the coefficients of the Hicks–Henne function determine a new airfoil shape using the leading- and trailing edges of the baseline airfoil. XFOIL uses a linear-vorticity stream-function panel method coupled with viscous integral boundary-layer equations for the analysis of airfoils with free or fixed transition, separation bubbles, blunt trailing edges, and limited trailing-edge separation over a range of Re values and Mach numbers. The validity, versatility and ease of use of the XFOIL programme for the preliminary design of subsonic airfoils and quick estimation of transition locations has been extensively documented. The direct search optimization algorithm EXTREM used for aerodynamic shape optimization can be used effectively for solving the multi-variable constrained optimization problem, converges quickly and requires no derivation. Once an extremum of optimization is attained, the optimization process is completed using parabolic extrapolation. Multi-objective aerodynamic optimization of airfoil design is a complex engineering problem that requires a combination of common multi-objective optimization methods and empirical evidence for specific problems to achieve a satisfactory design. The use of genetic algorithms underlies the functioning of multi-objective optimization. A typical optimization problem is expressed mathematically by the algorithm:

$$\textit{Find } x = \left\{x_1, x_2, \ldots, x_m\right\}^T$$

$$\textit{to maximize } f_1\left(x\right) = c_l$$

$$\textit{to maximize } f_2\left(x\right) = c_l / c_d$$

$$\textit{to minimize } f_3\left(x\right) = c_{m,1/4}$$

$$\textit{subjected to } x_i^L \le x_i \le x_i^U \textit{ for } i = 1 \sim 13.$$

where, $c_{m,¼}$ is the pitching moment about the quarter chord coefficient, and x_i^L and x_i^U are the upper and lower bounds of the design variables (Park *et al.*, 2008).

Numerical methods reduce the cost and effort for the preliminary configuration of aircraft and enable the expansion of the flight regime to be analysed well beyond the technical capabilities of a wind tunnel. Calculated results can be validated to confirm their accuracy. For example, Tornado is an open-source three-dimensional vortex-lattice program written in MATLAB® that can be used to model 3D wing surfaces and calculate aerodynamic forces and coefficients and flight-stability derivatives for UAV design optimization. The vortex-lattice method represents the wing as a planar surface broken into quadrilateral panels on which a horseshoe vortex is superimposed. The law of Biot–Savart is used to calculate the velocities induced by each horseshoe vortex and the vortex strength at a specific control point. Vortex strength is related to the circulation around the wing and the pressure differential between the upper and lower surfaces. Aerodynamic forces are obtained by integration of the pressure differentials. Longitudinal flight stability derivates computed with Tornado have been shown to be in good agreement with experimental data, and lateral-directional derivates are acceptable (Cardenas *et al.*, 2009).

Sanjay *et al.* (2012) conducted a CFD study on a UCAV configuration consisting of a moderately swept, tailless, flying wing with a cranked trailing edge. This was designed for low observability in hostile environments rather than high manoeuvrability, for reasons of survivability. CFD allows observation of the vortex flow in the wake of the flying UCAV. Aerodynamic studies were also carried out in a wind tunnel using different angles of attack. CFD data obtained using STAR CCM+ CFD code was compared to experimental wind tunnel results. Flow visualization provides insights into UCAV manoeuvrability, since it allows analysis to be made of vortex bursting in relation to aerodynamic lift under steady and unsteady conditions. The CFD analysis was carried out over 2–3000 iterations to attain convergence, and steady pressure and velocity distributions and lift and drag forces. In the computational domain with a tetrahedral mesh, the walls were positioned to effectively capture the turbulence flow around the UCAV and generate a realistic solution. To effectively capture the turbulence above the UCAV model, five prism layers of 18 mm were used. A tetrahedral surface mesher was used to mesh the entire geometry. An element size of 0.5 mm was maintained over the UCAV model to capture the exact geometry and critical regions. The outer domain was meshed with a tetrahedral mesh of an element size of 55 mm.

Successful application of turbulence models has greatly increased the quality of numerical simulations. Turbulence in a flow becomes dominant over all other flow phenomena. The complexity of turbulence models varies considerably depending on the amount of detail required in the numerical simulation. Complexity is due to the nature of the Navier–Stokes equation, which is a non-linear, time-dependent, three-dimensional partial differential equation.

One of the most prominent turbulence models, the (k-epsilon) model, is implemented in most general purpose CFD codes. The (k-epsilon) model is stable, numerically robust and accurate, and has excellent predictive capability. The first transported variable is turbulent kinetic energy (k) and the second is turbulent dissipation (ε), which determines the scale of the turbulence.

1.1.14 Further Reading

1 Azlin MA, Mat Taib CF, Kasolang S and Muhammad FH (2011). CFD analysis of winglets at low subsonic flow. *Proceedings of the World Congress on Engineering – Vol. I*. 6–8 July. London, United Kingdom.

2 Barbarino S, Bilgen O, Ajaj RM, Friswell MI and Inman DJ (2011). A review of morphing aircraft. *Journal of Intelligent Material Systems and Structures*. **22**: 823–877.

3 Boschetti PJ, Cárdenas EM and Amerio A (2012). Induced drag reduction of an airplane using local twist, panel method verification. *50th AIAA Aerospace Sciences Meeting*. 9–12 January. Nashville, Tennessee, pp. 1–10.

4 Cetinsoy E, Dikyar S, Hancer C, Oner KT, Sirimoglu E, Unel M and Aksit MF (2012). Design and construction of a novel quad tilt-wing UAV. *Mechatronics*. In press.

5 Cook MV, Buonanno A and Erbsloeh SD (2008). A circulation control actuator for flapless flight control. *The Aeronautical Journal*. **112**(1134): 483–489.

6 Gamboa P, Aleixo P, Vale J, Lau F and Suleman A (2007) Design and testing of a morphing wing for an experimental UAV. *Platform Innovations and System Integration*

for Unmanned Air, Land and Sea Vehicles (AVT-SCI Joint Symposium). Neuilly-sur-Seine, France, pp. 17–1–17-30.
7 Kroo I, Prinz F and Eaton J (2003). UAV aeroelastic control using redundant micro-actuators. Techical report, AFOSR. Arlington, VA.
8 Park K, Han J, Lim H, Kim B, and Lee J (2008). Optimal design of airfoil with high aspect ratio in unmanned aerial vehicles. *International Journal of Aerospace and Mechanical Engineering.* **2**(1): 66–72.
9 Phillips WF (2005). New twist on an old wing theory. *Aerospace America.* **January**: 27–30.
10 Reneaux J, Thibert JJ and Rodde AM (1997). Airfoil design for sailplanes and ultralight aircraft. *XXV OSTIV Congress.* 3–11 July. Saint-Auban sur Durance. France, pp. 1–7.
11 Ma R, Zhong B, Liu P and Drikakis D (2010). Multi-objective optimization design of low Reynolds-number airfoils S1223. *27th International Congress of the Aeronautical Sciences.* 19–24 September. Nice. France.
12 Sanchez A, García-Carrillo LR, Rondon E, Lozano R and Garcia O (2011). Hovering flight improvement of a quad-rotor mini UAV using brushless DC motors. *Journal of Intelligent and Robotic Systems.* **61**(1–4): 85–101.
13 Shyy W, Klevebring F, Nilsson M, Sloan J, Carroll B and Fuentes C (1999). Rigid and flexible low Reynolds number airfoils. *Journal of Aircraft.* **36**(3): 523–529.
14 Syed HM, Hameed MS and Manarvi IA (2011). A review of swept and blended wing body performance utilizing experimental, FE and aerodynamic techniques. *International Journal of Research and Reviews in Applied Sciences.* **8**(3): 371–385.
15 Tianyuan H and Xiongqing Y (2009). Aerodynamic/stealthy/structural multidisciplinary design. *Chinese Journal of Aeronautics.* **22**: 380–386.
16 Wilbur ML and Wilkie WK (2005). Active-twist rotor control applications for UAVs. *Proceedings of the 24th Army Science Conference.* 29 November–2 December. Orlando, Florida, pp. 1–8.
17 Wisnoe W, Nasir RE, Kuntjoro W and Mamat AM (2009). Wind tunnel experiments and CFD analysis of blended wing body (BWB) unmanned aerial vehicle (UAV) at Mach 0.1 and Mach 0.3. *13th International Conference on Aerospace Sciences and Aviation Technology.* 26¯28 May. Cairo, Egypt.
18 Xiaoping X, Xiaoping Z, Zhou Z and Ruijun F (2011). Application of active flow control technique for gust load alleviation. *Chinese Journal of Aeronautics.* **24**: 410–416.
19 Yang S, Han J and Lee I (2006). Characteristics of smart composite wing with SMA actuators and optical fiber sensors. *International Journal of Applied Electromagnetics and Mechanics.* **23**: 177–186.

1.1.15 Problems

Problem 1

Using the data in Table 1.1, plot the mean camber line for the NACA 2412 airfoil. Identify the chordwise location of the maximum camber.

Table 1.1 Mean camber line ordinates for the NACA 2414 airfoil.

x coordinate	Mean camber line
0.000	0.000
0.013	0.003
0.025	0.004
0.050	0.006
0.075	0.008
0.100	0.009
0.150	0.013
0.200	0.015
0.250	0.017
0.300	0.019
0.400	0.020
0.500	0.020
0.600	0.018
0.700	0.015
0.800	0.011
0.900	0.006
0.950	0.003
1.000	0.000

Problem 2

Using the Kutta–Joukowski theorem, calculate the circulation (Γ) and the lift per unit span (L') for a thin symmetric airfoil of chord (c) of 0.70 m, at angles of attack (α) of 5° and 10° in a standard atmosphere with air density (ρ_∞) of 1.225 kg/m^3 and airflow velocity (V_∞) of 40 m/s; see Eqs. (1.5) and (1.6).

$$\Gamma = \pi \alpha c V_\infty \tag{1.5}$$

$$L' = \rho_\infty V_\infty \Gamma = \pi \alpha c \rho_\infty V_\infty^2 \tag{1.6}$$

Problem 3

Calculate Γ and L' for a thin cambered airfoil of $c = 0.70$ m, at α of 5° and 10°, in standard atmosphere and V_∞ of 50 m/s. The Fourier coefficients are $A_0 = \alpha - 0.004517$ and $A_1 = 0.08146$; Eqs. (1.7) and (1.8).

$$\Gamma = cV_\infty \left(\pi A_0 + \frac{\pi}{2} A_1 \right) \tag{1.7}$$

$$L' = \rho_\infty V_\infty \Gamma = \rho_\infty V_\infty^2 c\pi = \left(A_0 + \frac{1}{2} A_1 \right) \tag{1.8}$$

Problem 4

Estimate the laminar boundary-layer thickness (δ_L) at the trailing edge for a chord length of 0.9 m, the net laminar skin-friction drag coefficient ($C_{f\mathrm{L}}$), the turbulent boundary layer thickness (δ_T) at the trailing edge, and the net turbulent skin-friction drag coefficient ($C_{f\mathrm{T}}$), for a thin airfoil. The chord-length Reynolds number ($\mathrm{Re_c}$) and chord-length Reynolds number based on distance x from the leading edge (Re_x) are (2×10^6); Eqs. (1.9)–(1.12).

$$\delta_\mathrm{L} = \frac{5c}{\sqrt{\mathrm{Re_c}}} \tag{1.9}$$

$$C_{f\mathrm{L}} = 2\frac{1.328}{\sqrt{\mathrm{Re_c}}} \tag{1.10}$$

$$\delta_\mathrm{T} = \frac{0.37x}{\mathrm{Re}_x^{1/5}} \tag{1.11}$$

$$C_{f\mathrm{T}} = 2\frac{0.074}{\mathrm{Re_c^{1/5}}} \tag{1.12}$$

Problem 5

A wing of $\mathrm{AR}_1 = 5$ and $\alpha_1 = 4°$ has a lift coefficient (C_L) of 1.0. Calculate the α_2 at which two wings of $\mathrm{AR}_2 = 8$ and 9 will have to operate to generate the same C_L; Eq. (1.13).

$$\alpha_2 = \alpha_1 + \frac{C_L}{\pi}\left(\frac{1}{\mathrm{AR}_2} - \frac{1}{\mathrm{AR}_1}\right) \tag{1.13}$$

Problem 6

An airfoil has $c_l = 0.0$ at $\alpha = -1°$ and $c_l = 0.8$ at $\alpha = 6°$. Calculate the airfoil lift slope (a_0) and the wing lift slope (a) for a wing of AR = 7.5. The lift efficiency factor (τ) is 0.09; Eqs. (1.14) and (1.15).

$$a_0 = \frac{c_{l2} - c_{l1}}{\alpha_2 - \alpha_1} \tag{1.14}$$

$$a = \frac{a_0}{1 + \left(\frac{a_0}{\pi \mathrm{AR}}\right)(1+\tau)} \left(\text{a and a}_0 \text{ are per radian}\right) \tag{1.15}$$

Problem 7

Calculate a for a low-AR straight wing of AR = 3.5; Eq. (1.16).

$$a = \frac{a_0}{\sqrt{1 + \left(\frac{a_0}{\pi \mathrm{AR}}\right)^2} + a_0/(\pi \mathrm{AR})} \tag{1.16}$$

Problem 8

The zero-lift angle of attack (α_{L0}) for the NACA 23012 airfoil is given by Eqs. (1.17) and (1.18). Compute α_{L0}.

$$\alpha_{L0} = -\frac{1}{\pi}[-2.8683\theta + 3.0576\sin\theta - 2.1843\sin\theta\cos\theta + 0.665\sin\theta(cos^2\theta + 2)]_0^{0.9335} \quad (1.17)$$

and

$$-\frac{1}{\pi}\left[0.022080\theta - 0.02208\sin\theta\right]_{0.9335}^{\pi} \quad (1.18)$$

1.2 UAV Flight Stability and Control

'The balancing of a gliding or flying machine is very simple in theory. It merely consists in causing the center of pressure to coincide with the center of gravity. But in actual practice there seems to be an almost boundless incompatibility of temper which prevents their remaining peaceably together for a single instant, so that the operator, who in this case acts as peacemaker, often suffers injury to himself while attempting to bring them together.'

Wilbur Wright

CONTENTS

1.2.1 Introduction – Flight Stability

Stability is defined as the tendency of the UAV to return to a condition of equilibrium when subjected to a disturbance in flight. This is typically caused by gusts or flight-control input (Nelson, 1998). *Static stability* is the initial response of the aircraft to regain equilibrium upon a disturbance. It can take be positive, negative, divergent or neutral. *Dynamic stability* refers to the response of the system over time. Even when an aircraft has inherent positive static stability, it usually tends to overshoot the equilibrium condition, producing a series of oscillations. If the magnitude of successive oscillations becomes smaller over time, the system has positive dynamic stability. When the time history of dynamic oscillations is short, the system is said to be highly damped. When oscillations take longer, the system is considered lightly damped. An aircraft with positive static stability but no damping continues to oscillate without experiencing any decrease in the amplitude of the oscillations; such an aircraft has neutral dynamic stability. The location of the centre of gravity (CG) that would cause the aircraft to have neutral static longitudinal stability is called the *neutral point* (NP). Moving the CG forward of the NP gives positive static stability, and moving the CG aft of the NP makes the aircraft statically unstable. A necessary condition for positive static longitudinal stability is, therefore, that the CG is forward of the NP. This criterion is stated in terms of the aircraft's *static margin* (SM), which is a non-dimensional measure of the aircraft's stability. A large SM implies a very stable but not particularly manoeuvrable aircraft. Highly manoeuvrable UAVs have a short positive SM. Vehicles with zero or negative SM require a computer fly-by-wire flight control system for flight control.

1.2.2 Control Surfaces

Conventional aircraft control surfaces produce the moments that cause rotations in pitch, roll and yaw. *Elevators* and *stabilators* generate a lift force at the aircraft's tail. The moment arm of this force from the aircraft's CG generates a pitching moment (Nelson, 1998). *Ailerons* deflect differentially and create more lift on one wing than the other, and therefore a net rolling moment. The *rudder* deflects to generate a lift force in a sideways direction. This force acts at a distance from the aircraft's CG, producing a moment about the vertical axis that causes the aircraft to yaw.

UAVs represent unique aircraft configurations, in that they employ novel types of hinged control surfaces. Often, the functions of two control surfaces are combined in a single surface. The names of these controls are a combination of the names of the two surfaces. A stabilator combines the functions of the horizontal stabilizer and the elevator in an 'all-moving' control surface. *Canard stabilators* are placed forward of the wings. A *flaperon* combines the functions of a plain flap and an aileron. *Elevons* combine elevators and ailerons, deflecting together to function as elevators for pitch control and also moving differentially, like ailerons, to provide roll control. In a V-tail UAV, the moveable control surfaces of the tail are called *ruddervators* (Figure 1.8). These function as rudders when moving differentially and as elevators when moving together.

1.2.3 Principles of Aircraft Stability

During flight, moments on a UAV are created by the aerodynamic load distribution and the thrust force not acting through the CG. Aerodynamic moments are expressed in terms of the dimensionless coefficients for pitching moment (C_m), rolling moment (C_ℓ) and yawing moment (C_n). For airplane configurations, the characteristic length (l) is the mean chord (c) for the pitching moment and the wing span (b) for the rolling and yawing moments. The values of C_m, C_ℓ and C_n depend on the angle of attack (α), Reynolds number (Re), Mach number (M) and sideslip angle (β) and are functions of the angular rates and sideslip of the aircraft (Nelson, 1998). Figure 1.9 shows the definition of velocities, forces and moments.

UAVs designed with inherent aerodynamic stability possess satisfactory handling or flying qualities. In contrast, UCAVs are purposely designed with limited stability in the interest of manoeuvrability in combat operations and require complex control management and electromechanical stability augmentation systems.

A UAV, particularly if in a single-main-rotor helicopter configuration, can be statically stable but dynamically unstable (Austin, 2010). However, to be dynamically stable

Figure 1.8 Down-angled ruddervators in an RQ-1 Predator. Photo: US Navy.

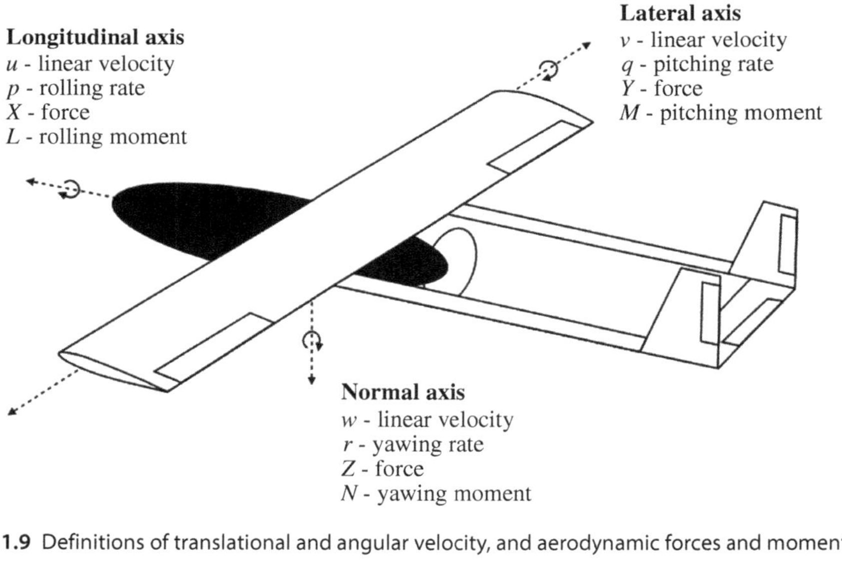

Figure 1.9 Definitions of translational and angular velocity, and aerodynamic forces and moments.

it is a necessary condition that the aircraft is statically stable. Dynamic stability entails damped oscillatory motion, whereby the oscillations caused by the disturbance are gradually dissipated by inertial and aerodynamic forces that oppose the motion of the vehicle. The measures of dynamic stability are the frequency and period of the oscillations, and the time it takes to damp the undesirable oscillatory motion to half of

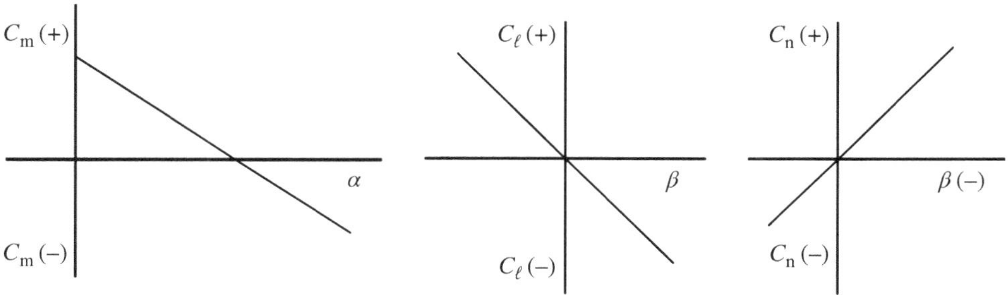

Figure 1.10 Stable longitudinal (in pitch), lateral (in roll) and directional (in yaw) conditions.

its amplitude (Nelson, 1998). A necessary condition for longitudinal static stability of the UAV is that the pitching moment curve has a negative slope through the equilibrium point. Also, to trim the vehicle at a positive α and produce useful lift, the C_m at $\alpha = 0°$ must be greater than zero. The slope must be negative for lateral static stability and positive for directional static stability (see Eq. (1.19) and Figure 1.10).

$$\text{pitch}\,\frac{dC_m}{d\alpha} < 0 \quad roll\,\frac{dC_\ell}{d\beta} < 0 \quad \text{yaw}\,\frac{dC_n}{d\beta} > 0 \tag{1.19}$$

1.2.4 UAV Configurations

In an orthodox tail-aft airplane configuration, the principal contributor to stability in pitch is the horizontal stabiliser in a *longitudinal dihedral* arrangement. However, the wing, fuselage and propulsion thrust also play a part (Nelson, 1998). A wing constructed using a conventional airfoil is destabilising in pitch, since such airfoils exhibit a negative nose-down moment. A positive pitching moment about the AC of the wing is achieved using a reflexed airfoil. The major contributor to stability in roll is the wing dihedral. Wing sweep, wing position on the fuselage and the vertical fin also influence the restoring roll moment.

The wing and fuselage are destabilizing in yaw, and directional stability is primarily attained by the vertical tail. Nonetheless, horizontal-takeoff-and-landing (HTOL) UAVs come in diverse airframe configurations, including, for example, canard, tailless, flying wing and delta wing (Austin, 2010). VTOL vehicles can have co-axial rotor, tandem rotor and quad rotor designs, and hybrid UAVs can use tilt-rotor, tilt-wing, tilt-wing-body, ducted fan and other designs. Flight stability in pitch in orthodox and unconventional vehicle configurations is achieved by methods that are specific to each category.

Orthodox layout: The aircraft's CG is located at the AC of the wing. The wing's camber produces a nose-down pitching moment. The tailplane generates a download to balance the wing's moment. An additional lift force must be developed by the wing to compensate for the tailplane download. For highly cambered wings, the tail's download must be larger. The horizontal stabiliser operates in the vortex-induced downwash of the wing and should consequently be set at a slightly negative angle relative to the local airflow in a *longitudinal dihedral* configuration, so that the tail generates the required downforce. The elevator modifies the tailplane's camber and adjusts the tailplane's download for different flight conditions.

Canard configuration: A canard foreplane is free from interference from the main wing downwash and the propeller slipstream. Unlike the aft-tail configuration, which yields a download, the canard foreplane produces lift to counteract the nose-down moment and adds to the total lift generated by the wing. Canard control works effectively to trim the additional nose-down moment originated by high-lift devices in the wing, such as flaps. However, a canard configuration reduces the vehicle's static stability. This is because any lifting surface located ahead of the CG moves the NP forward. In a canard, the foreplane causes the NP to lie ahead of the wing's AC. To enhance stability, a canard UAV must have the CG forward, and therefore the foreplane carries the camber-induced pitching load and an additional load caused by the forward position of the CG. The canard lies in the vortex-induced upwash of the wing. However, the forewing produces downwash that influences the α of the wing. The downwash can be reduced using a small forewing of large AR.

Tailless aircraft: Flying wing, delta, double delta, and lambda UAV configurations achieve longitudinal stability using wing sweepback and washout (Figure 1.11). The aircraft's CG is positioned ahead of the AC and the wingtips produce a download. A wing with reflex camber helps give a neutral pitching moment.

In the orthodox layout, stability in roll is primarily achieved using the wing dihedral, which in UAVs takes different forms such as elliptical, polyhedral, gull, and inverted gull dihedral. Typically, high-wing UAVs have no dihedral and the wing is therefore neutrally stable in roll (Nelson, 1998). As the aircraft rolls, the down-going wing meets the freestream at a greater α. At the same time, the α of the up-going wing is reduced. The rolling movement is therefore damped out, but there is no tendency to correct the banked attitude of the vehicle once the roll has stopped. The increase in lift force on the

Figure 1.11 Northrop Grumman X-47B UCAS-D. Photo: Rob Densmore. (*See insert for color representation of this figure.*)

down-going wing also increases the strength of the wingtip vortex creating a drag force that tends to yaw the aircraft. Thus there is cross-coupling of the rolling and yawing motions. The fin surface area (S) and its distance from the aircraft's CG are important considerations in obtaining adequate stability in yaw.

1.2.5 Relaxed Stability

For use in windy and harsh flight environments, it is desirable to reduce the response of some UAVs to turbulence in order to maintain payload sensors on the target (Austin, 2010). Strong inherent aerodynamic stability, large S and high AR in relation to the vehicle's mass cause UAVs to exhibit high responses to atmospheric turbulence. Near-neutral flight stability ensures the minimum disturbance from air turbulence. However, an automatic flight control and stability system (AFCS) is often required to ensure that the vehicle has positive spatial stability that will prevent the aircraft wandering off course.

1.2.6 Case Study

Cardenas *et al.* (2009) evaluated the static stability and open-loop dynamic stability of a small airplane designed for ecological conservation. The small UAV is a twin-boom, pusher-propeller configuration with maximum takeoff mass of 182 kg and a payload of 40 kg consisting of a high-technology camera that is used to find oil leakages. The wing is of rectangular planform, straight with local twist, and has no dihedral. The b is 5.2 m, AR is 8.57 and S is 3.13 m^2. The wing uses an NACA 4415 airfoil along the complete wingspan and Re is 1.41×10^6. Flight stability derivatives were computed using Tornado, an open-source 3D vortex lattice program with application of inviscid CFD.

Longitudinal stability was evaluated as a function of α, elevator deflection angle (δ_e) and q. The wing lift coefficient (C_L) for $\alpha = 0°$ and $\delta_e = 0°$ is 0.51, which increases to $C_L = 0.62$ for $\delta_e = 10°$. The wing lift slope (a) is 5.46 rad^{-1}. The C_L varies with δ_e by 0.68 rad^{-1}. Longitudinal stability was assessed using C_m curves at various δ_e. The C_m at $\alpha = 0°$ is 0.015 and the C_m slope is −2.73 rad^{-1}, whereby the negative sign indicates that the aircraft is inherently statically stable in pitch. The C_m slope increases to −2.87 rad^{-1} for $\delta_e = 10°$. The aircraft's CG is located at 0.25 c and the NP is positioned at 0.75 c.

The lateral-directional static stability coefficients were studied as a function of sideslip angle, rudder and aileron deflections, and yaw and roll rates at $\alpha = 0°$. The variation of side force coefficient with sideslip angle ($C_Y\beta$) has a derivative of $C_Y\beta = -0.56$ rad^{-1} and the slope is −0.017 rad^{-1}. Thus, the negative sign indicates that the airplane is inherently laterally stable. The aircraft is also directionally stable with a positive slope of 0.155 rad^{-1}. The variation in side force coefficient with rudder angle ($C_Y\delta_r$) has a derivative $C_Y\delta_r = -0.27$ rad^{-1}; the yawing ($C_n\delta_r$) and roll moment ($C_\ell\delta_r$) coefficients due to the deflection of the rudder were determined and their derivatives were $C_n\delta_r = 0.149$ rad^{-1} and $C_\ell\delta_r = -0.006$ rad^{-1} at the zero sideslip condition. The variation in rolling coefficient with aileron angle ($C_\ell\delta_a$) is $C_\ell\delta_a = -0.206$ rad^{-1}. Increasing the aileron angle increases the roll moment. The variation in yawing coefficient with aileron angle ($C_n\delta_a$) has a derivative $C_n\delta_a = 0.017$ rad^{-1} and the side force variation with aileron angle ($C_Y\delta_a$) is $C_Y\delta_a = -0.001$ rad^{-1}. The rolling (I_ℓ), pitching (I_M) and yawing (I_n) moments of inertia of the UAV are $I_\ell = 150$ kgm^2, $I_M = 400$ kgm^2 and $I_n = 400$ kgm^2, based on mass distribution.

Table 1.2 Characteristics of dynamic stability in pitch.

	Short-period mode	Phugoid mode
Period (T), s	2.05	26.04
Time to half amplitude ($t_{1/2}$), s	0.34	191.67
Number of cycles to half amplitude ($N_{1/2}$), cycles	0.17	7.36
Damping ratio (ζ)	0.551	0.015
Undamped natural frequency (ω_n), rad/s	3.68	0.24

The static stability data obtained using the vortex-lattice method are used to determine the airplane dynamic stability response characteristic (Table 1.2). The two longitudinal modes of motion – the short-period mode ($\lambda_{1,2}$) and the phugoid mode ($\lambda_{3,4}$) – were $\lambda_{1,2} = -2.025 \pm 3.070i$) and $\lambda_{3,4} = -0.004 \pm 0.241i$. The negative values of the longitudinal modes indicate that the UAV is dynamically stable. On a sudden disturbance, the response would decay sinusoidally over time.

1.2.7 Further Reading

1 Austin R (2010). *Unmanned Aircraft Systems: UAV Design, Development and Deployment.* John Wiley & Sons, Chichester.
2 Cárdenas EM, Boschetti PJ and Amerio A (2009). Stability and flying qualities of an unmanned airplane using the vortex-lattice method. *Journal of Aircraft.* ***46***(4): 1461–1464.
3 Johnson EN, Chowdhary G, Chandramohan R and Calise AJ (2011). UAV flight control using flow control actuators. *AIAA Flight Mechanics Conference,* Portland Oregon. pp. 1–6.
4 Lungu M (2012). Stabilization and control of a UAV flight attitude angles using the Backstepping method. *World Academy of Science, Engineering and Technology.* ***61***: 290–297.
5 Nelson RC (1998). *Flight stability and Automatic Control.* McGraw Hill, New York.
6 Salih AL, Moghavvemi M, Mohamed HA and Gaeid KS (2010). Flight PID controller design for a UAV quadrotor. *Scientific Research and Essays.* **5**(23): 3660–3667.
7 Uhlig DV and Selig MS (2011). Stability characteristics of micro air vehicles from experimental measurements. *29th AIAA Applied Aerodynamics Conference.* 27–30 June. Honolulu. Hawaii, pp. 1–13.

1.2.8 Problems

Problem 1

Figure 1.12 depicts a low-wing UAV configuration and the contributions of the wing and tail to static stability. The geometry and aerodynamics of the vehicle are shown in Table 1.3. A necessary condition for static stability is that $C_{m\alpha} < 0$, and $C_{m0} > 0$ is required to obtain useful lift and longitudinal trim at $\alpha > 0°$. Calculate the contribution

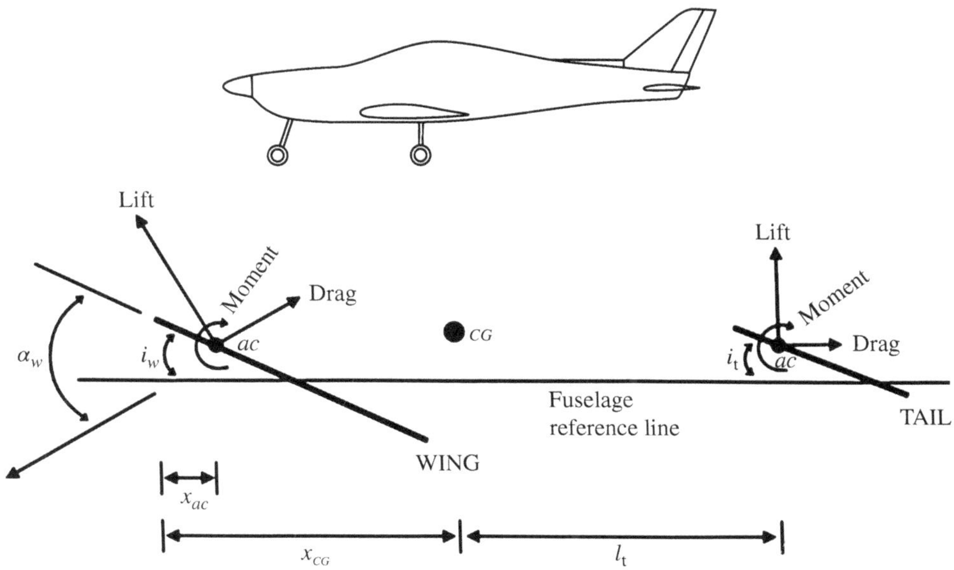

Figure 1.12 Low-wing UAV configuration and stability definitions.

Table 1.3 Vehicle geometry and aerodynamic characteristics.

Aircraft	Geometry	Wing airfoil	Tail airfoil
Mass = 200 kg	$b = 5\,\text{m}$	$i_w = 1.0°$	$i_t = -1°$
$V_{max} = 40\,\text{m/s}$	mean $c = 0.62\,\text{m}$	$c_{l\alpha w} = 0.092/\text{deg}$	$c_{l\alpha t} = 0.015/\text{deg}$
	$AR_w = 8.6$	$c_{macw} = -0.115$	$c_{mact} = 0$
	$S_w = 3.1\,\text{m}^2$	$x_{ac} = 0.25c$	
	$S_t = 0.72\,\text{m}^2$	$\alpha_0 = -3°$	
	$l_t = 2.4\,\text{m}$		

of the *wing* to static stability of the vehicle ($C_{m\alpha w} < 0$); Eqs. (1.4)–(1.7). Is the moment stabilizing or destabilizing?

The notation used is as follows:

Aircraft

C_{m0} = aircraft moment coefficient at $\alpha = 0°$
$C_{m\alpha}$ = aircraft moment curve slope
$V_{\max}$ = aircraft maximum cruise velocity

Wing

AR_w = aspect ratio of the wing planform
b = wing span

$\bar{c}$ = wing mean aerodynamic chord
C_{macw} = wing moment about the aerodynamic centre
c_{macw} = wing airfoil moment about the aerodynamic centre
C_{mcgw} = wing contribution to moment about aircraft's centre of gravity
C_{m0w} = wing moment at $\alpha = 0°$
$C_{m\alpha w}$ = wing moment curve slope
C_{Lw} = wing lift coefficient
C_{L0w} = wing lift coefficient at $\alpha = 0°$
$C_{L\alpha w}$ = wing lift curve slope
$c_{L\alpha w}$ = wing airfoil lift curve slope
$C_{l\alpha w}$ = wing lift curve slope (per radian)
i_w = wing incidence angle (mean chord to fuselage)
S_w = wing surface area
x_{ac} = distance from wing leading edge to aerodynamic centre
α = angle of attack
α_0 = zero-lift angle of attack of wing airfoil
α_{0w} = wing zero-lift angle relative to fuselage reference line

Tail

AR_t = aspect ratio of the tail planform
$C_{L\alpha t}$ = tail lift curve slope
$c_{L\alpha t}$ = tail airfoil lift curve slope
C_{mact} = tail moment about the aerodynamic centre
c_{mact} = tail airfoil moment about the aerodynamic centre
$C_{m\alpha t}$ = tail moment curve slope
C_{m0t} = tail moment at $\alpha = 0°$
i_t = tail incidence angle
l_t = tail moment arm
S_t = horizontal tail surface area
V_H = horizontal tail volume ratio
α = tail angle of attack
ε = downwash incidence angle
ε_0 = downwash angle at tail when wing is at $\alpha = 0°$
η = ratio of tail to wing dynamic pressures (tail efficiency)

Fuselage

$C_{m\alpha f}$ = fuselage moment curve slope
C_{m0f} = fuselage moment at $\alpha = 0°$
$k_2 - k_1$ = correction factor for body fineness ratio
l_f = fuselage length
w_f = average width of fuselage sections
x = coordinate
Δx = length of fuselage increments
ε_u = local induced angle due to upwash or downwash at each fuselage segment

Neutral point

x_{NP} = neutral point
The equations used are as follows:

Wing lift curve slope:

$$C_{L\alpha w} = \frac{C_{l\alpha w}}{1 + C_{l\alpha w} / (\pi AR_w)} \text{ in radians} \tag{1.20}$$

Wing moment coefficient at $\alpha = 0°$:

$$C_{m0w} = C_{macw} + C_{L0w}\left(\frac{x_{CG}}{\bar{c}} - \frac{x_{ac}}{\bar{c}}\right) \tag{1.21}$$

Wing moment curve slope:

$$C_{m\alpha w} = C_{L\alpha w}\left(\frac{x_{cg}}{\bar{c}} - \frac{x_{ac}}{\bar{c}}\right) \tag{1.22}$$

Wing lift coefficient at $\alpha = 0°$:

$$C_{L0w} = C_{L\alpha w}\left|\alpha_0\right| \tag{1.23}$$

Problem 2
The contribution of the tail to the static stability of the vehicle ($C_{m\alpha t} < 0$) is achieved by careful adjustment of V_H and $C_{L\alpha t}$. The $C_{L\alpha t}$ can be adjusted by modifying the tail AR_t. For the UAV in Problem 1, calculate the contribution of the *tail* to static stability using Eqs. (1.24)–(1.29). Is the moment stabilizing or destabilizing?

The following equations are used:

Tail moment coefficient at $\alpha = 0°$ (intercept):

$$C_{m0t} = \eta V_H C_{L\alpha t}\left(\varepsilon_0 + i_w - i_t\right) \quad \eta \text{ assumed to be unity} \tag{1.24}$$

Tail moment curve slope:

$$C_{m\alpha t} = -\eta V_H C_{L\alpha t}\left(1 - \frac{d\varepsilon}{d\alpha}\right) \tag{1.25}$$

Tail volume ratio:

$$V_H = \frac{l_t S_t}{S_w \bar{c}} \tag{1.26}$$

Downwash angle in radians:

$$\varepsilon = \frac{2C_{Lw}}{\pi AR_w} \tag{1.27}$$

Downwash angle at tail when wing is at $\alpha = 0°$:

$$\varepsilon_0 = \frac{2C_{L0w}}{\pi AR_w} \tag{1.28}$$

Downwash slope:

$$\frac{d\varepsilon}{d\alpha} = \frac{2C_{L\alpha w}}{AR_w} \quad C_{L\alpha w} \text{ is in radians} \tag{1.29}$$

Problem 3

For the UAV in Problem 1, calculate the contribution of the *fuselage* to static stability ($C_{m\alpha f}$) using Eqs. (1.30)–(1.34) (Multhopp's method). Is the moment stabilizing or destabilizing?

$$C_{m0f} = \frac{k_2 - k_1}{36.5 S_w \bar{c}} \sum_{x=l_f}^{x=0} w_f^2 \left(\alpha_{0w} + i_f\right) \Delta x \tag{1.30}$$

$$C_{m\alpha f} = \frac{1}{36.5 S_w \bar{c}} \sum_{x=l_f}^{x=0} w_f^2 \frac{\partial \varepsilon_u}{\partial \alpha} \Delta x \quad \left(deg^{-1}\right) \tag{1.31}$$

given the following equations:

Correction factor for body fineness ratio:

$$k_2 - k_1 = 0.84 \tag{1.32}$$

Summation of individual fuselage segments:

$$\sum_{x=l_f}^{x=0} w_f^2 \left(\alpha_{0w} + i_f\right) \Delta x = -400 \tag{1.33}$$

Summation of individual fuselage segments:

$$\sum_{x=l_f}^{x=0} w_f^2 \frac{\partial \varepsilon_u}{\partial \alpha} \Delta x = 35 \tag{1.34}$$

Problem 4

Compute the stick-fixed NP for the vehicle in Problem 1 using Eq. (1.35).

$$\frac{x_{NP}}{\bar{c}} = \frac{x_{ac}}{\bar{c}} - \frac{C_{m\alpha f}}{C_{L\alpha w}} + \eta V_H \frac{C_{L\alpha t}}{C_{L\alpha w}} \left(1 - \frac{d\varepsilon}{d\alpha}\right) \tag{1.35}$$

Part I

Novel Concepts in Unmanned Aircraft Aerodynamics

1.1 Fixed-wing (airplanes)

3

Aerodynamic Performance Analysis of Three Different Unmanned Re-entry Vehicles

Giuseppe Pezzella[1] *and Antonio Viviani*[2]

[1] *Italian Aerospace Research Centre, Capua, Italy*
[2] *University of Campania 'Luigi Vanvitelli', Aversa, Italy*

3.1 Introduction

This chapter deals with the aerodynamic performance analysis of three reusable and unmanned aerial vehicles conceived as flying laboratories to perform experimental flights in low Earth orbit (LEO). Each vehicle concept is an orbital re-entry vehicle (ORV), with re-entry energy of the order of 25 MJ/kg. They are flying test beds (FTBs) that will re-enter the Earth's atmosphere, thus allowing tests of critical re-entry technologies to be performed. The primary objective is to test in real flight conditions various thermal protection systems (TPSs) and hot structures that are potential candidates for next-generation re-entry vehicles. The secondary objective is to provide system-design tests of such re-entry vehicles, to address controlled gliding re-entry and to validate know-how related to in-flight measurement techniques.

Besides these objectives, they will also gather the aerothermal data needed to improve wind tunnel test facilities (including plasma wind tunnel facilities) and computational fluid dynamics (CFD) predictions and transpose them to flight. In particular, the vehicle will provide aerodynamic and aerothermodynamic flight data to correlate with ground testing results, such the 'Scirocco' plasma wind tunnel at CIRA[1]), thus providing new insights into the complex aerothermodynamic phenomena and improving prediction methodologies and their extrapolation to flight capabilities.

Based on experience in experimental vehicles, a progressive flight demonstration approach is preferable, since this limits the risks, allows progressive investment efforts, and ensures that more challenging developments benefit from results and findings obtained through a systematic approach.

So far, Europe has undertaken the development of three very different FTBs: namely ARD (Atmospheric Re-entry Demonstrator), EXPERT (European Experimental Reentry Testbed) and IXV (Intermediate Experimental Vehicle).

1 Centro Italiano Ricerche Aerospaziali, the Italian Aerospace Research Centre.

Advanced UAV Aerodynamics, Flight Stability and Control: Novel Concepts, Theory and Applications,
First Edition. Edited by Pascual Marqués and Andrea Da Ronch.

Companion website: http://www.wiley.com/go/marques/advanced_UAV_aerodynamics

ARD was a scaled-down version of an Apollo capsule. It was launched by ARIANE 5 V503 on October 21, 1998 [1]. After a successful sub-orbital and re-entry flight, it was recovered in the Pacific Ocean. ARD allowed an assessment of the aerodynamics of this kind of capsule, which is still a very attractive design solution for manned high-energy re-entry, such as return from Mars or Moon missions [2].

EXPERT, which has not yet flown, is a small sphere-cone FTB designed for in-flight testing of advanced TPSs, wall catalyticity, flow transition assessments, and so on [3].

Finally, the Intermediate Experimental Vehicle is shown in Figure 3.1. It is a rather blunt FTB that features a lifting-body configuration [4]. It was tested in re-entry flight conditions on 11 February 2015 at the end of a sub-orbital flight that had an energy level very close to that of an orbital re-entry [5].

IXV performed several in-flight experiments, such as the guidance, navigation and control (GNC) of a flapped lifting-body aeroshape and TPS catalyticity. A description of the aerodynamic and aerothermodynamic characterization of IXV can be found in the literature [4, 5].

The FTBs developed so far are capsules and lifting-body vehicles only. Therefore, in a step-by-step approach with increasing levels of complexity, only winged-body configurations have so far been explored, as discussed in this chapter.

A reusable ORV operates in different flight regimes from subsonic to hypersonic speeds. A typical mission profile includes:

- ascent phase, where the spacecraft is attached to a launch vehicle and flown to altitude;
- orbit phase, where the vehicle orbits in space until completion of its mission;
- descent phase, where the ORV re-enters the atmosphere and lands like a conventional airplane for subsequent use.

During the descent the spacecraft encounters several flowfield regimes (rarefied, transitional, and continuum flow) and speeds of flow (hypersonic, supersonic, transonic, and subsonic). Therefore, the choice of vehicle aeroshape and its aerodynamic

Figure 3.1 The IXV.

characterization in different flight conditions is fundamental for its safe return and mission success. Usually, the vehicle configuration is continuously adapted throughout the design phase by means of a trade-off study (a 'multidisciplinary design optimization') involving several competing ideas for meeting the mission requirements under the design constraints. For example, the vehicle concept will have to be carried to low-Earth orbit using a small launcher. It must then re-enter the atmosphere, allowing a number of experiments on critical re-entry technologies to be performed, before descending from a hypersonic regime down to landing. Figure 3.2 shows some wing–body configurations that might meet these requirements [6].

The configurations differ in terms of features such as planform shape, cross section, nose camber, wing-swept angle and vertical empennages. Of course, the winning configuration from the aerothermal point of view is the one with simultaneously the best aerodynamic and aerothermodynamic performance. The current most promising vehicle configurations according to the trade-off analyses, are shown in Figures 3.3–3.6. The different shapes of the side views are shown in Figure 3.3, ranging from blunt to sharp.

Figure 3.4 shows the sharp vehicle configuration, named ORV-WSB. It is also shown docked with a service module with its solar panels deployed [6, 7].

The blunt configuration, ORV-WBB, is shown in Figure 3.5 [7].

Finally, Figure 3.6 displays the spatular-body (SB) configuration, ORV-SB [7].

This shape is most attractive, being the only viable way to integrate scramjet propulsion (see Box 3.1) in the vehicle's aerodynamic configuration (see Figure 3.7, where the ORV-SB features a scramjet engine on the belly side), thus evolving towards a waverider aeroshape [8].

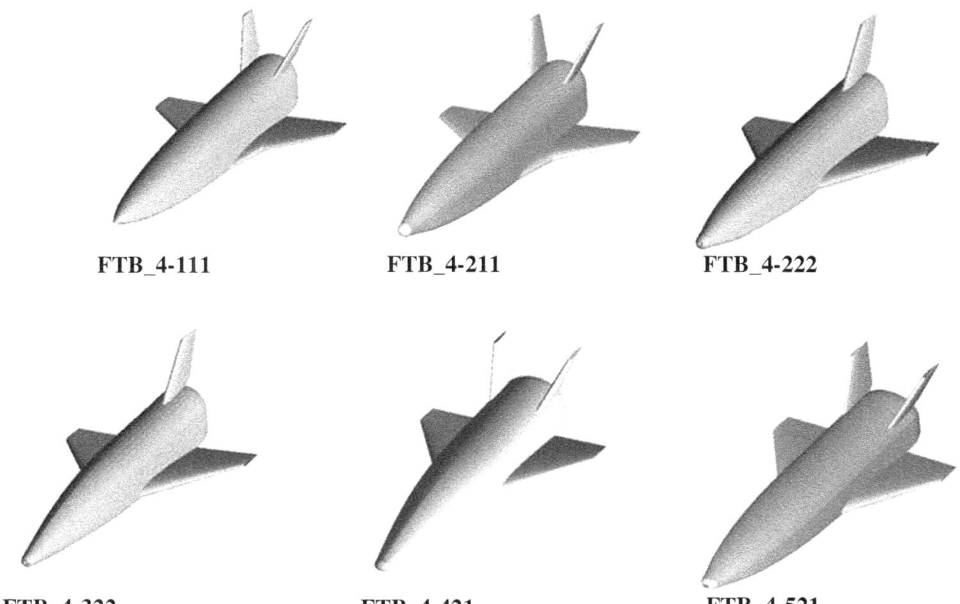

Figure 3.2 FTB trade-off configurations.

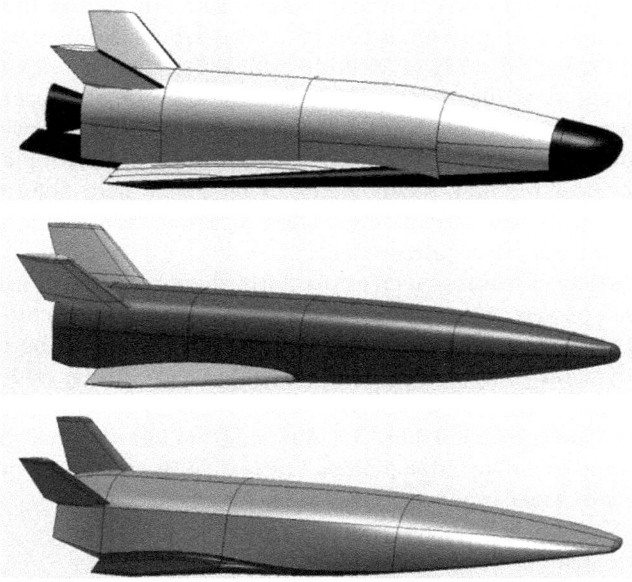

Figure 3.3 ORV-WSB body configurations: top, rather blunt; middle, sharp; bottom, spatulared. (*See insert for color representation of this figure.*)

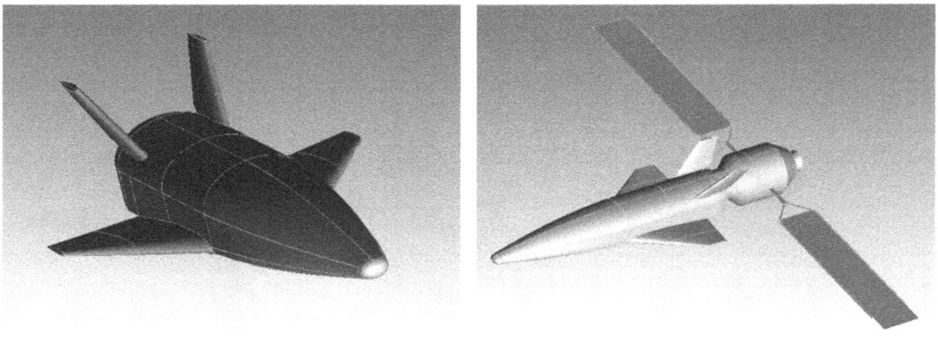

Figure 3.4 ORV-WSB configuration in flight and docked to a service module with solar panels.

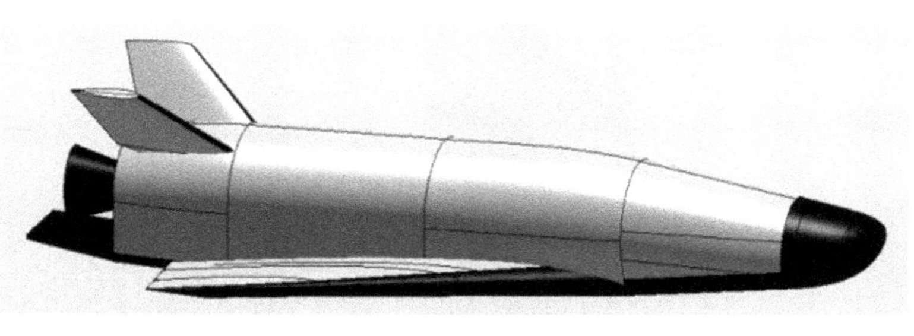

Figure 3.5 Rather blunt vehicle configuration.

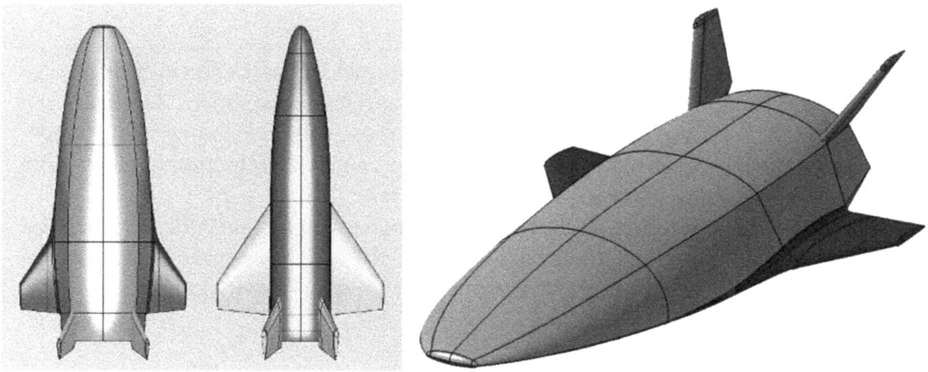

Figure 3.6 The spatular body configuration.

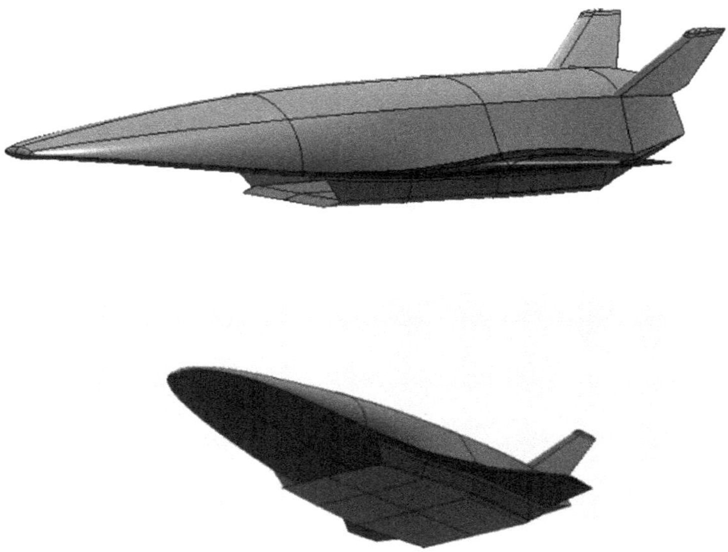

Figure 3.7 The spatular body configuration with scramjet.

So, all ORV concepts belong to the wing–body vehicle class. Such configurations, however, differ in terms of nose camber, planform shape, cross section, wing-swept angle and vertical empennages. These differences in aeroshape can be seen in Figure 3.11, where each configuration is superimposed.

These aeroshapes are conceived in such a way as to allow the ORV to re-enter the Earth's atmosphere with an aerodynamic efficiency that is better than the Space Shuttle, which had a maximum L/D ratio of about 1. The ORV will fly a re-entry trajectory with

Box 3.1 Scramjets.

Since the beginning of aviation, the trend in aircraft design has been towards greater speed. The next frontier of speed envelope is travel at hypersonic speeds [8]. One of the most practical and efficient approaches to reaching these speeds is known as the waverider. Figure 3.8 shows that high-performance flight vehicle architecture is converging with the technology of airbreathing configuration.

The waverider configuration provides high aerodynamic efficiency [8]. For instance, an infinitely thin flat plate is theoretically the most efficient hypersonic lifting surface. Its lift-to-drag ratio (L/D) is the highest that can be achieved at hypersonic speeds:

$$L/D = \cot \alpha$$

where α is the angle of attack (AoA). A flat plate, however, is obviously not practical, especially since it cannot contain any volume for payload, engines, fuel, and so on. Because of this, a more realistic design converges to a spatular vehicle architecture. The characteristics of this design are: a very small frontal area and highly streamlined shape to minimize total surface area, very little wing area but a fuselage shaped to generate additional lift; and a propulsion assembly that is highly integrated into the vehicle.

The waverider configuration is also dictated by the needs of the scramjet engine. In order to assess whether a ramjet engine can work at very high speeds, namely hypersonic Mach numbers, consider the schematic of a conventional subsonic-combustion ramjet engine in Figure 3.9 [9]. Let us suppose that the vehicle is flying at Mach 8 at a standard altitude of 30 km, where the air pressure and temperature are 1197 Pa and 226.51K, respectively.

A ramjet engine is an airbreathing propulsion device with essentially no rotating machinery. The flow, moving from left to right, enters the inlet, where it is compressed

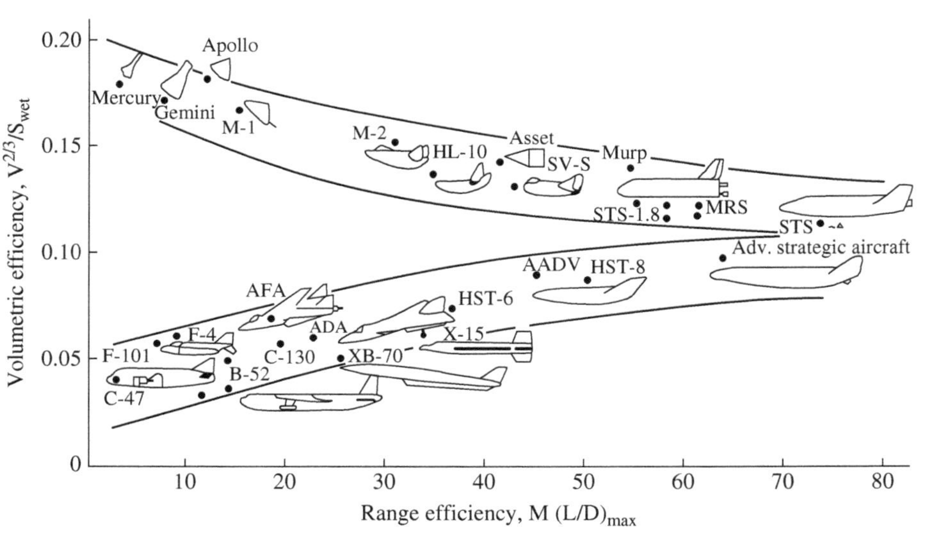

Figure 3.8 Space and atmospheric vehicle development, with the technology of high-performance vehicles converging with that of airbreathing aircraft. M, Mach number [8].

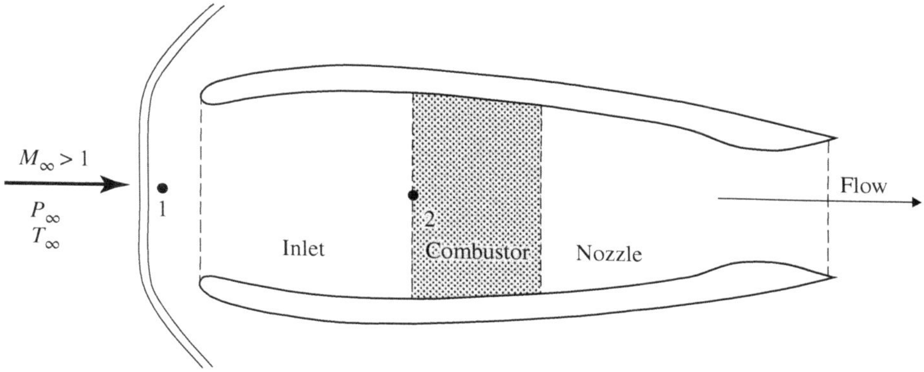

Figure 3.9 Schematic of a conventional subsonic-combustion ramjet engine.

and slowed down. The compressed air then enters the combustor at a very low subsonic speed, where it is mixed and burned with the fuel. The hot gas then expands through a nozzle, thus producing thrust. When traveling at supersonic speed, a local normal shock wave detaches in front of the ramjet. So, after passing through the shock wave, the flow from point 1 to 2, located at the entrance of the combustor, is isentropic.

Let us calculate the air temperature and pressure at point 2 when the Mach at that point is 0.2. The total pressure and total temperature of the freestream at $M_\infty = 8$ reads (isentropic flow properties):

$$\left(\frac{P_{0,\infty}}{P_\infty}\right)P_\infty = \left(0.9763\times10^4\right)\left(1.1970\times10^3\right) = 1.1690\times10^7 \text{ Pa}$$

$$\left(\frac{T_{0,\infty}}{T_\infty}\right)T_\infty = \left(0.138\times10^2\right)\left(226.510\right) = 31.260\times10^2 \text{ K}$$

At point 1 behind the normal shock, the total pressure is (normal shock properties for $M_\infty = 8$):

$$P_{0,1} = \left(\frac{P_{0,1}}{P_{0,\infty}}\right)P_{0,\infty} = \left(0.8488\times10^{-2}\right)\left(1.1690\times10^7\right) = 0.9922\times10^5 \text{ Pa}$$

The total temperature is constant across the shock, hence

$$T_{0,1} = T_{0,\infty} = 31.260\times10^2 \text{ K}$$

The flow is isentropic between points 1 and 2, hence P_0 and T_0 are constant between these points. Therefore,

$$P_{0,2} = 0.9922\times10^5 \text{ Pa}$$

$$T_{0,2} = 31.260\times10^2 \text{ K}$$

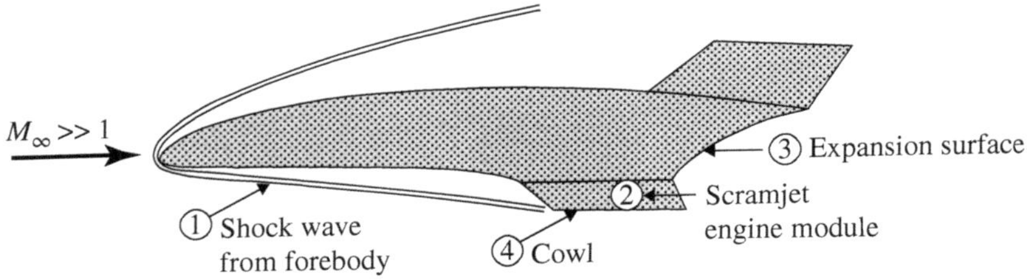

Figure 3.10 Schematic of a conventional scramjet-propelled aeroshape.

At point 2, where $M_2 = 0.2$, the ratios of the total-to-static pressure and total-to-static temperature (isentropic flow properties) are $P_{0,2}/P_2 = 1.028$ and $T_{0,2}/T_2 = 1.008$. Hence,

$$P_2 = \left(\frac{P_2}{P_{0,2}}\right) P_{0,2} = \frac{0.9922\times10^5}{1.028} = 0.9652\times10^5\ Pa = 0.95\ \text{atm}$$

$$T_2 = \left(\frac{T_2}{T_{0,2}}\right) T_{0,2} = \frac{31.260\times10^2}{1.008} = 31.012\times10^2\ \text{K}$$

These are extremely severe conditions. The temperature is so hot that the fuel injected into the combustor will decompose rather than burn, with little or no thrust being produced. Moreover, the structural design of the combustor would have to be extremely heavy, assuming in the first place that some special heat-resistant material could be found that could handle the high temperature. As a result, a conventional ramjet, where the flow is slowed down to a subsonic Mach number before entering the combustor, will not work at hypersonic Mach numbers.

The solution to this problem is not to slow the flow inside the engine to low subsonic speeds, but rather to slow it only to a lower but still supersonic speed [9]. In this way, the temperature and pressure increase inside the engine will be smaller and can be made tolerable. In such a ramjet, the entire flowpath through the engine remains supersonic, including inside the combustor. This type of ramjet is called a supersonic combustor ramjet or *scramjet* for short.

The concept requires modification of the intake to give an oblique shock wave instead of a strong normal shock (see Figures 3.9 and 3.10) [9]. The entire bottom surface of the vehicle is in essence an integrated portion of the scramjet engine.

a long-endurance leg in the upper atmosphere, using 'low' (with respect to the Shuttle guidance approach) and modulated AoA guidance. This brings a number of benefits, such as:

- overcoming the critical heat flux region in a quasi-thermal equilibrium condition, ensuring that the internal temperature remains less than the peak temperature that a steep re-entry can produce
- exposing small parts of the vehicle to a high heat flux (these are protected with an advanced TPS such as ultra-high-temperature ceramics), but leaving most of it exposed to a lower heat flux, allowing the use of simpler, cheaper, and lighter materials.

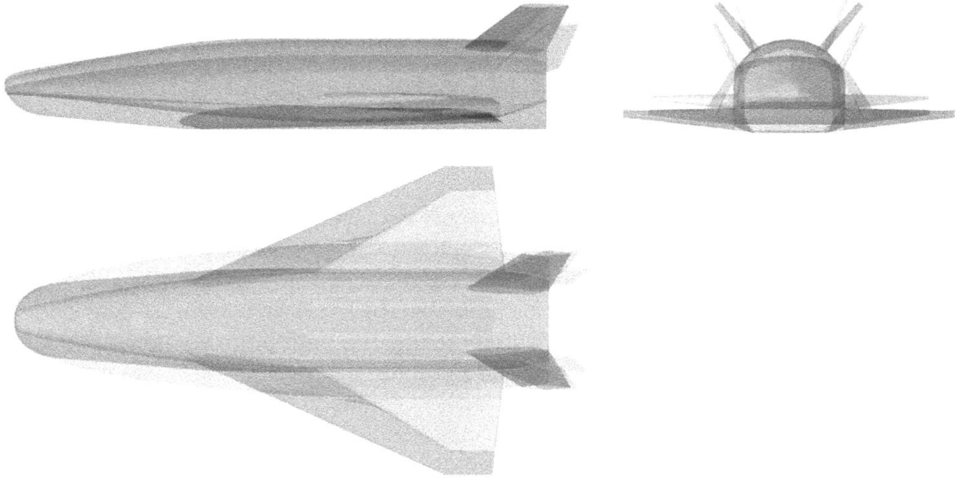

Figure 3.11 ORV aeroshapes comparison.

The aerodynamic performance of these ORVs, from the hypersonic down to subsonic regime, can be assessed. With this in mind, both low-order methods (hypersonic panel methods) and CFD design analysis have been considered in accordance with a phase-A design level.

The work is currently at an early design stage, so low-order methods are extensively used. CFD simulations are performed only to address the reliability of the low-order method design results and to investigate complex flowfield phenomena that are not predictable with simplified tools [10, 11, 12].

Both perfect gas and thermochemical non-equilibrium CFD simulations are performed at several points of the flight scenario in the range between Mach 0.3 and Mach 25. In the reacting gas computation, the air is modelled as a mixture of five species (O_2, N_2, NO, N, and O). An analysis of the longitudinal- and lateral-directional stability was also provided for one concept, together with the main interesting features of the flowfield past the vehicles at Mach numbers where real-gas effects occur. In fact, it is well known that the pitching moment can be highly modified by high-temperature real gas effects, thus affecting vehicle's stability and trimming conditions [13].

Box 3.2 The aerospace vehicle designer's perspective.

From the perspective of aerospace vehicle designers, the emphasis is on aerodynamic performance, stability, control, vehicle surface pressure, and shear and heating loads [14]. Fluid dynamics and physical processes are closely coupled to aerodynamic forces, moments, and aeroheating, and provide insights into the phenomena observed at the surface. Study can give an improved understanding of the flowfield about the vehicle. Therefore, the aerodynamic/aerothermodynamic challenge is to provide the optimum shape (the outer mold lines; OML) to safely satisfy mission requirements, including abort scenarios, and to reduce design conservatism, risk and cost. In other word, designers are seeking the optimum flying vehicle with the minimum structural (surface pressure/shear) and heating loads, which translates into reduced weight and reduced operational costs. As a result, aerothermodynamics provides crucial information to other key areas involved in the design, such as choice of structures, materials (including TPS), and the avionics, guidance, navigation, control and propulsion systems.

The vehicle aerothermodynamic design process is highly iterative, as shown in Figure 3.12 for the X-33 vehicle concept [14].

Aerothermodynamicists and systems analysis engineers responsible for the initial vehicle concept and for generating flight trajectories should work closely together from the beginning to the end of the program. Experienced aerothermodynamicists will often identify deficiencies in initial concepts, thereby allowing system analysis engineers to adjust the concept prior to its entering the aerothermodynamic process (Figure 3.12), specifically the screening and optimization. Such an exchange can save considerable time and resources. Systems analysis engineers should be involved in the development of wind tunnel test matrices and decisions relating to the output of data reduction procedures. Personnel working on guidance, navigation and control should also be brought into the aerothermodynamic process early, as they are the ultimate customer for aerodynamic measurements and predictions. Likewise, designers of the TPS, who are customers for the aeroheating measurements and predictions, should also be included [15].

Initial concepts of advanced aerospace vehicles are generally developed via systems analysis studies in which the various components (propulsion system, tankage, payload, and so on) are sized, weighed and packaged. Mission requirements, as expected, have a

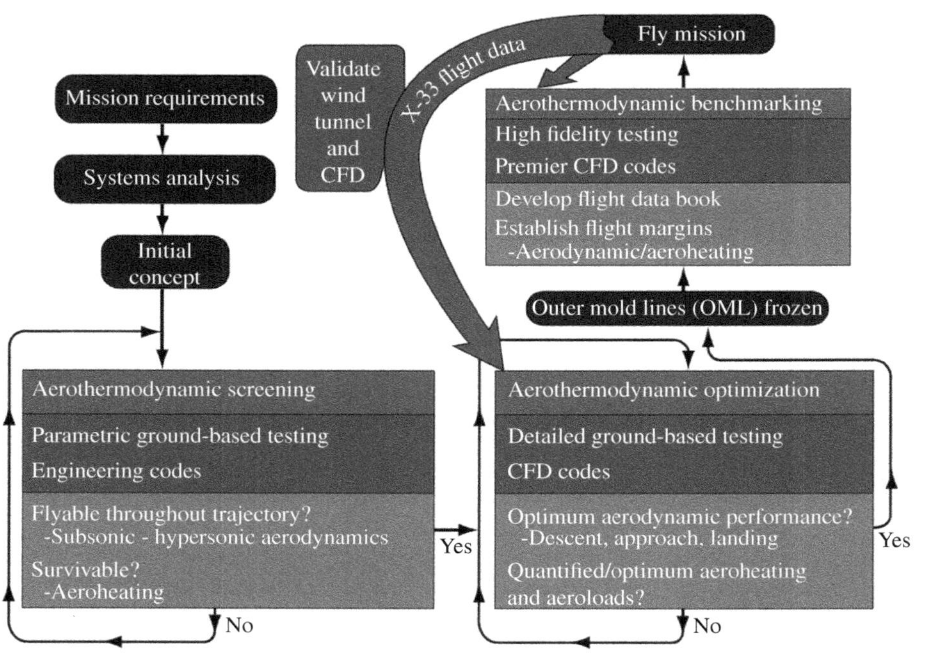

Figure 3.12 Aerothermodynamic design process.

major impact on the shape of the proposed vehicle. Aerodynamic performance characteristics for the vehicle concept are generally approximated using relatively simple engineering codes. Having generated an initial concept that might satisfy the mission requirements, the aeroshape (the OML) is provided to the aerothermodynamic community for screening/assessment of aerodynamic performance and aeroheating characteristics.

Aerothermodynamic screening begins with tests in wind tunnels using models with removable/replaceable components and control surfaces with different deflections. Since the purpose of the first series of tests is to provide a quick look at the aerothermodynamic characteristics, the model materials and fidelity are selected to expedite the fabrication process and to minimize the costs. The first aerodynamic tests performed are generally for subsonic conditions, to assess the approach and landing characteristics, and for hypersonic conditions, to assess trim and lift-to-drag characteristics. Forces and moments are measured for various stages of configuration buildup:

- fuselage only
- fuselage with wing/fin
- canards
- vertical tail
- engine module.

They are also measured over a range of attitudes (AoAs and sideslip) and control-surface deflections. If subsonic aerodynamic characteristics are unacceptable, an iterative process is initiated whereby the experimentalists, often in concert with the systems analysts, modify the OML of the concept. The concept is modified and retested until acceptable subsonic aerodynamic characteristics are achieved. These tests are generally performed in relatively small, low-cost tunnels. Hypersonic testing is generally performed in parallel with the subsonic testing thereby necessitating close communication between the two tests. Forces and moments are measured over a range of Reynolds number, attitudes and control-surface deflections.

The effects of configuration buildup are usually examined at hypersonic conditions, as are support interference effects. Thermal mappings to identify regions of high heating on the concept are measured using global thermography techniques. If required, and in concert with findings from subsonic testing, the vehicle OML is modified to achieve acceptable hypersonic aerodynamic performance and aeroheating characteristics. If these changes are extensive, an iterative process is initiated via additional tests performed at subsonic conditions to determine if approach and landing characteristics have been compromised. These tests represent the first phase of the aerothermodynamic process: the screening phase, in which wind tunnel results are generally complemented with predictions from relatively simple engineering codes.

Having refined the aeroshape to provide acceptable aerodynamic performance at subsonic and hypersonic conditions ('bounding the problem'), the aerothermodynamic optimization phase of the process is initiated. In this phase, the subsonic and hypersonic aerodynamic database is enriched via more detailed testing in these speed regimes. Most importantly, aerodynamic tests at transonic and supersonic conditions are performed. Depending primarily on the shape and the location of the centre of gravity (CoG) of the vehicle concept, the previously discussed approach – achieving acceptable aerodynamic performance at subsonic and hypersonic conditions and hoping that nothing bad happens in-between – may not be successful. Often, so-called 'show stoppers' are revealed via testing at transonic conditions, necessitating modifications, sometimes extensive, of the aerolined CoG location and retesting at subsonic and hypersonic conditions. Another difference between this and the screening phase is that quantitative values of aeroheating are measured in hypersonic wind tunnels and predicted via CFD codes. The measured aeroheating levels are extrapolated to flight and complemented with flight predictions via CFD to initiate TPS material selection (driven by peak heating), determination of the split line between different TPS materials (based on worst-case heating distributions), and sizing or thickness determination (based on total heat load). CFD is also applied tip-to-tail to predict aerodynamic performance, complementing the database generated via wind tunnel testing. Testing is continued in this optimization phase until closure is achieved on an aeroshape that provides:

- the optimum aerodynamic performance across the subsonic-to-hypersonic speed regime
- acceptable pressure and heating loads during the entire flight trajectory
- volumetric efficiency for effective packaging of the required components.

Once these criteria have been satisfied and, most importantly, the vehicle OML is frozen, benchmark testing (the third phase of process) is initiated in well-calibrated, high-flow quality wind tunnels. Depending on the risks associated with the program, this is done in the best available subsonic/transonic/supersonic/hypersonic tunnels. Emphasis is placed on the simulation of flight values of the Reynolds number throughout the trajectory. This is especially important at transonic-low supersonic conditions for many classes of aerospace vehicles. Some redundancy is desirable: testing in different facilities but at the same or similar flow conditions. High-precision, calibrated, quality assurance certified, highly instrumented models are used.

Model attitude is accurately set/maintained and monitored during a tunnel run, and support-system effects are quantified. In this final phase, steps are taken to provide highly accurate measurements of

- aerodynamic forces and moments
- detailed surface pressures (to extract loads)
- detailed heat transfer measurements.

The emphasis is on the accuracy and reliability of the data, as these will be used for final vehicle design, development of the aerodynamic flight data book and trajectory tailoring to ensure aeroheating levels in flight do not exceed TPS design limits. Tests are performed during this phase to determine the performance of the reaction control system (RCS), to simulate multibody aerodynamic separation if applicable, to examine aerodynamic performance and aeroheating characteristics during abort manoeuvres, and so on, to complete the flight data book. Aerodynamicists work closely with system analysts, using six-degree-of-freedom codes to provide inputs and margins for guidance, navigation and control. It is during this benchmarking phase of ground-based testing that predictions of aerothermodynamic characteristics obtained via CFD computer codes (high level, Navier–Stokes solvers) provide significant contributions. Having been calibrated against wind tunnel data, CFD codes are applied for flight conditions that often cannot be accurately simulated in wind tunnels.

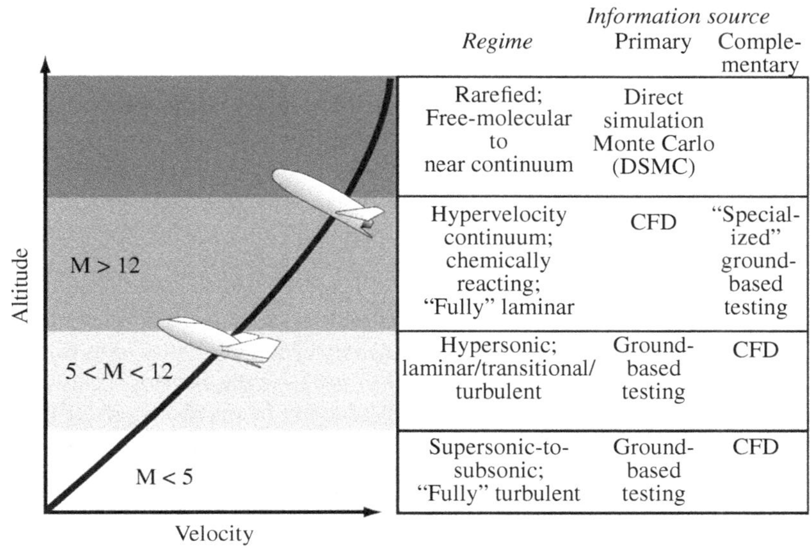

Figure 3.13 Aerothermodynamic design methodology.

The last phase of the aerodynamic/aerothermodynamic process is the extraction of aerothermodynamic data from the vehicle in flight and the comparison of wind tunnel simulations and CFD predictions to flight data. This comparison of predictions and flight results completes the aerothermodynamic triad of ground-based facilities, computer codes and flight experiments.

From the perspective of the aerospace vehicle designer, the approximate balance of experimental aerothermodynamic information required is 75–80% aerodynamic performance and 20–25% aeroheating, with 'trace' amounts relating to aerodynamic loads and fluid dynamic phenomena. For this reason, the aerothermodynamic methodology shown in Figure 3.13 is heavily weighted towards aerodynamic performance. To illustrate the methodology, a hypothetical trajectory is shown in Figure 3.13 for an aerospace vehicle descending into the Earth's atmosphere from LEO [14, 15].

The vehicle will encounter the classical hypersonic ($M_\infty > 5$), supersonic ($1.2 < M_\infty < 5$), transonic ($0.8 < M_\infty < 1.2$) and subsonic ($M_\infty < 0.8$) flow regimes. Within the hypersonic regime, three subregimes are shown in Figure 3.13.

At the highest altitudes, the *rarefied flow regime* extends from free-molecular behaviour to near-continuum behaviour as defined by the Knudsen number. At still relatively high altitudes, there is the *hypervelocity continuum* regime, characterized by laminar flow everywhere within the shock layer surrounding the vehicle, and chemically reacting in portions or throughout the flowfield. At still lower altitudes, the hypersonic continuum regime is characterized by vibrational excitation within the shock layer, but no (or very low levels of) dissociation, and by laminar/transitional/turbulent boundary layers and shear layers.

Aerothermodynamic information for aerospace vehicle concepts in the rarefied flow regime is provided by the direct simulation Monte Carlo (DSMC) method or by simple bridging relations between the free-molecular and continuum regimes. DSMC capabilities have advanced significantly over the last decade and DSMC is widely used because of its many successful predictions of aerothermodynamic performance for a wide spectrum of configurations for flight around the Earth and Mars.

There are several sources for aerothermodynamic information in the hypervelocity continuum regime. One is high-enthalpy ground-based facilities, such as

- piston-driven/combustion-driven shock tunnel
- expansion tube
- arc-driven *hotshot* tunnel
- ballistic, free-flight range.

The advantages and disadvantages of these specialized facilities are well recognized. A major disadvantage for the measurement of aerodynamic forces and moments on relevant aerospace vehicle shapes is the very short test times. Typically, but not exclusively, test times at these facilities ranges from a few milliseconds to microseconds at high enthalpy levels. Although short run times may prove advantageous for aeroheating measurements (they minimise conduction effects in transient techniques), this is not the case for force and moment measurements.

A second source of information is the simulation of certain aspects of a real gas via testing with a heavy gas (such as tetrafluoromethane, CF_4) at relatively low enthalpy levels. The use of a heavy gas in a conventional blowdown-to-vacuum hypersonic wind tunnel provides all the advantages of a conventional, as opposed to impulse, facility while simulating the high normal shock-density ratio and/or low values of specific heat ratios within the shock layer experienced in flight due to real-gas effects.

Testing with CF_4, which is three times heavier than air, provides higher values of density ratio than hypersonic air tunnels and lower values of the ratio of specific heats (gamma) within the shock layer about the model. These values are representative of those encountered by moderately-to-very blunt vehicles in hypervelocity flight, for which the gas within the shock layer of the vehicle is dissociated. For example, values of gamma over the windward portion of the orbiter flowfield at Mach 20 in flight were simulated via CF_4 and this simulation revealed that the pitch-up anomaly experienced on the maiden flight of the orbiter was due to real-gas effects; in other words, real-gas effects resulted in a greater expansion of the flow over the aft portion of the orbiter. This more rapid expansion decreased the surface pressure, as compared to levels inferred from tests in ideal-air wind tunnels and used as inputs to the original flight aerodynamic data book, and thereby increased the pitching moment (more nose up). Ground-based simulation of high-temperature real-gas effects on aerodynamic performance is important for all aerospace vehicles flying at Mach numbers above 12 or so, and for planetary probes [14, 15].

3.2 Vehicle Description

The vehicle concepts considered here feature a compact wing-body configuration equipped with a rounded-edge delta-like fuselage cross section, a delta wing, and a V-tail. The vehicle architecture shows a blended wing body (BWB) interface and a flat-bottomed surface to increase its overall hypersonic performance. The fuselage was designed to be longitudinally tapered, in order to improve aerodynamics and lateral-directional stability, and with a cross section large enough to accommodate all the vehicle subsystems such as, for example, the large propellant tanks for long missions.

This has a large impact on vehicle performance. From the aerodynamic point of view, the lift and the aerodynamic efficiency are mainly determined by the fuselage fineness and by the shape of the vehicle cross section [16]. The forebody is characterized by a rather simple cone–sphere geometry with smooth streamlined surfaces on the upper and lower side of fuselage, and by the nose drop-down configuration, typical of winged hypersonic vehicles. The nose camber is low enough to reduce the elevon size in order to provide the desired trim range with elevon deflections, and also to improve internal packaging of the vehicle's subsystems. The wing size and location were defined using trade-off studies so as to improve vehicle aerodynamics and to provide static stability and controllability during flight [10, 17].

The wing is swept back to optimise performance with respect to supersonic drag and aerodynamic heating. A properly designed strake could be added in the future, depending on the specific landing requirements. A wing dihedral angle of 5° is also provided to enhance lateral-directional stability. The wing also features a high length-to-width ratio to minimise drag, a section shape that is maintained from root to wing tip, and a nearly flat-bottomed surface to efficiently dissipate aeroheating. The vertical tails' sweep angle is 45°. Control power for the vehicle is provided by two wing-mounted elevon surfaces (which must serve as ailerons and elevators), and rudders. Used symmetrically, the elevons are the primary controls for pitch motion. Roll control is obtained through asymmetrical usage of the elevons. Rudders help to provide directional control, while vertical tails allow for sideslip stability. During entry, when the vehicle is flying at high AoAs, the rudders should be augmented by the RCS. For example, Figure 3.14 shows the ORV-WSB concept with RCS pods on the base.

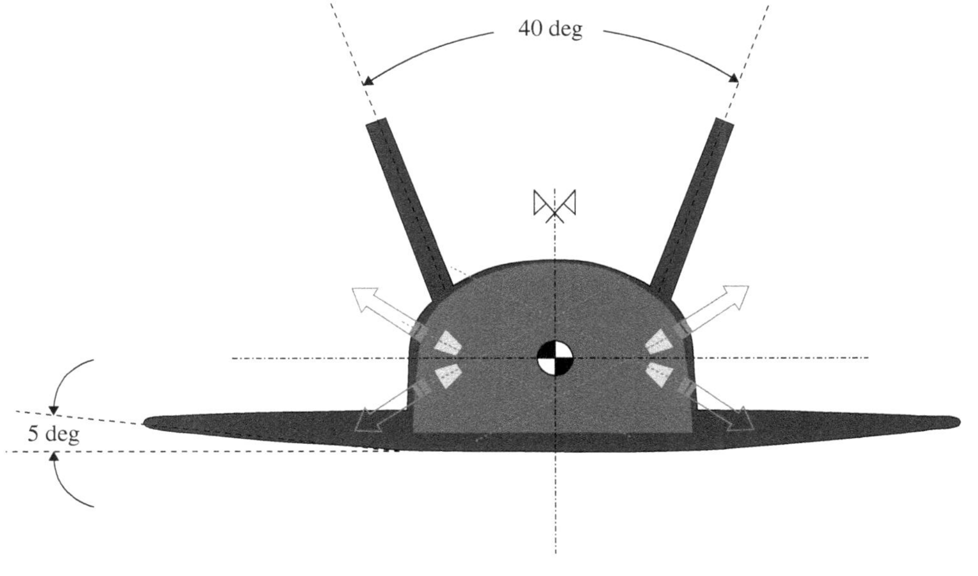

Figure 3.14 ORV-WSB dihedral angles with RCS pods on base.

Arrows indicate the thrust provided by each pod. Therefore, by appropriately combining the thrust vectors it is possible to control pitch, roll and yaw.

The vehicle may be provided by a body flap located at the trailing edge of the fuselage in order to augment pitch control and to shield the nozzle of the propulsion subsystem (see the image of ORV-WBB in Figure 3.5). Trim capability to relieve elevon loads is obtained by body-flap deflection. At hypersonic speeds, a surface behind the vehicle CG balances the nose-up pitching moment typical of this configuration at hypersonic speeds.

Finally, the aerodynamic control surfaces are designed to be large enough to provide stability without sacrificing too much lift.

3.3 Flight Scenario and Flow-regime Assessment

It is planned that the ORV should perform a complete re-entry flight from LEO orbit at 200 km. The return trajectory is based on a different strategy, from the guidance point of view. The vehicle will be held as long as possible in the highest level of the atmosphere that is compatible with the thermal constraints of the descent.

It has better gliding re-entry and a high-manoeuvring capabilities than the reference re-entry of the Space Shuttle, characterized by moderate AoAs (up to 20°). Longer flight durations are possible, allowing for more extended in-flight testing capabilities in high-energy hypersonic flight conditions.

In the framework of flight mechanics trade-off analyses, several re-entry trajectories have been computed, thus defining the vehicle flight envelope needed to address vehicle aerodynamics. For example, Figure 3.15 and Figure 3.16 show a number of re-entry trajectories evaluated supposing that the vehicle heat flux constraint ranges between 1.1 and 2 MW/m^2 and that the dynamic pressure limit is equal to 12 kPa. The Mach–Reynolds numbers grid in the altitude–velocity map of Figure 3.16 characterizes the vehicle aerodynamics according to the space-based design approach [18].

The time at which re-entry is ended – close to about 5000 s – is larger than for a conventional re-entry in, for example, the Space Shuttle. Note that for each trajectory shown in Figure 3.15 and Figure 3.16 the guidance technique is the same, but the AoA profiles are different (see Figure 3.15). Typically, the more common approach to the definition of the guidance strategy makes use of a predefined AoA profile, characterized by high values for the most critical re-entry phase, during which the heat flux is critical for vehicle safety. The bank angle control is usually applied in such a way that the flight path can follow a predefined drag–velocity profile and create the required cross-range.

An alternative approach has been adopted here, in order to be compliant with the specific ORV project requirements. This involves modulating the AoA profile during the re-entry critical phases. The AoA will be set as close as possible to the one corresponding to the maximum efficiency (at least in the 'dense' atmosphere range) that is compatible with the maximum heat flux the vehicle (TPS and cold structure) can sustain. This kind of AoA guidance profile should allow for greater vehicle manoeuvrability along the trajectory and will expose a smaller part of the vehicle to the higher heat flux, thus allowing for a TPS mass reduction and increasing the mission's safety level. For instance, the optimized long-endurance trajectories are characterized (for a relevant part of the trajectory) by a quasi-steady thermal equilibrium condition near the

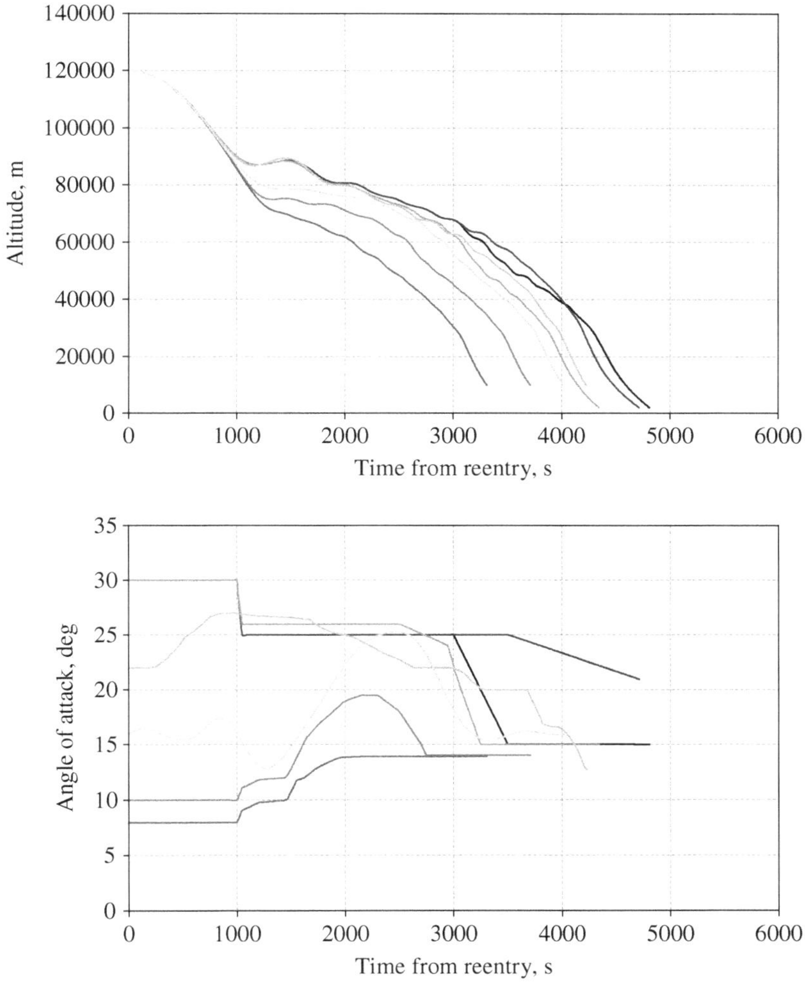

Figure 3.15 ORV re-entry: top, time history of light envelope; bottom, AoA profiles.

maximum admissible heat flux value at the stagnation point (see Figure 3.16, bottom), thus taking advantage of radiative cooling.

As far as flight-regime assessment is concerned, it is worth noting that along with the atmospheric re-entry trajectory, going from upper to lower altitudes, the vehicle experiences three main flow regimes. They are:

- the hypersonic–supersonic regime, in the upper part of the descent
- the transonic and subsonic regimes, in lower atmospheric layers.

At hypersonic-speed flow, three main (and heavily overlapping) regimes apply. They are:

- the rarefied-transitional regime, on the one hand, and
- the viscous-interaction regime and
- the real-gas regime, on the other hand, which lie in the continuum flow regime.

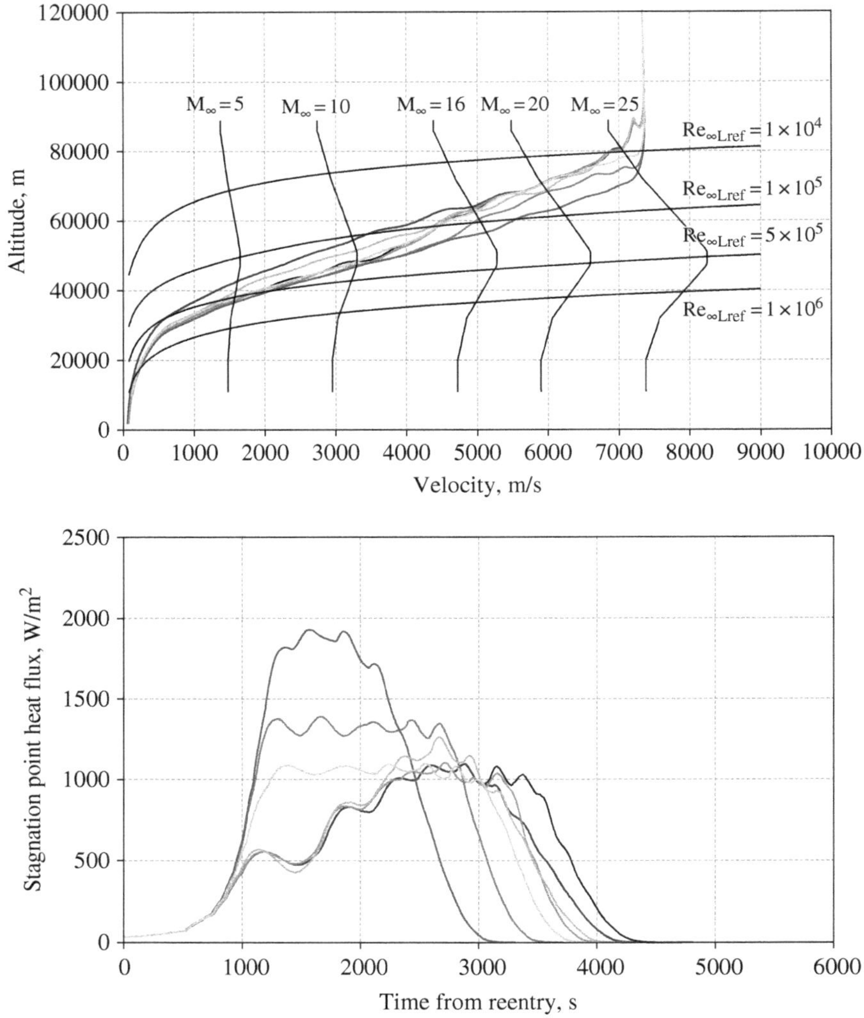

Figure 3.16 ORV re-entry: top, velocity–altitude map; bottom, time history of stagnation point heat flux profiles.

Consequently, knowing where in the re-entry trajectory each of these flight regimes is established is extremely important in order to address vehicle performance. For example, it is well-known that real-gas and viscous effects play a significant role in the vehicle's aerodynamics (drag rise and trim conditions) and aerothermodynamics. This means that vehicle design demands different

- flow models, ranging from a perfect gas to a reacting gas mixture
- flowfield solution approaches, ranging from DSMC to Navier–Stokes with and without slip flow conditions

to provide a full description of key flow phenomena affecting the vehicle's aerodynamic and aerothermodynamic characteristics.

3.4 Rarefied and Transitional Regimes

Once the vehicle has started its descent, the atmospheric density is low enough that the continuum assumption breaks down. Since individual molecular collisions are important, one must consider the general microscopic mass, force and energy-transfer problem at the vehicle surface. In these conditions, there are two distinct subregimes: the re-entry through the upper part of the atmosphere and that through the lower part of the high atmosphere. In the former, the free molecular flow (FMF) regime is completely established, while in the latter the transitional flow regime (TFR) applies. Of course, the limit between these two flow conditions depends on altitude and vehicle dimensions. For instance, the similarity parameter that governs these different flow regimes is the Knudsen number (rarefaction parameter), defined as:

$$\mathrm{Kn}_{\infty L_{ref}} = \frac{\lambda}{L_{ref}} = 1.25\sqrt{\gamma}\,\frac{M_\infty}{\mathrm{Re}_{\infty L_{ref}}} \tag{3.1}$$

where λ is the molecular mean-free path, L_{ref} is the characteristic length of the body and γ is the specific heat ratio [13, 17]. Indeed, when the air density becomes rarefied enough the molecular mean-free path can become as large as the scale of the body itself. This condition is indeed known as FMF regime. Here, the aerodynamic characteristics of the vehicle are determined by individual, scattered molecular impacts, and must be analysed on the basis of kinetic theory. Because of this, several particle simulations, such as DSMC analyses, are mandatory. FMF conditions result in an abrupt loss of vehicle aerodynamic efficiency due to a suddenly increase of drag and lift drop. Further down in the atmosphere (in other words, higher λ) the TFR is established. The DSMC approach is still valid but demands high computational effort to simulate the increasing number of molecules. Fortunately, in these flow conditions the slip conditions and temperature jump can be introduced in the continuum approach (in other words, a Navier–Stokes approximation) to take into account for the rarefaction effects.

The rarefied and transitional regimes for ORV concepts are shown in Figure 3.17.

The ORV re-entry trajectory is reported in the Mach–Reynolds numbers map together with iso-Knudsen curves, which bound the different flow regimes, according to the Bird regime classification. As one can see, the region for $10^{-3} < \mathrm{Kn}_\infty < 10$ is the transitional flow region. Therefore, above about 200 km altitude the ORV is in the FMF (in fact, $\mathrm{Kn}_{\infty L_{\mathrm{ref}}} \approx 70$); while from the entry interface (in other words 120 km, the red square in Figure 3.17) to about 87 km (green rhombus in Figure 3.17) the vehicle is in the TFR.

Finally, after that, continuum flow conditions are established.

3.5 Viscous-interaction Regime

It is the flow regime where, due to low Reynolds number and high Mach number effects, the boundary layer and the shock layer merge into a viscous shock layer enveloping the vehicle. This results in a loss of ORV aerodynamic performance assuming that the vehicle drag in this regime rises, thus lowering spacecraft aerodynamic efficiency.

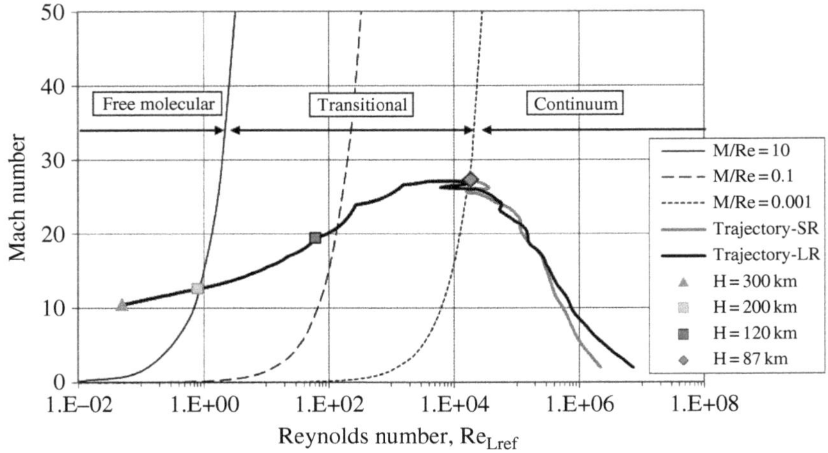

Figure 3.17 ORV re-entry scenario in the Mach–Reynolds map with iso-Knudsen curves.

The viscous interaction regime (VIR) is defined by the hypersonic viscous interaction parameter (VIP):

$$\bar{V}'_{\infty} = M_{\infty}\sqrt{\frac{C'_{\infty}}{Re_{\infty_{Lref}}}} \tag{3.2}$$

where

$$C'_{\infty} \propto f\left(\frac{T'}{T_{\infty}}\right) \tag{3.3}$$

and

$$\frac{T'}{T_{\infty}} = 0.468 + 0.532\frac{T_w}{T_{\infty}} + 0.195\frac{\gamma-1}{2}M_{\infty}^2 \tag{3.4}$$

By using the reference enthalpy approach to account for compressibility effects, an estimation of the VIR extension for the ORV can be found (Figure 3.18). This shows the vehicle re-entry trajectory both by means of an altitude-vs-Mach curve and a Mach-vs-VIP curve. As shown, two boundaries are also reported in the figure according to the criterion chosen in the past to address the VIR for the Space Shuttle. Therefore, very roughly one can conclude that for the ORV the VIR ranges from about 60 to 83 km altitude, close to the boundary of 86 km highlighted above as the beginning of the transitional flow regime.

3.6 High-temperature Real-gas Regime

The equilibrium/non-equilibrium real-gas regime is characterized by relatively high-density and high-velocity flow conditions. For instance, the hottest thermal environment is encountered and many chemical reactions occur. Therefore, the modelling of such

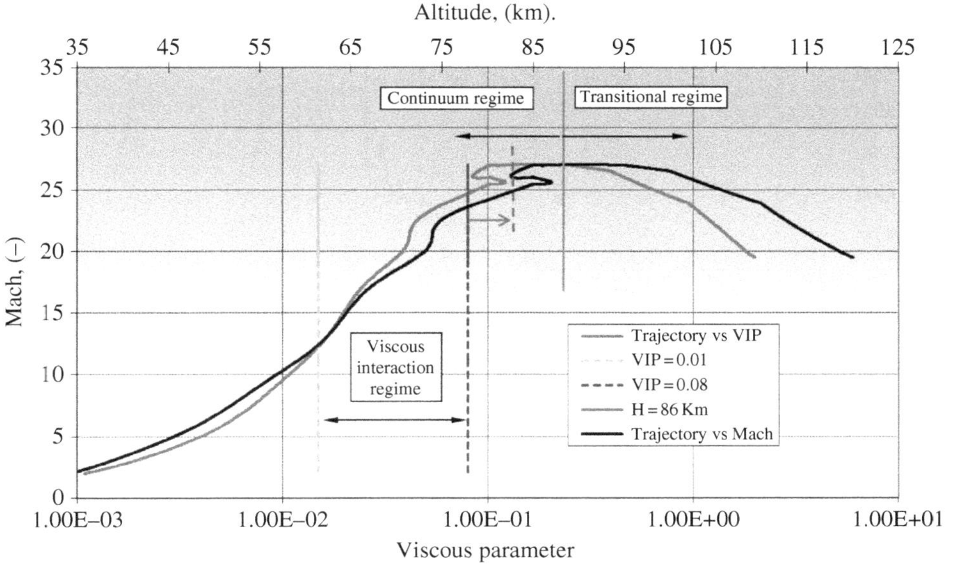

Figure 3.18 ORV re-entry trajectory in the Mach/altitude-viscous parameter map.

phenomena requires enforcing the Navier–Stokes equations with input data such as chemical equilibrium constants and reaction rates for the reaction mechanisms involving all the species that compose the gas mixture at the specified flight conditions.

In this framework, the current ORV nominal re-entry trajectory is illustrated in Figure 3.19 in an altitude–velocity map, where real-gas phenomena that occur during re-entry and which are relevant for assessing the vehicle aerodynamics and aerothermodynamics are shown [13].

For instance, Figure 3.19 highlights that when the ORV vehicle is flying at 6 km/s at about 70 km altitude, oxygen is completely dissociated because the velocity is larger than that corresponding to the O_2-dissociated domain. On the other hand, nitrogen begins to dissociate because we are near the green dashed line which represents the N_2 10% dissociation boundary. Indeed, the high-energy of re-entry flows leads to strong heating of the air in the vicinity of vehicle. Depending on the temperature level behind the bow shock wave (in other words, depending on the flight velocity), the vibrational degrees of freedom of the air molecules are excited and dissociation reactions of oxygen and nitrogen molecules may occur.

The high-temperature real-gas effects described here are enabled by energy transfer from the translational energy stored in the random motion of the air particles, which is increased by the gas heating up, to other forms of energy. Because this energy transfer is brought about by air particle collisions, it requires a certain time to happen. The time required to reach equilibrium is defined by the local temperature and density. Therefore, depending on the ratio of the relaxation time to a characteristic timescale of the flow, the chemical and thermal relaxation processes can be non-equilibrium, in equilibrium or frozen flow, thus influencing the vehicle's aerodynamic and aerothermodynamic performance. For example, flow thermochemical conditions influence vehicle pitch trim and aeroheating loads along with the re-entry flight.

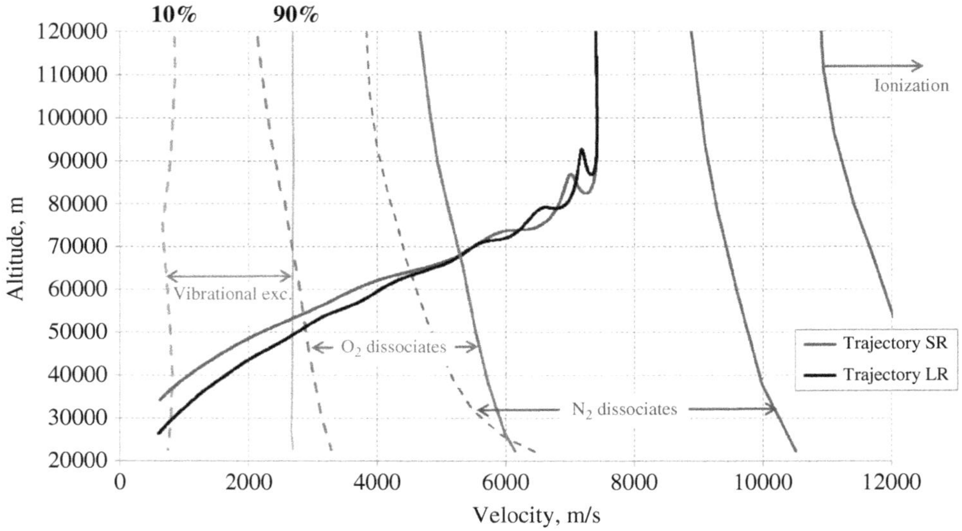

Figure 3.19 ORV re-entry scenario in the altitude–velocity plane with real gas map.

The similarity parameter that governs the thermochemical regimes is the Damkohler (Da) number, defined as:

$$\text{Da} = \frac{t_c}{t_r} = \frac{\textit{characteristic flow time}}{\textit{characteristic relaxation time}} \tag{3.5}$$

where the characteristic flow time of particles passing a region of the flow t_c is usually the ratio of the characteristic dimension L (the shock stand-off distance for instance) and the velocity u; the relaxation time refers to both chemical reactions (in other words chemical relaxation) and internal degrees of freedom of the flow molecules: vibration, rotation and so on (in other words thermal relaxation) [19]. Therefore,

$$\text{Da} = \frac{L}{ut_r} \tag{3.6}$$

So, for chemical reaction the Damkohler number reads:

$$\text{Da} = \frac{Ln_k}{u\dot{\omega}_k} \tag{3.7}$$

where n_k and $\dot{\omega}_k$ represent the number density and the chemical production rate of species k, respectively [19].

For thermal relaxation we can define the Damkohler as

$$\text{Da} = \frac{L}{u\tau_v} \tag{3.8}$$

where τ_v is the vibrational relaxation time, usually described by the Landau–Teller approximation for harmonic oscillators, which is valid to about 5000 K [19]. As a result, the chemical and thermal regimes may be loosely defined as follows:

- $Da = 0$ chemically/thermally frozen
- $10^{-2} < Da < 10^{3}$ chemical/thermal non-equilibrium flow
- $Da = \infty$ chemical/thermal equilibrium.

At very high altitudes where the densities are very small or at low speeds where temperatures are low, reaction rates are slow compared to the hydrodynamic timescales ($Da << 1$). This enables the frozen flow assumption to be used and solutions are restricted to those that model the flow of a multi-component fluid as a continuum (or free-molecular) problem with an appropriate equation of state and real (in other words, temperature-dependent) thermodynamic data.

On the other hand, at low altitudes the densities are high and the flow is likely to be characterized by fast chemical reactions with short timescales compared to the fluid velocity ($Da >> 1$). In this regime the assumption of equilibrium chemistry in the shock layer is more appropriate.

In the mid-range density regime, the assumption of either frozen or equilibrium flow, whilst accurate in some portions of the shock layer, would lead to over- (frozen) or under- (equilibrium) prediction of temperature in other areas [13, 17]. In reality, between these two regimes the density is such that the flow cannot be accurately characterized as being either frozen or in chemical equilibrium. It is in this flight region that non-equilibrium chemical kinetics must be considered [19].

Figure 3.19 also suggests that there will be an increasing number of species in the different domains along the descent trajectory. Therefore, accurate CFD simulations must rely on a flow thermochemical model appropriately tuned on the base of the freestream conditions of the trajectory point to simulate.

This discussion suggests that, due to the continual exchange of energy between the transitional and internal degrees of freedom of the flow molecules, the air at hypersonic speeds results in a mixture in thermal and/or chemical non-equilibrium in the different domains of the altitude–velocity map, as shown in Figure 3.20 and Tables 3.1 and 3.2.

This figure provides several insights into the trends expected in the air chemistry of the flight stagnation region of the ORV forebody. For instance, Figure 3.20 shows three different regions:

- A is the region of chemical and thermal equilibrium.
- B is the region of chemical non-equilibrium and thermal equilibrium.
- C is the region of chemical and thermal non-equilibrium.

Using Table 3.1, one can rapidly assess the most likely thermochemical model for the high-fidelity CFD simulation once trajectory freestream conditions are known.

3.7 Laminar-to-turbulent Transition Assessment

During descent from suborbit or orbit, re-entry vehicles experience transition from fully laminar to fully turbulent flow conditions. In the latter case, both aerodynamic drag and an aeroheating increase must be accounted for in the vehicle design [20].

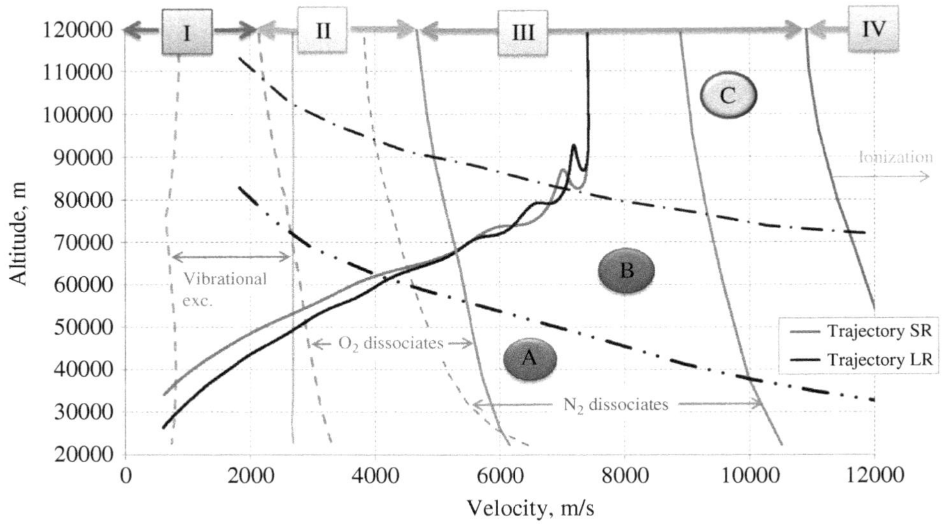

Figure 3.20 Re-entry scenario in the altitude–velocity plane with the stagnation point flow regimes and thermochemical phenomena. Refer to main text for key.

Table 3.1 Regions with their expected aerothemal phenomena.

Region	Aerothermal phenomena
A	Chemical and thermal equilibrium
B	Chemical non-equilibrium with thermal equilibrium
C	Chemical and thermal non-equilibrium

Table 3.2 Chemical species in high-temperature air.

Region	Model	Species
I	2 species	O_2, N_2
II	5 species	O_2, N_2, O, N, NO
III	7 species	O_2, N_2, O, N, NO, NO^+, e^-
IV	11 species	O_2, N_2, O, N, NO, NO^+, O^+_2, N^+_2, O^+, N^+, e^-

Therefore, a critical aeroheating design issue for all aerospace vehicles is to assess in what region of the descent flight trajectory the boundary-layer transition (BLT) occurs and to determine the corresponding increase in heat transfer/surface temperature.[2]

To avoid the worst case scenario from an aeroheating perspective, BLT needs to occur well past the region of peak heating on the trajectory (in other words, the region where

2 Typically, the windward heat transfer rate increases by a factor of 3–5 over the BLTs from laminar to turbulent.

the product of freestream density and velocity to the third power is a maximum) [14, 15]. Thus, a reasonably accurate determination of flight conditions where BLT occurs is essential for most aerospace vehicles, in order to ensure that aeroheating levels and loads remain within TPS design limits. Ideally, such information would be incorporated into the design of the TPS, including material selection, the split line definition, and choice of material thickness.

Although progress has been made in developing computational techniques for predicting hypersonic BLT onset, aerothermodynamicists still rely primarily on semi-empirical methods based on local flow conditions, such as local Mach and Reynolds numbers. For example, for the Shuttle Orbiter and other vehicles, an empirical correlation for hypersonic transition – the transition criterion based on the parameter Re_θ/M_e:

$$\frac{Re_\theta}{M_e} = 100 \tag{3.9}$$

was developed [13,14,15]. Here, Re_θ is the momentum thickness Reynolds number and M_e is the boundary-layer edge Mach number. Advances in numerical (in other words CFD) flowfield solutions and the phosphor thermography technique in wind tunnel experiments have given such approaches a boost.[3]

In practice, CFD results provide local conditions of interest about the model, specifically boundary layer thickness, M_e, and Re_θ at the location of the surface disturbance(s). Through experimental investigations, designers try to correlate the value assumed by the parameter Re_θ/M_e to the transition front detected in the wind tunnel.

Box 3.3 Transition characteristics of the X-33 concept.

Studies to deduce BLT characteristics for the X-33 concept (see Figure 3.21), performed at NASA Langley Research Center, have been illustrated the benefits of a synergistic experimental and computational capability [14, 15].

A large number of phosphor thermography aeroheating tests of the X-33 forebody were performed in the Langley 20-inch Mach 6 air tunnel for a range of AoAs, Reynolds numbers, and discrete boundary-layer trips of various heights (K) and locations on the model surface. Wind tunnel tests provided several complex BLT patterns measured on X-33 models, such as the one provided in Figure 3.22.

Those surface patterns were complemented with predictions from inviscid/boundary layer and CFD codes and the results were extrapolated to flight. For each tunnel run, the boundary-layer thickness (δ), momentum Reynolds number (Re_θ) and boundary-layer edge Mach number (M_e) were predicted via CFD code (LATCH), as shown in Figure 3.23.

Thus, applying the Re_θ/Me transition criterion to the X-33 experimental test campaign yielded values between 250 and 325, and an average value of Re_θ/Me = 285 was chosen for this vehicle [14, 15].

3 Thermography allows accurate determination of BLTs in hypersonic wind tunnel models, with and without discrete trip elements, over a wide range of attitude and flow conditions (primarily Reynolds number).

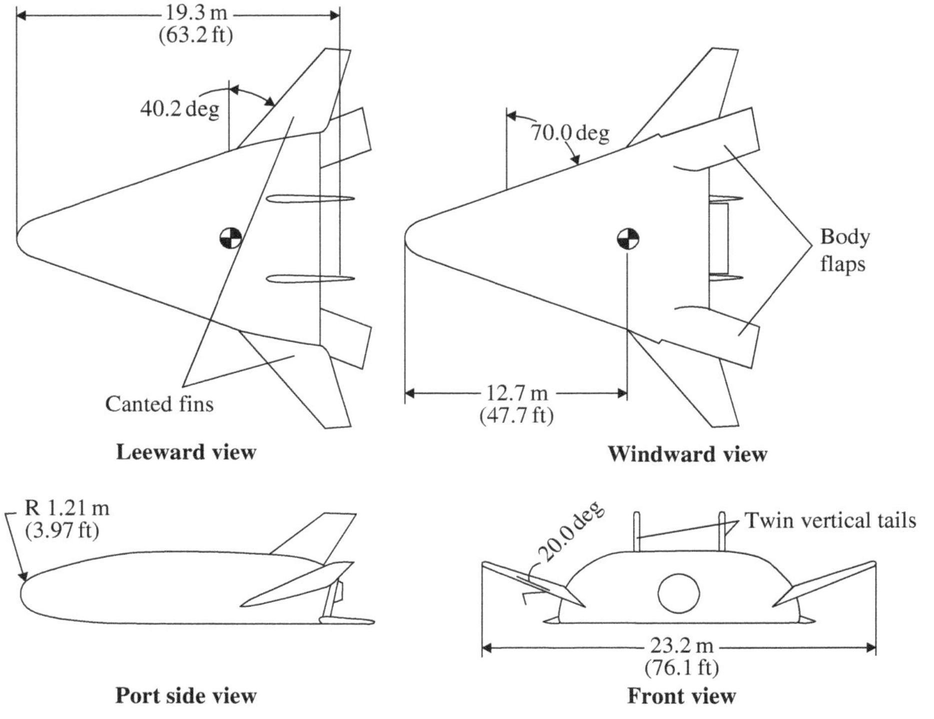

Figure 3.21 The Lockheed-Martin X-33 aeroshape.

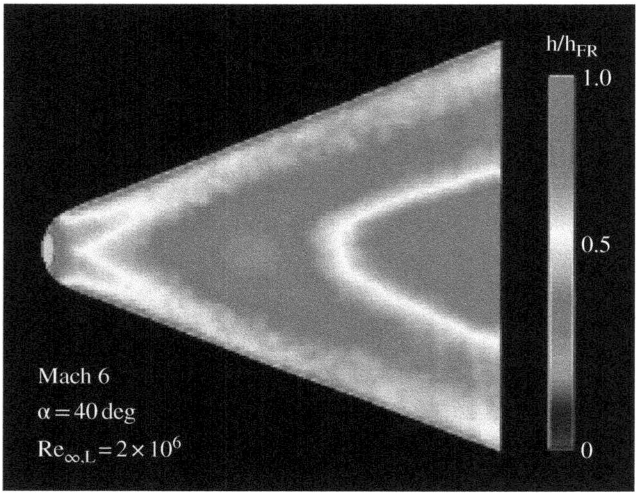

Figure 3.22 X-33 forebody BLT pattern at $M_\infty = 6$.

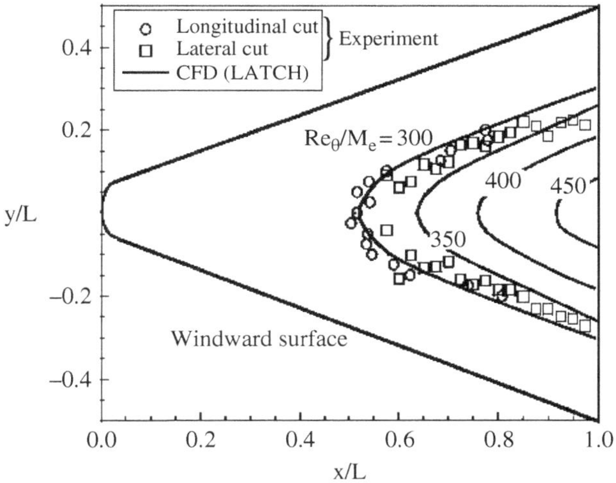

Figure 3.23 Correlation of BLT on X-33 forebody.

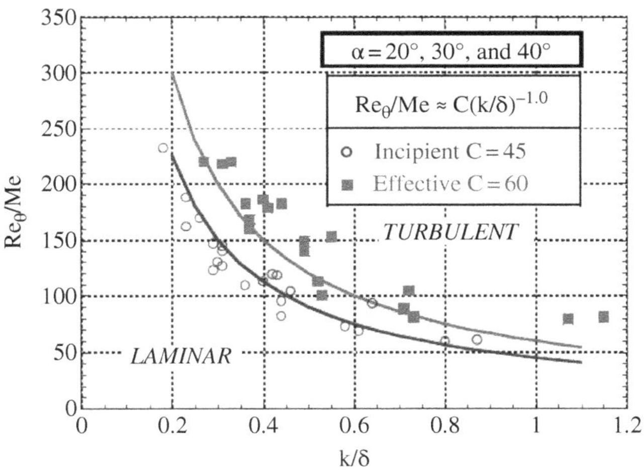

Figure 3.24 Correlation of BLT on X-33 forebody. Roughness transition correlation.

This large data set, involving several hundred tunnel runs and corresponding predictions, was also correlated in terms of Re_θ/Me vs K/δ. Results are provided in Figure 3.24.

In this framework, a relationship for the transition at a body point was derived by interpolation of the LATCH database to locate the altitudes where transition occurs for all combinations of velocity and AoA. For instance, Figure 3.25 shows this relationship using the criterion Re_θ/M_e referred to as a point at $x/L = 0.8$ on the vehicle windward centreline.

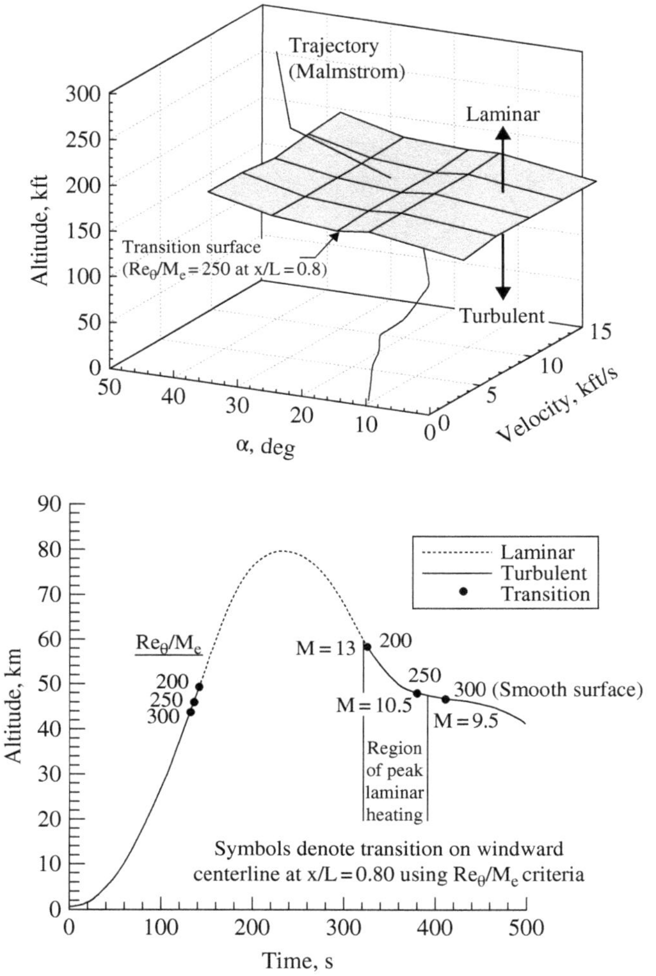

Figure 3.25 Transition criterion in trajectory space **X-33**.

The result (see Figure 3.25 top) is a nearly planar surface in a three-dimensional trajectory space (in other words, altitude, velocity, AoA) that separates laminar flow above from turbulent flow below.

In the same way, this criterion can also be applied to determine if laminar flow would indeed be achieved during ascent and the first portion of descent, as recognized on the right-hand side of Figure 3.25.

In practice, this analysis has proven to be extremely valuable in tailoring flight trajectories (both the ascent and descent legs) so to not exceed limits on the metallic TPS yet satisfying mission requirements.

In conclusion, if proven accurate, as is expected, this approach will be recommended for all aerospace vehicles expected to experience BLT near the region of peak heating during descent, and should be performed as early in the program as appropriate. That said, a major concern with this approach is the measurement of transition onset in a conventional-type hypersonic wind tunnel for which the freestream flow is acoustically contaminated. Freestream disturbances are expected to promote early transition on the model, hence the findings as applied to flight should be conservative. It is primarily because of these acoustic disturbances in the freestream of the wind tunnel, which result from the turbulent boundary layer on the nozzle wall, that flight measurements of transition onset are needed to substantiate the approach [14, 15].

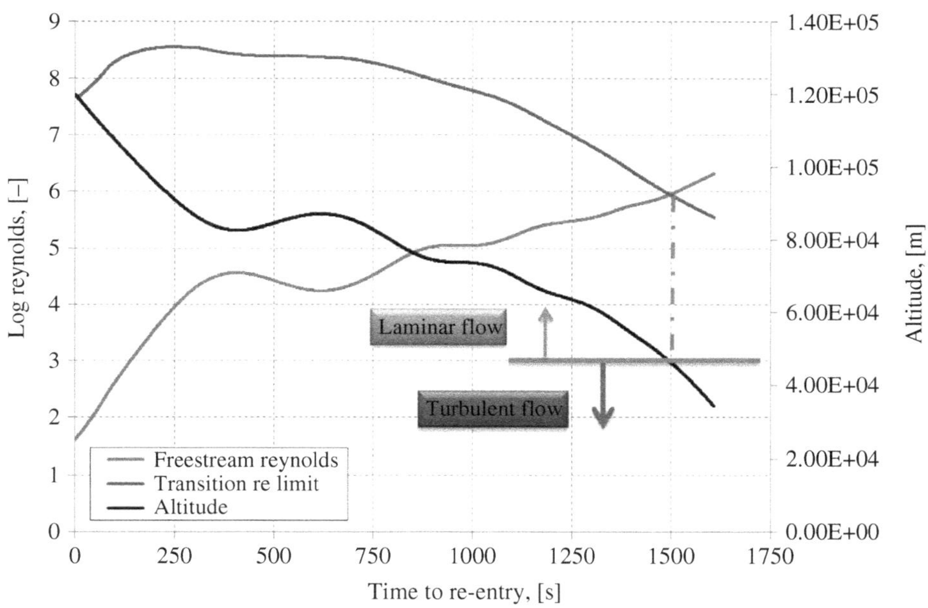

Figure 3.26 Assessment of laminar-to-turbulent transition.

Because the assessment of the local flow conditions demands accurate CFD computations, which are, of course, not compliant with a phase-A design level, a transition method based on freestream Reynolds (Re_∞) and Mach (M_∞) numbers has been adopted. For example, Figure 3.26 reports the transitional Reynolds limit evaluated by means of the following transition criterion:

$$Log\,Re_\infty > \left[Log\,Re_T + Cm(M_\infty) \right] \quad \text{Turbulent flow} \tag{3.10}$$

where Re_T and C_m depend on the type of flow, flying AoA, leading-edge sweep angle, and leading-edge nose bluntness [21]. As shown, the transition criterion highlights that, below about 46 km, altitude turbulent-flow conditions are expected.

3.8 Design Approach and Tools

A review of the aerodynamic analyses needed for the development of the aerodynamic database (AEDB) of the vehicle concept is now performed. These evaluations are aimed only at creating a preliminary AEDB, compliant with a phase-A design level [22,23,24]. The goal is to provide aerodynamic characteristics for flight-mechanics and thermal-shield design analyses. It must be verified that vehicle, after the de-orbit manoeuvre, is able to stay within the load constraints (in other words, entry corridor) when flying trimmed during descent to a conventional runway landing.

The AEDB is prepared as a function of Mach number, AoA, sideslip angle, aerodynamic control surface deflections, and Reynolds number, according to the space-based design approach [18]. This design approach dictates the generation of a complete dataset as function of a number of independent parameters (in other words, M_∞, Re_∞, α, β) as shown in Figure 3.27.

An accurate aerodynamic analysis of all these flight conditions, however, is very complex and time consuming, and is not compatible with a Phase-A design study, in which fast prediction methods are mandatory.

In the preliminary design phases, the evaluation of the vehicle AEDB is mainly accomplished by means of engineering tools, and a limited number of more reliable CFD computations (continuum regime only), according to the workflow shown in Figure 3.28 [22,23,24,25].

CFD computations are performed in order to verify the accuracy of the design results and to focus on some critical design aspects not predictable with simplified tools. This overall process is referred to as 'anchoring' of the engineering-level methods. The anchoring process permits a few select CFD solutions to be used beyond the specific flight conditions in which they were original run.

The anchoring process also allows for the cost-effective use of high fidelity, and computationally expensive, CFD solutions early in the design process, when the vehicle trajectories are often in a constant state of change [26, 27]. The CFD anchoring 'space' is defined by a small number of CFD solutions in Reynolds-Mach-AoA space, as shown in Figure 3.29.

Note that CFD analysis is nevertheless essential in preliminary design studies, keeping in mind the limited capability of an engineering-based approach in modelling complex flow interaction phenomena and aerodynamic interference, such as shock–shock and shockwave boundary-layer interactions.

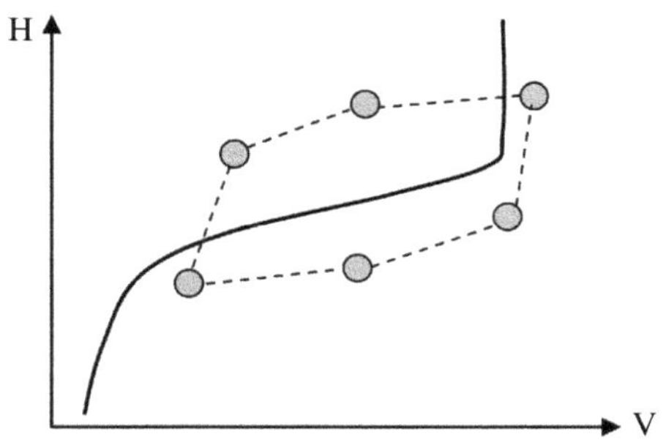

Figure 3.27 Space-based design approach in the altitude–velocity map.

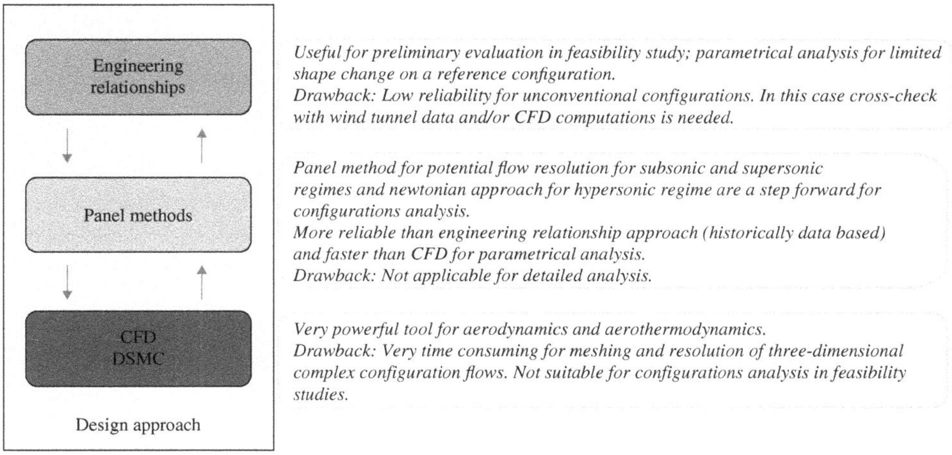

Figure 3.28 Tools and methods.

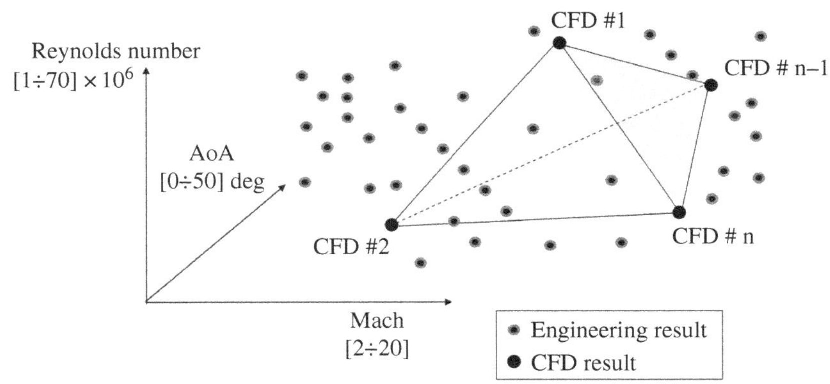

Figure 3.29 Hypothetical CFD anchoring mesh in Reynolds-Mach-AoA space.

Experience shows that an aerodynamic configuration that seems to be promising when evaluated by simplified methods is not always a feasible solution after performing more detailed calculations. But every layout deemed infeasible by the preliminary aerodynamic examination has no reasonable chance of realization. Therefore, the approach implemented in the engineering-based approach is fully justified, as it is much easier and more rapid, and hence the assessment of rearrangements can be done extremely quickly.

As a result, ORV aerodynamics in FMF conditions have also been provided with low-order methods. In the transitional flow regime (the one bridging the continuum and the free molecular regimes) the bridging relationships approach is applied.

Finally, in the continuum flow regime, vehicle aerodynamic appraisal is performed with both low-order and CFD methods.

Box 3.4 A look to the CFD benefits and future goals.

CFD contributes significantly to the development of aerothermodynamic databases for re-entry vehicles, addresses specific, localized phenomena such as shear layer impingement on different vehicle parts, and plays a complementary role in the development of aerodynamic/aerothermodynamic databases that require full tip-to-tail solutions [14, 15]. Through these applications to complex configurations over wide ranges of flow conditions and attitudes, CFD capabilities have increased considerably, as codes have been modified to enhance accuracy, increase speed, and provide new capabilities, such as RCS interactions, full wake solutions, and so on.

CFD predictions duplicate vehicle scale, velocity, attitude and atmospheric (free-stream) conditions, thereby avoiding the uncertainties associated with extrapolation of ground-based data to flight. For this reason, CFD calibrated against ground-based and flight data is the preferred source of aerothermodynamic information in the hypervelocity continuum regime. Wind tunnels provide the majority of aerothermodynamic information in the hypersonic continuum regime. Flight simulation via ground-based facilities has been tried and proven for this regime over the last 40 years. For a given interval of time, wind tunnels can provide several orders of magnitude more aerothermodynamic information than CFD codes. This discrepancy is larger when aerodynamic performance and aeroheating characteristics are addressed for a range of angles of sideslip. Use of both ground-based testing techniques and CFD codes is problematic in accurately predicting flight boundary-layer/shear-layer transition phenomena. The advantage in providing meaningful, relevant transition information resides with ground-based facilities, even in the absence of a hypersonic, low-disturbance/quiet testing capability. The hypersonic continuum regime affords the vehicle designer accurate/credible aerothermodynamic information via both experimental and computational capabilities, unlike the rarefied flow and hypervelocity continuum regimes where only computational capabilities are available/viable.

As the approach to generating aerodynamic and aeroheating information is developed, an important factor to be considered in ground-based testing is the basic shape of the aerospace vehicle concept(s), since the shape dictates which fluid dynamic phenomena are most important and which simulation parameters will dominate. From a hypersonic perspective, the test approach for a very slender configuration may be quite different from that for a blunt configuration. For example, a hypersonic airbreathing (in other words, scramjet) vehicle will most likely be slender, have sharp leading edges, and fly at low incidence and relatively high dynamic pressures. For a slender shape and low incidence, the flowfield about the vehicle will principally be supersonic/hypersonic. The effects of compressibility (Mach number), viscosity (Reynolds number), gas properties (such as ratio of specific heats), and thermal drivers (potential ratio of wall-to-adiabatic wall temperatures) are all expected to be important; that is, to have a first order influence on aerodynamic/aeroheating characteristics. Of particular importance is the state of the boundary layer (in other words, whether the boundary layer is laminar, transitional, and/or turbulent). This may influence the level of control effectiveness and will definitely influence the heating. Because of the small shock inclinations associated with sharp leading-edged slender bodies at small incidence, finite-rate chemistry effects on aerodynamics (for example through variation in the centre of pressure) and aeroheating (for example through thermochemical non-equilibrium heating including surface catalytic effects) are usually small, except in local regions of flow stagnation, shock–shock interactions and shock impingement.

At the other end of the shape spectrum, the flow over the forebody of a very blunt configuration is principally subsonic, becoming supersonic as it expands around the comers. For very blunt shapes, the most important hypersonic simulation parameter for aerodynamics is the density ratio across the normal portion of the bow shock or corresponding value of ratio of specific heats within the shock layer for continuum flow. In the continuum flow regime, Mach number effects for Mach numbers in excess of five are generally negligible (the Mach number independence principle), as are viscous effects on the forebody where the boundary layer is quite thin. The detachment distance of the bow shock from the forebody surface and the location of the sonic line separating the subsonic and supersonic flow regions are a strong function of the density ratio/post-shock gamma and the influence of density ratio/gamma on these parameters may have a first-order influence on the aerothermodynamic characteristics.

The AEDBs across the subsonic through hypersonic speed regimes, come essentially exclusively from wind tunnel testing. The role of CFD in the generation of vehicle aerodynamic databases is complementary in nature. There are two primary reasons for the dominance of wind tunnel testing over CFD. Once models and associated hardware are available, wind tunnels provide huge quantities of aerodynamic performance information over wide ranges of attitude (alpha, beta), control surface deflections, individually and in combination, and flow conditions in a relatively short period of time and with a high degree of credibility based on decades of previous experience. The second reason is that available wind tunnels cover the flight regimes for the spacecraft nicely, in that the maximum flight Mach number for the vehicle is expected to be around 7–10. As seen from Figures 3.13 and 3.30, the contribution of CFD increases significantly above Mach 10 or so, where reacting flowfields influence aerodynamic characteristics [14, 15].

In the generation of aeroheating databases, CFD is an equal and synergistic partner with wind tunnel testing. Because of the major role played by wind tunnels in establishing aerothermodynamic databases a significant portion of the subsequent lessons learned are associated with ground-based testing.

The ultimate credibility of aerodynamic and aeroheating information is achieved when independently performed wind tunnel measurements and CFD predictions for wind tunnel cases are compared and found to be in agreement over a range of attitude and flow conditions. Wind tunnel measurements and CFD predictions are highly complementary and together provide accurate aerothermodynamic information throughout the flight trajectory; in other words, across the subsonic to hypersonic regimes. There is a tendency of computationalists to bypass comparing CFD predictions to wind tunnel measurements and to apply CFD only to flight conditions. In the framework of entry-vehicle design programs, an aerothermodynamicist capable of performing both high-level experimental work and high level-computational (via CFD) work proved to be extremely valuable. Knowledge of the strengths and weaknesses of both disciplines has allowed the strengths to be systematically combined and optimized. The demonstration of these advantages has led to the development of more aerothermodynamicists with both skills. It is important to have experimental and computational aerothermodynamicists working on aerodynamic issues and those working

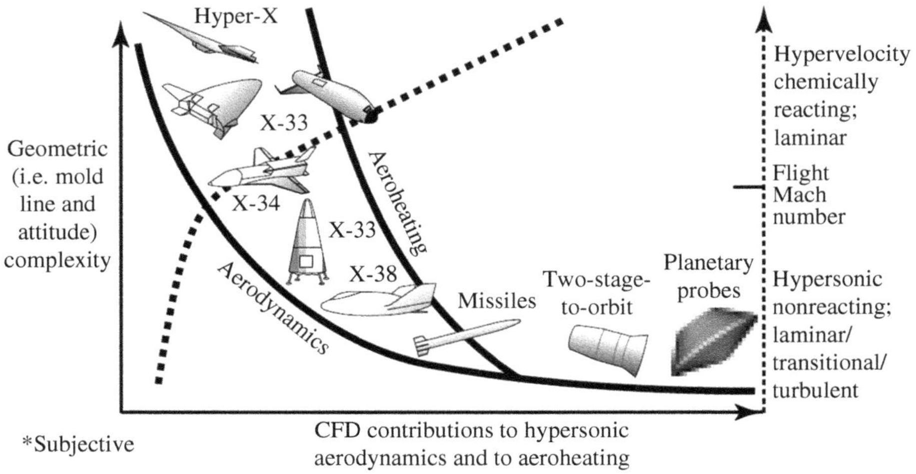

Figure 3.30 Relative contribution of CFD.

on aeroheating issues establish strong lines of communication. Often, anomalies observed in aerodynamic force and moment measurements can be explained by the detailed surface information from experimental aeroheating studies. Detailed studies of shock–shock interactions, flow separation–reattachment phenomena, BLTs, and so on, via aeroheating measurements, are beneficial to aerodynamicists in explaining force and moment trends. Naturally, inputs from computationalists who generate detailed surface and flowfield information are extremely valuable in this process. Working aerodynamic and aeroheating issues together, OMLs may be varied to both enhance the aerodynamic performance and to minimize aeroheating levels.

The same team of experimental aerodynamicists should test across the subsonic-to-hypersonic speed regimes, as opposed to different teams testing at subsonic, transonic, supersonic, and hypersonic conditions. The continuity and flexibility provided by a single team testing across the speed regimes is believed to outweigh the collective outputs of specialists in each regime, which must be coordinated and assembled into one story.

Configuration OMLs are modified/varied in an iterative manner to provide acceptable hypersonic aerodynamic performance, generally at relatively high AoAs, and subsonic approach and landing characteristics. Vehicle designers may be approaching closure on OMLs via this approach before testing at transonic conditions reveals significant aerodynamic performance problems. It is imperative for aerospace vehicle concepts that transonic aerodynamic information be obtained early in the program, so that subsonic, transonic, and hypersonic information can be used concurrently in the optimization of OMLs to achieve desired flying characteristics across the entire speed regime, from high altitude hypersonic conditions to approach and landing.

In most cases, the credibility of the experimental AEDB is enhanced considerably with the simulation of flight values of Reynolds number based on appropriate full-scale vehicle dimensions.

Generally, existing hypersonic wind tunnels simulate flight values of Reynolds number, for a given Mach number, and provide a sufficient Reynolds number range to produce

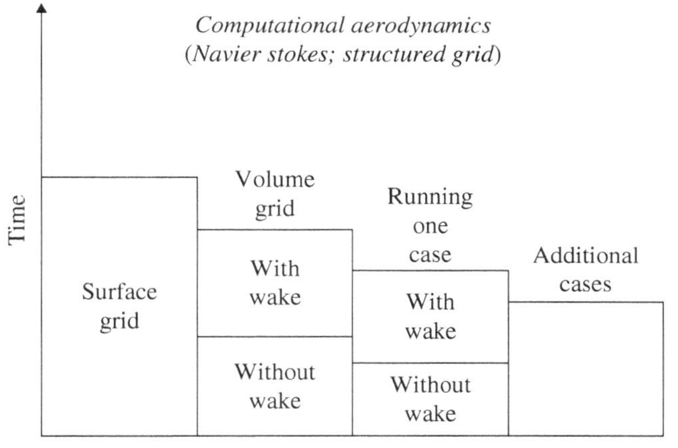

Figure 3.31 Relative times associated with typical CFD applications.

fully laminar and equilibrium turbulent boundary-layer/shear-layer flow about the test article. However, the simulation of flight Reynolds numbers at subsonic, transonic, and low supersonic conditions is a formidable task and can be accomplished in relatively few facilities. The vehicle designer is confronted with a trade of reduced risk for increased time and money. As a minimum, limited tests at transonic flight values of Reynolds number on the baseline configuration are recommended to determine if the major portion of the subsonic-to-supersonic AEDB obtained at lower-than-flight values of Reynolds number is credible. An alternative approach is to apply CFD to wind tunnel conditions corresponding to relatively low values of Reynolds number and, if prediction compares well with measured aerodynamic forces and moments, apply CFD to flight values of Reynolds number to determine the aerodynamic performance. Thus, subsonic-to-low-supersonic wind tunnel testing at flight values of Reynolds number or the use of CFD to extend wind tunnel conditions to flight should be used to decrease the risk level associated with the development of the aerodynamic flight data book.

The efficiency and effectiveness of surface and volume grid generation have been enhanced, although generation of these grids remains the major part of the total time required to generate a solution (see Figure 3.31) [14, 15].

Advantage has been taken of massive parallelization techniques to significantly reduce the running time required for solution convergence. Most importantly, confidence in using CFD to provide aerodynamic and aeroheating data for aerospace vehicles has increased appreciably by making comparisons of CFD predictions to wind tunnel measurements and code-to-code. These comparisons provided a better understanding of what physical and numerical models to use in future applications. Indeed, the strengths of CFD and ground-based testing are complementary and, used together, provide a better product.

Due to expected advances in CFD in the future, particularly in the reduced times required to run full Navier–Stokes solvers tip-to-tail the time will come when CFD plays a dominant role in the development of aerothermodynamic databases. However, it is believed that vehicle designers will rely primarily on wind tunnels for subsonic-to-low-hypersonic (in other words, Mach 0.1 to 10) aerothermodynamic information and on CFD

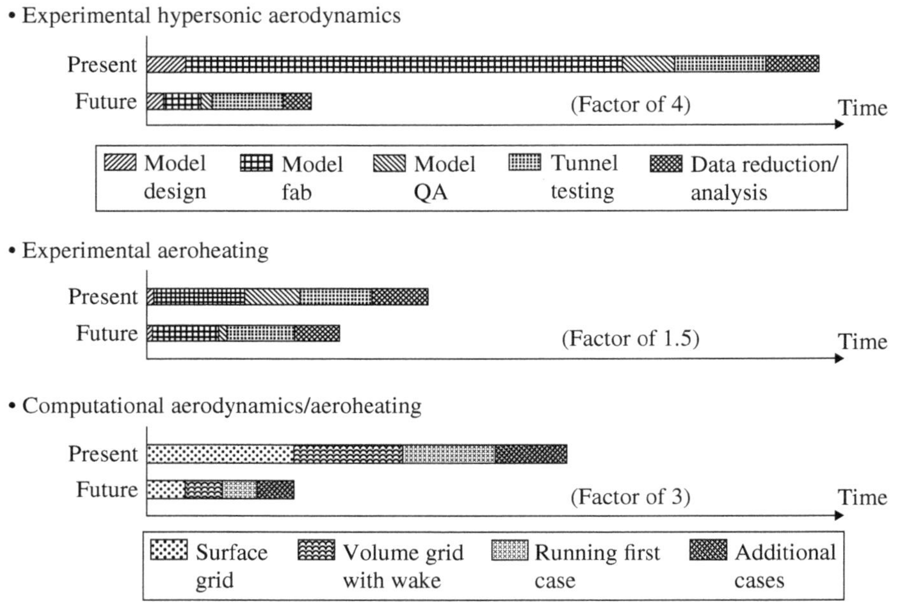

Figure 3.32 Future aerothermodynamic goals.

predictions for hypersonic-hypervelocity flows (in other words, Mach numbers in excess of 10). The emphasis on future enhancements is to provide aerothermodynamic information faster (see Figure 3.32) for all phases of the aerothermodynamic process: development of initial concepts, assessment, optimization, and benchmarking [14, 15].

Factors in computational and experimental aerothermodynamics that now require relatively long periods of time to perform are being worked on in an effort to make them quicker. Computationally, an ambitious goal was attained by the development of unstructured hypersonic viscous solvers. These allowed a reduction, of an order of magnitude or more, in the total time to generate grids and obtain a full viscous, tip-to-tail solution. This was because significant advances have been made in reducing the time required to generate structured surface/volume grids. This breakthrough greatly accelerated the development of comprehensive vehicle aerodynamic/aerothermodynamic databases. It is also an important ingredient in a process recently developed to rapidly predict peak heating and heating loads for a given trajectory and 'automatically' select optimum TPS materials, determine split lines between different TPS materials, and size the TPS. Additional computational plans include:

- the continued development and validation of advanced turbulence models for hypersonic flows
- enhancement of jet plume-flowfield-surface interaction (in other words, RCS interaction) capabilities
- possible revival of equilibrium radiation codes to address aerothermodynamic issues associated with very high velocity return to Earth missions

- continued advancements in DSMC by coupling to continuum Navier–Stokes solvers to predict RCS phenomena in the rarefied flow regime and extension of DSMC capabilities lower in the atmosphere
- the exploiting of boundary-layer and triple-deck theory to compute the effects of global and local changes to surface catalysis on computed heating rates
- development of the laminar wall function approximation to help reduce grid requirements and accelerate convergence of CFD solutions for hypersonic viscous flow.

All aerodynamic data are provided in a format that will allow a build-up from a basic configuration, by means of contributing elements to each force or moment component such as control surface effectiveness, and so on. Data are presented in a manner that treats each force and moment separately to facilitate the build-up procedure. Engineering codes are used for initial screening of aerospace vehicle concepts, trade studies and database construction.

In the framework of low-order methods codes, vehicle aerodynamics have been addressed by means of the HPM code, while CFD analyses for sub-transonic and hypersonic speed – both Euler and Navier–Stokes – have been carried out with the commercial code FLUENT. The HPM code is a 3D supersonic–hypersonic panel method code, developed at CIRA, that computes the aerodynamic characteristics of complex arbitrary 3D shapes using surface inclination methods (SIM), typical of Newtonian aerodynamics, including control surface deflections and pitch dynamic derivatives [25,26,27,28].

In HPM simulations, the surface is approximated by a system of planar panels, the lowest level of geometry used in the analysis being a quadrilateral element. Figure 3.33 shows a typical surface mesh used in HPM for design analysis.

The mesh differs from the finer ones that are typically considered for CFD analysis; only surface inclination variations are accounted for in HPM analysis (in other words, Newtonian aerodynamics), so flat surfaces have a coarse mesh. In the presence of curvature, a more dense mesh is used. Therefore, surface mesh generation is an important part of HPM design analysis.

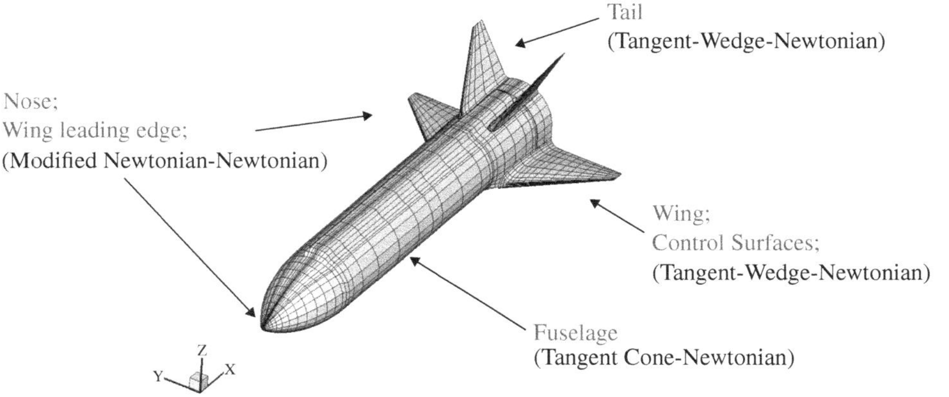

Figure 3.33 Typical hypersonic panel method mesh and representative flow models.

Table 3.3 Methods for inviscid Newtonian aerodynamic analysis.

Impact flow	Shadow flow
Modified Newtonian	Newtonian
Modified Newtonian/Prandtl-Meyer	Modified Newtonian/Prandtl-Meyer
Tangent wedge	Prandtl-Meyer empirical
Tangent wedge empirical	OSU blunt body empirical
Tangent cone empirical	Van Dyke unified method
OSU blunt body	High Mach no. BASE pressure
Van Dyke unified method	Shock expansion method
Blunt body skin friction	Input pressure coefficient
Shock expansion method	Free molecular flow
Free molecular flow	
Input pressure coefficient	
Hankey flat surface empirical	
Delta wing empirical	
Dahlem–Buck empirical	
Blast wave	
Modified tangent cone	

As shown in the figure, the pressure acting on each panel is evaluated by user-specified compression-expansion and approximate boundary-layer methods. The methods to be used in calculating the pressure in the impact and shadow regions of the vehicle may be specified independently and can be selected by the user; several methods are available, as reported in Table 3.3. More information about these theories can be found in the literature [13, 17, 25].

As shown in Figure 3.34, the generic vehicle configuration can be divided into a combination of simple shapes: cones, cylinders, flat plates, spheres, and wedges and so on, for which analytical solutions are available (see also Figure 3.33). For example, the wing leading edge was represented by a swept cylinder in order to obtain estimates for the leading-edge heating rates (outside of the shock–shock interaction region).

This design approach is viable, especially for vehicle aeroheating analyses.

In normal hypersonic vehicles applications, Prandtl–Meyer expansion flow theory and tangent cone/wedge methods are widely applied, together with the modified Newtonian method. Note that the flow separation is not accounted for, so the results obtained are not reliable for cases where significant flow separations exist.

The different parts of the vehicle (fuselage, wing and vertical tail) are analysed separately, and global vehicle aerodynamic coefficients are obtained by appropriate summation of the different component contributions, following a typical 'build-up' approach, according to Figure 3.33.

As far as the base drag is concerned, the following simple formula is used:

$$\Delta C_{Dbase} = -c_{Pbase} \frac{S_{base}}{S_{ref}} \tag{3.11}$$

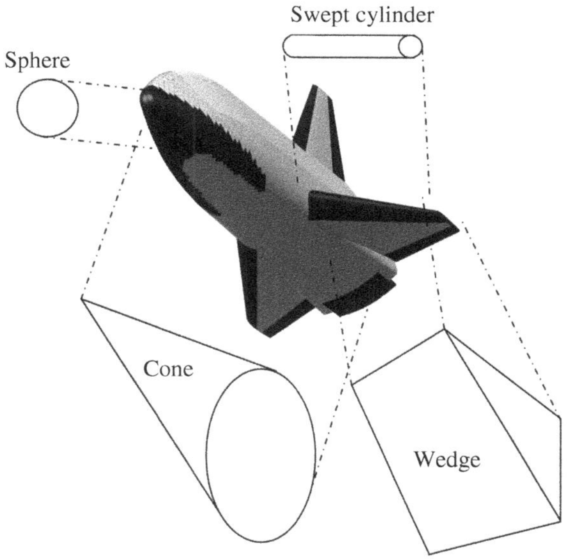

Figure 3.34 Design methodology. Representative vehicle part for flow models.

where

$$c_{P\text{base}} \cong -\frac{1}{M_\infty^2} \tag{3.12}$$

in hypersonic conditions and S_{base} and S_{ref} are the base and reference vehicle's surfaces, respectively [13, 17, 29].

In order to predict viscous contribution to aerodynamic forces and moments the shear force is determined on each vehicle panel. The skin friction is estimated based on the assumption of a laminar or turbulent flat plate as:

$$\Delta C_{D,\text{friction}} = C_f \frac{S_{wet}}{S_{ref}} \tag{3.13}$$

where C_f is the global skin friction coefficient, S_{wet} is the panel wetted area and S_{ref} is the reference vehicle's surfaces. Reference temperature and reference enthalpy methods are available for both laminar and turbulent flows [13, 17, 23, 24]. The viscous calculation is performed along with streamlines, and the results are then interpolated to the panel centroids. The streamlines are traced on the configuration, described by quadrilateral elements, using the Newtonian steepest-descent method, which uses only the element inclination angle relative to the velocity vector to determine the streamline trace. The only information required to generate streamlines is a quadrilateral description of the geometry and the flight attitude of the vehicle.

The HPM code is also able to perform detailed viscous analyses on 3D configurations [30].

The viscous analysis needs evaluation of the streamlines on vehicle's surface since it is performed along each streamline using a simple 1D boundary-layer method, and the results are then interpolated at each element centroid. To do this, the generic vehicle component is modelled as either a flat plate or a leading edge by selecting the appropriate boundary-layer model. The flat-plate boundary-layer model includes both laminar and turbulent methods as well as the cone correction. It sets the Mangler factor (M_f), which transforms the solution of the 2D boundary layer to the axially symmetrical case [23, 24]. The two available laminar skin friction and aeroheating correlations in the plate laminar method option are the Eckert and the ρ-μ methods. Both are based on the classic Blasius flat-plate boundary-layer solution corrected with the reference enthalpy compressibility factors. If the plate turbulent method is chosen, four turbulent methods are available: the Schultz–Grunow, the ρ-μ, the Spalding–Chi and the White methods.

In particular, for both laminar and turbulent flows, the reference temperature (T_{ref}), based on Eckert's method, reads:

$$T_{ref} = 0.5 \cdot (T_e + T_w) + 0.11 \cdot R_f \cdot (\gamma - 1) \cdot M_e^2 \cdot T_e \tag{3.14}$$

where the recovery factor, R_f, is calculated as the square root of the Prandtl number for laminar flows, and the cube root of the Prandtl number for turbulent flows. The Prandtl number is evaluated at the reference temperature.

For laminar flows, the skin friction coefficient and Stanton number are calculated as:

$$\begin{aligned} C_f &= 0.664/\sqrt{\mathrm{Re}/M_f} \cdot T_e / T_{ref} \\ S_t &= 0.5 \cdot C_f \cdot \mathrm{Pr}^{-2/3} \end{aligned} \tag{3.15}$$

For turbulent flows the relations for skin friction and Stanton number are:

$$\begin{aligned} C_f &= 0.37/\left[\log_{10}\left(\mathrm{Re}/M_f\right)\right]^{2.584} \cdot T_e / T_{ref} \\ S_t &= 0.5 \cdot C_f \cdot \mathrm{Pr}^{-2/3} \end{aligned} \tag{3.16}$$

In the above relations, M_f is the Mangler factor. For laminar flows, M_f is equal to 3 and for turbulent flows it is set to 2. The enthalpy-based film coefficient, recovery enthalpy and convective heating are defined as:

$$\begin{aligned} C_H &= \rho_{ref} \cdot U_e \cdot St \\ H_{rec} &= H_e + 0.5 \cdot R_f \cdot U_e^2 \\ \dot{Q}_{conv} &= C_H \cdot (H_{rec} - H_w) \end{aligned} \tag{3.17}$$

In the above relations, H_e is the edge static enthalpy, H_{rec} is the recovery enthalpy, U_e the edge velocity, and H_w is the wall enthalpy. H_w can be evaluated both at cold and radiative cooling wall boundary conditions. In the latter case, a Newton–Raphson technique is used to assess the wall temperature, with the energy radiated from the surface equal to the sum of the convective and incident shock-radiative heatings:

$$\left(\dot{q}_{\text{convective}} + \dot{q}_{\text{radiative}} = \sigma \cdot \varepsilon \cdot T_{\text{wall}}^4\right) \tag{3.18}$$

Indeed, since the convective heating depends on the final wall temperature, this nonlinear relationship must be solved iteratively at each panel centroid.

With the leading-edge boundary-layer model, the vehicle nose and leading edge may be modelled as either a sphere, cylinder or a swept cylinder. For instance, the stagnation point convective heat transfer for spherical and unswept cylinder leading edges, according to Fay and Riddell, reads:

$$\dot{q}_{co} = 0.57\left(\frac{4}{3}\right)^k Pr^{-0.6}\left(\rho_w \mu_w\right)^{0.1}\left(\rho_e \mu_e\right)_s^{0.4} \left[1+\left(Le^{\phi}-1\right)\left(\frac{h_d}{h_e}\right)\right]\left(h_e - h_w\right)\left(\frac{du_e}{dx}\right)_s^{0.5} \tag{3.19}$$

where the index $k = 0$ for 2D flow and k = 1 for axisymmetric flow. Here the subscripts w, e, and s denote conditions at the wall, the external flow, and the stagnation point. $\phi = 0.52$ for the equilibrium boundary layer and $\phi = 0.63$ for the frozen boundary layer and a fully catalytic wall.

The term in square brackets represents the effects of equilibrium chemical reactions occurring in the stagnation region and:

$$\begin{aligned} \mathrm{Pr} &= \frac{\mu c_p}{k} \\ Le &= \frac{\rho D_{12} c_p}{k} \\ h_D &= \sum_{i=1}^{n} c_i \Delta h_{f,i} \end{aligned} \tag{3.20}$$

The Prandtl number Pr and the Lewis number Le are nondimensional parameters, like the Reynolds number, and measure the relative importance of friction to heat conduction and of species diffusion (mixing) to conduction, respectively. The gas considered is air, which, for the purposes of mixing, can be considered to be a binary mixture of two species: atoms (O or N) and molecules (O_2 and N_2). The quantity D_{12} is the binary diffusion coefficient that measures the ability of species 1 to mix with species 2. The quantities c_i and $\Delta h_{f,i}$ are the molar concentrations of the individual species present (O, O_2, N, and N_2) and the chemical heat of formation of each of the species, respectively.

The Lewis number for air-like mixtures is close to unity: Le~1.4, so that the quantity $(Le^{0.52}-1) \approx 0.19$, and the contribution of the chemical reaction term can often be safely neglected in preliminary studies.

The velocity gradient along the x-axis – that is, along the body surface – at the stagnation point is:

$$\frac{du_e}{dx} = \frac{1}{R_b}\sqrt{\frac{2\left(p_e - p_\infty\right)}{\rho_e}} \tag{3.21}$$

Next, the methods used for spherical and unswept cylinder leading edges include Lee's method for laminar flow and the Detra–Hildalgo method for turbulent flow. These provide heat flux distributions around the leading edge.

As far as the influence of leading edge sweep angle is concerned, the analysis uses Lee's method with the addition of the sweep angle effect (K_{sweep}). According to the swept cylinder method, this reads:

$$\frac{\dot{q}_{\Lambda}}{\dot{q}(\Lambda=0)} = K_{sweep} \tag{3.22}$$

where K_{sweep}, according to Cato–Johnson, is:

$$K_{sweep} = \cos^{1.25}(\Lambda) + 0.05\sin(\Lambda) \tag{3.23}$$

while for Beckwith–Gallagher it reads:

$$K_{sweep} = 1 + \Lambda^2.(-1.875 + 1.097 \cdot \Lambda) \tag{3.24}$$

where Λ is the leading edge sweep angle.

The CFD code FLUENT solves the full Reynolds-averaged Navier–Stokes (RANS) equations in a finite-volume approach, with a cell-centred formulation on a multi-zone block-structured grid. In the present research effort, the thermal and chemical non-equilibrium flow field governing equations are integrated in a density-based approach with a flux difference splitting second-order upwind numerical scheme for the spatial reconstruction of the convective terms. For the diffusive fluxes, a cell-centred scheme is applied. In some computations, however, the flux vector is computed by using a flux-vector splitting scheme called the advection upstream splitting method. This provides an exact resolution of contact and shock discontinuities and it is less susceptible to carbuncle phenomena.

An implicit solver formulation was considered in the computations. Indeed, due to broader stability characteristics of the implicit formulation, a converged steady-state solution can be obtained much faster using the implicit formulation rather than the explicit formulation.

Calculation of global transport properties of the gas mixture relied on semi-empirical rules such as Wilke's mixing rule for viscosity and thermal conductivity. The viscosity and thermal conductivity of the ith species was obtained from the kinetic theory of gases. For the diffusion coefficient of the ith species in the mixture, the multi-component diffusion coefficient was applied, with species mass diffusivity evaluated by kinetic theory. Flowfield chemical reactions proceed with forward rates that are expressed in the Arrhenius form, while reaction-rate parameters are due to Park [37]. In particular, a number of in-house modifications (user defined functions; UDF) for the thermal non-equilibrium were used, since vibrational non-equilibrium conditions are not basic code features. In the UDF, vibrational relaxation is modelled using a Landau–Teller formulation, where relaxation times are obtained from Millikan and White, assuming simple harmonic oscillators.

Finally, the k-ω SST model was used to account for turbulence effects and only steady state computations have been carried out so far.

Box 3.5 Development of the aerothermodynamic database at NASA Langley Research Center.

At NASA's Langley Research Center Aerothermodynamics Branch, the format used in the development of aerodynamic/aerothermodynamic databases for several FTBs, such as the X-33 and X-34 vehicles, is that of the *aerothermodynamic chain*, the links of which are personnel, facilities, models/test articles, instrumentation, test techniques, and CFD [14, 15].

Aerothermodynamics is the foundation of the design, development, and flight of advanced space transportation vehicles and is on the critical path to success for such vehicles. It is defined here as encompassing:

- *aerodynamics:* involving forces, moments and pressure loads on the vehicle across the speed regime from take-off to orbit or beyond, and entry to landing (Mach 0.1–40)
- *aeroheating:* which includes convective and radiative heat-transfer rates for a configuration at flight conditions
- *fluid dynamics and physical processes:* which involve complex flow phenomena from the free molecular regime throughout the continuum regime (boundary-layer/shear-layer transition to turbulence, shock–shock interaction, shock impingement, flow separation and reattachment, plume-flowfield-surface interactions, and so on) and processes associated with high-temperature gases (chemical reactions, transport processes, thermodynamic non-equilibrium, gas-surface interaction, and so on).

This three-part definition of the aerothermodynamics is applied across the subsonic-to-hypersonic speed regimes for the full spectrum of aerospace vehicle configurations.

The three sources of aerothermodynamic information are as follows.

Ground-based facilities: Ground-based facilities have provided and continue to provide the majority of fundamental aerothermodynamic information for aerospace vehicle concepts. As is well recognized, duplication of all flight conditions in not possible via ground-based testing. Experimental aerothermodynamicists must resort to the simulation of important flight parameters, such as Mach number; Reynolds number and ratio of specific heat γ for subsonic to high-hypersonic conditions and to the duplication of certain physical aspects of flight (primarily velocity and the product of density and characteristic length-scale factor). Although no one facility can provide all the aerothermodynamic information required for the design of a vehicle, the combination of several facilities collectively providing ranges of Mach number, Reynolds number and γ can simulate a major portion of the flight trajectory.

CFD and/or engineering computer codes: Validated CFD may be used to predict surface and flow field conditions for the full-scale vehicle at atmospheric conditions (density, temperature and molecular weight) for points along the trajectory. The highest confidence in ground-based data and/or the pre-flight data book occurs when experimental and computational results are in full agreement.

Flight experiments: Flight experiments represent the third source of aerothermodynamic information. These experiments are generally performed with sufficient instrumentation to measure local phenomena (catalytic versus non-catalytic heating at hypersonic conditions), but not sufficient to accurately model spatially and temporally complex global phenomena such as laminar-to-turbulent BLTs and shear-layer reattachment. The advantages of flight experiments are well recognized; the primary disadvantage is that they require considerable time and cost to perform.

Utilization of this aerothermodynamic triad provides the optimum aerodynamic design to safely satisfy mission requirements while reducing design conservativism, risk and cost [14, 15].

3.9 Aerodynamic Characterization

Usually the vehicle aerodynamic characterization is provided in terms of force and moment coefficients, and control surfaces effectiveness. The force coefficients are lift (C_L), drag (C_D), and side (C_Y) and the moments refer to rolling ($C_l = C_{mx}$), pitching ($C_m = C_{my}$), and yawing ($C_n = C_{mz}$) coefficients, according to the following equations.

$$\begin{aligned}
C_i &= \frac{F_i}{\frac{1}{2}\rho_\infty V_\infty^2 S_{ref}} \quad i = L, D, Y \\
C_l &= C_{mx} = \frac{M_x}{\frac{1}{2}\rho_\infty V_\infty^2 b_{ref} S_{ref}} \\
C_m &= C_{my} = \frac{M_y}{\frac{1}{2}\rho_\infty V_\infty^2 L_{ref} S_{ref}} \\
C_n &= C_{mz} = \frac{M_z}{\frac{1}{2}\rho_\infty V_\infty^2 b_{ref} S_{ref}}
\end{aligned} \tag{3.25}$$

where S_{ref} is the reference surface (usually the vehicle planform area), L_{ref} is the longitudinal reference length (usually the fuselage length, L), and b_{ref} is the lateral-directional reference length (usually the wing-span).

3.9.1 ORV Aerodynamic Reference Parameters

For the ORV concept the geometric reference parameters (see Figure 3.35) that have been chosen in order to make aerodynamic forces (F_L, F_D, and F_Y) and moments (M_x, M_y, and M_z) non-dimensional coefficients are:

- L_{ref} (mean aerodynamic chord)
- b_{ref} (wing span)
- S_{ref} (wing area).

The pole coordinates are (x_{CoG}/L,0,0) m.

3.9.2 Reference Coordinate System and Aerodynamic Sign Conventions

In Figure 3.36 the reference frames adopted are shown. The subscript b indicates the body reference frame (BRF), s refers to the stability reference frame (SRF); while w indicates the wind reference frame (WRF). The origin of both reference systems is in the CoG of vehicle. The pole for the calculation of the moment coefficients is assumed as the CoG; the positive x-axis for the ORV body-axis system is shown parallel to the fuselage reference line.

The aerodynamic reference axis systems are sets of conventional, right-hand, orthogonal axes with the x- and z-axes in the plane of symmetry and with the positive x-axis directed out of the nose (in the body-axis system) or pointing into the component of the wind (in the stability-axis system) that lies in the plane of symmetry.

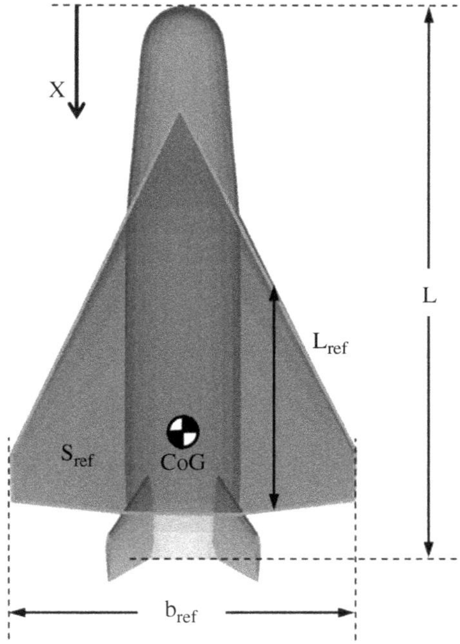

Figure 3.35 Aerodynamic reference parameters.

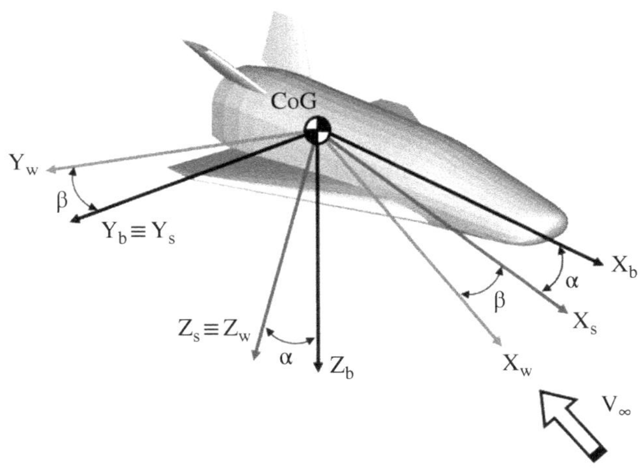

Figure 3.36 Reference frames.

The reference system for the aerodynamic data is a body-fixed axis system, compliant with the ISO 1151 standard (see Figure 3.37, in which the aerodynamic coefficient convention and sign rules are also provided).

Therefore, normal force (C_N), axial force (C_A), side force (C_Y), rolling moment (C_l), pitching moment (C_m), and yawing moment (C_n) coefficients refer to the BRF. Lift force (C_L) and drag force (C_D) coefficients are provided in the WRF.

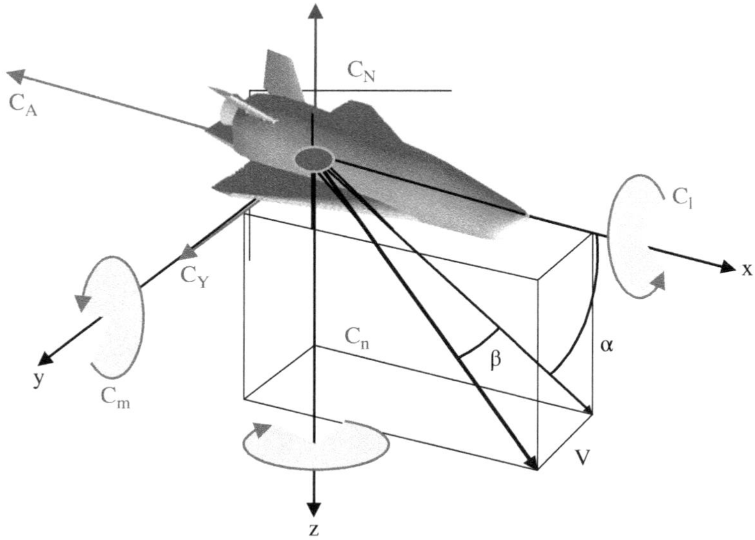

Figure 3.37 Reference frames and aerodynamic sign conventions.

The following aerodynamic sign convention (see Figure 3.37 where directions are positive as shown) for forces, moments, and velocity is adopted:

- AoA (α) is positive when free stream arrives from down of the pilot.
- Sideslip angle (β) is positive when free stream arrives from right of the pilot.
- Aileron deflection angle (δ_a) is positive when trailing edge is down.
- Elevon deflection angle (δ_e) is positive when trailing edge is down.
- Body flap deflection angle (δ_{bf}) is positive when trailing edge is down.
- Rudder deflection angle (δ_r) is positive when trailing edge turns on the left of the pilot.
- Axial force coefficient (C_A) is positive when force is pushing in front of vehicle towards the base.
- Normal force coefficient (C_N) is positive when force is pushing upwards on the belly side of vehicle.
- Side force coefficient (C_Y) is positive when force is pushing on the left-hand side of the vehicle towards the right.
- Rolling moment coefficient (C_l) is positive when the right wing is down.
- Pitching moment coefficient (C_m) is positive when the aircraft is nose up.
- Yawing moment coefficient (C_n) is positive when the right wing is backward.

This is the convention usually adopted in flight mechanics. As a result, the static stability conditions for the vehicle are the following:

- longitudinal static stability: $C_{m\alpha} < 0$
- lateral-directional static stability: $C_{n\beta} > 0$; $C_{l\beta} < 0$.

ORV aerodynamic control surface deflections, forces, and hinge moments are illustrated in Figure 3.38 and summarized in Table 3.4. In general, control surface deflection

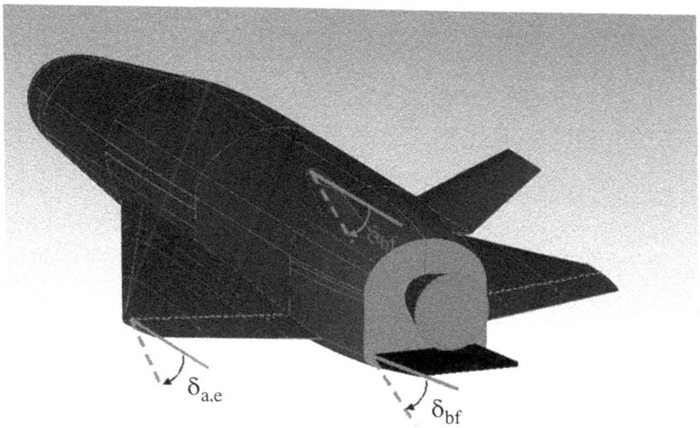

Figure 3.38 ORV aerodynamic control surface deflections, forces, and hinge moments.

Table 3.4 ORV aerodynamic control surface deflections, forces, and hinge moments.

Positive deflection of	Aero forces and moments
Rudder, δ_r	$+C_Y$, $-C_n$
Elevon, δ_e	$-C_m$
Right, $\delta_{e,R}$	$-C_l$
Left, $\delta_{e,L}$	$+C_l$

angles are measured in a plane perpendicular to the control surface hinge axis. An exception is the rudder control surface deflections, which are measured in a plane parallel to the fuselage reference plane.

3.9.3 Inputs for ORV AEDB Generation

Based on the re-entry flight scenario in Figures 3.15 and 3.16 the aerodynamic data set has been generated for the flight envelope bounded by the ranges shown in Table 3.5.

It must be noted that the range of Reynolds number was chosen to cover a wide part of the re-entry, based on a preliminary re-entry trajectory, as shown in Figure 3.39, which shows the ORV preliminary reference flight envelope, together with the iso-Mach and iso-Reynolds curves, in an altitude–velocity map.

3.9.4 ORV Aerodynamic Model

This section contains the ORV aerodynamic model (AM) for the development of the vehicle AEDB, used for flight mechanics analysis, subsystem design and analyses, as well as for flight control analysis.

An important aspect of developing an AEDB is the formulation of an aerodynamic model. For example, the accuracy of the database depends on the degree to which the AM represents the physics of the problem. Therefore, it is important that all the aerodynamic and control variables that may have influences on the given aerodynamic

Table 3.5 Flight envelope ranges.

Parameter	Value	Comment
Subsonic to supersonic flow conditions		
M_∞	[0.3, 0.5, 0.7, 0.9, 1.1, 1.3, 1.5, 1.7]	
$Re_{\infty/m}$	$1 \times 10^7\ m^{-1}$	
α	[0, 2, 4, 6, 8, 10, 12, 14, 16, 18, 20]	
β	[−8, −4, −2, 0, 2, 4, 8]	
δ	[−30, −20, −10, 10, 20, 30]	For aileron, elevon and rudder
For the remaining part of the re-entry trajectory		
M_∞	[2, 3, 4, 6, 8, 12, 16, 20, 25]	
$Re_{\infty/m}$	[2×10^4, 10^5, 5×10^5, 2×10^6]	
α	[0, 2, 5, 7, 10, 12, 14, 15, 17, 20, 22, 25, 27, 30, 35, 40, 45, 50]	
β	[−4, −2, 0, 2, 4]	
δ	[−30, −20, −10, 10, 20, 30]	For aileron, elevon and rudder

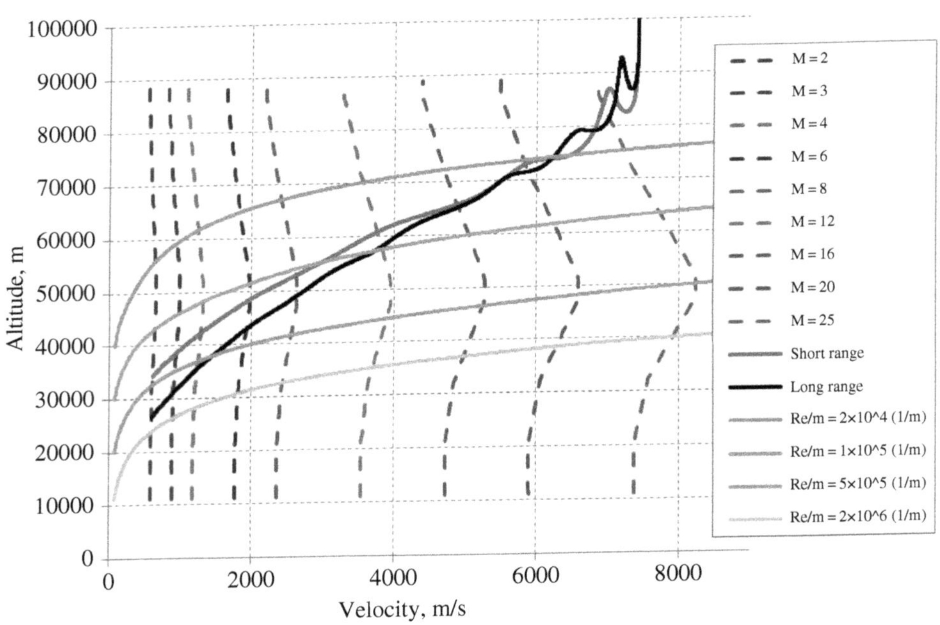

Figure 3.39 ORV re-entry scenario in an altitude–velocity map with iso-Mach and Reynolds curves.

coefficient be included in the aerodynamic model. To this end, a number of hypotheses have been assumed in developing the ORV AM. The independent variables that have been recognized as having influence on the ORV aerodynamic state are the set: $\{M, Re, \alpha, \beta, \delta_e, \delta_a, \delta_r, q, \dot{\alpha}\}$

The couple (M, Re)[4] identifies the aerodynamic environment, while the remaining variables completely describe the flowfield direction. Therefore, the functional structure of the AM of the ORV is based on these independent variables. Note, however, that the AEDB under development at this phase of the project does not consider the contribution of dynamic effects because low-order methods, widely used in this framework, are unable to reliably determine this contribution.

As done in the past for the US Orbiter and X-34 vehicles, the ORV AM development relies on the following assumptions:

- No RCS effects are considered.
- Only rigid body aerodynamic coefficients are evaluated; in other words, no aeroelastic deformations are accounted for.
- No Reynolds and Knudsen numbers effects on aerodynamic control surfaces are assumed.
- No sideslip effects on aerodynamic control surfaces are assumed, except for rudder efficiency.
- No effects of protrusions, gaps and roughness are considered.
- No Knudsen numbers effects on side force and aerodynamic moment coefficients are assumed (except for pitching moment coefficient).
- No mutual aerodynamic interference between control surfaces is considered.

Finally, it is worth noting that, as typically done in a classical approach, each aerodynamic coefficient can be derived by supposing that each contribution to the single global coefficient is independent from the others. This means, from an operational point of view, that each aerodynamic coefficient is described by a linear summation over a number of incremental contributions (in other words, a build-up approach). Each contribution is based on a small number of parameters.

3.9.5 Formulation of the Aerodynamic Database

The aerodynamic characteristics pertaining to the longitudinal (in other words, C_L, C_D, and C_m) and lateral (C_Y, C_l, and C_n) degrees of freedom are presented as full-scale rigid force and moment coefficients. They are presented in a form that allows a build-up to any desired configuration and/or flight condition by increments to the basic coefficient.

Each aerodynamic coefficient has been considered separately using the appropriate equation in which all the required contributions for obtaining the total coefficient for any selected flight condition appear. Following the formulation used for the Space Shuttle and assuming that the vehicle is operating at a combined AoA (α), and AoS (β), the total lift coefficient is given by:

$$\begin{aligned} C_{L,total} = & C_{L,b}(\alpha, \mathrm{M}, \mathrm{Re}) + \Delta C_{L,\delta_e} + \Delta C_{L,\delta_a} + \Delta C_{L,\delta_{bf}} \\ & + \Delta C_{L,\delta_r} + \Delta C_{L,b,\beta} + \Delta C_{L,\delta_r,\beta} \end{aligned} \tag{3.26}$$

where $C_{L,total}$ is the total lift coefficient of the vehicle for a given flight condition as expressed by the flight Mach number M, AoA is α, sideslip is β, elevon deflection is δ_e,

4 Note that the couple (M, Re) is also able to define the rarefied flow regime conditions since the Knudsen number is proportional to the Mach-to-Reynolds ratio; see Eq. (3.1).

ailerons deflections are δ_a, body flap deflections are δ_{bf}, and rudder deflection is δ_r. The parameter $C_{L,b}(\alpha, M, \mathrm{Re})$ is the baseline lift coefficient at zero sideslip and zero control surface deflections (in other words, in a clean configuration). It also takes into account rarefaction effects through a bridging relationship. The parameter $\Delta C_{L,\delta_e}$ represents the incremental lift coefficient due to elevon deflections above the baseline and is given by:

$$\Delta C_{L,\delta_e} = C_L(\alpha, M, \delta_e) - C_{L,b}(\alpha, M) \tag{3.27}$$

The parameter $\Delta C_{L,\delta a}$ represents the incremental lift coefficient due to aileron deflections above the baseline and can be evaluated using the data on elevons as follows:

$$\Delta C_{L,\delta a} = \left(\frac{\Delta C_{L,\delta_e = \delta_e, L} + \Delta C_{L,\delta_e = \delta_e, R}}{2} \right) - \Delta C_{L,\delta_e} \tag{3.28}$$

Here, we use the elevon data twice; once assuming $\delta_e = \delta_{e,L}$, obtaining $\Delta C_{L,\delta_e = \delta_e, L}$ and then assuming $\delta_e = \delta_{e,R}$ to determine $\Delta C_{L,\delta_e = \delta_e, R}$. As a check, when the aileron deflection is zero, in other words, $\delta_e = \delta_{e,L} = \delta_{e,R}$, the value of $\Delta C_{L,\delta_a}$ vanishes as expected.

The parameter $\Delta C_{L,\delta_{bf}}$ represents the incremental lift coefficient due to body flap deflections above the baseline and is given by:

$$\Delta C_{L,\delta_{bf}} = C_L(\alpha, M, \delta_{bf}) - C_{L,b}(\alpha, M) \tag{3.29}$$

The incremental lift coefficient $\Delta C_{L,\delta_r}$ due to the rudder is defined as follows:

$$\Delta C_{L,\delta_r} = C_L(\alpha, M, \delta_r) - C_{L,b}(\alpha, M) \cong 0 \tag{3.30}$$

The incremental lift coefficients due to the baseline and rudder in sideslip are given by:

$$\Delta C_{L,b,\beta} = C_L(\alpha, \beta, M) - C_L(\alpha, M) \tag{3.31}$$

$$\Delta C_{L,\delta r,\beta} = \left[C_L(\alpha, \beta, M, \delta_r) - C_L(\alpha, \beta, M) \right] - \Delta C_{L,\delta r}$$

Note that the first term in square brackets on the right-hand side of the last equation gives the combined incremental coefficient due to the rudder at an AoA and sideslip over the baseline at the same values of AoA and AoS. To get the incremental coefficient due only to sideslip β, we have to subtract the incremental due to AoA as shown by second term on the right-hand side of equation.

Those contributions represent aerodynamic cross-coupling effects, and they have been found to be significant, especially at higher values of AoA.

In a similar fashion, we assume that the drag and pitching moment coefficients are given by:

$$\begin{aligned} C_{D,total} = {} & C_{D,b}(\alpha, M, \mathrm{Re}) + \Delta C_{D,\delta_e} + \Delta C_{D,\delta_a} + \Delta C_{D,\delta_{bf}} \\ & + \Delta C_{D,\delta_r} + \Delta C_{D,b,\beta} + \Delta C_{D,\delta_r,\beta} \end{aligned} \tag{3.32}$$

$$\begin{aligned} Cmy_{total} = {} & Cmy_b(\alpha, M, \mathrm{Re}) + \Delta Cmy_{\delta_e} + \Delta Cmy_{\delta_a} + \Delta Cmy_{\delta_{bf}} \\ & + \Delta Cmy_{\delta_r} + \Delta Cmy_{b,\beta} + \Delta Cmy_{\delta_r,\beta} \end{aligned} \tag{3.33}$$

The change in pitching moment coefficient due to rudder deflection, $\Delta Cmy_{,\delta_r}$, is assumed to be zero. The side force coefficient is assumed to be given by:

$$\begin{aligned} C_{Y,total} &= C_{Y,b}(\alpha,M) + \Delta C_{Y,\delta_a} + \Delta C_{Y,\delta_r} + \Delta C_{Y,b,\beta} + \Delta C_{Y,\delta_r,\beta} \\ &= \Delta C_{Y,\delta_a} + \Delta C_{Y,\delta_r} + \Delta C_{Y,b,\beta} + \Delta C_{Y,\delta_r,\beta} \end{aligned} \tag{3.34}$$

since the vehicle configuration is symmetric; in other words, $C_{Y,b}(\alpha,M)=0$. Furthermore,

$$\Delta C_{Y,b,\beta} = C_{Y,b}(\alpha,\beta,M) - C_{Y,b}(\alpha,M) = C_{Y,b}(\alpha,\beta,M) \tag{3.35}$$

Similarly,

$$\begin{aligned} \Delta C_{Y,\delta_a} &= C_Y(\alpha,M,\delta_a) \\ \Delta C_{Y,\delta_r} &= C_Y(\alpha,M,\delta_r) \end{aligned} \tag{3.36}$$

Then,

$$C_{Y,total} = C_{Y,b}(\alpha,\beta,M) + C_Y(\alpha,M,\delta_a) + C_Y(\alpha,M,\delta_r) + \Delta C_{Y,\delta_r,\beta} \tag{3.37}$$

where the incremental side force coefficient due to rudder is sideslip is defined as

$$\Delta C_{Y,\delta_r,\beta} = \left[C_Y(\alpha,\beta,M,\delta_r) - C_{Y,b}(\alpha,\beta,M)\right] - \Delta C_{Y,\delta r} \tag{3.38}$$

Proceeding in a similar way, the rolling and yawing moment coefficients are assumed to be given by:

$$Cmx_{total} = Cmx_b(\alpha,\beta,M) + Cmx(\alpha,M,\delta_a) + Cmx(\alpha,M,\delta_r) + \Delta Cmx_{\delta_r,\beta} \tag{3.39}$$

$$Cmz_{total} = Cmz_b(\alpha,\beta,M) + Cmz(\alpha,M,\delta_a) + Cmz(\alpha,M,\delta_r) + \Delta Cmz_{\delta_r,\beta} \tag{3.40}$$

Therefore, it is assumed that the sideslip has effect only on the baseline, and when the rudder is deflected, but it has no effect when the elevons, body-flap or ailerons are deflected.

Moreover, as stated above, no dependency on Reynolds number is assumed for the elevon contribution. Indeed, even though this effect exists it is small and difficult to model. It has been considered as part of the uncertainty of the aerodynamic coefficients.

3.9.6 Process of Development of the Aerodynamic Database

The above formulation of the AM for free flight provides a framework for building the ORV AEDB. It consists of aerodynamic data tables in the form of total and incremental coefficients. These are provided in a convenient form so that the user can evaluate each of the terms in the free-flight aerodynamic models, and then sum them to get the desired aerodynamic coefficient.

In particular, the ORV AEDB relies on following segments:

- the free-molecular flow conditions
- the transitional flow conditions
- the continuum flow conditions

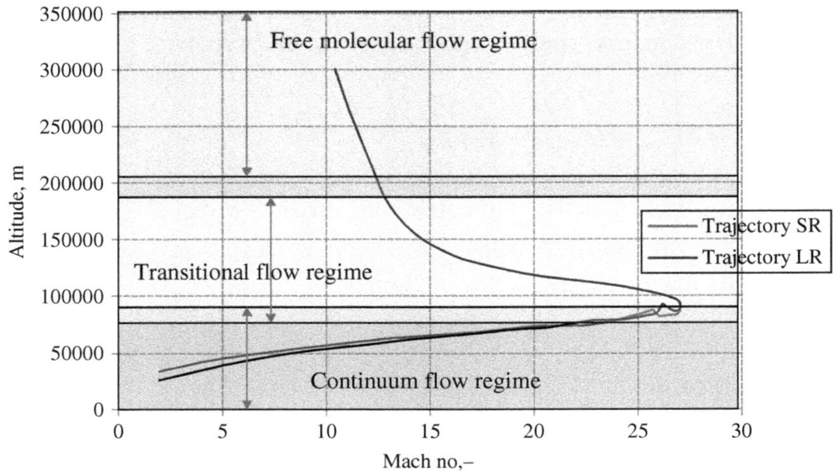

Figure 3.40 ORV flow regime in the altitude-Mach map.

- hypersonic flow
- supersonic flow
- transonic flow
- subsonic flow.

Figure 3.40 is a guideline used in developing vehicle AEDBs.

It shows the ORV reference flight scenario in an altitude-Mach map together with the altitude limits that bound the different flow regimes.

Finally, it is worth noting that in transitional flow conditions a very simple relationship to bridge the FMF regime to continuum flow (see Figure 3.40) can be given as:

$$C_{i\text{Transitional}} = C_{i\text{Continuum}} + \left(C_{i\text{FM}} - C_{i\text{Continuum}}\right) \cdot \bar{C}_i \tag{3.41}$$

where the normalized coefficient $\bar{C}_i$ uses the Knudsen number as the independent parameter:

$$\bar{C}_i = \frac{C_i - C_{i\text{Continuum}}}{C_{i\text{FM}} - C_{i\text{Continuum}}} = F\left(Kn_\infty\right) = \sin^2\left[\frac{\pi}{8}\left(3 + Log_{10} Kn_\infty\right)\right] \tag{3.42}$$

where $10^{-3} < Kn_\infty < 10$ and $C_{i\text{Continuum}}$ and $C_{i\text{FM}}$ are the aerodynamic coefficients in the continuum and FMF regimes, respectively. This formula was used for the US Orbiter aerodynamics assessment [31].

The results of each AEDB segment are described in detail in the next sections.

3.10 Low-order Methods Aerodynamic Results

Simplified aerodynamic analyses for supersonic and hypersonic speeds were performed on panel meshes similar to those shown in Figure 3.41.

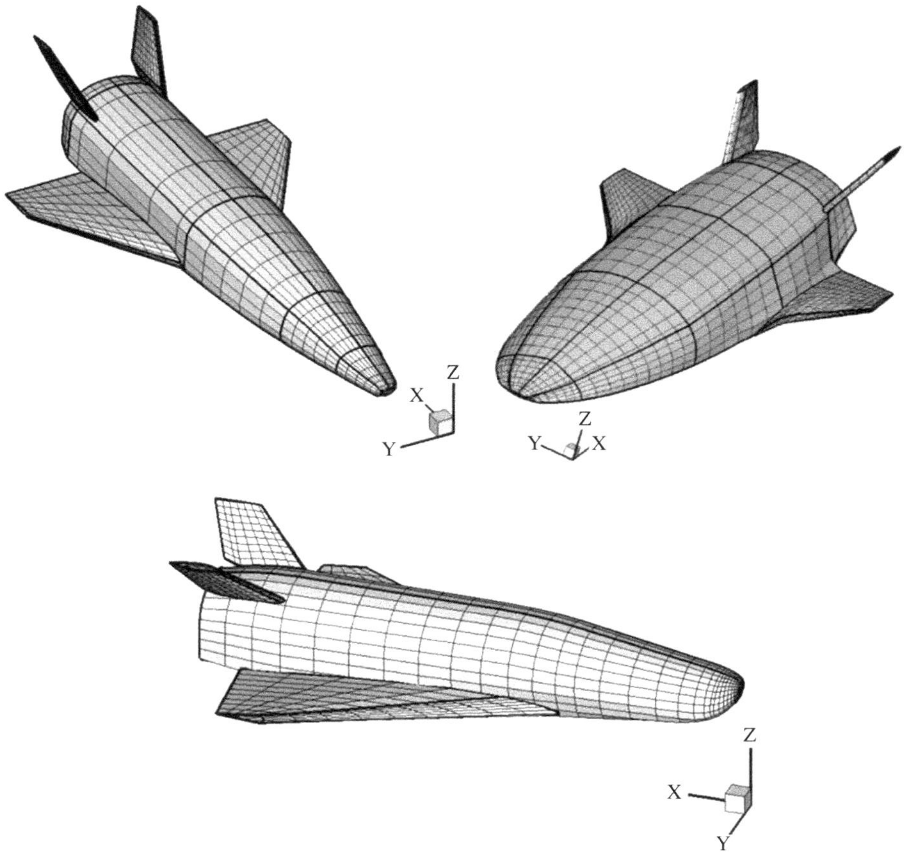

Figure 3.41 Examples of the ORV panel mesh.

3.10.1 HPM Results for Rarefied and Transitional Flow Conditions

As shown in Figures 3.17 and 3.40, above 200 km altitude the ORV experiences FMF conditions. This means that particle simulations, such as DSMC, should be performed to address the vehicle aerodynamics. At an early design stage, however, only low-order analyses, such as those obtained from SIM computations, can be carried out. The aerodynamic analysis was performed on the vehicle in a clean configuration, and at a free-stream velocity of 7330 m/s, considered constant with altitude. In all computations the wall temperature was set equal to 300K (with a body temperature ratio equal to 0.351), and the aerodynamic forces were evaluated on the assumption of a fully accommodated Maxwell model. Free-stream thermodynamic parameters were provided by the US Standard Atmosphere, 1976.

The ORV aerodynamics in FMF conditions are summarized in Figures 3.42–3.44.

These figures highlight that FMF conditions result in an abrupt loss of vehicle aerodynamic efficiency due to a sudden increase in drag and lift drop.

As far as transitional-flow aerodynamics are concerned, Figures 3.17 and 3.40 show that the ORV flies in that regime from about 200–87 km altitude. Lift, drag and

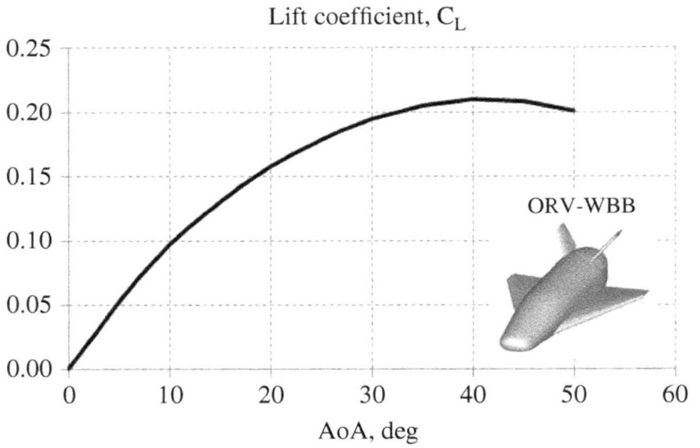

Figure 3.42 ORV-WBB lift coefficient vs AoA in FMF.

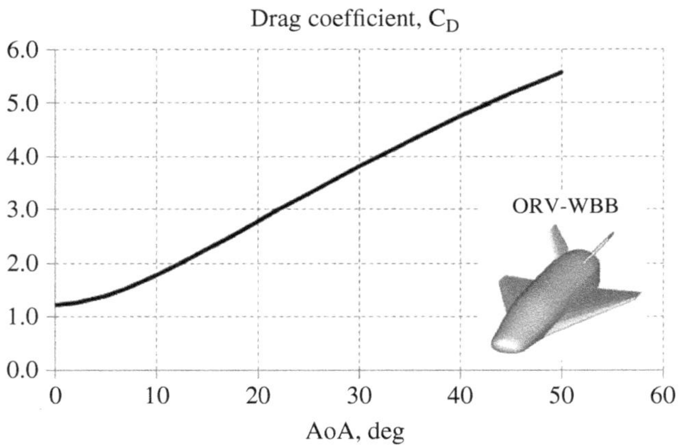

Figure 3.43 ORV-WBB drag coefficient vs AoA in FMF.

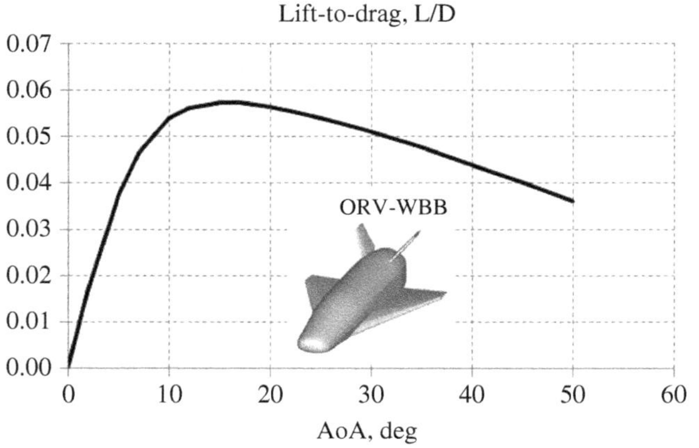

Figure 3.44 ORV-WBB aerodynamic efficiency vs AoA in FMF.

pitching moment coefficients versus the Knudsen ($Kn_{Lref\infty}$) number are reported in Figures 3.45–3.48 for AoAs of 10, 20, 30, and 40°. The behaviour of the lift coefficient at 10 and 20° and at 30 and 40° is shown in Figures 3.45 and 3.46, respectively.

The variation of drag and pitching moment coefficient of ORV-WBB at 30 and 40° AoA, along with the Knudsen number is reported in Figures 3.47 and 3.48, respectively.

The classical *S-shape* of the bridging formula is clearly recognized, ranging from continuum to FMF. Furthermore, note that in each figure the variation of the altitude with Knudsen number is also provided to help the reader understand with what flight conditions the aerodynamic results must be associated. The effect of rarefaction on the aerodynamic lift results in an abrupt loss of C_L. For example, at AoA = 40° C_L decreases by about 87.5% from 80 km to 200 km. The same consideration applies for C_m. As far as aerodynamic drag is concerned, Figure 3.47 shows that C_D at AoA = 30° increases by

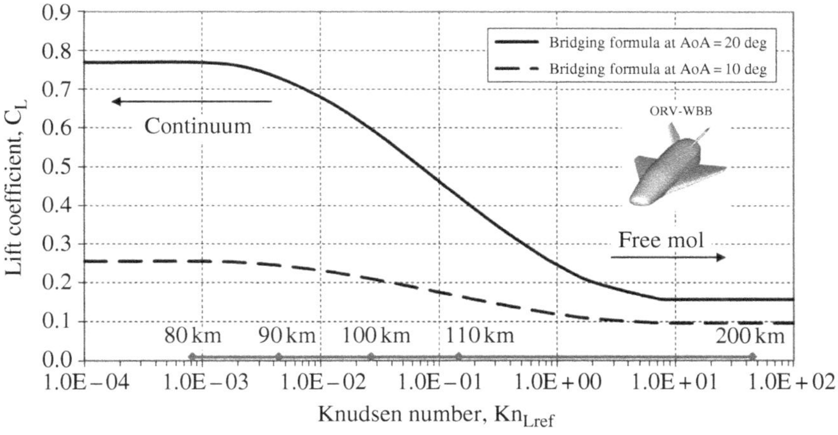

Figure 3.45 Lift coefficient vs Knudsen number for ORV-WBB at AoA = 10 and 20°.

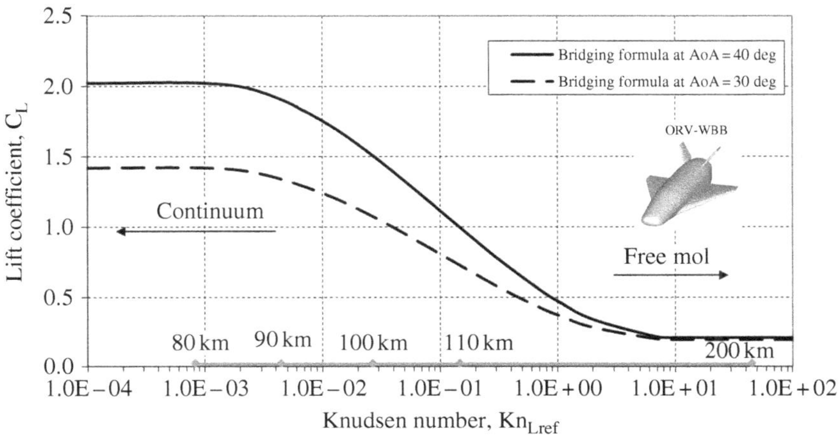

Figure 3.46 Lift coefficient vs Knudsen number for ORV-WBB at AoA = 30 and 40 deg.

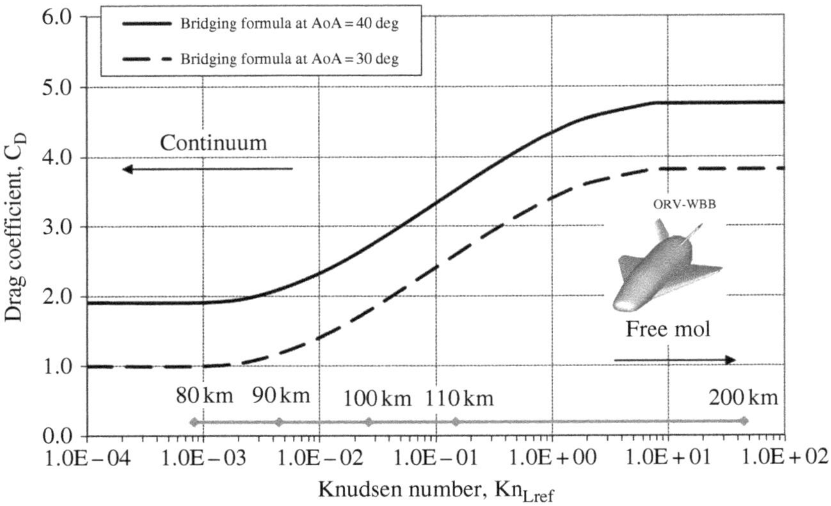

Figure 3.47 Drag coefficient vs Knudsen number for ORV-WBB at AoA = 30 and 40°.

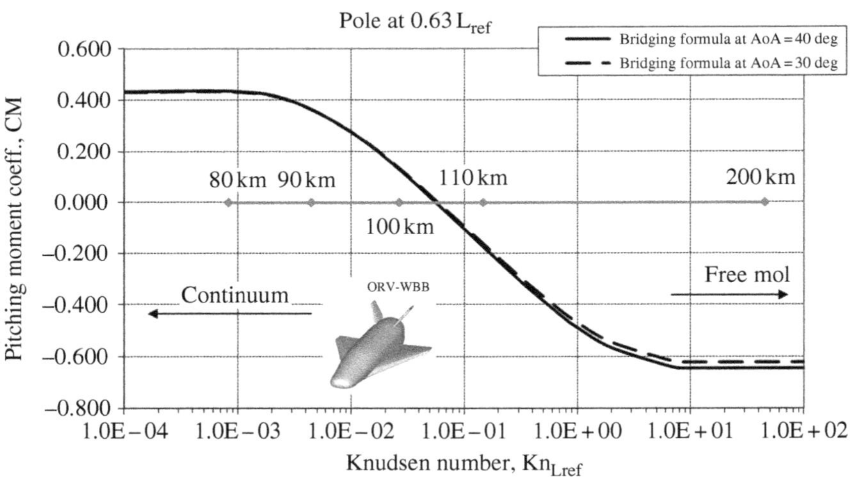

Figure 3.48 Pitching moment coefficient vs Knudsen number for ORV-WBB at AoA = 30 and 40°.

about 200% from 90 km to 110 km, whereas the drag at H = 200 km is 250% higher than at 90 km. Therefore, the strong reduction of the aerodynamic efficiency due to rarefaction effects is very clear in Figure 3.49.

3.10.2 HPM Results for Continuum Flow Condition

HPM results refer to both clean and flapped configurations. Trade-off design analyses highlight that the best surface inclination methods (see Table 3.3) to consider in assessing vehicle aerodynamic performance in continuum flow conditions are the empirical

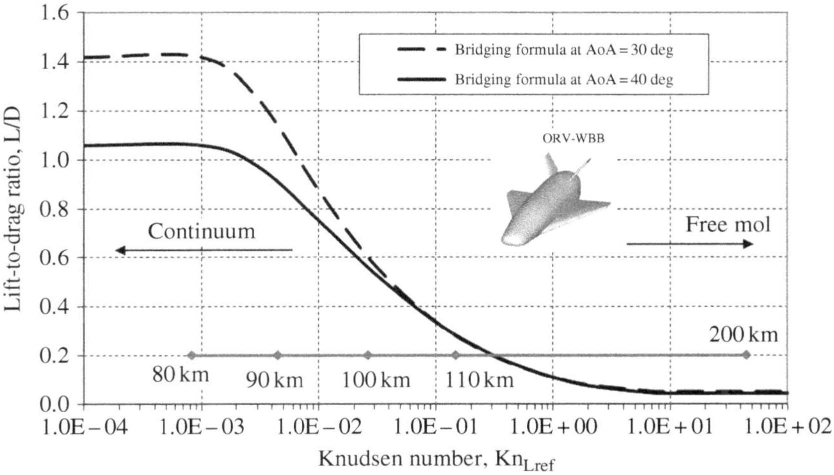

Figure 3.49 ORV-WBB aerodynamic efficiency vs Knudsen number for AoA = 30 and 40°.

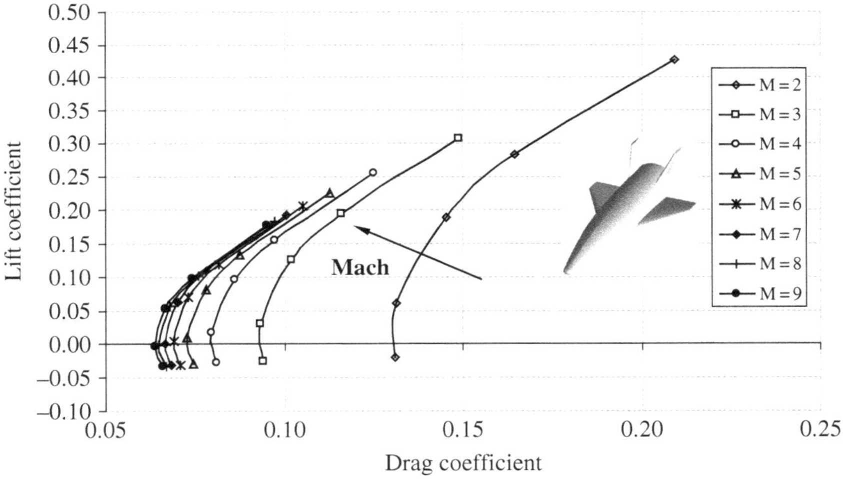

Figure 3.50 ORV-WSB aerodynamic polars for $2 \leq M_\infty \leq 9$.

tangent cone and tangent wedge, for the fuselage[5] and wing windside, respectively. A modified Newton–Prandtl–Meyer approach is used for the leading edges and the Newtonian method (in other words, $c_p = 0$) applies at the vehicle leeside [32, 33].

Some of main results obtained for the ORV-WSB for clean-configuration aerodynamics (in other words, no aerodynamic surfaces deflected) are shown in Figures 3.50–3.53.

Figures 3.50 and 3.51 show the aerodynamic polars and Figures 3.52 and 3.53 the pitching moment coefficients for Mach numbers ranging from 2 to 25 and α from 0 to 40°.

5 A tangent wedge empirical is applied at the fuselage windside of the ORV-SB.

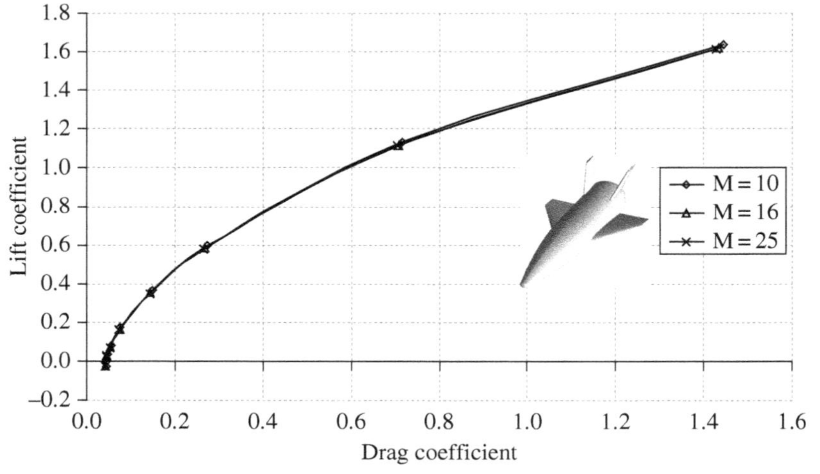

Figure 3.51 ORV-WSB aerodynamic polars for $M_\infty = 10$, 16, and 25.

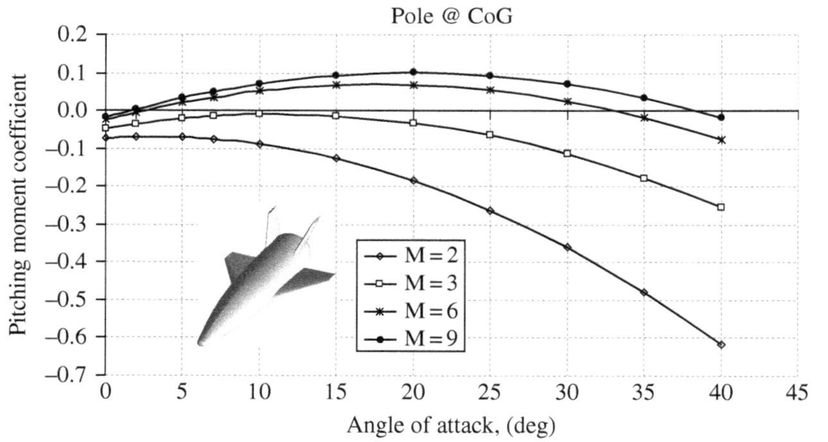

Figure 3.52 ORV-WSB pitching moment coefficient for $M_\infty = 2$, 3, 6, and 9.

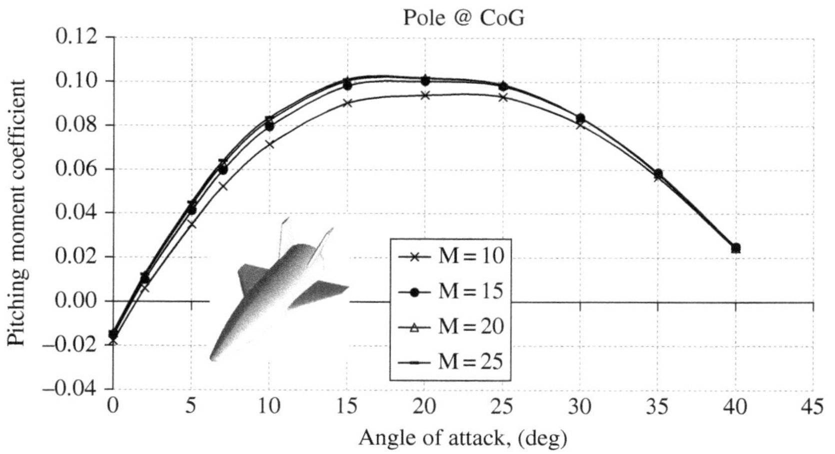

Figure 3.53 ORV-WSB pitching moment coefficient for $M_\infty = 10$, 15, 20, and 25.

As shown, ORV-WSB drag and lift decrease as Mach number increases, ultimately reaching a value that does not change even if Mach still rises, according to the Oswatich principle (in other words, independence of aerodynamic coefficients from M_∞) starting from $M_\infty = 10$ [13, 17]. Figure 3.52 also shows that at $M_\infty = 2$ and $M_\infty = 3$ the pitching moment derivative is negative for α larger than 5° and 15°, respectively. At hypersonic conditions, the configuration is statically stable (in other words, $C_{m\alpha} < 0$) for α higher than 20°. In particular, the concept in clean configuration features a natural trim point ($C_m = 0$) at about 33–38° AoA for $M_\infty = 6$ and 9. At higher Mach number trims, AoA ranges from about 42 to 44° (see Figure 3.53).

Box 3.6 An example of aerodynamic trade-off design.

As an example of the aerodynamic trade-off design results, Figures 3.54 and 3.55 show the aerodynamic polars (α = 0, 2, 5, 7, and 10°) and the pitching moment coefficients of all the competing FTB configurations of Figure 3.2, for $M_\infty = 6$ and $M_\infty = 7$, respectively [26, 34].

Note that the ORV-WSB concept, during trade-off analysis has been given the code string FTB_4-521. This code allows configurations to be distinguished from each other. The first four digits indicate the programme (for example, FTB_4 indicates the fourth series of the flying test bed); while the first, second and third digits of the remainder refer to fuselage, wing and tail, respectively. Therefore, the configurations FTB_4-211 and FTB_4-222 feature the same fuselage but different wings and tails (see Figure 3.2).

The differences are due to the combined effects of nose-up forebody and different wings: both the planform area and the sweep angle of the wing number 2 are higher

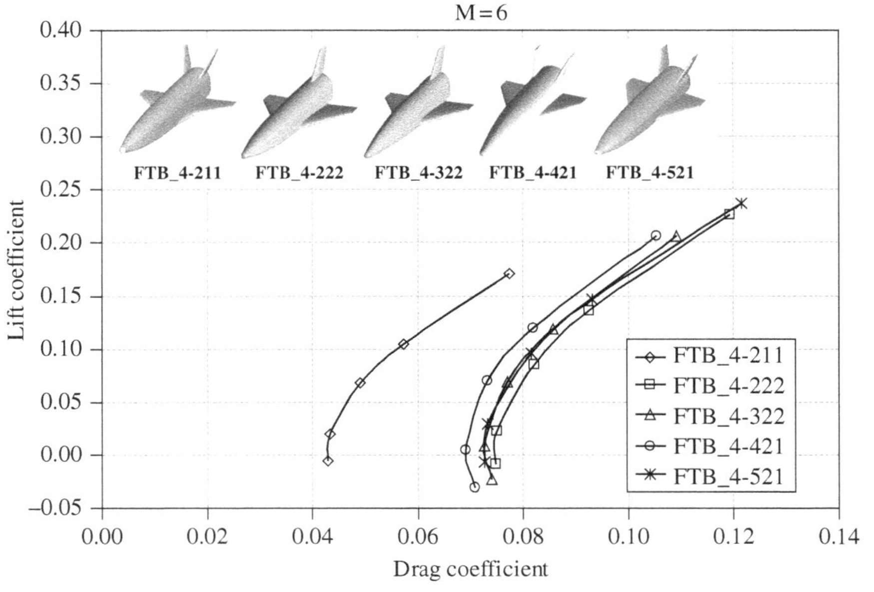

Figure 3.54 FTB_4 aerodynamic polars at $M_\infty = 6$.

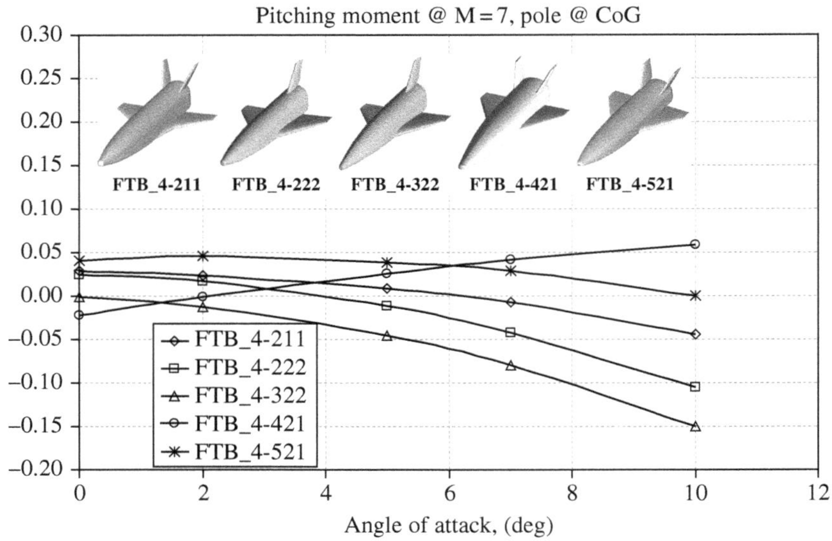

Figure 3.55 FTB_4 pitching moment coefficients at $M_\infty = 7$.

than those of wing number 1. Both FTB_4-322 and FTB_4-421 have the same lift coefficient since their planform area is the same.

For $\alpha \geq 5°$ the FTB_4-211 configuration shows lower aerodynamic lift force because its planform area is the lowest. As far as drag is concerned, all the configurations have the same C_D except for FTB_4-211, which is the sharp aeroshape. Figure 3.55 also shows that all the configurations are statically stable for α higher than 2° except for FTB_4-421 [26]. This configuration is characterized by static stability for AoAs larger than 15°. In particular, the concepts 222, 211 and 521 are trimmed respectively at about 4, 6 and 10° in their clean configurations.

FTB_4-322 can be trimmed through negative flap deflections for positive AoAs, whereas for FTB_4-222 α must be greater than 4°. FTB_4-521 is able to perform re-entry flight at a rather low AoA, thus flying like an airplane and not at a high AoA, as in the classical re-entry flight of the US Space Shuttle.

As far as the lateral-directional stability is concerned, Figures 3.56 and 3.57 show for $\alpha = 5°$ the effect of sideslip on rolling and yawing moment coefficients along with Mach number, respectively.

As shown, the FTB_4-421 configuration is statically stable in lateral-directional flight.

Recall that the safe flight of an airplane depends on the static directional stability (the weather vane effect) and on the dihedral effect (roll due to yaw). For directional stability, $C_{n\beta} > 0$. For the dihedral effect, $C_{l\beta} < 0$.

The effect of the wing flaps on vehicle aerodynamic coefficients as a function of aileron deflection and AoA is shown in Figures 3.58 and 3.59 for Mach number equal to 6 and 7 and $\alpha = 5°$. For all the cases, the magnitude of the increments increases with AoA, with a linear trend. In particular, from Figure 3.58 it can be seen that, at $M_\infty = 7$, for both the FTB_4-211 and FTB_4-521 configurations, the wing flap deflection needed to trim the

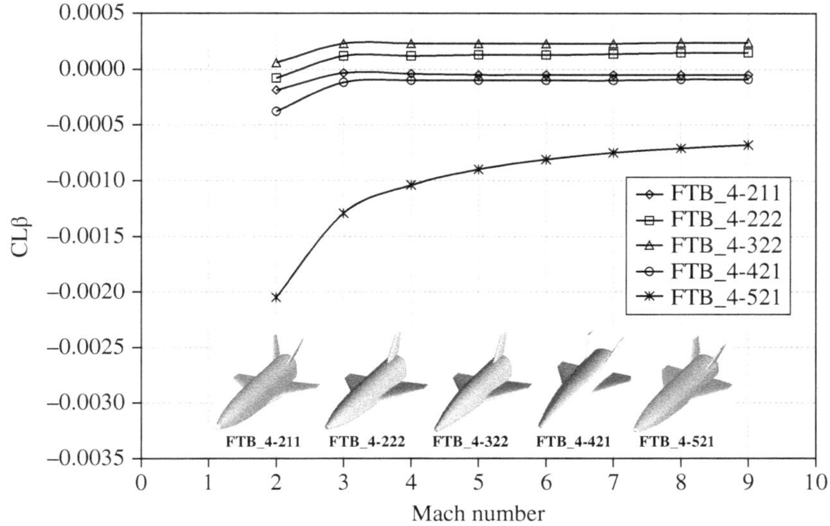

Figure 3.56 FTB_4: effect of sideslip on rolling moment coefficients along with Mach at $\alpha = 5°$.

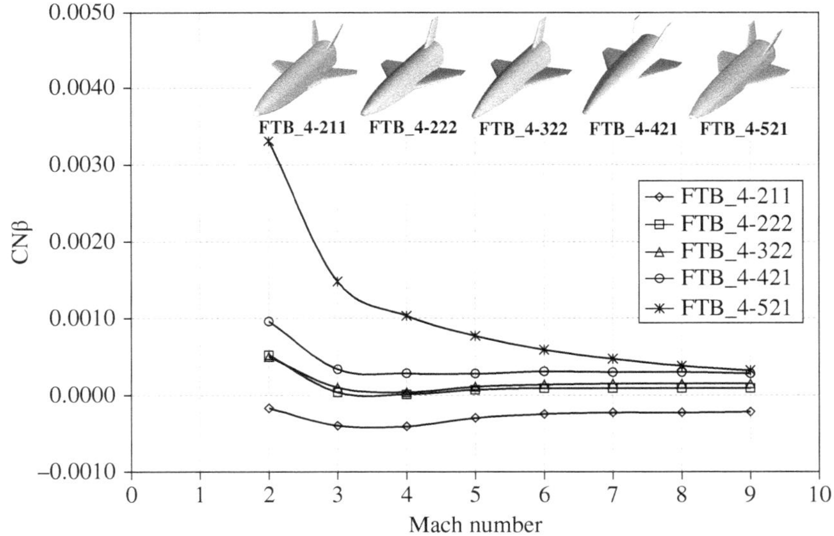

Figure 3.57 FTB_4: effect of sideslip on yawing moment coefficients along with Mach at $\alpha = 5°$.

vehicle is equal to about 10°, whereas at $M_\infty = 6$ the latter concept can be trimmed at a lower flap deflection (say about 7°).

For low deflections, Figure 3.59 highlights that the aerodynamic efficiency slightly increases. However, it must be stressed that this result is a consequence of the C_m trend, which depends on the final real position of the CoG. The current design is made to realize

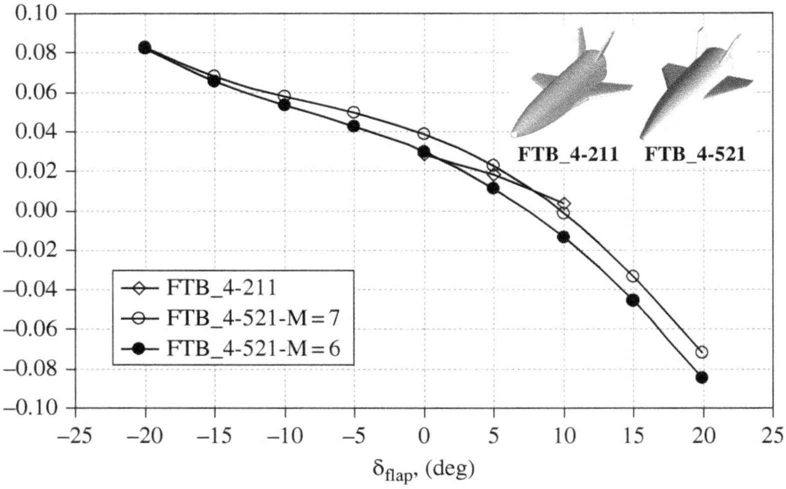

Figure 3.58 Effect of wing flap deflection on C_m at $M_\infty = 6, 7$ and $\alpha = 5°$. Pole at CoG.

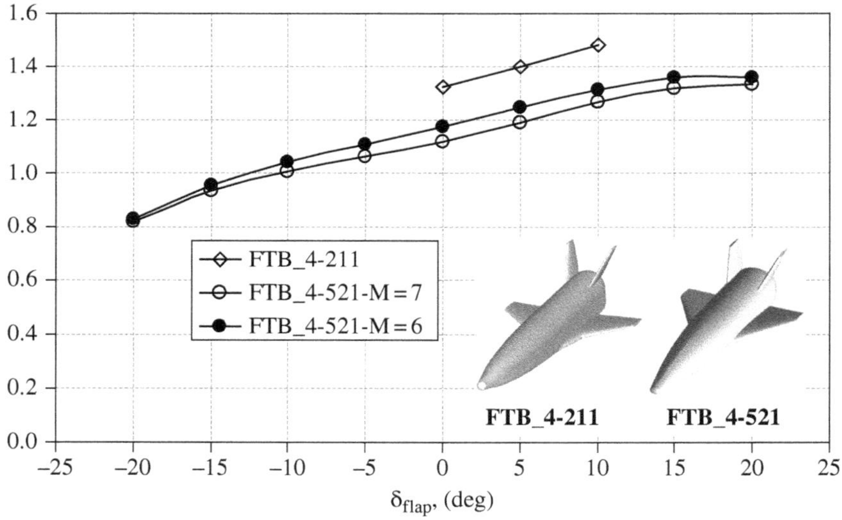

Figure 3.59 Effect of wing flap deflection on L/D at $M_\infty = 6, 7$ and $\alpha = 5°$.

small positive values of C_m in the flight conditions of interest, but the trim and stability analysis should also guarantee a sufficient margin in order to avoid negative values of C_{my}.

Finally, Figures 3.60 and 3.61 show the pressure coefficient (c_p) contours on the FTB_4-211 vehicle surface, flying at $\alpha = 5°$ at $M_\infty = 2$ and $M_\infty = 7$, respectively.

Another design issue dealt with in the trade-off analysis is the shock–shock interaction (SSI) phenomenon, detected by means of CFD simulations. For example, the SSI that

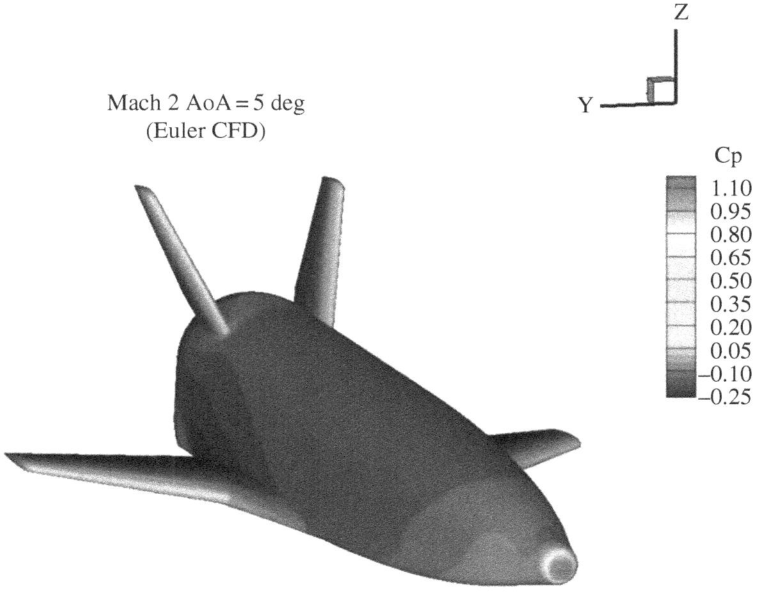

Figure 3.60 Pressure coefficient contours on vehicle surface (FTB_4-211) at $M_\infty = 2$ and $\alpha = 5°$.

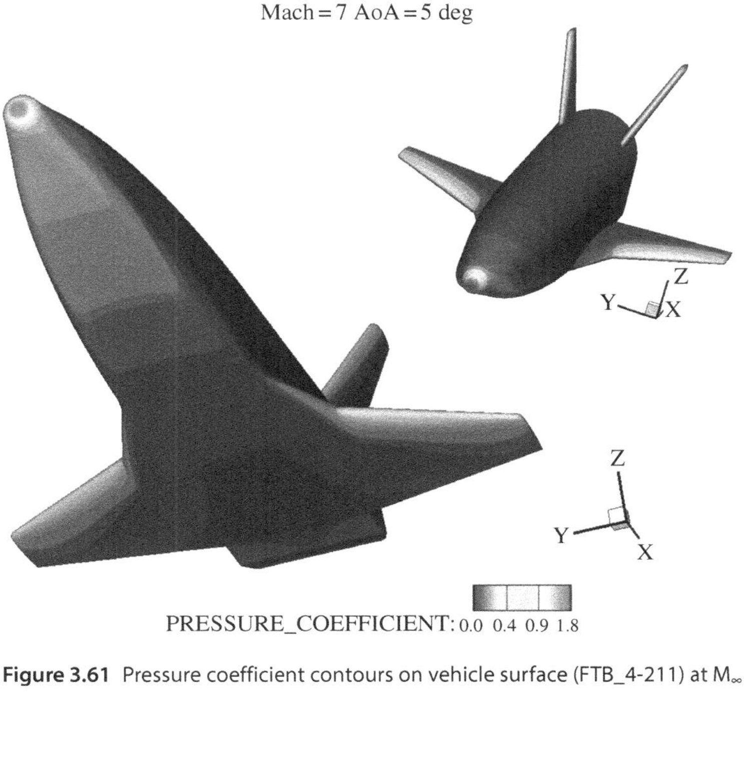

Figure 3.61 Pressure coefficient contours on vehicle surface (FTB_4-211) at $M_\infty = 7$ and $\alpha = 5°$.

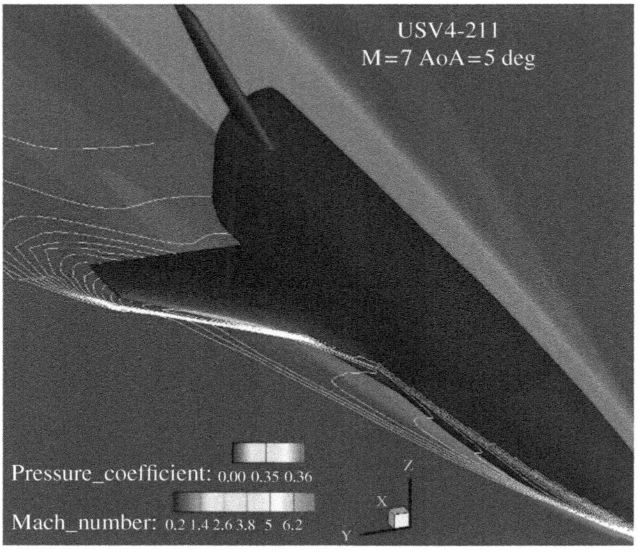

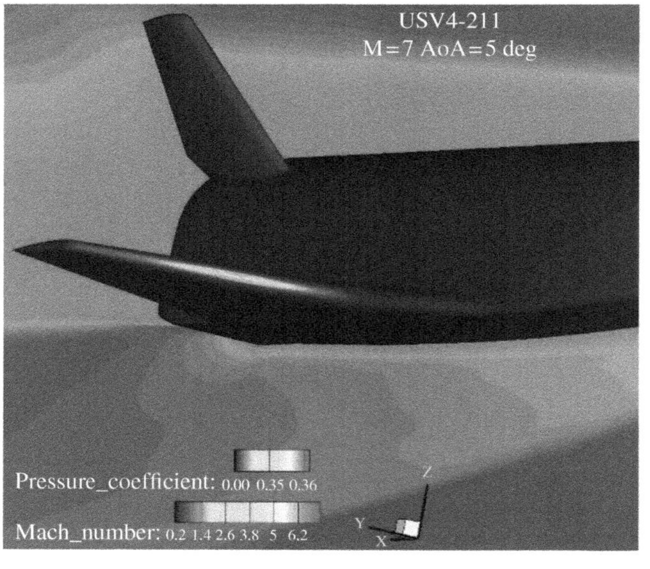

Figure 3.62 Contours of pressure coefficient over the FTB_4-211 surface. Mach number contours on vehicle symmetry plane and wing plane. (*See insert for color representation of this figure.*)

takes place on the FTB_4-211 at $M_\infty = 7$ and $\alpha = 5°$ can be seen in Figure 3.62, where contour plots of pressure coefficient on the vehicle surface and of Mach number on the FTB_4-211 symmetry plane are shown. The left-hand part of Figure 3.62 shows the bow shock cut in the wing plane, where the interaction with bow and wing shocks is clearly seen. This interaction results in pressure and heat flux overshoots where the SSI impinges the wing leading edge, as clearly shown in Figure 3.62. Therefore, for a reliable wing aeroheating estimation Navier–Stokes computations will be mandatory as the FTB_4 design matures.

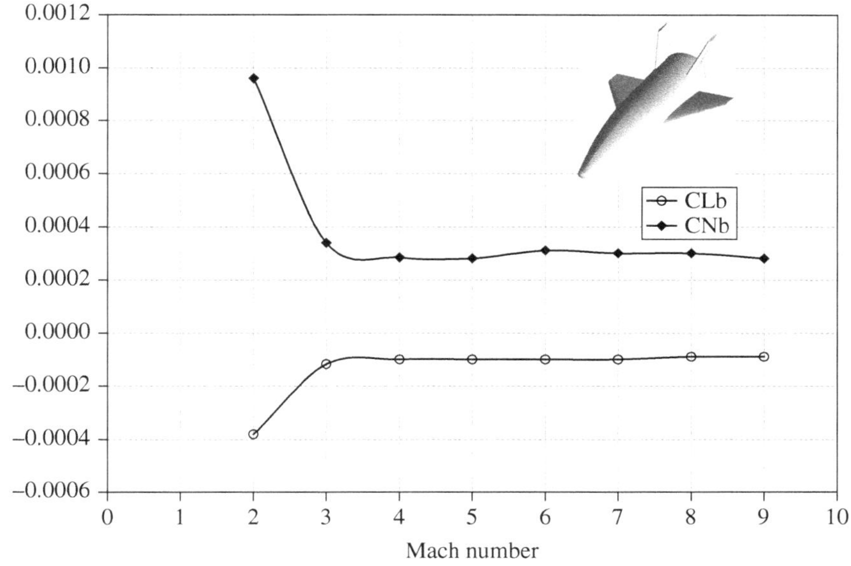

Figure 3.63 ORV-WSB: effect of sideslip on $C_{l\beta}$ and $C_{n\beta}$ up to $M_\infty = 9$, at $\alpha = 5°$.

As far as the lateral-directional stability is concerned, Figure 3.63 shows for $\alpha = 5°$ the effect of sideslip on the rolling (C_l) and yawing moment (C_n) coefficients, along with Mach number.

As shown, the configuration is statically stable in lateral-directional flight at $\alpha = 5°$. Note that the body flap is advantageous for both longitudinal and lateral-directional stability by providing margins on the CoG location. In fact, the body flap, located on the rear lower portion of the aft fuselage, allows the vehicle to be pitch trimmed, while the elevons provide roll control. The lift-to-drag ratio of ORV-WSB and ORV-SB concepts is shown in Figure 3.64 for Mach numbers ranging from 2 to 9 at $\alpha = 5°$.

The ORV-SB concept has enhanced aerodynamic efficiency due to its highly streamlined aeroshape compared with ORV-WSB, as expected.

Figure 3.65 shows the aerodynamic lift coefficients for ORV-WSB, ORV-WBB and ORV-SB. ORV-SB and ORV-WBB feature the same lift slope, which is larger than that of ORV-WSB.

The comparison of aerodynamic drag is shown in Figure 3.66. In this case, and in that of lift, discussed above, differences in aerodynamic performance at low AoAs are dictated by differences in fuselage forebody slopes and cross-sectional areas (see Figure 3.11). Indeed, the larger the vehicle's cross-section the greater the drag coefficient.

On the other hand, at high AoAs, differences in aerodynamic coefficients are due to the different planform shapes that characterize each concept (see Figure 3.11). The ORV-WSB has the lower planform surface.

The slope of the lift and drag coefficients is close respectively to $\sin^2\alpha \cdot \cos\alpha$ and $\sin^3\alpha$, as stated by impact flow theory [13, 17, 34, 35].

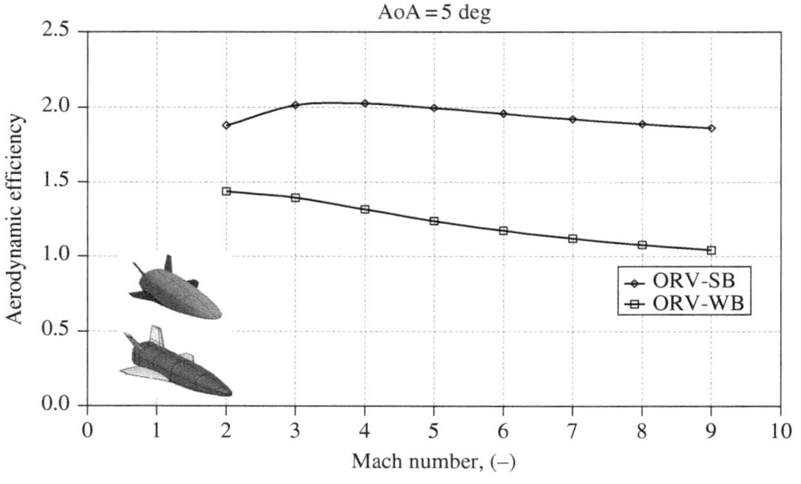

Figure 3.64 *L/D* versus Mach at α = 5°. Comparison between ORV-WSB and ORV-SB concepts.

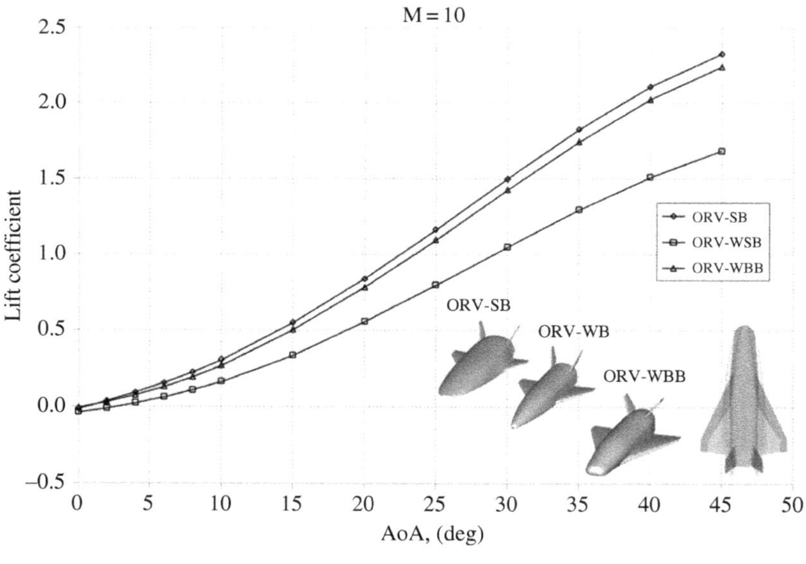

Figure 3.65 Lift coefficients at Mach 10 for ORV-WSB, ORV-WBB and ORV-SB.

Differences in aeroshape, discussed above, result in different *L/D* profiles[6] versus AoA, of course. For instance, *L/D* sharply increases with α in the case of streamlined configurations, but $(L/D)_{max}$ decreases as vehicle bluntness increases (the larger the vehicle bluntness the lower the peak aerodynamic efficiency) see Figure 3.67. It is worth

6 The lift-to-drag ratio is one of the most important features of vehicle aerodynamic performances. In fact, it has a direct impact on range and cross-range capabilities of a re-entry vehicle that has to reach its nominal landing site at the end of a space mission by unpowered flight.

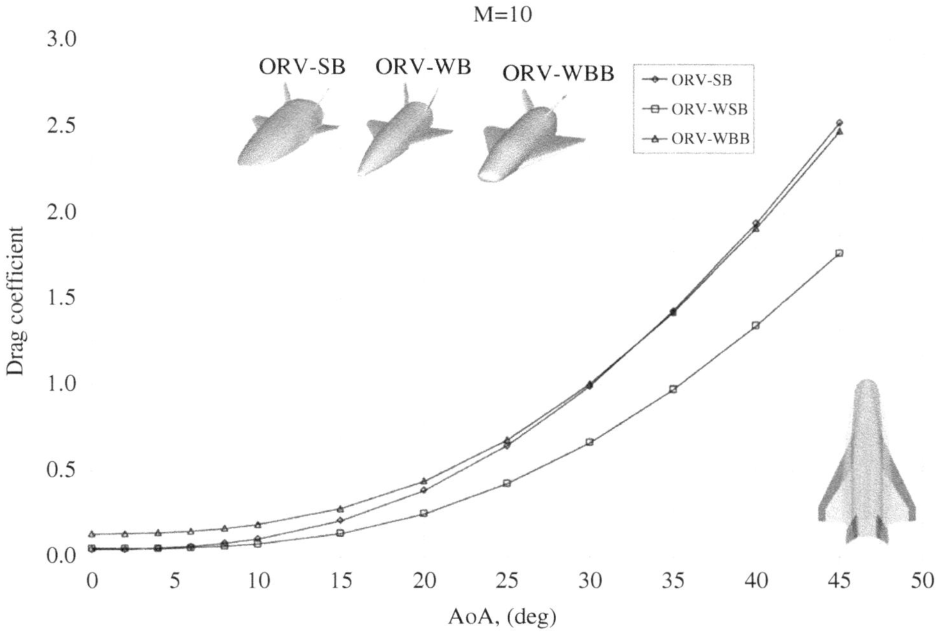

Figure 3.66 Drag coefficients at Mach 10 for ORV-WSB, ORV-WBB and ORV-SB.

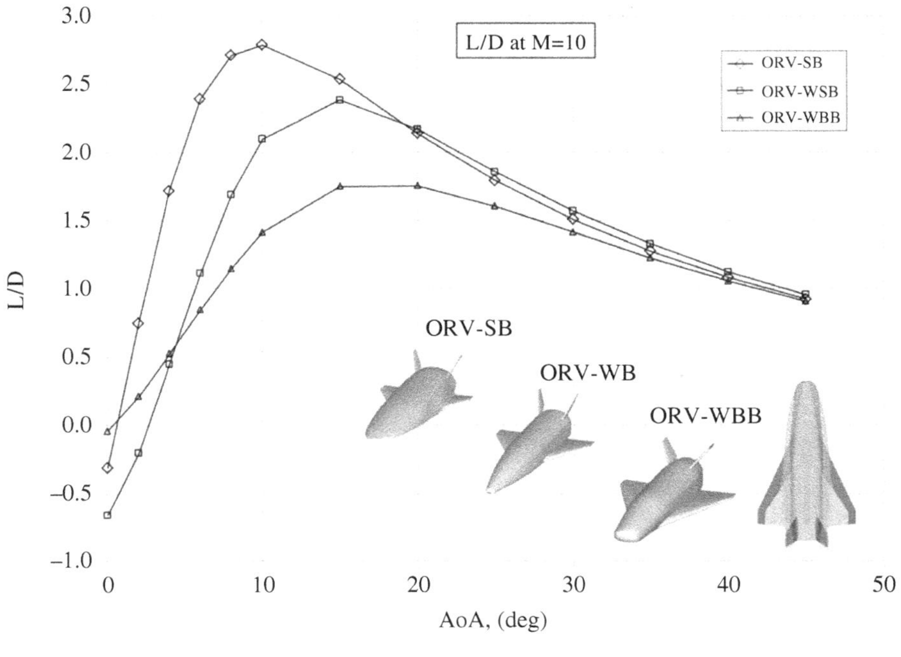

Figure 3.67 Lift-to-drag ratio at Mach 10 for ORV-WSB, ORV-WBB and ORV-SB.

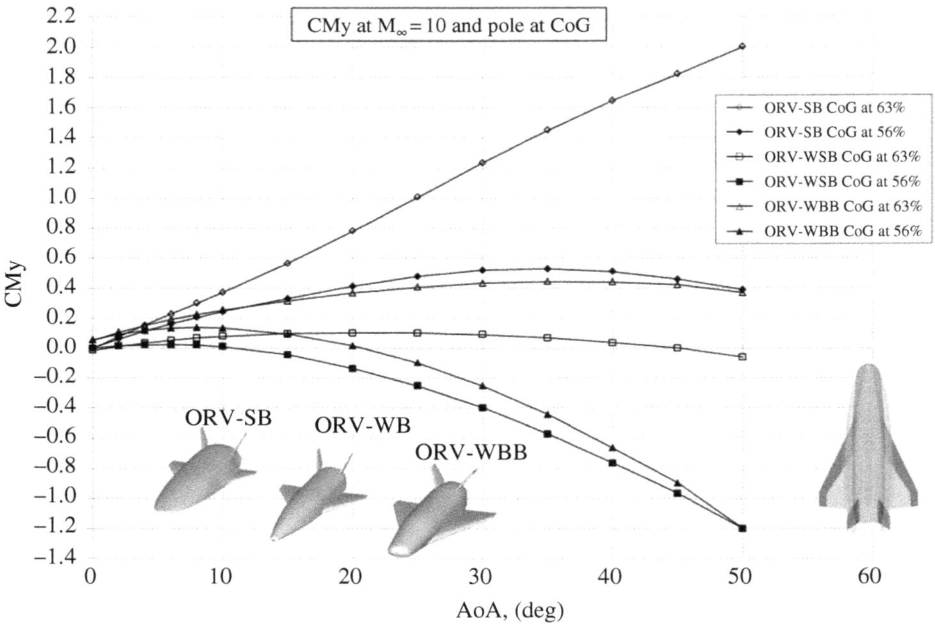

Figure 3.68 Effect of CoG position on C_m at Mach 10 for ORV-WSB, ORV-WBB and ORV-SB.

noting that this high (L/D) slope of very streamlined aeroshapes is extremely important for scramjet-propelled vehicles, since this kind of spacecraft must fly at small AoAs in order to assure proper airflow conditions at the scramjet intake and to minimize aerodynamic drag. In particular, the ORV-SB concept features the best lift-to-drag ratio of the three concepts up to about $\alpha = 20°$ and it has an $(L/D)_{max}$ equal to about 2.8 at $\alpha = 10°$. On the other hand, the maximum aerodynamic efficiency of ORV-WB and ORV-WBB is reached at about $\alpha = 15°$ at about 2.4 and 1.5, respectively.

For AoAs larger than 20°, differences in aerodynamic efficiency decrease as α increases, and they vanish for $\alpha > 35°$. As a result, in the framework of re-entry at high AoAs (35–40°; close to the AoA of the US Orbiter), differences in aeroshape do not significantly affect the descent flight. In fact, as said before, at hypersonic speeds and at high AoAs, vehicle aerodynamics are dictated essentially by planform shapes.

As far as the pitching moment is concerned, the effect of the CoG position (with respect to the fuselage length) on the C_m for each vehicle concept is summarized in Figure 3.68. When the CoG is at 63% of the fuselage length, ORV-SB features a strong static instability in longitudinal flight, highlighting that the centre of pressure is well ahead of the CoG (in other words, there is a negative static margin); ORV-WBB is static stable in pitch for $\alpha > 40°$ and can be trimmed by positive flap deflections; ORV-WSB is static stable in pitch for $\alpha > 30°$ and also has a natural trim point at about 45° AoA.

However, for a CoG at 56% of the fuselage length, the ORV-SB concept becomes statically stable in longitudinal flight for $\alpha > 35°$ and trim AoA can be attained by positive flap deflections; the other two concepts (WSB and WBB) are statically stable in pitch for $\alpha > 5°$ and feature a natural trim point at 10 and 20° AoA, respectively. In particular, they can be trimmed at high AoA by means of negative (in other words, trailing-edge up) flap deflections.

Note that, in statically stable and trimmed flight the attitude of the vehicle is aligned in such a way that the total external moment acting on the vehicle is zero. This means that after a disturbance the vehicle tends always to move back towards the trimmed state.

Pitching moment behaviour versus AoA indicates that vehicle subsystem arrangements (in other words, CoG position) must be carefully addressed. Indeed, in order to have a static stable and trimmable vehicle concept, the CoG has to be carefully chosen with reference to the vehicle shape.

3.11 CFD-based Aerodynamic Results

For the numerical flowfield computations, on the basis of the flight envelope of Figure 3.15 a number of flight conditions have been chosen for CFD computations in steady-state conditions, according to the space-based design approach [13, 17, 34,35,36].

Numerical results aim to anchor engineering analyses in order to increase their accuracy, and to focus on some critical design aspects that are not predictable with simplified tools, for example SSI and shock-wave–boundary-layer interactions (SWIBLI) and high-temperature real-gas effects. SSI determines pressure and heat flux overshoots on the leading edges of both wing and tail, which must be accounted for in TPS design; SWBLI influences control surface effectiveness. The CFD test matrix is shown in Table 3.6.

It is worth nothing that at $M_\infty = 8$, 10, 16, and 20, non-equilibrium CFD computations are also carried out. Real-gas effects can be important because, during atmospheric re-entry, dissociation processes take place in the shock layer, which can have an influence on the aerodynamic coefficients. High-temperature real-gas effects are expected to influence stability and control derivatives of the vehicle, in particular its pitching moment, as highlighted by the first Space Shuttle re-entry (STS-1), where an nose-up pitching moment required a body-flap deflection twice than that predicted by the pre-flight analyses in order to trim the vehicle [36].

Furthermore, real-gas effects cause a shock closer to the vehicle than a perfect gas theoretically would (in other words, there is a thin shock layer) [35,36,37]. This means that a strong SSI phenomenon arises. These effects obviously occur only at high Mach numbers [37].

Numerical non-equilibrium investigations allow assessment and characterization of the wall catalyticity in order to reduce the design margins, preventing use of the overly conservative hypothesis of a fully catalytic wall, as usually used in TPS design.

CFD simulations also help in finding the effect of laminar-to-turbulent transitions, which must be accounte for in vehicle TPS design. It is well known that these transiations can cause strong and even dangerous overheating of the vehicle skin [38,39,40].

Numerical CFD computations have been carried out on both multi-block structured and hybrid unstructured grids, similar to those shown in Figures 3.69 and 3.70. Figure 3.69 shows the sub-transonic computational domain that were considered when addressing ORV-WSB aerodynamics at sub-transonic speeds.

The external boundary of the half-body grid has been built as a Cartesian block, and far-field surfaces are located at about ten body lengths away from the body (upstream and downstream) to ensure that the flow at the boundary is close to free stream. Indeed, the farther we are from the vehicle, the less effect it has on the flow and so the more accurate is the farfield boundary condition.

Table 3.6 CFD test matrix.

CFD test matrix										
	AoA @ AoS=0 deg							**AoS @ AoA=5 deg**		
Mach	0	5	10	20	30	40	45	2	4	8
0.3	X	X	X	X						
0.5	X	X	X	X						
0.8	X	X	X	X						
0.95	X	X	X	X						
1.1	X	X	X	X						
1.25	X	X	X	X						
2	X	X	X	X				X		
3	X	X	X	X				X	X	X
4		X						X		
5	X	X	X	X				X		
6	X	X	X					X		
7	X	X	X					X		
8		X	X							
8		Y	Y							
10			Y	Y	Y					
16			Y	Y	Y					
20			X	X	X	X				
20			X	X	X	X				
25					Y		Y			

X, perfect gas; Y, reacting gas. Note that each cell identifies a CFD run (in other words, check point).

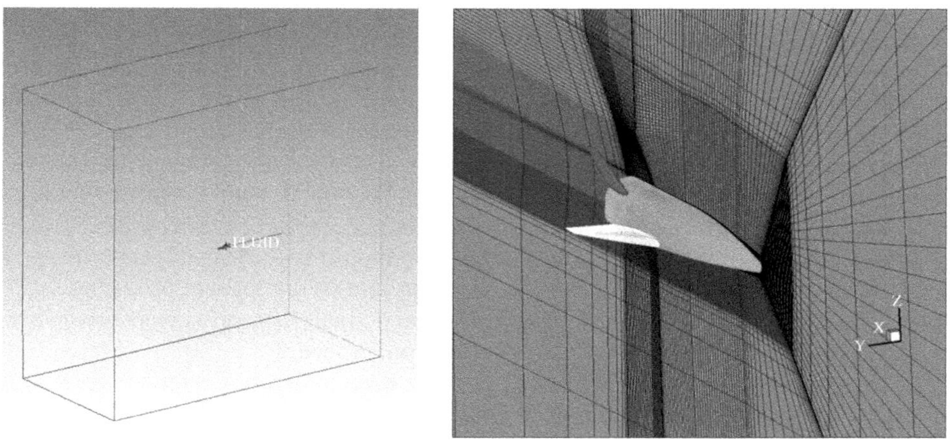

Figure 3.69 Sub-transonic computational domain: mesh on symmetry plane and ORV-WSB surface.

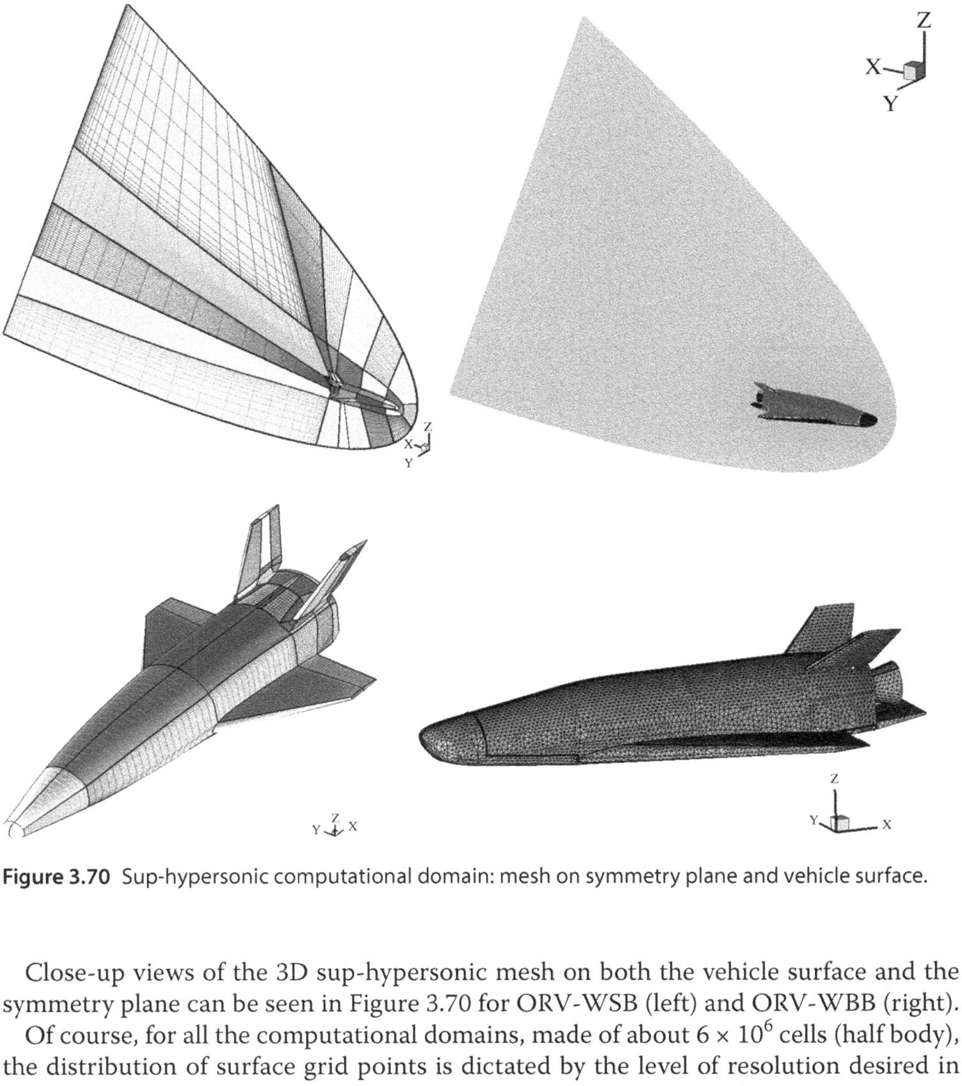

Figure 3.70 Sup-hypersonic computational domain: mesh on symmetry plane and vehicle surface.

Close-up views of the 3D sup-hypersonic mesh on both the vehicle surface and the symmetry plane can be seen in Figure 3.70 for ORV-WSB (left) and ORV-WBB (right).

Of course, for all the computational domains, made of about 6×10^6 cells (half body), the distribution of surface grid points is dictated by the level of resolution desired in various areas of vehicles, such as the stagnation region. Grid refinements in strong gradient regions of the flowfield are addressed by means of a solution-adaptive approach. The coordinate y^+ of the first cell adjacent to the surface is about 1.

As far as numerical results are concerned, it is worth noting that they refer to both converged and grid-independent computations. Figures 3.71 and 3.72 show the ORV-WSB aerodynamics at Re = 10^6, for Mach numbers ranging from 0.3 to 5 and for α = 0, 5, 10, and 20°. As shown, lift and drag coefficients rise in the transonic region, where the presence of the shock wave produces a large increase of the aerodynamic forces on the vehicle: the increase of wave drag and base drag that in this region reach their maximum values, as expected. On the other hand, when Mach number increases, aerodynamic coefficients tend to reach a limit value, according to the Mach number independence principle (the Oswatich principle).

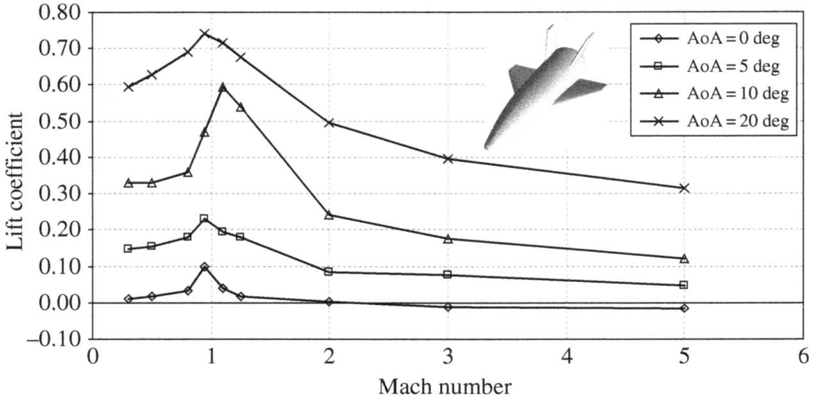

Figure 3.71 Lift coefficient versus Mach number for different AoAs; ORV-WSB concept.

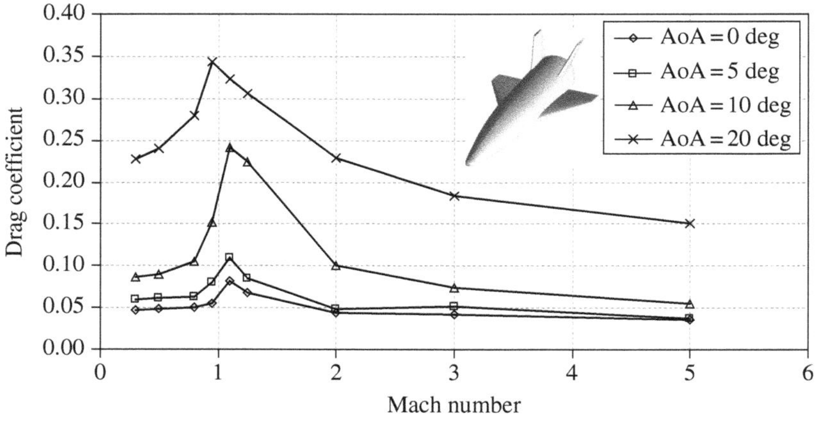

Figure 3.72 Drag coefficient versus Mach number for different AoAs; ORV-WSB concept.

In particular, this behaviour is similar for all the considered AoAs, and the strong dependence of the drag on the AoA is essentially due to the large increases of the induced drag with α [41,42,43,44].

Figure 3.73 shows the aerodynamic lift, drag and *L/D* versus α at $M_\infty = 0.3$. Aerodynamic results at these flight conditions are extremely important considering that $M_\infty = 0.3$ represents the nominal landing condition at the end of an unpowered re-entry flight. Figure 3.73 shows that the lift coefficient at $M_\infty = 0.3$ steadily increases from $\alpha = 0$ to 20° indicating that the vehicle does not stall up to 20° at this Mach number. Therefore, the drag coefficient at zero AoA is about 0.047 and it continues to rise as AoA increases, and is expected to reach about 0.22 at α = 20°. The maximum *L/D*, close to 3.5, is attained at α = 10°. Of course, extra lift can be provided by means of positive flap deflections, thus also providing static stable trim conditions at landing AoA.

In Figure 3.74 vehicle polars for Mach numbers ranging from 0.3 to 1.25 and for α = 0, 5, 10, and 20° are also reported, thus showing ORV-WSB's aerodynamics in sub-transonic flow conditions.

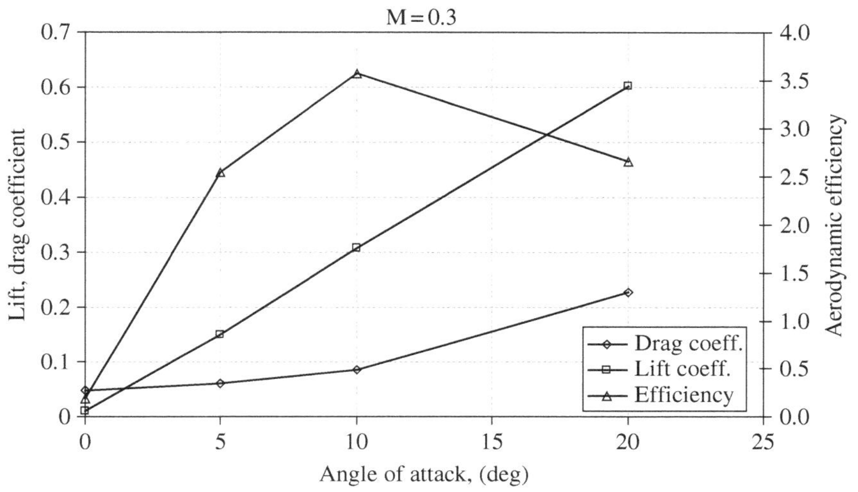

Figure 3.73 C_L, C_D and *L/D* versus AoA for $M_\infty = 0.3$; ORV-WSB concept.

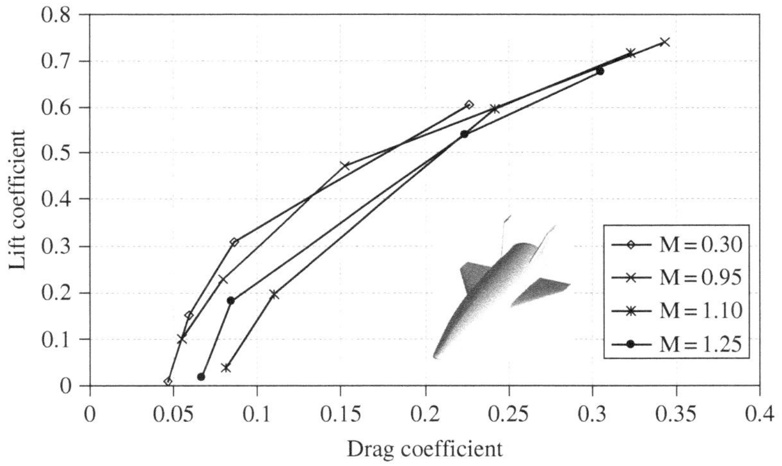

Figure 3.74 Drag polars for $0.3 \leq M_\infty \leq 1.25$ and for $\alpha = 0, 5, 10$, and 20°; ORV-WSB concept.

The variation of the pitching moment coefficient of ORV-WSB versus α is presented in Figure 3.75.

At low subsonic speeds (Mach 0.3 and 0.8), the ORV-WSB vehicle is statically unstable in longitudinal flight (in other words, $C_{m\alpha} > 0$) with the given CoG for all the considered AoAs. Concept stability changes for Mach numbers larger than 0.95. This means that at these flight conditions a perturbation that produces an increased angle of attack causes a nose-down (negative) pitching moment, which decreases the AoA, restoring the vehicle to its original position. In particular, C_m in the sub-transonic regime follows a fairly linear trend. Furthermore, the AoA at which the moment coefficient about the centre of gravity equals zero (in other words, trim condition) is approximately 5° for $M_\infty = 0.30$ and $M_\infty = 0.95$, 12° at $M_\infty = 0.80$, and $\alpha = 2°$ for $M_\infty = 1.25$.

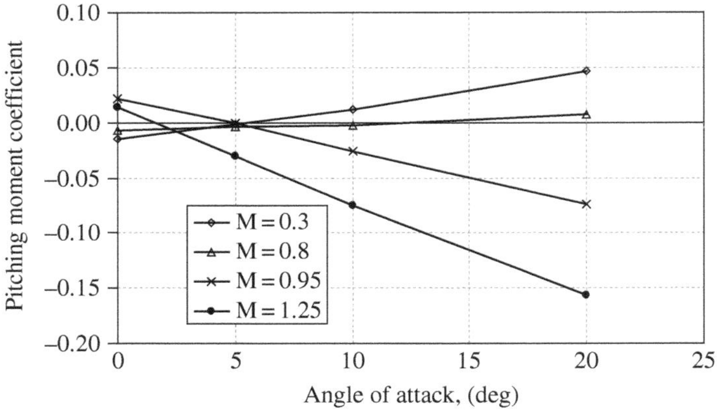

Figure 3.75 C_m vs α at different Mach numbers. ORV-WSB concept.

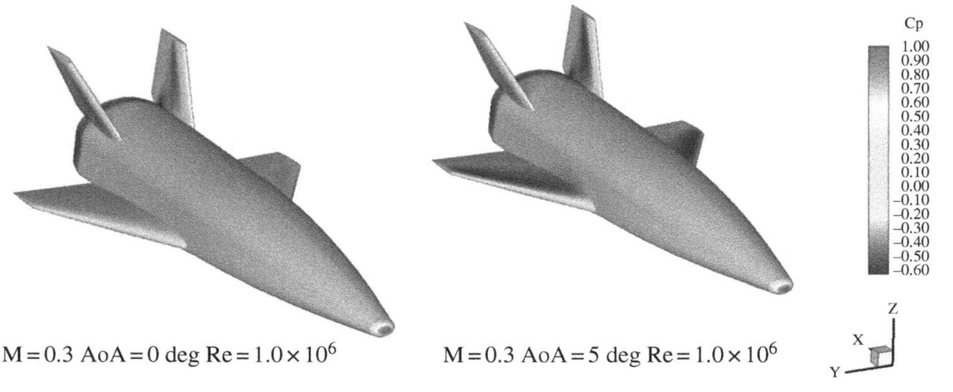

Figure 3.76 Pressure coefficients at $M_\infty = 0.3$ and for $\alpha = 0$ and 5°; ORV-WSB concept.

In order to demonstrate the subsonic aerodynamic performance, Figures 3.76–3.78 show the pressure coefficient contour maps over the ORV-WSB surface for all the analyzed AoAs at $M_\infty = 0.3$.

On the leading edge, we see a stagnation point where the velocity of the flow is nearly zero. The fluid accelerates on the upper surface (in other words, leeside) as highlighted by the negative value of C_p (in other words, pressure less than that of the surrounding atmosphere). On the other hand, at the trailing edge, the flow on the upper surface decelerates and converges with the flow on the lower surface (in other words, windside) according to the Bernoulli equation (in other words, whenever there are high velocities, we have low pressures and vice versa).

With the pressure distributions as sketched, the effect on the upper surface is the larger, and there is a resultant upwards force on the section, that is, lift.

In particular, the pressure reduction on the upper surface increases both in intensity and extent at high incidence (see Figures 3.77 and 3.78); the stagnation point moves progressively further back on the lower surface, and the increased pressure on the lower surface covers a greater proportion of the surface, as shown in Figure 3.78.

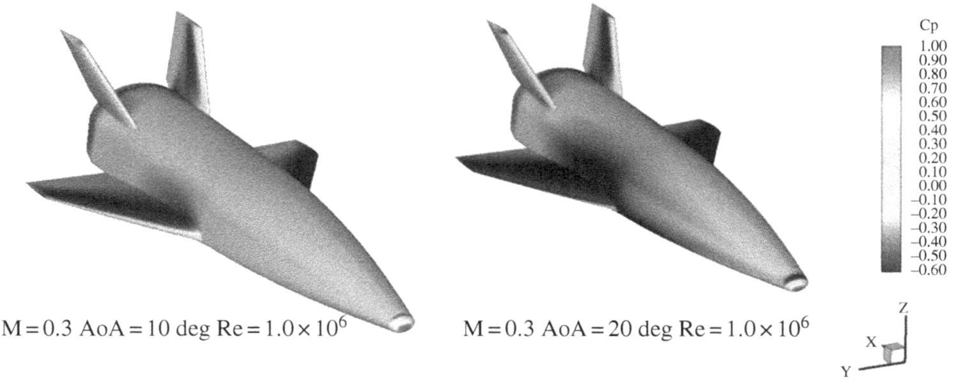

Figure 3.77 Pressure coefficients at $M_\infty = 0.3$ and for $\alpha = 10$ and 20°; ORV-WSB concept.

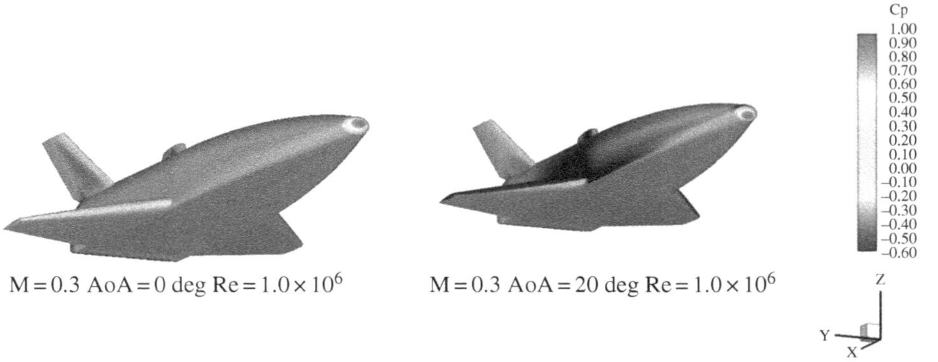

Figure 3.78 Pressure coefficient at $M_\infty = 0.3$ and for $0 \leq \alpha \leq 20°$; ORV-WSB concept.

Therefore, the following conclusions may be drawn: at low incidence the lift is generated by the difference between the pressure reductions on the upper and lower surfaces; at higher incidences the lift is partly due to pressure reductions on the upper surface and partly due to pressure increases on the lower surface. Figure 3.79 shows the effect of AoA on the flowfield streamlines in the vehicle symmetry plane at $M_\infty = 0.3$ and for α ranging from 0 to 20°. Streamtraces are displayed in order to highlight the flowfield structure surrounding the vehicle, especially the complexity of the wake flow. There is a rather strong base expansion together with a vortex feature. Indeed, for all the AoAs, a pattern of two swirling vortices is clearly visible in the wake flow. This kind of pattern is close to those known as von Karman vortices. These swirling vortices characterize the unsteady[7] downstream separation of flow over the blunt bodies (in this case the vehicle base). This brings about an increase in drag (in other words, pressure drag).

7 When large flow separation arises at the vehicle base, for example at high angles of incidence, large oscillations of global aerodynamic loads can be observed. In this case, unsteady CFD analyses are mandatory to address the vehicle aerodynamic performance. A non-dimensional frequency of Strouhal number must be chosen as a target value for the simulation time step.

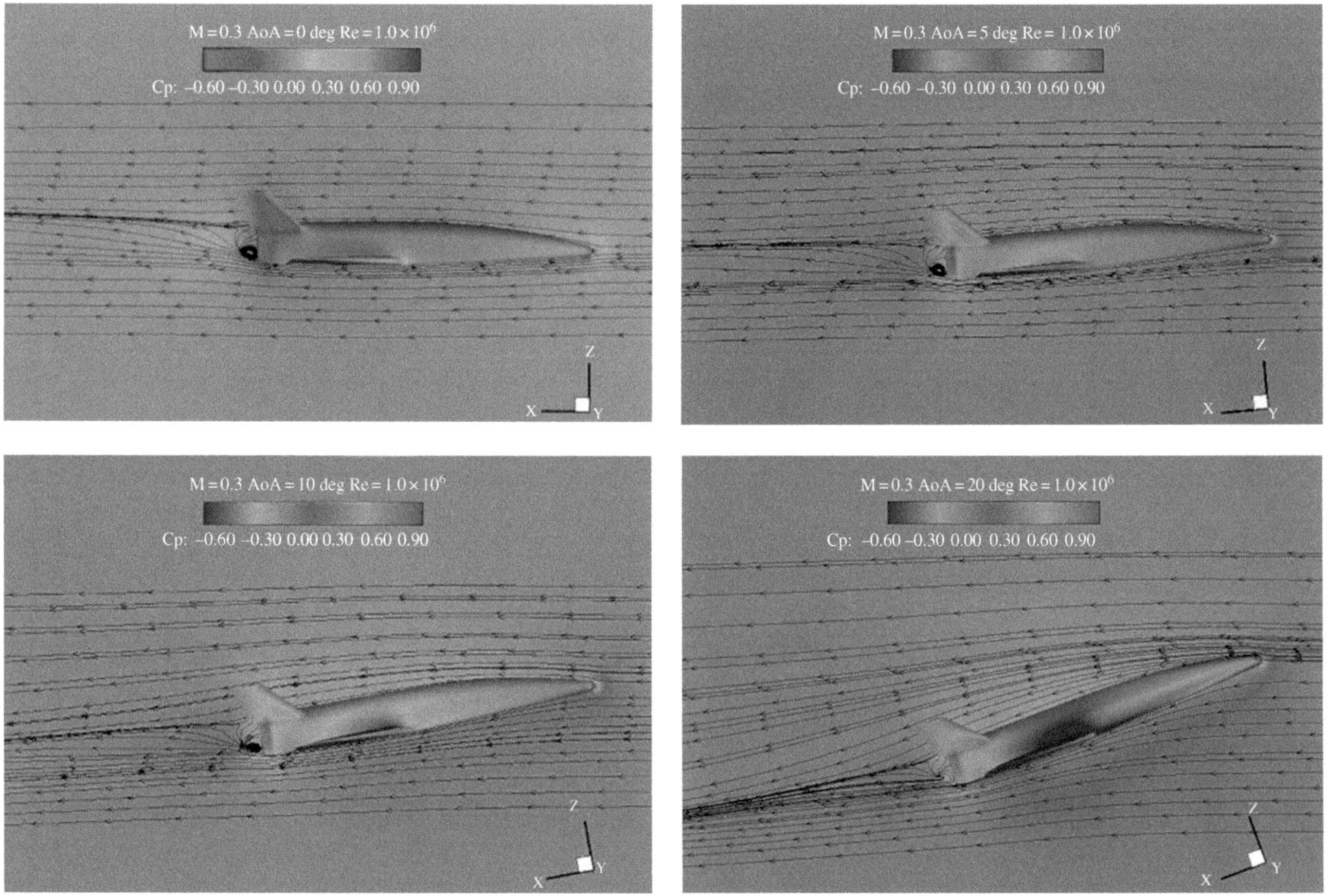

Figure 3.79 Pressure field past ORV-WB, with streamtraces on the vehicle symmetry plane at $M_\infty = 0.3$ and for $\alpha = 0, 5, 10$ and $20°$.

It is worth noting that space vehicles need drag to decelerate during the atmospheric descent, in order to lose as much internal energy (kinetic plus potential) as possible so that they can land safely. Therefore, space vehicles typically have a truncated afterbody (or blunt base), which results in base drag being a significant component of the total vehicle drag.

The blunt base of the re-entry vehicle is also needed to allocate the nozzles of the rocket motor (as in the Space Shuttle Orbiter) for the boost needed in the ascent phase and the de-boost necessary to start the de-orbiting manoeuvre before atmospheric re-entry (see nozzle at the base of ORV-WBB in Figure 3.5).

In Figure 3.80 an example of the 3D flowfield streamtraces predicted by CFD computations is reported for the ORV-WSB concept. It is possible to see the formation of the leading-edge vortex, typical of delta-wing configurations, and its interaction with the wing-tip vortex.

Note that leading-edge vortex lift is an important source of vehicle lift. The pressure on the upper surface of the wing is lower than that of the surrounding atmosphere, while the pressure on the windside is greater than that on the leeside. Thus, over the upper surface, air will tend to flow inwards towards the root from the tips, being replaced by air that was originally outboard of the tips. Similarly, on the undersurface, air will either tend to flow inwards to a lesser extent, or may tend to flow outwards. Where these two streams combine at the trailing edge, the difference in spanwise velocity will cause the air to roll up into a number of small streamwise vortices, distributed along the whole span. These small vortices roll up into two large vortices just inboard of the wing tips.

In the following figures, some of main interesting hypervelocity flowfield features obtained for the ORV-WSB concept are shown. For example, the Mach number and flow species fields for M_∞ = 10 and 20 at AoA = 10, 30, and 40° are reported from Figures 3.81–3.85. Looking at the contour fields on the vehicle symmetry plane, one can appreciate the rather strong bow shock that occurs ahead of the vehicle concept during descent at these flight conditions, especially at higher AoAs. Figures 3.81 and 3.82 show the Mach number field past the rather sharp WBV for M_∞ = 10 and for 10 and 30° AoA, respectively.

Figure 3.83 shows a blow-up of the flowfield past the vehicle for both free-stream conditions. Stream traces are also displayed.

As shown, even if the CFD computations are carried out for perfect gas flow, the bow shock is very close to the vehicle because of its streamlined configuration.

The effect of AoA on the flowfield past ORV-WSB flying at M_∞ = 10 can be appreciated as well. Figure 3.84 displays a blow-up of the flowfield surrounding the rather sharp wing body vehicle for M_∞ = 20 and α = 40°. The Mach number and the static temperature contour field both on the vehicle symmetry plane and the concept outer surface are provided. Streamtraces are also displayed in order to highlight the flowfield structure surrounding the vehicle, especially the complexity of the wake flow. The flow stagnates downstream, thus increasing the local temperature distribution and a strong base expansion together with a vortex feature can be seen at α = 40°.

In this case, air was considered to be made up of five chemical species (O_2, N_2, O, N and NO) and in thermochemical non-equilibrium conditions. The corresponding oxygen (O) and nitrogen oxide (NO) mass fraction contour fields are shown in Figure 3.85.

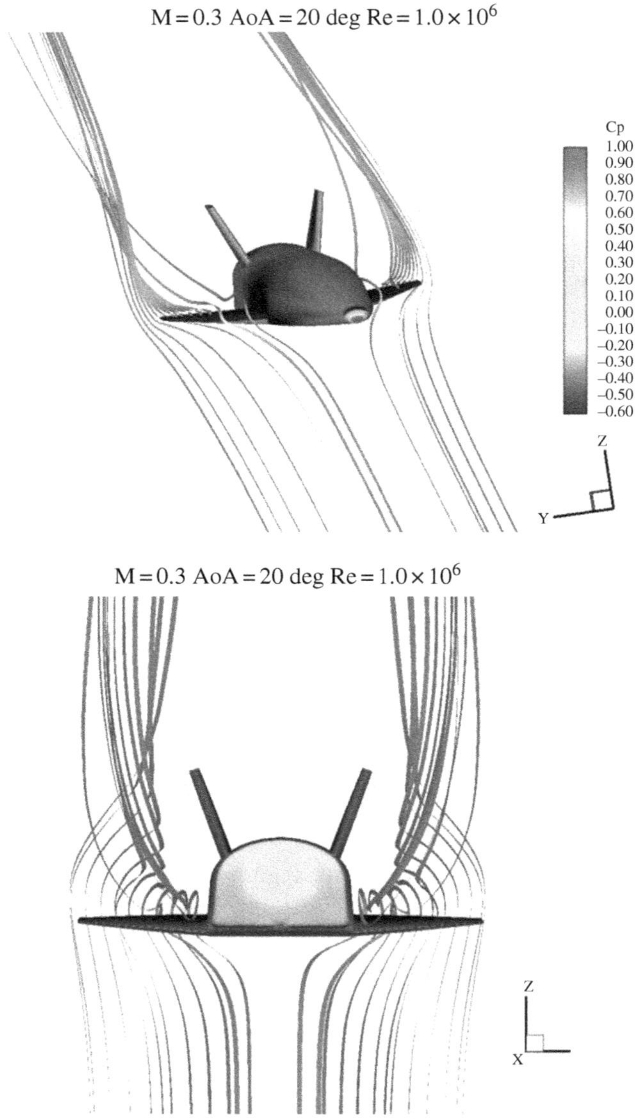

Figure 3.80 C_p on ORV-WSB surface with flowfield streamtraces at $\alpha = 20°$ and $M_\infty = 0.3$.

As far as the AoS effect is concerned, Figure 3.86 shows the Mach number contours on a fuselage cross plane and those of the static pressure on the vehicle surface at $M_\infty = 6$, $\alpha = 5°$ and $\beta = 2°$. Figure 3.87 shows the same flowfield features as Figure 3.86 but for $M_\infty = 3$, $\alpha = 5°$ and $\beta = 4$ and 8°.

A different line of sight for $M_\infty = 3$, $\alpha = 5°$ and $\beta = 8°$ is provided in Figure 3.88.

The effect of sideslip angle on flowfield features is clearly shown in the non-symmetrical distribution of the contour fields. This determines, apart from lift, drag and pitching moment, the lateral-directional aerodynamic forces and moments acting on the vehicle concept.

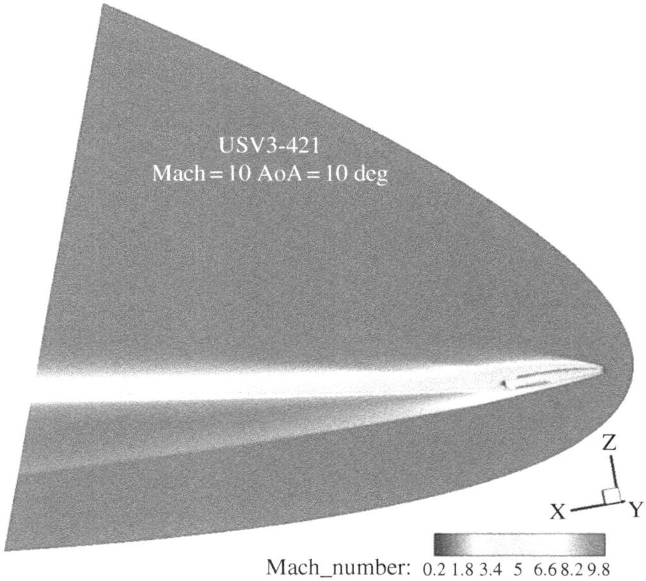

Figure 3.81 Mach contours on symmetry plane and ORV-WSB surface at $M_\infty = 10$ and $\alpha = 10°$.

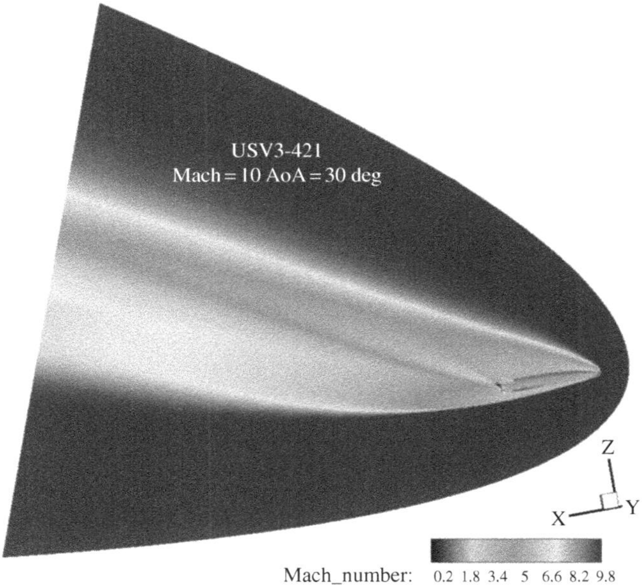

Figure 3.82 Mach contours on symmetry plane and ORV-WSB surface at $M_\infty = 10$ and $\alpha = 30°$.

Finally, Figures 3.89 and 3.90 show the pressure spike due to SSI that takes place on the ORV-WSB at $M_\infty = 6$, $M_\infty = 7$ and $\alpha = 5°$, respectively. Of course, the point of the wing leading edge where this interaction takes place depends on the free-stream conditions.

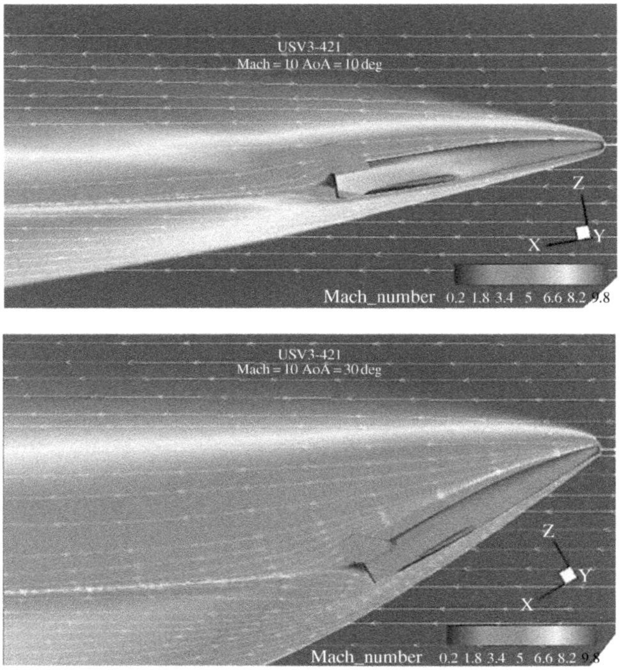

Figure 3.83 Mach number contours on symmetry plane and vehicle surface at $M_\infty = 10$ and $\alpha = 10$ and 30°.

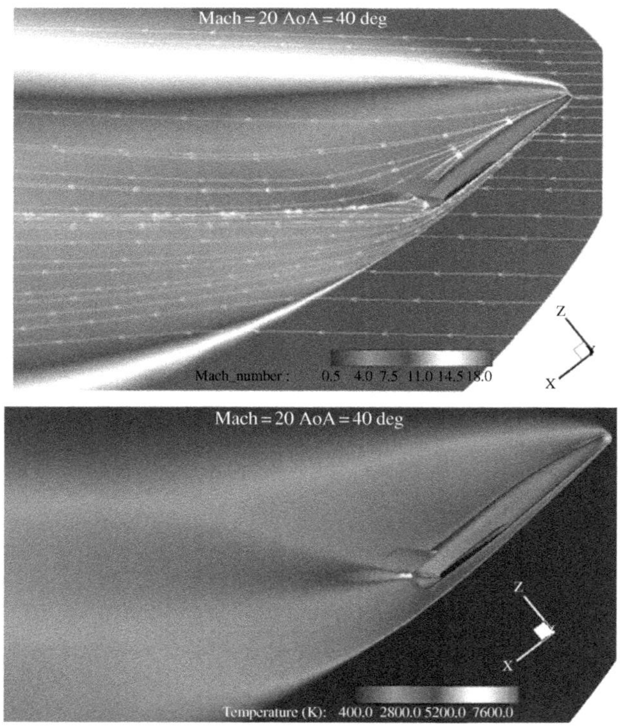

Figure 3.84 Non-equilibrium CFD computation at $H_\infty = 70$ km, $M_\infty = 20$ and $\alpha = 40°$. Mach and temperature contour fields; ORV-WSB.

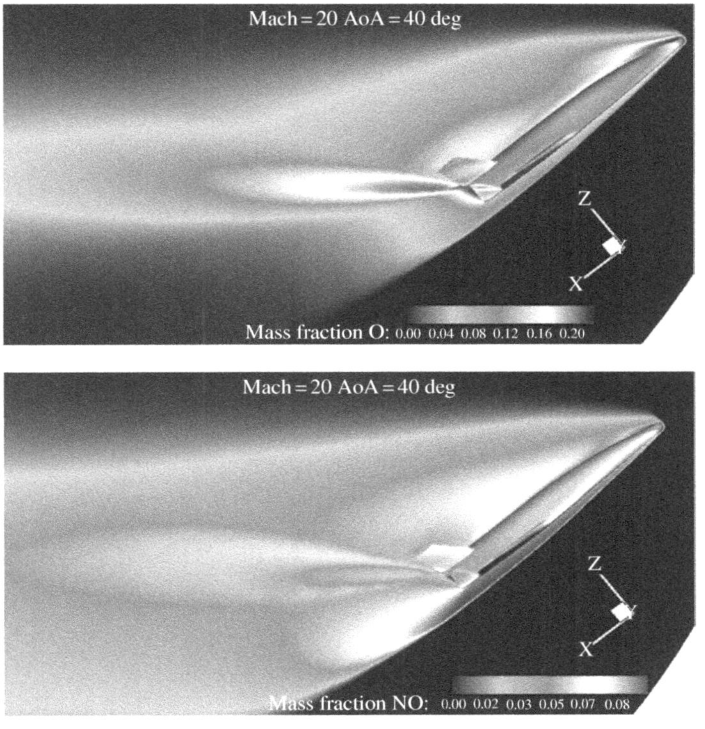

Figure 3.85 O and NO mass fractions. CFD computation at $H_\infty = 70$ km, $M_\infty = 20$ and $\alpha = 40°$; ORV-WSB concept.

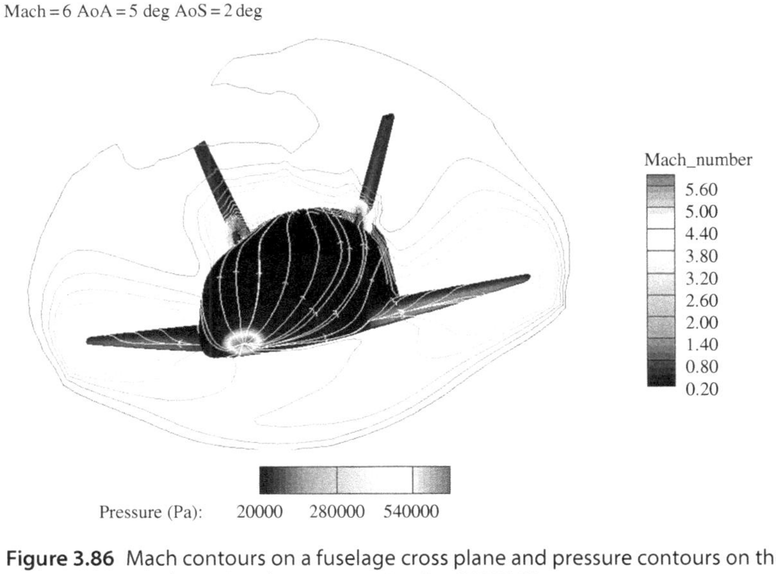

Figure 3.86 Mach contours on a fuselage cross plane and pressure contours on the vehicle surface at $M_\infty = 6$, $\alpha = 5°$ and $\beta = 2°$. Euler computation for ORV-WSB concept.

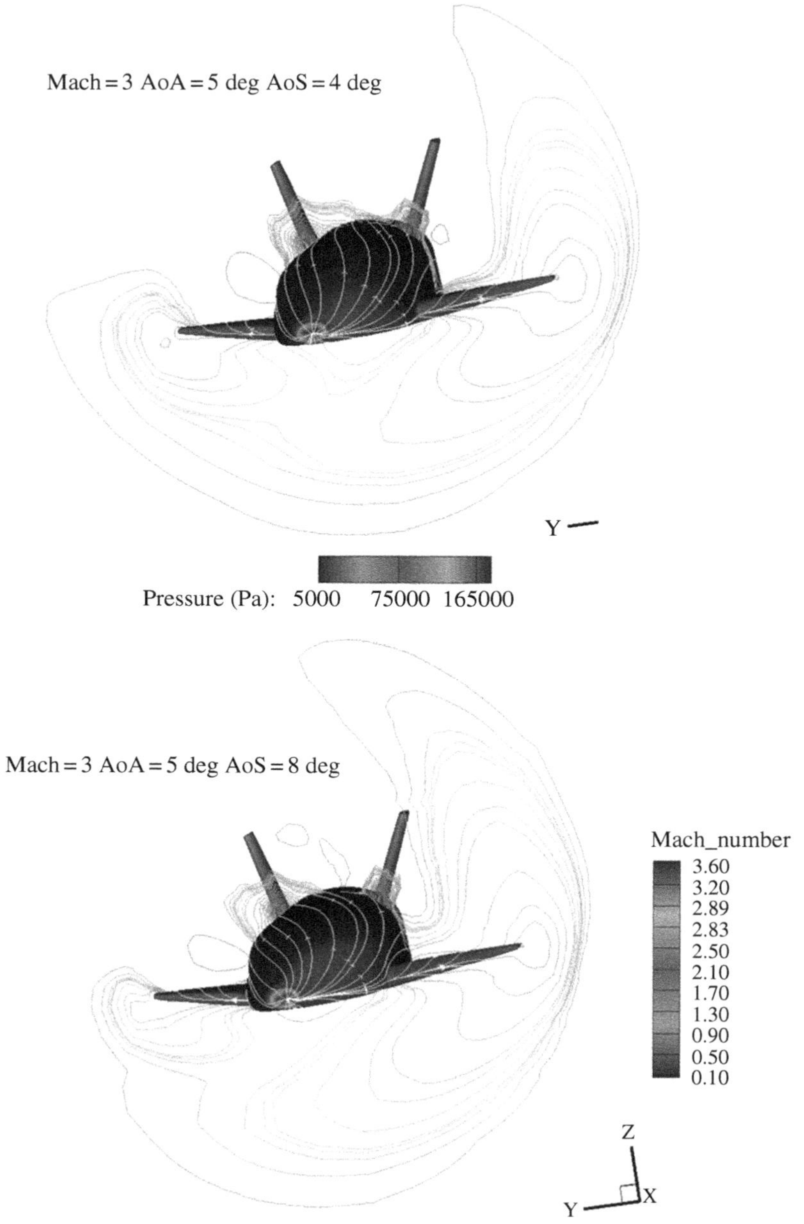

Figure 3.87 Mach contours on a fuselage cross plane and pressure contours on the vehicle surface at $M_\infty = 3$, $\alpha = 5°$ and $\beta = 4$ and $8°$. Euler computation for ORV-WSB concept.

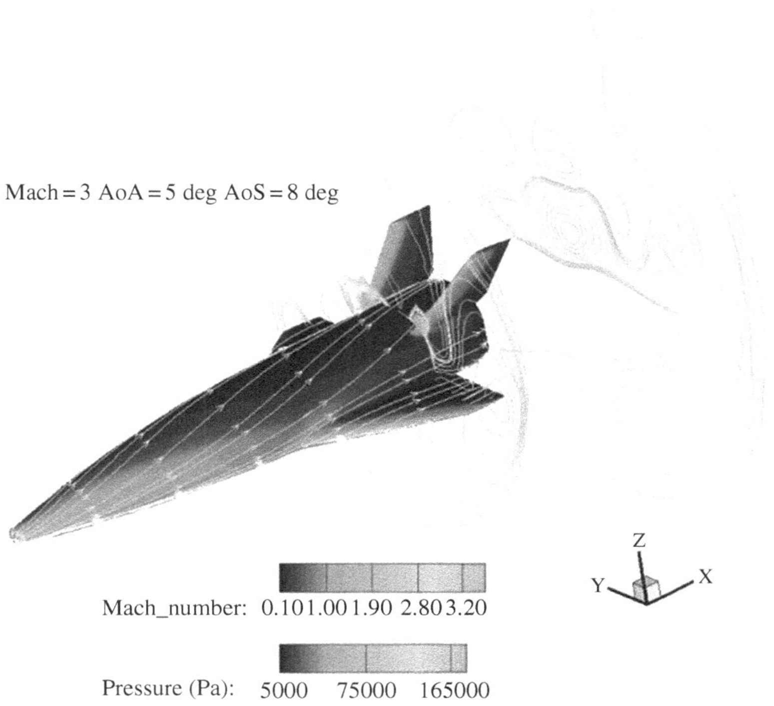

Figure 3.88 Mach contours on two fuselage cross planes and pressure contours on the vehicle surface at $M_\infty = 3$, $\alpha = 5°$ and $\beta = 8°$. Euler computation for ORV-WSB concept.

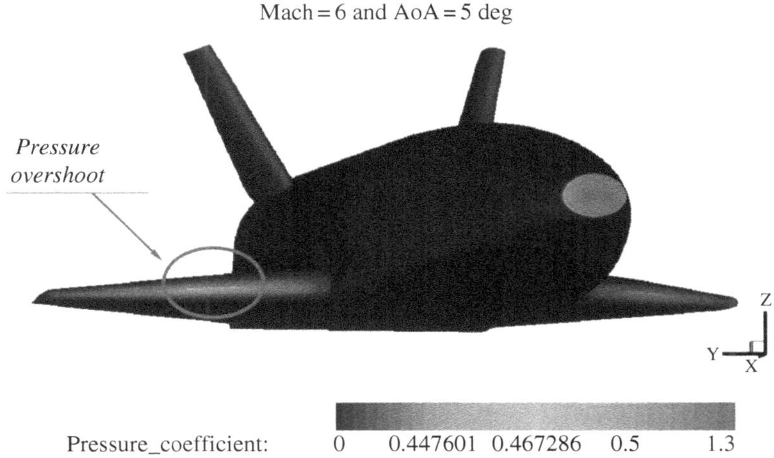

Figure 3.89 Pressure coefficient distribution on the ORV-WSB surface at $M_\infty = 6$ and $\alpha = 5°$.

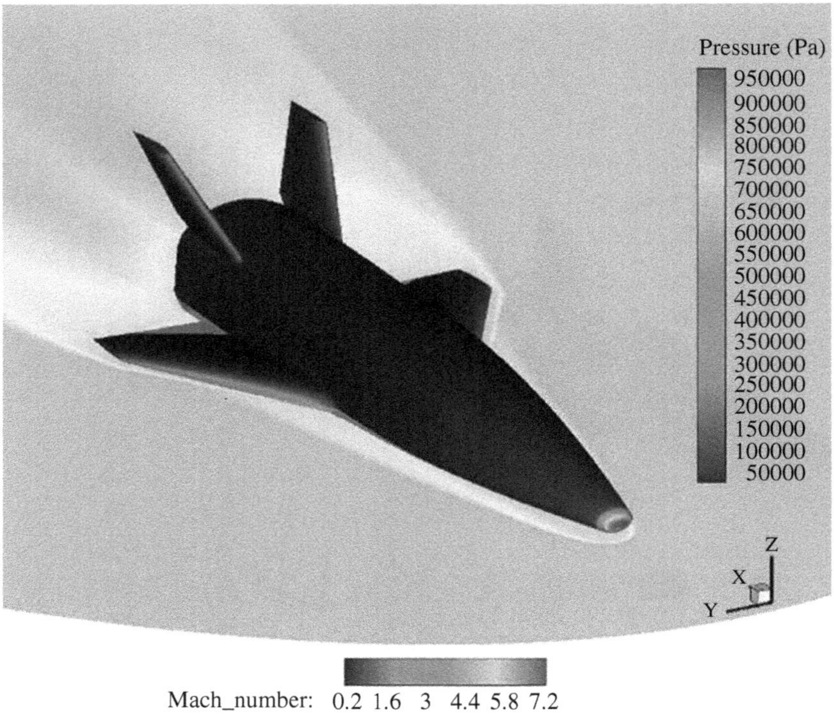

Figure 3.90 Pressure and Mach number distributions on ORV-WSB surface and wing plane at $M_\infty = 7$ and $\alpha = 5°$.

Figure 3.89 shows the contour plot of the pressure coefficient on the vehicle surface and Figure 3.90 displays the Mach number on the wing plane. The pressure overshoot where SSI impinges the wing leading edge is clearly seen, as is the trace of the bow shock on the wing plane.

Numerical flowfield results for the ORV-WBB concept are shown in Figures 3.91 and 3.92 for $M_\infty = 25$ and $\alpha = 30°$. Figure 3.91 highlights the Mach number and the N_2 mass fraction contour fields on the vehicle symmetry plane and c_p on the concept outer surface.

Side and bottom views of the surface distribution of c_p at these flight conditions are shown in Figure 3.92. Here, three flowfield cross-sections are also shown to give an idea of the 3D bow shock that envelops the ORV-WBB.

Regarding the flowfield past the ORV-SB concept, Figure 3.93 shows the Mach number contour field in the vehicle symmetry plane and the pressure distribution on the concept surface at $M_\infty = 25$ and $\alpha = 30°$. Streamlines are also provided.

3D streamtraces at these flight conditions together with a cross-flow section coloured by Mach number are provided in Figure 3.94. A very narrow shock layer is also expected for this concept.

To compare the aerodynamic results of numerical and engineering analyses, results for $M_\infty = 10$, 20 and for $0 < \alpha < 40°$ are provided in Figures 3.95 and 3.96 for the ORV-WSB concept.

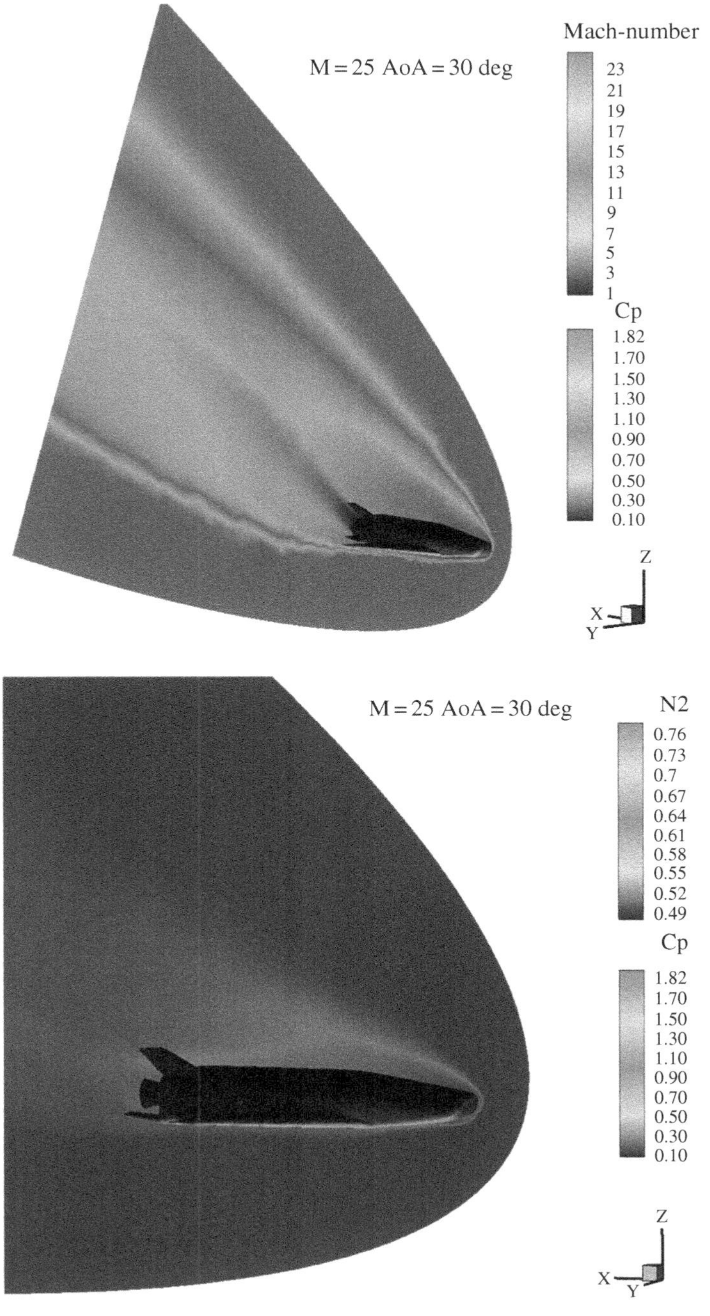

Figure 3.91 Mach (top) and N_2 (bottom) contours on the symmetry plane and c_p on the vehicle surface at $M_\infty = 25$ and $\alpha = 30°$; ORV-WBB concept.

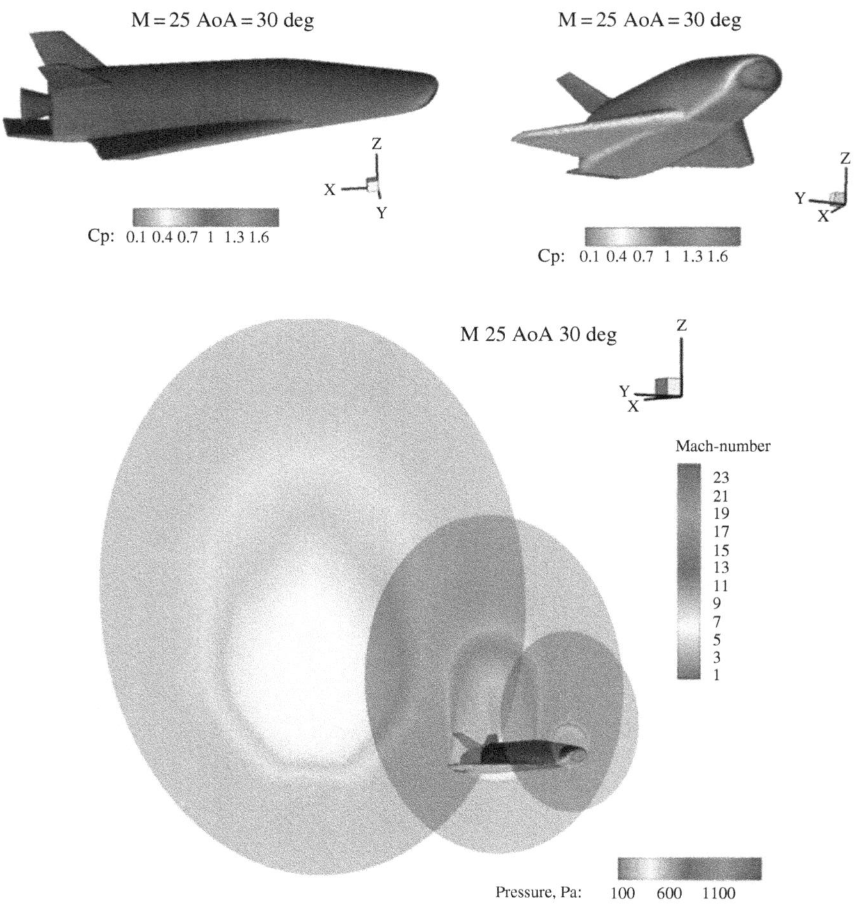

Figure 3.92 c_p on vehicle surface at $M_\infty = 25$ and $\alpha = 30°$; ORV-WBB concept.

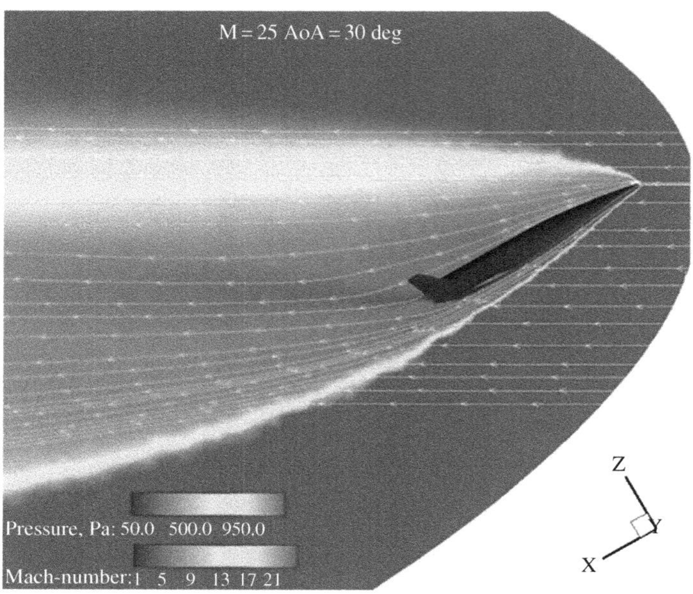

Figure 3.93 c_p on the ORV-SB surface at $M\infty = 25$ and $\alpha = 30°$.

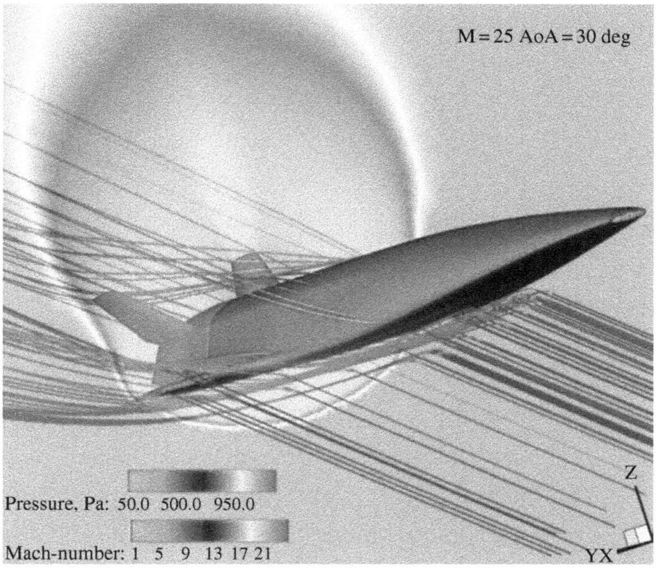

Figure 3.94 c_p on the ORV-SB surface at $M_\infty = 25$ and $\alpha = 30°$ with 3D streamtraces and flowfield cross-section coloured by Mach number.

Figure 3.95 C_L vs α; HPM and CFD results comparison at $M_\infty = 10$ and 20; ORV-WSB concept.

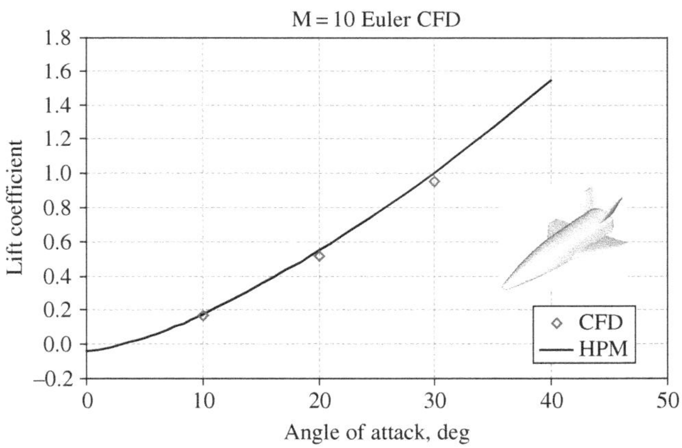

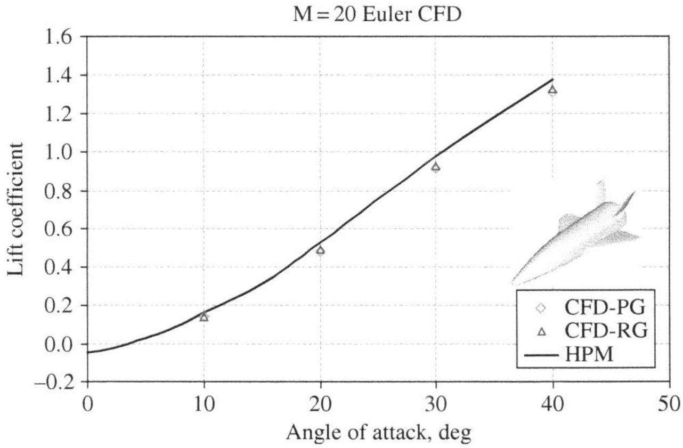

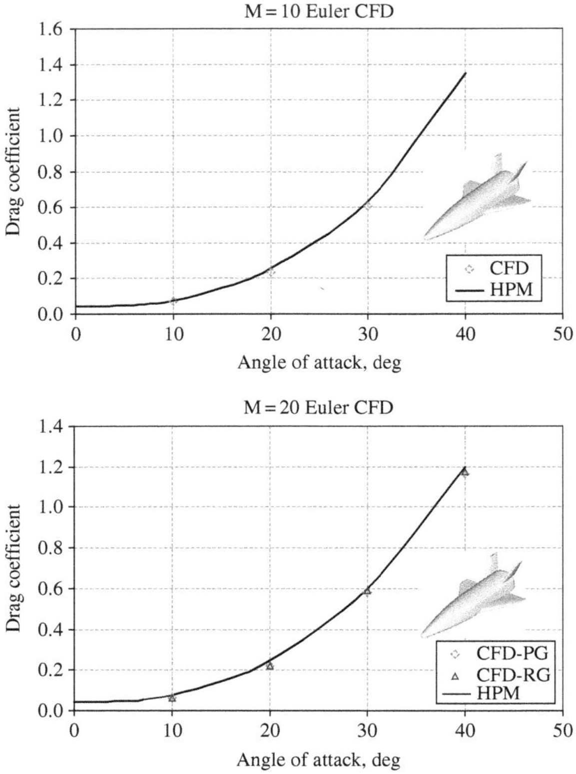

Figure 3.96 C_D vs α; HPM and CFD results comparison at $M_\infty = 10$ and 20; ORV-WSB concept.

Table 3.6 Real-gas effects on ORV-WSB aerodynamics.

		CL	
AoA (°)	**PG**	**RG**	**Error (%)**
10	0.1386	0.1402	−1.15
20	0.4875	0.4809	1.37
30	0.9310	0.9130	1.97
40	1.3270	1.3079	1.46
		CD	
AoA (°)	**PG**	**RG**	**Error (%)**
10	0.0633	0.0610	3.80
20	0.2228	0.2185	2.00
30	0.5917	0.5817	1.72
40	1.1755	1.1645	0.94
		CM	
AoA (°)	**PG**	**RG**	**Error (%)**
10	0.0358	0.0432	−17.01
20	0.0864	0.0939	−7.98
30	0.0818	0.0958	−14.55
40	0.0248	0.0476	−47.82

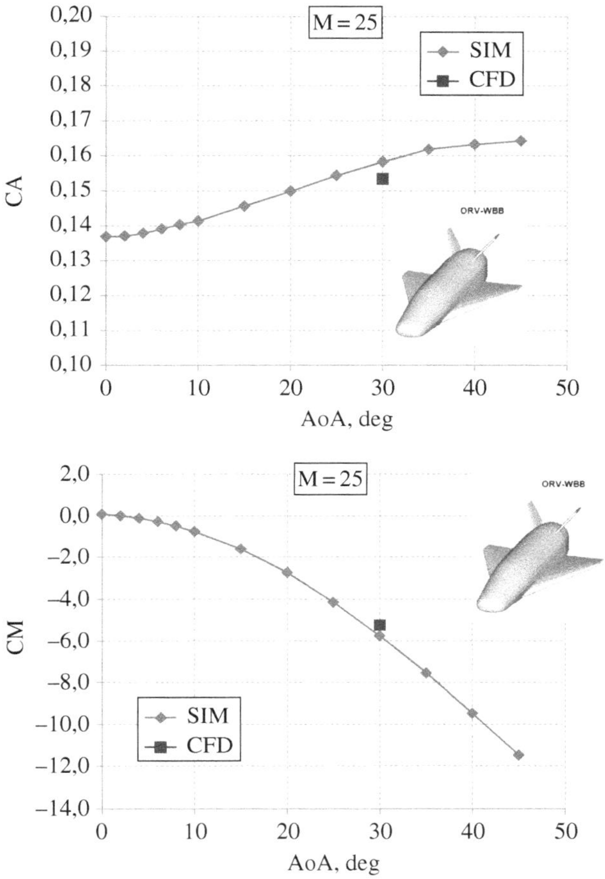

Figure 3.97 C_A and C_M at $M_\infty = 25$ and $\alpha = 30°$; CFD and HPM comparison; ORV-WBB concept.

This confirms that engineering-based (in other words, HPM) and numerical data match very well. Moreover, the differences between real-gas and frozen-gas coefficients, shown in Table 3.6 for $M_\infty = 20$ and $10<\alpha<40°$., suggest that high-temperature gas effects are negligible for lift and drag; whereas a significant effect on vehicle pitching moment is expected along the re-entry trajectory varying from 8 to 50%.

Finally, numerical and engineering-based aerodynamic results for ORV-WBB and ORV-SB are shown in Figures 3.97 and 3.98, respectively.

Figure 3.97 compares the aerodynamic results for the axial force coefficient (C_A) and pitching moment coefficient (C_M) at $M_\infty = 25$ and $\alpha = 30°$ (pole at vehicle nose); Figure 3.98 compares the results for the normal force coefficient (C_N) at $M_\infty = 10$ and $\alpha = 30°$ and for the axial force coefficient (C_A) at $M_\infty = 25$ and $\alpha = 30$ and $45°$. Here too, HPM provides reliable results in the framework of a phase-A design.

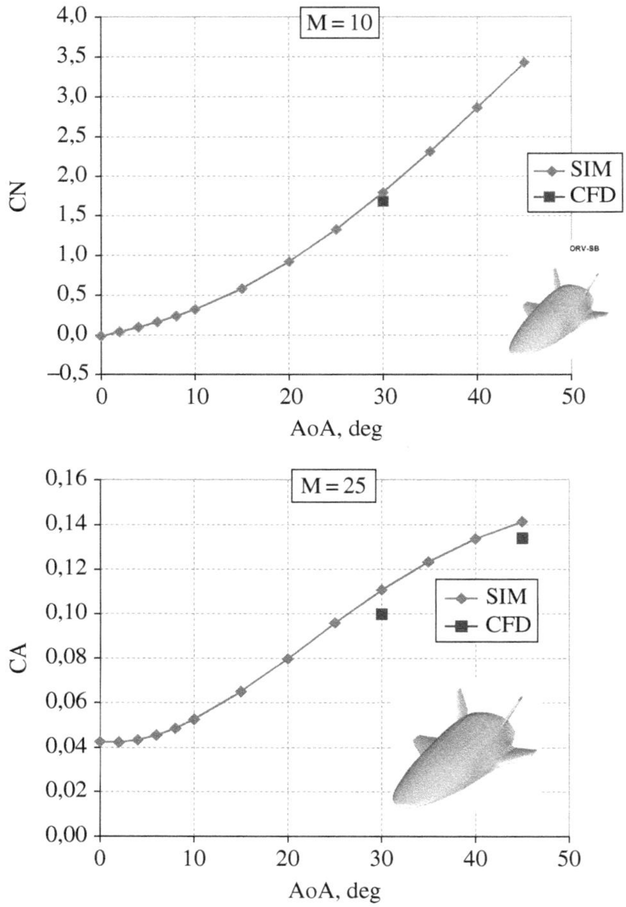

Figure 3.98 C_A and C_N at $M_\infty = 25$ and $\alpha = 30°$; CFD and HPM comparison; ORV-SB concept.

References

1 The Atmospheric Reentry Demonstrator. ESA Bulletin. BR-138, October 1998

2 P. Tran, J.C. Paulat, P. Boukhobza, 'Re-entry flight experiments lessons learned – the atmospheric reentry demonstrator ARD'. Technical report RTO-EN-AVT-130, NATO RTO.

3 A. Schettino, R. Votta, P. Roncioni, et al. 'Aerodynamic and Aerothermodynamic database of expert capsule'. Proceedings of the West-East High Speed Flow Field Conference; 19–22 Nov 2007, Moscow, Russia.

4 S. Paris, D. Charbonnier, D. Tran, 'Experimental and numerical investigation of aerothermal characteristics of the IXV hypersonic vehicle'. Paper presented at the 7th European Symposium on Aerothermodynamics for Space Vehicles. 9–12 May 2011, Brugge.

5 G. Pezzella, G. Marino, G. Rufolo, 'Aerodynamic database development of the ESA intermediate experimental vehicle'. *Acta Astronautica* 2014, Volume 94, Issue 1, pp. 57–72.
6 A. Viviani, G. Pezzella, 'Aerodynamic performance analysis of an unmanned re-entry vehicle from hypersonic down to subsonic regime'. Proceedings of the 63rd International Astronautical Congress 2012, Naples, Italy.
7 A. Viviani, G. Pezzella, 'Aerodynamic performance analysis of three different vehicle concepts at hypersonic speed', Proceedings of the XXII Conference of Italian Association of Aeronautics and Astronautics, 9–12 Sept 2013. Naples, Italy.
8 'Aerodynamic problems of hypersonic vehicles.' AGARD Lecture Series No. 42; Volume 1, Advisory Group for Aerospace Research and Development, 1997.
9 J.D. Anderson, *Fundamental of Aerodynamics*, McGraw-Hill, 2001.
10 A. Schettino, G. Pezzella, et al., 'Mission trade-off analysis of the Italian USV reentry flying test bed'. Proceedings of the 14th AIAA-AHI Space Planes and Hypersonic Systems and Technologies Conference; 6–9 Nov 2006, Canberra. Paper AIAA-2006-8017.
11 A. Schettino, G. Pezzella, et al., 'Aero-thermal trade-off analysis of the Italian USV Re-entry flying test bed', Proceedings of the 14th AIAA-AHI Space Planes and Hypersonic Systems and Technologies Conference; 6–9 Nov 2006, Canberra. Paper AIAA-2006-8114.
12 G. Pezzella, F. Battista, A. Schettino, et al., Hypersonic aerothermal environment preliminary definition of the CIRA FTB-X Reentry Vehicle, Proceedings of the West-East High Speed Flow Field Conference; 19–22 Nov 2007, Moscow, Russia.
13 J.D. Anderson, *Hypersonic and High Temperature Gas Dynamics*, McGraw-Hill, 1989.
14 C.G. Miller, 'Development of X-33/X-34 Aerothermodynamic databases: lessons learned and future enhancements'. RTO AVT Symposium on 'Aerodynamic Design and Optimization of Flight Vehicles in a Concurrent Multi-Disciplinary Environment', 18–21 Oct 1999, Ottawa, Canada.
15 C.G., Miller, 'Aerothermodynamic flight simulation capabilities for aerospace vehicles'. Proceedings of the 20th AIAA Advanced Measurements and Ground Testing Technology Conference. 15–18 Jun 1998. Albuquerque, NM. AIAA Paper 98-2600.
16 G. Pezzella, M. Marini, P. Roncioni, J. et al. 'Preliminary design of vertical takeoff hopper concept of future launchers preparatory program'. *Journal of Spacecraft and Rockets* 2009. Volume 46, Issue 4, pp. 788–799.
17 J.J. Bertin, *Hypersonic Aerothermodynamics*', AIAA, 1994.
18 D.K. Prabhu, System design constraints – trajectory aerothermal environments, in: RTO AVT/VKI Lecture Series in Critical Technologies for Hypersonic Vehicle Development, May 10–14, 2004.
19 Advisory Group for Aerospace Research and Development (North Atlantic Treaty Organization) (1997) Capsule aerothermodynamics, AGARD-R-808. AGARD, Neuilly sur Seine.
20 R.A. Thompson, 'Review of X-33 hypersonic aerodynamic and aerothermodynamic development'. ICAS 2000 Congress, 2000.
21 L. Gomg, W.L. Ko, R.D. Quinn, 'Thermal response of space shuttle wing during re-entry heating', Technical report TM 85907, NASA, 1984.
22 G. Pezzella, R. Gardi, G. Guidotti, C. Richiello, 'Aerodynamic and aerothermodynamic trade-off analysis of the Italian USV2 flying test bed in the framework of an hypersonic flight test'. Proceedings of 3rd International ARA Days. 2–4 May 2011, Arcachon. France. Paper AA-1-2011-54.

23 D.J. Kinney 'Aero-thermodynamics for conceptual design', Proceedings of the 42nd AIAA Aerospace Sciences Meeting and Exhibit, Reno, NV, USA, 5–8 Jan, 2004, Paper AIAA-2004-31.
24 D.J. Kinney, 'Aerodynamic shape optimization of hypersonic vehicles', Proceedings of the 44th AIAA Aerospace Sciences Meeting and Exhibit, Reno, NV, USA, 9–12 Jan 2006. Paper AIAA-2006-239.
25 E. Bonner, W. Clever, K Dunn, 'Aerodynamic preliminary analysis system II Part I – Theory', NASA Contractor Report 182076. April 1991.
26 G. Pezzella, 'Aerodynamic and aerothermodynamic trade-off analysis of a small hypersonic flying test bed'. *Acta Astronautica* 2011, Volume 69, Issue 3–4, pp. 209–222.
27 G. Pezzella, 'Aerodynamic and aerothermodynamic design of future launchers preparatory program concepts'. *Aerospace Science and Technology* 2012, Volume 23, Issue 1, pp. 233–249.
28 G. Pezzella, 'Hypersonic aerothermal environment assessment of the CIRA FTB-X reentry vehicle'. *Aerospace Science and Technology* 2013, Volume 25, Issue 1, pp. 190–202.
29 F.G. Moore, T. Hymer, F. Wilcox, 'Base drag prediction on missile configurations', *Journal of Spacecraft and Rockets* 2014, Volume 31, Issue 5, pp. 759–765, 1994.
30 K. Burns et al., 'Viscous effects on complex configurations' Software User's Manual WL-TR-95-3060, McDonnell Douglas Aerospace, 1995.
31 C.R. Blanchard, K.T. Larman, C. Moats, 'Rarefied-flow shuttle aerodynamics flight model'. *Journal of Spacecraft and Rockets*, 1994, Volume 31, Issue 4, pp. 550–556.
32 M. Maughmer, L. Ozoroski, D. Straussfogel, L. Long, 'Validation of engineering methods for predicting hypersonic vehicle control forces and moments', *AIAA Journal of Guidance, Control, and Dynamics* 1993, Volume 16, Issue 4, pp. 762–769.
33 M. Moore, J. Williams, 'Aerodynamic prediction rationale for analyses of hypersonic configurations', Proceedings of the 27th Aerospace Sciences Meeting, 9–12 Jan 1989, Reno, NV, Paper AIAA-89-0525.
34 A. Viviani, G. Pezzella, *Aerodynamic and Aerothermodynamic Analysis of Space Mission Vehicles*. Springer International Publishing. 2015.
35 C. Weiland, *Aerodynamic Data of Space Vehicles*. Springer, 2014.
36 'Aerodynamic problems of hypersonic vehicles.' AGARD Lecture Series No. 42; Volume 1, Advisory Group for Aerospace Research and Development, 1972.
37 C. Park, 'Review of chemical kinetic problems of future NASA missions: earth entries'. *Journal of Thermophysics and Heat Transfer* 1993, Volume 7, Issue 3, pp. 385–398.
38 D. Olynick, Trajectory-based thermal protection system sizing for an X-33 winged vehicle concept, *Journal of Spacecraft and Rockets*, Volume 35, Issue 3, pp. 249–257.
39 G. Pezzella, E. Filippone, M. Serpico, 'Re-entry aerodynamics and aerothermodynamics analyses of the flying test bed USV-X in the framework of a high lift return', Proceedings of 16th AIAA/DLR/DGLR International Space Planes and Hypersonic Systems and Technologies Conference. Paper AIAA 2009-7425, 2009.
40 A. Viviani, G. Pezzella, 'Computational flowfield analysis over a blunt-body reentry vehicle', *Journal of Spacecraft and Rockets* 2010, Volume 47, Issue 2, pp. 258–270.
41 M. De Stefano, A. Fumo, G. Pugliese et al. 'Small autonomous winged aeroshape trade-off through mission and system guidelines'. 65th International Astronautical Congress, 29 Sept–3 Oct 2014, Toronto, Canada. Paper IAC-14-D2.6.6.

42 F. Petrosino, M. De Stefano, G. Fumo et al., 'Aerodynamic performances of USV3 CIRA re-entry vehicle', 63rd International Astronautical Congress, 1–5 Oct 2012. Naples, Italy. Paper IAC-12-D2.3.7.
43 D. Cinquegrana, F. Petrosino, G. Pezzella et al. 'USV3 aerodynamics assessment: a step forward in the ongoing design phase'. 45th AIAA Plasmadynamics and Lasers Conference, 2014. Paper AIAA 2014-2379.
44 D. Cinquegrana, G. Pezzella, P. Catalano, 'Aerodynamic and aerothermodynamic of the USV3 re-entry vehicle', XXII Conference of Italian Association of Aeronautics and Astronautics (AIDAA 2013), 9–12 Sept 2013, Naples, Italy.

4

Nonlinear Reduced-order Aeroservoelastic Analysis of Very Flexible Aircraft

Nikolaos D. Tantaroudas[1] *and Andrea Da Ronch*[2]

[1] *European Dynamics, SA, Athens, Greece*
[2] *University of Southampton, Southampton, UK*

This chapter overviews the technical difficulties that aerospace engineers have to face during the design of environmentally friendly aerial vehicles. Present trends in aerospace design are driven by two factors: increasing the global aerodynamic efficiency and reducing the structural weight to a minimum. The failure in flight of some prototypes of very flexible aircraft, as described in Section 4.1, has shown that traditional (linear) methods are no longer adequate for the analysis and design of future aerial platforms. In Section 4.2, the reader will be introduced step by step to the mathematical models needed to predict the complex interactions that may occur between the aerodynamic, structural, flight mechanical, and control fields. The computational costs of these models, however, are high because of the large number of calculations needed to ensure structural integrity over the flight envelope, and such computational costs are not viable for industrial aircraft design. The latest developments in the field of model reduction will be discussed in Section 4.3, and the reader will have the opportunity to practice and use these mathematical models, utilising the programming codes associated with this chapter. Finally, active control techniques to enhance the performance and resilience of a very flexible aircraft test case to atmospheric gust and turbulence are discussed in Section 4.4. After the conclusions are given in Section 4.5, some exercises are provided in Section 4.6, which can be solved using the programming codes accompanying this chapter.

The chapter is intended to provide the reader with the essential knowledge required to design high-altitude long-endurance (HALE) vehicles, which have gained considerable attention in recent years. However, it is worth noting that the mathematical models developed within this chapter are applicable to any aircraft configuration and, in particular, to large transport aircraft with their increasingly larger aeroelastic effects. The generality of the models described here allows more traditional (rigid) aircraft configurations – a subset of very flexible aircraft – to be dealt with too.

Advanced UAV Aerodynamics, Flight Stability and Control: Novel Concepts, Theory and Applications,
First Edition. Edited by Pascual Marqués and Andrea Da Ronch.

Companion website: http://www.wiley.com/go/marques/advanced_UAV_aerodynamics

4.1 Introduction

The interest in HALE vehicles has increased steadily over recent years because they provide low-cost efficient platforms for a variety of applications. Their low structural mass and high aerodynamic efficiency allow fight at high altitudes and low speeds with minimal energy consumption. The range of applications of HALE aircraft varies from monitoring and collecting data of the atmospheric environment to rescue missions in biohazard environments. The advantage of unmanned HALE aircraft is the ability to operate at extreme conditions for long duration times without putting at risk human life.

The analysis and design of HALE aircraft, however, presents some unique challenges that are not critical for more rigid (and stiff) aircraft. The dynamic interactions between deformable wings, flow development and flight mechanics may cause structural failure, as occurred in 2003 in the NASA 'Helios' prototype. The Helios aircraft was developed under the Environmental Research Aircraft and Sensor Technology (ERAST) program as a HALE-class vehicle. It was a proof-of-concept solar-electric-powered flying wing designed to operate at high altitudes for long-duration flights. Two configurations were produced, the first designed to achieve high-altitude flight and the second to achieve long-endurance flight. On 13 August 2001, the first Helios configuration flew at a record altitude of 96,863 ft above sea level. The second configuration was lost in flight on 26 June 2003 after encountering low-level turbulence. Approximately 30 min into the flight, atmospheric turbulence caused larger than expected wing deformations and the aircraft began a slowly diverging pitch oscillation. The wing dihedral angles remained high and the oscillations never subsided. Instead, they grew with each period and this led the structure of the aircraft and the skin to pull apart, see Figure 4.1.

4.1.1 Challenges and Prospects

From a technical standpoint, the challenges to be overcome in the analysis and design of HALE aircraft can be summarised as follows:

- The development of a multidisciplinary framework to realistically model the nonlinear interactions occurring between the fluid, structure, flight dynamics, and control fields.
- The lack of an approach to systematically reduce large computational models to a smaller system for faster simulation times and for control synthesis design.
- The exploitation of advanced control design strategies to tame aeroelastic phenomena and improve the effectiveness of the controlled response to atmospheric gusts and extend the flight envelope.

To define the state of the art in the design of HALE vehicles and identify opportunities for progress, a brief review of recent developments in the field is first considered.

Nonlinear Time-domain Multidisciplinary Framework

There is a large body of work on the development of multidisciplinary frameworks to predict the dynamic response of free-flying flexible aircraft as they encounter atmospheric gusts and turbulence. Three representative references are those of Dillsaver and Cesnik (2011), Cook et al. (2013) and Tantaroudas et al. (2014). While a nonlinear formulation is generally used for the structural model, which is based on a beam stick representation of the aircraft components, and for the flight dynamics, the aerodynamic model in most studies is inferred from linear assumptions. It is now recognised that the ability to

(a)

(b)

Figure 4.1 The NASA Helios prototype. From Noll et al. (2004).

obtain realistic (nonlinear) flow predictions is a key remaining issue in the field of very flexible aircraft. Unsteady time-domain analyses of highly flexible aerial vehicles in atmospheric turbulence are still expensive despite modern computing power. The simulation costs become prohibitive when high-fidelity numerical models are used in an industrial environment with a very large number of simulations required. Parametric searches are performed to estimate the critical loads that the aircraft will encounter during its expected life cycle and these are used for structural sizing. Inaccuracies in the load estimates can jeopardise the entire project or result in a very conservative (and inefficient) design.

Nonlinear Model Order Reduction

The response of a dynamic system to given initial conditions or external forces may be obtained either in the frequency or time domain. The advantages of the analysis performed in the frequency domain are offset by the underlying assumption of linearity in the system response. The time-domain analysis, through use of numerical integration schemes, allows predictions to be made of the response of a nonlinear system. Unsteady time-domain analysis is, however, computationally expensive and particularly so in the case of a large-dimensional model. The objective of model order reduction is to produce a low-dimensional system that is computationally efficient yet accurate enough to approximate, to some desired threshold, the response characteristics of the original system (Antoulas 2005). The resulting reduced-order model can then be used to replace the original system for routine calculations or to develop a simple and fast controller suitable for real-time applications. An appropriate model-reduction methodology will ensure the following properties are met:

- The approximation error may be reduced to a desired threshold by increasing the size of the reduced order model. Hence convergence of the reduced model on the original system should be proved.
- The properties of the original system may be retained. Stability properties, in particular, are important for evaluating the system's dynamic response.
- The algorithm to perform model reduction must be systematic, computationally stable, and efficient.

As described in more detail in Section 4.3, there are two general approaches to reducing the complexity and cost of a large computational model.

Advanced Active Control Strategies

Another important aspect of the design of flexible aircraft is the flight control system design. The system can be active or passive. Here, we will focus on active control, which has the potential to increase aircraft performance and to extend flight envelopes to new limits. The design of a gust-tolerant vehicle needs an accurate mathematical model to realistically simulate the nonlinear interactions that dominate such aerial platforms. Nevertheless, the use of fairly large nonlinear physics-based models introduces complications in the design, synthesis and testing of control strategies. Two major difficulties arise when dealing with control: the design and the implementation.

The design becomes complicated when the system is of high order and includes many unobservable or unmeasurable states, especially when the system is nonlinear. More important, however, is the limited availability of computational resources.

As for implementation, it is difficult to scale applications in relation to available resources such as memory and power limitations and still guarantee the real-time response when implemented. As a result, the derivation of low-order controllers based on reduced models becomes of high importance. This is accomplished by model-order reduction techniques. In this way, not only is the problem of fast and accurate predictions of loads overcome, but also the design of the flight control system is simplified and the hardware implementation of the controller becomes feasible (Campos-Delgado et al. 2003).

4.2 Large Coupled Computational Models

The reference aircraft configuration around which this chapter is based shares many common aspects with the Global Hawk (Figure 4.2), built by Northrop Grumman and first flown in 1998. The design is unrestricted and is available on request from authors.

The reference aircraft is described in Section 4.2.1, and the mathematical models for the fluid, structural, and flight-dynamic fields are introduced in Sections 4.2.2–4.2.5.

(a)

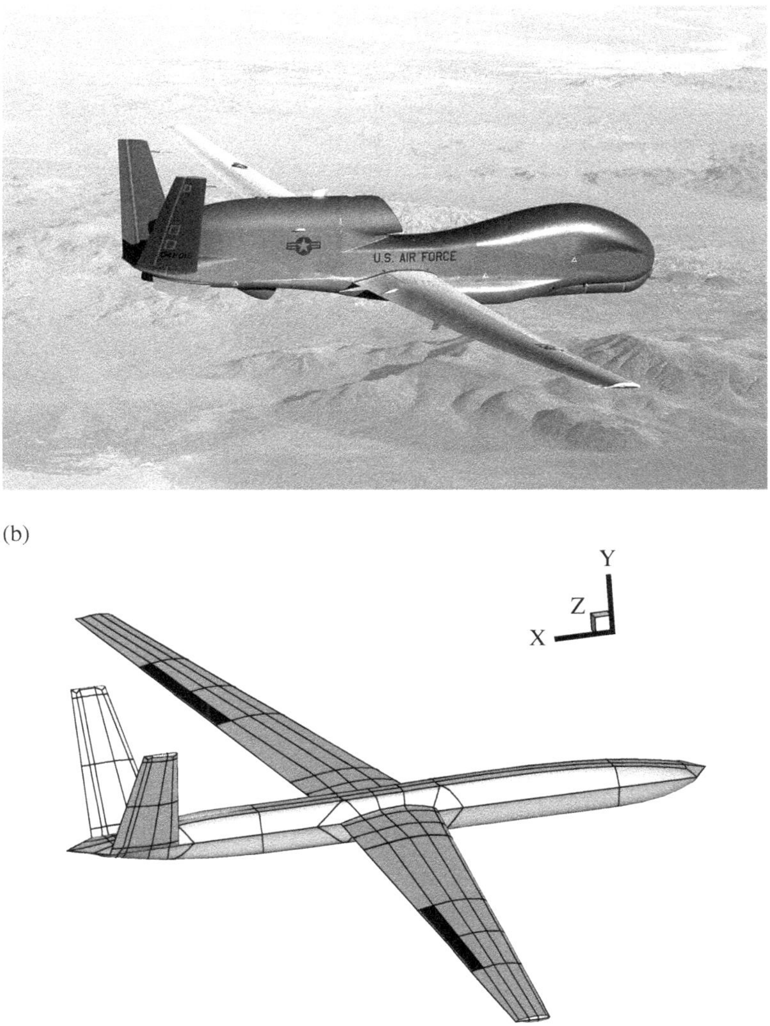

Figure 4.2 Examples of high–altitude UAVs: (a) RQ4 Global Hawk (courtesy U.S. Air Force); (b) the test case described in this chapter.

4.2.1 Aircraft Test Case

The test case is a flexible unmanned aerial vehicle (UAV) that generally resembles the RQ4 Global Hawk. Figure 4.2 is a three-dimensional view of the aircraft, which features high aspect-ratio wings, a fairly rigid streamlined fuselage, and a V-tail. A set of trailing-edge control surfaces is located on each semi-wing at a distance of 37–77% of the wing span from the wing root, and at 32% of the local chord from the wing trailing edge. The basic geometric characteristics are shown in Figure 4.3.

A detailed finite-element structural model of the airframe created in MSC/NASTRAN[1] was available for accurate stress calculations, and this was later used to create an equivalent beam model. The structure was built of composite material, and the structural model included a combination of various finite-element types. With fuel tanks on the wings between the front and rear spars accounting for over 4700 kg, the centre of gravity was 6.38 m from the nose of the aircraft.

The starting finite-element model of the structure was then reduced to an equivalent beam model. A beam stick representation of the aircraft follows easily, because lifting surfaces are of high aspect ratio. For the wings and tail, the beam model was located at the centre of the corresponding structural box, between the front and rear spars. The mass and stiffness properties of the beam model were iteratively refined to ensure a good agreement of the lowest modeshapes and frequencies with the original detailed structural model.

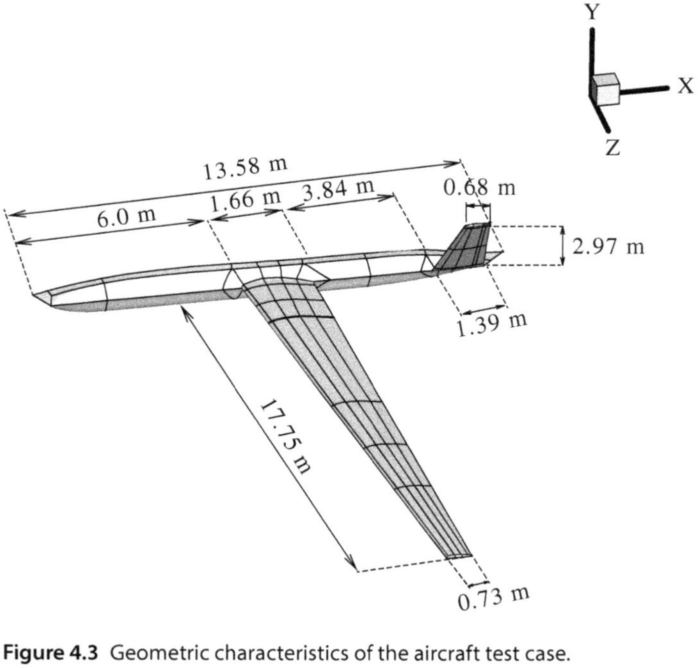

Figure 4.3 Geometric characteristics of the aircraft test case.

1 http://www.mscsoftware.com/product/msc-nastran.

Table 4.1 First five modeshapes and frequencies of the UAV test case main wing.

	Modeshape	MSC/NASTRAN	Beam model	
			17 elements	27 elements
Mode number		Hz	Hz	Hz
1	First bending	3.56	3.74	3.58
2	Second bending	7.75	7.92	6.84
3	First torsion	14.90	12.87	17.18
4	Third bending	15.70	10.83	11.98
5	Fourth bending	24.60	14.57	19.80

A comparison of the first five lowest modeshapes and frequencies determined in the original detailed model and the beam stick model is shown in Table 4.1. Tuning the mass and stiffness properties of the beam model reveals a reasonably good agreement for all the modeshapes shown, with increasing inaccuracies at higher frequencies. Following a study aimed at investigating the dependency of the frequencies on the number of beam elements used, it was found that 27 elements were adequate to discretise the aircraft wing, and 10 were used for the tail. The fuselage, on the other hand, is modelled as a rigid body.

4.2.2 Aerodynamic Model

Several options for the aerodynamics can be used. Using an engineering approach, we aim for an aerodynamic model that is as simple as possible yet is still accurate enough. In the most simple case, a two-dimensional linear aerodynamic model can be used on a representative two-dimensional section of the aeroelastically most-critical lifting surface. The unsteady flow is modelled by a frequency-domain expression for the incompressible two-dimensional potential flow over a flat plate in harmonic motion, originally formulated by Theodorsen (1935).

An extension to this approach, called strip theory, adapts the same two-dimensional unsteady flow model for a three-dimensional aeroelastic system by combining section aerodynamics with a beam model for the wing structure. Strip theory can provide fairly reliable, and usually conservative, results for divergence speed, critical flutter speed and aileron reversal. However, it requires that the physical characteristics of the aircraft configuration under analysis can be adequately reduced to a beam-type structure and that three-dimensional aerodynamic effects do not have a significant impact on the aerodynamics.

The total aerodynamic loads consist of contributions arising from the section motion, trailing-edge flap rotation, and the penetration into a gusty field, as illustrated in Figure 4.4. The aerodynamic loads due to an arbitrary input time-history are obtained through convolution against a kernel function. Since the assumption is of linear aerodynamics, the effects of the various influences on the aerodynamic forces and moments

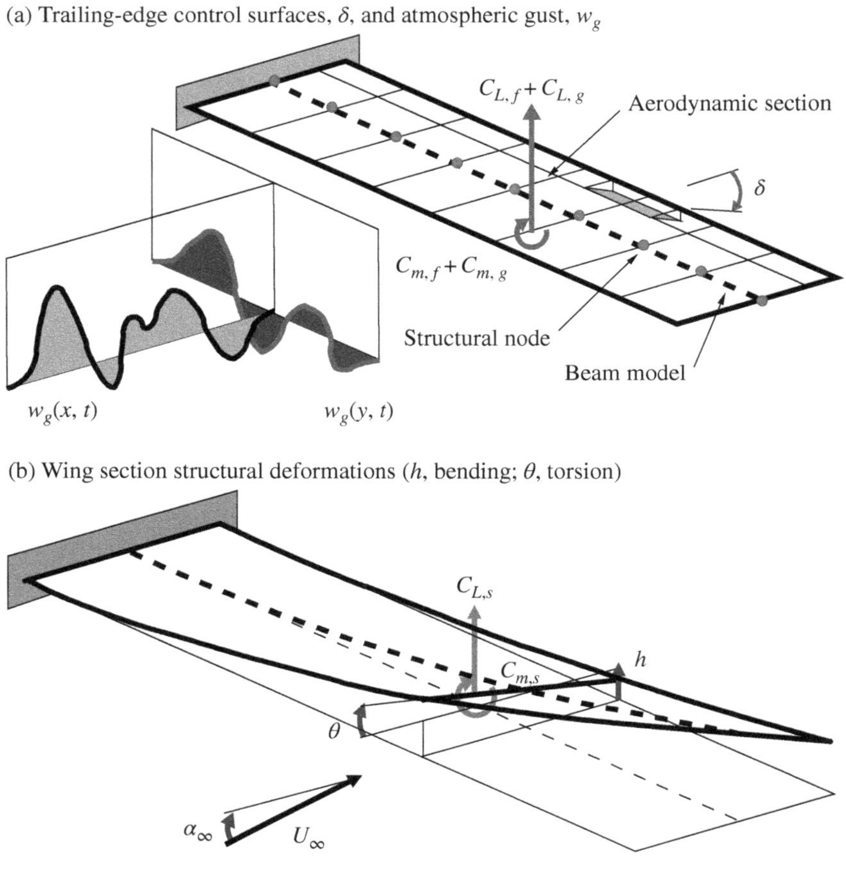

Figure 4.4 Schematic of a slender wing structure showing various contributions to the aerodynamic loads.

are added together to find the variation of the forces and moments in time for a given motion and gust. It follows that

$$C_i = C_{i,s} + C_{i,f} + C_{i,g} \tag{4.1}$$

where the dependence on time is not shown explicitly. The sub-index i is used to denote the lift coefficient, $i = L$, and pitch-moment coefficient, $i = m$, whereas s, f and g indicate the contributions from the section motion, flat rotation, and gust perturbation, respectively. A schematic representation of the various contributions to the aerodynamic loads is shown in Figure 4.4.

A brief description of each contribution to the total aerodynamic loads is summarised in the following three subsections. Note that, as is common in aerodynamics, the loads are formulated using non-dimensional time, $\tau = tU_\infty/b$, which is also adopted for the time derivative, $(\bullet)' = d(\bullet)/d\tau$. The subindex 0 will be used to indicate initial conditions.

Section Motion

The first term on the right-hand side of Eq. (4.1) indicates the increment in the aerodynamic loads caused by a generic motion of the wing section. Each structural node of the beam stick model (see Section 4.2.3), has six degrees of freedom: three rotations and three translations. As the aerodynamic model here is two-dimensional, the resulting motion of the wing section in the three-dimensional space is projected onto the plane defining the wing cross section. Referring to Figure 4.4b, the motion of the wing section contributing to the aerodynamic loads consists of the vertical displacement of the structural beam model, denoted by h, and a rotation around the elastic axis, denoted by θ. This information is readily available from the solution of the structural problem.

We denote by α the effective angle of incidence of the wing section, which includes the freestream angle of attack, α_∞, and the wing torsional deformation, θ. Scale the vertical displacement, h, by the semi-chord of the wing cross section, $\xi = h/b$. The resulting force and moment coefficients for any arbitrary section motion in pitch and plunge are formulated as

$$\begin{aligned} C_{L,s}(\tau) = {} & \pi\left(\xi''(\tau) - a_h\alpha''(\tau) + \alpha'(\tau)\right) + \\ & 2\pi\left(\alpha_0 + \xi_0' + (1/2 - a_h)\alpha_0'\right)\phi_w(\tau) + \\ & 2\pi\int_0^\tau \phi_w(\tau-\sigma)\left(\alpha'(\sigma) + \xi''(\sigma) + (1/2 - a_h)\alpha''(\sigma)\right)d\sigma \end{aligned} \tag{4.2}$$

$$\begin{aligned} C_{m,s}(\tau) = {} & \pi(1/2 + a_h)\left(\alpha_0 + \xi_0' + (1/2 - a_h)\alpha_0'\right)\phi_w(\tau) + \\ & \pi(1/2 + a_h)\int_0^\tau \phi_w(\tau-\sigma)\left(\alpha'(\sigma) + \xi''(\sigma) + (1/2 - a_h)\alpha''(\sigma)\right)d\sigma + \\ & \frac{\pi}{2}a_h\left(\xi''(\tau) - a_h\alpha''(\tau)\right) - (1/2 - a_h)\frac{\pi}{2}\alpha'(\tau) - \frac{\pi}{16}\alpha''(\tau) \end{aligned} \tag{4.3}$$

The Wagner function, φ_w, accounts for the influence of the shed wake, and is known exactly in terms of Bessel functions. For a practical evaluation of the integral, the exponential approximation of Jones (1940) is used:

$$\phi_w(\tau) = 1 - \Psi_1 e^{-\varepsilon_1\tau} - \Psi_2 e^{-\varepsilon_2\tau} \tag{4.4}$$

where the constants are $\Psi_1 = 0.165$, $\Psi_2 = 0.335$, $\varepsilon_1 = 0.0455$ and $\varepsilon_2 = 0.3$.

Trailing-edge Flap Rotation

The second term on the right-hand side of Eq. (4.1) represents the increment in the aerodynamic loads for any arbitrary trailing-edge rotation, see Figure 4.4a. The build-up in the loads not only depends on the instantaneous flap rotation but also on its time derivatives (velocity and acceleration). The relations between the control surface input, δ, and the load coefficients are

$$\begin{aligned} C_{L,f}(\tau) = {} & -T_4\delta'(\tau) - T_1\delta''(\tau) + \\ & 2\pi\left[\left(\frac{1}{\pi}T_{10}\delta_0 + \frac{1}{2\pi}T_{11}\delta_0'\right)\phi_w(\tau) + \right. \\ & \left. \int_0^\tau\left(\frac{1}{\pi}T_{10}\delta' + \frac{1}{2\pi}T_{11}\delta''\right)\phi_w(\tau-\sigma)d\sigma\right] \end{aligned} \tag{4.5}$$

$$C_{m,f}(\tau) = -\frac{(T_4 + T_{10})}{2}\delta(\tau) - \frac{\left(T_1 - T_8 - (c - a_h)T_4 + \frac{1}{2}T_{11}\right)}{2}\delta'(\tau) + \frac{(T_7 + (c - a_h)T_1)}{2}\delta''(\tau) + \pi(a_h + 1/2)\left[\left(\frac{1}{\pi}T_{10}\delta_0 + \frac{1}{2\pi}T_{11}\delta_0'\right)\phi_w(\tau)) + \int_0^\tau \left(\frac{1}{\pi}T_{10}\delta' + \frac{1}{2\pi}T_{11}\delta''\right)\phi_w(\tau - \sigma)d\sigma\right] \tag{4.6}$$

The coefficients T_1, T_4, T_7, T_8, T_{10} and T_{11} are geometric constants that depend on the size of the trailing-edge flap relative to the chord of the wing section. The reader can find the full expressions in Da Ronch et al. (2014).

Atmospheric Gust

The last term on the right-hand side of Eq. (4.1) describes the effect that atmospheric gusts and turbulence have on the build-up of aerodynamic loads. For an arbitrary gust time-history, the load coefficients are computed by the following relations:

$$C_{L,g}(\tau) = 2\pi\left(w_{g0}\Psi_k(\tau) + \int_0^\tau \Psi_k(\tau - \sigma)\frac{dw_g}{d\sigma}d\sigma\right) \tag{4.7}$$

$$C_{m,g}(\tau) = \pi(1/2 + a_h)\left(w_{g0}\Psi_k(\tau) + \int_0^\tau \Psi_k(\tau - \sigma)\frac{dw_g}{d\sigma}d\sigma\right) \tag{4.8}$$

where the gust intensity, w_g, is normalised by the freestream speed. The integration uses the exponential approximation of the Küssner function

$$\Psi_k(\tau) = 1 - \Psi_3 e^{-\varepsilon_3\tau} - \Psi_4 e^{-\varepsilon_4\tau} \tag{4.9}$$

where the coefficients $\Psi_3 = 0.5792$, $\Psi_4 = 0.4208$, $\varepsilon_3 = 0.1393$ and $\varepsilon_4 = 1.802$ are from Leishman (1994). Appropriate forms of w_g to model realistic atmospheric gust and turbulence time-histories are presented in some detail in Section 4.2.4.

4.2.3 Flexible-body Dynamics Model

For the structural model, the geometrically-exact nonlinear beam equations are used (Hesse and Palacios 2012). Results are obtained using two-noded displacement-based elements. In a displacement-based formulation, nonlinearities arising from large deformations are cubic terms, as opposed to an intrinsic description where they appear up to second order. The coupled flexible multibody nonlinear equations are expressed in the form

$$M(w_s)\begin{Bmatrix}\ddot{w}_s\\ \ddot{w}_r\end{Bmatrix}+Q_{\mathrm{gyr}}(\dot{w}_s, w_s, w_r)\begin{Bmatrix}\dot{w}_s\\ \dot{w}_r\end{Bmatrix}+Q_{\mathrm{stiff}}(w_s)\begin{Bmatrix}w_s\\ w_r\end{Bmatrix}=R_F \tag{4.10}$$

The subscripts s and r denote elastic and rigid-body degrees of freedom, respectively. The terms Q_{gyr} and Q_{stiff} indicate, respectively, gyroscopic and elastic forces, whereas R_F contains all external forces acting on the system, including aerodynamic contributions. More details of the structural modelling of multibody dynamics using finite elements can be found in Geradin and Cardona (2001).

Equation (4.10) is coupled with the linearised quaternion equations that propagate the orientation of the body frame with respect to the inertial frame.

$$\dot{\zeta}_i + C_{QR}\dot{w}_r + C_{QQ}\zeta_i = 0 \tag{4.11}$$

4.2.4 Atmospheric Gust and Turbulence Models

In flight, aircraft regularly encounter atmospheric turbulence. For linear analysis, turbulence is regarded as a set of component velocities superimposed on a steady background flow. The aircraft experiences rapid changes in lift and moment forces, which cause rigid and flexible dynamic responses of the entire aircraft. These responses can cause passenger discomfort and introduce large loads on the structure and they must be accounted for during the design stage to ensure aircraft safety. The models used for the prediction of the aircraft response have to incorporate events that are perceived as discrete, usually described as 'gusts', as well as the phenomena described as continuous, in other words turbulence.

A concise summary of the mathematical models used to approximate discrete and continuous turbulent events is given next. The reader may find a more extensive review in Etkin (1981).

Discrete Deterministic Gusts

Discrete events are isolated encounters with steep gradients in the speed of air, which typically occur at the edges of thermals and downdrafts, in the wakes of structures or mountains, or at temperature inversions. Discrete gusts may also appear as rare extremes of turbulence in clouds, and so on, possibly associated with organised structures embedded in the otherwise random background. These organised extremes are not adequately allowed for in the usual Gaussian models of continuous random turbulence, and specialised discrete models should then be used.

The most common discrete gust model, which has evolved over the years from the isolated sharped-edge gust function used in the earliest airworthiness assessments, is the 'one-minus-cosine' function:

$$w_g(x,t)=\begin{cases}\frac{1}{2}w_{g0}\left(1-\cos\left(\frac{\pi U_\infty}{L_g}\left(t-\frac{x}{U_\infty}\right)\right)\right) & x\in\left[tU_\infty-2L_g, tU_\infty\right]\\ 0 & \text{otherwise}\end{cases} \tag{4.12}$$

where w_{g0} is the gust intensity, L_g is the gust length, and x is the position of a point on the aircraft relative to an aircraft-attached frame of reference (see Figure 4.5).

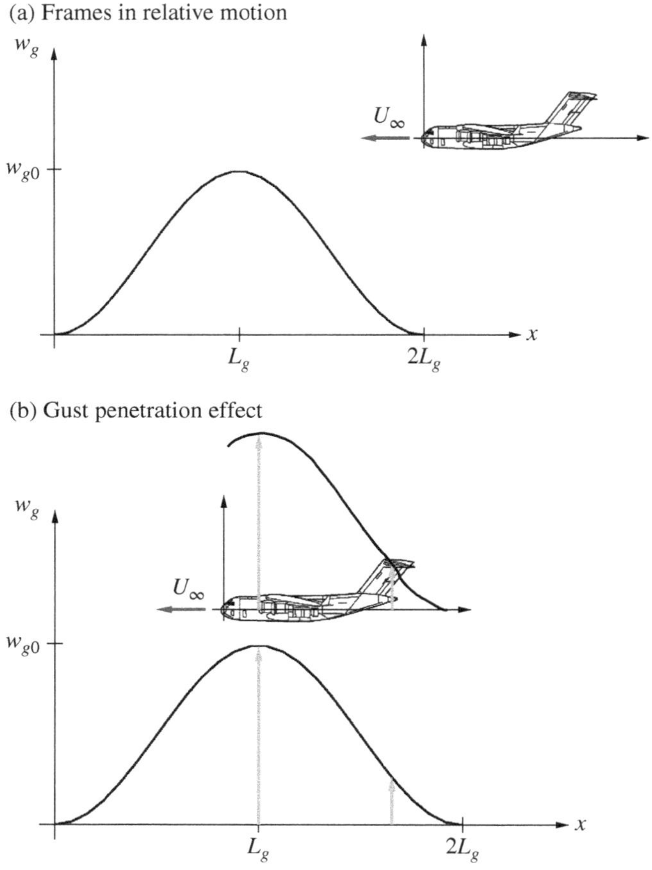

Figure 4.5 Discrete model of a one-minus-cosine gust.

The design gust velocity, w_{g0}, varies with the gust length and the altitude and speed of flight (Hoblit 1988). In the simple case of Eq. (4.12), the gust intensity depends on one spatial coordinate, x, in addition to the time coordinate, t. The rate of change of the gust intensity at different points on the aircraft – say the main wing and tailplane – largely depends on two ratios (see Figure 4.5). The first ratio describes the relative size of the gust compared to the aircraft characteristic length. The second relates to the time it takes for the aircraft to fly over the gusty field. As these two ratios decrease, the dependence on the spatial coordinate becomes more and more apparent and should be modelled appropriately when simulating the aircraft response to relatively short gusts.

Random Turbulence

Random turbulence is a chaotic motion of the air and is described by its statistical properties. The main statistical features that need to be considered are: stationarity, homogeneity, isotropy, time and distance scales, probability distributions, correlations and spectra. Atmospheric turbulence is a vector process in which the velocity vector is a random function of the position vector and of time. Because of the complexity introduced by this multi-dimensionality, the description of turbulence and the associated input/response problems are often simplified, whether justified or not, to a one-dimensional representation.

The engineering model of random turbulence at altitude has been developed over many years; see for example (Houbolt 1973). It is now widely accepted that it is satisfactory to treat atmospheric turbulence as frozen, homogeneous and isotropic in relatively large patches. The frozen-field assumption, closely allied to Taylor's hypothesis, is that turbulent velocities do not change during the time of passage of the airplane. This is a valid assumption in most cases. The Dryden and the von Kármán models are considered adequate engineering models to predict the correlation and spectra, with the weight of experimental evidence favouring the latter. Although there is much evidence that turbulence is not in fact a Gaussian process, with small and large values both occurring more frequently than in a normal distribution, the assumption that individual patches are Gaussian is widely used because of the great analytical advantage it offers.

A commonly used spectrum that matches experimental data is the von Kármán model. The power spectral density (PSD, in $[\mathrm{m}^2/(\mathrm{s}^2\,\mathrm{Hz})]$) for the vertical direction, Φ_z, according to Military Specification MIL-F-8785C, is given by (Moorhouse and Woodcock 1982):

$$\Phi_z(\Omega)=\frac{\sigma_z^2 2L_z}{U_\infty}\frac{1+8/3\left(aL_z\Omega\right)^2}{\left(1+\left(aL_z\Omega\right)^2\right)^{11/6}} \tag{4.13}$$

where $\Omega=\omega/U_\infty$ is the scaled frequency (in [rad/m]), σ_z is the root mean square turbulence velocity (in [m/s]), L_z is the characteristic scale wavelength of the turbulence (in [m]), and $a=1.339$ is the von Kármán constant. Figure 4.6 illustrates the PSD spectrum as function of the frequency. Whilst the system response in the frequency domain to a random turbulence can easily be calculated once the frequency-response function is known, this approach is linear and does not permit nonlinear effects to be included in the analysis. An alternative approach is to generate a random turbulence time signal with the required spectral characteristics defined in Eq. (4.13), and solve the nonlinear system of equations in the time domain.

A method to calculate the time-domain response of a nonlinear aeroelastic model to random turbulence is as follows. First, take the Fourier transform of a unit variance band-limited white noise signal, $X(\Omega)$, and pass it through a filter defined as the square root of the PSD spectrum in Eq. (4.13), $H_z(\Omega)$. Then calculate the output signal using the relation

$$W_g(\Omega)=H_z(\Omega)X(\Omega) \tag{4.14}$$

Take the inverse Fourier transform of $W_g(\Omega)$ to obtain the random turbulence in the time domain, $w_g(t)$. This method, which applies a Fourier transform twice, is preferred over an alternative method that does not make use of the Fourier transform. More details may be found in Gianfrancesco (2014).

The method described above is implemented in an open source MATLAB toolbox called the Von Kármán Turbulence Generator (VKTG). The VKTG toolbox implements the mathematical representations of random turbulence defined in Military Specification MIL-F-8785C and Military Handbook MIL-HDBK-1797, allowing for the dependence of the root mean square turbulent velocity and turbulence length scale on aircraft mission parameters and weather conditions. As shown in Figure 4.6, at higher frequencies,

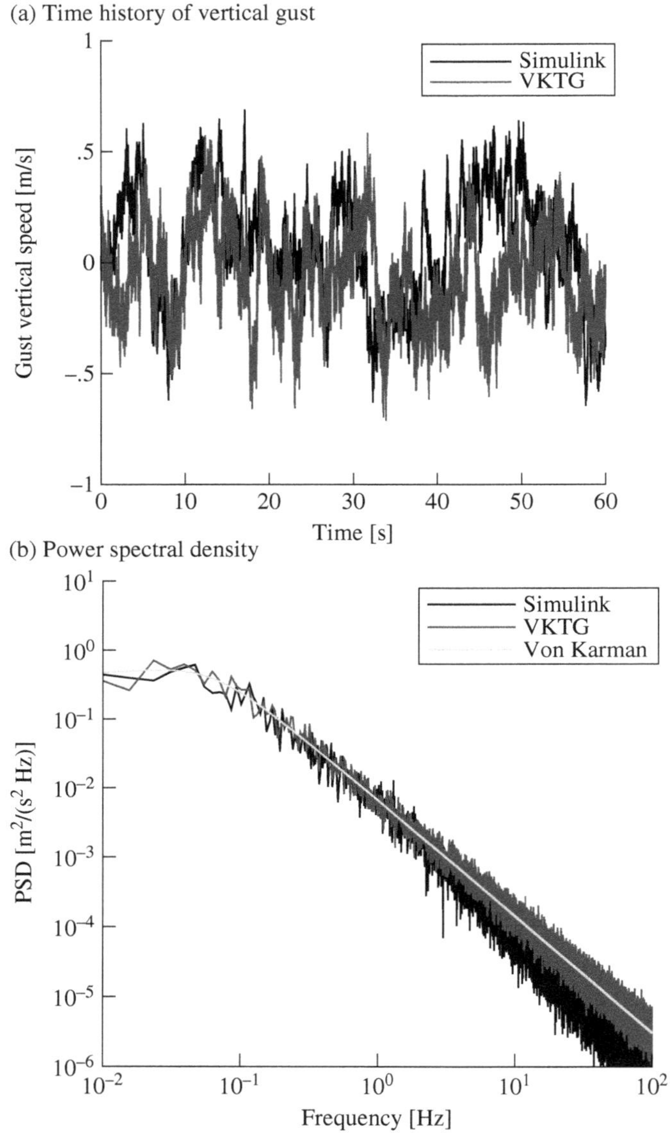

Figure 4.6 Random vertical gust intensity using the von Kármán spectral representation (Military Specification: MIL-F-8785C; flight speed: $V = 280$ m/s; altitude: $h = 10{,}000$ m; and turbulence intensity: 'light, 10^{-2}'); the terms 'Simulink' and 'VKTG' denote, respectively, the Von Kármán Wind Turbulence Model block of MATLAB and the Von Kármán Turbulence Generator implementation.

the PSD of the VKTG model correlates more closely with the von Kármán spectrum of Eq. (4.13) than an off-the-shelf MATLAB/SIMULINK model. Note that the VKTG toolbox accompanying this chapter is also available online.[2]

2 http://www.personal.soton.ac.uk/adr1d12/.

4.2.5 Numerical Implementation

With the previous sections as background, the coupled dynamics of a flexible flying aircraft encountering atmospheric gusts and turbulence requires a careful integration of each single discipline within a comprehensive simulation environment. It is generally possible to recast the complete system of equations in state-space form, this being convenient for the derivation of the reduced-order model, as discussed in Section 4.3.

4.3 Coupled Reduced-order Models

The numerical solution of a nonlinear system in the time domain requires the integration of the differential equations that govern its dynamics. Numerical schemes are often referred to as being explicit or implicit. For the conditional stability of explicit methods, a small time step is required to solve the small timescales that are always present in spatially detailed models but that are not needed in our analysis. Implicit methods, which are more complex to program and require more computational effort in each solution step, are therefore preferred because they allow for longer time steps. As most of the computational methods to solve nonlinear systems were developed for first-order ordinary differential equations (ODEs), it is convenient to reformulate the coupled system as a system of first-order ODEs.

Let us consider as a starting point the nonlinear system

$$\dot{\boldsymbol{x}} = \boldsymbol{F}\left(\boldsymbol{x}(t), \boldsymbol{u}(t), \boldsymbol{d}(t), t\right) \tag{4.15}$$

where the nonlinear operator $\boldsymbol{F}: \mathbb{R}^{\hat{n}} \rightarrow \mathbb{R}^n$ depends on the specific formulation used. For brevity, $(\dot{\bullet})$ denotes the time derivative, $d(\bullet)/dt$. Consider the coupled system partitioned as follows

$$\boldsymbol{x} = \begin{Bmatrix} \boldsymbol{x}_f \\ \boldsymbol{x}_s \\ \boldsymbol{x}_r \end{Bmatrix}, \quad F = \begin{Bmatrix} \boldsymbol{F}_f \\ \boldsymbol{F}_s \\ \boldsymbol{F}_r \end{Bmatrix} \tag{4.16}$$

where the subindices f, s and r indicate, respectively, the fluid, structural, and rigid-body (flight dynamics) degrees of freedom. Denote by u the vector of manipulable inputs, generally used for control purposes, and by d the vector of exogenous disturbances perturbing the system. Indicate the ith component of $\boldsymbol{F}$ by f_i. If the components of $\boldsymbol{F}$ are differentiable at $\boldsymbol{x} \in \mathbb{R}^n$, define the Jacobian matrix $\boldsymbol{A} \in \mathbb{R}^n \times \mathbb{R}^n$ as

$$a_{ij}(x) = \frac{\partial f_i}{\partial x_j}(\boldsymbol{x}) \tag{4.17}$$

or in matrix form as

$$\boldsymbol{A}(\boldsymbol{x}) = \frac{\partial \boldsymbol{F}}{\partial \boldsymbol{x}}(\boldsymbol{x}) \tag{4.18}$$

for any constant vector of manipulable inputs and exogenous disturbances, $\tilde{\boldsymbol{u}}$ and $\tilde{\boldsymbol{d}}$, respectively.

4.3.1 Approaches to Model Reduction

The difficulty with Eq. (4.15) is the size of the computational model which, for realistic applications, may include $\mathcal{O}(10^4)$ to $\mathcal{O}(10^7)$ degrees of freedom. To accelerate the time-domain analysis and allow design of a low-order control system, a reduced-order model of Eq. (4.15) is often needed in practice.

Two approaches to reduce the size of a system exist. The first is based on system identification techniques that attempt to replace the original large computational model, which is treated as a 'black box', with a system of smaller size. Often the model structure is simplified, allowing for some forms of nonlinearity to be included. The advantages are the easiness of the numerical implementation and the availability of various techniques (Volterra series, indicial functions, surrogate models, and so on). The disadvantages are the lack of robustness of the system parameters and the limited validity and certainty for conditions outside those used to generate the model. More details may be found in Ghoreyshi et al. (2013) and Da Ronch et al. (2011).

The second approach involves manipulation of the governing equations in Eq. (4.15). The advantages are the ability to retain nonlinear effects in a smaller system and that the system validity depends on the assumptions made to derive the model. The disadvantage is the added complexity in the model implementation. Two well-established reduced-order models for application in unsteady aerodynamics and aeroelasticity are based on the harmonic balance method (Da Ronch et al. 2013a) and nonlinear model projection (Da Ronch et al. 2012), respectively.

The approach to model reduction first presented in Da Ronch et al. (2012) and thereafter applied extensively to various problems and test cases (Da Ronch et al. 2013b, 2014, 2013c; Tantaroudas et al. 2014; Timme et al. 2013), is discussed in more detail in the following sections. The approach is based on the manipulation of the governing equations and consists of a systematic procedure to generate both linear and nonlinear reduced-order models, independently of the mathematical formulation used in the coupled system. The approach also satisfies the properties outlined in Section 4.1.1.

4.3.2 Stability Analysis

Neglect first the explicit dependence of the system in Eq. (4.15) on the time variable, and consider a constant vector of manipulable inputs and exogenous disturbances, $\tilde{\boldsymbol{u}}$ and $\tilde{\boldsymbol{d}}$. Here, for simplicity, $\tilde{\boldsymbol{d}} = \boldsymbol{0}$ is assumed. An equilibrium point, $\tilde{x}$, satisfies the relation

$$\boldsymbol{F}(\tilde{\boldsymbol{x}},\tilde{\boldsymbol{u}}) = \boldsymbol{0} \tag{4.19}$$

The stability of the system in the neighbourhood of the equilibrium point, $\tilde{\boldsymbol{x}}$, is studied through an eigenvalue analysis. Define a small increment with respect to the equilibrium point, $\boldsymbol{w} = \boldsymbol{x} - \tilde{\boldsymbol{x}}$. The linearised homogeneous system around the equilibrium point, $\tilde{\boldsymbol{x}}$, takes the form

$$\dot{\boldsymbol{w}} = \boldsymbol{A}\boldsymbol{w} \tag{4.20}$$

where $\boldsymbol{A}$ is the Jacobian matrix, Eq. (4.18). Let us consider a solution of Eq. (4.20) in the exponential form

$$\boldsymbol{w} = \phi e^{\lambda t} \tag{4.21}$$

where λ is a constant scalar and ϕ is a constant n-dimensional vector. The right and left eigenvalue problems associated with this ansatz, after substituting Eq. (4.21) into Eq. (4.20), are

$$\begin{aligned} &\text{right}: \quad \lambda_i \phi_i = A\phi_i \qquad i = 1,2,\ldots,n \\ &\text{left}: \quad \lambda_i \psi_i = A^T \psi_i \quad i = 1,2,\ldots,n \end{aligned} \tag{4.22}$$

For large computational models, the solution of the above eigenvalue problems using off-the-shelf algorithms is unfeasible. Because the subject is out of the scope of this chapter, the interested reader is referred to Badcock et al. (2011). Here, let us assume an appropriate eigenvalue solver is available for solving the right and left problems.

It is convenient to normalise the eigenvectors to satisfy the biorthonormality conditions. This will be of great benefit when deriving the reduced-order model.

$$\begin{aligned} &\phi_i^H \phi_i = 1 \qquad i = 1,2,\ldots,n \\ &\psi_j^H \phi_i = \delta_{ij} \quad i,j = 1,2,\ldots,n \end{aligned} \tag{4.23}$$

so that the relation holds

$$\psi_j^H A\phi_i = \lambda_i \delta_{ij} \quad i,j = 1,2,\ldots,n \tag{4.24}$$

where δ_{ij} is the Kronecker delta and the operator $(\bullet)^H$ indicates the Hermitian transpose: the transpose of the complex conjugate.

4.3.3 Model Projection

An attempt to solve Eq. (4.15) may not only require large computational resources but it is also not well suited to the design of a low-order control system. The idea of transforming the above set of n simultaneous nonlinear equations into one that lends itself to an easier solution arises naturally. As is common in structural dynamics, a transformation of coordinates defined in terms of the n orthonormal modal vectors $\phi_1, \phi_2, \ldots, \phi_n$, is an efficient choice (Meirovitch 1990). Let define w in the form of a linear combination of modal vectors

$$\begin{aligned} w &= \phi_1 z_1 + \bar{\phi}_1 \bar{z}_1 + \phi_2 z_2 + \bar{\phi}_2 \bar{z}_2 + \ldots + \phi_n z_n + \bar{\phi}_n \bar{z}_n \\ &= \sum_{i=1}^{n} \left(\phi_i z_i + \bar{\phi}_i \bar{z}_i \right) \end{aligned} \tag{4.25}$$

where the over-bar sign $\overline{(\bullet)}$ indicates the conjugate. It is worth noting that the solution of the eigenvalue problems in Eq. (4.22) provides the eigenvalues and the associated eigenvectors of the coupled system, describing the interactions between the fluid, structural, and flight mechanics fields. It follows that the transformation of coordinates defined in Eq. (4.25) generalises the well-known approach used in structural dynamics for coupled problems. The approach retains the high efficiency of the basis functions, created using the coupled modal vectors, to allow fast convergence even for non-homogeneous cases.

Modal analysis in structural dynamics is a powerful tool to ensure good approximations of the 'exact' solution obtained through time integration of Eq. (4.15) using a relatively small number of modal vectors. This property holds for the coupled model presented herein. Because the structural response is generally well represented by the low-frequency normal modes, it is not unexpected that a small basis of coupled mode-shapes dominated by the motion of the structure is critical for the aeroelastic response.

To achieve a significant reduction in the system size, a small basis of biorthonormal coupled eigenvectors and associated eigenvalues representative of the system dynamics defined in Eq. (4.15) should be identified. Let us collect m coupled eigensolutions in the modal matrices of the right and left eigenvectors

$$\begin{aligned}\boldsymbol{\Lambda} &= \text{Diag}\left(\lambda_1, \lambda_2, \ldots, \lambda_m\right) \\ \boldsymbol{\Phi} &= \left[\phi_1, \phi_2, \ldots, \phi_m\right] \\ \boldsymbol{\Psi} &= \left[\psi_1, \psi_2, \ldots, \psi_m\right]\end{aligned} \tag{4.26}$$

where the diagonal matrix $\boldsymbol{\Lambda}$ has dimension $(m \times m)$ and the matrices $\boldsymbol{\Phi}$ and $\boldsymbol{\Psi}$ have dimension $(n \times m)$.

Truncate the linear combination in Eq. (4.25) to include $m \ll n$ terms

$$w \approx \sum_{i=1}^{m}\left(\phi_i z_i + \bar{\phi}_i \bar{z}_i\right) = \boldsymbol{\Phi} z + \bar{\boldsymbol{\Phi}} \bar{z} \tag{4.27}$$

where the columns ϕ_i of the $(n \times m)$ matrix $\boldsymbol{\Phi}$ span the subspace within which the system motion is now restricted. The transformation of coordinates relates the state-space vector of the large-order coupled model, $w \in \mathbb{R}^n$, with the stat- space vector of the reduced-order coupled model, $z \in \mathbb{C}^m$. To ensure convergence of the results, it is common to start with a small basis of coupled eigenvectors, then expanding this until the results have fully converged. A study of the model convergence is described in Section 4.3.6.

4.3.4 Linear Reduced Order Model

Let us start with the linearised model of Eq. (4.15). For simplicity, consider a constant equilibrium point, $\tilde{\boldsymbol{x}}$, when no control inputs and disturbances are acting on the system, $\boldsymbol{u} = \boldsymbol{0}$ and $\boldsymbol{d} = \boldsymbol{0}$, respectively. Note that $\boldsymbol{u}$ and $\boldsymbol{d}$ are independent of the equilibrium point, whereas $\tilde{\boldsymbol{x}}$ depends on both control and disturbance vectors. It follows that the linearised system has the form

$$\dot{w} = \boldsymbol{A} w + \frac{\partial \boldsymbol{F}}{\partial \boldsymbol{u}}\left(\tilde{\boldsymbol{x}}\right)\boldsymbol{u} + \frac{\partial \boldsymbol{F}}{\partial \boldsymbol{d}}\left(\tilde{\boldsymbol{x}}\right)\boldsymbol{d} = \boldsymbol{A} w + \boldsymbol{B}_u \boldsymbol{u} + \boldsymbol{B}_d \boldsymbol{d} \tag{4.28}$$

As the above linearised system still retains the size of the coupled nonlinear system, the transformation of coordinates may be used to derive a linear reduced-order model. First, substitute Eq. (4.27) into Eq. (4.28), then pre-multiply each term by the Hermitian transpose of the left modal matrix, $\boldsymbol{\Psi}^H$. Recalling the biorthonormal properties in Eq. (4.23) and (4.24), it follows that a linear reduced-order model has the form

$$\dot{\boldsymbol{z}} = \text{Diag}\left(\lambda_i\right)\boldsymbol{z} + \boldsymbol{B}_{ru}\boldsymbol{u} + \boldsymbol{B}_{rd}\boldsymbol{d} \tag{4.29}$$

where $\boldsymbol{B}_{ru} = \Psi^H \boldsymbol{B}_u$ and $\boldsymbol{B}_{rd} = \Psi^H \boldsymbol{B}_d$. Solving the linear reduced-order model in Eq. (4.29) offers two important computational advantages. The first is that the system consists of a set of independent (uncoupled) equations that exploit the biorthonormal properties of the modal basis which makes the system much easier to solve. The second advantage is that we have achieved a reduction in the system size, $m \ll n$, by exploiting very efficient basis functions. As a result, a very efficient model of small size can be used to predict the time-domain response.

4.3.5 Nonlinear Reduced-order Model

Next, let us consider a way to incorporate nonlinear effects in the reduced-order model. There are two basic requisites to meet. The first relates to the difficulty of the implementation and the cost of model generation, which should be as low as possible to facilitate its exploitation. The second is that it is desirable to have a formulation that is independent of the equations used to create the model, so that the approach to nonlinear model reduction is systematic and applicable, in principle, to any coupled system. The method presented in Da Ronch et al. (2012) meets both requirements.

An approach to systematically derive nonlinear reduced-order models indeed exists. Expand the nonlinear system in Eq. (4.15) in a Taylor series retaining terms up to third order in the perturbation. It follows that

$$\dot{\boldsymbol{w}} = \boldsymbol{A}\boldsymbol{w} + \boldsymbol{B}_u \boldsymbol{u} + \boldsymbol{B}_d \boldsymbol{d} + \frac{1}{2!}\boldsymbol{B}(\boldsymbol{w},\boldsymbol{w}) + \frac{1}{3!}\boldsymbol{C}(\boldsymbol{w},\boldsymbol{w},\boldsymbol{w}) \tag{4.30}$$

where the additional terms, compared to Eq. (4.28), indicate the second- and third-order Jacobian operators. It is immediate seen that the operators $\boldsymbol{B}$ and $\boldsymbol{C}$ are, respectively, bi-linear and tri-linear with respect to the arguments and are analytically obtained as

$$\boldsymbol{B}(\boldsymbol{w},\boldsymbol{w}) = \sum_{i=1}^{n}\sum_{j=1}^{n} \frac{\partial^2 \boldsymbol{F}}{\partial x_i \partial x_j} w_i w_j \tag{4.31}$$

$$\boldsymbol{C}(\boldsymbol{w},\boldsymbol{w},\boldsymbol{w}) = \sum_{i=1}^{n}\sum_{j=1}^{n}\sum_{k=1}^{n} \frac{\partial^3 \boldsymbol{F}}{\partial x_i \partial x_j \partial x_k} w_i w_j w_k \tag{4.32}$$

The difficulty with substituting the transformation of coordinates, Eq. (4.27), into Eq. (4.30) is the treatment of the quadratic and cubic terms, $\boldsymbol{B}$ and $\boldsymbol{C}$, respectively. After some manipulations and recalling the linearity of the high-order operators, a relatively simple relation is found. For brevity, the form of the quadratic term is reported

$$\begin{aligned}\boldsymbol{B}(z,z) = \sum_{r=1}^{n}\sum_{s=1}^{n}\Big(&\boldsymbol{B}(\phi_r,\phi_s) z_r z_s + \boldsymbol{B}(\phi_r,\bar{\phi}_s) z_r \bar{z}_s + \\ &\boldsymbol{B}(\bar{\phi}_r,\phi_s) \bar{z}_r z_s + \boldsymbol{B}(\bar{\phi}_r,\bar{\phi}_s) \bar{z}_r \bar{z}_s\Big)\end{aligned} \tag{4.33}$$

The interested reader is invited to refer to Da Ronch et al. (2012) for the relation for the cubic term. Here it is sufficient to note that the double sum may be further simplified by taking advantage of the bi-linearity of $\boldsymbol{B}$, that is, $\boldsymbol{B}(\phi_r,\phi_s) = \boldsymbol{B}(\phi_s,\phi_r)$, reducing the total number of calculations required to compute $\boldsymbol{B}(z,z)$.

The final step, as already done for the linear reduced-order model, is to pre-multiply each term of Eq. (4.30), once expressed as a function of z and not of w, by the Hermitian transpose of the left modal matrix, $\boldsymbol{\Psi}^H$. The nonlinear reduced-order model then has the form

$$\dot{\boldsymbol{z}} = \mathrm{Diag}(\lambda_i)\boldsymbol{z} + \boldsymbol{B}_{ru}\boldsymbol{u} + \boldsymbol{B}_{rd}\boldsymbol{d} + \boldsymbol{f}_{\mathrm{nln}}(\boldsymbol{z}) \tag{4.34}$$

where $\boldsymbol{f}_{\mathrm{nln}}(\boldsymbol{z})$ contains the nonlinear terms of the quadratic and cubic operators.

Numerical Implementation

A final note on the numerical approach used to calculate the higher-order terms of the nonlinear reduced-order model. It is possible to calculate all the contributions without having to resort to complex arithmetic or calculating all the second- and third-order partial derivatives analytically. Because it is only their action on vectors that is required, matrix-free products may be used. The evaluation of the finite differences suffers from the truncation error for values of the step size that are too large, and from the rounding error for values that are too small. The latter effect is more significant for the coefficients that include a third Jacobian product. In cases where convergence of the finite differences for various step sizes is not found, it is possible to resort to the MATLAB libraries supporting extended-order arithmetics; see Da Ronch et al. (2013c).

The approach to the generation of the reduced-order model detailed above leads to the set of equations described next. The equations of the reduced-order model are independent of the specific formulation used for the full order model, and are always expressed in a state-space form.

$$\begin{cases} \dot{\boldsymbol{z}} = \mathrm{Diag}(\lambda_i)\boldsymbol{z} + \boldsymbol{B}_{ru}\boldsymbol{u} + \boldsymbol{B}_{rd}\boldsymbol{d} + \boldsymbol{f}_{\mathrm{nln}}(\boldsymbol{z}) \\ \boldsymbol{w} = \boldsymbol{\Phi}\boldsymbol{z} + \bar{\boldsymbol{\Phi}}\bar{\boldsymbol{z}} \end{cases} \tag{4.35}$$

The dynamics of the nonlinear reduced-order model in Eq. (4.35) are given in terms of a complex-valued state vector, z, which is of small size. The output equation defines how the physical degrees of freedom of the original full-order model can be retrieved if necessary. The control inputs and disturbances are in the vectors u and d, respectively. For the linear aerodynamics described in Section 4.2.2, it is apparent that the control surface rotation, angular velocity and acceleration are treated as commanded inputs. Keeping these quantities as separate inputs is not convenient, as the three quantities are all linked by a time integration/derivation relationship. Referring to Da Ronch et al. (2014), it is possible to see that the actual commanded control input is the angular acceleration of the control surface, and that the angular velocity and rotation are easily deduced by numerical time integration.

4.3.6 Aircraft Test Case Gust Response

Having presented a set of mathematical models for the description of flexible aircraft dynamics and a model-reduction strategy to reduce costs, a demonstration of these tools is now performed for the flexible unmanned aircraft test case of Section 4.2.1. The freestream conditions are $U_\infty = 59$ m/s, $\alpha_\infty = 4°$ and $\rho_\infty = 0.0789\,\mathrm{kg/m^3}$. In flight, the aircraft exhibits large wing deformations. The deformed shape is computed from a static aeroelastic solution and taken as the equilibrium point for the reduced model

Table 4.2 Basis of coupled eigenvalues used for the model projection; see Eq. (4.26).

Mode number	Modeshape	Real part	Imaginary part
		Hz	Hz
1	First bending	-8.82×10^{-1}	±1.97
2	Second bending	-8.04×10^{-1}	$\pm9.81\times10^{1}$
3	First torsion	-1.71×10^{-1}	$\pm1.45\times10^{1}$
4	Third bending	-7.03×10^{-1}	$\pm2.17\times10^{1}$
5	Fourth bending	-6.04×10^{-1}	$\pm4.19\times10^{1}$
6	Gust mode	-9.90	0.00
7	Gust mode	-1.01×10^{1}	0.00
8	Gust mode	-1.01×10^{1}	0.00

generation. The coupled large-order model, hereafter referred to as the full-order model (FOM), consists of 540 degrees of freedom, 324 associated with the structural model and 216 with the aerodynamic model.

First, the right and left eigenvalue problems are solved around the static aeroelastic deformed shape. As the identification of an adequate basis for the model projection is critical for the analysis, a preliminary study is done to ensure convergence by increasing the size of the modal basis, Eq. (4.26). A reasonable approach is to initially include a number of coupled modes that are dominated by the structural response. These modes are associated with the normal modes of the structure when the surrounding fluid is removed. In addition to this clear choice, the inclusion of the so called 'gust modes' is needed to enrich the modal basis for gust load prediction. In linear aerodynamics, these modes are easily identified, being related to the smallest Küssner constant, $\epsilon_3 = 0.1393$. The eigenvalues of the 'gust modes' are $\lambda_i = -\epsilon_3 U_\infty / b_i$ (in [Hz]). Tests to ensure convergence of the modal basis were done using up to eight coupled eigenvalues, as summarised in Table 4.2. The first five coupled modes are mainly dominated by the structural response and are traced for these flight conditions from the corresponding normal modes of the structure. The remaining modes are 'gust modes' and provide the mechanisms to describe the influence of an atmospheric gust on the structural response. The variation of the frequencies of the coupled modeshapes with respect to the freestream speed is shown in Figure 4.7.

The convergence of a linear reduced-order model for increasing size of the modal basis is shown in Figure 4.8. The open-loop response is computed for a random turbulence with statistical properties defined by the von Kármán spectrum. The reduced-order model predictions are compared with those of the original full-order model, with dimension 540 degrees of freedom. Good agreement is observed, with as low as eight coupled modes for the reduced-order model. To emphasise when a model is nonlinear, 'N' is appended to the shorthand notation.

Next, the reduced-order model is demonstrated for the efficient search of the worst-case gust. The search is conducted for the one-minus-cosine gust family considering gust wavelengths between 0 and 776 aircraft mean aerodynamic chords (with a step size

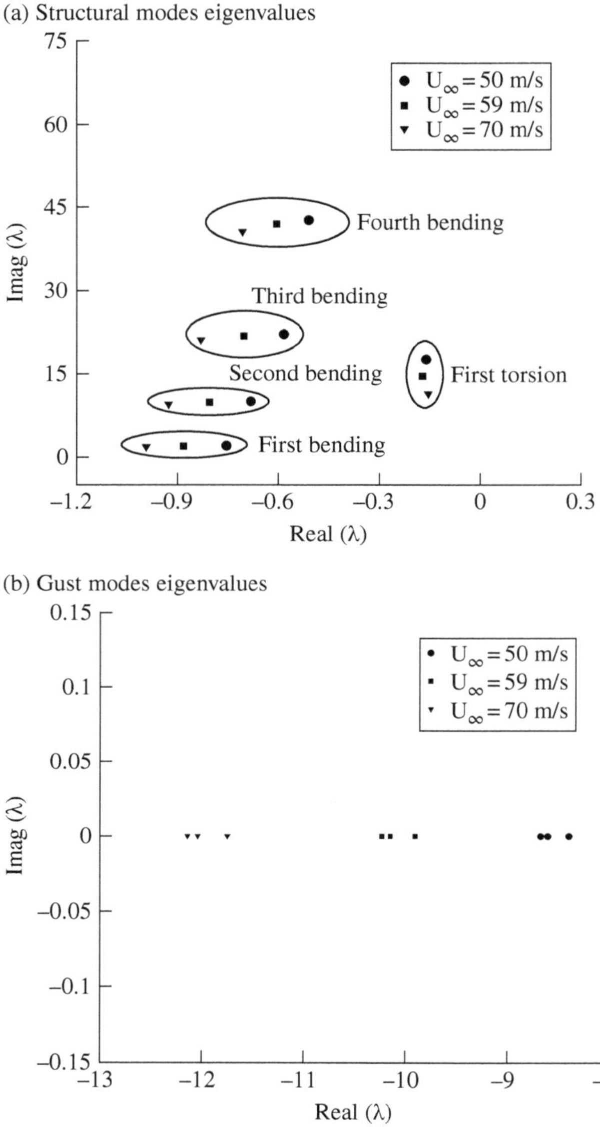

Figure 4.7 Dependence on freestream speed of coupled aeroelastic frequencies for: (a) structurally-dominated modeshapes and (b) gust modeshapes.

of 9.7). A strong gust intensity, 14% of the freestream speed, causes large wing structural deformations. In addition to the linear reduced model above, a nonlinear reduced-order model was generated with the same modes but including terms up to second order. The inclusion of higher-order terms did not modify the convergence properties of the model. The search was performed using both the full- and reduced-order models and 80 calculations were performed in total. Figure 4.9 illustrates the largest upward and

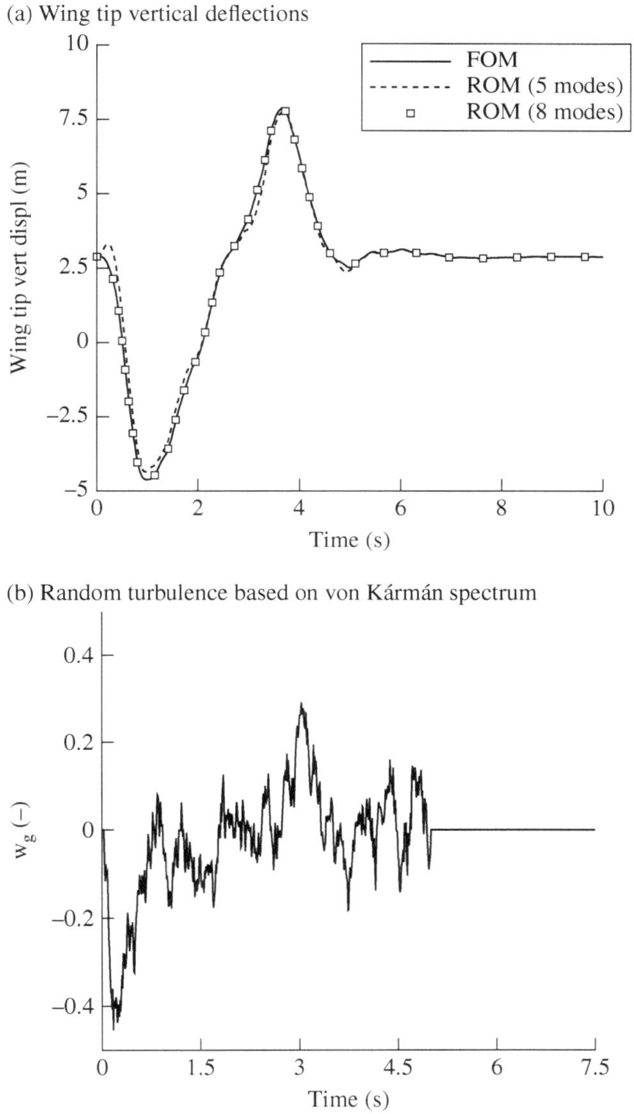

Figure 4.8 Gust response of the aircraft test case: (a) convergence for increasing number of coupled modes and (b) vertical gust intensity normalised by $U_{\circ}$. Military specification: MIL-F-8785C. $U_{\infty} = 59$ m/s, $\alpha_{\infty} = 4°$, $\rho_{\infty} = 0.0789\,\text{kg/m}^3$ and turbulence intensity: 'severe 10^{-5}'. FOM, full-order model; ROM, reduced order model.

downward structural deflections of the wing tip for various gust wavelengths. These are reported along the horizontal axis. The worst-case gust causing the largest structural deformation is seen to have a 4 s duration, corresponding to a length of 197 mean aerodynamic chords at the flying speed of 59 m/s. The dynamic response to the worst-case gust is compared for linearised and nonlinear models, and also for the

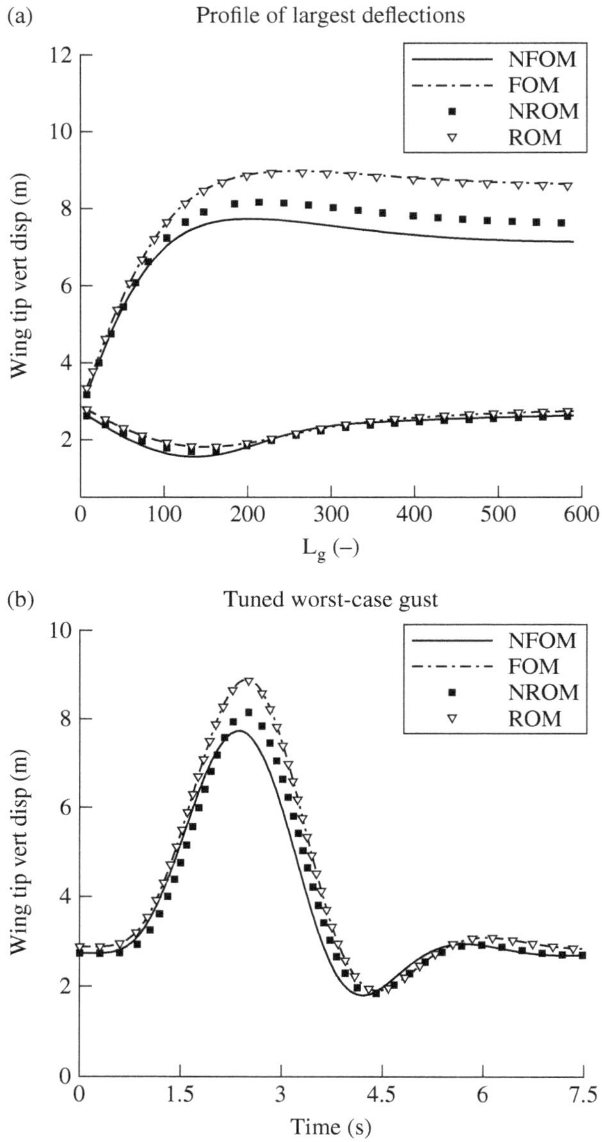

Figure 4.9 Worst-case gust search for the one-minus-cosine gust family: (a) profile of largest structural deflections at the wing tip and (b) dynamic response for the tuned worst-case gust. NFOM, nonlinear full-order model; FOM, full-order model; NROM, nonlinear reduced-order model; ROM, reduced-order model.

full- and reduced-order models. Deformations of 9 m are considered large, because the wing span is 17.75 m, and it is not unexpected in this case that the linearised (full and reduced) models over-predict the deformations. The computational cost to obtain the gust profiles in Figure 4.9 with the reduced models was a fraction of that needed for the original full model: for the linear case, the reduced model was 10 times as fast; for the nonlinear case, an increased performance of about 30 times was recorded. These indicative values are expected to increase considerably as the size of the original model increases (Da Ronch et al. 2013c), demonstrating the practical use and advantage of our approach to model reduction.

To conclude, it is demonstrated through application to a realistic HALE test case that the reduced-order models significantly reduce the computational cost for parametric worst-case gust searches. Section 4.4.3 will also show that the reduced-order models are adequate for a variety of control designs for gust load alleviation.

4.4 Control System Design

Previous work by the authors has considered the systematic generation of nonlinear reduced models that not only capture the system nonlinearities but also are parameterised with respect to flow conditions such as freestream speed and density (Da Ronch et al. 2012, 2013b; Papatheou et al. 2013). The reduced models were used for H_∞ control design for gust load alleviation. In general, the H_∞ control provides strong disturbance rejection but, in principle, a controller that can adapt during changes of the flow conditions, such as airspeed or density, is desired. Recently, an adaptive controller was found efficient for gust load alleviation for a flexible wing (Tantaroudas et al. 2014). That study, in addition to a comparison of control design performance, emphasised the fact that inherently different control methodologies, from robust controllers to nonlinear adaptive controllers, can be designed based on the very same nonlinear reduced-order model in Eq. (4.35).

This section continues with an overview of the theory behind two control strategies: H_∞ synthesis in Section 4.4.1 and model reference adaptive control in Section 4.4.2. Finally, Section 4.4.3 presents a comparison of the two control strategies in gust load alleviation in the aircraft test case.

4.4.1 H_∞ Synthesis

Starting from Eq. (4.35), a manipulation is needed to recast the system of equations for H_∞ control synthesis. First, observe that the equations of the reduced-order model dynamics are complex-valued and this is incompatible for control design. The system of equations is easily recast in real form, splitting the real and imaginary parts into separate equations and doubling its original dimension. Second, recall that for linear aerodynamics, the rotation δ, angular velocity δ' and acceleration δ'' of a control surface contribute to the build-up of the aerodynamic loads. As an interdependency exists between these three quantities, the actual commanded input is chosen to be the control surface angular acceleration, from which the deflection and angular velocity can be calculated numerically. It follows that it is more convenient to perform control design synthesis around the following set of equations rather than Eq. (4.35):

$$\begin{cases} \dot{\boldsymbol{x}} = \boldsymbol{A}\boldsymbol{x} + \boldsymbol{B}\boldsymbol{u} + \boldsymbol{D}\boldsymbol{w}_e + \boldsymbol{F}_{\text{nln}}\left(\boldsymbol{x}\right) \\ \boldsymbol{y}_{\text{ctrl}} = \boldsymbol{C}_1\boldsymbol{x} + \boldsymbol{D}_{11}\boldsymbol{w}_e + \boldsymbol{D}_{12}\boldsymbol{u} \\ \boldsymbol{y}_{\text{msrd}} = \boldsymbol{C}_2\boldsymbol{x} + \boldsymbol{D}_{21}\boldsymbol{w}_e + \boldsymbol{D}_{22}\boldsymbol{u} \end{cases} \tag{4.36}$$

where $w_e = (d, w_d)^T$, with w_d indicating an artificial disturbance used in tuning the variables of the control synthesis, and $x = (\text{Real}(z), \text{Imag}(z), \delta, \delta')^T$. The output is distinguished by what the controller is aiming to control, y_{ctrl}, and what the controller has information about, y_{msrd}.

The H_∞ control problem with additional input-shaping techniques for control tuning purposes for the classical H_∞ problem formulation is written as follows (Zhou and Doyle 1998),

$$\begin{Bmatrix} \dot{x} \\ y_{\text{ctrl}} \\ y_{\text{msrd}} \end{Bmatrix} = \begin{bmatrix} A & D & B \\ C_1 & D_{11} & D_{12} \\ C_2 & D_{21} & D_{22} \end{bmatrix} \begin{Bmatrix} x \\ w_e \\ u \end{Bmatrix} \tag{4.37}$$

The resulting controller has the linear form

$$u(s) = K(s) y_{\text{msrd}}(s) \tag{4.38}$$

where $K(s)$ is the H_∞ controller transfer function in the Laplace domain. The aim is to minimise the transfer of the disturbance signal from d to y_{ctrl} by creating a controller that uses information from y_{msrd} to change the input u. This can be written as

$$\frac{\sup \int_0^\infty || y_{\text{msrd}}(t) ||^2 \, dt}{\sup \int_0^\infty || d(t) ||^2 \, dt} \le \gamma \tag{4.39}$$

where γ represents the ratio of the maximum output energy to the maximum input energy.

The problem is expanded to include a weight on inputs, K_c, which carries over to an additional element on controlled output and a weight on measurement noise, K_d, which carries over to an additional element on measured output. The H_∞ control is derived on the basis of the linearised model and is applied directly to the nonlinear full-order model by using the reduced matrices from the nonlinear model order-reduction framework.

4.4.2 Model Reference Adaptive Control

Assume the nonlinear reduced-order model in the form of Eq. (4.35), and consider an ideal model reference in the form

$$\dot{x}_m = A_m x_m + B_m u + B_d d + \bar{f}_{\text{nln},m}(x_m) \tag{4.40}$$

Matrix A_m is a stable Hurwitz matrix that meets the desired properties of the reference system. This could mean eigenvalues with increased damping compared to the actual aeroelastic system. Matrix B_m is user defined and describes the influence of the control inputs on the states of the reference model. The states of the reference model due to the increased damping in matrix A_m will decay to zero faster under the same disturbances or flap actuation while their magnitude will be in general smaller as well. The physical displacements of the system can be retrieved using the output equation.

The goal is to find a dynamic control input, $\boldsymbol{u}$, related to the flap actuation that satisfies the condition $\lim_{t\to\infty} \| y(t) - y_m(t) \|$. The exact control feedback for the model matching conditions is defined as

$$\boldsymbol{u} = \boldsymbol{K}_x^* \boldsymbol{x} + \boldsymbol{K}_r^* \boldsymbol{r} \tag{4.41}$$

where $\boldsymbol{r}$ is a reference signal applied to both systems, as shown in Figure 4.10, representing in our case the flap angle, and $\boldsymbol{K}_x^*, \boldsymbol{K}_r^*$ are the exact gains acting on the states and control input to match the two models. Substituting Eq. (4.41) in Eq. (4.36) and satisfying the model matching conditions yields

$$\begin{cases} \boldsymbol{A} + \boldsymbol{B}_c \boldsymbol{K}_x^* = \boldsymbol{A}_m \\ \boldsymbol{B}_c \boldsymbol{K}_r^* = \boldsymbol{B}_m \end{cases} \tag{4.42}$$

Since $\boldsymbol{A}$ and $\boldsymbol{B}_c$ are considered to be unknown to the controller, the values of $\boldsymbol{K}_x^*, \boldsymbol{K}_r^*$ in Eq. (4.41) are also unknown at the initial time and the actual control signal applied at the current time-step is defined as

$$\boldsymbol{u} = \boldsymbol{K}_x \boldsymbol{x} + \boldsymbol{K}_r \boldsymbol{r} \tag{4.43}$$

$\boldsymbol{K}_x$ and $\boldsymbol{K}_r$ in Eq. (4.43) are dynamic gains that need to be solved and at the end will be required to converge to the values that provide a solution to Eq. (4.42).

However, in adaptive control systems there is a major uncertainty about the convergence of the adaptive gains, even in deterministic ideal situations. In the presence of stochastic disturbances, this issue becomes even more challenging and complicated and this topic has remained open for many years in the field of nonlinear adaptive control design. There are many cases where the adaptive gains converge to values that are different to the actual analytical pre-calculated ideal gains, even without the presence of

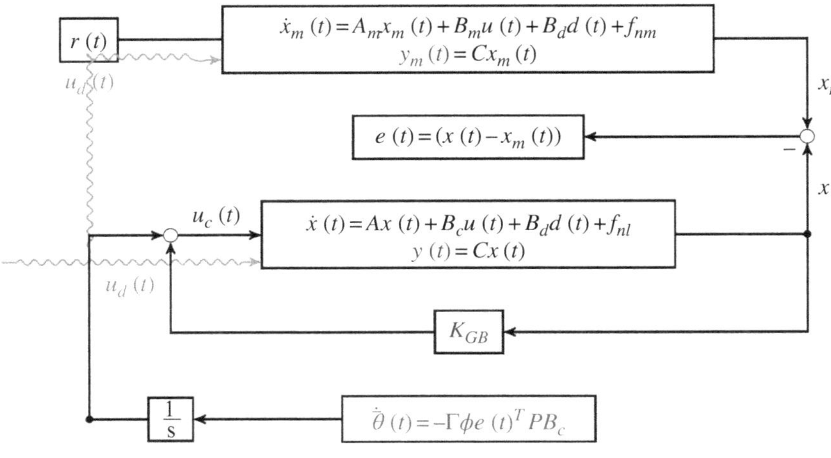

Figure 4.10 Block diagram of a nonlinear adaptive control algorithm.

disturbances. Recently, Barkana (2005) showed that in cases where the adaptive gains do not reach the unique solution that the preliminary design suggests, it is not because there is something wrong with the control design. This is because the adaptive controller only needs a specific set of gains that correspond to a particular input command, rather than the unique solution of gains for all inputs that an exact design suggests.

The closed-loop dynamics of the nonlinear reduced model at this point may be expressed as

$$\dot{\boldsymbol{x}} = \left(\boldsymbol{A} + \boldsymbol{B}_c\boldsymbol{K}_x\right)\boldsymbol{x} + \boldsymbol{B}_c\boldsymbol{K}_r\boldsymbol{r} + \boldsymbol{B}_d\boldsymbol{d} + \boldsymbol{f}_{\text{nln}}\left(\boldsymbol{x}\right) \tag{4.44}$$

Let $\theta^* = \left\{\boldsymbol{K}_x^*\boldsymbol{K}_r^*\right\}^T$ and $\theta = \left\{\boldsymbol{K}_x\boldsymbol{K}_r\right\}^T$. The estimation error between the instantaneous and the ideal gains is defined as

$$\bar{\theta} = \theta^* - \theta = \left(\bar{\theta}_x\bar{\theta}_r\right)^T \tag{4.45}$$

with $\bar{\theta}_x = \boldsymbol{K}_x^* - \boldsymbol{K}_x$ and $\bar{\theta}_r = \boldsymbol{K}_r^* - \boldsymbol{K}_r$. Now define $\phi = (\boldsymbol{x},\boldsymbol{r})^T$. In this case, the closed-loop system dynamics in Eq. (4.44) are expressed as

$$\begin{aligned}\dot{\boldsymbol{x}} &= \left(\boldsymbol{A} + \boldsymbol{B}_c\boldsymbol{K}_x^*\right)\boldsymbol{x} + \boldsymbol{B}_c\boldsymbol{K}_r^*\boldsymbol{r} - \boldsymbol{B}_c\bar{\theta}_x\boldsymbol{x} - \boldsymbol{B}_c\bar{\theta}_r\boldsymbol{r} + \boldsymbol{B}_d\boldsymbol{d} + \boldsymbol{f}_{nl}\left(\boldsymbol{x}\right)\\ &= \boldsymbol{A}_m\boldsymbol{x} + \boldsymbol{B}_m\boldsymbol{r} - \boldsymbol{B}_c\phi^T\bar{\theta} + \boldsymbol{B}_d\boldsymbol{d} + \boldsymbol{f}_{nl}\left(\boldsymbol{x}\right)\end{aligned} \tag{4.46}$$

For the purpose of the stability proof of the closed-loop system one needs to define the error dynamics between the two systems (Barkana 2013).

$$\boldsymbol{e} = \boldsymbol{x} - \boldsymbol{x}_m \tag{4.47}$$

The derivative of this expresses the rate of change between the two systems and can be written as

$$\begin{aligned}\dot{\boldsymbol{e}} &= \dot{\boldsymbol{x}} - \dot{\boldsymbol{x}}_m\\ &= \boldsymbol{A}_m\left(\boldsymbol{x} - \boldsymbol{x}_m\right) - \boldsymbol{B}_c\phi^T\bar{\theta} + \left(\boldsymbol{f}_{nl}\left(\boldsymbol{x}\right) - \bar{\boldsymbol{f}}_{nm}\left(\boldsymbol{x}_m\right)\right)\\ &= \boldsymbol{A}_m\boldsymbol{e} - \boldsymbol{B}_c\phi^T\bar{\theta} + \left(\boldsymbol{f}_{nl}\left(\boldsymbol{x}\right) - \bar{\boldsymbol{f}}_{nm}\left(\boldsymbol{x}_m\right)\right)\\ &= \boldsymbol{A}_m\boldsymbol{e} - \boldsymbol{B}_c\phi^T\bar{\theta} + \boldsymbol{F}_{Df}\end{aligned} \tag{4.48}$$

At this point, the Lyapunov equation needs to be solved for the reference model because its solution will be part of the steady part of the Lyapunov candidate function that we define and that will lead to the stability proof of the nonlinear reduced model (Ioannou and Sun 1996).

$$\boldsymbol{P}\boldsymbol{A}_m + \boldsymbol{A}_m{}^T\boldsymbol{P} = -\boldsymbol{Q}, \qquad \boldsymbol{Q} = \boldsymbol{Q}^T \geq 0 \tag{4.49}$$

where in Eq. (4.49) $\boldsymbol{Q}$ is a semi-definite positive user-defined matrix. A scalar quadratic Lyapunov function $\boldsymbol{V}$ in e and $\bar{\theta}$ may be defined, such that the system becomes asymptotically stable by satisfying $\boldsymbol{V} > 0$. Its time derivative is semi-definite negative $\boldsymbol{V}' \leq 0$ Ioannou and Sun (1996). This function will provide insight on the selection of the parameter update law of the time-varying gains in Eq. (4.43). The Lyapunov function

$$V\left(\boldsymbol{e},\theta\right) = \boldsymbol{e}^T\boldsymbol{P}\boldsymbol{e} + \bar{\theta}^T\Gamma^{-1}\bar{\theta} > 0 \tag{4.50}$$

is considered, where $\boldsymbol{P} = \boldsymbol{P}^T > 0$ is the solution of the algebraic Lyapunov equation (4.49) for a particular selection of $\boldsymbol{Q}$ while $\Gamma = \Gamma^T \geq 0$ is a user-defined semi-definite positive matrix. Note that the positiveness of the above Lyapunov function is guaranteed only if the system under examination is a minimum-phase system, which is enforced in the reduced-order model generation. Differentiating the above equation with respect to time yields

$$\dot{V}(\boldsymbol{e},\theta) = \dot{\boldsymbol{e}}^T(\boldsymbol{P}+\boldsymbol{P}^T)\boldsymbol{e} + 2\bar{\theta}^T\Gamma^{-1}\dot{\bar{\theta}} + \boldsymbol{ePF}_{Df} \tag{4.51}$$

By substitution of the error dynamics and by using Eq. (4.49), Eq. (4.51) is expanded as follows

$$\dot{V}(\boldsymbol{e},\theta) = -\boldsymbol{e}^T\boldsymbol{Qe} + 2\bar{\theta}^T\Gamma^{-1}\left(\Gamma\phi\boldsymbol{e}^T\boldsymbol{PB}_c + \dot{\bar{\theta}}\right) + \boldsymbol{ePF}_{Df} \tag{4.52}$$

From thus, one can determine the adaptation parameter to satisfy the semi-definite negativeness of the derivative of the Lyapunov function as

$$\dot{\bar{\theta}} = -\Gamma\phi\boldsymbol{e}^T\boldsymbol{PB}_c \tag{4.53}$$

which leads to

$$\dot{V}(\boldsymbol{e},\theta) = -\boldsymbol{e}^T\boldsymbol{Qe} + +\boldsymbol{ePF}_{Df} \tag{4.54}$$

The term $-\boldsymbol{e}^T\boldsymbol{Qe}$ in Eq. (4.54) is negative-definite with respect to $\boldsymbol{e}$ and this is enforced by the semi-definitive positive matrix $\boldsymbol{Q}$. The derivative of the Lyapunov function remains negative definite in both $\boldsymbol{x}(t)$ and $\boldsymbol{e}(t)$ if additionally the second term in Eq. (4.54) is not too large, or alternatively if the following inequality is satisfied (Torres and Mehiel 2006):

$$\|\boldsymbol{F}_{Df}\| = \|\boldsymbol{f}_{nl}(\boldsymbol{x}) - \bar{\boldsymbol{f}}_{nm}(\boldsymbol{x}_m)\| \leq \frac{\|\boldsymbol{Q}\|}{\|\boldsymbol{P}\|}\|\boldsymbol{x}-\boldsymbol{x}_m\| \tag{4.55}$$

However, it is impossible to come up with a general mathematical proof that ensures the stability of the nonlinear adaptive control scheme of flexible aircraft for all types of nonlinearities. Instead, the efficiency of the control design is demonstrated on the nonlinear system for realistic amplitudes of external disturbances. The dynamic time-varying gains in Eq. (4.43) are updated by the adaptive law so that the time derivative of the Lyapunov function decreases along the error dynamic trajectories as in Eq. (4.54). By using Barbalat's lemma this translates in boundness of the error dynamics with respect to the time evolution, and as a result the model-matching conditions are satisfied.

In general, this control approach is limited to minimum phase systems. Thus, when applied in unstable non-minimum phase systems, unstable zero-pole cancellation may occur and the error between the two assumed models slowly diverges to infinity. However, a simple feedback based on the Bass–Gura formula (Ogata 2010) can be applied to the reduced-order model to place any unstable zeros in the left half plane. The implementation of the computational algorithm is summarised in the block diagram shown in Figure 4.10.

4.4.3 Aircraft Test Case Gust-load Alleviation

H_∞ Synthesis

The design of a controller for load alleviation was performed on the linear reduced model considering the tuned worst-case gust. The good performance of the controller in suppressing the vibrations of the linear model induced by the worst-case gust is not unexpected, as the controller was designed specifically for that scenario. However, its performance will be shown on the nonlinear full model. The question addressed in this section is whether good alleviation can be achieved for a different shape of the gust, but using the same controller. The responses shown in Figure 4.11 are for the discrete worst-case gust and for a continuous-gust model based on the von Kármán spectrum. The aeroelastic vibrations of the closed-loop system are significantly reduced when

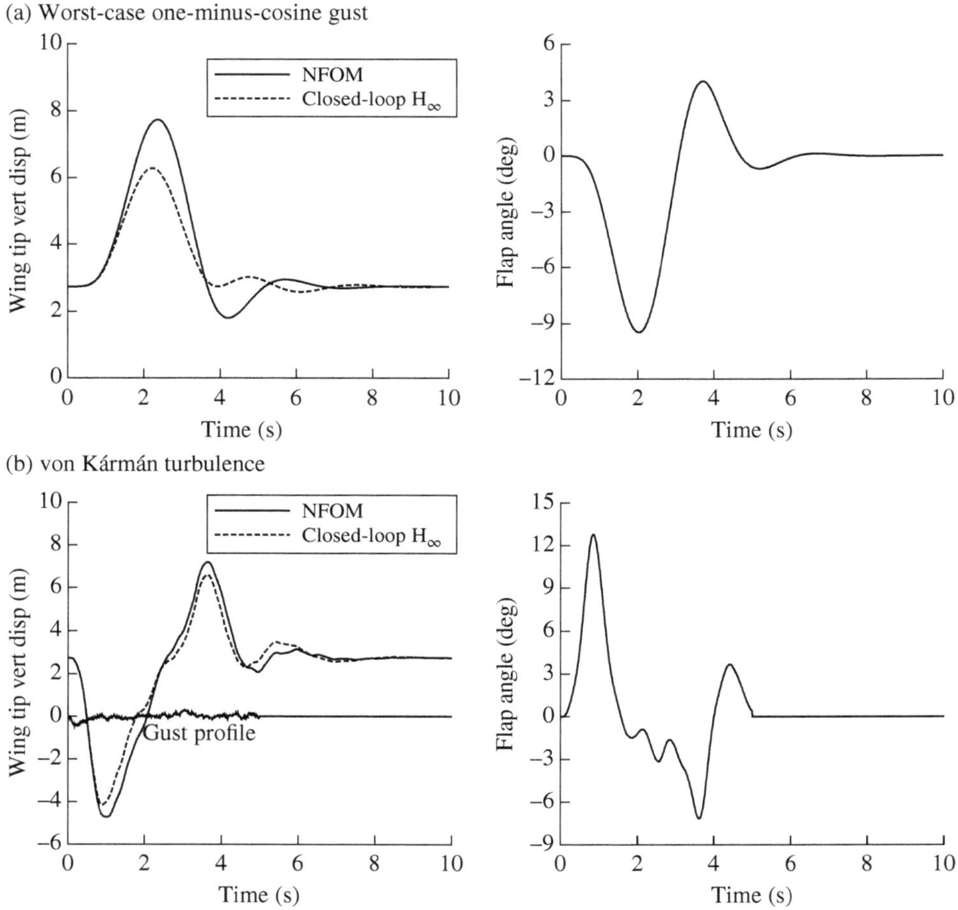

Figure 4.11 Gust-load alleviation response using the H_∞ controller compared to the open-loop response: (a) worst-case one-minus-cosine gust from Figure 4.9; (b) von Kármán turbulence model. NFOM, nonlinear full-order model.

compared to the open-loop response. However, the performance of the optimal robust controller is seen to degrade when applied to the nonlinear system for very strong stochastic disturbances.

The efficiency of the optimal control approach using the reduced models for gust-load alleviation can be demonstrated in a case with noticeable differences between the linear and nonlinear full-order models; see Figure 4.9. However, the performance of the optimal robust controller is reduced when applied to the nonlinear system for very strong stochastic disturbances.

Model Reference Adaptive Controller

The nonlinear reduced model was implemented to simplify and speed up the calculation of the adaptive model reference control framework. The computed control surface deflection was applied to the nonlinear full-order model, which is under external disturbances. The selection of the reference model is of critical importance; a bad choice could potentially lead the flap to undergo unrealistic rotations. In this case, a reference model was created with additional damping added to the first bending and torsional modes. As a result, the aeroelastic vibrations of the reference system die out more quickly than those of the plant to be controlled. The eigenvalues of the linearised reference system are summarised in the Table 4.3, which should be comprated to Table 4.2 for the uncontrolled system. Note that damping is added to the first five complex conjugate eigenvalues. The eigenvalues and a comparison of the plant model and the selected reference model for the worst-case gust are shown in Figure 4.12.

The selection of the semi-definite-positive matrix $\boldsymbol{Q}$ that provides a solution to the Lyapunov equation given a stable Hurwitz matrix of a reference model $\boldsymbol{A}_m$ is also critical. In this case, $\boldsymbol{Q}$ was chosen to be a diagonal matrix with elements $Q_{ii} = 10^{-4}$. As shown in Eq. (4.53), the selection of the reference model will affect how $e(t)$ will evolve during the time integration, which is part of the adaptation parameter. The reference model in that case needs to be stable so that the error decreases asymptotically. Finally, observe that the adaptation parameter is affected by $\boldsymbol{P}$ and, as a result, by matrices $\boldsymbol{Q}$ and $\boldsymbol{\Gamma}$.

Table 4.3 Eigenvalues of the reference model.

Mode number	Real part	Imaginary part
1	-9.53×10^{-1}	± 2.01
2	-8.53×10^{-1}	$\pm 1.02\times10^{1}$
3	-1.71×10^{1}	$\pm 2.79\times10^{1}$
4	-5.73×10^{-1}	$\pm 4.74\times10^{1}$
5	-1.21×10^{1}	$\pm 6.55\times10^{1}$
6	-9.90	0.00
7	-1.01×10^{1}	0.00
8	-1.01×10^{1}	0.00

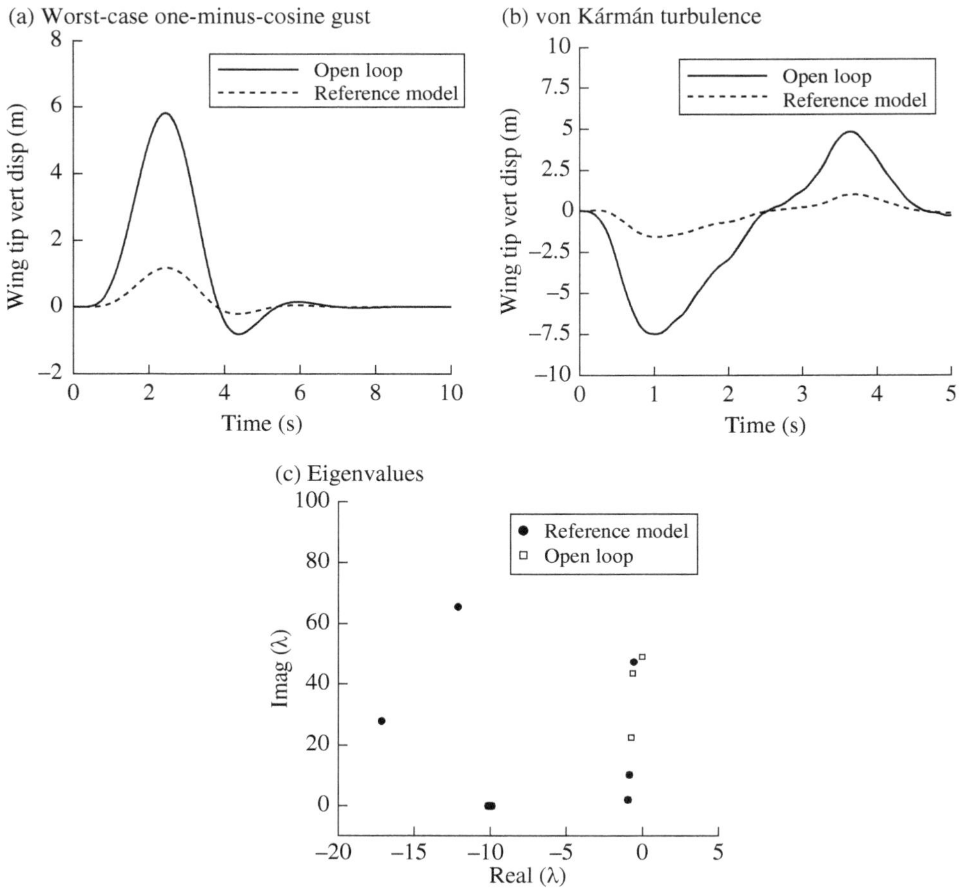

Figure 4.12 Ideal reference model for the MRAC controller design compared to the open-loop response for: (a) worst-case one-minus-cosine gust from Figure 4.9; (b) von Kármán turbulence model; (c) eigenvalues of the system to be controlled and the reference system.

Table 4.4 Adaptation parameter selection.

	Discrete gust case	Continuous gust case
$\boldsymbol{\Gamma}$	$0.01\boldsymbol{Q}$	$0.01\boldsymbol{Q}$
$\boldsymbol{\Gamma}$	$0.10\boldsymbol{Q}$	$0.10\boldsymbol{Q}$
$\boldsymbol{\Gamma}$	$1.00\boldsymbol{Q}$	$1.00\boldsymbol{Q}$

The effect of the adaptation matrix $\boldsymbol{\Gamma}$ is therefore investigated for the performance of the closed-loop system. The discrete selection of the semi-definite-positive matrix $\boldsymbol{\Gamma}$ is shown in Table 4.4 for both discrete and continuous gust-load alleviation.

The derived controller based on the reduced model is directly applied to the full-order nonlinear aeroelastic system. The wing-tip vertical displacements for different

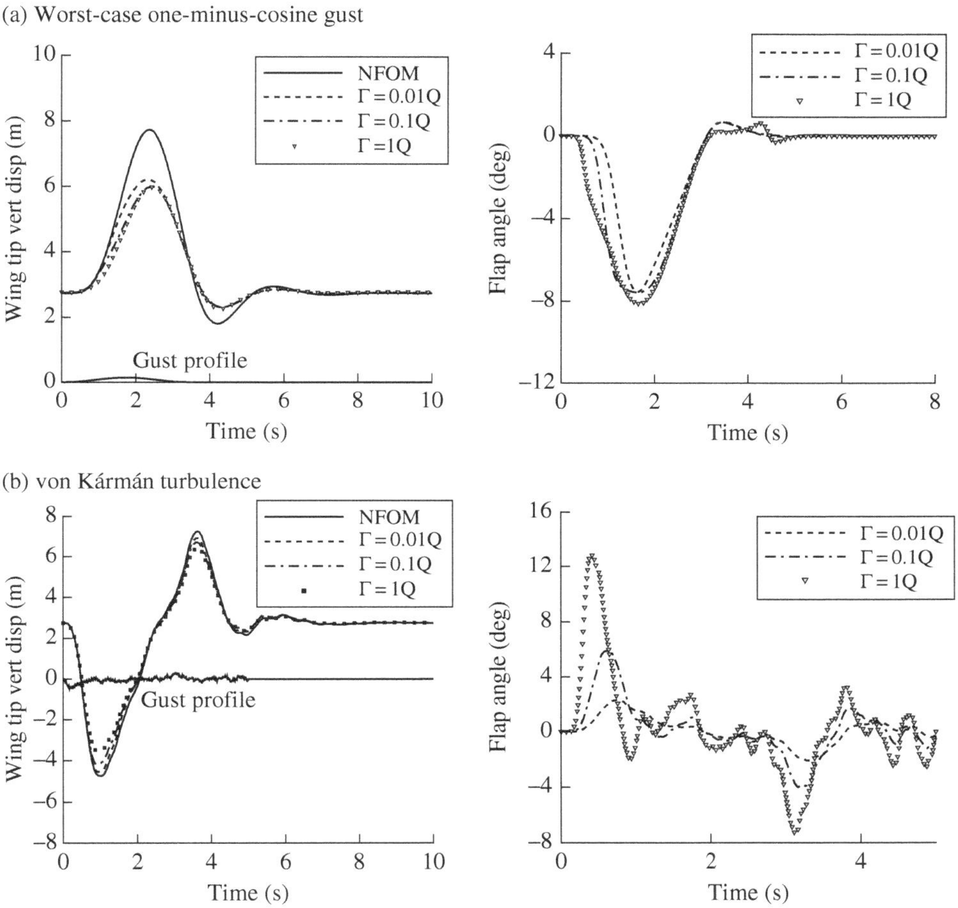

Figure 4.13 Gust-load alleviation response using the MRAC controller for various adaptation gains compared to the open-loop response for: (a) worst-case one-minus-cosine gust from Figure 4.9; (b) von Kármán turbulence model. NFOM, nonlinear full-order model.

adaptation rates for the worse-case one-minus-cosine gust and for a continuous gust are shown in Figure 4.13.

The results show a significant reduction of the wing-tip deflections for the closed-loop system for both linear and nonlinear cases, with realistic flap deflections in all cases. It can be seen that for the particular selection of the semi-definite-positive matrix $\boldsymbol{Q}$, a larger adaptation gain $\boldsymbol{\Gamma}$ is required during the fluid–structure–gust interaction to alleviate the disturbances. A further increase of the adaptation gain may lead to a non-realistic flap rotation with a flap angle of over 15°, which is a common constraint for the flap maximum rotation. Therefore, very large adaptation rates are not recommended, because the flap might overshoot during the fluid–structure–gust interaction.

Control Design Comparison

Both control designs were found adequate for gust-load alleviation of a very flexible aircraft. However, a 'good' controller not only guaranteed that the closed-loop structural deformations are smaller than those of its open-loop counterpart, but also that this is

Table 4.5 Comparison of control performance for a discrete one–minus–cosine gust.

Controller design	Reduction in wing-tip deflection [%]	Maximum flap rotation [deg]
H_∞	23.15	−9.47
MRAC, $\Gamma = 10^{-2}\boldsymbol{Q}$	24.45	−7.54
MRAC, $\Gamma = 10^{-1}\boldsymbol{Q}$	28.89	−7.56
MRAC, $\Gamma = 1\boldsymbol{Q}$	29.45	−8.11

Table 4.6 Comparison of control performance for a stochastic gust.

Controller design	Reduction in wing-tip deflection [%]	Maximum flap rotation [deg]
H_∞	10.26	12.79
MRAC, $\Gamma = 10^{-2}\boldsymbol{Q}$	4.73	2.31
MRAC, $\Gamma = 10^{-1}\boldsymbol{Q}$	8.00	5.89
MRAC, $\Gamma = 1\boldsymbol{Q}$	12.68	12.83

achieved with a realistic, optimal and minimum control effort. The performance of the H_∞ and MRAC controllers for the discrete one-minus-cosine gust is reported in Table 4.5. It is found that the adaptive control methodology achieves a better performance in reducing wing-tip deflection than the H_∞ control strategy, and the performance in gust-load alleviation increases for increasing adaptation rates. The reduction in wing-tip deflection is also achieved with a smaller control effort.

Finally, the performance of the two controllers is summarised in Table 4.6 for random turbulence based on the von Kármán spectrum. The gust-load alleviation proves more challenging in this case because of the larger frequency content compared to the one-minus-cosine gust. The choice of the adaptation rate is critical, because it affects the capability of the control system to follow rapid changes in gust loads. It is not unexpected, therefore, that the performance of the MRAC controller degrades for smaller adaptation rates. For larger adaptation rates, the adaptive control design achieves about the same level of gust-load alleviation, than the H_∞ controller, but with a smaller control effort.

The comparison of the performance of the two control strategies indicates that, in general, gust-load alleviation for random turbulence is more challenging than for discrete gusts, and may result in degraded performance, at least to some degree.

Note that the ability to investigate the two control strategies is enabled by the model-reduction technique presented in this chapter, demonstrating its practical applicability.

4.5 Conclusion

A unified methodology to facilitate control synthesis design starting from arbitrarily large computational models of flexible flying aircraft was presented in this chapter. The methodology requires the accurate calculation of the coupled eigenvalues and modeshapes to form an efficient basis for model projection, and a Taylor series expansion to retain some of the nonlinearities affecting the system dynamics. The methodology is demonstrated for an aircraft test case in various flight conditions and atmospheric gusts and turbulence, highlighting the benefits of the proposed approach. The methodology is found to be effective for practical use, and its generality allows application to any large computational model.

4.6 Exercises

1 Investigate the impact that the altitude has on the statistics of the random vertical gust intensity, Eq. (4.13). Assume a flight speed $U_\infty = 250$ m/s and turbulence intensity 'moderate 10^{-3}'. Refer to the Military Specification MIL-F-8785C and use the MATLAB von Kármán Turbulence Generator (VKTG) toolbox that accompanies this chapter.

2 A process model that describes the relation between the velocity and displacement is given by the following dynamic equation.

$$\dot{x}(t) = \alpha x(t) + bu(t) \tag{4.56}$$

However, the desired dynamic response is given by a model with dynamics of the form

$$\dot{x}_m(t) = \alpha_m x_m(t) + b_m u_c(t) \tag{4.57}$$

Assume a controller of the form $u(t) = \theta_1(t)u_c(t) - \theta_2(t)x(t)$ to assess the problem of tracking between the two given systems.

a) Calculate the derivative of the error, e (error dynamics), between the two systems in the closed-loop solution.
b) Assuming a Lyapunov candidate function $V = V\left(e(t), \theta_1(t), \theta_2(t)\right)$

$$V = \frac{1}{2}\left[e^2(t) + \frac{1}{b\gamma}\left(\left(b\theta_2(t) + \alpha - \alpha_m\right)^2 + \left(b\theta_1(t) - b_m\right)^2\right)\right] \tag{4.58}$$

calculate:

i) the set of b, γ that satisfies $V \geq 0$
ii) the derivative of the Lyapunov function in e, θ_1, θ_2
iii) the adaptation parameters θ_1, θ_2 such that the closed loop solution is asymptotically stable, performing this simulation in MATLAB/SIMULINK and investigating the effect of the adaptation parameters in the closed-loop solution.

References

Antoulas AC 2005 *Approximation of Large-scale Dynamical Systems*. Siam. doi: 10.1137/1.9780898718713.

Badcock KJ, Timme S, Marques S, Khodaparast H, Prandina M, Mottershead JE, Swift A, Da Ronch A and Woodgate M 2011 Transonic aeroelastic simulation for instability searches and uncertainty analysis. *Progress in Aerospace Sciences* **47**, 392–423.

Barkana I 2005 Gain conditions and convergence of simple adaptive controls. *International Journal of Adaptive Control and Signal Processing* **19**, 13–40. doi: 10.2514/6.2013-1485.

Barkana I 2013 Simple adaptive control – a stable direct model reference adaptive control methodology – brief survey. *International Journal of Adaptive Control and Signal Processing* **28**, 567–603.

Campos-Delgado DU, Femat R and Velasquez ER 2003 Design of reduced-order controllers via h_∞ and parametric optimisation: Comparison for an active suspension system. *European Journal of Control* **9**, 48–60. doi: 10.3166/ejc.9.48-60.

Cook RG, Palacios R and Goulart P 2013 Robust gust alleviation and stabilization of very flexible aircraft. *AIAA Journal* **51**(2), 330–340.

Da Ronch A, Badcock KJ, Wang Y, Wynn A and Palacios RN 2012 Nonlinear model reduction for flexible aircraft control design. AIAA Paper 2012–4404 *AIAA Atmospheric Flight Mechanics Conference*, Minneapolis, MN. doi: 10.2514/6.2012-4404.

Da Ronch A, McCracken A, Badcock KJ, Ghoreyshi M and Cummings RM 2011 Modeling of unsteady aerodynamic loads, AIAA Paper 2011–6524 *AIAA Atmospheric Flight Mechanics Conference*, Portland, OR. doi: 10.2514/6.2011-6524.

Da Ronch A, McCracken A, Badcock KJ, Widhalm M and Campobasso MS 2013a Linear frequency domain and harmonic balance predictions of dynamic derivatives. *Journal of Aircraft* **50**(3), 694–707. doi: 10.2514/1.C031674.

Da Ronch A, Tantaroudas ND and Badcock KJ 2013b Reduction of nonlinear models for control applications. AIAA Paper 2013–1491 *54th AIAA/ASME/ASCE/AHS/ASC Structures, Structural Dynamics, and Materials Conference*, Boston, MA. doi: 10.2514/6.2013-1491.

Da Ronch A, Tantaroudas ND, Jiffri S and Mottershead JE 2014 A nonlinear controller for flutter suppression: from simulation to wind tunnel testing. AIAA Paper 2014–0345 *AIAA SciTech 2014, 55th AIAA/ASMe/ASCE/AHS/SC Structures, Structural Dynamics, and Materials Conference*, National Harbor, MD. doi: 10.2514/6.2014-0345.

Da Ronch A, Tantaroudas ND, Timme S and Badcock KJ 2013c Model reduction for linear and nonlinear gust loads analysis. AIAA Paper 2013-1492 *54th AIAA/ASME/ASCE/AHS/ASC Structures, Structural Dynamics, and Materials Conference*, Boston, MA. doi: 10.2514/6.2013-1492.

Dillsaver M and Cesnik C 2011 Gust load alleviation control for very flexible aircraft. AIAA Paper 2011-6368 *AIAA Atmospheric Flight Mechanics Conference*, Portland, OR. doi: 10.2514/6.2011-6368.

Etkin B 1981 Turbulent wind and its effect on flight. *Journal of Aircraft* **18**(5), 327–345.

Geradin M and Cardona A 2001 *Flexible Multibody Dynamics: A Finite Element Approach*. John Wiley & Sons.

Ghoreyshi M, Cummings RM, Da Ronch A and Badcock KJ 2013 Transonic aerodynamic loads modeling of X-31 aircraft pitching motions. *AIAA Journal* **51**(10), 2447–2464. doi: 10.2514/1.J052309.

Gianfrancesco M 2014 *Functional modelling and design of energy harvesters for slender wing structures* Master's thesis Department of Management and Production Engineering, Politecnico di Torino, Italy & Faculty of Engineering and the Environment, University of Southampton, UK.

Hesse H and Palacios R 2012 Consistent structural linearisation in flexible-body dynamics with large rigid-body motion. *Computers& Structures* **110–111**(0), 1–14.

Hoblit FM 1988 *Gust Loads on Aircraft: Concepts and Applications.* American Institute of Aeronautics & Astronautics. doi: 10.2514/4.861888.
Houbolt JC 1973 Atmospheric turbulence. *AIAA Journal* **11**, 421–437.
Ioannou PA and Sun J 1996 *Robust Adaptive Control.* Prentice-Hall.
Jones R 1940 The unsteady lift of a wing of finite aspect ratio. Technical Report 681, NACA.
Leishman JG 1994 Unsteady lift of a flapped airfoil by indicial concepts. *Journal of Aircraft* **31**(2), 288–297.
Meirovitch L 1990 *Dynamics and Control of Structures.* John Wiley & Sons.
Moorhouse DJ and Woodcock RJ 1982 Background information and user guide for MIL-F-8785C, military specification – flying qualities of piloted airplanes. Technical report, DTIC Document.
Noll TE, Brown JM, Perez-Davis ME, Ishmael SD, Tiffany GC and Gaier M 2004 Investigation of the helios prototype aircraft mishap. Technical Report Volume I – Mishap Report.
Ogata K 2010 *Modern Control Engineering.* Prentice-Hall.
Papatheou E, Tantaroudas ND, Da Ronch A, Cooper JE and Mottershead JE 2013 Active control for flutter suppression: An experimental investigation. IFASD Paper 2013-8d *International Forum on Aeroelasticity and Structural Dynamics*, Bristol, UK.
Tantaroudas ND, Da Ronch A, Gai G, Badcock KJ and Palacios R 2014 An adaptive aeroelastic control approach by using non linear reduced order models. AIAA Paper 2014-2590 *14th AIAA Aviation Technology, Integration, and Operations Conference*, Atlanta, GA. doi: 10.2514/6.2014-2590.
Theodorsen T 1935 General theory of aerodynamic instability and the mechanism of flutter. Technical Report NACA–496, National Advisory Committee for Aeronautics.
Timme S, Badcock KJ and Da Ronch A 2013 Linear reduced order modelling for gust response analysis using the dlr-tau code. IFASD Paper 2013-36a *International Forum on Aeroelasticity and Structural Dynamics*, Bristol, UK.
Torres S and Mehiel E 2006 Nonlinear direct adaptive control and disturbance rejection for spacecraft. AIAA Paper 2006-6038 *AIAA Guidance, Navigation, and Control Conference and Exhibit*, Keystone, CO. doi: 10.2514/6.2006-6038.
Zhou K and Doyle JC 1998 *Essentials of Robust Control.* Prentice Hall.

5

Unmanned Aircraft Wind Tunnel Testing

R. Bardera Mora

Low-speed Experimental Aerodynamics Laboratory, National Institute for Aerospace Technology, Madrid, Spain

5.1 Introduction

Many techniques are involved in the design of a new aircraft, but wind tunnel testing is traditionally considered as a key part of the assessment of an aerodynamic design. This chapter describes wind tunnel testing's role in the development of the Diana aircraft, a new unmanned aerial vehicle (UAV) designed by the National Institute for Aerospace Technology (INTA) in Spain. In a first phase, the airfoils forming the wings and tail were selected in a preliminary design after definition of the configuration requirements. In a subsequent phase, both analytical and computational methods were applied to estimate the aerodynamic characteristics of the aircraft. Aircraft tests were performed in wind tunnels in order to provide experimental aerodynamic insights that would validate the results obtained by the analytical and computational fluid dynamics (CFD) methods that were used in the aerodynamic aircraft design process. Aerodynamic force and moment measurements were carried out by means of a six-component internal balance. Dimensionless coefficients were calculated directly from raw data in centre-balance body axes, and were subsequently transferred to wind axes in the usual manner. Finally, characteristic curves of the aerodynamic coefficients were determined and compared with those obtained by CFD methods. Flow visualization techniques were used to complete the aerodynamic study. An oil-film technique was used to visualize the flow on the aircraft surface at angles of attack close to the aerodynamic stall, and the results were compared those from CFD surface-flow visualization. The effect of turbojet exhaust gases on the aircraft tail was investigated by wind tunnel testing, and infrared (IR) thermography was used to obtain temperature measurements of the exhaust gases while the turbojet was running, in a simulation of cruise conditions.

Advanced UAV Aerodynamics, Flight Stability and Control: Novel Concepts, Theory and Applications,
First Edition. Edited by Pascual Marqués and Andrea Da Ronch.

Companion website: http://www.wiley.com/go/marques/advanced_UAV_aerodynamics

5.2 The Diana UAV Project

The aim of the Diana Unmanned Aircraft project is the design and development of a high-speed fixed-wing aerial target drone, which will be used to simulate threats from current and future weapons. In addition, it is intended to be used as a platform for research and development of high-speed UAV technologies (see Figure 5.1). The distinctive features of Diana are its high-speed flight, high manoeuvrability and a modular design with moderate system and operational costs. The maximum take-off weight is 1600 N and the manufacturer's empty weight is 900 N. Aircraft endurance is more than 1 h, flight ceiling is 6000 m and range is 90 km. Diana is a turbojet-powered aircraft with 1100 N of thrust and a maximum speed of 200 m/s. The take-off system is based on launching by catapult.

Figure 5.2 shows the Diana aircraft's configuration. It has a slightly swept mid-wing, resulting in a trapezoidal planform, and a V-tail design. The main external dimensions are 3.467 m length, 1.844 m wingspan and 0.354 m diameter for the cylindrical fuselage. The nacelle of the turbojet can be seen below the fuselage.

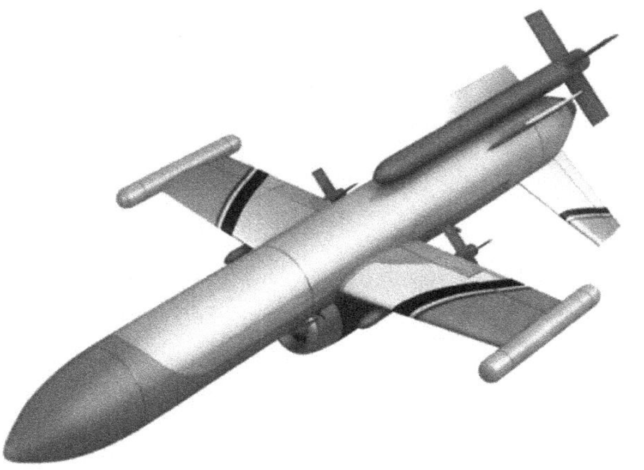

Figure 5.1 Diana unmanned aircraft.

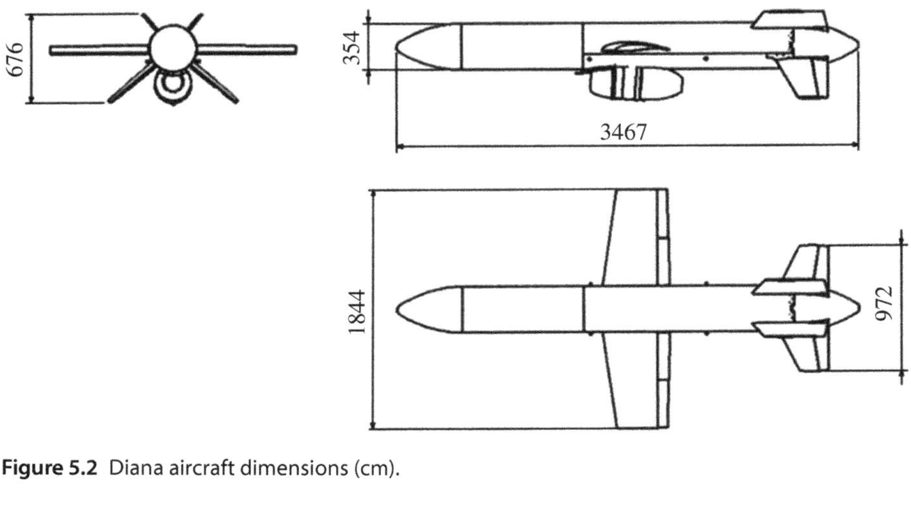

Figure 5.2 Diana aircraft dimensions (cm).

Figure 5.3 Diana aircraft model supported by sting. (*See insert for color representation of this figure.*)

5.3 Experimental Facility

The testing was carried out in a low-speed subsonic wind tunnel at INTA, Madrid. The wind tunnel is a continuous-flow closed-circuit type with an open test section 2.8 × 1.9 m^2 in size, producing velocities of up to 50 m/s with a turbulence intensity of 0.27%. A 1:2 scale model of Diana was built in aluminium alloy for the tests. The nose ogive was fabricated in carbon fibre, in order to produce a rear displacement of the centre of mass, because carbon fibre is a lighter material than aluminium. During the test campaign the aircraft model was supported by a rear sting system attached to the aircraft tail cone, as shown in Figure 5.3. Different wind angle aircraft orientations were achieved by moving the sting with an automatic positioning system.

5.4 Force and Moment Measurements

Aerodynamic forces and moments were measured at the maximum airspeed of 50 m/s and a Reynolds number based on the mean aerodynamic chord of 7.5×10^5. The aircraft model was tested at angles of attack (AOAs) ranging from −6– + 18° with an increment of 2° of pitch angle. No rolling angle was applied while testing. The total force and moment on the complete aircraft model were measured in order to obtain the dimensionless coefficients for different flight configurations. Similarity laws were invoked to apply the dimensionless coefficients from the wind tunnel tests on the model to full scale aircraft.

5.4.1 Reference Frames

The reference frame is a set of three orthogonal axes, in general in a right-hand sequence. The most used reference frames are the body axes and wind axes [1–2]. The body axes are fixed to the model (body) and move with it. The x_b–z_b plane is usually a plane of symmetry of the model. Figure 5.4 shows the body axes frame.

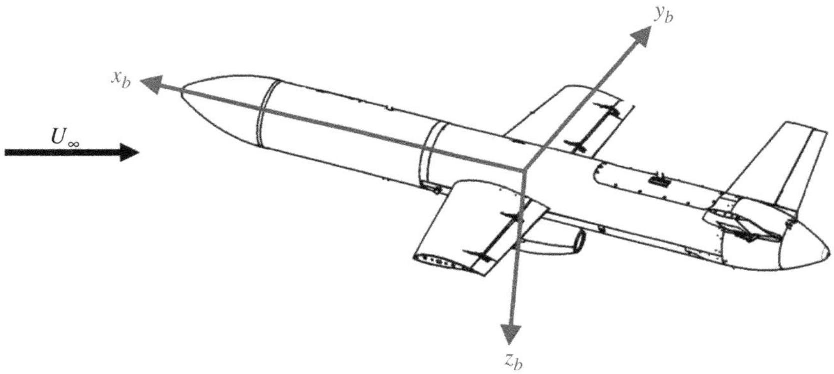

Figure 5.4 Body axes frame.

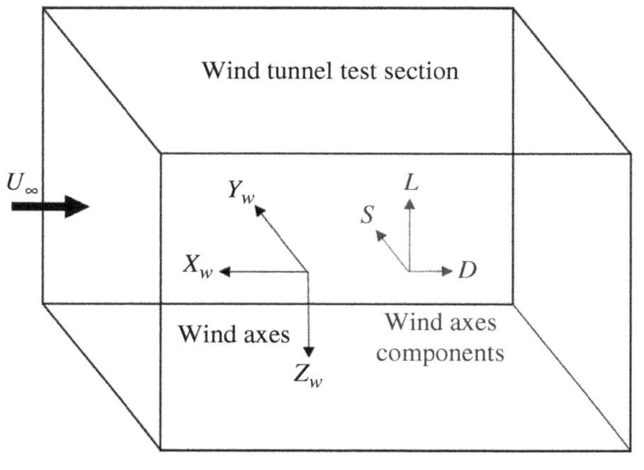

Figure 5.5 Wind axes frame.

Force components on body axes are referred as axial (**A**), side (**Y**) and normal (**N**) for the x_b, y_b and z_b force components, respectively. Body force vector $\mathbf{F}_b$ with this convention is given by,

$$[F_b]=\begin{bmatrix} X_b \\ Y_b \\ Z_b \end{bmatrix}=\begin{bmatrix} -A \\ Y \\ -N \end{bmatrix} \tag{5.1}$$

The wind axes system has the x_w axis pointing in the wind direction, y_w pointing to the right looking into the wind and z_w pointing down. With this convention, drag force (**D**) is in the negative x_w direction, side force (**S**) is in the positive y_w direction while lift force (**L**) is in the negative z_w direction, as shown in Figure 5.5.

The wind axes force vector $\mathbf{F}_w$ in this convention is as follows,

$$[F_w] = \begin{bmatrix} X_w \\ Y_w \\ Z_w \end{bmatrix} = \begin{bmatrix} -D \\ S \\ -L \end{bmatrix} \tag{5.2}$$

The relationship between the components on the body and wind frames is determined by a transformation matrix $[\mathbf{L}_{bw}]$, as follows,

$$[F_b] = [L_{bw}][F_w] \tag{5.3}$$

where $[\mathbf{L}_{bw}]$ is a matrix containing the expressions for converting wind axes forces to body axes forces. For the most common case, a roll angle of zero, $[\mathbf{L}_{bw}]$ matrix components are given by,

$$[L_{bw}] = \begin{bmatrix} \cos\alpha\cos\beta & -\cos\alpha\sin\beta & -\sin\alpha \\ \sin\beta & \cos\beta & 0 \\ \sin\alpha\cos\beta & -\sin\alpha\sin\beta & \cos\alpha \end{bmatrix} \tag{5.4}$$

where α and β are the attack and yaw angles, respectively.

The moment components on the x, y and z axes are referred to as rolling, pitching and yawing moments, respectively. The moment components in the body axes are the following:

$$[M_b] = \begin{bmatrix} M_x \\ M_x \\ M_x \end{bmatrix} = \begin{bmatrix} l \\ m \\ n \end{bmatrix} \tag{5.5}$$

where l, m and n denote moment components with no subscript for body axes.

Moment components on the wind axes are, with subscript w, as follows,

$$[M_w] = \begin{bmatrix} l_w \\ m_w \\ n_w \end{bmatrix} \tag{5.6}$$

A very common operation is the transfer of force and moment from one point of reference, such as the balance centre (bc), to another reference point, for example the centre of gravity (cg) of the aircraft model. In this case, if a system of force produces a resultant force F_{bc} and a resultant moment M_{bc} relative to the balance centre, statics rules indicate that an equivalent system acting at the centre of gravity produces the same force F_{cg},

$$F_{cg} = F_{bc} \tag{5.7}$$

but a different moment $\mathbf{M}_{bc}$, given by,

$$M_{cg} = M_{bc} - r_{bc-cg} \times F_{bc} \tag{5.8}$$

where $\mathbf{r}_{bc\text{-}cg}$ is the vector from the balance centre to the aircraft centre of gravity.

5.4.2 Instrumentation

Force measurements were carried-out by using an AB Rollab I6B128 six-component strain gauge internal balance. The balance is made of high-tensile steel and every load component is measured by means of a strain gauge bridge. The repeatability of this strain gauge balance is normally better than 0.1% of full scale, but the accuracy of the force measurements depends entirely on the quality of the calibration equipment and evaluation process.

The maximum force measurement range for this balance is,

$$\begin{bmatrix} F_x \\ F_y \\ F_z \end{bmatrix} = \begin{bmatrix} 800 \\ 2000 \\ 3000 \end{bmatrix} N \tag{5.9}$$

and the maximum moment measurement range is,

$$\begin{bmatrix} M_x \\ M_y \\ M_z \end{bmatrix} = \begin{bmatrix} 100 \\ 150 \\ 80 \end{bmatrix} Nm \tag{5.10}$$

This balance was selected because its range of measurements matches the expected loads on the aircraft model during the test campaign.

During the measurements, the balance is connected to an electric power supply HP 6624. The signal from each bridge of the balance is amplified by a NEC DC Strain Amplifier AS2101. Data acquisition is performed by a National Instruments PXI electronic system. Finally, a computer manages the whole measurement and data acquisition process. Figure 5.6 shows the measurement equipment.

Internal balances are usually located inside the aircraft model, so the model is joined to the balance at one end and to the sting at the other. Figure 5.7 shows the balance fixed to the sting support system before the model aircraft was assembled to the balance.

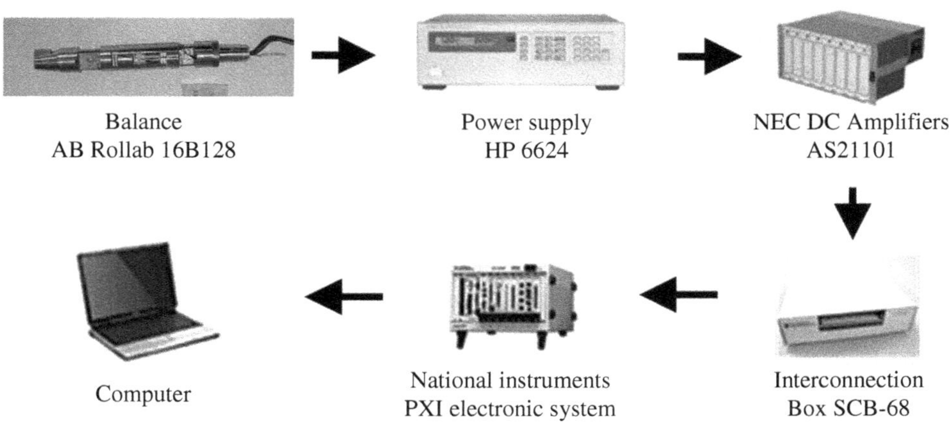

Figure 5.6 Balance-measurement instrumentation.

Figure 5.7 Balance fixed to sting support.

5.4.3 Balance Calibration

When subjected to a load, the balance provides a corresponding voltage signal. The relationship between balance output and load input (forces and moments) can be expressed by a second-order mathematical model as follows [1–4],

$$F_i = \sum_{j=1}^{6} K_{ij} S_j + \sum_{j=k}^{6} \sum_{k=1}^{6} K_{ijk} S_j S_k + R_i \tag{5.11}$$

where F_i is the 'i' component of the force matrix ($i = 1, 2 \ldots 6$), S_j are the balance voltage signals, K_{ijk} are the calibration coefficients and R_i are the residuals. Removing the residuals term, in compact matrix notation,

$$[F] = [K][S] \tag{5.12}$$

where [F] is the load vector with six components (three forces and three moments), [S] is the signals matrix with 27 elements and [K] is the calibration matrix (6 rows by 27 columns). The estimation of the calibration matrix is a multiple regression problem, usually solved by applying the least squares method, as follows,

$$[K] = [F][S]^T [E]^{-1} \tag{5.13}$$

where T and −1 indicate the transposed and inverse matrices, respectively and the matrix [E] is given by,

$$[E] = [S][S]^T \tag{5.14}$$

Prior to the Diana test campaign, the isolated balance was calibrated outside the wind tunnel using the classical load static method. This method is based on the measurement of balance responses when subjected to calibrated loads in order to determine the

Figure 5.8 Balance calibration layout.

unknown calibration matrix. For this purpose, a correct calibration requires several loading combinations. Assuming that n loading states are used to calibrate the balance, the corresponding expression is given by,

$$[F]_{6\times n} = [K]_{6\times 27}[S]_{27\times n} \tag{5.15}$$

where matrix subscripts indicate rows × columns.

Additionally, sting deflections are determined by measuring the support deformation when the balance is subjected to loads during the calibration process. Figure 5.8 shows the layout of the rig where the balance is to be calibrated by the classical load static method.

5.4.4 Force and Moment Balance Measurements

The results of the balance measurements are presented in this section as dimensionless coefficients based on a body balance axes system for a maximum wind tunnel airspeed of 50 m/s with zero yaw ($\beta = 0°$) and zero roll ($\phi = 0°$) angles. The Reynolds number based on the mean aerodynamic chord was 7.5×10^5.

Forces measured by the balance are represented as dimensionless coefficients in the usual manner, as follows,

$$C_i = \frac{F_i}{q\,A} \quad (i = x, y, z) \tag{5.16}$$

where F_i is the force along the i axis, A is a reference area taken as the wings' area and q is the dynamic pressure of the air flow defined as $q = 0.5\rho U_\infty^2$, where ρ is the air density and U_∞ is the free stream velocity. Aerodynamic coefficients are calculated as a function of the effective AOA, the angle of attack corrected for the sting deflections.

The moment coefficients are defined in analogue form as force coefficients,

$$Cm_i = \frac{M_i}{q\,A\,l} \quad (i = x, y, z) \tag{5.17}$$

where l is a reference length. Thus, l is the span when $i = x$, z and the mean aerodynamic chord when $i = y$.

Results of the force coefficients measurements are depicted in Figure 5.9. The curves show the coefficient C_x being slightly negative up to an AOA of 5°. Above this angle C_x is positive. Coefficient C_z is decreasing with AOA showing a negative slope. Finally, coefficient C_y is zero as expected.

Moment coefficients in the balance axes are plotted in Figure 5.10 for the complete aircraft. Coefficients Cm_x and Cm_z are approximately zero and the pitching moment coefficient Cm_y grows with the angle of attack, approximately in a linear form.

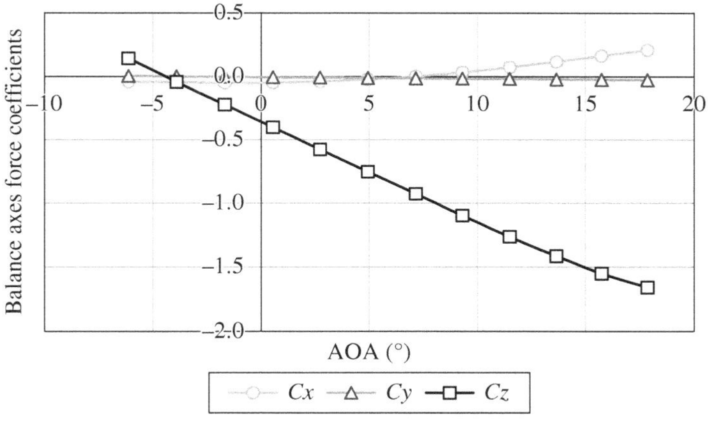

Figure 5.9 Force coefficients versus angle of attack.

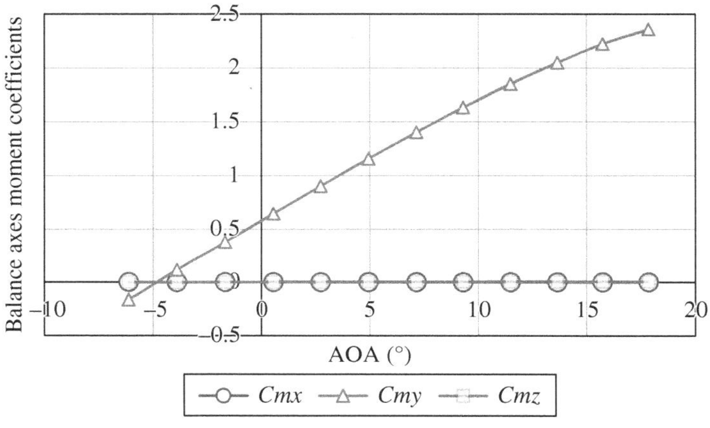

Figure 5.10 Moment coefficients versus angle of attack.

5.4.5 Force and Moment in Wind Axes

The results of the balance measurements were transferred from the body axes frame to a more usable wind axis frame. The relationship between the components in the wind and body frames is determined by the following expression,

$$[F_w] = [L_{bw}]^T [F_b] \tag{5.18}$$

where $[L_{bw}]^T$ is the transposition of matrix $[L_{bw}]$, and contains the expressions for converting from the body axes frame to the wind axes frame. For the case for which roll (ϕ) and yaw (β) angles are zero, $[L_{bw}]^T$ matrix components are given by,

$$[L_{bw}]^T = \begin{bmatrix} cos\alpha & 0 & sin\alpha \\ 0 & 1 & 0 \\ -sin\alpha & 0 & cos\alpha \end{bmatrix} \tag{5.19}$$

where α is the AOA.

When the wind axes frame is fixed to the centre of gravity of the model and the balance axes frame is fixed to the balance centre, the force relationship is as follows,

$$[F_{w-cg}] = [L_{bw}]^T [F_{w-bc}] \tag{5.20}$$

Following statics rules, if a system of forces produces a resultant $F_{w\text{-}bc}$ in the wind axes frame, an equivalent system acting at the centre of gravity produces the same force F_{w-cg} in the wind axes frame,

$$F_{w-cg} = F_{w-bc} \tag{5.21}$$

Finally, a forces system acting at the centre of gravity of the model expressed in the wind axes frame is given by,

$$[F_{w-cg}] = \begin{bmatrix} -D \\ S \\ -L \end{bmatrix}_{cg} = [L_{bw}]^T \begin{bmatrix} X_b \\ Y_b \\ Z_b \end{bmatrix}_{bc} \tag{5.22}$$

where cg and bc indicate the centre of gravity of the model and the balance centre, respectively.

The wind axes coefficient forces measured are defined as,

$$C_k = \frac{k}{q\,A} \quad (k = D, S, L) \tag{5.23}$$

where *D*, *S* and *L* are drag, side and lift force, respectively, *A* is a reference area taken as the wings' area and *q* is the dynamic pressure of the air flow defined as $q = 0.5\rho U_\infty^2$, where ρ is the air density and U_∞ is the free stream velocity.

Wind axes force coefficients were calculated after transferring the force between frames and the results are shown in Figure 5.11. Lift coefficient C_L is increasing with AOA whereas the drag coefficient C_D is slowly rising. The side coefficient C_S is approximately zero.

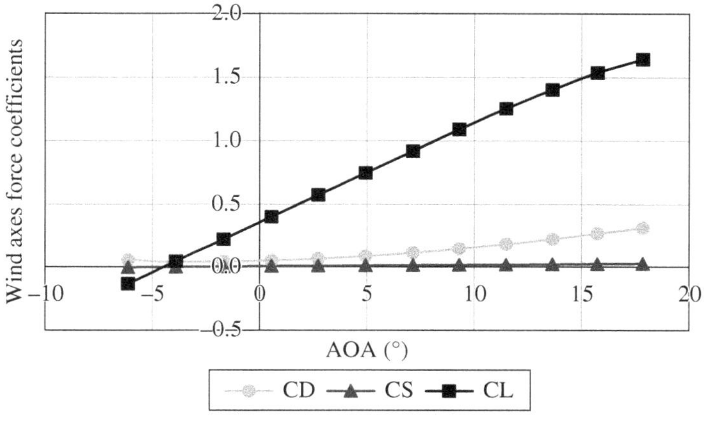

Figure 5.11 Force coefficients versus angle of attack (wind axes).

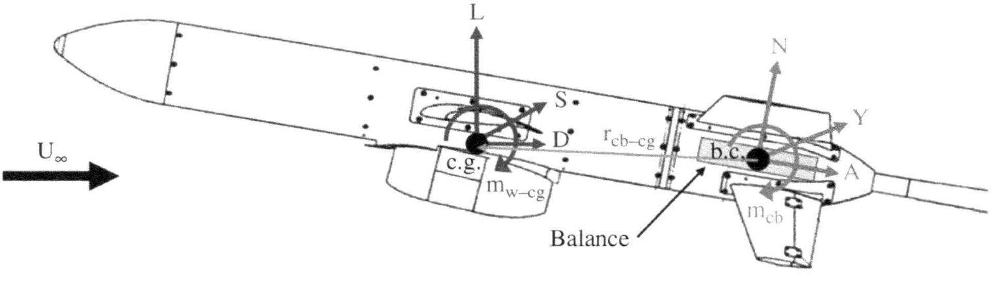

Figure 5.12 Transference of force and moment from balance centre to centre of gravity.

The moment at the centre of gravity is computed from force and moment measured at the balance centre, with all components in wind axes, as follows [1]:

$$\left[M_{w-cg}\right]=\begin{bmatrix} l_{w-cg} \\ m_{w-cg} \\ n_{w-cg} \end{bmatrix}=\begin{bmatrix} l_{w-bc} \\ m_{w-bc} \\ n_{w-bc} \end{bmatrix}-\begin{bmatrix} 0 & -z_{cg} & y_{cg} \\ z_{cg} & 0 & -x_{cg} \\ -y_{cg} & x_{cg} & 0 \end{bmatrix}\begin{bmatrix} -D \\ S \\ -L \end{bmatrix} \tag{5.24}$$

Figure 5.12 shows the transfer of force and moment from the balance centre to the aircraft model centre of gravity. Body axes force is represented on the balance centre (bc) while wind axes force is located at the model centre of gravity (cg).

The pitching moment around the model centre of gravity in wind axes is defined in analogue form as force coefficients:

$$Cm_{cg}=\frac{m_{w-cg}}{q\,A\,c} \tag{5.25}$$

where c is the mean aerodynamic chord.

Figure 5.13 shows the pitching moment coefficient, which is decreasing as AOA is increasing.

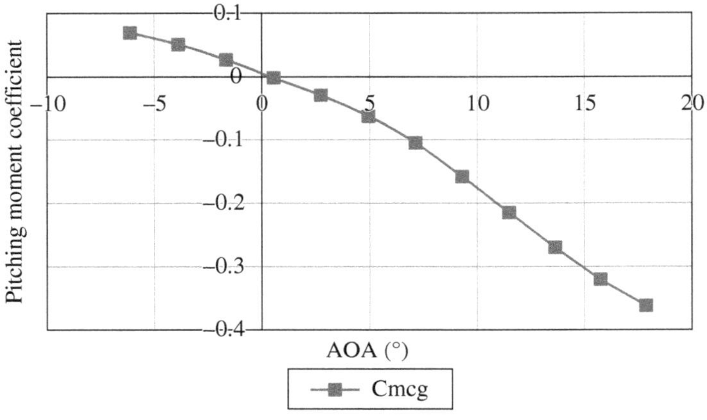

Figure 5.13 Pitching moment versus angle of attack (wind axes).

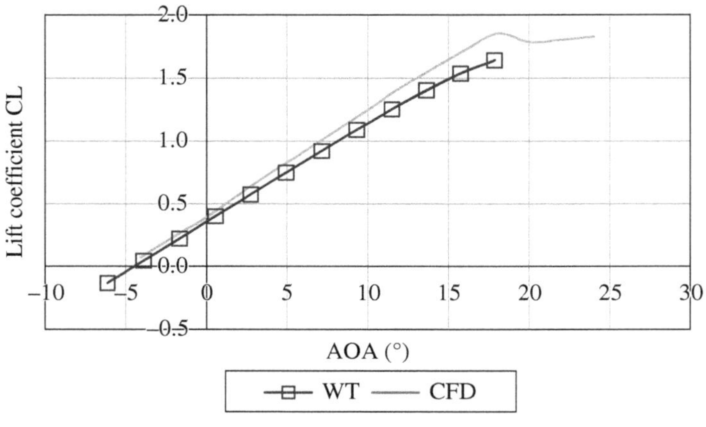

Figure 5.14 Lift curve versus angle of attack.

5.5 Wind Tunnel and CFD Comparisons

This section compares the complete aircraft results calculated using CFD methods and wind tunnel (WT) methods.

5.5.1 Lift Curve

Figure 5.14 shows the curve of the lift coefficient C_L as a function of the AOA of the complete aircraft. CFD and WT curves show a linear trend but with slightly different slopes. The maximum WT value of C_L is 1.64 at AOA = 18°, corresponding to an overestimated CFD value of $C_L = 1.85$. Lift drop out is observed in the CFD curve above 18° but with a slight recovery from 20°. AOAs higher than 18° were not tested in the wind tunnel due to vibrations of the model giving erroneous balance measurements.

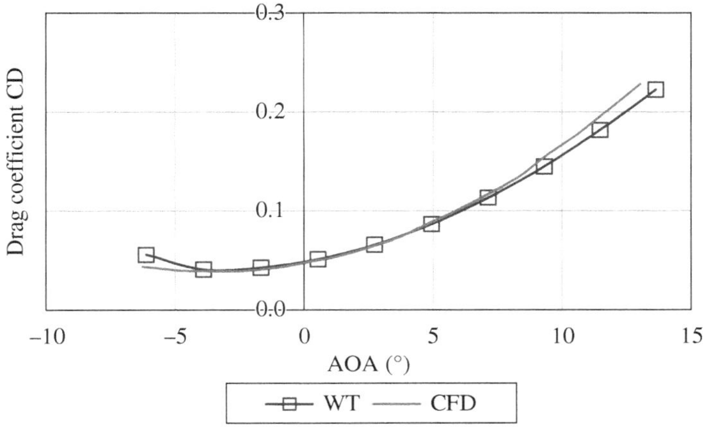

Figure 5.15 Drag curve versus angle of attack.

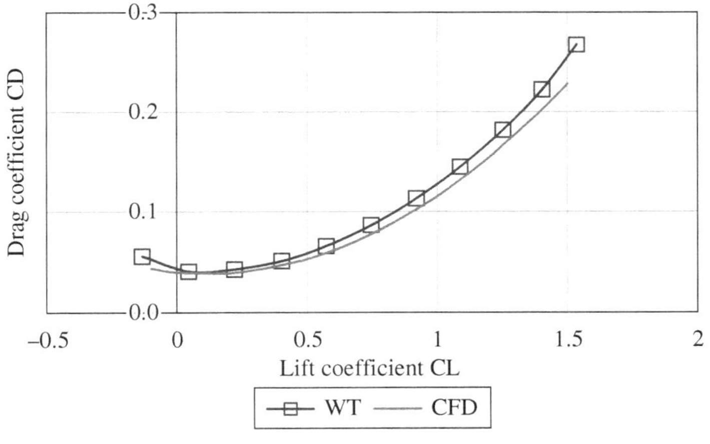

Figure 5.16 Drag polar curve.

5.5.2 Drag Curve

The drag curve represents the variation of drag coefficient C_D with AOA. Figure 5.15 shows the drag coefficient curve for the complete aircraft using CFD and WT. Both curves show almost the same trend. High coincidence is found when the minimum value of $C_D = 0.0407$ is reached at an AOA of −3.9°. At higher AOAs, the CFD method tends to overestimate the drag force.

5.5.3 Drag Polar Curve

Drag polar curve represents the drag force coefficient C_D versus the lift force coefficient C_L taking into account the dependence of both coefficients on the angle of attack. Figure 5.16 shows the drag polar curve of the complete aircraft obtained from CFD and WT. The minimum value of C_D can be observed to be coincident in both curves, but slightly different values are obtained when C_L is increasing, mainly due to differences between the lift coefficient curves.

5.5.4 Pitching Moment Curve

The variation of the pitching moment around the model centre of gravity as a function of AOA is shown in Figure 5.17 by the corresponding dimensionless pitching moment coefficient (Cm_{cg}) versus AOA. Curves from CFD estimates and WT data are sketched in Figure 5.17. They have the same trend with slight differences when the AOA increases. The CFD method underestimates the pitching moment compared to the wind tunnel data. Negative slope ($\partial Cm_{cg}/\partial\alpha < 0$) indicates positive static longitudinal stability [5].

Another typical curve, the pitching moment coefficient Cm_{cg} versus the lift coefficient C_L is shown in Figure 5.18. The trim point is $C_L = 0.40$ (AOA = 0.55°), corresponding to a zero pitching moment coefficient. This figure shows that $C_L = 0$ gives a pitching moment $Cm_{cg} = 0.051$. The CFD curve shows an elbow corresponding to stall (AOA = 18°).

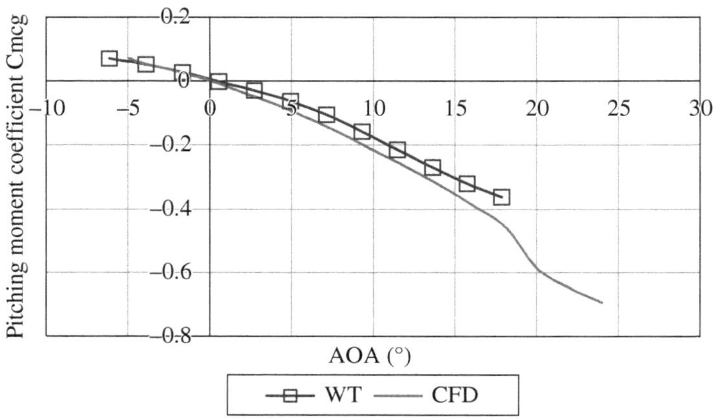

Figure 5.17 Pitching moment versus AOA.

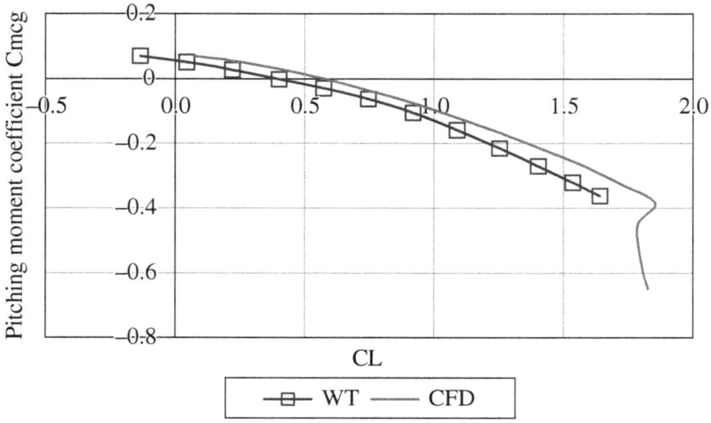

Figure 5.18 Pitching moment versus lift.

5.6 Flow Visualization

Flow visualization is a standard non-intrusive wind tunnel technique that provides qualitative information about the flow-field topology, giving an interesting picture of the whole-flow behaviour. In our research, we applied two flow visualization techniques – the surface oil-film technique and infrared (IR) thermography –to the investigation of the turbojet influence on the tail surfaces.

5.6.1 Oil-film Technique

The oil-film technique is based on observations of the flow pattern when a specifically prepared paint is applied to the model surface and when this is illuminated with a UV or similar light source [6–9]. The Diana aircraft model was placed in the wind tunnel test section and painted with a uniform and very thin layer of a mixture of gasoline, oil and a fluorescent pigment in the form of a fine powder. The tunnel was run at the maximum test air speed. The frictional force of the free stream moved the paint and the gasoline evaporated. After turning off, the paint dried into the flow pattern on the model surface, indicating the flow direction, transition, separation, and so on. Enhanced images were obtained, illuminating with black light tubes and digital image recording. The flow visualization was performed at an AOA of 20° and a wind tunnel free stream velocity of 50 m/s. Figure 5.19 shows a view of the fuselage flow visualization.

Figure 5.20 is a top view of the flow pattern over the left aircraft wing.

Figures 5.21 and 5.22 show a comparison of the flows as calculated from CFD codes and the oil-film technique. CFD simulations were implemented using the commercial software Fluent 14.5 (ANSYS). Figure 5.21 shows fuselage flow visualization images. The left-hand picture represents the flow streamlines computed by CFD and the right-hand side shows the flow from the oil-film technique. The greyscale version of the image was obtained after brightness and contrast modifications in order to enhance the image quality. The images show good agreement.

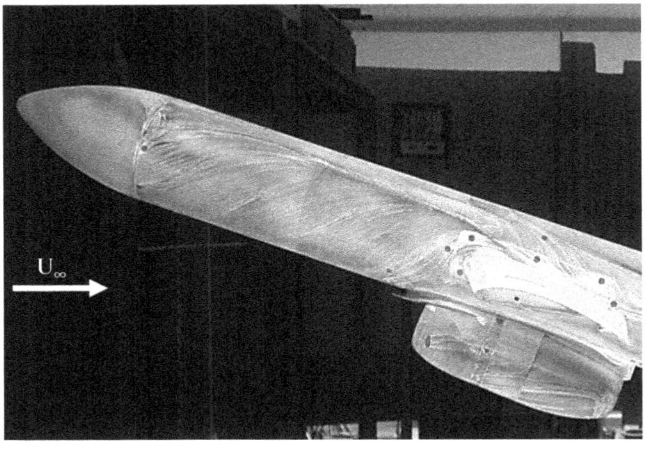

Figure 5.19 Oil-film flow visualization on the aircraft fuselage.

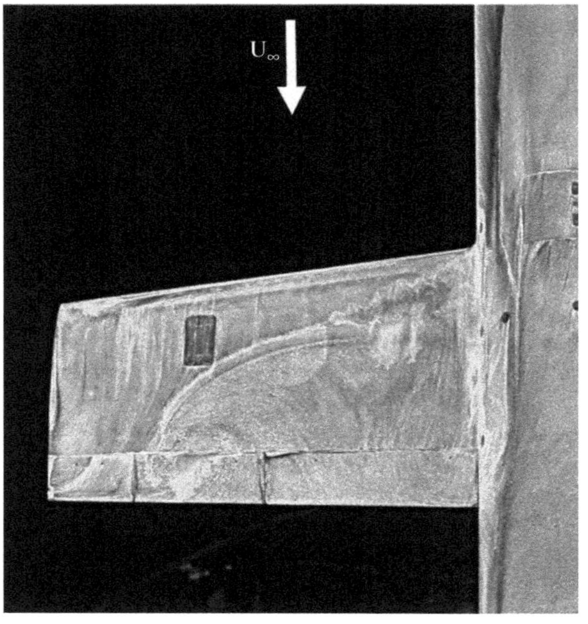

Figure 5.20 Oil-film flow visualization of to the aircraft wing.

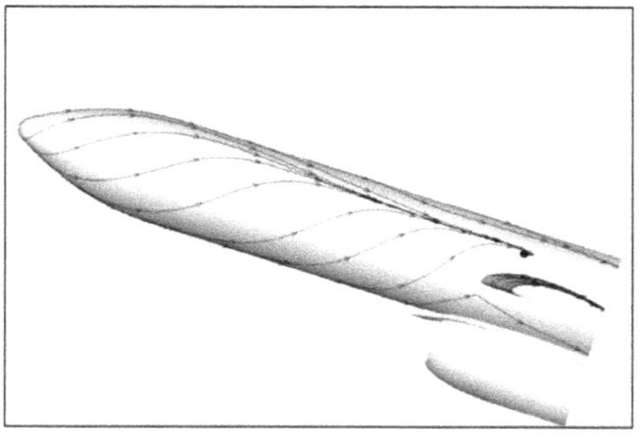

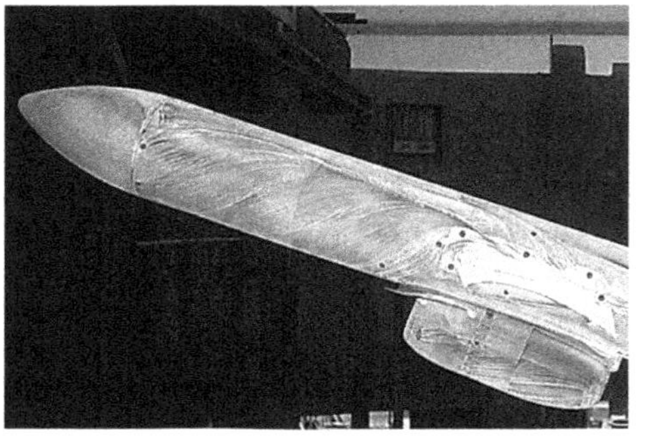

Figure 5.21 Fuselage flow visualization: left, CFD; right, oil-film.

Figure 5.22 shows the aircraft wing flow visualization images. The picture on the left represents the flow streamlines obtained by CFD and that on the right shows the flow visualization obtained by wind tunnel testing (in a greyscale version after brightness and contrast adjustments). Results provided by CFD are close to those given by the oil-film technique, showing a similar flow pattern, including the separation line.

5.6.2 IR Thermography

Infrared thermography is an experimental technique used to determine the temperature distribution on the model surface. Infrared thermography is based on the remote detection of infrared radiation from objects. This radiation is captured by an IR camera that converts the thermal radiation into a signal video. This gives a measurement of temperature by representing it as a colour temperature map [10]. The objective of the

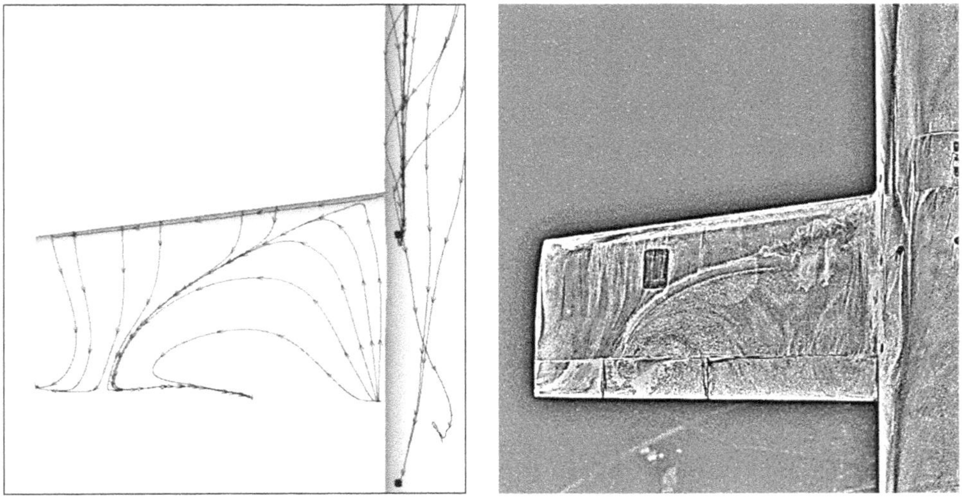

Figure 5.22 Flow visualization over the aircraft wing: left, CFD; right, wind tunnel.

Figure 5.23 Diana aircraft inside the wind tunnel test section.

thermography experiment was to investigate the influence of the turbojet exhaust gases over the tail control surfaces. For this purpose, another motorized and reduced-scale (1:2) Diana model was installed in the wind tunnel test section, supported by a bearing axle located at 1/4 of the wing chord. Figure 5.23 shows the layout of this experiment.

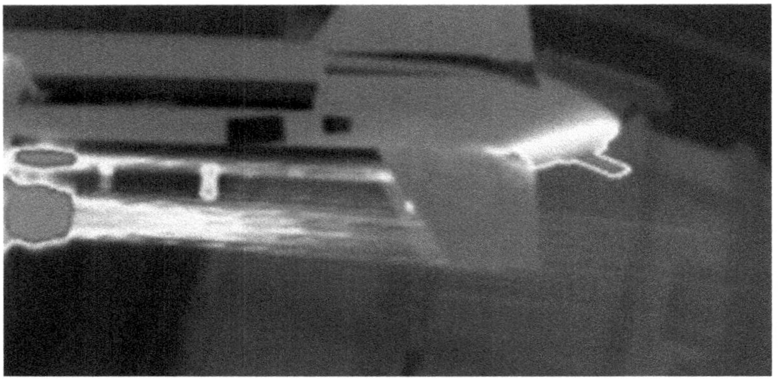

Figure 5.24 Thermography image of the turbojet running.

A Thermovision 900 LW camera from AGEMA Infrared Systems was used for thermal data acquisition. The thermography experiment was carried out at a free stream velocity of 50 m/s with the turbojet engine running to simulate aircraft flight conditions, with hot turbojet exhaust gases arriving at the aircraft tail surfaces. The aircraft was radio controlled during the wind tunnel tests in order to verify the correct behaviour of the aircraft controls.

Figure 5.24 shows the thermography image representing the aircraft in flight with the turbojet running. As can be observed on the image, the turbojet nacelle appears in red whereas orange zones indicate hot regions of the jet flow downstream of the turbojet. Colder regions of the jet are in blue. The IR thermography tests show the flow interaction of the jet with the tail control surfaces, which must be taken into account when determining the flight performance for aircraft operation. In addition, the analysis of thermography images provides information about the real variations of aircraft attitude in AOA as a function of tail control-surface deflection.

5.7 Summary and Conclusions

Experimental investigation of the aerodynamics of a new UAV is a classical part of its design process, and it is a fundamental tool in the validation of computational results. An experimental investigation of the Diana UAV was carried out in a low-speed wind tunnel of the National Institute for Aerospace Technology in Madrid. The results of aerodynamic force and moment measurements were presented by typical dimensionless coefficients. Flow visualizations by the oil-film technique and IR thermography were also obtained. A comparison of the results from the two methods was performed. This showed that the force computed by CFD generally agrees well with experimental wind tunnel data over a large range, but, the lift, drag, drag polar and pitching moment curves differ at high angles of attack. Flow visualizations of the fuselage and wing surfaces were performed by means of the oil-film technique. The results are in good agreement with CFD surface streamlines for the fuselage but slight differences appear in the flow over the wing, although the main lines of separation are well predicted by computational techniques. Infrared thermography was used in the wind tunnel testing of a motorized Diana aircraft model. The turbojet was running during the test experiments and the effect of the exhaust gases interacting with the tail surfaces was recorded with the thermography camera. Wind tunnel experiments have a fundamental role in the aerodynamic design process of a general purpose aircraft and can, in many UAV cases, take advantage of the reduced size of the vehicle to use motorized models of the airplane, allowing for a very realistic and accurate description of the aerodynamic characteristics of the UAV.

Acknowledgments

Author wish to acknowledge the support of J.M. Olalla in development of the theoretical and CFD methods that were applied in the aerodynamic design of the Diana UAV. He wish to thank especially S. Sor and M. Urdiales, who were in charge of the Diana wind tunnel experiments, for their collaboration. Last, but not least, our thanks go to all the people involved in the Diana project.

References

1 Barlow JB, Rae WH Jr, Pope A 1999 *Low-speed Wind Tunnel Testing*, 3rd edn. John Wiley & Sons,.
2 Tropea C, Yarin A, Foss JF 2007 *Handbook of Experimental Fluid Mechanics*. Springer.
3 *Calibration and use of internal strain-gauge balance with application to wind tunnel testing*. Technical report R-091-2003, AIAA, 2003.
4 Dubois M 1981 Six-components strain-gauge balances for large wind tunnels. *Experimental Mechanics*: Nov; 401–407.
5 Dole CE and Lewis JE 2000 *Flight Theory and Aerodynamics*, 2nd edn. John Wiley & Sons.
6 Merzkirch W 1987 *Flow Visualization*. 2nd edn. Academic Press.
7 Maltby RL 1962 Flow visualization in wind tunnels using indicators. AGARDograph 70. Advisory Group for Aeronautical Research and Development, Paris.
8 Goldstein RJ 1996 *Fluid Mechanics Measurements*. 2nd edn. Taylor & Francis.
9 Russo GP 2011 *Aerodynamic Measurements*. Woodhead Publishing.
10 Minkina W and Dudzik S 2009 *Infrared Thermography: Errors and Uncertainties*. John Wiley & Sons.

6

Chord-dominated Ground-effect Aerodynamics of Fixed-wing UAVs

Qiulin Qu[1] and Ramesh K. Agarwal[2]

[1] *Beijing University of Aeronautics and Astronautics, Beijing, China*
[2] *Washington University in St. Louis, St. Louis, MO, USA*

6.1 Introduction

The UAV is an acronym for 'unmanned aerial vehicle', an aircraft with no pilot on board. UAVs can be remote-controlled (flown by a pilot at a ground control station) or can fly autonomously based on pre-programmed flight plans or more complex dynamic automation systems. Currently, UAVs are widely used for military and civilian operations. UAVs can be classified into

- horizontal take-off and landing fixed-wing UAVs
- vertical take-off and landing (VTOL) rotor UAVs.

During take-off and landing on a runway or on an aircraft carrier deck, the aerodynamics and flow field of a fixed-wing UAV will be influenced by the ground effect (GE). Ever since the early days of aviation, pilots have experienced the GE phenomenon while operating very close to the ground. Either during take-off or landing, air vehicles experience improved efficiency near the ground in the form of increased lift.

When the aerodynamics are changed by GE during take-off and landing, the pilot of a crewed aircraft can adjust its attitude in real time. However, it is difficult for the remotely operated UAV or the automatic control system of a UAV to do the same thing in real time. Thus, an understanding of their GE aerodynamics will benefit the development of automatic takeoff and landing systems for UAVs.

Nomenclature

In this chapter, the following nomenclature is used:

A	area of the channel between the lower surface of the airfoil and the ground
C_D	drag coefficient
C_L	lift coefficient
C_p	pressure coefficient
c	chord length of the airfoil

Advanced UAV Aerodynamics, Flight Stability and Control: Novel Concepts, Theory and Applications, First Edition. Edited by Pascual Marqués and Andrea Da Ronch.

Companion website: http://www.wiley.com/go/marques/advanced_UAV_aerodynamics

DGE	dynamic ground effect
d	width of local stream tube
d_{in}	width of the stream tube on the inlet boundary
GE	ground effect
h	ride height; the distance between the lowest point on the airfoil surface and the ground/the top of the platform
$\mathbf{j}$	unit vector along y-axis
L	the distance between the leading edge of airfoil and the side of platform
MGE	mutational ground effect
Ma	Mach number
$\mathbf{n}$	unit vector normal to the airfoil surface
p	local static pressure
p_∞	freestream pressure
Re	Reynolds number based on airfoil chord length
S	surface area of the airfoil
SGE	static ground effect
T	oscillation period of unsteady aerodynamic forces
V_s	sink velocity
V_∞	freestream velocity
W	compression work done by the airfoil
x, y	Cartesian coordinates with the origin located at the leading edge of the airfoil
y^+	dimensionless wall distance of the first mesh layer
α	angle-of-attack (AOA)
β	velocity vector angle; the angle between a velocity vector and the positive direction of the x-axis
γ	flight path angle
θ	pitch angle
Γ	circulation around the airfoil
ΔC_L	increment in lift coefficient
$(\Delta C_L)_R$	dimensionless lift coefficient increment in the MGE
Δt	time step for unsteady simulation
$\Delta\alpha$	increment in angle-of-attack (AOA)
ρ	air density
$\boldsymbol{\tau}$	shear stress on the airfoil surface

Subscripts

h	ground effect with the ride height of h
low	lower surface of the airfoil
up	upper surface of the airfoil
∞	unbounded flow

6.2 Categories of Ground Effect

Based on flow physics, the GE can be categorized into the attached-flow GE of high-aspect-ratio wings for transport aircraft and the separated-flow GE of delta wings for fighters. The former can be further classified into the chord-dominated GE and the span-dominated GE (Rozhdestvensky 2006). For a 2D airfoil at positive angle of attack (AOA), ground proximity generally causes a high-pressure distribution on the lower surface of the airfoil leading to increase in lift, nose-down pitching moment and the lift-to-drag ratio; this phenomenon is called the 2D chord-dominated GE. For a 3D wing at positive AOA, ground proximity often pushes the wingtip vortices outward along the span leading to a decrease in downwash angle and induced drag; this phenomenon is called the 3D span-dominated GE.

Depending on the application, the GE can be categorized into the static ground effect (SGE, in which the ride height does not change with time), the dynamic ground effect (DGE, in which the ride height changes continuously with time) and the mutational ground effect (MGE, in which the ride height changes suddenly with time) (Figure 6.1; see also Qu et al. 2015b). Wing-in-ground (WIG) craft use the SGE to improve the aerodynamic efficiency while cruising on the water surface (Qu et al. 2014); conventional aircraft undergo the DGE during taking-off and landing (Qu et al. 2014b); fixed-wing carrier aircraft undergo the MGE effect during taking-off and landing on carrier decks (Coton 1998).

6.3 Chord-dominated Static Ground Effect

For the 2D chord-dominated GE, the aerodynamics and flow physics at low-to-moderate AOA have been extensively investigated. Hsiun and Chen (1996) studied the effect of camber and thickness on the aerodynamics of a 2D airfoil in the SGE by numerical simulation. They found that the aerodynamic forces on the airfoil were determined by the passage between the lower surface of the airfoil and the ground. Chun and Chang (2003) numerically simulated the flow fields around a Clark-Y airfoil at $\alpha = 5.92°$ and a NACA4412 airfoil at $\alpha = 5°$ in the SGE using fixed and moving grounds. Their investigation indicated that there were very small differences in lift and pitching moment predictions between the results with fixed and moving grounds, but the drag in the fixed-ground simulation was smaller than that in the moving-ground simulation. Ahmed et al. (2007) reported experimental results for a NACA4412 airfoil in GE at $\alpha = 0$–$10°$. They found that when the ride height was reduced, the drag increased, but the lift trend depended on the shape of the passage between the lower surface of the airfoil and the ground, which changed with the AOA. Zerihan and Zhang (2000) and Mahon and Zhang (2005) performed numerical simulations and wind tunnel experiments to study the negative GE of a Tyrrell-02 airfoil, which is a highly cambered inverted airfoil. When the ride height decreased, the downward force first increased gradually to a peak value and then decreased dramatically.

The archival information about GE aerodynamics at high AOAs is very limited. Hiemcke (1997) has reported experimental results for a NACA5312 airfoil in GE at high AOA. It was found that the attached flow on the airfoil evolved into separated flow when the ride height was reduced at high AOA. However, Hiemcke has not described the physical reasons behind the occurrence of flow separation and the influence of flow separation on the aerodynamic forces. Ying et al. (2010) conducted numerical studies on the stall characteristics of a NACA0012 airfoil in GE. They found that the stall AOA decreased with the reduced ride height, but did not show the flow structure or explain the reasons behind what they observed.

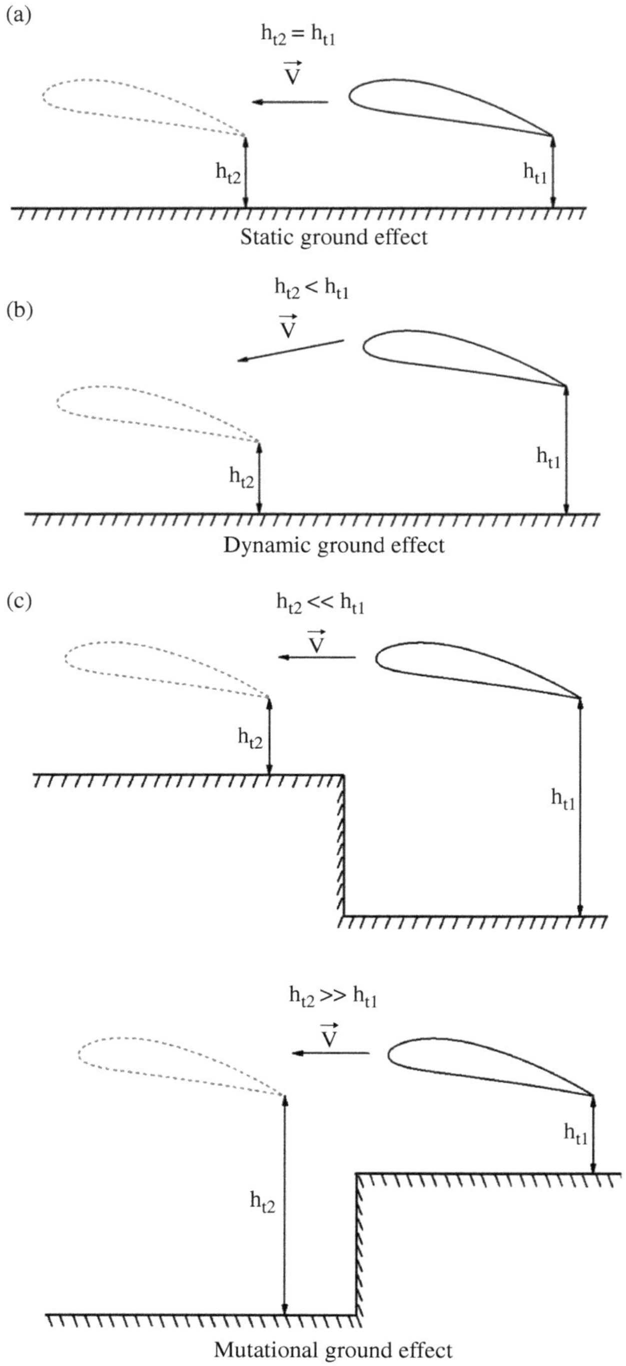

Figure 6.1 Category of GE based on application areas.

Here, the aerodynamics and flow physics of a NACA4412 airfoil in the SGE for a wide range of AOA (Qu et al. 2015a) are discussed.

6.3.1 Physical Model

A NACA4412 airfoil with $c = 1.0\,m$, $Ma = 0.26$ and $Re = 6 \times 10^6$ is investigated. A wide AOA range of $\alpha = -4$–$20°$ (with $\Delta\alpha = 2°$) is considered in order to study both the attached and separated flow patterns in GE. For 2D chord-dominated GE, several studies have concluded that the critical ride height (defined as the minimum height below which the aerodynamics is affected by the GE) is approximately $h/c = 1.0$ for low and moderate AOAs. Thus, relative ride heights of $h/c = \infty$, 1.0, 0.8, 0.6, 0.4, 0.2, 0.1 and 0.05 are considered. Values of $h/c = 5.0$ and 2.0 are additionally considered for cases of $\alpha = 4°$ and $8°$.

Here the relative motion principle is adopted. The airfoil is fixed in space; the ground and the inflow from upstream move opposite to the flight velocity of the airfoil.

6.3.2 Lift Characteristics

For convenience in explaining the results, the following definitions and notations related to lift coefficient are introduced. The total lift coefficient of an airfoil is defined as follows:

$$C_L = \frac{1}{0.5\rho V_\infty^2 c} \oint_{S_{up}+S_{low}} \left((p - p_\infty)\vec{n} + \vec{\tau}\right) \cdot \vec{j} dS \tag{6.1}$$

The upper surface lift coefficient of an airfoil is defined as follows:

$$C_{L,up} = \frac{1}{0.5\rho V_\infty^2 c} \oint_{S_{up}} \left((p - p_\infty)\vec{n} + \vec{\tau}\right) \cdot \vec{j} dS \tag{6.2}$$

The lower surface lift coefficient of an airfoil is defined as follows:

$$C_{L,low} = \frac{1}{0.5\rho V_\infty^2 c} \oint_{S_{low}} \left((p - p_\infty)\vec{n} + \vec{\tau}\right) \cdot \vec{j} dS \tag{6.3}$$

The total lift is written as:

$$C_L = C_{L,up} + C_{L,low} \tag{6.4}$$

For the GE flow field with ride height h, the lift increments in total, upper surface and lower surface lifts due to GE are defined as follows:

$$\Delta C_L = (C_L)_h - (C_L)_\infty \tag{6.5}$$

$$\Delta C_{L,up} = (C_{L,up})_h - (C_{L,up})_\infty \tag{6.6}$$

$$\Delta C_{L,low} = (C_{L,low})_h - (C_{L,low})_\infty \tag{6.7}$$

The lift curves and lift increment curves of the NACA4412 airfoil at various ride heights are shown in Figure 6.2. The total lift curve in unbounded flow has an expected

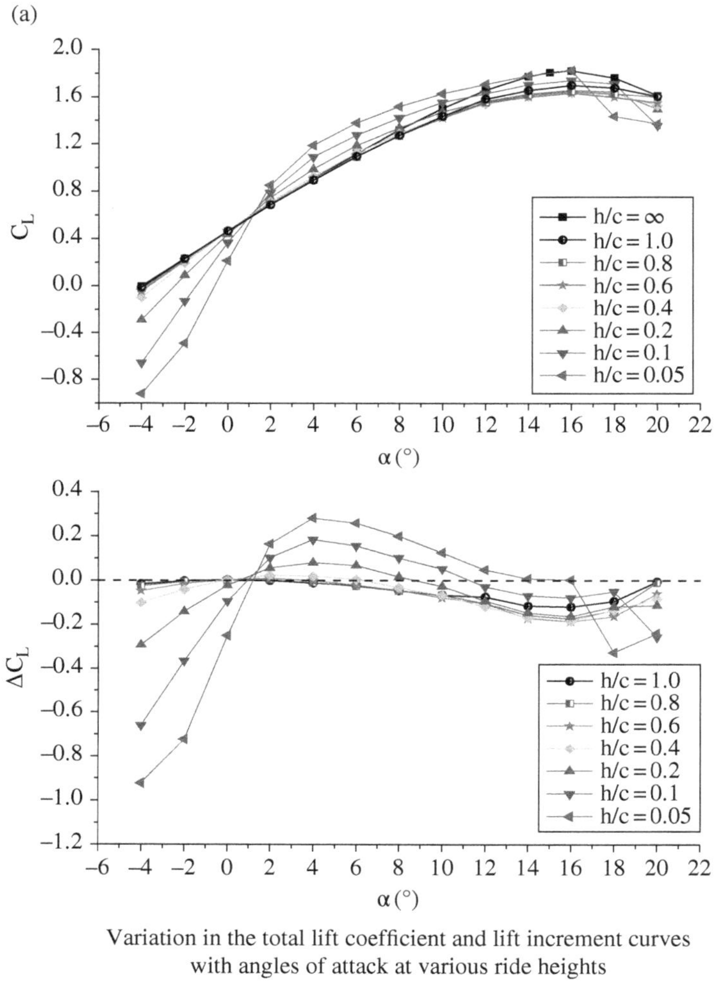

Figure 6.2 Lift curves and lift increment curves for the NACA4412 airfoil at various ride heights.

linear portion in the AOA range from −4–8°. However, the total lift curve in GE has a convex portion instead of a linear portion in this range. Therefore, there are two intersection points between the two lift curves: the lift curve in GE and the lift curve in unbounded flow. The smaller the ride height, the bigger are the curvatures of the convex portion of the airfoil and the AOAs corresponding to the intersection points. In the two regions outside the two intersection points, the lift decreases with the reduced ride height. In the region between the two intersection points, the lift increases with decreasing ride height; this is the region where traditional GE aerodynamics research has focused so far. All the cases in unbounded flow and the GE flow field for this airfoil have the same stall AOA of 16°. However, compared to the unbounded flow, in GE the maximum lift is smaller for $h/c = 1.0–0.1$ and is larger for $h/c = 0.05$. The stall characteristics become sharper for h = 0.1 and 0.05.

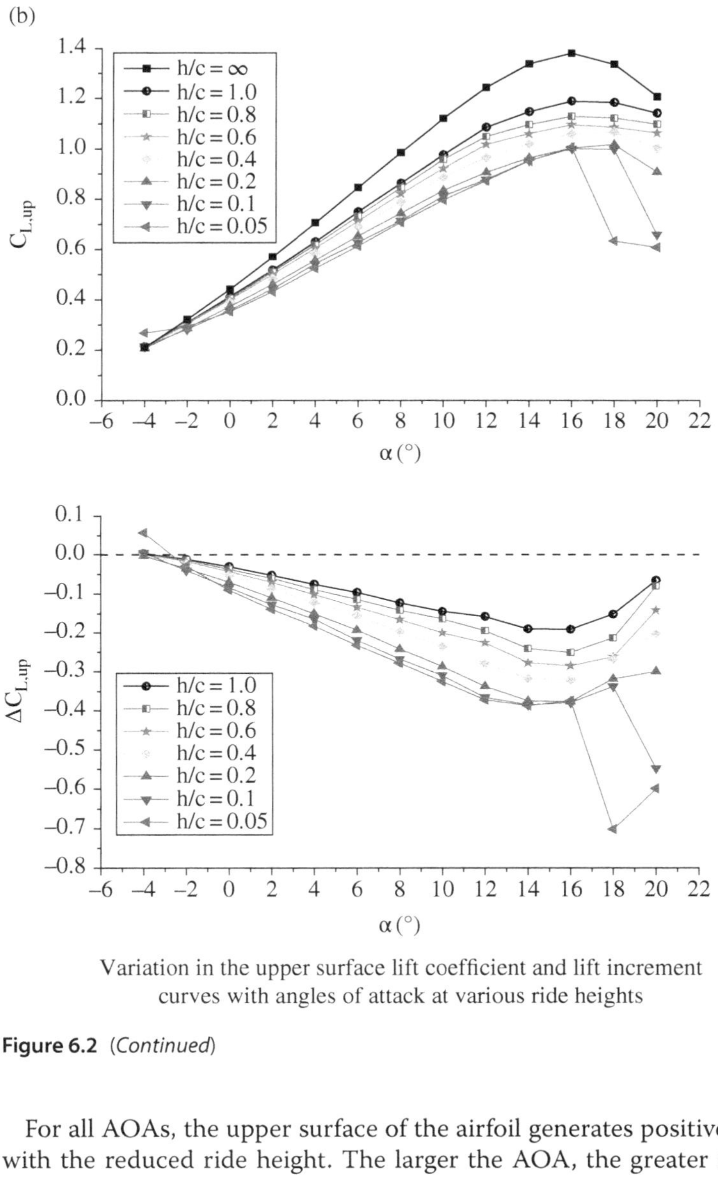

Variation in the upper surface lift coefficient and lift increment curves with angles of attack at various ride heights

Figure 6.2 *(Continued)*

For all AOAs, the upper surface of the airfoil generates positive lift, which decreases with the reduced ride height. The larger the AOA, the greater is the decrease in the upper surface lift. For low, moderate and high AOA, the lower surface produces positive lift, which increases with the reduced ride height. For negative AOA, the lower surface produces negative lift, which decreases with the reduced ride height.

Previous investigations have determined that the critical ride height is approximately $h/c = 1.0$ for the 2D chord-dominated GE at low and moderate AOAs. From Figure 6.2a, one can see that the critical ride height changes with AOA. For negative to moderate AOA, say $-4° < \alpha \le 8°$, the critical ride height can be seen as approximately one chord; but for high AOA, say, $8° < \alpha \le 20°$, the critical ride height is larger than one chord. Furthermore, when h/c decreases from ∞ to 1.0, the total lift does not change

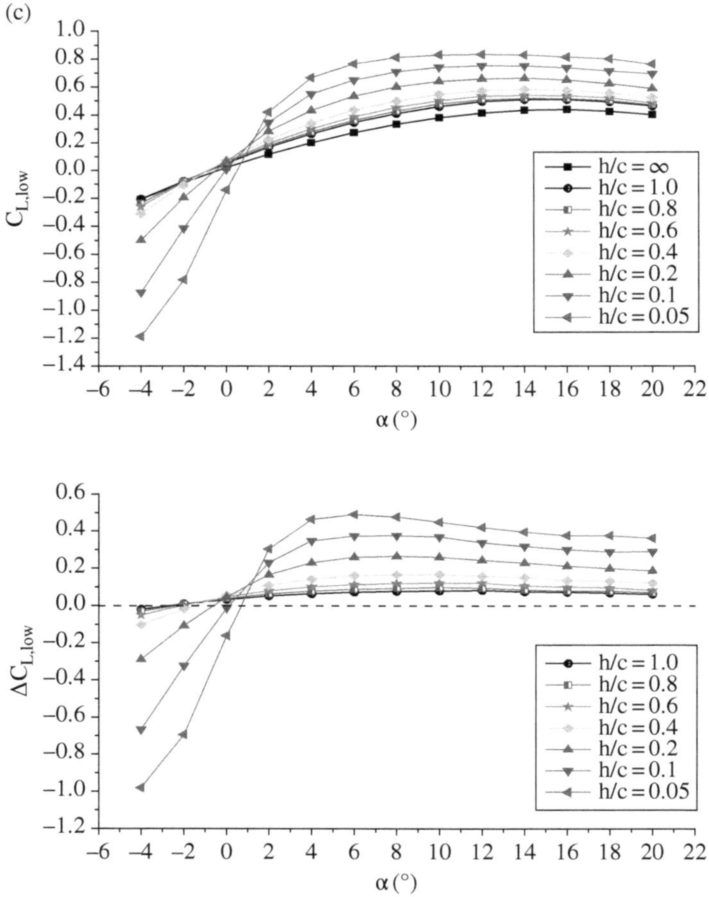

Variation in the lower surface lift coefficient and lift increment curves with angles of attack at various ride heights

Figure 6.2 *(Continued)*

appreciably for negative to moderate AOA, but the variations in both the upper and lower surface lifts are obvious.

The lift increment contours of the NACA4412 airfoil in the AOA–ride-height plane are shown in Figure 6.3. Based on the sign of the total lift increment value, the AOA–ride-height plane can be divided into three regions (Figure 6.3a):

- Region I of positive GE
- Region II of negative GE
- Region III of negative GE.

The lift characteristics and flow physics for the typical AOAs in the three regions are analyzed in detail.

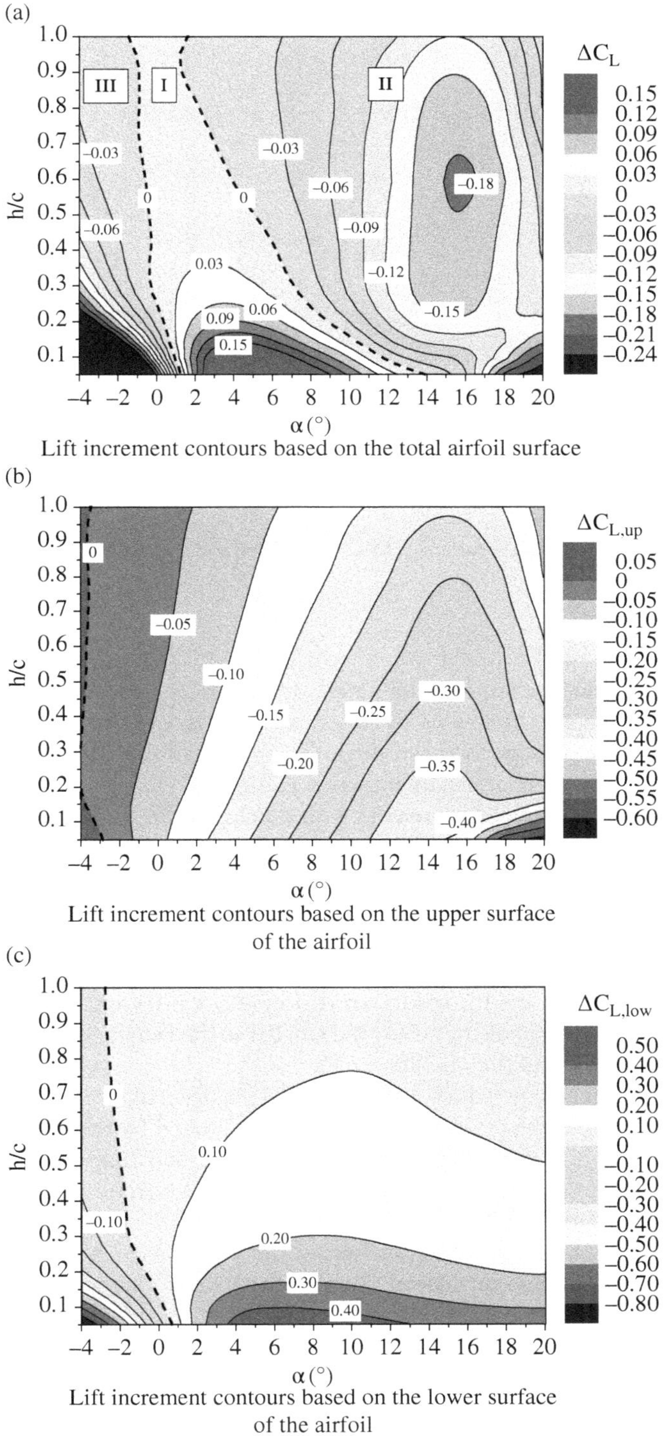

Figure 6.3 Lift increment contours of the NACA4412 airfoil in the AOA–ride-height plane. (*See insert for color representation of this figure.*)

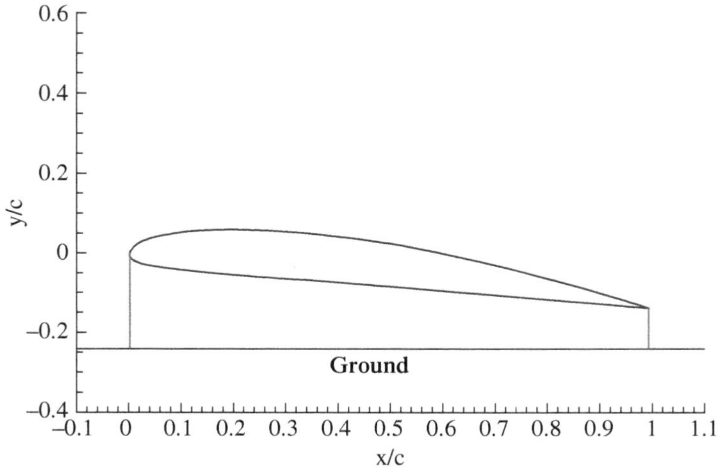

Figure 6.4 Convergent passage between the lower surface of NACA4412 airfoil and the ground at $\alpha = 8°$.

6.3.3 Effect of Low-to-moderate AOA

Lift and Pressure Distribution

For the low-to-moderate AOA range, when the ride height decreases, the airfoil enters Region I from Region II (Figure 6.3a). In this AOA range, we analyze the two typical cases with $\alpha = 4°$ and $8°$, where the passage between the lower surface of the airfoil and the ground forms a typical convergent channel, as shown in Figure 6.4.

Figure 6.5 shows the variations in lift and lift increment with ride height for $\alpha = 4°$ and $8°$. In the following analysis, the ride height of $h/c = 10$ represents the unbounded flow.

In the unbounded flow, $C_{L,up}$ is much greater than $C_{L,low}$. When the ride height decreases, $C_{L,up}$ decreases gradually and $C_{L,low}$ increases gradually. The smaller the ride height, the faster is the increase in $C_{L,low}$. Therefore, $C_{L,low}$ becomes larger than $C_{L,up}$ at very small ride height.

At some value of AOA, there is a ride height that results in $\Delta C_L = 0$, which divides the lift increment curve into two regions: the lift reduction region (Region II) and the lift enhancement region (Region I). The higher the AOA, the smaller is the ride height that gives $\Delta C_L = 0$; it is $h/c = 0.6$ for $\alpha = 4°$ and $h/c = 0.3$ for $\alpha = 8°$.

From $h/c = \infty$ to the ride height for which $\Delta C_L = 0$, ΔC_L is less than 0 since $|\Delta C_{L,up}| > |\Delta C_{L,low}|$. In this region, the trend of total lift is dominated by the upper surface. From the ride height at which $\Delta C_L = 0$ to $h/c = 0.05$, ΔC_L is greater than 0 since $|\Delta C_{L,up}| < |\Delta C_{L,low}|$, and it increases dramatically. In this region, the trend of total lift is dominated by the lower surface.

Figure 6.6 shows the pressure coefficient distribution on the NACA4412 airfoil at various ride heights. For the two AOAs considered, there is hardly any change in the trend of pressure distribution with ride height. When the ride height is reduced, $C_{p,up}$ increases first quickly and then slowly, resulting in a decrease in $C_{L,up}$; $C_{p,low}$ increases first slowly and then quickly, resulting in an increase in $C_{L,low}$.

From $h/c = \infty$ to $h/c = 0.2$, the suction peak decreases continually to a minimum. When the ride height decreases, the airfoil suction peak increases slightly, creating a larger adverse pressure gradient in the chord-wise direction. This is because the stagnation point moves backwards.

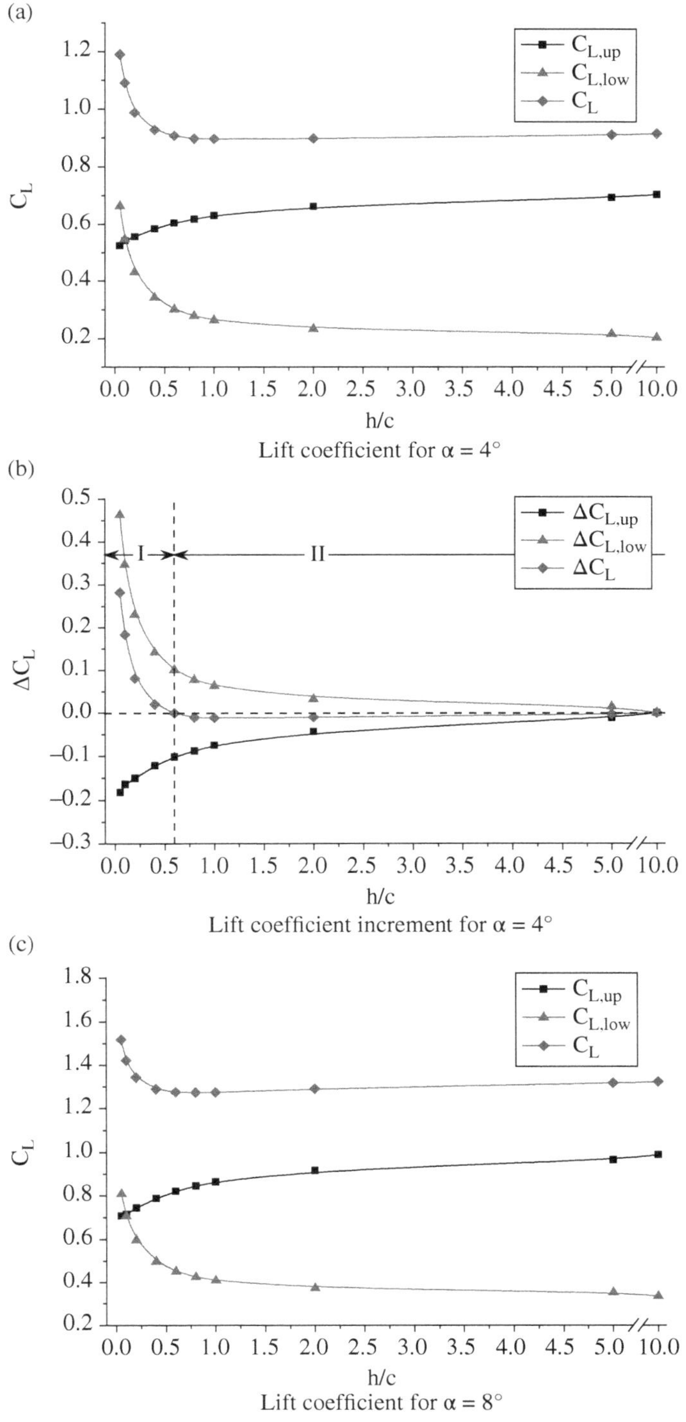

Figure 6.5 Variations in lift and lift increment with ride height.

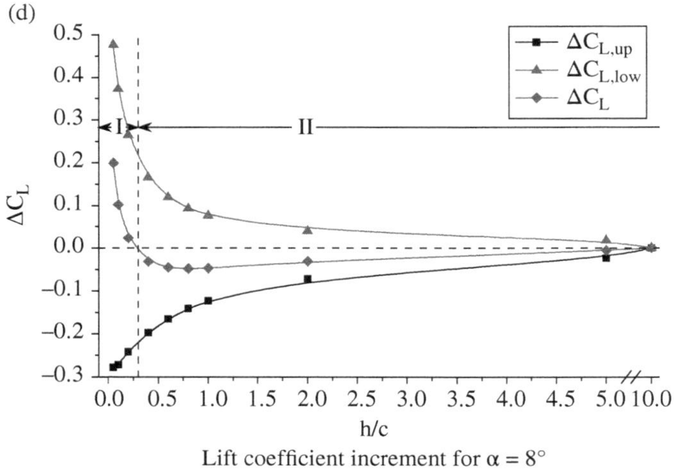

Figure 6.5 (*Continued*)

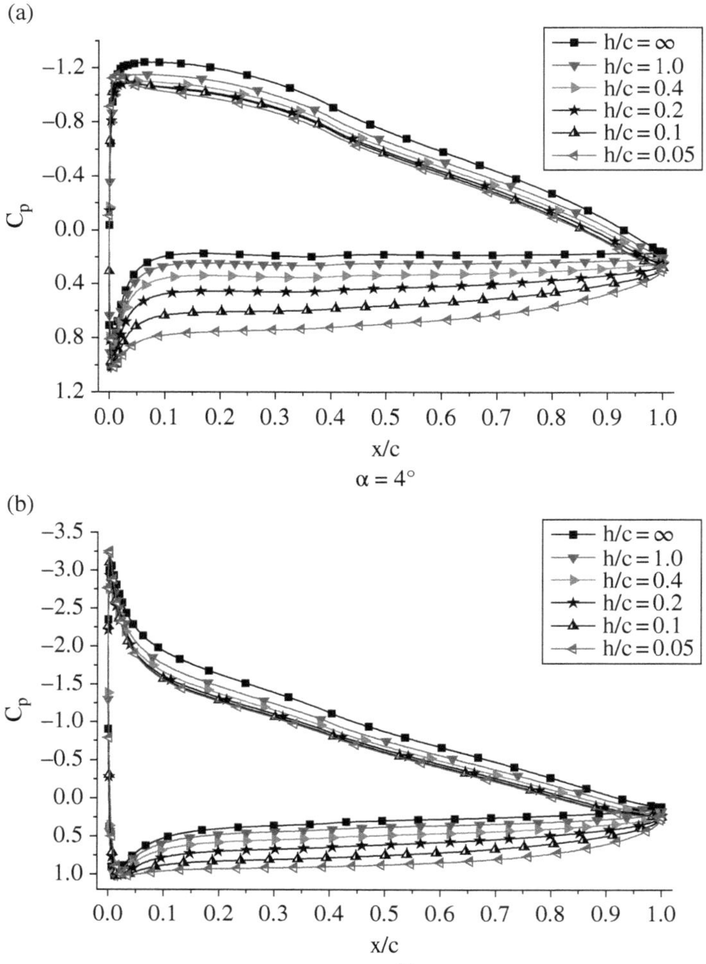

Figure 6.6 Pressure coefficient distributions on the NACA 4412 airfoil at various ride heights.

Flow Physics

Figure 6.7 shows the streamlines around the NACA4412 airfoil in the unbounded flow field and in the GE flow field with $h/c = 0.1$. There are two special streamlines in the flow field around an airfoil. One terminates at the stagnation point on the airfoil and the other begins from the trailing edge of the airfoil. The two streamlines are collectively referred to as the 'stagnation streamline', which divides the upstream inflow into two branches: one goes over the airfoil and the other goes under the airfoil. In the unbounded flow field and in the GE flow field, the stagnation streamlines are labeled as F1 and G1 respectively.

The stagnation streamline extends upstream to the location $x/c = -40$. Another two streamlines located above and under the stagnation streamline at an equal distance of $d_{in} = 0.1c$ from it are released from the location of $x/c = -40$. These are labeled as (F0, F2) and (G0, G2) for the unbounded flow field and the GE flow field respectively (Figure 6.7). Note that the streamline G2 is coincident with the ground for the GE flow field of $h/c = 0.1$.

For the AOA producing positive lift, the airfoil induces the surrounding streamlines to deflect upwards. In the unbounded flow field, the streamline deflection region is very large in the whole flow field. However, in the GE flow field, this region is much smaller and is only in the vicinity of the airfoil. The streamlines G0–G2 remain straight until they approach the leading edge of the airfoil.

Figure 6.8 shows the deflection of the stagnation streamline with ride height. When the ride height decreases, the position where the deflection begins moves towards the airfoil.

In order to explain the reduction in streamline defection in GE flow field, the mirror-image model (Mook and Nuhait 1989) for positive GE is introduced. This is shown in Figure 6.9. The clockwise circulation around the airfoil is defined as $-\Gamma_I$ and the anticlockwise image circulation below the ground is defined as Γ_I. In the upstream and downstream regions, the image circulation Γ_I induces downwash and upwash flow respectively, which makes the streamlines approach the ground.

In GE, the reduction in streamlines' upward deflection produces two results: one of reducing the effective AOA and the other of blocking the airflow below the airfoil.

Figure 6.10 shows the velocity vector angle (defined as the angle between the velocity vector and the positive direction of the x-axis.) distribution on the stagnation streamline at $\alpha = 8°$. When the ride height is reduced, the velocity vector angle decreases, resulting in the reduction of effective AOA. The smaller the ride height, the smaller is the effective AOA.

In both the unbounded flow field and the GE flow field, the streamlines (F1 and F2) and (G1 and G2) respectively form a typical stream tube below the NACA4412 airfoil. Figure 6.11 shows the width of the stream tube along the flow direction. The tube width almost retains the same value from the inlet to $x/c = -1$, and then increases sharply to a maximum at $x/c \approx 0$ (corresponding to the leading edge). After that, the width decreases quickly until $x/c = 1$ (corresponding to the trailing edge), and then decreases slowly to the inlet value at $x/c = 2$.

It is clear that the width of the stream tube below the airfoil in the GE flow field is much larger than that in the unbounded flow field. This is because the airflow is blocked in the convergent passage below the airfoil, so the velocity decreases and the pressure rises. The smaller the ride height, more airflow is blocked.

In GE, the effective AOA decreases, which would make $C_{p,up}$ increase and $C_{p,low}$ decrease. The airflow is blocked, which would make $C_{p,low}$ increase. The combination of the two effects makes both $C_{p,up}$ and $C_{p,low}$ increase.

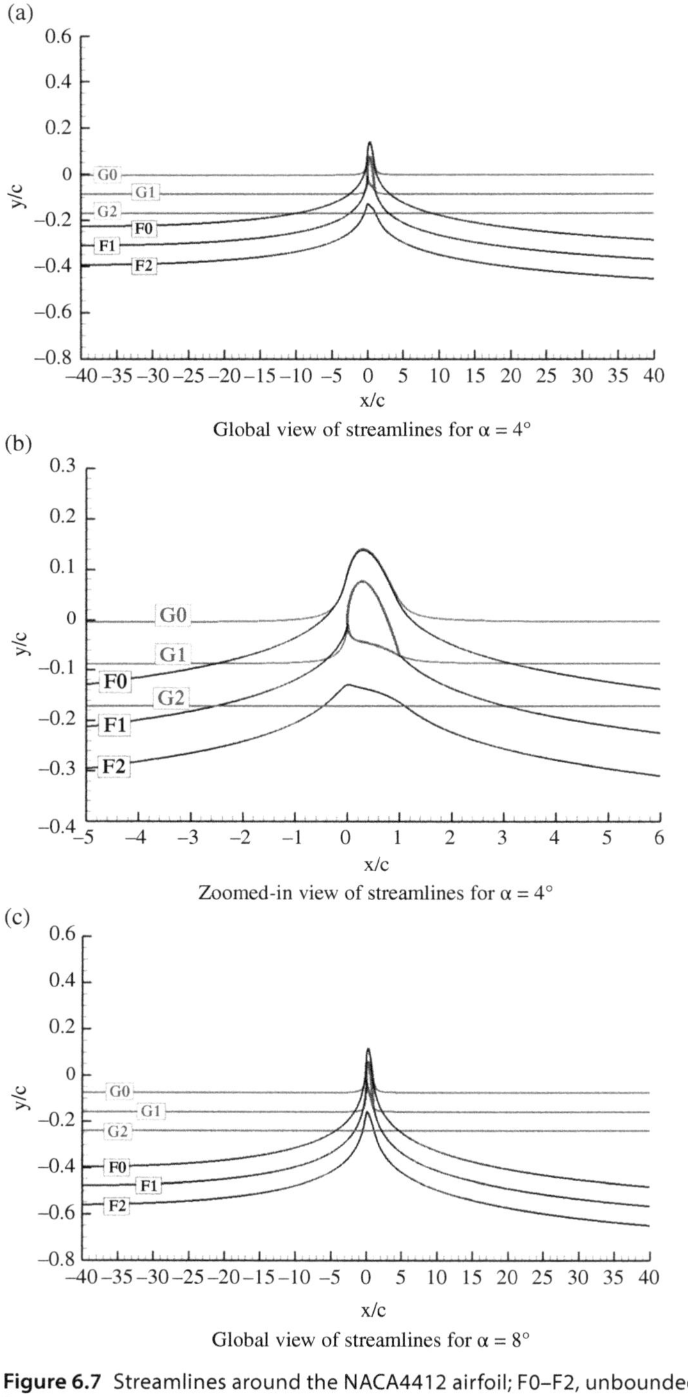

Figure 6.7 Streamlines around the NACA4412 airfoil; F0–F2, unbounded flow, G0–G2, GE with $h/c = 0.1$.

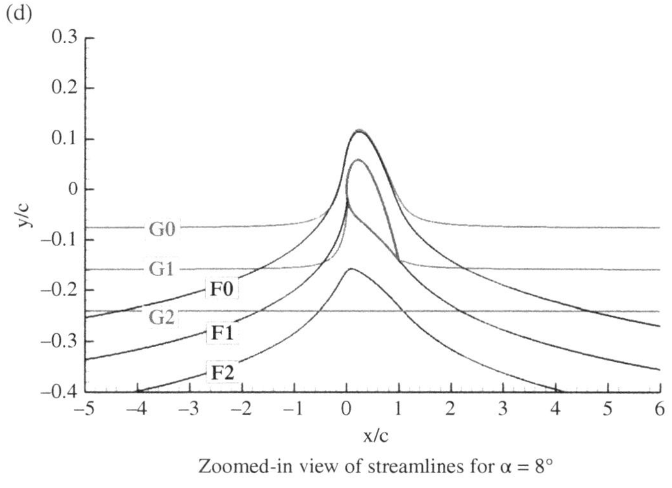

Figure 6.7 (*Continued*)

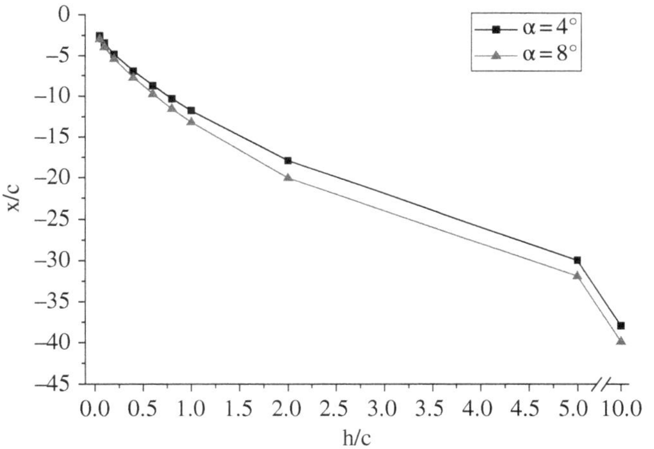

Figure 6.8 Deflection position of the stagnation streamline with ride height.

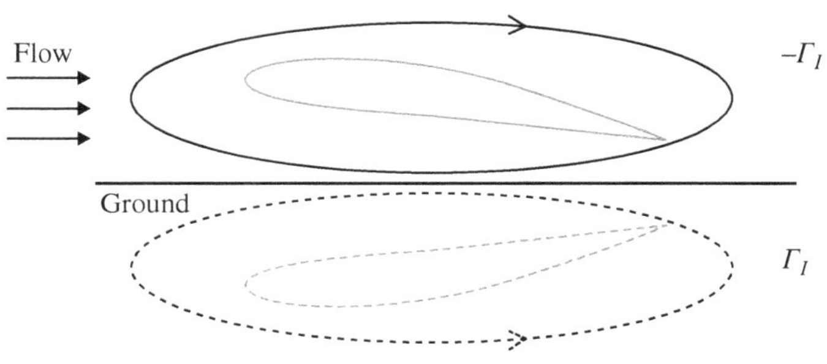

Figure 6.9 Mirror-image model of positive GE.

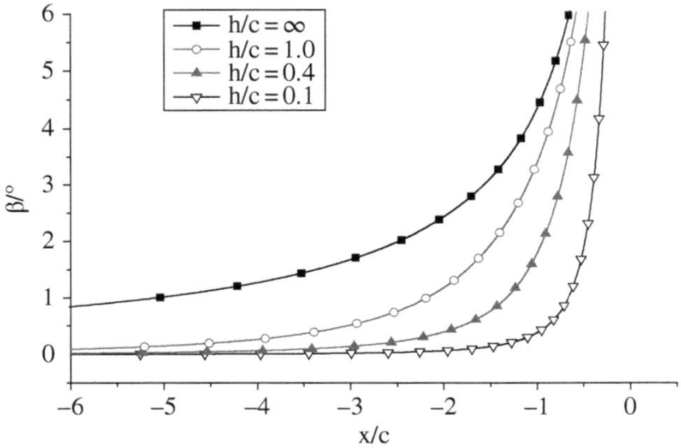

Figure 6.10 Velocity vector angle distribution on the stagnation streamline at $\alpha = 8°$.

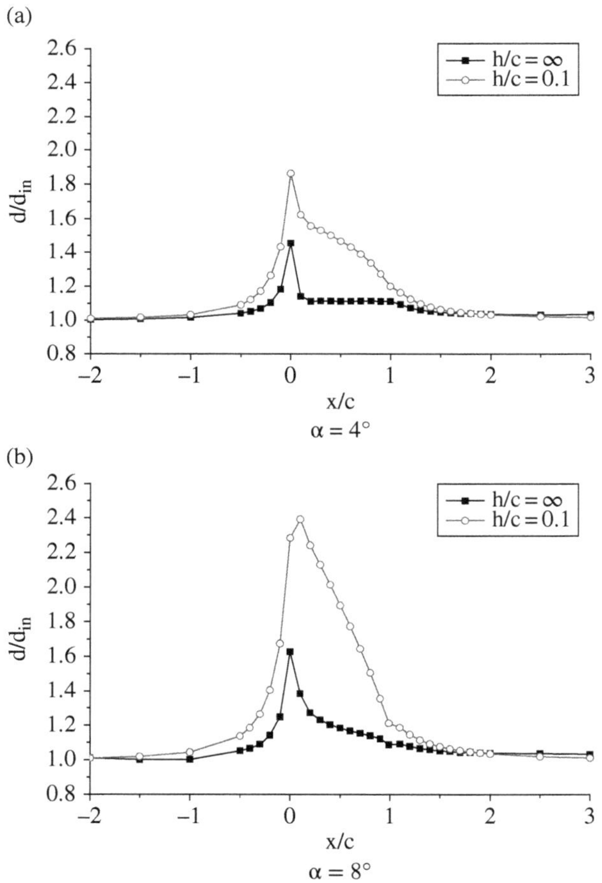

Figure 6.11 Width of the stream tube below the NACA4412 airfoil along the flow direction.

When the ride height decreases from $h/c = \infty$ to the ride height for which $\Delta C_L = 0$, the reduction in effective AOA dominates, thus $|\Delta C_{L,up}| > |\Delta C_{L,low}|$ and the total lift decreases, resulting in negative GE. When the ride height decreases from the ride height for which $\Delta C_L = 0$ to $h/c = 0.05$, the airflow-blocking effect dominates, thus $|\Delta C_{L,up}| < |\Delta C_{L,low}|$ and the total lift increases, resulting in positive GE.

6.3.4 Effect of High AOA

Lift and Pressure Distribution

In the high-AOA range located in Region II (Figure 6.3a), we analyze the two typical cases with $\alpha = 16°$ and $18°$, for which the passage between the lower surface of the airfoil and the ground is still a typical convergent channel. Figure 6.12 shows the variation of lift increment with ride height. When the ride height is reduced, $\Delta C_{L,up}$ decreases and

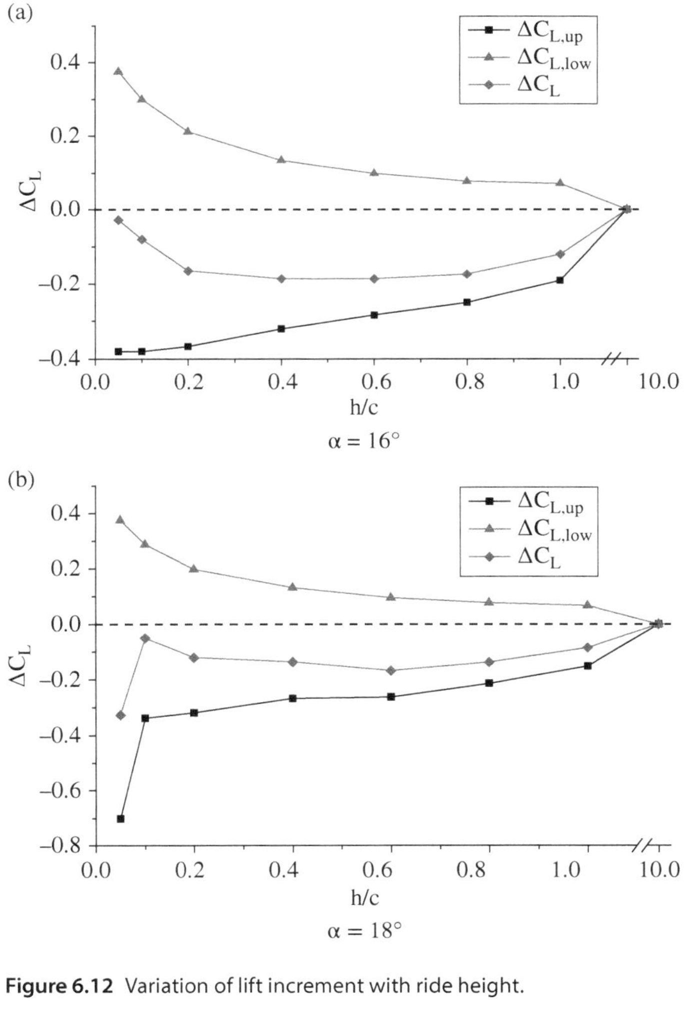

Figure 6.12 Variation of lift increment with ride height.

$\Delta C_{L,low}$ increases. $|\Delta C_{L,up}| > |\Delta C_{L,low}|$ in the entire ride height region and therefore the total lift increment is negative, resulting in negative GE. Note that $\Delta C_{L,up}$ decreases sharply from $h/c = 0.1$ to 0.05 for $\alpha = 18°$; this is because the separated flow suddenly extends to the leading edge of the airfoil (see Figure 6.18).

Figure 6.13 shows the pressure coefficient distribution on the NACA4412 airfoil at various ride heights. When the ride height is reduced, $C_{p,up}$ increases quickly and $C_{p,low}$ increases slowly.

In the unbounded flow, when AOA increases, $C_{p,low}$ increases and $C_{p,up}$ decreases. The maximum value of $C_{p,low}$ is limited to 1, but the minimum value of $C_{p,up}$ is unrestricted. Thus the change in $C_{p,up}$ is larger than that in $C_{p,low}$.

In GE, when AOA increases, the pressure increment of the lower surface due to the airflow-blocking effect decreases and the suction decrement of the upper surface due to

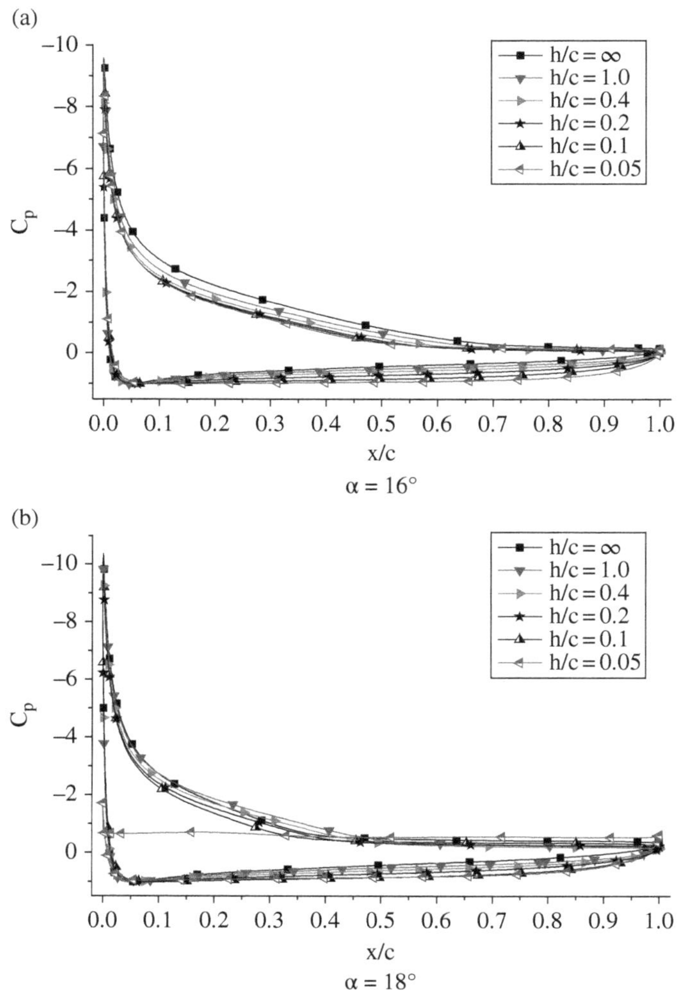

Figure 6.13 Pressure coefficient distributions on the NACA 4412 airfoil at various ride heights.

the reduction of effective AOA increases. Thus the positive GE in Region I evolves into the negative GE in Region II.

There is an obvious pressure plateau region near the trailing edge of the upper surface of the airfoil indicating flow separation. When the ride height is reduced, the pressure plateau region extends towards the leading edge. For the case of $\alpha = 18°$ and $h/c = 0.05$, nearly the entire upper surface is occupied by the pressure plateau region, which results in a steep reduction in upper surface lift and the total lift due to sharp stall characteristics.

Flow Physics

Figure 6.14 shows the streamlines around the NACA4412 airfoil in the unbounded flow field and the GE flow field at $h/c = 0.1$. Figure 6.15 shows the width of the stream tube below the airfoil along the flow direction. The main characteristics of the streamline deflection and the stream tube width are similar to that for the low-to-moderate AOA cases. However, it is a qualitative analysis (based on less accurate quantitative information) due to separated flow at high AOA.

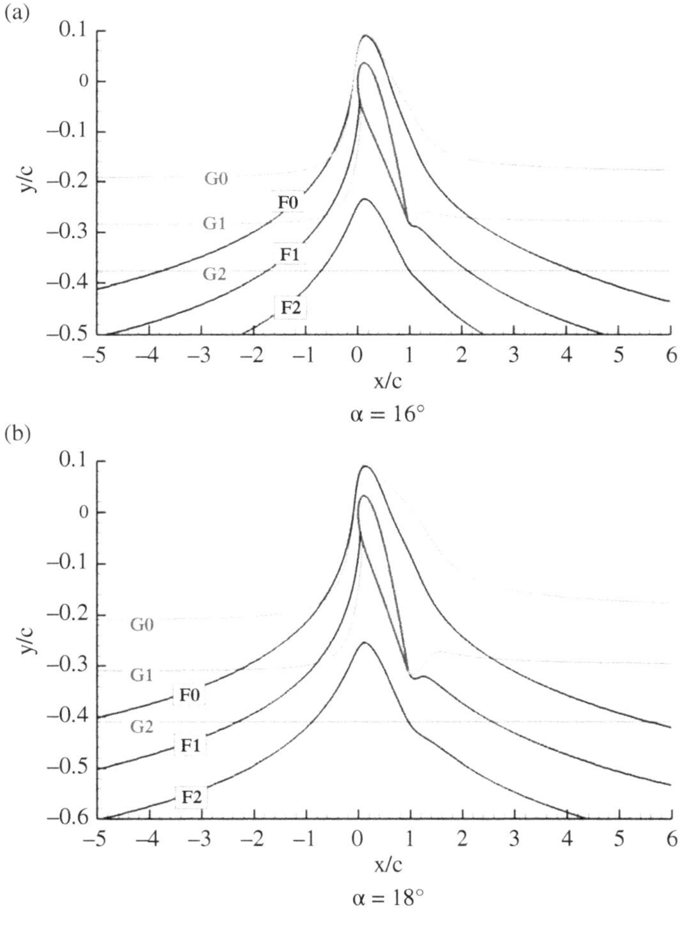

Figure 6.14 Streamlines around the NACA4412 airfoil; Fo–F2, unbounded flow, G0–G2, GE with $h/c = 0.1$.

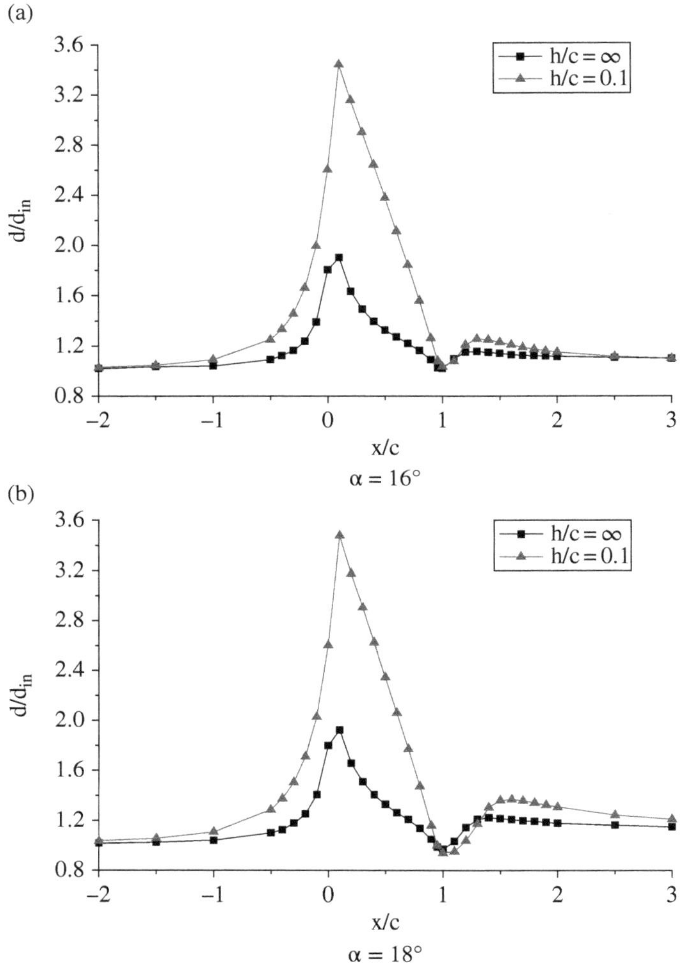

Figure 6.15 Width of the stream tube below the NACA4412 airfoil along the flow direction.

For low-to-moderate AOA, the upper surface flow remains attached in both the unbounded flow field and the GE flow field, so the streamlines G0 and F0 are almost coincident above the airfoil (see Figure 6.7). For high AOA, however, the upper surface separated flow region in the GE flow field is greater than that in the unbounded flow field, so the streamline G0 is located above F0 in the vicinity of the airfoil.

Figure 6.16 shows the steady separated flow pattern on the upper surface of the NACA4412 airfoil. When the ride height is reduced, the separation point moves towards the leading edge (Figure 6.17) and the separation region is enlarged since the adverse pressure gradient along the chord direction increases.

From Figure 6.17, it can be noted that the change in the location of the separation point at $\alpha = 18°$ is very sharp as the ride height changes from $h/c = 0.1$ to 0.05. At $h/c = 0.05$, the separation point is located close to the leading edge and the separated flow is periodic. Figure 6.18 shows the unsteady separated flow in one period of oscillation. In the

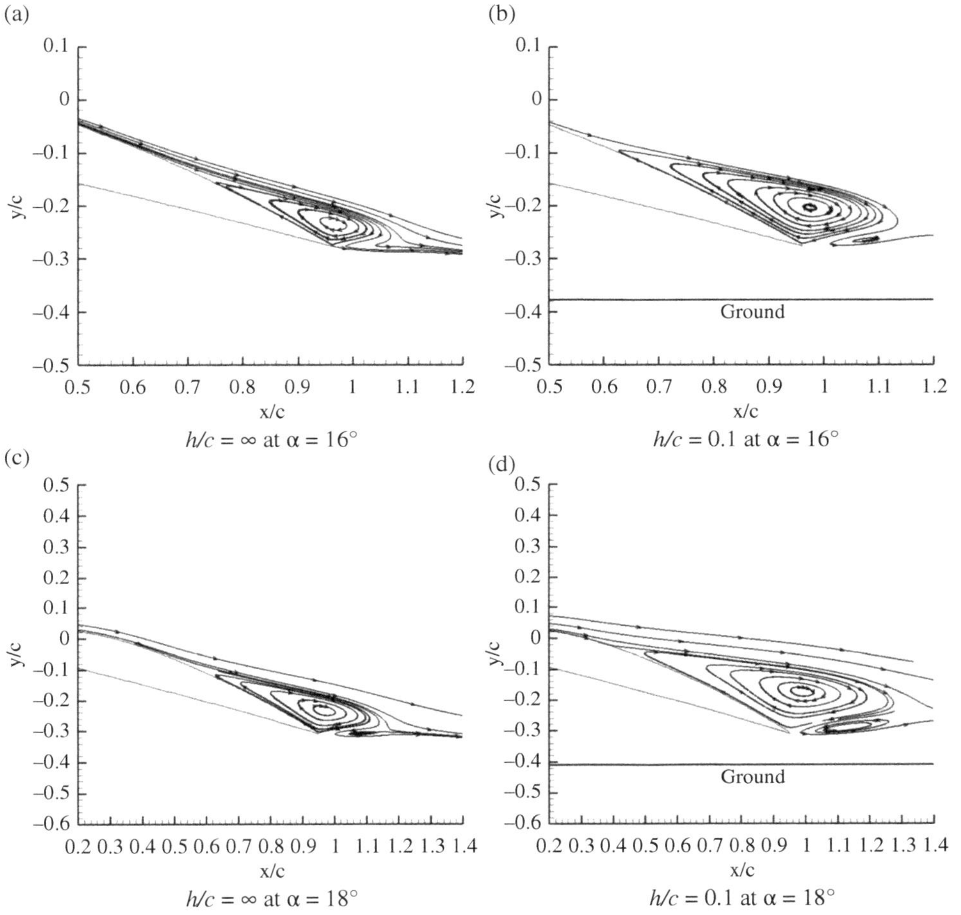

Figure 6.16 Steady separated flow region on the upper surface of the NACA4412 airfoil.

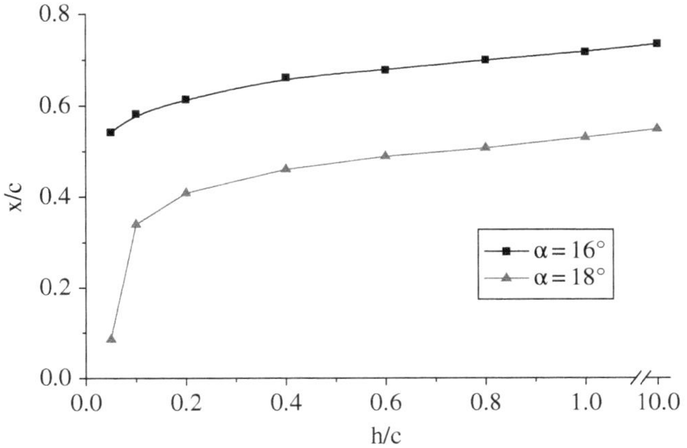

Figure 6.17 Position of separation point as a function of ride height.

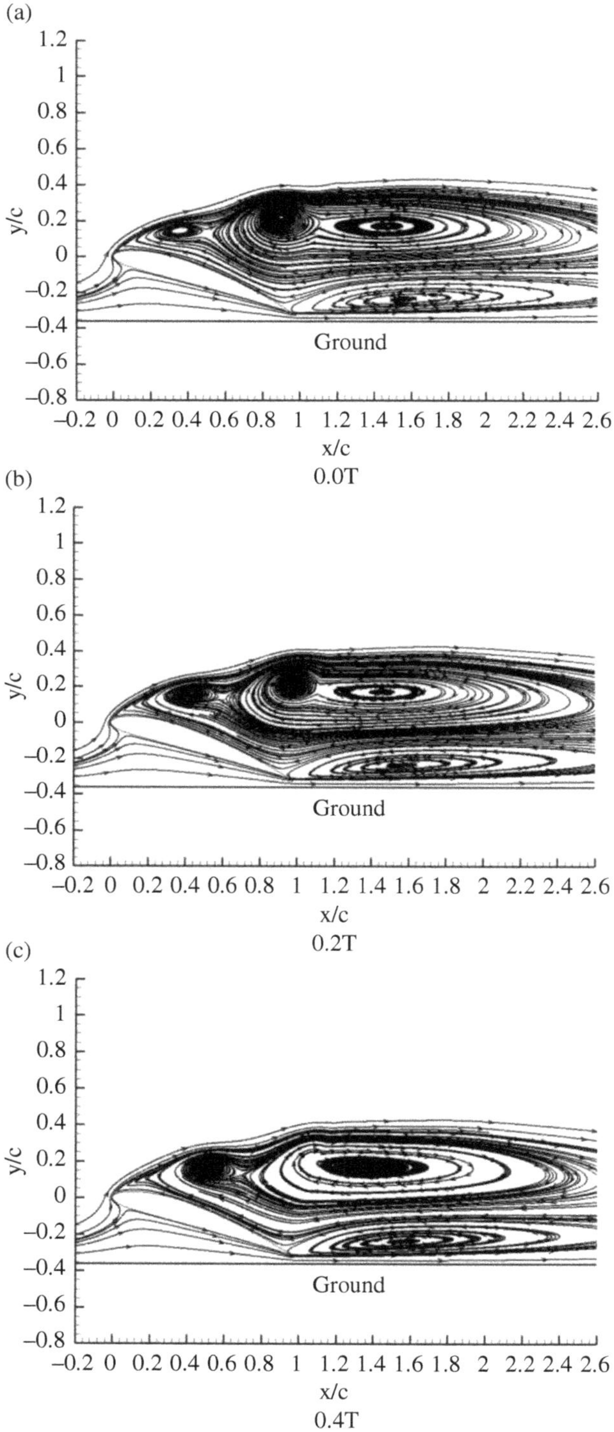

Figure 6.18 Unsteady separated flow on the upper surface of NACA4412 airfoil at $\alpha = 18°$ and $h/c = 0.05$.

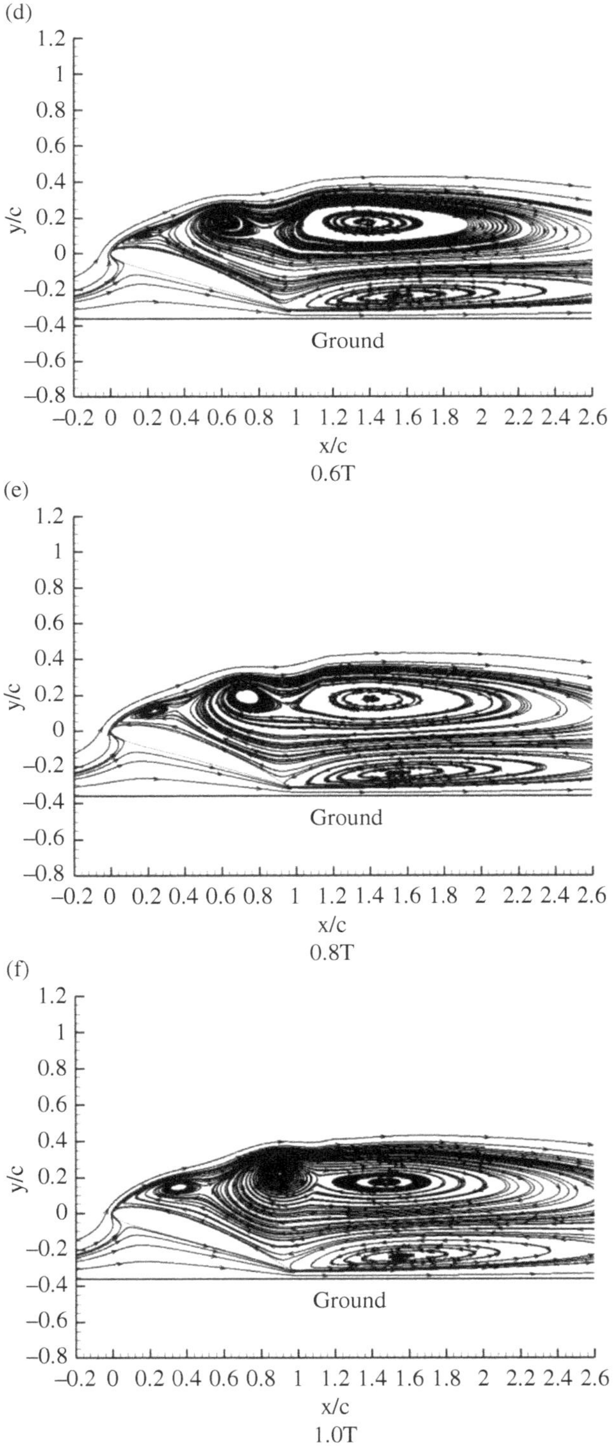

Figure 6.18 *(Continued)*

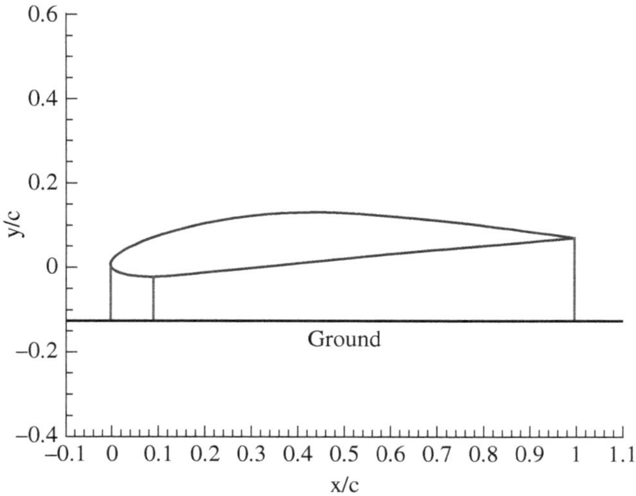

Figure 6.19 Convergent-divergent passage between the lower surface of the NACA4412 airfoil and the ground at $\alpha = -4°$.

separation region, vortices are constantly shed from the leading edge and then merge downstream. The entire upper surface is occupied by the separated flow, resulting in a pressure plateau region on the entire upper surface of the airfoil (Figure 6.13b).

6.3.5 Effect of Negative AOA

Lift and Pressure Distribution

In the negative AOA range located in Region III (Figure 6.3a), we analyze the typical case of $\alpha = -4°$ where the passage between the lower surface of the airfoil and the ground is a typical convergent-divergent channel like a 2D Venturi tube (as shown in Figure 6.19). Thus the airflow first accelerates to a maximum velocity at the throat section and then decelerates gradually, resulting in a greater suction peak at the throat section.

Figure 6.20 shows the variations in lift and lift increment with ride height. When the ride height is reduced, $C_{L,up}$ initially does not change and then increases slightly ($h/c \leq 0.1$), and $C_{L,low}$ decreases first gradually and then dramatically. Thus the trend in total lift is determined by the lower surface, so C_L decreases with reduced ride height.

Figure 6.21 shows the pressure coefficient distribution on the NACA4412 airfoil at various ride heights for $\alpha = -4°$. It is interesting to note that the suction region on the lower surface is apparent due to the Venturi effect. When the ride height is reduced, $C_{p,up}$ initially does not change and then decreases slightly, and thus the suction peak on the lower surface increases and the suction region is enlarged.

Flow Physics

Figure 6.22 shows the streamlines around the NACA4412 airfoil at $\alpha = -4°$. The lift is negative in the entire ride height range (see Figure 6.20a), so the airfoil induces the surrounding streamlines to deflect downwards, the opposite of what was seen in the above cases producing positive lift at positive AOA. The reduction in streamlines' downward deflection in GE increases the effective AOA.

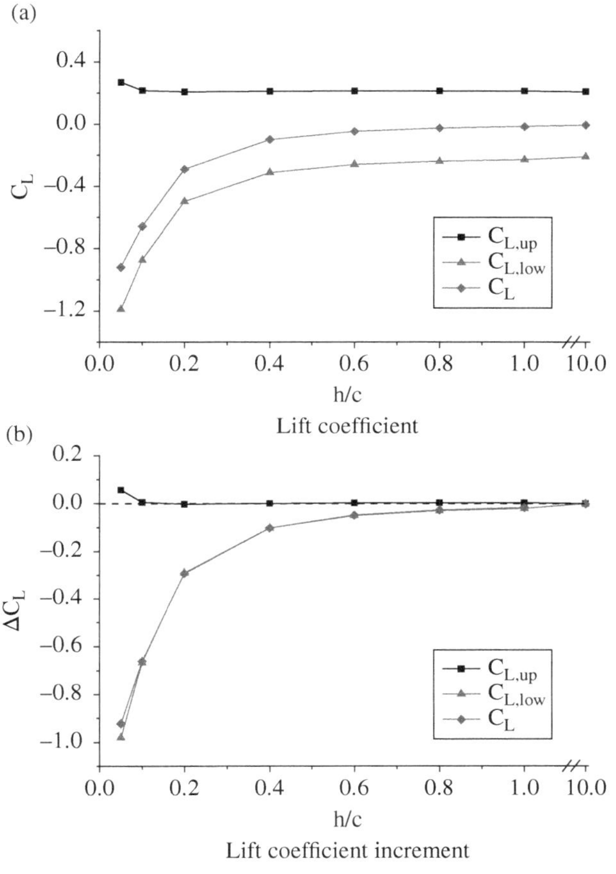

Figure 6.20 Variations in lift and lift increment with ride height at $\alpha = -4°$.

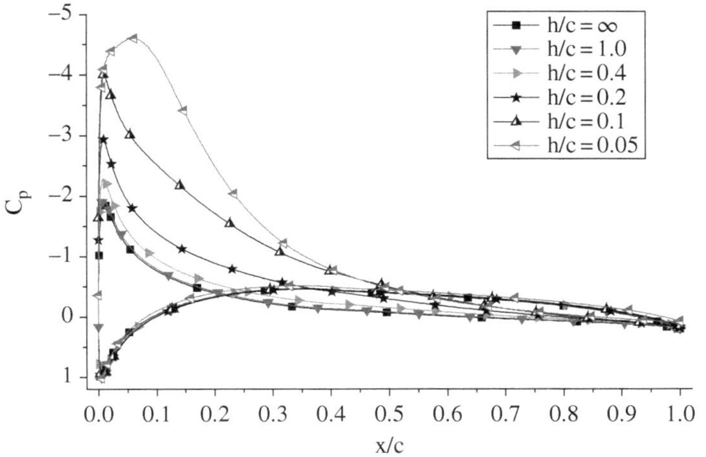

Figure 6.21 Pressure coefficient distributions on the NACA 4412 airfoil at various ride heights for $\alpha = -4°$.

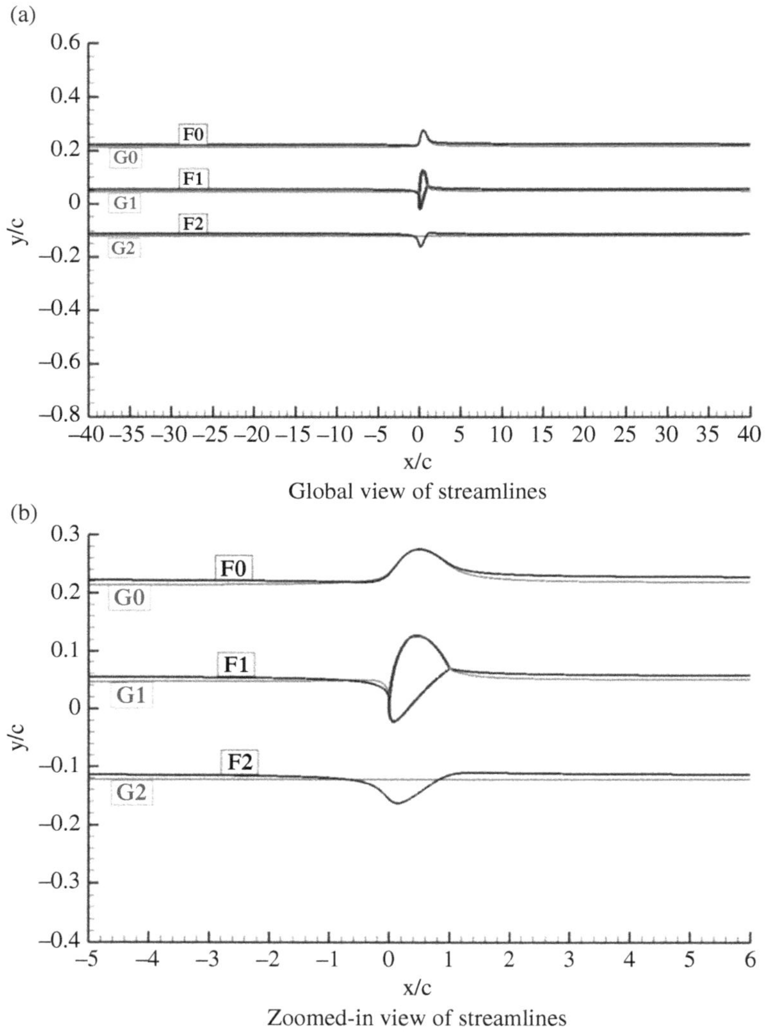

Figure 6.22 Streamlines around the NACA4412 airfoil at $\alpha = -4°$; F0–F2, unbounded flow, G0–G2, GE with $h/c = 0.1$.

The mirror-image model of negative GE for $\alpha = -4°$ is shown in Figure 6.23. Since the case of $\alpha = -4°$ in the unbounded flow generates negative lift, the circulation around the airfoil is anticlockwise and is defined as Γ_{III}; the image circulation below the ground is clockwise and is defined as $-\Gamma_{III}$. In the upstream and downstream regions, the image circulation induces upwash and downwash flow respectively, which makes the streamlines approach the ground.

Figure 6.24 shows the width of the stream tube below the airfoil along the flow direction at $\alpha = -4°$. Because the airfoil generates negative lift, the stream tube becomes narrow in the vicinity of the airfoil. The tube width in GE is much smaller than in the unbounded flow due to the Venturi effect, so the velocity increases and the pressure decreases. The smaller the ride height, more apparent is the Venturi effect.

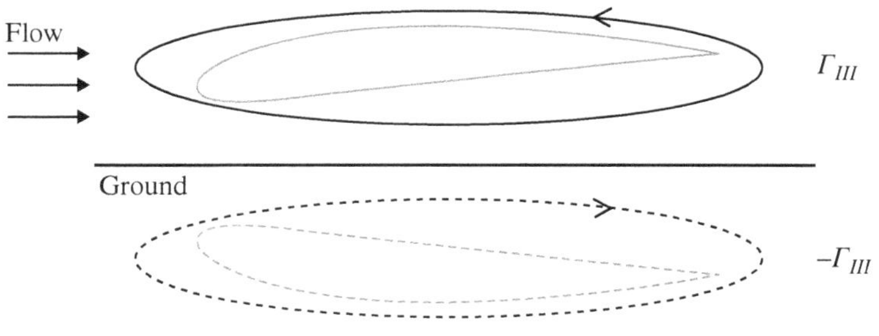

Figure 6.23 Mirror-image model of negative GE.

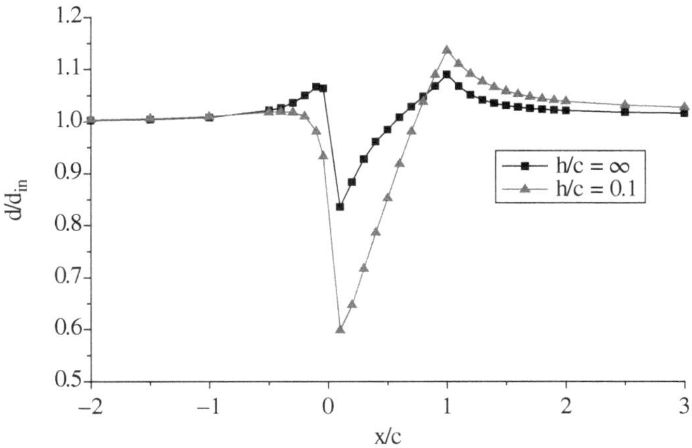

Figure 6.24 Width of the stream tube below the NACA4412 airfoil along the flow direction at $\alpha = -4°$.

In GE at $\alpha = -4°$, the effective AOA increases, which makes $C_{p,up}$ decrease (here the upper surface is the pressure side), especially at small ride height, and $C_{p,low}$ increase (here the lower surface is the suction side). The Venturi effect is also enhanced, which makes $C_{p,low}$ decrease dramatically. Since the increment effect of effective AOA is much weaker than the Venturi effect, the trend of total lift is decided by the Venturi effect.

6.3.6 Conclusions on Chord-dominated Static Ground Effect

The AOA–ride-height plane can be divided into three regions based on the sign of the lift increment value: Region I of positive GE and Regions II and III of negative GE.

For low-to-moderate AOA, when the ride height decreases, the airfoil enters Region I from Region II, and the pressure on the lower surface of the airfoil increases due to the airflow-blocking effect from the convergent passage between the lower surface and the ground and the pressure on the upper surface increases due to the reduction of effective AOA as a result of reduction in streamlines' upward deflection. In Region II, the airflow-blocking effect is weaker than the reduction effect in effective AOA resulting in a negative lift increment; in Region I, the airflow-blocking effect is stronger than the reduction effect in effective AOA resulting in a positive lift increment.

For high AOA located in Region II, when the ride height is reduced, the separation point moves towards the leading edge of the airfoil and the separated flow region is enlarged since the adverse pressure gradient along the chord direction increases. The pressure increment on the upper surface due to the enlarged flow separation region is greater than the pressure increment on the lower surface due to the airflow-blocking effect; therefore the lift increment is negative.

For negative AOA located in Region III, when the ride height is reduced, the suction on the lower surface increases due to the Venturi effect from the convergent-divergent passage between the lower surface and the ground. The Venturi effect is stronger than the increment effect of effective AOA due to reduced downward deflection of streamlines; therefore the lift increment is negative.

6.4 Chord-dominated Dynamic Ground Effect

During take-off and landing, an aircraft respectively moves away or towards the ground and its ride height above the ground changes continuously with time, so the air flow around the aircraft is unsteady. In the past, the majority of investigators have treated the DGE as quasi-steady, but recent studies (Matsuzaki et al. 2008, Molina and Zhang 2011) indicate that the unsteady flow can be considered as quasi-steady only when the sinking rate or the climbing rate of the aircraft are very small. However, the sink rate is usually large during emergency landings. Therefore, the study of the DGE at higher sink rates is of considerable interest in efforts to ensure flight safety during take-off and landing.

Regarding chord-dominated DGE, Chen and Schweikhard (1985) and Nuhait and Zedan (1993) studied a two-dimensional flat plate moving towards the ground using the vortex-lattice method. Chen and Schweikhard assumed that the wake was straight along the flight path, while Nuhait and Zedan allowed the wake to deform freely. Both studies have indicated that the lift in the DGE is larger than that in the SGE, and the difference becomes larger with larger sink rate. Matsuzaki et al. (2008) conducted a wind tunnel test of a NACA6412M airfoil in the DGE using a fixed ground plane. The airfoil was set to move perpendicular to the free stream, upward as well as downward, at a constant velocity. The results showed that, compared with the steady-state case at the same height and the same effective AOA, the lift in the sinking case was smaller. Contrary results were found in climb. However, Matsuzaki et al. did not measure the flow field in detail, so these differences could not be easily explained by changes in the flow physics.

Here, the DGE of a NACA4412 airfoil moving towards the ground (Qu et al. 2014a; Qu et al. 2014b) is discussed.

6.4.1 Physical Model

A NACA4412 airfoil with $c = 0.15\,m$, $V_\infty = 30.8\,m/s$ and $Re = 3 \times 10^5$ is investigated. Other parameters are shown in Table 6.1. The airfoil moves towards the ground with a vertical component of the flight velocity; the ground and the inflow from upstream moves at the opposite of the horizontal component of the flight velocity.

The formula for AOA in the DGE is:

$$\alpha = \theta - \gamma \tag{6.8}$$

Table 6.1 Parameters used in the SGE and DGE.

SGE	
Angle of attack α (°)	0, 2, 4, 6, 8, 10
Ride height *h/c*	1.00, 0.80, 0.60, 0.40, 0.30, 0.20, 0.15, 0.10, 0.05
DGE	
Pitch angle θ (°)	2, 4, 6, 8
Sink rate velocity Vs (m/s)	1.075, 2.154, 3.237
Flight path angle γ (°)	−2, −4, −6

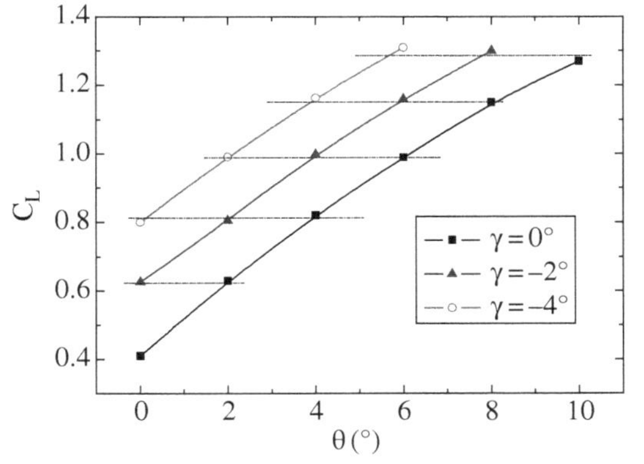

Figure 6.25 Lift coefficients at different sink rates in an unbounded flow field.

6.4.2 Lift Characteristics

Intuitively, the DGE is the coupling of two phenomena: the SGE and the dynamic effect due to sinking. For the dynamic effect in an unbounded flow field, in other words the airfoil descending perpendicular to the direction of the free-stream, the lift coefficients at different sink rates are plotted in Figure 6.25. It can be seen from this graph that C_L vs. θ curves for $\gamma = -2°$ and $-4°$ are similar and parallel to the static C_L curve ($\gamma = 0°$) with lateral displacement of 2° and 4° to the left, respectively. Thus the lift of the descending airfoil is equal to that of a static airfoil with a same AOA. Therefore, the dynamic effect can be explained by the airfoil incidence effect in an unbounded flow field.

Figures 6.26 and 6.27 show the variation in lift coefficient with ride height at different sink rates. The lift coefficients in both the SGE and DGE increase as the ride height *h/c* decreases (except for the case with $\alpha = 0°$ in the SGE, where the lift decreases as *h/c* decreases). This is because of the Venturi tube that forms between the ground and the lower surface of the airfoil.

As can be seen from Figures 6.26 and 6.27, for $h/c \geq 0.5$, C_L in the DGE is equal to C_L in the SGE for the same AOA. When the ride height is below $h/c = 0.5$, the lift coefficient in the DGE is larger than in the SGE and the difference between the two becomes larger with the decrease in ride height and the increase in sink rate.

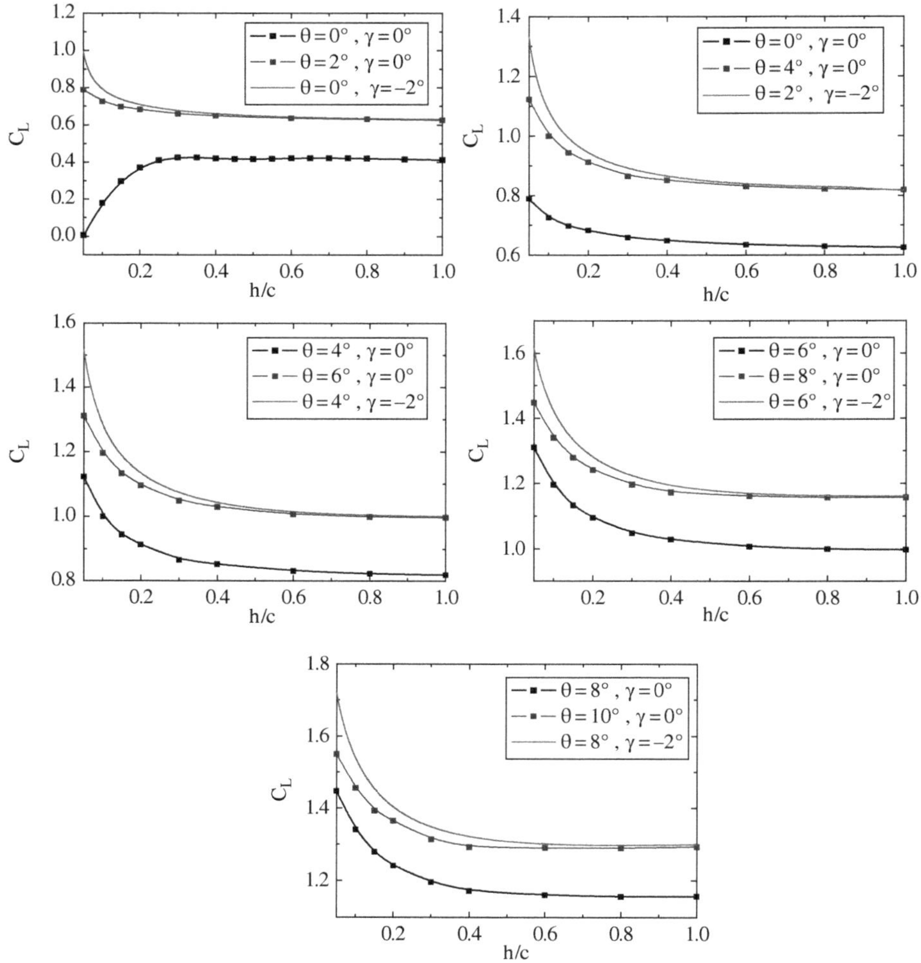

Figure 6.26 Variation of C_L with ride height for various values of θ and $\gamma = 0$ and $-2°$.

The relationship between the lift coefficient and the ride height in the DGE can be divided into three regions (see Figure 6.28).

1) **DGE 1:** $h/c > 1.0$, the lift in the DGE remains unchanged with decrease in h/c and is the same as that in the SGE for the same AOA. The major physics that governs the flow in this region is due to the incidence effect.
2) **DGE 2:** $0.5 < h/c < 1.0$, the lift in the DGE increases with decrease in h/c, and is nearly the same as that in the SGE for the same AOA (the former is slightly larger). The main physics governing the flow is due to both the SGE and incidence effect.
3) **DGE 3:** $h/c < 0.5$, the lift in the DGE increases dramatically with the decrease in h/c, and is much larger than that in the SGE for the same AOA. As h/c decreases, this difference becomes more significant. Three physical factors control the flow: the SGE, the incidence effect and the compression work effect (described below).

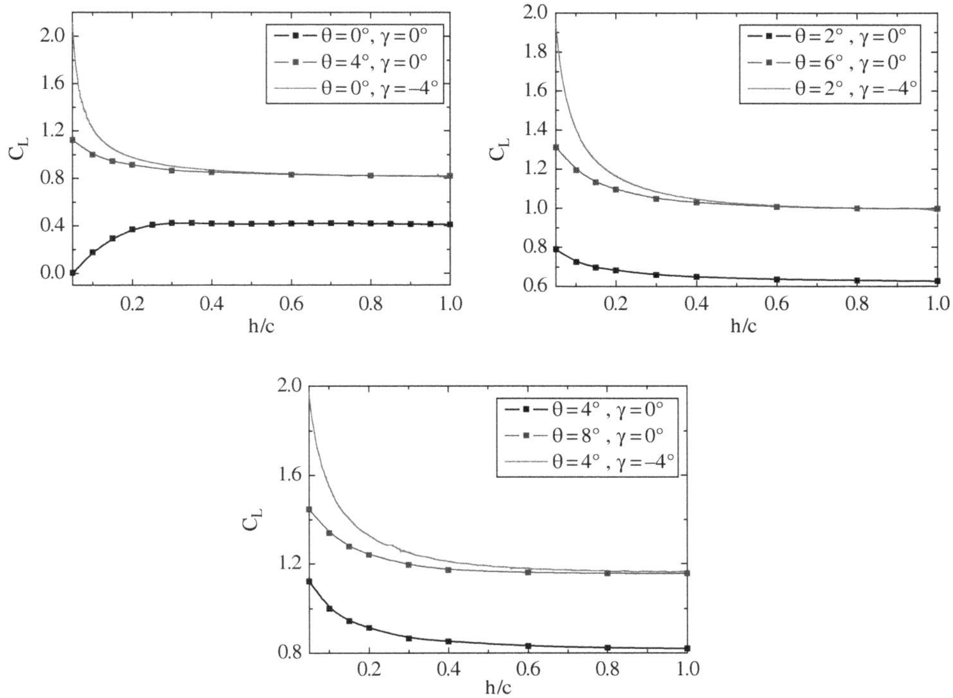

Figure 6.27 Variation of C_L with ride height for various values of θ and γ = 0 and −4°.

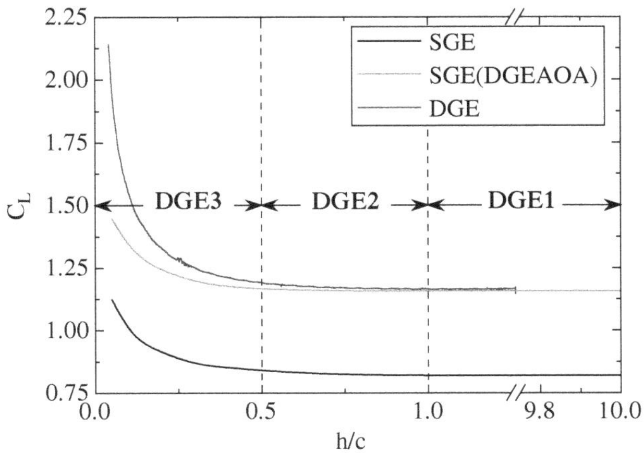

Figure 6.28 DGE regions defining the variation of lift curve with ride height SGE (DGEAOA) shows the lift curve of the SGE with the same AOA as DGE.

From the above discussion, we can conclude that when the sink rate is small, the DGE can be considered to be quasi-steady SGE. Furthermore, when the sink rate increases (the largest sink rate corresponds to $\gamma = -6°$), as long as $h/c > 0.5$, the DGE can still be treated as quasi-steady SGE with high accuracy.

6.4.3 Pressure Distribution

The pressure coefficients of NACA4412 with θ = 4°, different ride heights and different sink rates are shown in Figure 6.29. As the airfoil moves closer to the ground, C_p increases dramatically on the high-pressure surface of the airfoil. $C_p > 1$ is observed on almost the entire lower surface of the airfoil at small heights (see Figure 6.29c,d); the suction peak exhibits an increase while suction decreases slightly over other parts of the suction surface resulting in a larger adverse pressure gradient. Another aspect to note is that the increase in suction peak and lower surface pressure increases with increasing sink rate.

In many aerodynamic problems where one is primarily interested in lift, the work done by external forces such as gravity and viscous forces is neglected. For incompressible flow, the pressure coefficient at the stagnation point is 1. In this investigation, C_p is much larger than 1 on the pressure surface of the airfoil when the ride height in the DGE is very small; the explanation for this result is given below.

As shown later, in Figure 6.33, in the DGE there are two branches of airflow: the uniform inflow from upstream and the downwash flow due to the moving airfoil. These

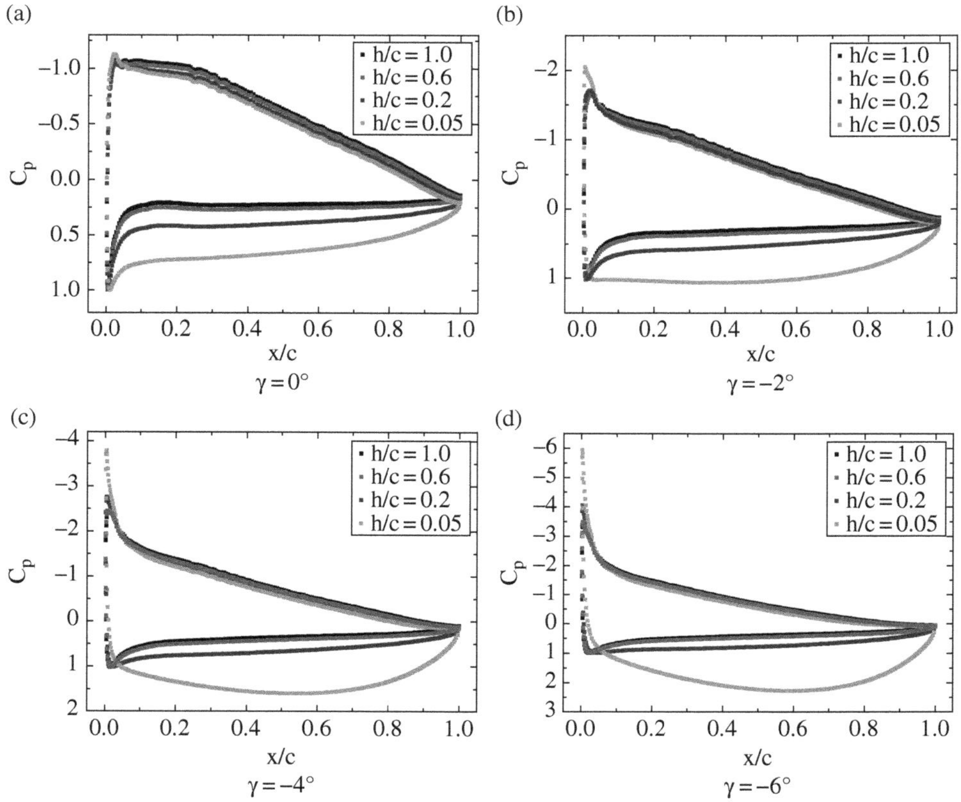

Figure 6.29 Pressure coefficient C_p for NACA4412 airfoil in the DGE at θ = 4° with different sink rates γ and ride heights *h/c*.

two streams of airflow couple into one composite flow. In the unbounded flow field, the downwash flow moves freely, while in the DGE the downwash flow is blocked by the ground and moves towards the leading edge and trailing edge of the airfoil to escape. The channel between the lower surface of the airfoil and the ground becomes narrower with decreasing *h/c*, so the downwash flow does not have sufficient space to escape and therefore is compressed. As a result, the mechanical energy of the air increases; in other words, the airfoil does compression work on the fluid below it.

In order to analyze the compression work on the air done by the airfoil, we define an area A of the channel between the lower surface of the airfoil and the ground as shown in Figure 6.30.

The total pressure p_t of an arbitrary point in A is defined as:

$$p_t = \frac{1}{2}\rho V^2 + p \tag{6.9}$$

The averaged total pressure coefficient $\overline{p_t}$ of the fluid in A with a ride height of *h/c* and flight path angle γ is defined as:

$$\overline{p_t}\left(h/c,\gamma\right) = \frac{1}{A}\iint_A p_t \mathrm{d}A \tag{6.10}$$

Figure 6.31 shows the variation of the averaged total pressure of air in A with ride height. $\overline{p_t}$ remains constant at different ride heights in DGE 1. In DGE 2, the total pressure increases slightly with decrease in h, indicating that the airfoil begins to

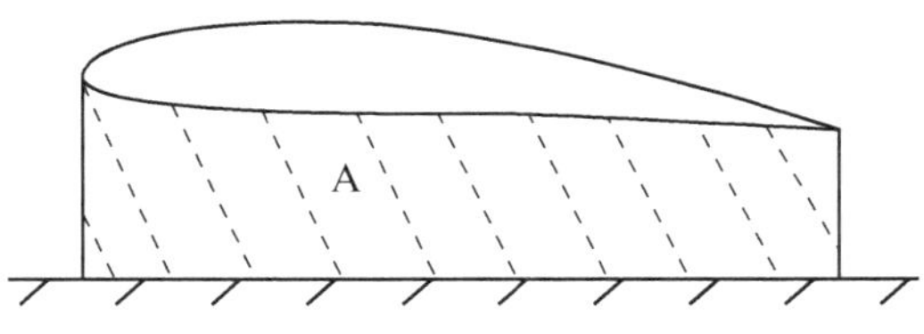

Figure 6.30 Sketch of area A between the airfoil and the ground.

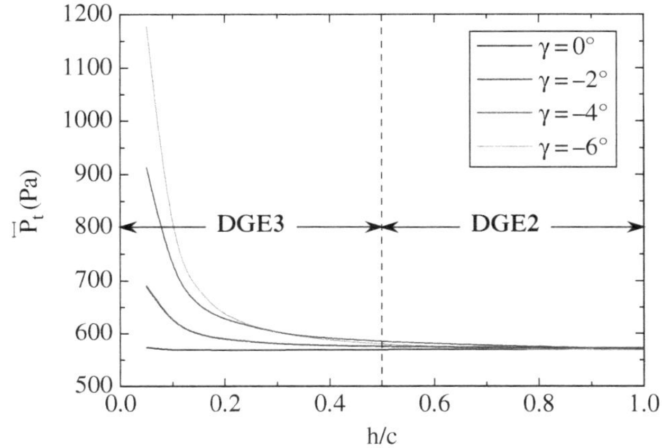

Figure 6.31 Variation of the averaged total pressure in area A of Figure 6.30 with ride height at θ = 4°.

compress the air, but the compression effect is very weak. In DGE 3, the total pressure increases rapidly and gets larger with larger sink rate indicating that the compression effect has become a significant factor.

Neglecting viscosity, the energy conservation equation for the air below the airfoil in the DGE can be approximately written as:

$$p_\infty + \frac{1}{2}\rho V_\infty^2 + W(h,\gamma) = \frac{1}{2}\rho V^2 + p \tag{6.11}$$

where $W(h, \gamma)$ is the compression work done by the airfoil and is a function of both h and γ. When the ride height decreases, the moving airfoil does more work. When the sink rate increases, the compression ratio of the air increases, resulting in larger compression forces and more compression work.

The pressure coefficient C_p on the lower surface of the airfoil in the DGE can be obtained as:

$$C_p = \frac{p - p_\infty}{\frac{1}{2}\rho V_\infty^2} = 1 - \left(\frac{V}{V_\infty}\right)^2 + \frac{W(h,\gamma)}{\frac{1}{2}\rho V_\infty^2} \tag{6.12}$$

From Eq. (6.12), it can be seen that C_p on the lower surface of the airfoil in the DGE can be greater than 1. This is because the air below the airfoil is subject to compression and thus its mechanical energy increases, resulting in an increase in static pressure.

Figure 6.32 shows C_p on the airfoil with $\theta = 4°$ and $\gamma = -4°$ in the DGE and at $\alpha = 8°$ in the SGE. The pressure coefficients in the DGE and the SGE agree quite well at $h/c = 0.6$; this figure clearly demonstrates that the major factors that govern the flow in DGE 2 are the combination of the incidence effect and the SGE. At $h/c = 0.2$, the pressure on the lower surface of the airfoil and the suction peak in the DGE are significantly larger than those in the SGE. As the ride height decreases, for example at $h/c = 0.05$, the difference becomes much greater, which indicates that in DGE 3 compression work done by the airfoil is a major factor in addition to the former two factors.

6.4.4 Flow Physics

Figure 6.33 depicts the streamlines and pressure contours around a NACA 4412 airfoil in an unbounded flow field and in ground effect at $\theta = 4°$. All streamlines are created using a velocity that is relative to the initial position of the airfoil. Figure 6.33a shows a static airfoil in uniform inflow. Compared to the unbounded flow field, the streamlines below the airfoil in the SGE are straighter, the stream tubes are broader, and thus the flow velocity is slower and the pressure is larger. This is the reason that the lift in the SGE is larger than that in the unbounded flow field.

In Figure 6.33b, there is no inflow from upstream and the airfoil descends at a constant velocity. The airfoil drives the air below it and produces a downwash flow, simultaneously forming the leading-edge and trailing-edge vortices. In the unbounded flow field, the downwash flow can develop freely and is vertically directed.

In the DGE, there is insufficient space for the downwash flow to develop and therefore the flow is compressed greatly. As a result, the pressure increases dramatically. Furthermore, the downwash flow is separated into two streams to escape from the

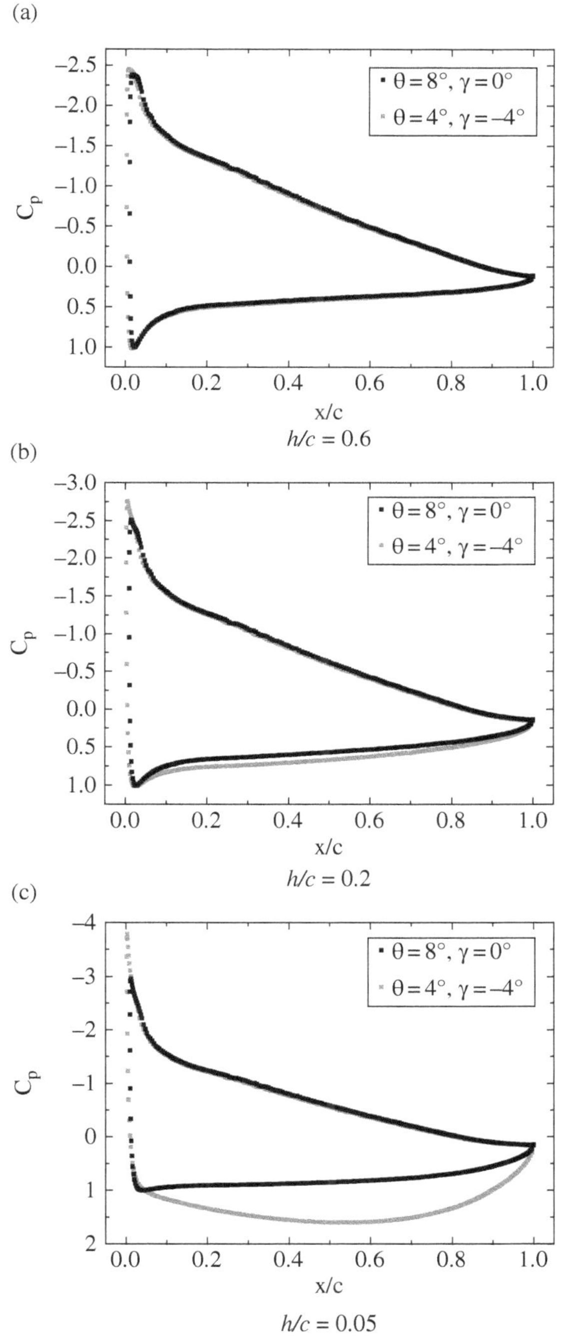

Figure 6.32 Comparison of pressure coefficients of NACA 4412 airfoil for the DGE and the SGE at various ride heights.

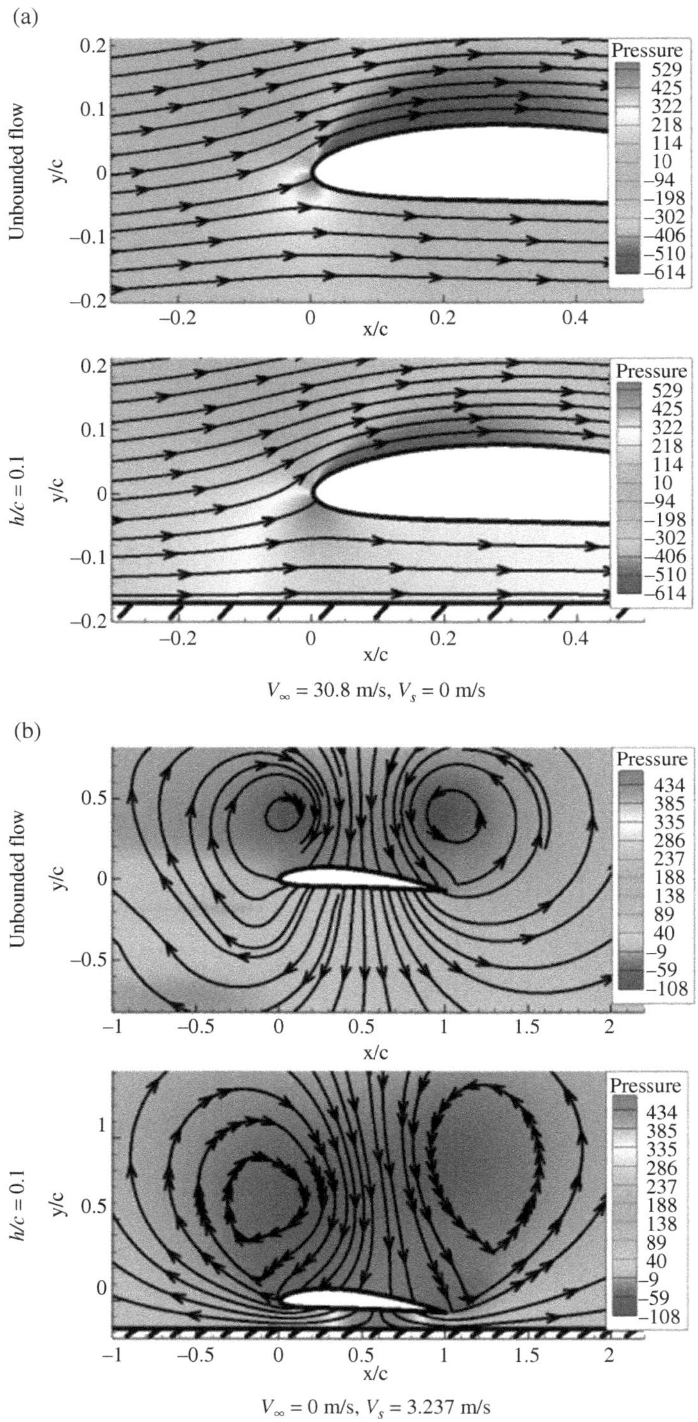

Figure 6.33 Streamlines and pressure contours around NACA 4412 at $\theta = 4°$. (*See insert for color representation of this figure.*)

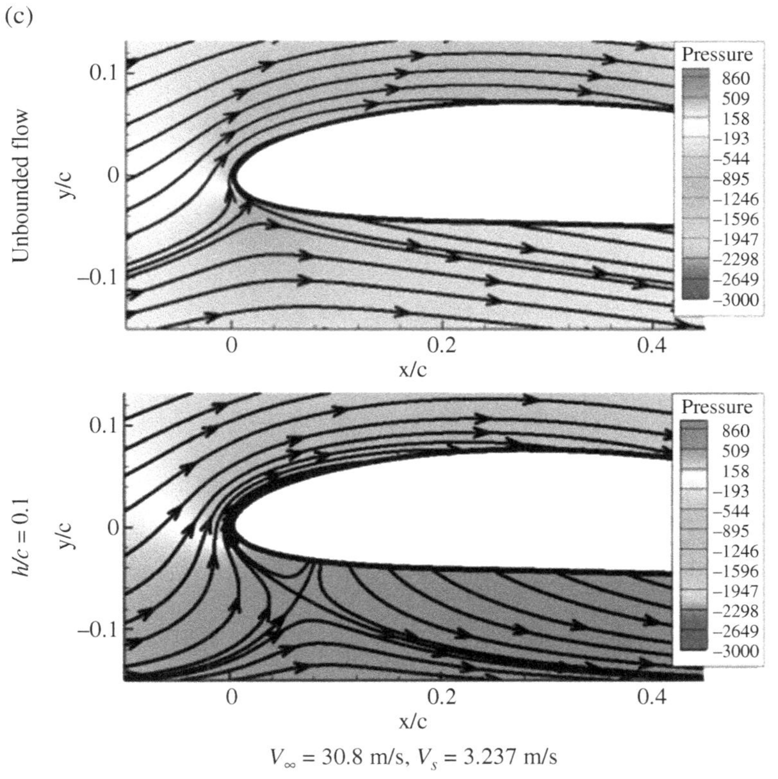

Figure 6.33 (*Continued*)

leading and trailing edges. This escaping airflow moves almost parallel to the ground, so the distance between the cores of two vortices and the upper surface of the airfoil increases.

In Figure 6.33c, there is uniform inflow from upstream and the airfoil descends at a constant velocity. In this case, the upstream inflow and the downwash flow streams couple into one composite flow. Because of the velocity reference point selected here, there is only a saddle point in the flow field instead of the stagnation point on the airfoil surface. The composite flow is divided into four streams by the saddle point. The inflow from upstream is separated into an upward stream and a downward stream respectively passing from the upper and lower surfaces of the airfoil. The downwash flow is separated into a forward flow and a backward flow respectively escaping from the leading edge and the trailing edge.

Figure 6.34 depicts the streamlines and pressure contours around NACA 4412 airfoil in the DGE at $\theta = 4°$ and $\gamma = -4°$.

In the unbounded flow field (DGE 1), the saddle point is very close to the leading edge of the lower surface of the airfoil. Because the airfoil does not do compression work in this region and the static pressure of the air below the airfoil is so small, it cannot resist the upstream inflow. As a result, almost all the downwash flow passes through the trailing edge.

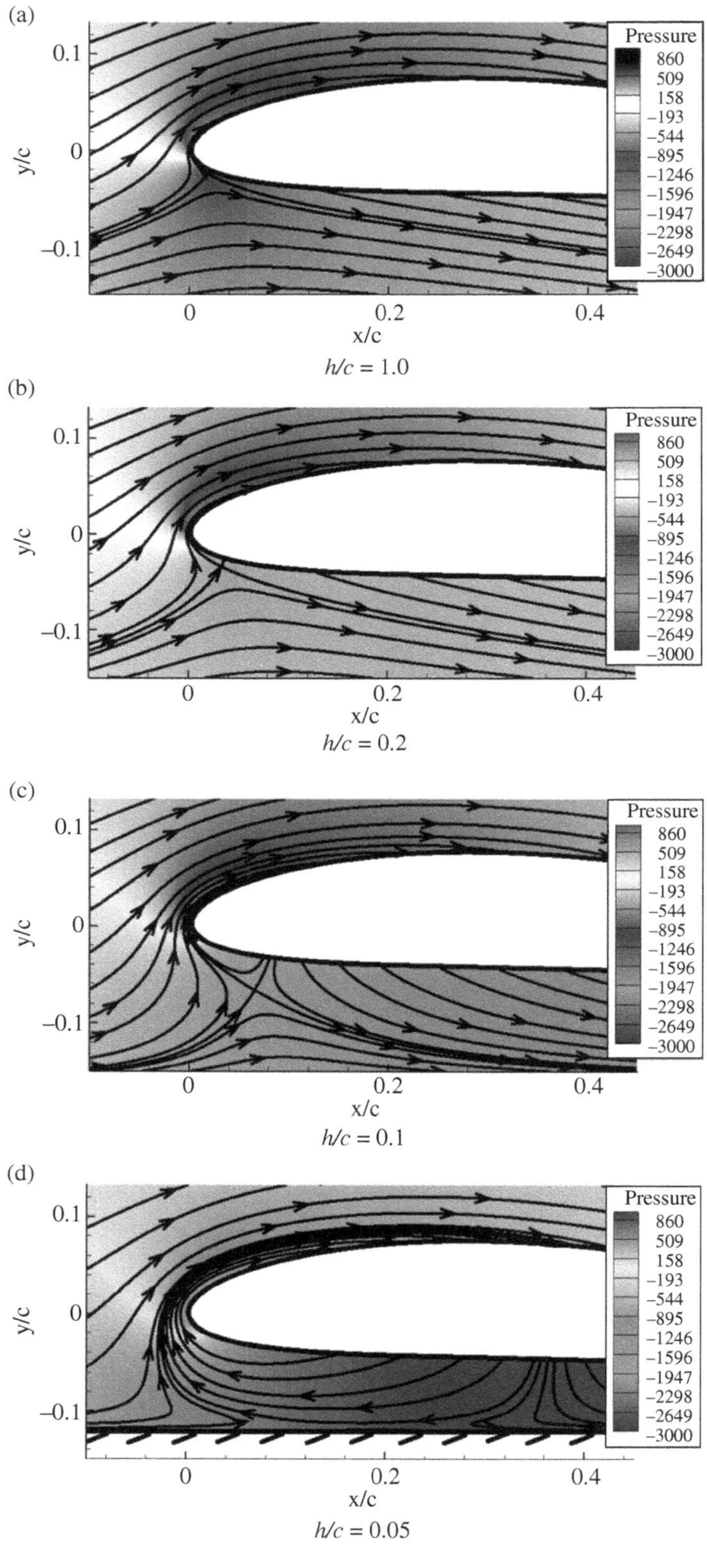

Figure 6.34 Streamlines and pressure distributions around NACA 4412 airfoil at $\theta = 4°$ and $\gamma = -4°$ for various *h/c*.

In DGE 2 (see Figure 6.34a), the position of the saddle point is nearly the same as that in DGE 1 since the compression work is still very small; the static pressure below the airfoil does not increase significantly, so it cannot resist the upstream inflow.

In DGE 3 (see Figures 6.34b–d), the saddle point moves toward the trailing edge and the ground when h/c decreases. It is important to note here that the moving velocity of the saddle point increases with decrease in h/c. For example, at $h/c = 0.05$, the saddle point is almost on the ground. This is because the compression work done by the airfoil increases significantly as the airfoil moves closer to the ground, resulting in a large increase in static pressure of the air below the airfoil. This high-pressure air can ultimately resist the upstream inflow, forcing more upstream inflow to go over the upper surface. At the same time, more downwash flow is diverted from the leading edge. At $h/c = 0.05$, the increase of static pressure due to the compression work is so large that nearly all the upstream inflow is diverted from the upper surface. The channel below the lower surface is primarily occupied by the downwash flow.

It can be concluded from the above analysis that in the DGE, more upstream inflow and downwash flow are forced to go over the upper surface of the airfoil with the reduced ride height. As a result, the fluid velocity near the leading edge increases including the suction peak (see Figures 6.29 and 6.32).

6.4.5 Conclusions

From the analysis of the flow properties and aerodynamic forces, the DGE can be divided into three regions based on the ride height:

DGE 1 ($h/c > 1.0$): the lift in the DGE is equal to that in the SGE for the same AOA and does not change with the ride height. It is because DGE 1 is out of ground effect and the main physics that governs the flow is the incidence effect induced by the sinking movement.

DGE 2 ($0.5 < h/c < 1.0$): the lift in the DGE is almost equal to that in the SGE for the same AOA and increases with decreasing ride height. This is because of increase in pressure on the lower surface of the airfoil. The main factor that governs the flow is the coupling of the SGE and the incidence effect.

DGE 3 ($h/c < 0.5$): when the ride height decreases significantly, the lift in the DGE gradually becomes larger than that in the SGE for the same AOA. The lift difference between the DGE and SGE becomes more significant with larger sink rate. Furthermore, in addition to the SGE and the incidence effect, the compression work effect becomes an important factor that governs the flow. The air below the airfoil does not have sufficient space to escape as the airfoil moves towards the ground, so this air is compressed and the static pressure increases significantly, resulting in an increase in lift. Compression work effect becomes quite apparent at relatively small ride heights.

Furthermore, in the study of the chord-dominated DGE during landing, a dynamic case can be treated as a static case with the same AOA in the following two circumstances:

1) The sink rate is very small.
2) The relative ride height is larger than 0.5 ($h/c > 0.5$), even if the sink rate is very large.

6.5 Chord-dominated Mutational Ground Effect

So far, the literature about the MGE is very sparse. Coton (1998) conducted a catapult experiment on a model aircraft to measure the aerodynamic forces during its flight over an aircraft carrier, and found that the unsteady aerodynamic characteristics are very important in the MGE. Yang and Luh (1998) calculated the unsteady aerodynamic forces on a 2D ellipse leaping a platform using potential flow theory. Shipman et al. (2008) numerically investigated the flow coupling between an F/A-18 and a CVN class aircraft carrier during approach and recovery to the flight deck using the dynamic mesh technology. They focused on the unsteady aerodynamic forces, but did not try to focus on the flow physics resulting in variations in the aerodynamic forces.

6.5.1 Physical Model

Here the take-off and landing of a NACA4412 airfoil on a rectangular platform are investigated. The conditions are as follows:

$c = 1.0\,\text{m}$
$V_\infty = 60\,\text{m/s}$
$\text{Re} = 4.1 \times 10^5$
$\gamma = 0°$
$\alpha = 4°, 8°, 10°$ and $14°$
$h/c = 1.0, 0.5, 0.2,$ and $0.05.$

Here, for the calculations, the principle of relative motion is adopted. The airfoil is fixed in space; the platform and the inflow from upstream move in the opposite direction to the flight velocity of the airfoil (see Figure 6.35).

6.5.2 Typical Landing Process

A typical landing, with $\alpha = 8°$ and $h/c = 0.2$, is analyzed in detail. Figure 6.36 shows the variation in lift and drag coefficients with L/c for a typical landing. Before the leading

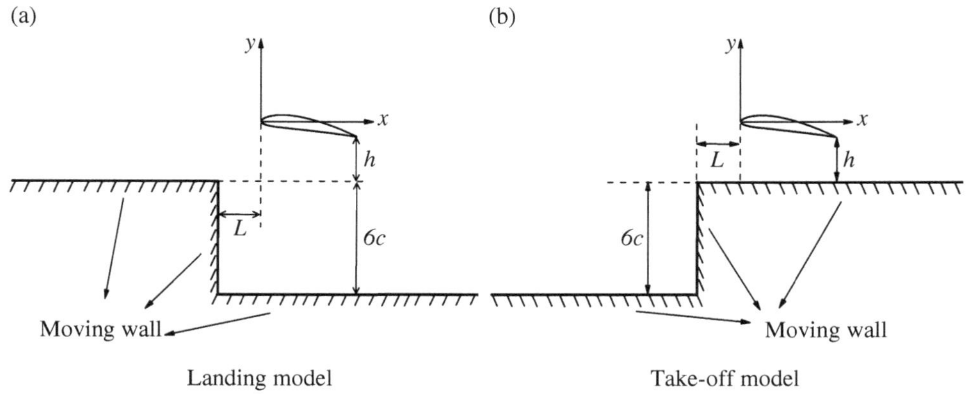

Figure 6.35 Take-off and landing models used in the application of relative motion principle.

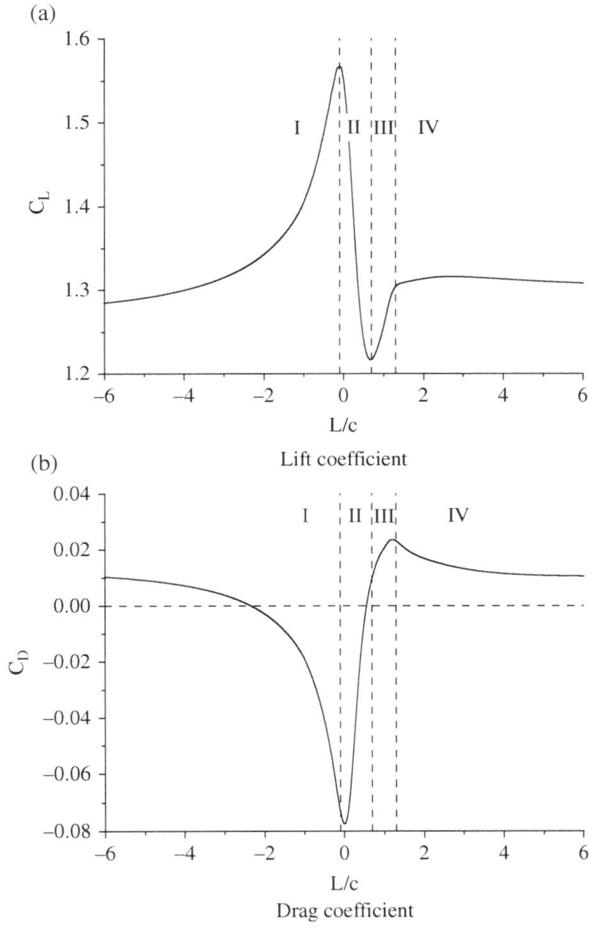

Figure 6.36 Lift and drag curves of NACA4412 airfoil during a typical landing process.

edge of the airfoil reaches the side of the platform, L/c has a negative value; afterwards, L/c becomes positive. Note that in Figure 6.35, the origin is at leading edge of the airfoil, with the x-axis pointing towards the right.

In the region $L/c << 0$, the lift and drag have stable values in the unbounded flow field. At $L/c \approx -6$, the airflow around the airfoil begins to be affected by the platform and the aerodynamic forces begin to change. With L/c increasing from approximately −6 to a positive value of around 0.1, the lift first gradually increases by 22.9% to the maximum value at $L/c \approx 0.1$, then decreases to the minimum value at $L/c \approx 0.7$, and then has a sharp increase until $L/c \approx 1.3$, and finally it gradually tends to the stable value of lift in the SGE.

With L/c increasing from approximately −6 to zero, the drag coefficient first gradually decreases to the lowest value at $L/c \approx 0$ (when the resistance due to drag becomes thrust), then rapidly increases to the maximum value at $L/c \approx 1.2$, and later gradually decreases to the stable value in the SGE. The thrust occurs in the region $-2.4 \leq L/c \leq 0.4$. In Figure 6.36, square symbols between the vertical dotted lines show the four regions from left to right.

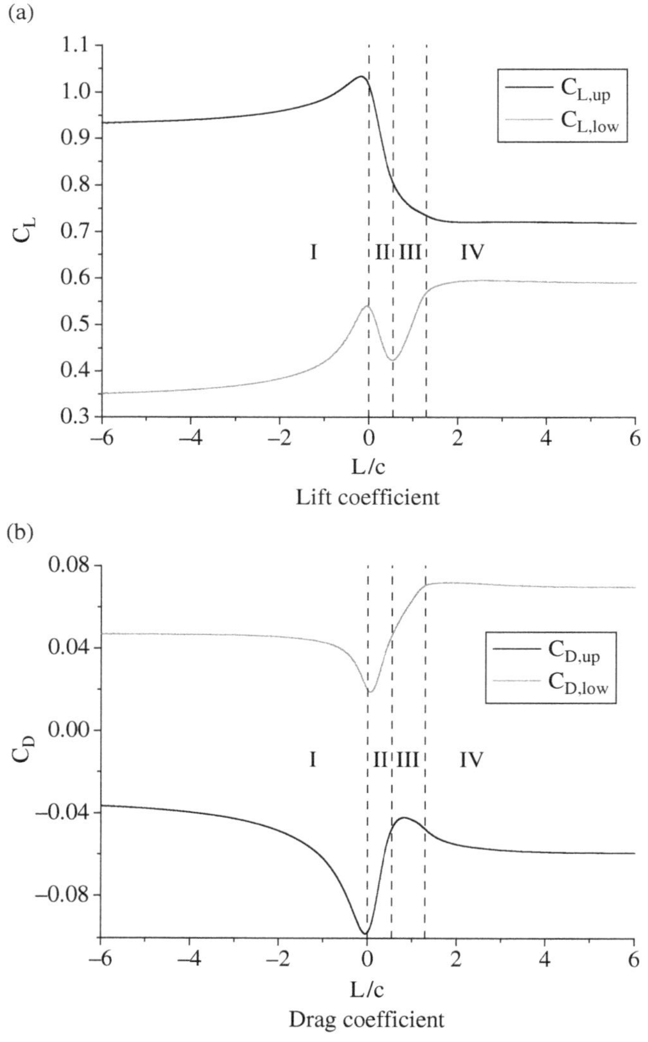

Figure 6.37 Aerodynamic force history curves of the upper and lower surfaces during a typical landing process.

Figure 6.37 shows the aerodynamic force history curves of the upper and lower surfaces of a NACA 4412 airfoil during a typical landing process. The changes in the histories of lifts of the upper and lower surfaces are not identical; the lift of the upper surface is apparently larger than that of the lower surface. According to the history characteristics of the lift of the lower surface, the MGE during a landing process can be divided into four phases.

- In Phase I ($-6 \le L/c \le 0$), the lift of the lower surface monotonically increases and the lift of the upper surface first increases to a maximum and then slightly decreases; thus the total lift increases.

- In Phase II ($0 \leq L/c \leq 0.55$), the lift of the upper and lower surfaces and the total lift all decrease.
- In Phase III ($0.55 \leq L/c \leq 1.3$), the lift of the lower surface increases, the lift of the upper surface decreases, and the lift increment of the lower surface is greater than the lift decrement of the upper surface, so the total lift increases.
- In Phase IV ($L/c \geq 1.3$), the lifts of the upper and lower surfaces and the total lift all tend to the stable values in the SGE.

Figure 6.37b shows that the upper and lower surfaces produce negative drag (thrust) and positive drag respectively. The thrust is caused by the suction peak on the leading edge of the upper surface.

- In Phase I, the drags of the upper and lower surfaces and the total drag all decrease.
- In Phase II, the drags of the upper and lower surfaces and the total drag all sharply increase.
- In Phase III, the drag of the lower surface continues to increase and the drag of the upper surface first increases and then decreases with small amplitude, so the total drag slowly increases.
- In Phase IV, the drags of the upper and lower surfaces and the total drag all gradually tend to the stable values in the SGE.

Figure 6.38 shows the streamlines, velocity vector angle contours and pressure contours around a NACA4412 airfoil during a typical landing process on an aircraft carrier. The velocity vector angle is defined as the angle between the velocity vector and the positive direction of the x-axis. The streamlines are drawn in the coordinate system fixed to the airfoil.

Figures 6.38a–c show the process of an airfoil approaching the platform in Phase I. Above the platform, the streamlines are parallel to the top of platform; on passing the platform, the streamlines significantly deflect upward, resulting in an upwash flow around the airfoil. The airfoil thereby experiences an increment in effective AOA. Thus, the suction on the upper surface and the pressure on the lower surface all increase, and thus the lift of both the upper and lower surfaces increases.

Figures 6.38d and 38e show the leading edge of the airfoil passing the platform in Phase II. The top of the platform weakens the upwash flow around the leading edge of the airfoil, but the upwash flow still exists in the middle and at trailing edge of the airfoil. At the same time, the leading part of the airfoil starts to form the SGE due to the top surface of the platform. The SGE is significantly weaker than the reduced effect of the upwash flow, so the lift of the upper and lower surfaces decrease.

Figures 6.38f–h show the process of the airfoil passing the whole platform in Phase III. The upwash flow completely disappears and the SGE gradually becomes the main controlling factor. The SGE makes the pressure on both the lower and the upper surface increase, so the lift of the upper surface decreases but the lift of the lower surface increases.

6.5.3 Typical Take-off Process

Here the typical take-off process at $\alpha = 10°$ and $h/c = 0.2$ is analyzed in detail. Figure 6.39 shows the aerodynamic force history curves for the typical take-off process. Before the leading edge of the airfoil reaches the side of the platform, L/c has a negative value; afterwards, L/c becomes positive.

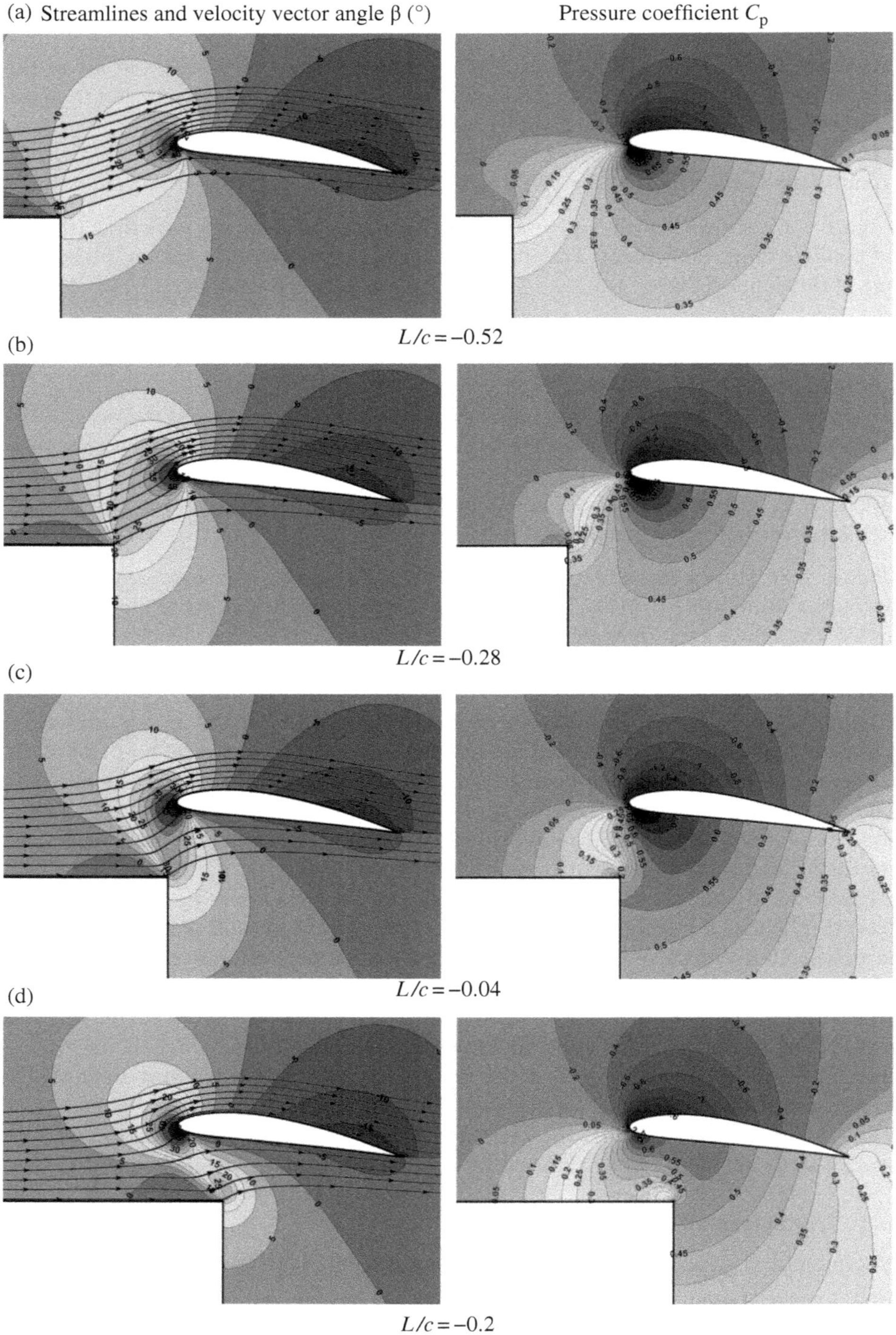

Figure 6.38 Streamlines, velocity vector angle contours and pressure contours around NACA4412 during a typical landing process.

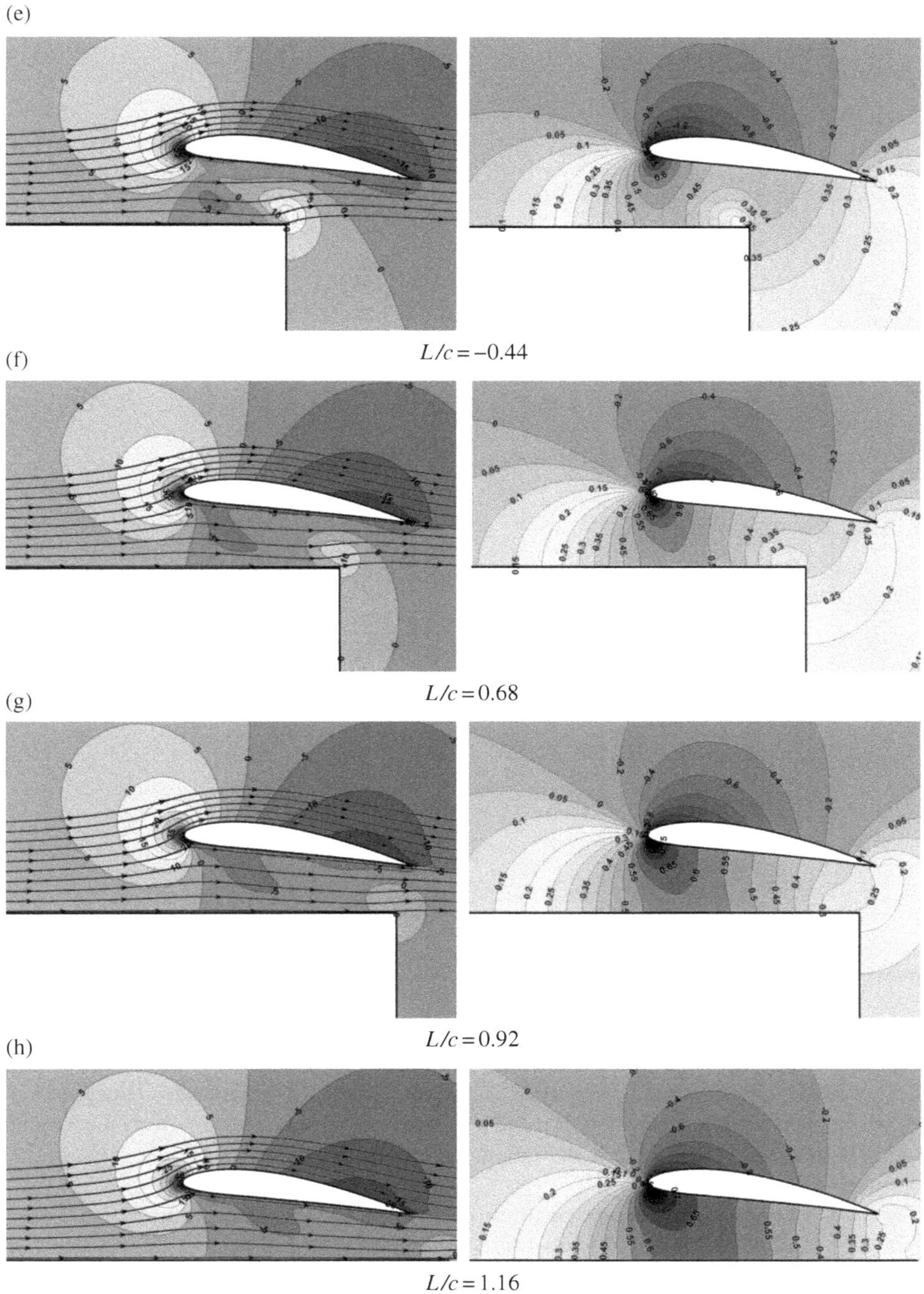

Figure 6.38 (*Continued*)

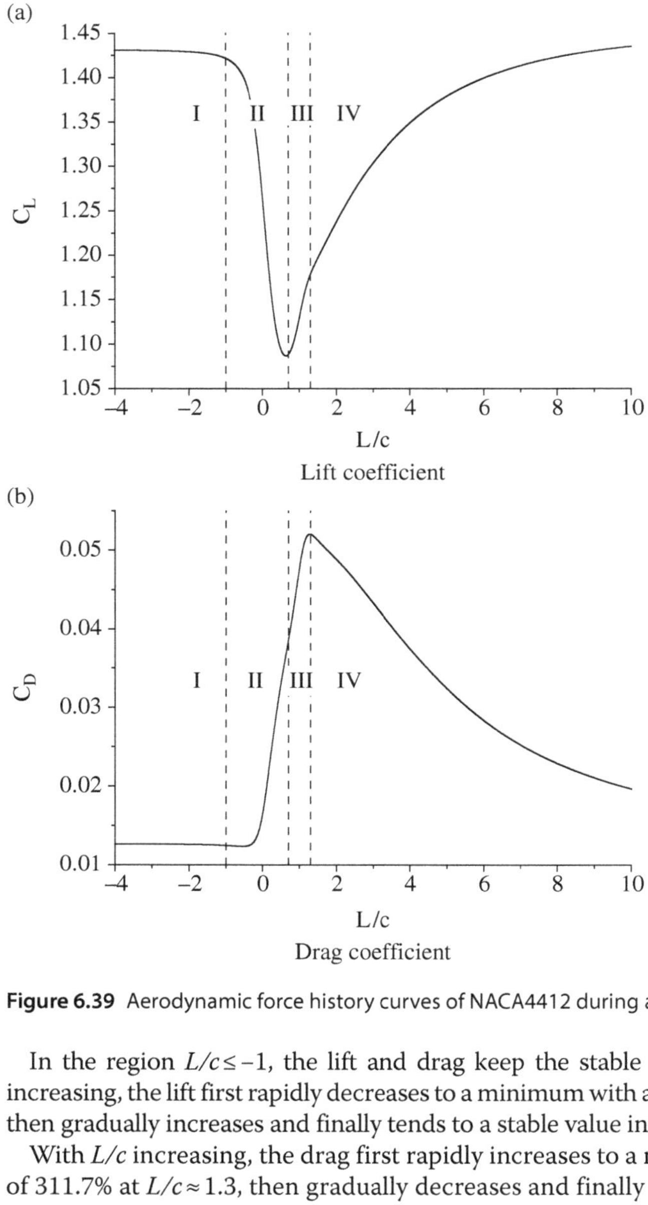

Figure 6.39 Aerodynamic force history curves of NACA4412 during a typical take-off process.

In the region $L/c \leq -1$, the lift and drag keep the stable values of the SGE. With L/c increasing, the lift first rapidly decreases to a minimum with a decrease of 23.9% at $L/c \approx 0.7$, then gradually increases and finally tends to a stable value in the unbounded flow field.

With L/c increasing, the drag first rapidly increases to a maximum with an increment of 311.7% at $L/c \approx 1.3$, then gradually decreases and finally tends to a stable value in the unbounded flow field.

Figure 6.40 shows the aerodynamic force history curves of the upper and lower surface of NACA4412 airfoil during a typical take-off. The changes in the histories of the lift of the upper and lower surfaces are identical: the lifts first decrease, then increase and finally tend to stable values in the unbounded flow field. Additionally, the lift of the upper surface is apparently larger than that of the lower surface. Based on the history characteristics of the lift of the lower surface, the MGE during take-off can be divided into four phases.

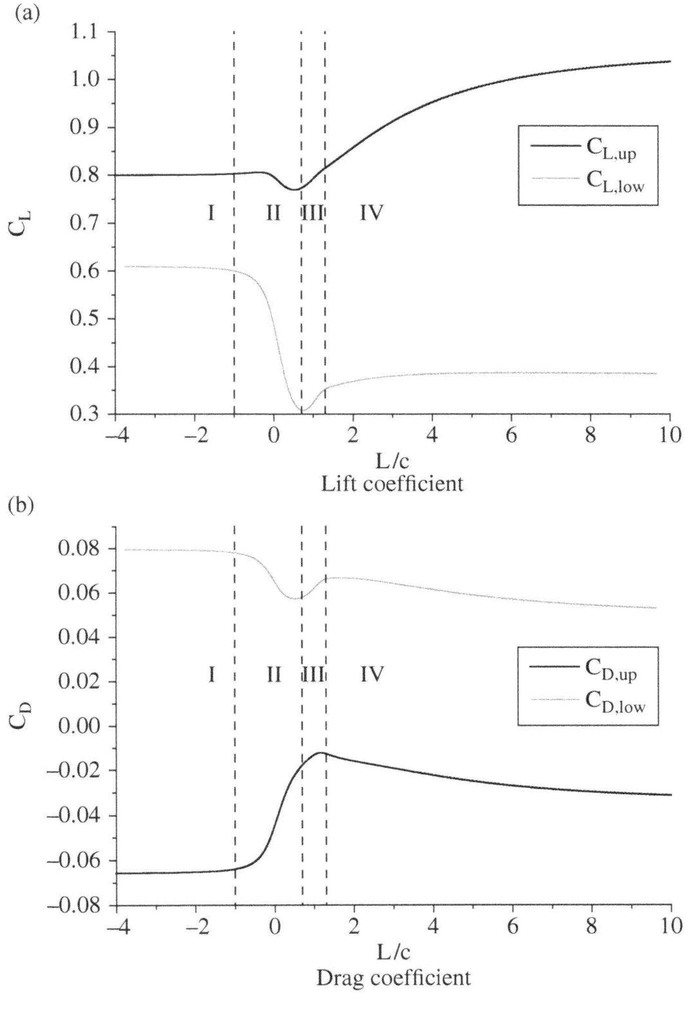

Figure 6.40 Aerodynamic force history curves of the upper and lower surfaces during a typical take-off.

- In Phase I ($L/c \leq -1$), the flow around airfoil is not affected by the side of the platform, so the total lift, and the lift of the lower and upper surfaces all remain constant in the SGE.
- In Phase II ($-1 \leq L/c \leq 0.7$), the lift of the upper and lower surfaces and the total lift all decrease, and the decrease in lift of the lower surface is greater than that of the upper surface.
- In Phase III ($0.7 \leq L/c \leq 1.3$), the lift of the upper and lower surface and the total lift all rapidly increase.
- In Phase IV ($L/c \geq 1.3$), the lift of the lower surface first increases over a short time and then tends to a constant value; the lift of the upper surface gradually increases over a longer timescale and then tends to a constant value. Thus the total lift gradually tends to the stable value in the unbounded flow field.

In Phase I, the drag of the upper and lower surface and the total drag all have the stable values in the SGE. In Phase II, the drag of the lower surface decreases, the drag of the upper surface increases and the increment of the upper surface is larger than decrease in drag of the lower surface in the latter half period, so the total drag first keeps a constant value and then sharply increases. In Phase III, the drag of the upper and lower surface and the total drag all increase. In Phase IV, the drags of the upper and lower surfaces and the total drag all gradually decrease over a long time and finally tend to stable values in the unbounded flow field.

Figure 6.41 shows the streamlines, velocity vector angle contours and pressure contours around NACA4412 during a typical take-off. Figure 6.41a shows the flow field in the SGE in Phase I. The streamlines are parallel to the top surface of the platform. The airflow is not affected by the side of the platform but is blocked in the channel between the lower surface of airfoil and the top surface of the platform, which leads to a high-pressure region under the airfoil.

Figures 6.41b–e show the process of the leading edge of airfoil passing the platform in Phase II. Due to the disappearing of the top of the platform, the streamlines significantly deflect downward behind the side of platform, resulting in a downwash flow around the leading edge of the airfoil. Additionally, the lower boundary of the airflow, being blocked under the airfoil, gradually disappears, so the SGE gradually weakens. Both these two effects make the lift of both the upper and lower surface decrease.

Figures 6.41f–h show the process of the trailing edge of the airfoil leaving the side of the platform in Phase III. The SGE completely disappears and the downwash flow gradually weakens. The latter is the main controlling factor. Therefore, the lifts of both the upper and lower surfaces increase.

6.5.4 Effects of AOA and Ride Height on the MGE

First, four cases with $h/c = 0.2$ and $\alpha = 4°$, $8°$, $10°$ and $14°$ are compared to study the effect of AOA on the MGE. Figures 6.42 and 6.43 show the lift history curves for different AOAs, during landing and take-off, respectively. The dimensionless lift coefficient increment is defined as follows:

$$(\Delta C_L)_R = \frac{C_L - (C_L)_{L/c=-\infty}}{(C_L)_{L/c=-\infty}} \tag{6.13}$$

For take-off and landing, the lift history curves for different AOAs include four phases (see Figures 6.42a and 6.43a). With the AOA increasing, the dimensionless lift increment during landing decreases in Phase I, stays almost constant in Phase II, and decreases in Phase III. The peak of the lift decrement during take-off decreases.

Compared to the lifts in the unbounded flow field, the lifts of $\alpha = 4°$ and $8°$ in the SGE with $h/c = 0.2$ is larger, indicative of a positive GE; the lifts of $\alpha = 10°$ and $14°$ in the SGE with $h/c = 0.2$ is smaller indicative of a negative GE.

Next, four landing cases with $\alpha = 8°$ and $h/c = 0.05$, 0.2, 0.5 and 1.0, and four take-off cases with $\alpha = 10°$ and $h/c = 0.05$, 0.2, 0.5 and 1.0, are compared to study the effect of ride height on the MGE. Figures 6.44 and 6.45 show the lift history curves for different ride heights, during landing and take-off respectively. The fluctuations in the unsteady lift decrease with increasing ride height.

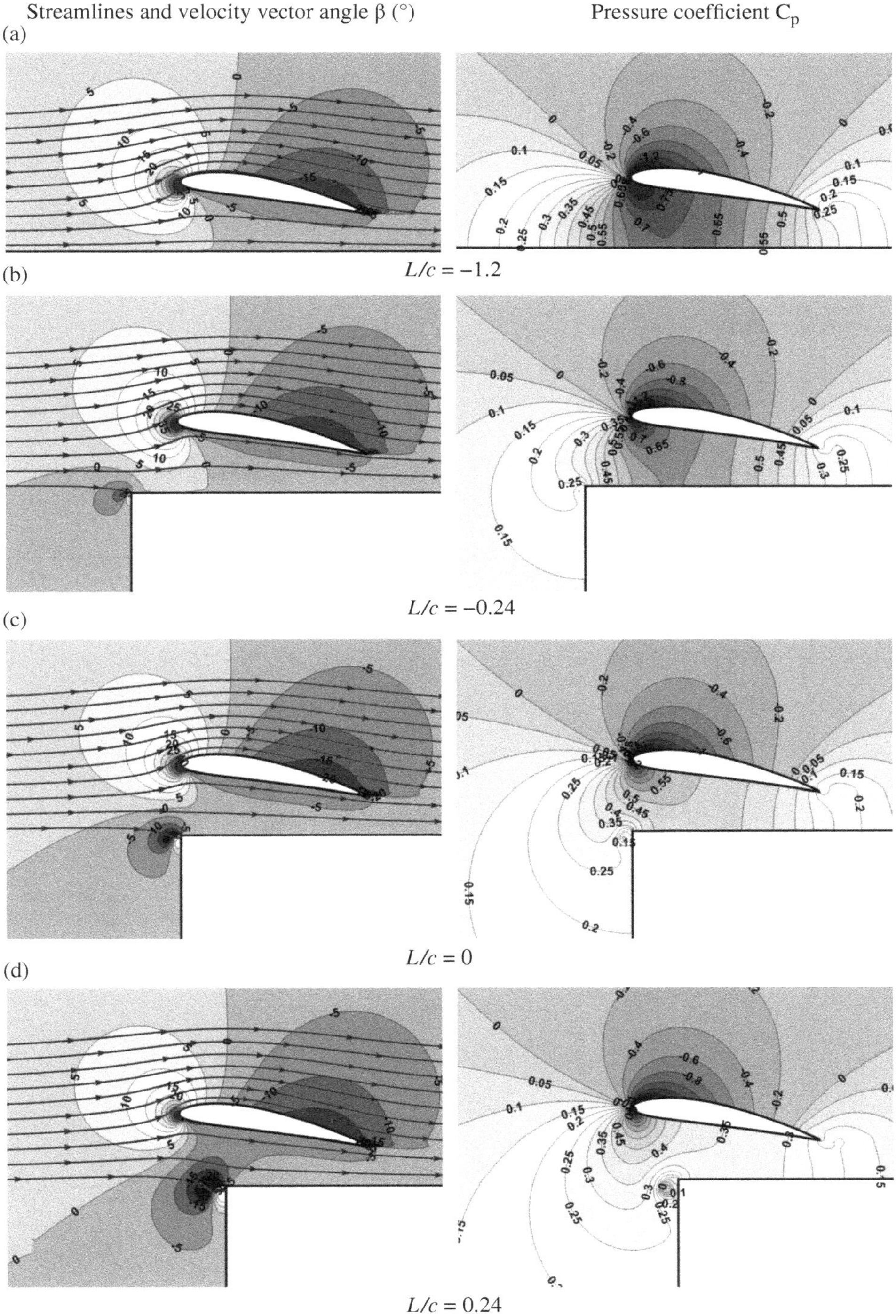

Figure 6.41 Streamlines, velocity vector angle contours and pressure contours around NACA4412 during a typical take-off.

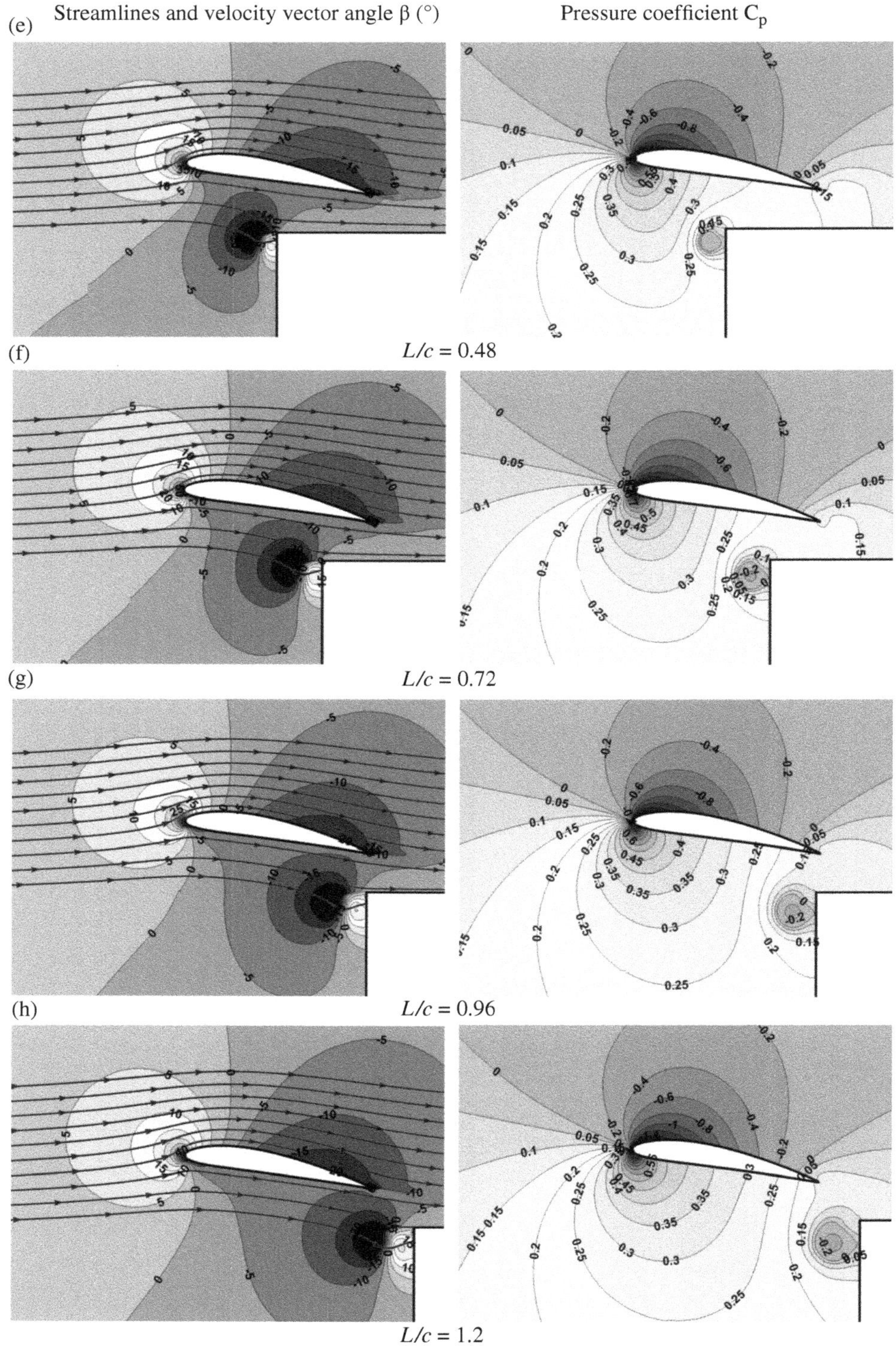

Figure 6.41 (*Continued*)

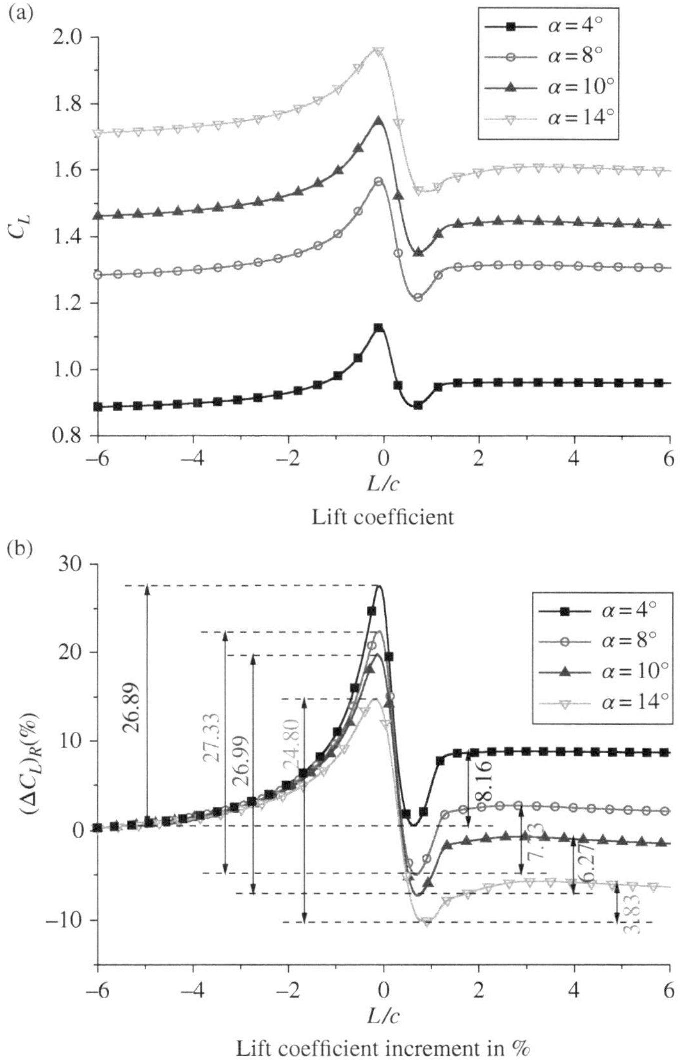

Figure 6.42 Lift history curves of NACA4412 during landing for different AOAs.

Compared to the unbounded flow field, the cases of $h/c = 0.2$ and 0.05 show a positive GE, but the cases of $h/c = 0.5$ and 1.0 show a negative GE.

6.5.5 Conclusions

The process for the landing of a NACA4412 airfoil on an aircraft carrier deck can be divided into four phases. In Phase I, the airfoil gradually approaches the side of platform; the upwash flow increases the effective AOA and thus the lift increases. In Phase II, the leading half of the airfoil enters the space above the platform; the upwash flow weakens and the SGE begins to appear, but the former is the main controlling

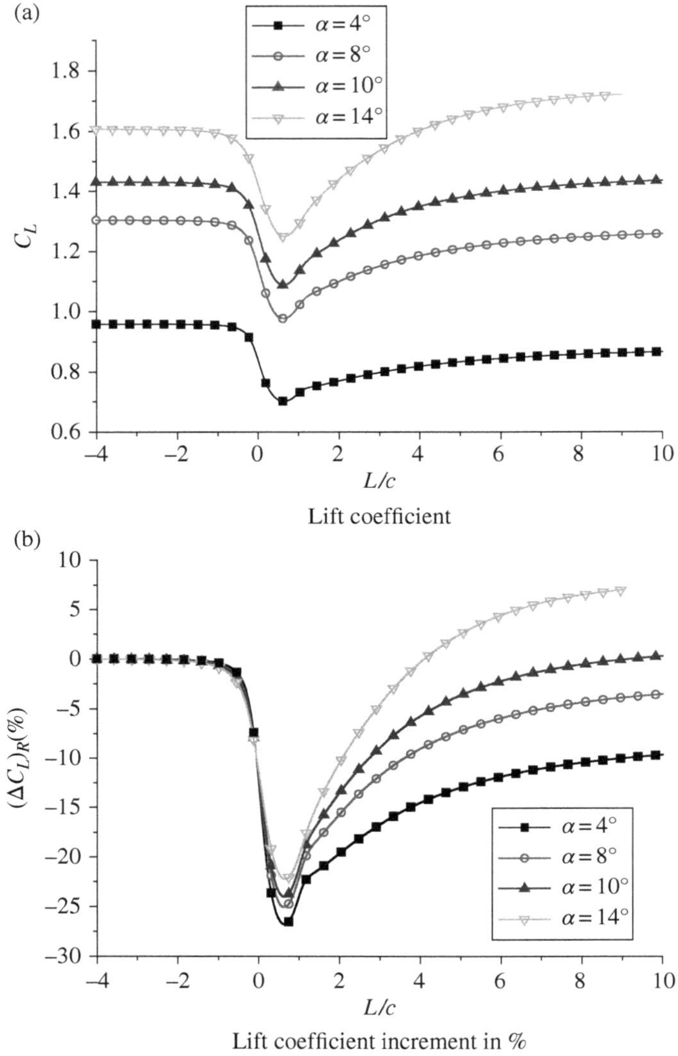

Figure 6.43 Lift history curves of NACA4412 during take-off for different AOAs.

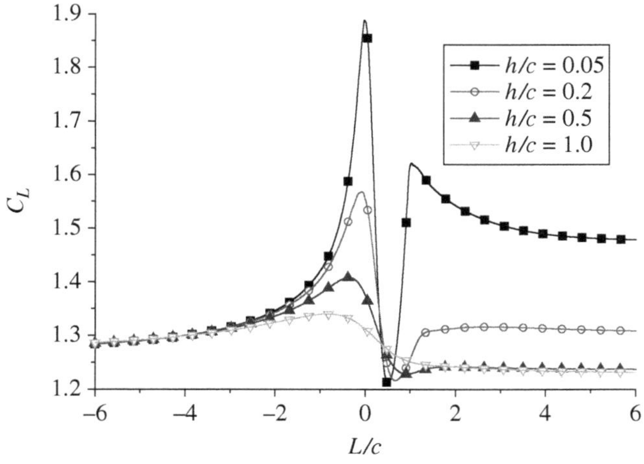

Figure 6.44 Lift history curves of NACA4412 during landing process for different *h/c*.

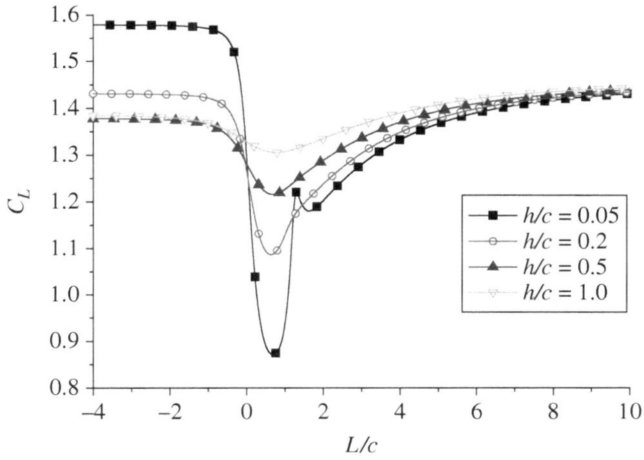

Figure 6.45 Lift history curves of NACA4412 during take-off process for different *h/c*.

factor and thus the lift decreases. In Phase III, the airfoil is completely above the platform; the upwash flow entirely disappears and the SGE becomes the main controlling factor, so the lift increases. In Phase IV, the lift and flow around the airfoil all gradually tend to the situation in the SGE.

The take-off process of NACA4412 airfoil on a platform can be divided into four phases too. In Phase I the airfoil gradually approaches the side of platform and the lift keeps the stable value of the SGE. In Phase II, the leading half of the airfoil leaves the platform space, the SGE gradually disappears and the downwash flow decreases the effective AOA, so the lift sharply drops. In Phase III, the airfoil completely leaves the platform space, the SGE completely disappears and the downwash flow weakens but the remains the main controlling factor, so the lift increases. In Phase IV the flow around the airfoil and lift gradually tend to the situation in the unbounded flow field.

The phase characteristics of the lift history curves during take-off and landing do not change with increase in AOA and the fluctuating amplitude of the unsteady lift decreases with increasing ride height.

Acknowledgments

This work was partially supported by the National Natural Science Foundation of China (No. 11302015).

References

Ahmed MR, Takasaki T, and Kohama Y (2007). 'Aerodynamics of a NACA4412 airfoil in ground effect,' *AIAA Journal*, **45**(1): 37–47.

Chen YS, and Schweikhard WG (1985). 'Dynamic ground effects on a two-dimensional flat plate,' *Journal of Aircraft*, **22**(7): 638–640.

Chun HH, and Chang RH (2003). 'Turbulence flow simulation for wings in ground effect with two ground conditions: fixed and moving ground,' *International Journal of Maritime Engineering*, **145**(A3): 1–18.
Coton P (1998). 'Study of environment effects by means of scale model flight tests in a laboratory,' *21st ICAS Congress*, Melbourne: pp. 13–18.
Hiemcke C (1997). 'NACA 5312 in ground effect – Wind tunnel and panel code studies,' AIAA Paper 1997–2320, *15th Applied Aerodynamics Conference.*
Hsiun CM, and Chen CK (1996). 'Aerodynamic characteristics of a two-dimensional airfoil with ground effect,' *Journal of Aircraft*, **33**(2): 386–392.
Mahon S, and Zhang X (2005). 'Computational analysis of pressure and wake characteristics of an aerofoil in ground effect,' *Journal of Fluids Engineering-Transactions of the ASME*, **127**(2): 290–298.
Matsuzaki, Yoshioka TS, Kato T, and Kohama Y (2008). 'Unsteady aerodynamic characteristics of wings in ground effect,' *Proceedings of the 40th JAXA Workshop on Investigation and Control of Boundary-Layer Transition*: pp. 53–56.
Molina J, and Zhang X (2011). 'Aerodynamics of a heaving airfoil in ground effect,' *AIAA Journal*, **49**(6): 1168–1179.
Mook D, and Nuhait A (1989). 'Numerical simulation of wings in steady and unsteady ground effects,' *Journal of Aircraft*, **26**(12): 1081–1089.
Nuhait AO, and Zedan MF (1993). 'Numerical simulation of unsteady flow induced by a flat plate moving near ground,' *Journal of Aircraft*, **30**(5): 611–617.
Qu QL, Jia X, Wang W, Liu PQ, and Agarwal RK (2014a). 'numerical simulation of the flowfield of an airfoil in dynamic ground effect,' *Journal of Aircraft*, **51**(5): 1659–1662.
Qu QL, Jia X, Wang W, Liu PQ, and Agarwal RK (2014b). 'Numerical study of the aerodynamics of a NACA 4412 airfoil in dynamic ground effect,' *Aerospace Science and Technology*, **38**: 56–63.
Qu QL, Lu Z, Liu PQ, and Agarwal RK (2014). 'Numerical study of aerodynamics of a wing-in-ground-effect craft,' *Journal of Aircraft*, **51**(3): 913–924.
Qu QL, Wang W, Liu PQ, and Agarwal RK (2015a). 'Airfoil aerodynamics in ground effect for wide range of angles of attack,' *AIAA Journal*, **53**(10): 3144–3154.
Qu QL, Zuo PY, Wang W, Liu PQ, and Agarwal RK (2015b). 'Numerical investigation of the aerodynamics of an airfoil in mutational ground effect,' AIAA Paper 2015–0256, *53rd AIAA Aerospace Sciences Meeting.*
Rozhdestvensky KV (2006). 'Wing-in-ground effect vehicles,' *Progress in Aerospace Sciences*, **42**(3): 211–283.
Shipman JD, Arunajatesan S, Cavallo PA, Sinha N, and Polsky SA (2008). 'Dynamic CFD simulation of aircraft recovery to an aircraft carrier,' AIAA Paper 2008-6227, *26th AIAA Applied Aerodynamics Conference.*
Yang SA, and Luh PA (1998). 'A numerical simulation of hydrodynamics forces of ground-effect problem using Lagrange's equation of motion,' *International Journal for Numerical Methods in Fluids*, **26**(6): 725–747.
Ying CJ, Yang W, and Yang ZG (2010). 'Numerical simulation on stall of wing in ground effect,' *Flight Dynamics*, **28**(5): 9–12.
Zerihan J, and Zhang X (2000). 'Aerodynamics of a single element wing in ground effect,' *Journal of Aircraft*, **37**(6): 1058–1064.

1.2 Rotary-wing (helicopter)

7

Dynamics Modelling and System Identification of Small Unmanned Helicopters

Cunjia Liu and Wen-Hua Chen

Department of Aeronautical and Automotive Engineering, Loughborough University, Loughborough, UK

7.1 Introduction

Small unmanned helicopters are versatile and are promising platforms for tasks such as surveillance and reconnaissance in confined and dynamic surroundings, due to their abilities in hovering and low-altitude cruising, and their manoeuvrability.

To this end, the flight dynamics of helicopters need to be understood so that their operational capabilities can be fully exploited. However, in order to apply modern control strategies and to design aircraft that work in a broad range of flight conditions, a precise mathematical model is a necessary. The model should not only have an accurate reflection of the main dynamic characteristics of the helicopter, but should also involve some simplification, to allow control design and real-time prediction. Due to the complexity of the dynamics of helicopters, developing such a model is a challenge.

In the control community, there are two well-established approaches for modelling system behaviour:

- first-principles modelling, where models describing dynamics are built from scratch based on underlying physical laws
- system identification, where algorithms are used to find the relationship between the inputs and outputs of a given system using data collected from experiments.

A typical first-principles mathematical model for a general helicopter is nonlinear and contains many states, since it has to account for interactions between each component of the aircraft: the motor, the rotors and the fuselage (Padfield 2007). The advantage of this kind of model is that it can cover a wide range of flight conditions. However, the quality may be degraded by inappropriate simplifications and assumptions. Moreover, there are dozens of inherent physical parameters that need to be determined. Some physical parameters can be directly measured, but the others may have to be estimated through experiments or calculated from theory, often including assumptions or the results of experience. All of these factors will cause inaccuracy and uncertainty in

Advanced UAV Aerodynamics, Flight Stability and Control: Novel Concepts, Theory and Applications,
First Edition. Edited by Pascual Marqués and Andrea Da Ronch.

Companion website: http://www.wiley.com/go/marques/advanced_UAV_aerodynamics

modelling a small helicopter, although this issue can be addressed to some extent by arduous experimental validation and refinement of the model.

A good example of using the first-principles method to model a helicopter is the Minimum-Complexity Helicopter Simulation Math Model (Heffley & Mnich 1988). A dedicated model for small-scale helicopters comes from Gavrilets et al. (2001), with dynamics equations developed using basic helicopter theory, and nonlinear expressions provided for external forces and moments. The model was successfully used for aerobatic flight control design (Gavrilets et al. 2002). In recent years, a number of projects involving autonomous helicopters have further developed first-principles nonlinear models, including those at Aalborg University (Hald et al. 2005) and the National University of Singapore (Cai et al. 2012).

The system identification approach naturally focuses on input–output data, trying to specify a model that can interpret such experimental phenomena. This method is more direct and more effective since it integrates validation into the modelling process. System identification is not a standalone process as it requires a good understanding of helicopter dynamics, so the identification and first-principles approaches are essentially complementary and should be integrated to develop a new model (Mettler 2003).

An early attempt to identify a small helicopter model can be found in the Caltech experiment (Morris et al. 1994), where a linear model for attitude motion was derived and applied. In order to guarantee the linear behaviour of a Kyosho EP Concept 30 helicopter, it was mounted on a stand, allowing only angular movements. The authors used the linearised rigid-body equation of motion as a parametrised model, and identified these parameters by using a prediction error method (PEM). The resulting model was successfully used for control design for attitude movements. However, helicopter dynamics on a stand are not representative of a free helicopter, so there was still a big gap between the model and the real flight response. Another notable study was carried out by Kim & Tilbury (2009). The authors developed a parametrised model using the first-principles method and this explicitly accounted for the flybar (stabilizer bar) dynamics. However, their identification was based on single-input–single-output transfer functions, and the coefficients in the transfer functions had no physical meanings.

There was no significant progress in the system-identification approach to modelling a small-scale helicopter until Mettler and his co-workers completed their remarkable work (Mettler et al. 1999, 2001). Their modelling and identification were based on Carnegie Mellon's Yamaha R-50 helicopter. First, a linear parametrised model was developed, mainly based on the rotor–fuselage equations, extended by the inclusion of the stabilizer-bar dynamics. With a proper simplification, the main rotor was modelled through first-order tip-path-plane dynamics, and the stabilizer bar was regarded as a secondary rotor. All the coupled components were connected by cross-coupling derivatives. The final model resulted in a linear MIMO state-space model of 13 orders. The authors then applied a frequency-domain identification tool called CIFER (Comprehensive Identification from Frequency Response) to identify these unknown parameters in the state-space model. The results showed that the model had high fidelity despite its relatively simple structure. This parametrised model appears to be general enough to describe other small-scale rotorcraft; the same methodology has since been successfully applied to the MIT X-Cell 60 helicopter (Mettler et al. 2004).

A number of other modelling projects have been carried out using the model structure proposed by Mettler. Others used this model structure but relied on the linear time-domain identification tool PEM to estimate the parameters (Kim & Shim 2003,

Shin et al. 2005). Although there were difficulties reported in converging to the globally optimal values of the model parameters during the identification, the control designs based on these models were successful. A more comprehensive study was carried out by La Civita et al. (2002). The authors combined nonlinear modelling with linear system identification to produce a high-fidelity model that can cover a wider range of operating conditions. More recently, Grauer et al. (2009) adopted a two-step equation-error/output-error process in both time and frequency domains to identify the model for a miniature helicopter. A general summary on the system identification of small unmanned aircraft can be found in Hoffer et al. (2014).

The studies just described are a good starting point for the further development of a mathematical model for miniature helicopters. There are still open problems in modelling and system identification, especially some of the practical aspects. Firstly, from a control point of view, a helicopter model that is simple but can accurately predict the dynamic response is desired. Thus a complicated model specified only from first principles is unsuitable; it should be validated using real flight data to guarantee its fidelity. Secondly, in system-identification experiments, input signals to the helicopter should have sufficiently large magnitudes so as to guarantee that the helicopter is fully actuated and the corresponding flight modes are excited. Therefore, the model needs to account for the nonlinearity caused by manoeuvres. Thirdly, since the model structure needs nonlinear terms, the selected identification algorithm should be able to process nonlinear models, which implies that frequency-domain methods may not be directly appropriate.

To solve these difficulties and to obtain a practical control-oriented model, in this chapter a helicopter model is derived with nonlinear terms to account for helicopter dynamics in a broad range of flight conditions. However, it retains the first-order approximation of external forces and moments, so as to reduce the complexity to facilitate control design. The unknown parameters in the model structure are identified from flight data using the PEM algorithm in the Matlab System Identification Toolbox (Ljung 2009). This algorithm supports nonlinear grey-box identification (with unknown structure and unknown parameters), which makes it well suited for our purpose. Furthermore, to improve the accuracy of the parameter estimation, the entire nonlinear model is broken down into several small identification processes. The corresponding flight tests are also carefully designed and performed to support the identification process. All these efforts guarantee the fidelity of the model and the accuracy of the identification. The resulting parameterised nonlinear model shows a consistent match with flight data, even when both lateral and longitudinal channels are excited. Of particular interest is that the prediction of translational speeds is better than previous studies. As there is no dedicated software involved in the identification process, the proposed method can easily be picked up by other research groups to accelerate the development of rotorcraft-based autonomous systems.

7.2 Model Development

7.2.1 Overview

The miniature helicopter in this study is a Trex-250 electronic helicopter, as shown in Figure 7.1. It is a 200-sized helicopter with a main-rotor diameter of 460 mm and a trail-rotor diameter of 108 mm. The miniature size and 3D aerobatic ability make it

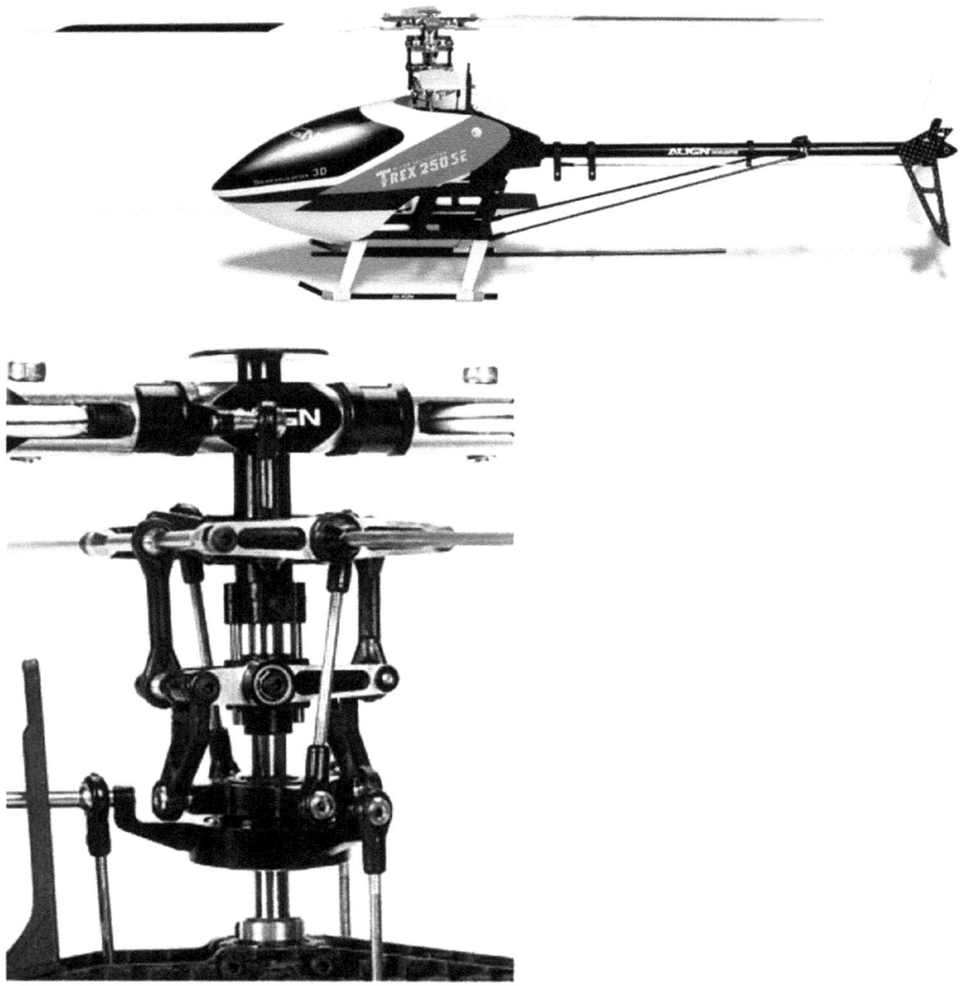

Figure 7.1 Trex-250 helicopter.

well-suited for indoor flight tests. Trex-250 has a collective pitch rotor and a well-designed Bell-Hiller stabilizer mechanism that makes it representative of most widely used small-scale helicopters. In addition, the Trex family features a series of helicopters, from the Trex-250 micro-size to the Trex-700 small-size, all with a similar structures and handing qualities. This means that the research outcomes can be easily applied quite widely.

A helicopter's versatile flight ability is supported by its effective yet complicated rotor system. The main rotor blades can change their pitch angle simultaneously and cyclically, while rotating around the main shaft. This movement is achieved via a swashplate mechanism. When the swashplate moves vertically along the rotor shaft under the control of a collective input δ_{col}, the pitch angles of all blades are altered at the same time, so that the thrust is changed. When the swashplate is tilted, the blade pitching follows the swashplate's perimeter cyclically to produce different lifts at different angles. Then the whole rotor disc exhibits a flapping motion to generate torques on the helicopter fuselage. The cyclic control inputs can be divided into lateral (roll) and longitudinal (pitch) components δ_{lat} and δ_{lon}, respectively. On the other hand, the tail rotor only has a collective input to alter its thrust, which is used to counteract the main rotor moment and to change the heading. As in a full-size helicopter, where this control is accessible via pedals, it is denoted as δ_{ped}. In addition, there is a Bell-Hiller stabiliser bar in the rotor system, which is used to improve the flyability of model helicopters, by providing a mechanical rate feedback (damping).

To study the behaviour of a helicopter, a body-fixed reference coordinate is established and attached to the helicopter as shown in Figure 7.2. Its origin is at the helicopter's centre of gravity (CG), the x-axis is along the forward direction of the helicopter, the z-axis is downward and the y-axis is determined from the 'right-hand' rule. The other principal variables are also shown on the x, y and z body axes. They include: the projections of inertial velocity in the three body axes u, v, w; the Euler angles ϕ, θ, ϕ representing the helicopter orientation; and the body angular rates p, q, r. The main rotor is represented by a disc, which can tilt about the rotor hub in both longitudinal and lateral directions. This motion is described through the angles β_{1c} and β_{1s} measured with respect to a plane perpendicular to the rotor shaft.

The transformation between the inertial coordinates and the body-fixed coordinates can be parameterised in terms of Euler angles. The Euler angles refer to a specific sequence of rotations about the vehicle body axes. A widely used definition in flight dynamics is a sequence of yaw-pitch-roll, which are angle ψ about the z-axis, angle θ about the current y-axis, and φ about the current x-axis (Cook 2007), respectively. In this way, the kinematic relationship is

$$[\dot{x} \quad \dot{y} \quad \dot{z}]^T = \mathbf{R}_b^i [u \quad v \quad w]^T \tag{7.1}$$

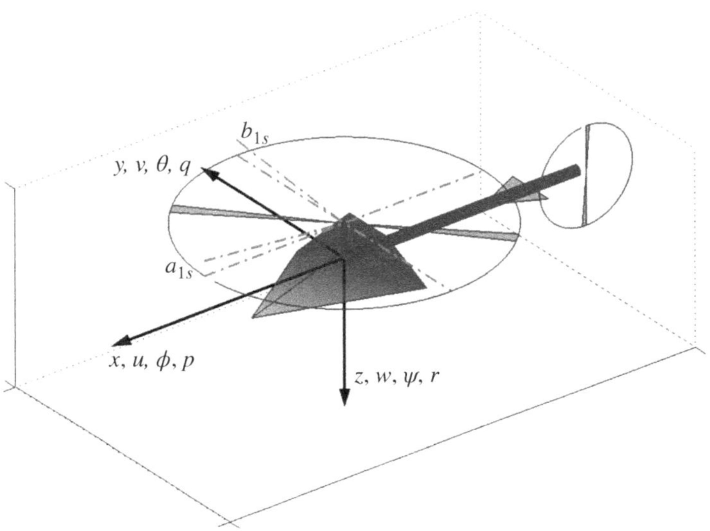

Figure 7.2 Body coordinates of the helicopter.

where the transformation matrix is (with s denoting sin(·) and c for cos(·))

$$\mathbf{R}_b^i = \begin{bmatrix} c\theta c\psi & s\phi s\theta c\psi - c\phi s\psi & c\phi s\theta c\psi + s\phi s\psi \\ c\theta s\psi & s\phi s\theta s\psi + c\varphi c\psi & c\phi s\theta s\psi - s\phi c\psi \\ -s\theta & s\phi c\theta & c\phi c\theta \end{bmatrix} \tag{7.2}$$

In addition, the relationship between the angular rate (p, q, r) and the Euler angle rate $(\dot{\phi}, \dot{\theta}, \dot{\psi})$ can be expressed as

$$\begin{bmatrix} \dot{\phi} \\ \dot{\theta} \\ \dot{\psi} \end{bmatrix} = \begin{bmatrix} 1 & \sin\phi\tan\theta & \cos\phi\tan\theta \\ 0 & \cos\phi & -\sin\phi \\ 0 & \sin\phi\sec\theta & \cos\phi\sec\theta \end{bmatrix} \begin{bmatrix} p \\ q \\ r \end{bmatrix} \tag{7.3}$$

7.2.2 Rigid-body Model

Rigid-body equations of motion are extensively used in aerospace engineering to describe the dynamics of a vehicle, in air or in outer space, which is free to translate and rotate in all six degrees of freedom (6DOF). The rigid-body equations of motion can be developed from Newton–Euler equations of motion expressed in the inertial reference coordinates. For a constant vehicle mass m and moment of inertia (inertial tensor) $\mathbf{I}$, they are:

$$m\frac{d\mathbf{v}_I}{dt} = \mathbf{F} \tag{7.4}$$

$$\mathbf{I}\frac{d\omega_I}{dt} = \mathbf{M} \tag{7.5}$$

where $\mathbf{F} = [X \;\; Y \;\; Z]^T$ is the vector of external forces acting on the vehicle's CG, and $\mathbf{M} = [L \;\; M \;\; N]^T$ is the vector of external moments. For a helicopter, the external forces and moments are produced by the main rotor and the tail rotor, the gravitational force, and the aerodynamic forces produced by the fuselage and the tail surfaces.

Using the kinematic principle of moving reference coordinates, the equations of motion can be expressed with respect to the helicopter body-fixed reference coordinates:

$$m\dot{\mathbf{v}} + m(\omega \times \mathbf{v}) = \mathbf{F} \tag{7.6}$$

$$\mathbf{I}\dot{\omega} + (\omega \times \mathbf{I}\omega) = \mathbf{M} \tag{7.7}$$

where $\mathbf{v} = [u \;\; v \;\; w]^T$ and $\omega = [p \;\; q \;\; r]^T$ are the fuselage velocities and angular rates in the body coordinates, respectively, and × denotes the cross-product. For a 6DOF rigid-body system, Eq. (7.6) produces the three differential equations describing the helicopter's translational dynamics about its three body axes:

$$\begin{aligned} \dot{u} &= (-wq + vr) + X/m \\ \dot{v} &= (-ur + wp) + Y/m \\ \dot{w} &= (-vp + uq) + Z/m \end{aligned} \tag{7.8}$$

Similarly, Eq. (7.7) produces the three ordinary differential equations describing the vehicle's rotational dynamics, with the assumption that the cross-products of the inertia are small:

$$\begin{aligned} \dot{p} &= -qr\left(I_{yy} - I_{zz}\right) / I_{xx} + L / I_{xx} \\ \dot{q} &= -pr\left(I_{zz} - I_{xx}\right) / I_{yy} + M / I_{yy} \\ \dot{r} &= -pq\left(I_{xx} - I_{yy}\right) / I_{zz} + N / I_{zz} \end{aligned} \tag{7.9}$$

After the rigid-body dynamics of the helicopter have been expressed in Eqs. (7.8) and (7.9) as a function of the external forces **F** and moments **M**, the next step is to find the components that construct the external forces and moments, and express them as functions of control inputs and/or vehicle states. This is the most difficult step in the modelling of a helicopter due to the complicated nature of rotor aerodynamics and interactions. However, the principal components of **F** and **M** are generated by the main and tail rotors, the gravitational forces, and the aerodynamic forces caused by the fuselage (including tail boom and vertical and horizontal fins), such that:

$$\begin{aligned} X &= X_M + X_F + G_x \\ Y &= Y_M + Y_F + Y_T + G_y \\ Z &= Z_M + Z_Z + G_z \\ L &= L_M + Y_M h_M + Z_M y_M + Y_T h_T \\ M &= M_M + M_T - X_M h_M + Z_M l_M \\ N &= N_M + Y_M l_M - Y_T l_T \end{aligned} \tag{7.10}$$

where X_*, Y_*, Z_* and L_*, M_*, N_* denote, respectively, the forces and moments from different sources, with the subscripts $_M$, $_T$, $_F$ indicating the main rotor, tail rotor, and fuselage, respectively; (l_M, y_M, h_M) and (l_T, y_T, h_T) denote the coordinates of the main-rotor and tail-rotor shafts relative to the CG in the body coordinates; and (G_x, G_y, G_z) denote the projections of gravitational force in the body coordinates in the form of:

$$[G_x \quad G_y \quad G_z]^T = \mathbf{R}_i^b \cdot [0 \quad 0 \quad g]^T \tag{7.11}$$

where g is acceleration due to gravity.

Note that in the construction of Eq. (7.10), it is assumed that the tail rotor only generates force and torque along the y direction and that torques arising from the aerodynamic forces of the fuselage are negligible. This is reasonable because among all the components in Eq. (7.10), the forces and moments produced by the main rotor and the thrust of the tail rotor have the most significant influences on a small helicopter. Moreover, they can be altered by the control inputs so that the helicopter can be controlled. To account for their mechanism and complete the helicopter model accordingly, the rigid-body model needs to be extended to include the main-rotor's behaviour.

7.2.3 Extension of the Rigid-body Model

The objective of extending the rigid-body model discussed in the previous subsection is to take into account the higher-order effects that exist in helicopter dynamics, and to investigate their interactions with the rigid-body dynamics. Except for the rigid-body motion, the most obvious and important movements in helicopter dynamics are the tilting or flapping movements of the main rotor. A combination of these two parts gives a hybrid model that is sufficiently simple, but captures the main helicopter characteristics.

Simplified Rotor Dynamics

The rotor dynamics are naturally complicated due to the high number of degrees of freedom of the motion of the rotor blades plus the elusive aerodynamic characteristics. Consequently, detailed rotor equations of motion can be extremely complex (Padfield 2007), and they may not be suitable for system identification and control design. However, earlier work on modelling of small-scale rotorcraft has shown that the main rotor system with a stabiliser bar (flybar) can be lumped and represented by tip-path-plane (TPP) flapping dynamics with only two states. In the following discussion, this chapter will describe the development of a highly simplified TPP rotor model for system identification and later for control design.

The flapping motion quantified by a flapping angle β is a 2π periodic function as it associated with blades rotating about the rotor shaft (speed Ω, position Ψ). Thus the general solution to the flapping equation can be approximated by a first-order Fourier series

$$\beta(\Psi) \approx \beta_0(t) - \beta_{1c}(t)\cos\Psi - \beta_{1s}(t)\sin\Psi \tag{7.12}$$

This type of motion results in a cone-shaped rotor, as shown in Figure 7.3. The top of the cone is the so-called tip-path-plane. The constant term β_0 describes the cone angle and the coefficients of the first harmonic β_{1c} and β_{1s} describe the tilting of the rotor TPP in the longitudinal and lateral directions, respectively. By using the TPP presentation, the rotor-flapping dynamics can be further simplified, because a tilting movement of the plane replaces the detailed rotation movements of the blades. The complete derivation and simplification can be found in Mettler (2003). The solution in that paper is directly adopted here to extend the rigid-body model.

To investigate the dynamic modes of the rotor TPP, simplified notation – a instead of β_{1c}, b instead of β_{1s} and a_0 instead of β_0 – are used to formulate a tip-path-plane state vector $\mathbf{a} = [a_0 \quad a \quad b]^T$. Note that for the Trex-250 helicopter, the coining angle β_0 is exactly equal zero due to its hingeless rotor head. For system identification, the rotor dynamics need to be further simplified, because not all effects can be identified from flight data. By focusing on the low-frequency flapping dynamics, the longitudinal and lateral flapping dynamics can be approximated by first-order equations (Mettler 2003):

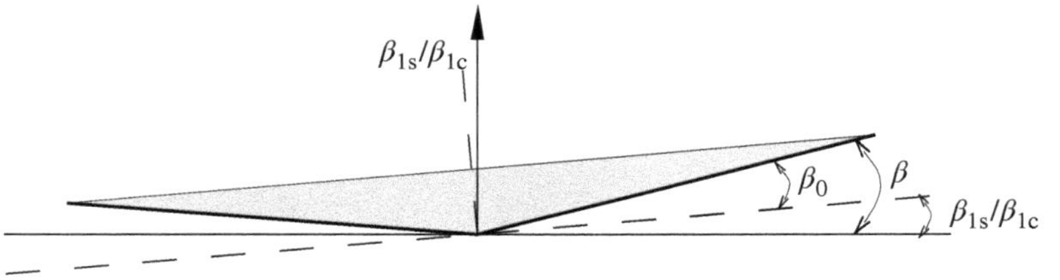

Figure 7.3 Tip-path-plane rotor representation.

$$\frac{16}{\gamma\Omega}\dot{a} = -a - \frac{16}{\gamma\Omega}q + \frac{p}{\Omega} + \frac{8}{\gamma\Omega^2}\frac{k_\beta}{I_\beta}b - B_1 \tag{7.13}$$

$$\frac{16}{\gamma\Omega}\dot{b} = -b - \frac{16}{\gamma\Omega}p + \frac{q}{\Omega} + \frac{8}{\gamma\Omega^2}\frac{k_\beta}{I_\beta}b + A_1 \tag{7.14}$$

where Ω is the rotor speed, k_β is the spring constant of the flapping restraint, I_β is the blade moment of inertia about the flapping hinge, A_1 and B_1 are, respectively, the lateral and longitudinal cyclic blade-pitch angles, and γ is the blade Lock number, which represents the ratio between the aerodynamic and inertial forces acting on the blade:

$$\gamma = \frac{\rho c C_{la} R^4}{I_\beta} \tag{7.15}$$

where R is the rotor radius, ρ is the air density, c is the blade chord length, and C_{la} is the lift curve slope.

These equations capture the key TTP responses from control inputs and helicopter motion. Moreover, these equations reflect rotor dynamics in a low-frequency range within which they will be coupled into fuselage dynamics. In Eqs. (7.13) and (7.14), one important coefficient is $\frac{16}{\gamma\Omega}$, which is defined as the rotor time constant τ. Note that it depends on the rotor Lock number γ and rotor angular speed Ω. By using the notation $\tau = \frac{16}{\gamma\Omega}$, we have two terms τq and $-\tau p$, which are important pitch and roll damping terms in the lateral and longitudinal directions, respectively. They can also be seen as a rate feedback mechanism, which is actually provided by the flybar. In contrast, $-\frac{q}{\Omega}$ or $+\frac{p}{\Omega}$ is the lateral (or longitudinal) flapping produced by a body roll rate p (or pitch rate q). This is one of the main rotor cross-coupling effects. $\frac{8}{\gamma\Omega^2}\frac{k_\beta}{I_\beta}$ is another cross-coupling effect arising in the presence of a flapping restraint. It can be denoted by the derivatives A_b and B_a for longitudinal and lateral flapping, respectively. Finally, the blade-pitch angles A_1 and B_1 in the lateral and longitudinal directions are decided by the cyclic movement of the swashplate, which can be modelled to be proportional to control inputs δ_{lat} and δ_{lon}:

$$A_1 = A_{\text{lat}}\delta_{\text{lat}} + A_{\text{lon}}\delta_{\text{lon}} \tag{7.16}$$

$$B_1 = -B_{\text{lat}}\delta_{\text{lat}} - B_{\text{lon}}\delta_{\text{lon}} \tag{7.17}$$

where A_{lat}, A_{lon}, B_{lat} and B_{lon} are control derivatives, in which A_{lat} and B_{lon} are used to account for unmodelled cross-coupling effects.

The simplified rotor dynamic equations that are suitable for system identification can be summarised as:

$$\tau\dot{a} = -a - \tau q + A_b b + B_{\text{lat}}\delta_{\text{lat}} + B_{\text{lon}}\delta_{\text{lon}} \tag{7.18}$$

$$\tau\dot{b} = -b - \tau p + B_a b + A_{\text{lat}}\delta_{\text{lat}} + A_{\text{lon}}\delta_{\text{lon}} \tag{7.19}$$

where $-\frac{q}{\Omega}$ or $+\frac{p}{\Omega}$ are dropped, because they are one order of magnitude smaller than $\frac{16}{\gamma\Omega}$, when the latter is determined from its physical definition and the final identification results.

After the flapping dynamics of the rotor have been determined, it becomes possible to investigate the basic function of the main rotor and its by-products: the main rotor thrust T and torque Q_M. The former can be modelled as

$$T = K_T P_M \Omega^2 \tag{7.20}$$

where K_T denotes the aerodynamic constant of the rotor's blade, and P_M is the collective pitch angle of the blade, which is controlled by input δ_{col}. The main rotor's torque has a more complicated form, which according to Gavrilets et al. (2001) can be modelled as:

$$\begin{aligned} I_{\text{rot}}\dot{\Omega} &= Q_M - Q_M^R \\ Q_M^R &= c\Omega^2 + dP_M^2\Omega^2 \\ Q_M &= P_{\text{e}}/\Omega \\ P_{\text{e}} &= \bar{P}_{\text{e}}\delta_{\text{thr}} \end{aligned} \tag{7.21}$$

where I_{rot} is the moment of inertia of the rotor blades about the main shaft, Q_M^R is a reaction torque due to the aerodynamic resistance of blade (with physical parameters c and d depending on blade characteristic), and P_{e} denotes the engine power assumed to be proportional to throttle input δ_{thr} with coefficient $\bar{P}_{\text{e}}$. Although the main rotor torque varies in such a complicated way, modern unmanned helicopters are equipped with onbard heading gyros, which help the tail rotor to generate the correct thrust to cancel the main rotor torque.

It can be seen from the above discussion that many of the coefficients in rotor dynamics depend on the rotor speed. Fortunately, in most modern unmanned rotorcraft, the rotor speed Ω is kept constant by an onboard electronic governor. As a result the corresponding coefficients can be considered as constant at the current modelling level.

Rotor–Fuselage Interaction

The interaction between the main rotor and the fuselage is reflected by the relation between the rotor tip-path-plane angle, the thrust vector, and the forces and moments produced by the rotor, which is shown in Figure 7.4.

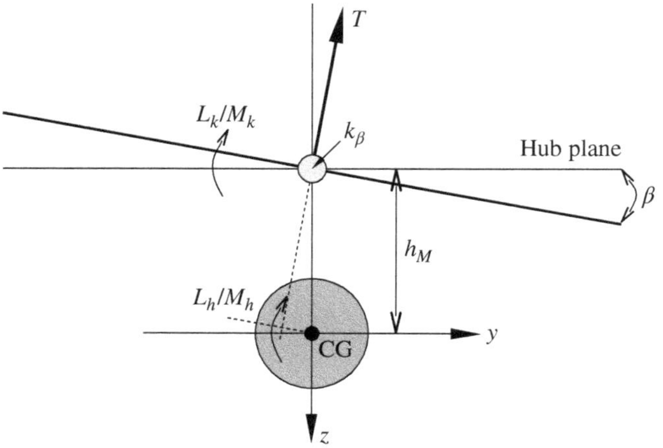

Figure 7.4 Rotor force and moments acting on the helicopter fuselage.

The forces produced by the main rotor are considered to be the result of the thrust vector tilting. With the assumption that the direction of the thrust vector is perpendicular to the rotor TPP (a common assumption for hovering and low-speed flight), X_M, Y_M and Z_M in Eq. (7.10) are the projections of the thrust vector on the x, y and z body axes, such that

$$\begin{aligned} X_M &= -T\sin a\cos b \approx -Ta \\ Y_M &= T\sin b\cos a \approx Tb \\ Z_M &= -T\cos a\cos b \approx -T \end{aligned} \tag{7.22}$$

Note that here we use the fact that the magnitude of flapping angle is small.

The moment produced by rotor flapping acting on the fuselage has two parts. The first contribution results from the restraint in the blade attachment to the rotor hub. This restraint can be approximated by a linear torsional spring model. Using the TPP lateral and longitudinal flappings b and a, the corresponding torsional moments are

$$\begin{aligned} L_k &= k_\beta b \\ M_k &= k_\beta a \end{aligned} \tag{7.23}$$

The second moment contribution is from the tilting of the thrust vector. As the thrust vector tilt is proportional to the rotor flapping angles, its projection in the hub plane produces a moment with an arm of length equal to the distance h_M. Taking into account Eq. (7.22), the resulting roll and pitch moments are

$$\begin{aligned} L_T &= h_M Y_M \approx h_M Tb \\ M_T &= h_M\left(-X_M\right) \approx h_M Ta \end{aligned} \tag{7.24}$$

The total moments due to rotor flapping action on the fuselage are obtained by summing the hub restraint and thrust tilting contributions.

Complete Helicopter Model

After the rotor dynamics and their coupling with the rigid-body dynamics have been determined, the complete helicopter model for system identification can be formulated by combining every force and moment component in Eq. (7.10). In the first-principles modelling of helicopters, every noticeable contribution to external forces and moments needs to be taken into account. However, in the development of a model for system identification, some terms in Eq. (7.10) need to be selectively dropped or further simplified. Firstly, this is due to their limited contributions, which cannot be identified from experimental data. Effects from trivial contributions can be easily corrupted by disturbances in flight test or submerged in sensor noise. Secondly, the extra terms need more free parameters to describe them, which will increase the degrees of freedom of the parameters, thereby potentially causing difficulties in identification.

It is a recursive process to refine the model structure, determining which terms in the equations should remain or be dropped by assessment of the final identification results. According to Mettler (2003) and initial system identification trails, it was found that it is adequate to approximate some forces (or moments) using a first-order Taylor series. To this end, we express forces and moments using stability derivatives and control derivatives, which are first partial derivatives in Taylor series with respect to vehicle states and control inputs, respectively. Note that in the translational and rotational equations, the forces are normalised by the helicopter mass, and the moments are normalised by the related moment of inertia.

In hover and slow flight conditions, the main rotor thrust is approximately equal to the vehicle weight, so Eq. (7.20) is approximated as

$$\frac{T}{m} = -Z_{\mathrm{col}}\delta_{\mathrm{col}} + g \approx g \tag{7.25}$$

where Z_{col} represents the collective pitch derivatives. Note that a more comprehensive thrust model can be adopted, (see, for example, Cai et al. 2011), but its parameters need to be obtained through a complicated experimental process. By following the approximation on Eq. (7.25) and combining Eqs. (7.11) and (7.22), the forces in Eq. (7.10) can be reformulated as:

$$X/m = X_u u - g\sin\theta - ga \tag{7.26}$$

$$Y/m = Y_v v + g\cos\theta\sin\phi + gb \tag{7.27}$$

$$Z/m = Z_w w + g\cos\theta\cos\phi + Z_{\mathrm{col}}\delta_{\mathrm{col}} - g \tag{7.28}$$

where, X_u, Y_v and Z_w are velocity damping derivatives, which account for the relative fuselage drag forces. The approximation $\frac{T}{m} = g$ is only used in the horizontal forces of the main rotor as their effects are reduced by small flapping angles. Note that the lateral force Y_T produced by the tail rotor is dropped, as it is compensated by a partition of the main thrust generated by a 'sitting angle' in flight and this effect can be eliminated by removing the mean values of the roll angle in the identification process.

Next, it follows from Eqs. (7.23) and (7.24) that the pitch and roll moment derivatives with respect to flapping angles can be expressed by

$$M_a = \left(k_\beta + h_M T\right)/I_{yy} \tag{7.29}$$

$$L_b = \left(k_\beta + h_M T\right)/I_{xx} \tag{7.30}$$

Thereby, the normalised moments are

$$L/I_{xx} = L_a a + L_b b \tag{7.31}$$

$$M/I_{yy} = M_a a + M_b b \tag{7.32}$$

$$N/I_{zz} = N_r r + N_{\mathrm{ped}}\delta_{\mathrm{ped}} + N_{\mathrm{col}}\delta_{\mathrm{col}} \tag{7.33}$$

where L_a and M_b are cross-coupling derivatives used to capture unmodelled coupling effects, N_r is the yawing damping derivative accounting for the rate feedback from the onboard heading gyro, N_{col} is collective control derivative, and N_{ped} is the yawing control derivative. Note that the torque N_M produced by the main rotor is ignored, because it can be compensated by the anti-spinning torque $Y_T l_T$ from the tail rotor.

So far, the model structure used for describing the helicopter dynamics has been completed by substituting Eqs. (7.26)–(7.28) and (7.31)–(7.33) into Eq. (7.10). The schematic of the model structure is illustrated in Figure 7.5. It shows that the control signals first alter

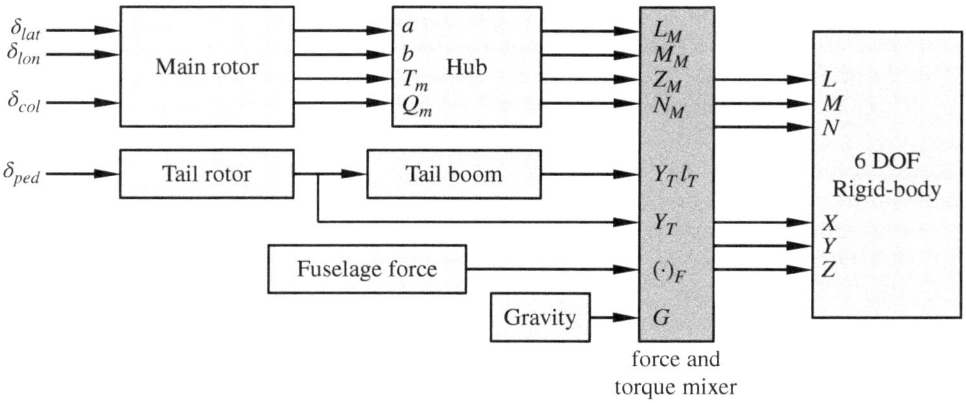

Figure 7.5 Schematic of the model structure.

Table 7.1 Equations in the model schematic.

Blocks	Corresponding equations
Main rotor	(7.18), (7.19), (7.20), (7.21)
Tail rotor	Y_T, $Y_T l_T$
Hub	(7.22), (7.23), (7.24)
Fuselage force	$X_u u$, $Y_v v$, $Z_w w$
Gravity	(7.11)
Mixer	(7.10), (7.26)–(7.28), (7.31)–(7.33)
Rigid-body	(7.8), (7.9), (7.1), (7.3)

the states of the main and tail rotors, which produce corresponding thrusts, torques and flapping angles, then these elements act on the main rotor hub and tail boom to produce different forces and torques on the rigid body of the helicopter. These forces and torques, combining fuselage forces and gravity through a mixer block, complete the helicopter model. The equations contained in each block are summarised in Table.7.1. The question remaining in the modelling is how to decide the specific values of the unknown parameters in the model structure.

7.3 System Identification

This section introduces the identification procedure including the design of flight experiments, identification algorithm, model breakdown and parameter estimation.

7.3.1 Preparation

Flight experiments play an important role in the system identification. Ideally, the flight experiments are executed in an open-loop style. In order to excite the desired mode and obtain the dynamic response over a large frequency range, the pilot should input sweep frequency signals and try not to add too much control effort to remedy the stability of the helicopter (so long as the helicopter does not crash). Moreover, the input signals in the channels of interest should have sufficient magnitude to suppress the interference from disturbances. This is critical since the helicopter is small and light, and the response to disturbances may submerge the real control responses. In the identification process, the model needs to be broken down into small blocks, and corresponding flight tests need to be performed. During the flight tests, at least two sets of data need to be collected for each pattern of flight. One is for identification and the other is for validation.

The algorithm for parameter estimation used in this work is PEM, which is a conventional identification technique and has been included in the Matlab System Identification Toolbox (Ljung 2009). PEM needs a parameterised model and determines the unknown parameter so that the prediction based on the model matches measured data as accurately as possible. It can be considered an optimization process, where the optimisation variables are the unknown parameters of the model, the constraints are the parameter ranges and model equations, and a cost function is used to penalise any deviation between model prediction and measured helicopter response. The cost function that PEM minimises is defined as follows:

$$V_N\left(\theta,Z^N\right)=\frac{1}{N}\sum_{k=1}^{N}l\left(\varepsilon\left(k,\theta\right)\right) \tag{7.34}$$

where θ is the vector of parameters to be estimated, $Z^N=[\mathbf{u_k},\mathbf{y_k}]$, $k=1,2,\cdots,N$ is the given experiment data set, $l(\cdot)$ is a positive defined scalar function, and ε is the error between measured data and predicted responses from the model which depended on θ.

Since essentially there is a nonlinear optimisation involved in the estimation, there is a chance that the algorithm may trap in a local minimum that is not the actual value for a specific system, and the results can be sensitive to the initial guesses of the parameters. To overcome these problems, the full helicopter is partitioned into a number of subsystems to describe its specific movements. The flight tests for each kind of movement are carried out in order to serve the corresponding subsystem identification.

7.3.2 Model Breakdown and Identification

As stated above, identifying all the parameters simultanously is very difficult or even impossible, because there are 16 unknown parameters in the model structure. However, by breaking down the full helicopter model into several parts, it is possible to find suitable initial values for the unknown parameters. The helicopter dynamics can be divided into four channels: roll, pitch, yaw and heave. Thereby, the identification and the corresponding flight tests are implemented according to this sequence. When one channel is excited, the other channel should be kept as calm as possible. To investigate the coupling effect between roll and pitch, those two channels also need to be excited simultaneously. The identification flowchart is shown in Figure 7.6. Note that in the identification process, first the inner-loop attitude dynamics are identified, and then the outer-loop translational movements are investigated.

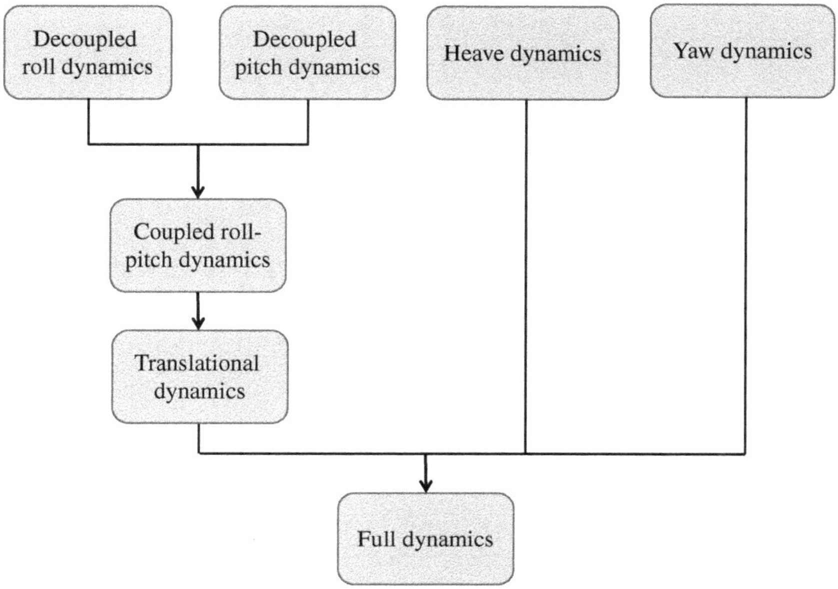

Figure 7.6 Flowchart of identification process.

After the system identification process is broken down, the relevant model in each step is also simplified, because there are fewer unknown parameters to estimate. Furthermore, the results obtained in the current step is set as the initial value for the following identification procedure to avoid sensitivity to the initial guess for the parameter. The detailed steps and results for Figure 7.6 are given in the following subsections.

Decoupled Roll and Pitch dynamics

The roll and pitch dynamics are initially identified in this step. Cross-coupling effects are avoided in the flight experiments, so these terms in the equations can be dropped. Specifically, in identifying the roll channel, the coefficients for flapping angle a and pitch rate q in Eqs. (7.19) and (7.31) are ignored to yield the following equations:

$$\begin{aligned} \dot{p} &= L_b b \\ \dot{b} &= -p - b/\tau + B_{\text{lat}}\delta_{\text{lat}} + B_{\text{lon}}\delta_{\text{lon}} \end{aligned} \tag{7.35}$$

where L_b, B_{lat} and B_{lon} are unknown parameters to be estimated. Similarly, the equations for identifying the pitch channel can be derived from Eqs. (7.18) and (7.32) as:

$$\begin{aligned} \dot{q} &= M_a a \\ \dot{a} &= -q - a/\tau + A_{\text{lat}}\delta_{\text{lat}} + A_{\text{lon}}\delta_{\text{lon}} \end{aligned} \tag{7.36}$$

where M_a, A_{lat} and A_{lon} need to be identified. The results in this step are given in Figure 7.7, where the estimated responses and experimental responses are compared.

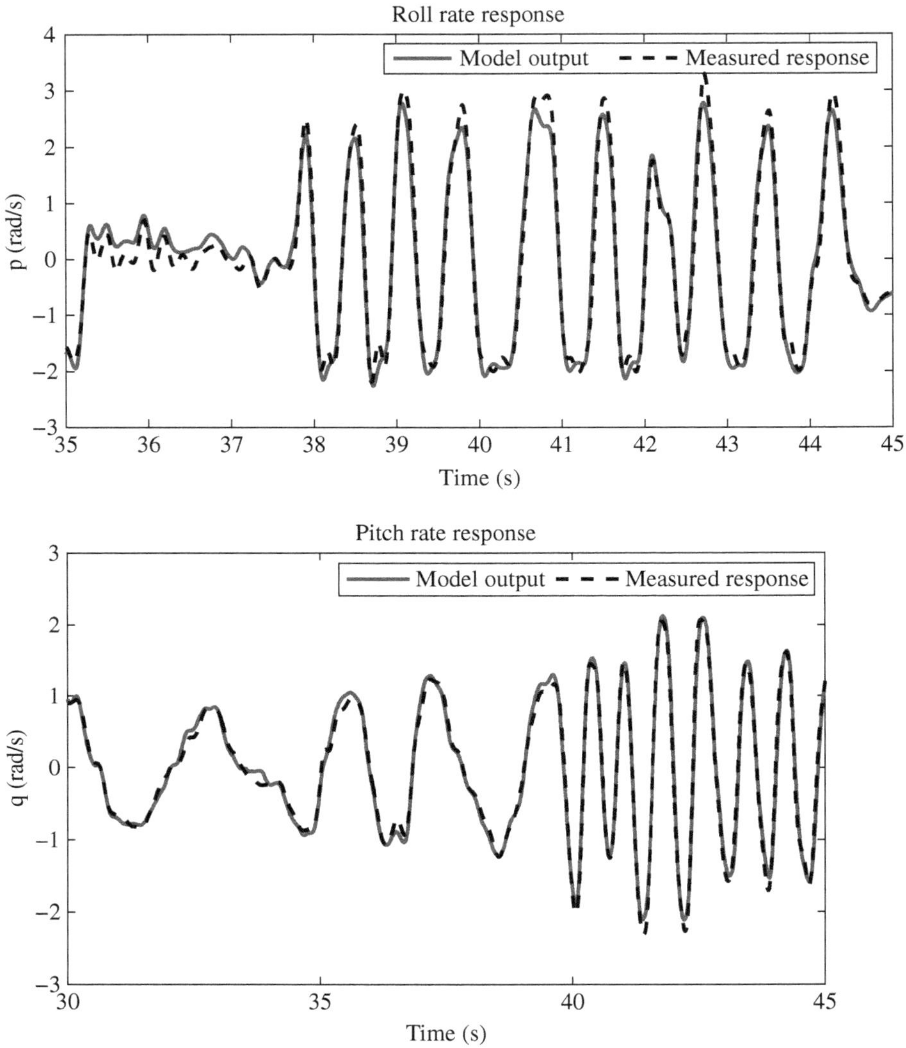

Figure 7.7 Decoupled roll and pitch identification.

Coupled Roll–Pitch Dynamics

In this step, roll and pitch dynamics are integratd through cross-coupling terms L_a and M_b, such that

$$\begin{aligned}\dot{p} &= L_a a + L_b b \\ \dot{q} &= M_a a + M_b b\end{aligned} \tag{7.37}$$

The corresponding flight test should excite both channels in an interactive way to identify L_a and M_b and refine the other parameters from the previous step. The control signals and identification results are shown in Figure 7.8.

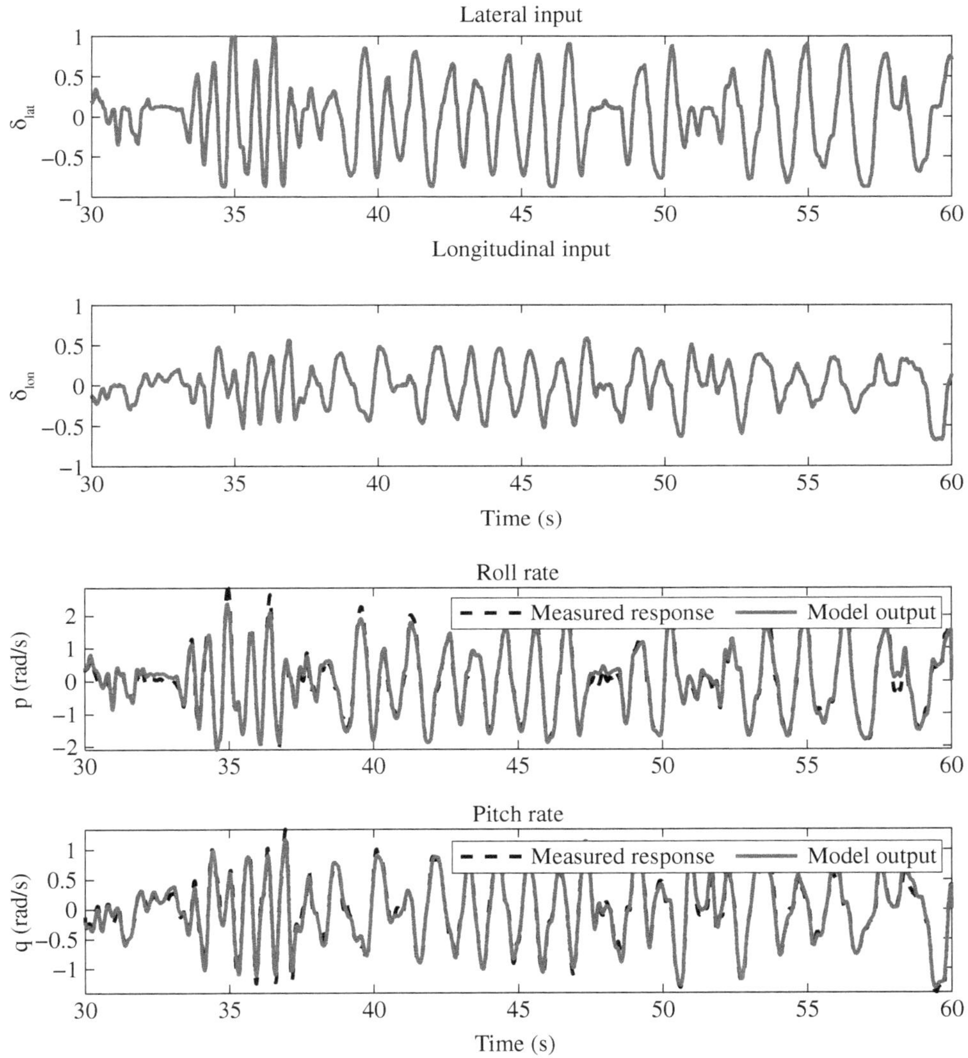

Figure 7.8 Coupled roll and pitch identification.

Translational Equations of Motion

In this step, roll–pitch dynamics from Eq. (7.37) and translational dynamics from Eqs. (7.26)–(7.27) and (7.8) are considered. To reduce the complexity, two variables are defined in Eq. (7.38) to substitute for the corresponding nonlinear terms in Eqs. (7.26) and (7.27). Since the newly defined variables can be measured in the flight test, they can be treated as the control inputs. The key derivatives to be identified in this step are the translational damping terms X_u and Y_v.

$$\begin{aligned} termX &= -wq + vr - g\sin\theta \\ termY &= -ur + wp + g\sin\phi\cos\theta \end{aligned} \tag{7.38}$$

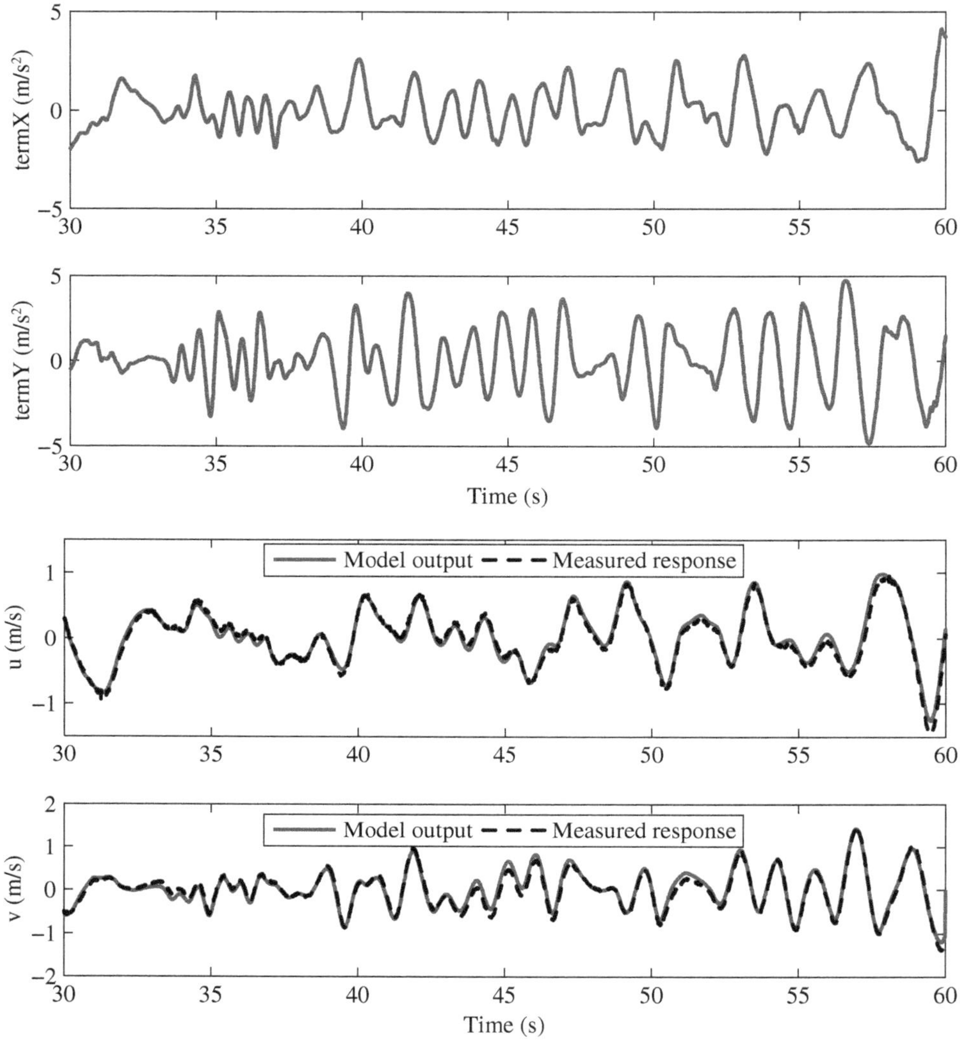

Figure 7.9 Translational dynamic identification.

The same set of data used for the coupled roll–pitch dynamics is used to identify the translational movements. Comparing with the existing state-space helicopter model of Mettler et al. (2001), the model proposed in this paper introduces the two nonlinear terms to increase the fidelity. The values of *termX*, *termY* and identification results are shown in Figure 7.9.

Yaw Dynamics

The Trex-250 helicopter is equipped with a head holding gyro, which makes the yaw channel quite stable and decoupled from the other channels. Hence, a first-order system

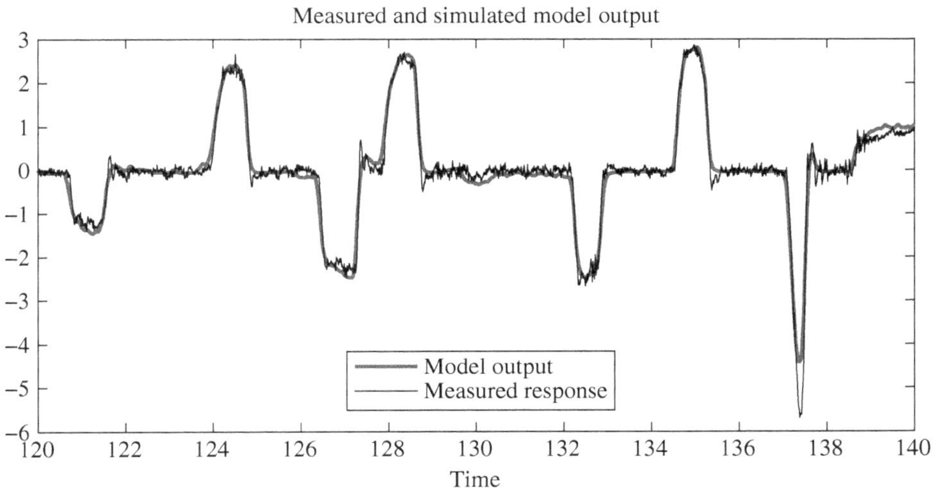

Figure 7.10 Yaw dynamic identification.

is adequate to model the dynamics with derivatives of N_r, N_{ped} and N_{col} from Eq. (7.33). The control signal and identification results are given in Figure 7.10.

Heave Dynamics

In this step, the unknown parameters Z_w and Z_{col} in Eq. (7.28) are identified. Similar to the strategy for the translational equations of motion, the nonlinear term is defined as an input variable in Eq. (7.39).

$$termZ = -vp + uq + g\cos\phi\cos\theta - g \tag{7.39}$$

Full Helicopter Model

After the previous steps, there are initial estimates for all of the unknown parameters. The full helicopter model is now evaluated in this step, and all the parameters are further refined using the previous results as initial values. The parameters identified are given in Table 7.2.

7.3.3 Result Validation

The model validation is conducted by driving the identified model using another set of flight data that was not used in the identification process. The validation flight data includes excitations on all the channels to verify the model accuracy with respect to coupled helicopter dynamics. The validation results are presented in Figure 7.11. There is good agreement between the flight data and the predictions, especially for longitudinal and lateral movements.

To examine the results quantitatively, a measurement of the best fit Ljung (2009) is introduced, which is defined as the percentage of the output that the model reproduces:

$$\text{BestFit} = \left(1 - \frac{|\mathbf{y} - \hat{\mathbf{y}}|}{|\mathbf{y} - \bar{\mathbf{y}}|}\right) \times 100 \tag{7.40}$$

Table 7.2 Identified parameters for Trex-250.

Parameter (unit)	Identified value	Standard deviation	Parameter (unit)	Identified value	Standard deviation
X_u (s^{-1})	−0.233	0.006	Y_v (s^{-1})	−0.329	0.006
Z_w (s^{-1})	−0.878	0.022			
L_a (s^{-2})	83.98	11.21	L_b (s^{-2})	745.67	13.85
M_a (s^{-2})	555.52	9.89	M_b (s^{-2})	11.03	3.44
τ (s)	0.045	0.001	N_r (s^{-1})	−23.98	0.63
A_{lat} (rad)	0.196	0.003	A_{lon} (rad)	1.945	0.006
B_{lat} (rad)	2.120	0.007	B_{lon} (rad)	−0.38	0.010
Z_{col} (m/s^2)	−5.71	0.06			
N_{col} (rad/s^2)	8.89	2.9	N_{ped} (rad/s^2)	113.65	2.9

where $\mathbf{y}$ is the measured helicopter response, $\hat{\mathbf{y}}$ is the output predicted by the identified model and $\overline{\mathbf{y}}$ is the mean of $\mathbf{y}$. A value of 100% corresponds to a perfect fit, and 0% indicates that the fit is no better than approximating the output by a constant ($\hat{\mathbf{y}} = \overline{\mathbf{y}}$). The best fit values are also provided in Figure 7.11.

In the flight test for model validation, all channels in the helicopter dynamics are excited simultaneously, and the model predictions still match the experimental data with high accuracy. The identification process outlined here therefore captures the majority of the cross-coupling effects. In contrast, in most of the literature (Mettler et al. 2001 and Shin et al. 2005), time-domain validations are carried out with only one channel excited at a time.

There are some mismatches in both the heave and yaw channels. For the former, the differences are mainly due to the high nonlinearity inherent in the rotor system and excited by the aggressive flight. Effects such as inflow air instability and rotor-speed variation cannot be captured by the present thrust model. To this end, a more comprehensive thrust model must be adopted to improve the accuracy. The mismatch in the yaw channel mainly exists in the high-frequency range. This is due to the presence of the head holding gyro, which always tries to compensate for external disturbances. The identified model is able to capture the trend of the yaw movement and it is adequate for control design.

It is worth pointing out that although efforts have been made to obtain an accurate model from flight test data, there is still a modelling uncertainty with respect to real helicopter dynamics. The uncertainty, on one hand, is caused by the limitation of the current model structure and the identification method as discussed above. But it can also be introduced by physical changes in the helicopter. Because of the light structure of a miniature helicopter, its dynamics are vulnerable to mechanical changes, different payloads, component upgrades and repairs after crashes. Therefore, the control algorithm designed for a miniature helicopter requires a certain robustness to perform well in practice.

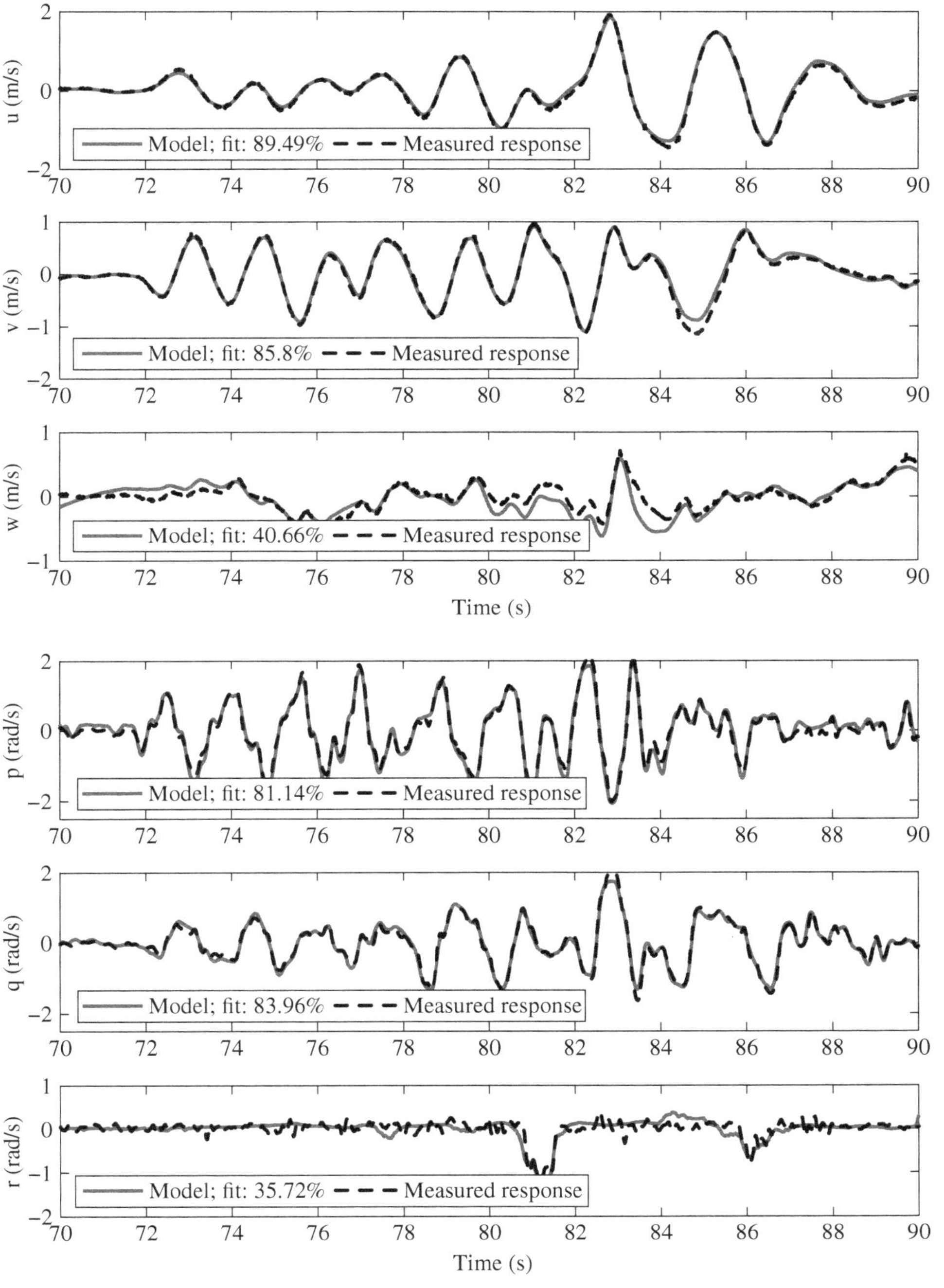

Figure 7.11 Model validation.

7.4 Basic Control Design

After a mathematical model of the helicopter has been obtained, a flight control system can be designed using model-based techniques. This section describes the controller design to support the non-aggressive flight of a small helicopter. Around the hover mode, helicopter dynamics can be considered as a linear system with multiple inputs and multiple outputs. Then, by decoupling the helicopter dynamics into several subsystems, a cascaded PID controller can be developed.

7.4.1 Model Analysis

A helicopter has four primary input commands δ_{lat}, δ_{lon}, δ_{col} and δ_{ped} for controlling the roll and pitch rates, vertical velocity and yaw rate, respectively. In piloting a helicopter, the cyclic inputs produce rotor moments and then change the fuselage attitude, but they cannot control the vehicle's position and velocity directly. After the fuselage's roll and pitch angles are changed, the rotor thrust is tilted accordingly to produce horizontal projections as propulsive forces. In this way, the helicopter's translational states can be controlled. On the other hand, the dynamics of a helicopter can be divided into slow translational movements (outer loop) and fast attitude movements (inner loop). Although this decomposition is only valid in non-aggressive flight, it provides a guideline for control analysis. So when a desired vehicle position is given, the controller first transfers a position requirement into an attitude command in a low bandwidth, and then controls the helicopter to track such a command in a high bandwidth.

7.4.2 PID Controller

By following the above principle, a cascaded control structure can be derived for a helicopter's low-speed flight. First, an inner-loop controller is used to stabilise the attitude dynamics; this follows attitude commands and generates cyclic controls. Then, an outer-loop controller is designed for tracking the position and heading reference. It generates the collective pitch and pedal controls directly based on height and heading errors and provides desired attitude commands for an inner-loop controller based on lateral and longitudinal errors. The control diagram is shown in Figure 7.12.

In this control structure, the outer-loop controller employs four PID controllers. The first two measure the lateral and longitudinal errors e_x^b and e_y^b in body-fixed coordinates and generate the desired attitude requirements ϕ_r and θ_r. Note that the position error is usually compared in the inertial coordinates, so a correction block is introduced to transfer errors $e_x^i = x_r - x$ and $e_y^i = y_r - y$ into body coordinates according to the heading angle ψ:

$$
\begin{aligned}
e_x^b &= e_x^i \cos\psi + e_y^i \sin\psi \\
e_y^b &= -e_x^i \sin\psi + e_y^i \cos\psi
\end{aligned}
\tag{7.41}
$$

The other two PID controllers in the outer-loop eliminate the height and heading errors directly. In the inner-loop, there are another two PID controllers used to track the attitude command.

An important feature of PID-based control is that it can be implemented without a specific model of the vehicle dynamics. All of the feedback gains in the control structure can be tuned empirically based on the observation of flight performance. However, this

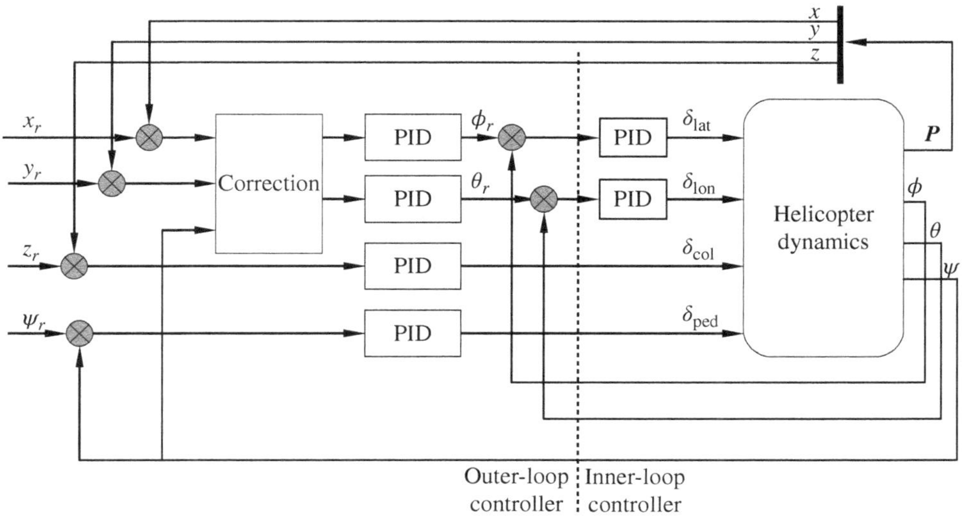

Figure 7.12 Cascaded control structure.

Table 7.3 PID gains.

	K_p	K_i	K_d
Lateral	0.415 2	0.002	0.592
Longitudinal	−0.313	−0.002	−0.450
Height	−0.455	−0.010	−0.470
Heading	0.750	0.010	0.500
Roll	0.507	0.010	0.318
Pitch	0.751	0.010	0.696

is a very tedious process and will be very risky if a 'trial and error' approach is adopted for a real helicopter. Since the helicopter model is available, the tuning of feedback gains can be carried out using model-based design. To this end, a Simulink model is created including both the helicopter model and the control structure. The overall control system is then decomposed into four single-input, single-output subsystems corresponding to four control channels: δ_{lat} to position y, δ_{lon} to position x, δ_{col} for height z and δ_{ped} for heading ψ. For each subsystem, we use Matlab's control design toolbox to optimise PID gains based on a step-response criterion. The optimised gains are given in Table 7.3.

Simulations are carried out to verify the performance of the basic PID control. The step-response results are shown in Figure 7.13. It can be seen that the cascaded PID has satisfactory performance for non-aggressive flight.

It can be seen that the cascaded PID has its performance limits, as it is designed based on a linearised and decoupled model. The decoupling means not only separation of translational and attitude dynamics – slow outer loop and fast inner loop – but also four separated control channels: lateral, longitudinal, heading and heave. In addition, PID control exhibits poor robustness in experiments. The control parameters tuned for a

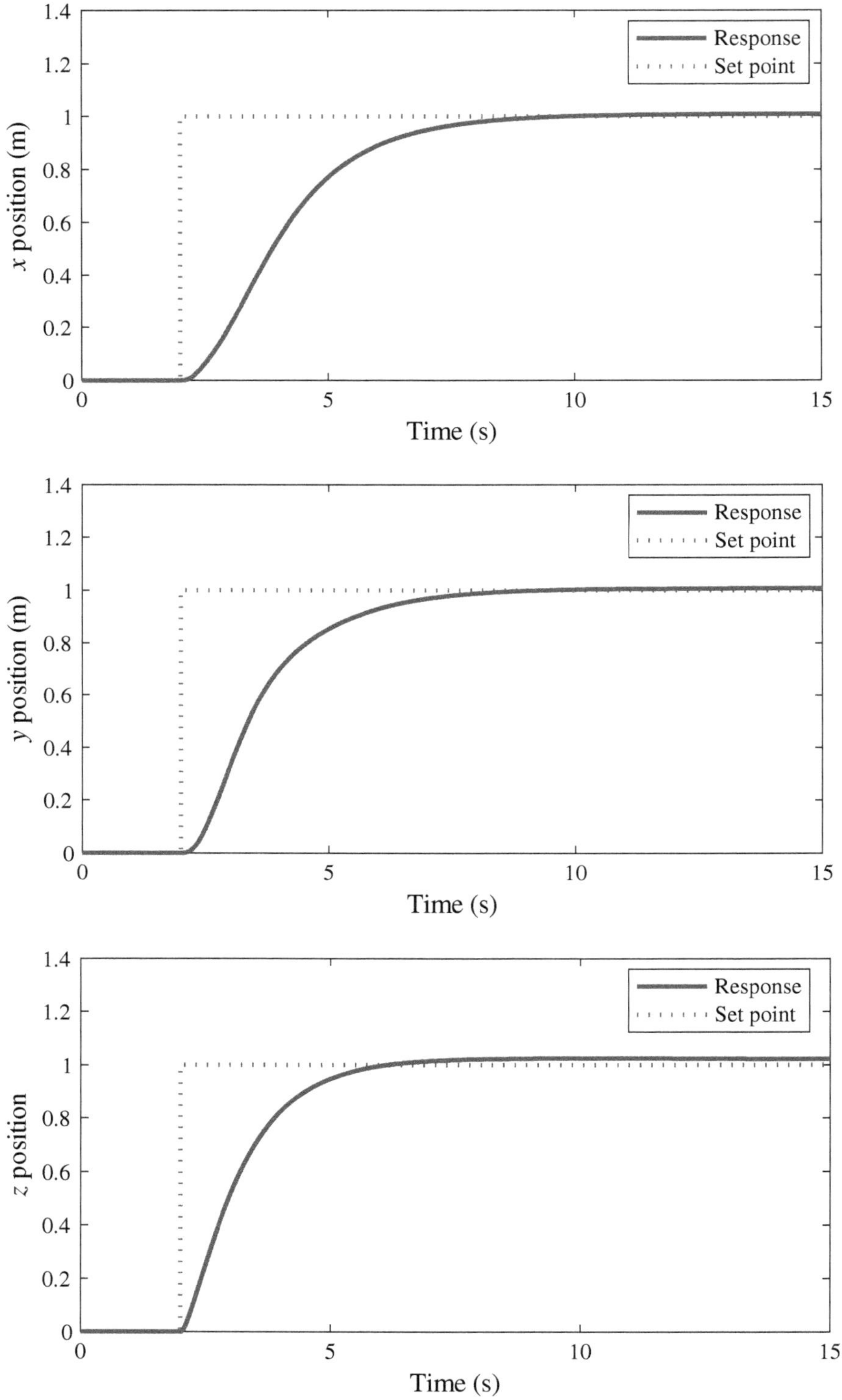

Figure 7.13 Step responses under PID control.

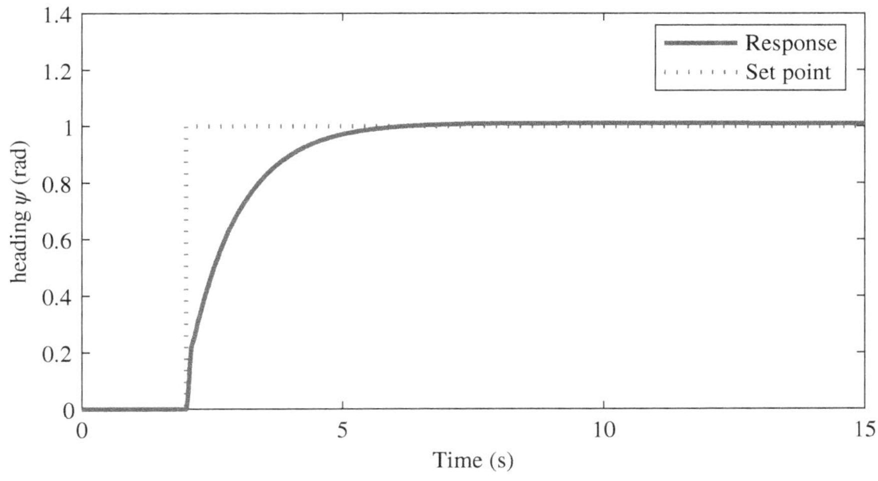

Figure 7.13 *(Continued)*

helicopter may not work for the same helicopter in the next experiment due to the trivial mechanical changes. However, the nonlinear dynamic model developed in this chapter paves the way to many advanced model-based control techniques that can improve the flight performance and increase robustness in various flight conditions. A good survey on unmanned helicopter control can be found in Kendoul (2012).

7.5 Conclusion

This chapter describes a control-oriented modelling and system identification procedure based on a Trex-250 miniature helicopter. This can be extended to other small helicopters with Bell-Hiller stabilizer bar mechanism. The model adopted in this work has a simple structure but contains the major nonlinear effects, so as to cover a larger flight region. To circumvent some difficulties in optimisation-based nonlinear system identifications, a practical process is detailed, including model breakdown and the corresponding flight manoeuvre design. Specific flight test patterns have been used in different identification steps to provide more deterministic estimations. The cross-coupling feature of the helicopter has also been captured during this process, with both lateral and longitudinal channels excited. The consequent identification results show a good match between the model predictions and real flight data. The identified helicopter model can be used for flight control design and online prediction in path-planning algorithms (Liu et al. 2011). In addition, simulation shows the cascade PID controller provides satisfactory performance; this could be used as a baseline for further studies on advanced flight control techniques.

Bibliography

Cai G, Chen B and Lee T 2011 *Unmanned Rotorcraft Systems* Advances in Industrial Control. Springer.

Cai G, Chen BM, Lee TH and Lum KY 2012 Comprehensive nonlinear modeling of a miniature unmanned helicopter. *Journal of the American Helicopter Society* **57**(1), 1–13.

Cook M 2007 *Flight Dynamics Principles* Elsevier Aerospace Engineering Series. Elsevier/Butterworth-Heinemann.

Gavrilets V, Martinos I, Mettler B and Feron E 2002 Control logic for automated aerobatic flight of miniature helicopter *AIAA Guidance, Navigation, and Control Conference and Exhibit, Monterey, CA*.

Gavrilets V, Mettler B and Feron E 2001 Dynamic model for a miniature aerobatic helicopter *AIAA Guidance, Navigation, and Control Conference and Exhibit, Montreal, Canada*.

Grauer J, Conroy J, Hubbard Jr J and Pines D 2009 System identification of a miniature helicopter. *Journal of Aircraft*.

Hald U, Hesselbæk M, Holmgaard J, Jensen C, Jakobsen S, Siegumfeldt M, la Cour-Harbo A and Thomsen J 2005 Autonomous helicopter – modelling and control. Technical report, Aalborg University.

Heffley R and Mnich M 1988 Minimum-complexity helicopter simulation math model. Technical report, NASA, Moffett,CA.

Hoffer N, Coopmans C, Jensen A and Chen Y 2014 A survey and categorization of small low-cost unmanned aerial vehicle system identification. *Journal of Intelligent & Robotic Systems* **74**(1–2), 129–145.

Kendoul F 2012 Survey of advances in guidance, navigation, and control of unmanned rotorcraft systems. *Journal of Field Robotics* **29**(2), 315–378.

Kim HJ and Shim DH 2003 A flight control system for aerial robots: algorithms and experiments. *Control Engineering Practice* **11**(12), 1389–1400.

Kim S and Tilbury D 2009 Mathematical modeling and experimental identification of an unmanned helicopter robot with flybar dynamics. *Journal of Robotic Systems* **21**(3), 95–116.

La Civita M, Messner W and Kanade T 2002 Modeling of small-scale helicopters with integrated first-principles and system-identification techniques *Proceedings of the 58th Forum of the American Helicopter Society*, vol. **2**, pp. 2505–2516.

Liu C, Chen WH and Andrews J 2011 Piecewise constant model predictive control for autonomous helicopters. *Robotics and Autonomous Systems* **59**(7–8), 571–579.

Ljung L 2009 MATLAB System Identification Toolbox, Version 7.3.

Mettler B 2003 *Identification Modeling and Characteristics of Miniature Rotorcraft*. Kluwer Academic Pub.

Mettler B, Dever C and Feron E 2004 Scaling effects and dynamic characteristics of miniature rotorcraft. *Journal of Guidance, Control, and Dynamics* **27**(3), 466–478.

Mettler B, Tischler M and Kanade T 1999 System identification of small-size unmanned helicopter dynamics *Proceedings of the 55th Forum of the American Helicopter Society*.

Mettler B, Tischler MB and Kanade T 2001 System identification modeling of a small-scale unmanned helicopter. *Journal of the American Helicopter Society*.

Morris J, van Nieuwstadt M and Bendotti P 1994 Identification and control of a model helicopter in hover *Proceedings of the 1994 American Control Conference*, vol. **2**, pp. 1238–1242.

Padfield G 2007 *Helicopter Flight Dynamics: the Theory and Application of Flying Qualities and Simulation Modeling*. American Institute of Aeronautics and Astronautics.

Shin J, Nonami K, Fujiwara D and Hazawa K 2005 Model-based optimal attitude and positioning control of small-scale unmanned helicopter. *Robotica* **23**(01), 51–63.

8

Aerodynamic Derivative Calculation Using Radial Basis Function Neural Networks

Ranjan Ganguli

Department of Aerospace Engineering, Indian Institute of Science, Bengaluru, India

8.1 Introduction

Aerodynamic stability and control parameters or derivatives are widely used in real-time simulation, handling qualities analysis and control system design. They play an important role in the development of robotic helicopter UAVs (Kendoul et al 2009; Mokhtari and Benallegue 2004; Mondrag et al 2010). Such helicopters are important for the monitoring of traffic, search and rescue operations, agricultural spraying, logistics and so on (Shakernia et al 1999a,b). Rotorcraft are important unmanned systems due to their capabilities of vertical and low-speed flight. Several methods of aerodynamic derivative calculation for rotorcraft have been proposed (Agard 1991; Padfield 1999; Prouty 1986). The aerodynamic parameters calculated using the system identification based method are more accurate than those derived from other methods, such as analytical and numerical differentiation (Agard 1991; Padfield 1999). System identification involves reconstructing a simulation model structure and model parameters from experimental flight data. Typical system identification techniques range from simple curve fitting algorithms to complex statistical error analysis.

System identification has become a significant tool for applications such as model validation, handling qualities evaluation, control law design, and flight-vehicle design and certification (Jategaonkar et al 2004). Estimation and system identification are important problems for rotorcraft micro air vehicles (MAVs). Shen et al (2013) point out that rotorcraft MAVs fly well in 3D environments and can hover in place and maneuver around obstacles. They focus on the problem of estimating the state of rotorcraft MAVs using only onboard cameras and an inertial measurement unit (IMU). They used the pelican quadrotor with an IMU (accelerometer, gyroscope). The vision-based state estimation onboard the rotorcraft MAV permits autonomous flight. Chowdhary and Jategaonkar (2010) mention that aerodynamic parameter estimation is integral to aerospace system design. They used the unscented Kalman filter (UKF) (Julier and Uhlmann 2004) for aerodynamic parameter estimation of a rotary-wing UAV from real flight data. The baseline vehicle was the Benda Genesis 1800 helicopter. They used a

Advanced UAV Aerodynamics, Flight Stability and Control: Novel Concepts, Theory and Applications, First Edition. Edited by Pascual Marqués and Andrea Da Ronch.

Companion website: http://www.wiley.com/go/marques/advanced_UAV_aerodynamics

linear estimation model, which is suitable for rotorcraft in hover (Lorenz and Chowdhary 2005; Mettler 2003). Sensor measurement for all velocities, angular rates and Euler angles were obtained. These measurements are typically corrupted by noise and vibration of the rotorcraft. They concluded that the UKF is a powerful system identification tool.

The methodology and significance of system identification for flight vehicles is available in the literature (Hamel and Kaletka 1997; Hamel and Jategaonkar 2005; Jategaonkar 2006; Morelli and Klein 2005; Tishler and Remple 2006). The system identification method uses techniques such as the maximum likelihood method, equation error method, output error method and filter error method. These methods require a mathematical model of the aircraft with a set of initial values for the parameters to initiate the algorithm (Iliff and Maine 1985, 1986). The identification methods are also affected by the presence of noise, such as state noise or measurement noise (Agard 1991). The identification process becomes more difficult with the increase in the number of degrees of freedom (DoFs) and model parameters. This can happen due to insufficient information content in real-time flight data (Fu and Kaletka 1993). These problems related to conventional parameter estimation techniques hinder the inclusion of higher-order DoFs in model development. However, higher-order modeling is essential for nap-of-earth flight, aerial combat, high-*g* maneuvers and the design of high-gain flight control systems (Fu and Kaletka 1993; Pavel 2001). All of these are important issues for unmanned rotorcraft.

New methods for aircraft parameter estimation aim to overcome the shortcomings of classical system identification methods. One such method is the application of artificial neural networks (ANNs) for parameter estimation (Raol and Jategaonkar 1995; Raisinghani et al 1998a,b; Vijaykumar et al 2006). The potential of ANNs has already been demonstrated in fields such as nonlinear flight control, system identification, structural damage detection and identification, flight certification, failure rate prediction and modeling complex phenomena such as ice accretion (Amin et al 1997; Al-Garni et al 2006; Ganguli et al 1997, 1998; Habib and Zaghloul 1996; Kim and Calise 2005; Malaek and Izadi 2006; Ogretim et al 2006; Reddy and Ganguli 2003; Suresh et al 2005, 2006; Tsou and Shen 1994). Neural networks have been applied to helicopter problems (Horn et al 2002; Kottapalli 2006; Sahani et al 2006). However, the applicability of recurrent neural networks in parameter estimation is limited because of the fixed number of neurons needed for state-space formulation (Raol and Jategaonkar 1995). In contrast, feed-forward neural networks (FFNNs) can approximate any measurable function to any desired level of accuracy (Horni et al 1989). This property of FFNNs makes them an ideal choice for aircraft parameter estimation.

To avoid the shortcomings of conventional methods, two new neural network based techniques, the delta method and the zero method, were proposed by Raisinghani and his co-workers (1998a; 1998b). These methods do not require initial estimates of the parameters; they can be directly extracted from the flight data. Raisinghani's approach complements existing methods and also has the potential for online parameter estimation. It is based on a type of FFNN known as the multi-layer perceptron (MLP) which is popular in many applications (Ganguli et al 1998; Suresh et al 2004). However, there are some drawbacks to using MLP, such as the slow convergence rate, computational memory requirements and the sensitivity to outliers. Such difficulties can be avoided by using a different type of FFNN known as radial basis function networks (RBFNs) (Baraldi et al 2000; Park et al 2002; Suresh et al 2004).

Raisinghani et al (1998a) estimated aerodynamic derivatives from simulated fixed-wing data generated from decoupled lateral and longitudinal modes. This modal decoupling is a reasonable assumption for a fixed-wing aircraft. For rotary-wing aircraft, lateral and longitudinal motions are strongly coupled, so the coupled equations of motion should be used to generate simulated data. Raisinghani et al (1998b) also suggested the removal of 25% of estimates from both ends of the ordered set of estimates because of practical problems in parameter estimation in that region This requirement has a negative impact in helicopter problems as it severely effects both transition and the characteristics of the high-speed flight regime. This drawback of the MLP-based delta method in derivative computation is overcome by application of RBFN in place of MLP (Kumar et al 2008).

The objective of the present chapter is to outline a methodology for computation of aerodynamic derivatives directly from flight data using RBFN. The proposed approach does not require a mathematical model of the rotorcraft. The delta method using RBFN is first applied to simulated data with added measurement and state noise. The simulated data is generated by a 6-DoF nonlinear helicopter simulation model, which includes the coupled longitudinal and lateral modes. The method is used to compute derivatives from real time flight test data of the Bo 105 helicopter. The results from the RBFN-based delta method are compared with the identified derivatives from 6-DoF and 9-DoF models available in the literature. The modified 3211 pilot control input is used for rotorcraft parameter estimation. The estimated parameters are then validated from the data generated by a frequency sweep pilot control input, allowing assessment of the predictive capability of the RBFN-based delta method.

8.2 Helicopter Aerodynamic Derivatives

The helicopter can be mathematically modeled by considering it as a number of subsystems: the main rotor, fuselage, powerplant, empennage, tail rotor and flight control systems (Padfield 1999). The helicopter dynamics are assessed about its center of gravity (cg). An orthogonal axes system is set up with the cg as the origin, as shown in Figure 8.1.

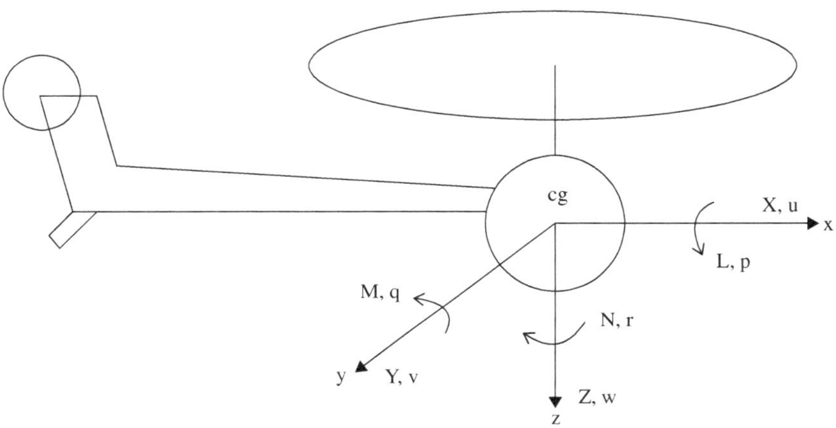

Figure 8.1 Helicopter coordinate system, velocity components and external forces and moments.

We assume that the cg is fixed, even though the rotor blades flap. The equations of motion for this helicopter are derived using physical laws of conservation of momentum and energy and can be expressed as,

$$\frac{d\mathbf{x}}{dt} = \mathbf{F}(\mathbf{x},\mathbf{u},t)$$

With initial conditions $\mathbf{x}(0) = \mathbf{x}_0$. Here $\mathbf{x}(t)$ is a vector of state variables, $\mathbf{u}(t)$ is a vector of control variables and f is a nonlinear function. For a 6-DoF rigid-body model,

$$\mathbf{x} = [u,w,q,\theta,v,p,\phi,r,\psi]$$

Here, u, v, w are three translational velocity components; p, q, r are three rotational velocity components and ϕ, θ and ψ are Euler's angles. The function $\mathbf{F}$ includes the applied forces and moments that typically emanate from aerodynamic gravitational, inertial and structural sources. The control vector is,

$$\mathbf{u} = [\theta_0, \theta_{1s}, \theta_{1c}, \theta_{tr}]$$

Here θ_0, θ_{1s}, θ_{1c} and θ_{tr} are the main rotor collective, longitudinal cyclic, lateral cyclic and tail rotor collective. The expanded form of the equations is (Padfield 1999):

$$\begin{aligned}
\dot{u} &= -(wq - vr) + \frac{X}{M_a} - g\sin\theta \\
\dot{v} &= -(ur - wp) + \frac{Y}{M_a} + g\cos\theta\sin\phi \\
\dot{w} &= -(vp - uq) + \frac{Z}{M_a} + g\cos\theta\cos\phi \\
I_{xx}\dot{p} &= (I_{yy} - I_{zz})qr + I_{xz}(\dot{r} + pq) + L \\
I_{yy}\dot{q} &= (I_{zz} - I_{xx})rp + I_{xz}(r^2 - p^2) + M \\
I_{zz}\dot{r} &= (I_{xx} - I_{yy})pq + I_{xz}(\dot{p} - qr) + N \\
\dot{\phi} &= p + q\sin\phi\tan\theta + r\cos\varphi\tan\theta \\
\dot{\theta} &= q\cos\phi - r\sin\phi \\
\dot{\psi} &= q\sin\phi\sec\theta + r\cos\phi\sec\theta
\end{aligned}$$

Here M_a is the aircraft mass and the moment and products of inertia are (Padfield 1999):

$$\begin{aligned}
&x-\text{axis}: \quad I_{xx} = \int_{\text{body}} (y^2 + z^2)\,dm \\
&y-\text{axis}: \quad I_{yy} = \int_{\text{body}} (x^2 + z^2)\,dm \\
&z-\text{axis}: \quad I_{zz} = \int_{\text{body}} (x^2 + y^2)\,dm \\
&xz-\text{axes}: \quad I_{xz} = \int_{\text{body}} xz \quad dm
\end{aligned}$$

The external forces X, Y, Z and moments L, M, N can be expanded as a Taylor series. For example,

$$X = X_e + \frac{\partial X}{\partial u}\delta u + \frac{\partial X}{\partial w}\delta w + \cdots + \frac{\partial X}{\partial \theta_0}\delta\theta_0 + \cdots \text{etc.}$$

Similar expansions can be obtained for Y, Z, L, M and N. The partial derivative implies that all other state and control variables are held fixed and only one is perturbed. By convention, the derivatives are written as

$$\frac{\partial X}{\partial u} = X_u, \quad \frac{\partial X}{\partial \theta_{1c}} = X_{\theta_{1c}}$$

Using the derivatives, the linearized equation of motion are written as,

$$\dot{\boldsymbol{x}} = \boldsymbol{A}\boldsymbol{x} + \boldsymbol{B}(\boldsymbol{t}) + \boldsymbol{f}(\boldsymbol{t})$$

where the function $\boldsymbol{f}(\boldsymbol{t})$ represents disturbances. The system matrix $\boldsymbol{A}$ and control matrix $\boldsymbol{B}$ can be written as

$$\boldsymbol{A} = \left(\frac{\partial \boldsymbol{F}}{\partial \boldsymbol{x}}\right)_{x=xe}, \quad \boldsymbol{B} = \left(\frac{\partial \boldsymbol{F}}{\partial \boldsymbol{x}}\right)_{x=xe}$$

There are 36 stability derivatives and 24 control derivatives for a 6-DoF model. Some of these are more important than others. Estimation of these derivatives allows us to form a model for developing control algorithms for the unmanned rotorcraft. This chapter presents a general approach for calculating these derivatives.

8.3 Radial Basis Function Neural Networks

Artificial neural networks (ANNs) are a nonlinear function approximation tool that can be used to model complex relationships between the input and output of a system. ANNs can be classified on the basis of the type of connectivity between neurons, type of network architecture and the number of layers in the network. Feed forward neural networks (FFNNs) are composed of several layers of neurons, namely input layer, hidden layer(s) and output layer. Each neuron is connected to others through weights. The training process of FFNNs involves changing the weights to obtain a desired input–output relationship. More details about ANNs of various types is available in Haykins (1994).

Radial basis function networks (RBFNs) are a curve-fitting method in high-dimensional space. The basic architecture of RBFNs involves three layers, as shown in Figure 8.2. The input layer contains source nodes. The second layer is the hidden layer, which transforms the input space to a high-dimensional hidden space. Generally the hidden layer performs a nonlinear transformation. The third layer is made of output neurons, which transforms the hidden space to the output space. Typically, the output neurons perform a linear transformation. The weights in RBFNs exist only between the hidden layer and the output layer. RBFNs are analogous to the Gaussian density function, which

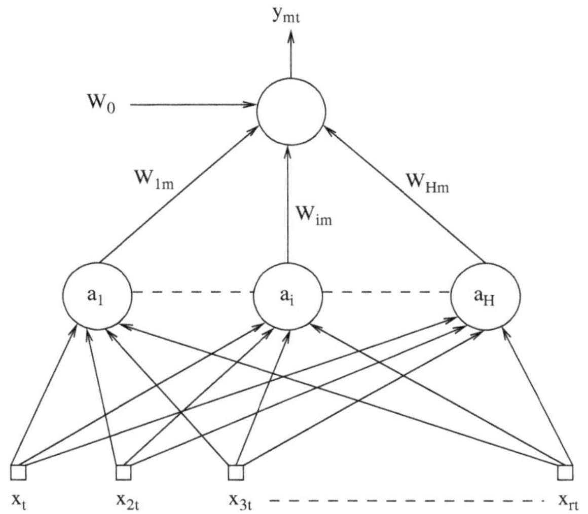

Figure 8.2 Architecture of radial basis function networks.

is defined by a center position and width parameter. The Gaussian density function is a maximum when the input variables are at the center and decreases monotonically as the distance of the variables from the center increases. The batch mode *k*-means clustering algorithm is applied for determining RBFN centers and the weights between the hidden layer and output layer are calculated by the linear least squares optimization algorithm (Haykins 1994). The *k*-means clustering algorithm uses the given tracing data to find a set of clusters with each dimensional center. The dimensions of each center are determined by the number of input nodes. This cluster center becomes the center of the RBFN unit.

The overall input–output mapping can be written as:

$$y_{mt} = \sum_{i=1}^{H} W_{im} a_i(x_t) + W_0 \tag{8.1}$$

where Gaussian function $a_i(x_t)$ is defined as

$$a_i(x_t) = \exp\left[-\sum_{j=1}^{\bar{r}} \frac{\left[x_{jt} - \hat{x}_{ji}\right]^2}{\sigma_i^2}\right] \tag{8.2}$$

8.4 The Delta Method

The delta method is an ANN-based approach for parameter estimation. The derivative calculated using this method considers all the data points, thus smoothing noisy data. Derivative calculation by neural network is given by

$$\frac{dy}{dx}\Big|_{x=x_i} = \frac{\hat{f}(x_1, \ldots.\, x_{i+1}, \ldots x_n) - \hat{f}(x_1, \ldots..\, x_{i-1}, \ldots x_n)}{2(x_{i+1} - x_{i-1})} \tag{8.3}$$

The neural network is first trained for a particular set of input–output values to create the approximate function $\hat{f}(x)$. Then modified input files are prepared in which the variable about which the derivative has to be calculated is perturbed and rest of the inputs are kept constant. These modified input files are then used to calculate derivatives using the trained neural network and the predicted values of outputs are used with the central difference technique. The aerodynamic derivative calculation procedures for fixed-wing aircraft are given by Raisinghani et al (1998a,b). The practical difficulties faced by those authors at the extremes can be overcome by using RBFNs in place of MLP. As mentioned before, this is necessary for rotorcraft.

8.4.1 A Simple Example

The delta method is illustrated by applying it to a simple, single-input, single-output, nonlinear function given by

$$y = 0.8 \sin x \qquad 0 \le x \le \pi \tag{8.4}$$

The data is generated for 100 points, in which 50 points are used for training and 50 points for testing of the neural network. Then the delta method is applied to both RBFN and MLP. The number of trainable parameters in both networks is kept identical, to ensure a fair comparison. The derivative is calculated by both RBFN and MLP and the results are compared with the analytical solution given by Eq. (8.5).

$$\frac{dy}{dx} = 0.8 \cos x \qquad 0 \le x \le \pi \tag{8.5}$$

It can be seen from Figure 8.3 that the MLP-based delta method gives an error at the extremes. On the other hand, the RBFN-based delta method computes the derivative

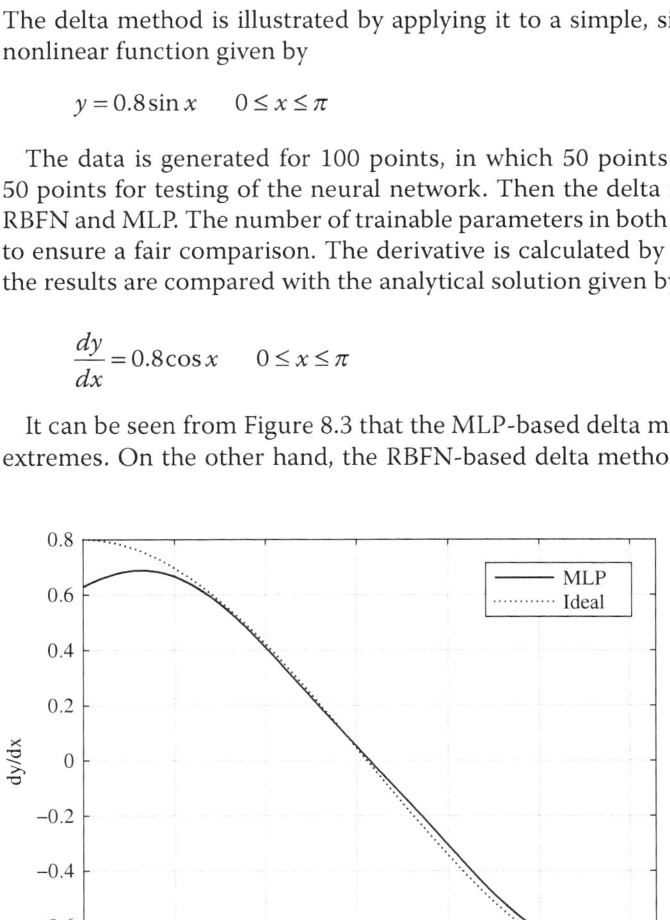

Figure 8.3 Derivative computation of $y = 0.8 \sin x$ using the delta method and multi-layer perceptron compared with ideal analytical values.

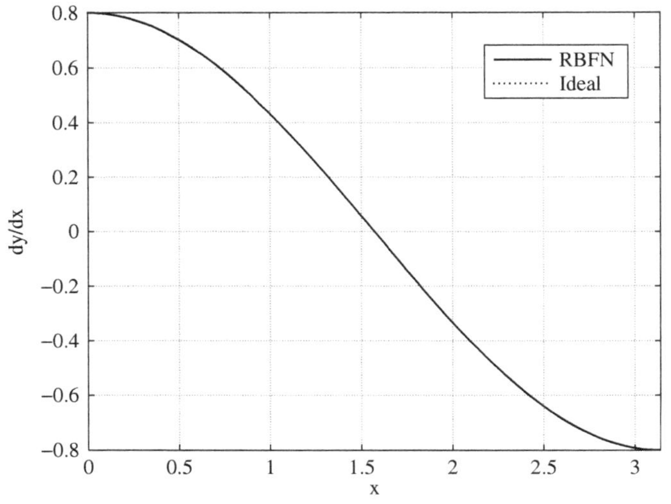

Figure 8.4 Derivative computation of $y = 0.8\sin x$ using delta method and radial basis function network compared with ideal analytical values.

accurately, over the whole range, as shown in Figure 8.4. The effect of noise in the data on derivative computation is also illustrated for this simple function. The noise in the simulated data is added as follows. Assume that we have to include noise in a quantity $\boldsymbol{Y}$, which is a vector of k elements. The $\boldsymbol{Y}_{\text{noise}}$ is obtained as

$$\boldsymbol{Y}_{\text{noise}} = \boldsymbol{Y}\left(1 + \left(2\alpha\boldsymbol{R} - \alpha\right)\right) \quad (8.6)$$

where α is the noise level and $\boldsymbol{R}$ is a vector of random numbers between 0 and 1 and has the same size as vector $\boldsymbol{Y}$.

The noise level is kept as 1%. The effect of noisy data on the estimated derivative is shown in Figure 8.5. The direct application of finite difference (FD) method gives substantially more error than derivative computation with the help of the RBFN-based delta method. Thus the illustrative example clearly shows that the RBFN-based delta method gives better results at the extremes of the dataset than the MLP-based delta method. The RBFN-based delta method is more suitable for rotorcraft parameter estimation, enabling the computation of aerodynamic derivatives in both the transition flight regime and the high-speed flight regime, which together define the extremes of forward-speed variation. Moreover, the RBFN-based delta method gives good results with noisy data, thus making it appropriate for application to real-time flight test data, which contains noise of varied levels and types. Chowdhary and Jategaonkar (2010) point out that rotorcraft system identification is complicated by noise in measurement and processing due to vibration of the rotorcraft frame and windy conditions. The effect of wind is amplified by the small size of most unmanned rotorcraft. All further results in this chapter are obtained using the RBFN-based delta method.

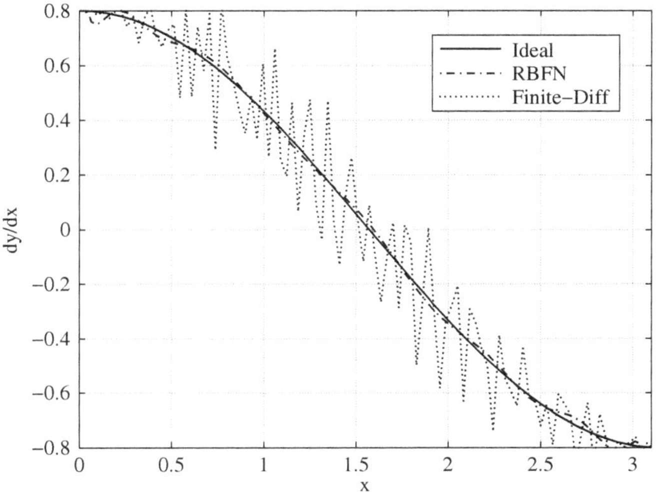

Figure 8.5 Derivative computation with delta method (using RBFN) and the FD method and comparison for noisy data.

8.5 Parameter Estimation Using Simulated Data

The 6-DoF nonlinear helicopter simulation model is based on the DLR Research Bo 105, S123 helicopter. Bo 105 is a lightweight, twin-engine, multi-role helicopter with a hingeless rotor. The simulation model used in this work is based on the Level-1 model and subsystem modeling equations given by Padfield (1981, 1999).

8.5.1 Simulated Data

Simulated flight data for computation of the quasi-steady aerodynamic derivatives are obtained from a 6-DoF nonlinear simulation model. The model is trimmed at various forward flight speeds and the forces and moments are obtained at these trim points. Trim involves the solution of the vehicle equilibrium equations of the helicopter. The quasi-steady aerodynamic derivatives are then obtained using the FD method by numerical perturbation of the state and control variables. The finite difference results are compared with the derivatives estimated using the delta method. The effect of state and measurement noise on both quasi-steady derivatives is investigated by adding Gaussian noise of varying proportions to the simulated data.

8.5.2 Delta Method to Simulated Data

The simulated data is generated for straight and level flight. The rotorcraft model is trimmed at 70 speed points in the interval (0, 70)m/s and the following trim parameters are computed.

- trim control angles $(\theta_0, \theta_{1c}, \theta_{1s}, \theta_{tr})$
- trim forces (X, Y, Z)

- trim moments (L, M, N)
- forward speed (u).

These trim parameters are then used for the derivative computation over a flight range of (0,70) m/s. The derivative computation is done by both the FD method and the delta method. The steps in the delta method are shown in Figure 8.6 and also given below using the Y_u stability derivative as an example:

1) The input file consists of trim control angles and forward speed.
2) The radial basis function network is then mapped with Y force as output.
3) Two different modified input datasets with $u \pm \Delta u$ are made, keeping the other variables constant.
4) The dataset for $u + \Delta u$ is presented to the previously trained network and the perturbed output is Y^+.
5) The dataset for $u - \Delta u$ is presented to the trained network and the perturbed output is Y^-.
6) The Y_u derivative is computed as $((Y^+) - (Y^-))/2\Delta u$.

The selection of the number of data points for training is an important aspect in the delta method. We use a training set of 35 data points. The testing set consists of 15 data points, different from the training set.

The delta method is now used to compute some stability and control derivatives. The derivatives are selected keeping in mind the nonlinearities and couplings involved to show that RBFN can capture these effects. The influence of the number of data points on derivative computation is illustrated using Y_u derivative computation in Figures 8.7

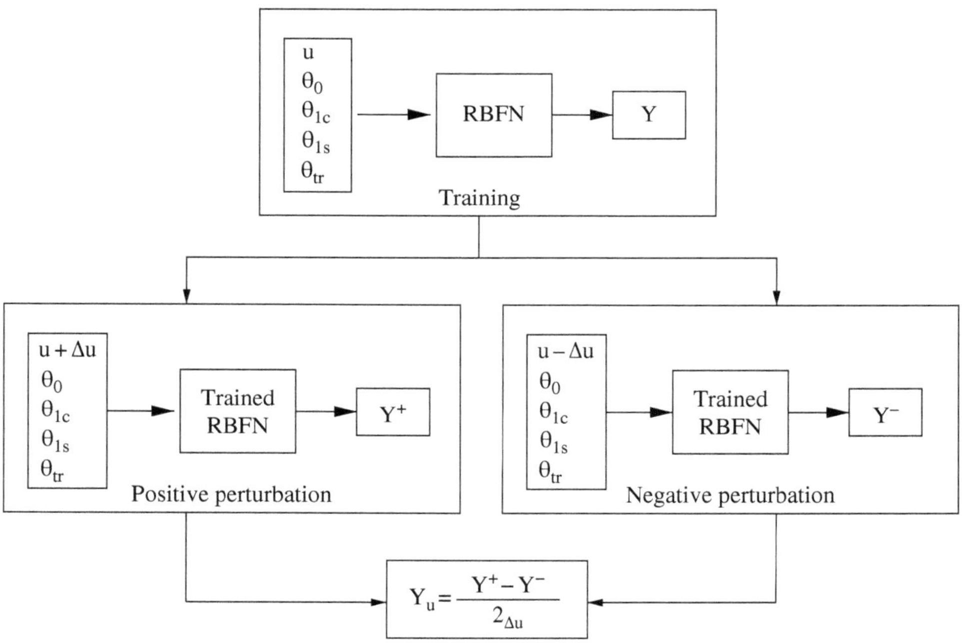

Figure 8.6 Y_u schematic representation of derivative computation using the delta method.

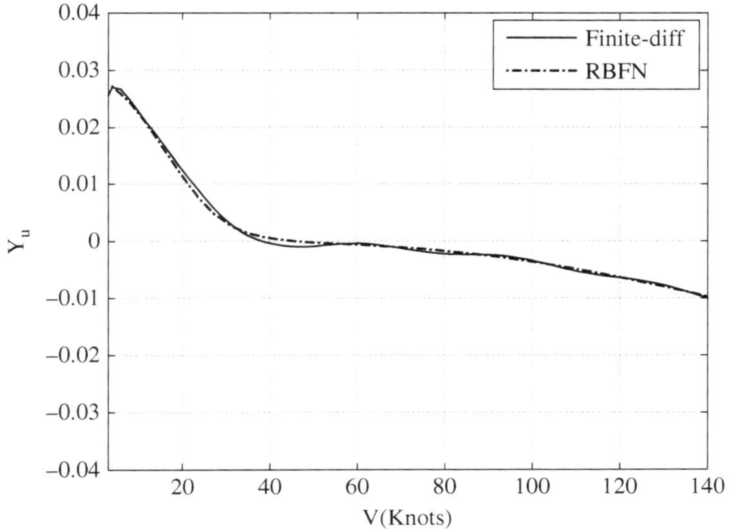

Figure 8.7 Y_u stability derivative calculated using delta method with RBFN and FD for 35 data points.

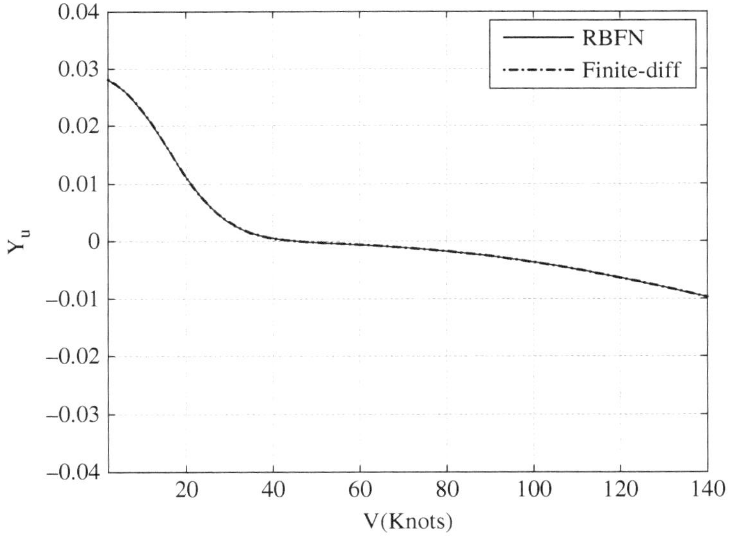

Figure 8.8 Y_u stability derivative calculated using RBFN-based delta method and FD for ideal (zero-noise) data.

and 8.8. In both cases, 35 data points are used for derivative computation using the delta method. In Figure 8.7, the FD technique uses the same 35 data points as used in the delta method. Figure 8.8 shows the result when the FD method uses all 70 points of the dataset. Comparing Figures 8.7 and 8.8, it is clear that the FD technique requires more data points for good results. However, the delta method gives good

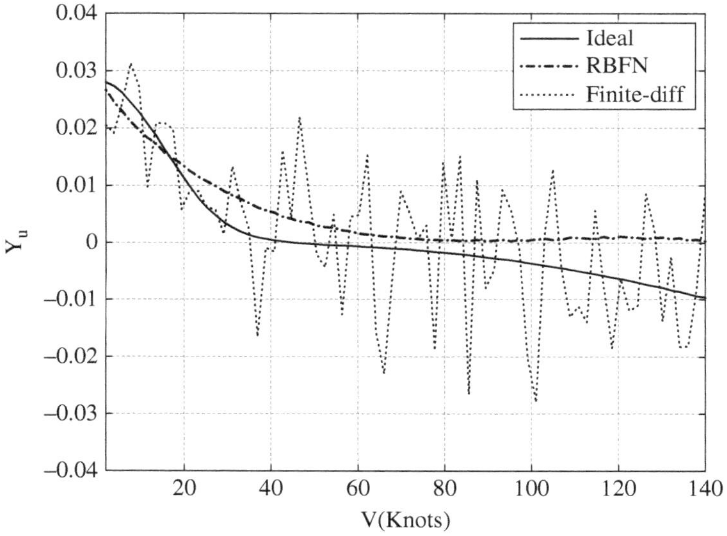

Figure 8.9 Y_u stability derivative calculated using ideal (zero-noise, finite difference) and noisy data.

results even with fewer data points. Since flight test data is expensive to obtain and available at few points, the delta method should be used to obtain accurate derivatives in such cases.

The derivatives are also plotted for 1% measurement noise and state-noise level for stability derivatives and 5% measurement noise and state-noise level for control derivatives. Figure 8.9 shows the Y_u stability derivative at different forward speeds for noise-free ideal data. Now Y_u is a coupled derivative type and the relatively high value of Y_u in low-speed flight is due to the strong non-uniform inflow effects in this flight region. This non-uniform inflow effect reduces with increase in speed, thus making the value of Y_u approach zero. In the absence of noise, both the delta and FD derivatives match closely, indicating that the neural network has been properly trained. The effect of measurement and state noise on the Y_u derivative is also shown in Figure 8.9. The delta method computes the derivative accurately, whereas the FD approach gives substantial errors. A similar methodology is used to compute the static speed derivative M_u, shown in Figure 8.10.

The Z_{θ_0} control derivative computation using RBFN is given in Figure 8.11. Now Z_{θ_0} is an important control derivative, which estimates the heave control sensitivity and quantifies the heave response to the pilot's main rotor collective input. RBFN successfully computes the control derivative and gives results similar to those obtained from FD for ideal data. As shown in Figure 8.12, the presence of noise in the data deleteriously effects the derivative computation using FD, while the RBFN derivatives closely match the ideal case.

The derivative values are also presented in Tables 8.1 and 8.2 to exemplify the effect of different noise levels on the derivative computation by both FD and RBFN methods. The same levels for both state and measurement noise is used in these comparisons. The comparison uses an error norm between the numerical FD and RBFN outputs both for simulated and noisy data. The derivatives obtained from noisy data by both FD and

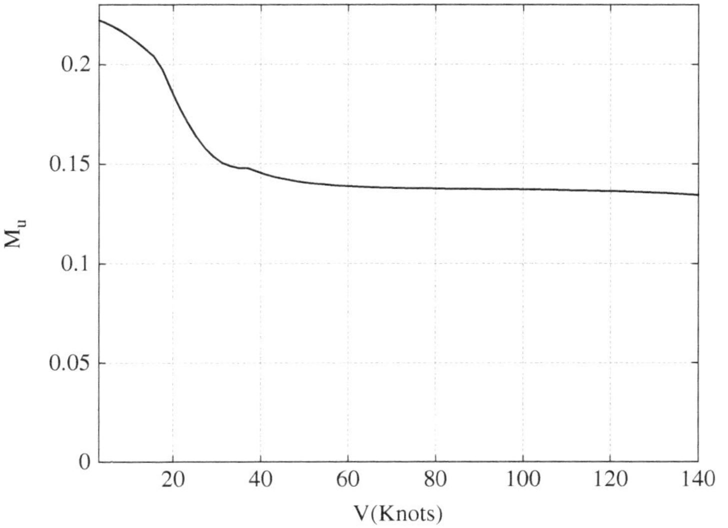

Figure 8.10 M_u stability derivative calculated using ideal (zero-noise) data.

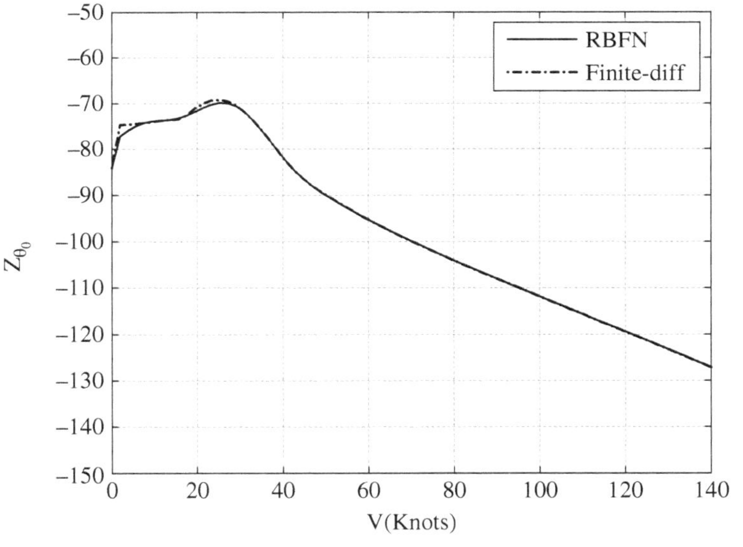

Figure 8.11 Z_{θ_0} control derivative calculated using radial basis function network (RBFN) based delta method and finite difference for ideal (zero noise) data.

RBFN techniques are compared with the derivative computed using the FD method from noise-free simulated data. The error norm is defined as

$$\varepsilon_{X_u} = \sqrt{\frac{\| \hat{X}_{u,\alpha} - X^{FD}_{u,\alpha=0} \|}{\| X^{FD}_{u,\alpha=0} \|}} \tag{8.7}$$

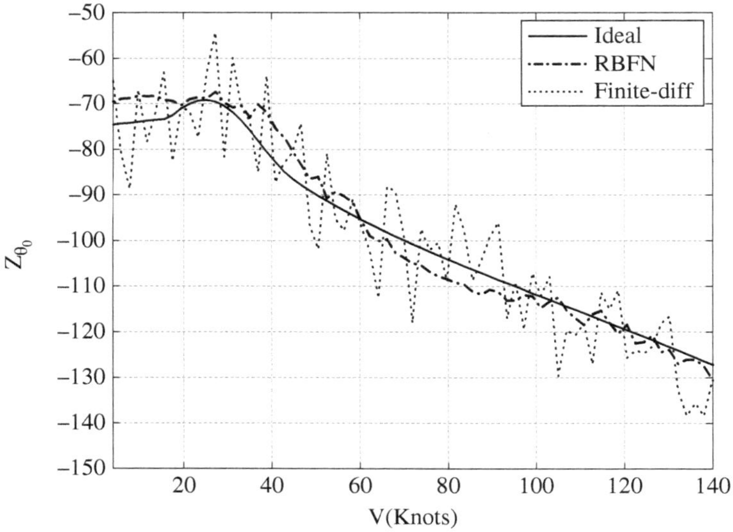

Figure 8.12 Z_{θ_0} control derivative calculated using ideal (zero noise, FD) and noisy data.

Table 8.1 Error in the Y_u derivative for simulated data with different noise levels.

Noise level α(%)	Delta	Method Finite difference
0	0.2260	0
1	0.3953	1.4016
3	0.6502	4.3485
5	0.7638	7.7113
7	0.8737	9.7445
10	1.0393	14.9452

where X denotes any general force and moment and u is a general state or control variable. Here $\hat{X}_{u,\alpha}$ is the derivative calculated from either the FD or RBFN methods at a noise level α. Table 8.1 shows the comparison between RBFN and FD for the Y_u stability derivative. The error in the derivative calculated by FD increases steadily with increase in the noise level in the simulated data, while the RBFN gives satisfactory output at those higher noise levels. Table 8.2 compares the RBFN and FD results for the Z_{θ_0} control derivative.

To summarize, we see that the quasi-static aerodynamic derivatives can be calculated accurately using the delta method based on the RBFN neural network. For the ideal or zero-noise case, the delta method results are almost identical to those obtained using the FD method. Furthermore, the delta method using RBFN does not give any deleterious effects at extreme flight conditions, a problem which haunts the MLP approach.

Table 8.2 Error in the Z_{θ_0} derivative for simulated data with different noise levels.

Noise level α(%)	Delta	Method Finite difference
0	0.1043	0
1	0.1159	0.1388
3	0.1485	0.2789
5	0.1611	0.3057
7	0.1981	0.3481
10	0.2172	0.4424

The strength of the delta method is clear when the simulated data is contaminated with state and measurement noise. In such cases, the delta method is robust to noise while the FD method amplifies it. Since rotorcraft data is noisy, the RBFN method is therefore very useful for such vehicles. Since MAV and UAVs are much lighter than full-scale helicopters, they are prone to noise due to wind gusts and turbulence.

8.6 Parameter Estimation Using Flight Data

Parameter estimation from real-time flight test data is a complicated process. Conventional parameter estimation methods such as the maximum likelihood estimator need a mathematical model of the system (Chowdhary and Jategaonkar 2010). Such a mathematical model is hard to develop and may not be accurate. The unknown parameters are estimated by exciting the system by a suitable input and then measuring the input and actual system response. The unknown parameters are then computed to ensure that the response of the model to a given input matches the actual response. Mathematical models emanating from system identification are useful for control system design. In general, system identification can be online or offline (Jategaonkar 2006); offline methods need access to the complete set of data. Online methods use data as it becomes available. Recursive system identification is an online method which handles flight data as it becomes available from onboard sensors and calculates the aerodynamic derivatives in real time. Most aerospace systems are nonlinear to some extent and this issue needs to be addressed in system identification.

Unfortunately, model-based methods become complicated in the presence of noise. Higher-order modeling also complicates the identification process because the number of parameters is large. The delta method is model-free and the parameters can be computed directly from flight data.

8.6.1 Flight Test Data

Flight tests with the Bo 105 S123 helicopter were conducted by DLR in 1987 to create a database for system identification. This flight test data is used in this chapter to test the delta method. For flight test data, the helicopter was trimmed in steady-state horizontal flight at

80 m/s at a density altitude of 915 m. The flight test data includes linear and angular acceleration components, angular rates, linear velocity components, control inputs at the pilot's stick and rotor blades, along with other variables. The pilot inputs used in the flight tests were:

- positive and negative doublet inputs for each control variable
- positive and negative modified 3211 inputs for each control variable
- pilot-generated frequency sweeps for each of the four controls.

These inputs were administered to each control individually while the other three controls were kept fixed at their trim values. Details about the flight test input specification are given by Fu and Kaletka (1993). The doublet test input can exite modes in either axis very well, but simple test signals of this type have limitations in a highly coupled rotorcraft model. Doublets are not ideal test inputs for helicopter identification although they may be useful when used in conjunction with other types of input (Agard 1991). In order to alleviate the shortcomings of doublet inputs, a second form of multi-state input known as the modified 3211 input was developed. The 3211 input excites a wide frequency band within a short period of time, which is suitable for unstable and coupled systems (Agard 1991; Fu and Kaletka 1993). Another form of pilot input known as the frequency sweep is also useful in frequency-domain rotorcraft identification. The rotorcraft identification problem is difficult to solve using conventional time-domain techniques due to a high level of measurement noise and the presence of inter-axis coupling (Fu and Kaletka 1993; Tischler et al 1992, 1985). Fu and Kaletka (1993) applied the modified 3211 input for identification and performed model validation using the frequency sweep input. Similarly, the present study uses data generated by the modified 3211 input for estimation of the parameter. The same input signals as used by Iliff and Maine (1986) are used for identification to ensure a proper comparison of results.

8.7 Delta Method with Flight Data

The flight data is available in terms of linear and angular accelerations. The data is converted to force and moment components by multiplying by the corresponding mass and moment of inertia values, respectively. The modified 3211 signal has a time length of 7 s. Therefore, the parameters are extracted from the data generated for the first 7 s after applying the modified 3211 pilot control. The training dataset is then created to train the RBFN. Each state and control variable is included in the training of the RBFN. This is needed for the mapping of coupling and cross-axis effects peculiar to rotorcraft. The components of the training set are:

- u, v, w
- p, q, r
- $\theta_0, \theta_{1c}, \theta_{1s}, \theta_{tr}$
- required force or moment component as output.

First, a training set for a particular stability or control parameter is selected. Then the network is mapped for the given force or moment with the training variables. The input variable about which the derivative has to be computed is perturbed. The rest of the variables are kept unchanged at their original values. The perturbation is made on both the positive and negative sides of the particular input variable. This results in two

Table 8.3 Computational parameters used in RBFN-based delta method.

Computational parameters	Value
Number of neurons	234
Spread of radial basis function	10–35
$\Delta\ u$	0.001 m/s
$\Delta\ v$	0.001 m/s
$\Delta\ w$	0.001 m/s
$\Delta\ p$	0.0001 rad/s
$\Delta\ q$	0.0001 rad/s
$\Delta\ r$	0.0001 rad/s
$\Delta\ \theta_0$	0.0001 rad
$\Delta\ \theta_{1c}$	0.0001 rad
$\Delta\ \theta_{1s}$	0.0001 rad

different datasets: one has a positively perturbed value of the selected state or control variable along with other constant variables and the other has a negatively perturbed value. These modified input datasets are now presented to the trained FBFN. The positive and negative perturbed datasets give the corresponding force or moment perturbed value. The RBFN parameters used are shown in Table 8.3.

The derivative is then calculated using the central difference formula. The process is illustrated in Figure 8.13 using the X_u derivative as an example. A longitudinal modified 3211 input is used as the pilot input. The steps are:

1) Input file is made of all state and control variables.
2) The radial basis function network is then mapped with X force as output.
3) Two different modified input datasets with $u \pm \Delta u$ are made, keeping the other variables constant.
4) The dataset with $u + \Delta u$ is presented to the previously trained network and the perturbed output is X^+.
5) Similarly, the dataset with $u - \Delta u$ is presented to the trained network and the perturbed output is X^-.
6) The X_u derivative can now be computed as $((X^+) - (X^-))/2\Delta u$.
7) The arithmetic mean of X_u computed at all time steps up to 7 s is taken as the final result and is used for comparison with results obtained from conventional techniques.

The methodology outlined above is used for computing of all the stability and control derivatives. The derivatives estimated using the delta method are compared with derivatives from the 14-order model computed by Fu and Kaletka (1993) (see Table 8.4). The parameters estimated by using the RBFN-based delta method are in the same range as those obtained by the 9-DoF model.

The computed stability and control parameters are also compared with the identified parameters of the Bo 105 helicopter obtained by researchers at other organizations (see Tables 8.5–8.8; Agard 1991). The delta method estimates most of the stability and

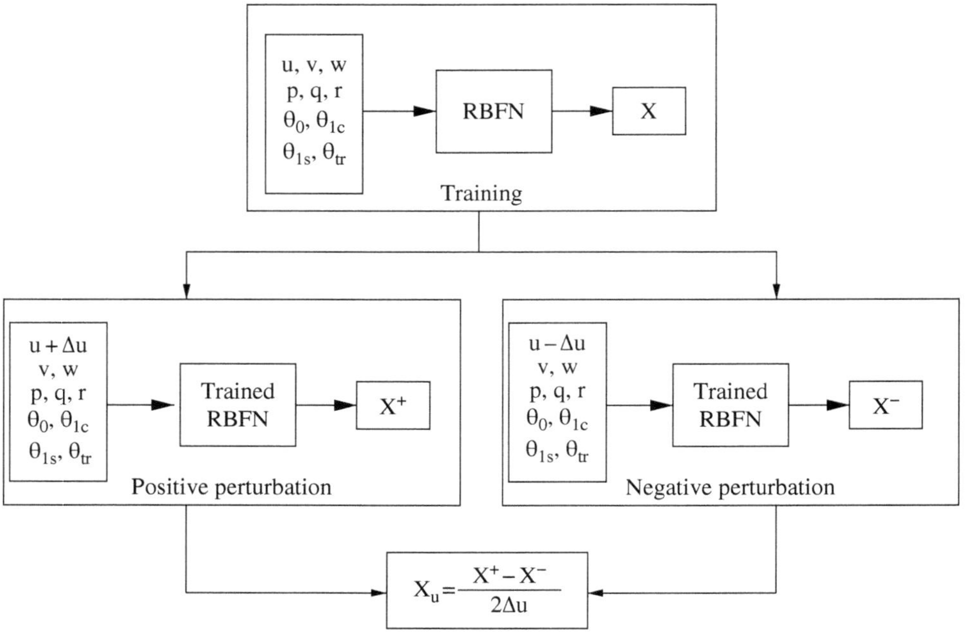

Figure 8.13 X_u schematic representation of derivative computation using the delta method.

Table 8.4 Identification results: comparison of delta method and 14-order model.

Derivative	Delta method	14-order model
X_u	−0.0340	−0.0352
X_w	−0.0709	−0.0992
Y_v	−0.1245	−0.1027
Y_p	−0.3419	1.3823
Z_u	−0.0185	−0.0067
Z_w	−0.2646	−0.2175
Z_p	−0.1257	−0.5669
L_u	−0.0833	−0.0059
L_v	−0.1393	−0.1080
L_w	−0.0861	−0.0683
L_q	0.7040	0.5287
M_u	−0.0161	−0.0081
M_v	0.0267	0.0281
M_w	−0.0181	−0.0196
M_p	−0.7393	−0.0906
N_u	−0.0086	−0.0039
N_v	0.0781	0.1276
N_w	0.0228	0.0141
N_p	−0.0784	0.2973
N_r	−1.0738	−0.8952

Table 8.5 Force derivatives with state variables from delta method and 6-DoF model.

Derivative	Delta method	AFDD	DLR	Glasgow	NAE
X_u	−0.0340	−0.0385	−0.059	−0.032	−0.050
X_v	−0.0190	0*	0*	0*	0.0043
X_w	−0.0709	−0.061	0.036	−0.0422	−0.017
X_p	0.4166	0.756	0*	0*	0.479
X_q	1.5740	2.548	0*	0*	1.206
Y_u	0.0307	0*	0*	0*	−0.061
Y_v	−0.1245	−0.221	−0.170	−0.131	−0.279
Y_w	−0.1356	−0.083	0*	0*	−0.0307
Y_p	−0.3419	−2.030	0*	0*	−2.993
Y_r	6.5181	0.950	1.332	0*	0.807
Z_u	−0.0185	0.2457	0.014	−0.0883	0.1016
Z_v	−0.1859	0*	0*	0*	−0.135
Z_w	−0.2646	−1.187	−0.998	−0.791	−1.106
Z_p	−0.1257	2.622	0*	0*	0.385

* Eliminated from model structure. Source: Agard 1991.

Table 8.6 Moment derivatives with state variables from delta method and 6-DoF model.

Derivative	Delta method	AFDD	DLR	Glasgow	NAE
L_u	−0.0833	−0.061	−0.081	−0.027	−0.099
L_v	−0.1393	−0.207	−0.271	−0.098	−0.270
L_w	−0.0861	0.168	0.116	0.130	0.116
L_p	−4.1502	−8.779	−8.501	−4.470	−1.895
L_q	0.7040	3.182	3.037	0*	4.454
L_r	0.9065	0.991	0.410	1.318	0.434
M_u	−0.0161	0*	0.029	0.0203	0.0078
M_v	0.0267	0.050	0.048	0*	0.0248
M_w	−0.0181	0.096	0.053	0.0491	0.0696
M_p	−0.7393	−0.998	−0.419	−1.367	−1.414
M_q	−2.3217	−4.493	−3.496	−2.217	−2.992
N_v	0.0781	0.082	0.117	0.0784	0.112
N_w	0.0228	−0.119	0.034	0.0281	−0.0634
N_p	−0.0784	−0.466	−1.057	−1.302	−0.692
N_r	−1.0738	−1.070	−0.858	−0.756	−1.017

* Eliminated from model structure. Source: Agard 1991.

Table 8.7 Force derivatives with control variables from delta method and 6-DoF model.

Derivative	Delta method	AFDD	DLR	Glasgow	NAE
X_{θ_0}	0.0319	−0.046	0*	0*	−0.0222
$X_{\theta_{1s}}$	−0.0268	−0.072	−0.028	−0.048	−0.05
$X_{\theta_{1c}}$	−0.0063	0*	0*	0*	−0.00176
Y_{θ_0}	−0.0309	−0.032	0*	0*	−0.0152
Z_{θ_0}	−0.2841	−0.388	−0.349	−0.259	−0.337
$Z_{\theta_{1s}}$	−0.1371	−0.103	−0.303	−0.140	−0.175

* Eliminated from model structure, Source: agard.

Table 8.8 Moment derivatives with control variables by delta method and 6-DoF model.

Derivative	Delta method	AFDD	DLR	Glasgow	NAE
L_{θ_0}	0.0130	0.058	0.032	0*	0.0142
$L_{\theta_{1s}}$	0.0243	0.073	0.024	0*	−0.005
$L_{\theta_{1c}}$	0.0392	0.179	0.185	0.0764	0.1361
M_{θ_0}	0.0292	0.073	0.057	0.039	0.048
$M_{\theta_{1s}}$	0.0470	0.098	0.093	0.0565	0.0787
$M_{\theta_{1c}}$	−0.017	0*	−0.009	0*	0.0106
N_{θ_0}	−0.0104	−0.051	0*	0*	−0.036
$N_{\theta_{1s}}$	−0.0436	−0.075	0*	0*	−0.051
$N_{\theta_{1c}}$	0.0318	0.033	0.026	0*	0.047

* Eliminated from model structure. Source: Agard 1991.

control parameters satisfactorily. Since rotorcraft parameter identification from flight test data is an abstruse problem, there is considerable variation in the predictions of different researchers.

Verification is an important step in the system identification process. Verification should be conducted with data not used in identification. Therefore, flight test data from frequency sweeps are used for the verification. Tables 8.9–8.11 compare estimated derivatives from modified 3211 and frequency sweep pilot control inputs. There is good agreement in direct and coupled stability derivatives, demonstrating the predictive capability of the delta method.

Table 8.9 Verification of force derivatives using frequency sweep input.

Derivative	3211	Frequency sweep
X_u	−0.0340	−0.0373
X_v	−0.0190	−0.0117
X_w	−0.0709	−0.0566
X_p	0.4166	1.09
X_q	1.5740	1.2188
Y_u	0.0307	0.0444
Y_v	−0.1245	−0.1143
Y_w	−0.1356	−0.1835
Y_r	6.5181	6.4246
Z_u	−0.0185	−0.0263
Z_v	−0.1859	−0.1987
Z_w	−0.2646	−0.3169

Table 8.10 Verification of moment derivatives using frequency sweep input.

Derivative	3211	Frequency sweep
L_u	−0.0833	−0.0425
L_v	−0.1393	−0.0179
L_w	−0.0861	−0.0210
L_q	0.7040	2.4766
M_u	−0.0161	−0.0327
M_v	0.0267	0.0527
M_w	−0.0181	−0.0133
M_p	−0.7393	−0.5531
N_u	−0.0086	−0.004
N_v	0.0781	0.1326
N_w	0.0228	0.0376
N_p	−0.0784	−0.5635
N_r	−1.0738	−1.0872

The RBFN approach is philosophically different from the other classical system identification approaches. The RBFN-based delta method is model-free and therefore provides a useful tool for rotorcraft parameter identification from flight test data. RBFN-based derivatives can also be used to compare to derivatives obtained using conventional system identification techniques.

Table 8.11 Verification of control derivatives using frequency sweep input.

Derivative	3211	Frequency sweep
X_{θ_0}	0.0319	0.0304
$X_{\theta_{1s}}$	−0.0268	−0.0237
Y_{θ_0}	−0.0309	−0.0407
Z_{θ_0}	−0.2841	−0.3778
$Z_{\theta_{1s}}$	−0.1371	−0.094
L_{θ_0}	0.0130	0.0172
$L_{\theta_{1s}}$	0.0392	0.583
$L_{\theta_{1c}}$	0.0243	0.0208
M_{θ_0}	0.0292	0.0317
$M_{\theta_{1s}}$	0.0470	0.0545
$M_{\theta_{1c}}$	−0.0107	−0.0093
$N_{\theta_{1c}}$	0.0318	0.0271

8.8 Summary

Rotorcraft parameter estimation based on the RBFN-based delta method has been discussed. This technique is model-free and is suitable for rotorcraft UAVs and MAVs. The method is first evaluated on simulated data generated by a nonlinear simulation model. Both ideal (noise free) and noise-contaminated data are considered. Both state and measurement noise are considered. Then the approach is used to compute the stability and control derivatives directly from real time flight test flight data of a Bo 105 helicopter.

The RBFN is able to successfully map the coupling and nonlinearities in the system and estimate satisfactory stability and control derivatives using the delta method, both for simulated ideal and noisy data. The delta method is robust to noise present in the data. The performance is satisfactory in case of state and measurement noise. The delta method identifies the parameters from flight data and the results have about the same level of accuracy as a 14-order model. The delta method gives satisfactory results for sparse data points. This helps in parameter estimation when the fewer data points are obtained from flight tests. Parameter estimation using RBFN has acceptable predictive capability: the verification results from frequency sweep input are in the same range as those estimated using a modified 3211 input. The RBFN approach provides an alternative to the conventional parameter identification approach when a mathematical model is not available. It can also be used as a method that gives an estimate of the derivatives, which can be used to guide conventional parameter estimation work and model development.

Acknowledgements

The author thankfully acknowledges Dr R. Jategaonkar and Dr W.V. Gruenhagen, DLR, Germany, for providing the Bo 105 flight data. He also acknowledges his research collaborator Dr S.N. Omkar and his former students Dr. M. Vijaya Kumar and Mr. Rajan Kumar.

References

Agard1991agard Agard 1991 Rotorcraft system identification, Technical Report AR 280, LS 178, Advisory Group for Aerospace Research & Development.

Amin SM, Gerhart V and Rodin EY 1997 System identification via artificial neural networks – applications to on-line aircraft parameter estimation, *SAE Transactions*, **106**, 1787–1808.

Al-Garni AZ, Jamal A, Ahmad AM, Al-Garni AM and Tozan M 2006 Failure-rate prediction for De Havilland Dash-8 tires employing neural network technique, *Journal of Aircraft*, **43** (2), 537–543.

Baraldi A, Blonda P, Satalino G, D'Addabbo A and Tarantino C 2000 RBF two stage learning networks exploiting supervised data in the selection of hidden unit parameters; an application to SAR data classification, *Proceedings in IGARSS*, Honolulu, HI, pp. 672–674.

Chowdhary G. and Jategaonkar R. 2010 Aerodynamic parameter estimation from flight data applying extended and unscented Kalman filter, *Aerospace Science and Technology*, **14** (2), 106–117.

Fu KH and Kaletka J 1993 Frequency domain identification of Bo105 derivative models with rotor degrees of freedom, *Journal of American Helicopter Society*, **38** (1), 73–83.

Ganguli R, Chopra I and Haas DJ 1997 Detection of helicopter rotor system simulated faults using neural networks, *Journal of American Helicopter Society*, **42** (2), 161–171.

Ganguli R, Chopra I and Haas DJ 1998 Helicopter rotor system fault detection using physics-based model and neural networks, *AIAA Journal*, **36** (6), 1078–1086.

Hamel PG and Kaletka J 1997 Advances in rotorcraft system identification, *Progress in Aerospace Sciences*, **33** (4), 259–284.

Hamel PG and Jategaonkar RV 2005 Evolution of flight vehicle system identification, *Journal of Aircraft*, **33** (1), pp. 9–28.

Habib S and Zaghloul ME, System identification using time dependent neural networks – a simpler approach, *AIAA Guidance, Navigation, and Control Conference*, San Diego, CA. AIAA Paper 1996–3803

Horn J, Calise AJ and Prasad JVR 2002 Flight envelope limit detection and avoidance for rotorcraft, *Journal of American Helicopter Society*, **47** (4), 253–262.

Horni KK, Stinchcombe M and White H 1989 Multi-layer feed forward networks are universal approximaters, *Neural Networks*, **2** (5), 359–368.

Haykin S 1994 *Neural Networks: A Comprehensive Foundation*, Macmillan.

Iliff KW and Maine RE 1985 More than you may want to know about maximum likelihood estimation, Technical report TM 85905, NASA.

Iliff KW and Maine RE 1986 Application of parameter estimation to aircraft stability and control the output error approach, Technical report RP 1168, NASA.

Jategaonkar RV, Fischenberg D. and Gruenhagen WV 2004 Aerodynamic modeling and system identification from flight data – Recent applications at DLR, *Journal of Aircraft*, **41** (4), 681–691.

Julier SJ and Uhlmann JK 2004 Unscented filtering and nonlinear estimation, *Proceedings of the IEEE*, **92** (3), 401–422.

Jategaonkar RV 2006 *Flight Vehicle System Identification: A Time Domain Methodology*, AIAA.

Kendoul F, Kenzo N, Isabelle F. and Rogelio L 2009 An adaptive vision-based autopilot for mini flying machines guidance, navigation and control, *Autonomous Robots*, **27** (3), 165–188.

Kim BS and Calise AJ 2005 Nonlinear flight control using neural networks, *Journal of Guidance, Control, and Dynamics*, **20** (1), 26–33.

Kottapalli S 2006 Neural network modeling of UH-60A pilot vibration, *Journal of American Helicopter Society*, **51** (2), 195–201.

Kumar R. and Ganguli R. and Omkar SN and Kumar MV 2008 Rotorcraft parameter identification from real time flight data, *Journal of Aircraft*, **45** (1), 333–341.

Lorenz S and Chowdhary G 2005 System identification for a miniature rotorcraft, Preliminary Results, *AHS 61st Forum*, Grapevine, TX.

Malaek SMB. and Izadi HA, 2006 Progressive certification process based on living systems architecture using learning capable controllers, *Journal of Aircraft*, **43** (1), 207–215.

Mettler B 2003 *Modeling Identification and Characteristics of Miniature Rotorcrafts*, Kluwer Academic Publishers.

Mokhtari A and Benallegue A 2004 Dynamic feedback controller of Euler angles and wind parameters estimation for a quadrotor unmanned aerial vehicle, *Proceedings of IEEE International Conference on Robotics and Automation*, Vol. **3**, pp. 2359–2366.

Mondragón IF, Olivares-Méndez MA, Campoy P, Martínez C and Mejias L 2010 Unmanned aerial vehicles UAVs attitude, height, motion estimation and control using visual systems, *Autonomous Robots*, **29** (1), 17–34.

Morelli EA and Klein V 2005 Application of system identification to aircraft at NASA Langley Research Center, *Journal of Aircraft*, **42** (1), 12–25.

Ogretim E, Huebsch W and Shinn A 2006 Aircraft ice accretion prediction based on neural networks, *Journal of Aircraft*, **43** (1), 233–240.

Padfield GD 1981 A theoretical model of helicopter flight mechanics for application to piloted simulation, Technical report 81048, RAE.

Padfield GD 1999 *Helicopter Flight Dynamics: The Theory and Application of Flying Qualities and Simulation Modeling*, AIAA.

Pavel M 2001 On the necessary degrees of freedom for helicopter and wind turbine low-frequency mode modeling, PhD Thesis, Technische Universiteit Delft.

Park JW, Harley RG and Venayagamoorthy GK 2002 Comparison of MLP and RBF neural networks using deviation signals for on-line identification of a synchronous generator, *IEEE Power Engineering Society Winter Meeting*, **1**, 274–279.

Prouty RW 1986 *Helicopter Performance, Stability, and Control*, Krieger Publishing.

Raol JR and Jategaonkar RV 1995 Aircraft parameter estimation using recurrent neural networks-a critical appraisal, *20th Atmospheric Flight Mechanics Conference*. AIAA Paper 95–3504.

Raisinghani SC, Ghosh AK and Kalra PK 1998a Two new techniques for parameter estimation using neural networks, *The Aeronautical Journal*, **102** (1011), 25–29.

Raisinghani SC, Ghosh AK and Khubchandani S 1998b Estimation of aircraft lateral-directional parameters using neural networks, *Journal of Aircraft*, **35** (6), 876–881.

Reddy RRK and Ganguli R 2003 Structural damage detection in a helicopter rotor blade using radial basis function neural networks, *Smart Materials and Structures*, **12** (2), 232–241.

Sahani NA, Horn JF, Jeram GJJ and Prasad JVR 2006 Hub moment limit protection using neural network prediction, *Journal of American Helicopter Society*, **51** (4), 331–340.

Shakernia OMY, Koo TJ and Sastry S 1999 Landing an unmanned air vehicle: Vision-based motion estimation and nonlinear control, *Asian Journal of Control*, **1** (3), 128–145.

Shakernia O, Ma Y, Koo TJ, Hespanha J and Sastry SS 1999 Vision guided landing of an unmanned air vehicle, *Proceedings of the 38th IEEE Conference on Decision and Control*, Vol. **4**, pp. 4143–4148.

Shen S, Mulgaonkar Y, Michael N and Kumar V 2013 Vision-based state estimation for autonomous rotorcraft MAVs in complex environments, *IEEE International Conference on Robotics and Automation*, pp. 1758–1764.

Suresh S, Omkar SN, Mani V and Sundararajan N 2005 Nonlinear adaptive neural controller for unstable aircraft, *Journal of Guidance, Control, and Dynamics*, **28** (6), 1103–1111.

Suresh S, Omkar SN, Mani V and Sundararajan N 2006 Direct adaptive neural flight controller for F-8 fighter aircraft, *Journal of Guidance, Control, and Dynamics*, **29** (2), 454–464.

Suresh S, Omkar SN, Ganguli R and Mani V 2004 Identification of crack location and depth in a cantilever beam using a modular neural network approach, *Smart Materials and Structures*, **13** (4), 907–915.

Tishler MB and Remple RK 2006 *Aircraft and Rotorcraft System Identification: Engineering Methods WITH Flight-Test Examples*, AIAA.

Tsou P and Shen MHH 1994 Structural damage detection and identification using neural networks, *AIAA Journal*, **32** (1), 176–183.

Tischler MB and Cauffman MG 1992 Frequency-response method for rotorcraft system identification: flight applications to Bo105 Coupled rotor/fuselage dynamics, *Journal of American Helicopter Society*, **37** (3), 3–17.

Tischler MB, Leung JGM and Dugan DC 1985 Frequency-domain identification of XV-15 tilt-rotor aircraft dynamics in hovering flight, *Journal of American Helicopter Society*, **30** (2), 38–48.

Vijaykumar M, Omkar SN, Ganguli R, Sampath P and Suresh S 2006 Identification of helicopter dynamics using recurrent neural networks and flight data, *Journal of American Helicopter Society*, **51** (2), 164–174.

9

Helicopter BERP Tip: Literature Review of Helicopter Blade Shape Optimisation Methods

Catherine S. Johnson, Mark Woodgate and George N. Barakos

School of Engineering, University of Liverpool, UK
School of Engineering, University of Glasgow, UK

9.1 Introduction

The design of rotor blades is complex, in that it involves many disciplines of engineering, such as aerodynamics, structural dynamics, aeroelasticity and flight control systems. These disciplines do not just play an individual part in the design of the rotor blade, but are coupled; some more strongly than others (Caradonna 1992). Even within a single discipline, such as the aerodynamics of the rotor, there are often conflicting design requirements: forward flight tends to have opposite requirements to hover, the blade on the advancing side has opposite requirements to those on the retreating side of a forward-moving helicopter and so on (Leishman 2005). Therefore, defining an optimum blade tends to be a compromise between these various conditions. The optimum is therefore determined by the objectives set for each particular rotor design. While the initial rotor design may be relatively easy to come up with, finding the optimum design parameters is not an easy task.

For the above reasons, computer codes to aid helicopter designers have been, and are still being, developed and used in the industry (Johnson 2010a,b). In the past, these methods were limited to using simple theories that modelled the aerodynamics of a rotor. This limitation was due to the high cost of obtaining sufficient data to make valid comparisons for a set of design parameters, either by experiments or by simulation. Both methods involve high costs. The computational costs, however, can be reduced by reducing the complexity of the models used to simulate the aerodynamics around a rotor. This, however, compromises the accuracy of the data. Nevertheless, over time, computing power has increased, allowing more advanced simulation models to be used. This has attracted a lot of interest in design and optimisation methods, as outlined in Section 9.2.

A variety of methods have been developed, and the majority of the applications have been for cases that require a small computational domain (such as aerofoils) or a simple aerodynamic analysis (such as cases where operation is optimised for a single static condition). The current challenge is to apply these methods to a complex design such as

Advanced UAV Aerodynamics, Flight Stability and Control: Novel Concepts, Theory and Applications, First Edition. Edited by Pascual Marqués and Andrea Da Ronch.

Companion website: http://www.wiley.com/go/marques/advanced_UAV_aerodynamics

a helicopter rotor blade, where the aerodynamics are complex and change, and the design space has a large number of dimensions, and to do this accurately and efficiently. The initial concept design is a well-established process and designers and engineers of rotor blades have substantial experience to build on, as well as the assistance of efficient codes to aid them in obtaining a preliminary rotor design. The optimisation methods become more applicable when optimisation of an existing design is required to obtain even better performance from a rotor.

Optimisation techniques usually require a starting design point, so the design stage is just as important as the optimisation. While it is possible for optimisation to lead to new designs, the time and effort involved would attract a high cost. Take the BERP tip blade as an example (Harrison et al. 2008). The BERP tip blade is not something that can easily be created simply by using optimisation methods. However, if the designer's ideas and experience in the field of aerodynamics are used to create an initial design, then the dimensions and extent of the BERP blade characteristics can be perfected to improve performance. When many characteristics of a rotor are adjusted in such a way, a considerable improvement can be made (Leishman2005). This is what makes optimisation so important and has led to the increase in the amount of research conducted and the number of academic papers written on this topic.

In addition to the added performance gained by using optimisation procedures, the use of numerical solutions for the optimisation problem removes some of the workload of obtaining the optimum design from the designer, while still giving the designer flexibility. In the specific case of rotors, there are many criteria and objectives that must be fulfilled simultaneously, a situation known as 'multiobjective optimisation'. Depending on the required performance, it is possible to program the optimiser to create a rotor tailored to expectations in many diverse conditions.

The challenge of optimisation for helicopters has been summarised in the following quote from Celi (1999):

> Researchers in helicopter applications of optimisation face a complex multidisciplinary problem, with several possible choices of design variables, objective functions, behavior and side constraints, analysis models, sensitivity formulations, approximation concepts, optimisation algorithms, not to mention the many types of results that can be generated and presented.

9.2 Literature Review

As mentioned in Section 9.1, the design of helicopters is a complex task and hence requires a number of iterations before the final design is obtained. Even obtaining a good initial concept can be an iterative process. For example, Pereira et al. (2002) described an upgrade of a helicopter. In their paper, they provide some basic insights into the design of helicopters, especially the rotor. For example, the preliminary sizing to determine the number of blades, the diameter, aspect ratio and the solidity of the rotors to achieve the required cruise speed at low cost is described. This process incorporates many variables, such as aerofoil sections, taper, twist and blade-tip design (which itself includes sweep, taper and blade-tip anhedral angle).

NASA has documented a design-from-scratch procedure for rotorcraft (Orr and Narducci 2009). Figure 9.1a is an image from this document showing a simplified overall design process for helicopters. Figure 9.1b is an expanded version of a part of that process, which involves the design of the rotor. As can be seen, the design of the rotor is an iterative process involving several disciplines. However, even within the aerodynamic discipline alone, the process requires a number of iterations that must fit in with constraints that may be due to sizing, material selection or manufacturing and

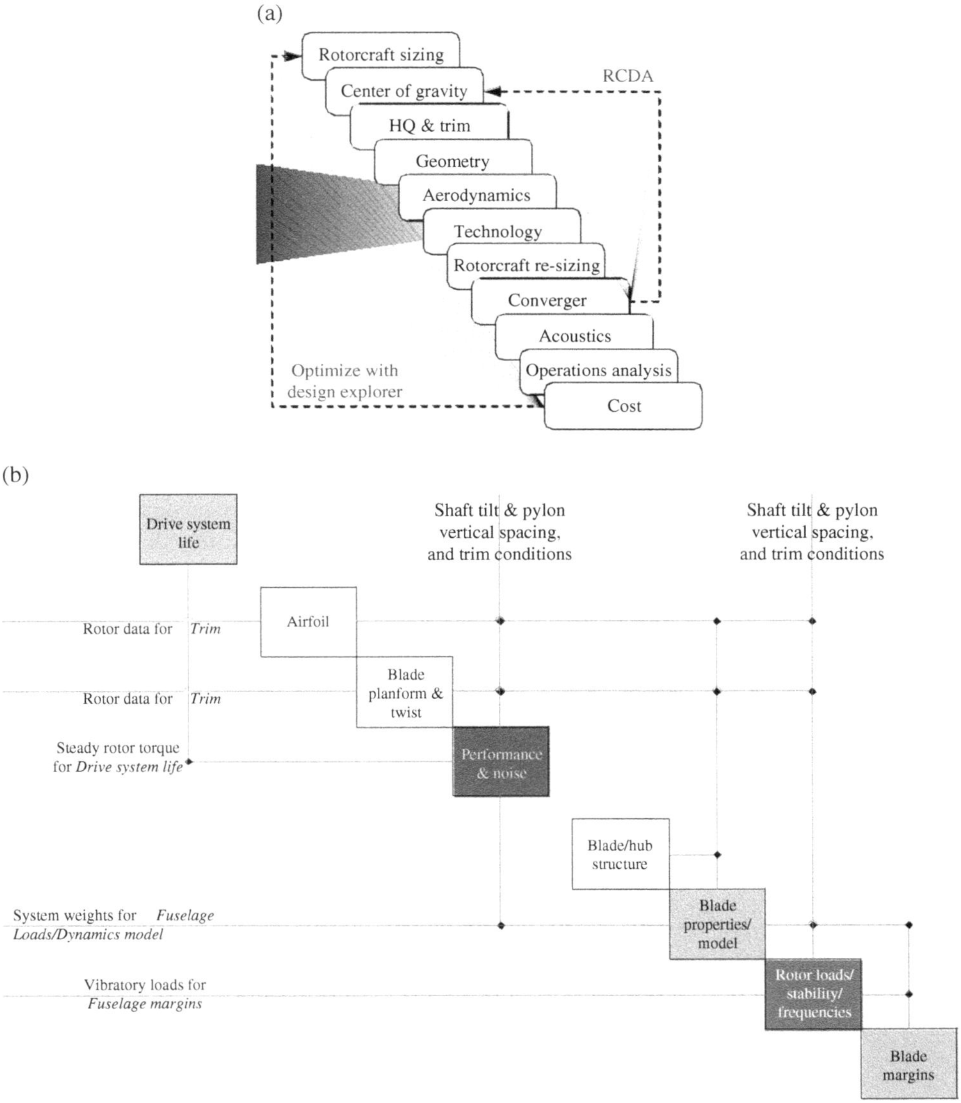

Figure 9.1 (a) General design process for helicopters, (b) The design process involving rotors. RCDA, rotorcraft conceptual design and analysis. Source: NASA/CR-2009-215563 (Orr and Narducci 2009).

other limiting factors, not necessarily related to aerodynamics. This narrows the aspects of a blade that can be optimised aerodynamically.

NASA also developed a more recent conceptual design tool (Johnson 2010a) that designs rotorcraft for specific requirements and then analyses the performance for a set of flight conditions and missions. It has the ability to calculate the weight, power and size of typical configurations but allows flexibility in introducing new concepts. Various components can be added and sized accordingly. The governing equations are simplified methods and most are analytical, although higher-order methods such as CAMRAD have been coupled with them (Acree, Jr. 2010).

Typically, however, optimisation for rotor aerodynamics is initiated from an already 'optimised' design, adjusting the values of certain design parameters so that more performance is obtained in hover and forward flight. Over time, many studies have been carried out on the effect of various design parameters for rotor blades. Leishman's book (Leishman 2005) gives some insights into the significance of each parameter. It also provides the basic theory behind the effect of the geometry of a rotor on its overall performance. In some cases, the significance is large, but a design parameter may have to be primarily optimised for a non-aerodynamic purpose. For example, in the preliminary design of the rotor, the rotor diameter is usually specified according to non-aerodynamic constraints such as storage, gear box ratio, torque limits, stiffness or drooping of the blades when stationary. However, the rotor diameter does affect the aerodynamic performance of the helicopter, since the bigger the diameter, the better the hover performance because of the lower disc loading and hence the lower the induced velocity and power. In addition, the autorotation capabilities are improved.

Another characteristic sometimes affected by non-aerodynamic parameters is solidity, σ. Aerodynamically, reducing the rotor solidity reduces the profile power but increases the disc loading and hence the induced power. Since there is less lifting area, the angle of attack of the blades may need to be increased to maintain the same lift. Therefore, σ is limited by the stall margin of the blade. In forward flight, a higher solidity is required to maintain the same lift, but the greater the σ, the lower the figure of merit (FM).

In his book, Leishman (2005) also gives a basic guideline on how to select aerofoils. For hover, the rotor FM that represents the ratio of power and ideal power required by the rotor is given as Eq. (9.1):

$$FM = \frac{\dfrac{C_T^{3/2}}{\sqrt{2}}}{\dfrac{\kappa C_T^{3/2}}{\sqrt{2}} + \dfrac{\sigma C_{d_o}}{8}} = \frac{1}{\kappa + \dfrac{2.6}{\sqrt{\sigma}}\left(\dfrac{C_L^{3/2}}{C_D}\right)^{-1}} \tag{9.1}$$

According to Eq. (9.1), the aerofoils must have high $C_l^{3/2}/C_d$. The general emphasis is typically on obtaining a high $C_{l_{max}}$. This will allow:

- lower solidity and high manoeuvre loads
- a higher drag divergence Mach number permitting higher flight speeds without too much power loss and noise

- a good lift-to-drag over a wide range of Mach numbers to maintain low profile power consumption
- low autorotative descent and a low pitching moment to minimise blade torsion moments, vibrations and to keep loads on the control system small.

Usually, aerofoil thickness is kept small to accommodate higher Mach numbers. To compensate for reduced lift, camber is added. In some cases, trailing-edge tabs and reflex angles are added to highly cambered aerofoils so that high lift and zero moments can be obtained. On the retreating side, where stall is likely to occur, thicker and more cambered aerofoils stall more gently than thin aerofoils with sharp leading edges. To obtain the best performance, a compromise is required.

Martin and Leishman (2002) performed a study on how different tip shapes (rectangular, sweep, taper and subwings) affect wake geometry in hover, which is a significant contributor to the performance of rotors – especially in hover (Caradonna 1992) – as it can result in velocity gradients that cause stall on the following blade, high induced velocities and radial flow. They came up with a number of findings. For example, in hover, tip sweep appeared to decrease both radial and axial convection of the vortex core while taper increased radial and decreased axial convection. Also, the primary effect of sweep is that the vortex core is pushed outwards past the tip plane path because a swept tip does not conform to the circular streamlines, but is 'sheared' so that it extends to outside $r/R = 100\%$. The vortex peak swirl velocity (scaled with the tip speed) is a function of vortex strength, core circulation normalised with the farfield value and vortex radius. It is also a function of the blade planform (twist, taper and sweep) since these parameters influence the peak bound circulation, which then influences the vortex strength.

Previously, Caradonna also carried out experiments on wakes and looked specifically at vortex strength (Caradonna and Tung 1981). His experiments used a hot-wire probe to measure the wake trajectories and velocities. He used a method similar to that used by Cook (1972) to separate the velocity induced by the wake and the velocity induced by the rest of the system. Even though there is quite a lot of variability between vortices, they all seem to have similar appearances to the classic Rankine vortex (Anderson, Jr. 1995). However, as rotor speed increases, the appearance diverges from that of the Rankine vortex. The non-dimensional vortex strength remains, however, independent of tip speed. Furthermore, the tip vortices seem to contain all the blade circulation.

Caradonna also wrote a second paper on the use of computational fluid dynamics (CFD) in rotorcraft design (Caradonna 1992). Some of the compromises in the design of the main rotor – such as the tip speed – are pointed out. The compromise here is between transonic flow on the advancing side and stall on the retreating side. However, this is strongly dependent on the aerofoil section, twist, planform and propulsive requirements. For hover, the single most important flow issue is the wake, because it determines the rotor induced power. For forward flight, it has slightly less significance, because the wake is convected away from the blade faster. However, its detailed prediction is still required for vibratory loadings, for example.

Very specific design parameters such as twist, aerofoil sections, tip sweep, anhedral and planform geometry begin to emerge as significant contributors to rotor blade performance as more design studies are looked at. Keys et al. (1987) also carried out wind tunnel studies on the effect of twist on rotor blades. Twist improves hover performance because of the better distribution of loads; more uniform downwash in the far wake leads to lower induced power. However, very low and even negative pitch angles on the tips of rotor blades can adversely affect forward flight performance and vibratory loads. Also, as twist is increased, there is more downwash inboard in hover, and so there is an increased downward push on the fuselage. In effect, one has a rotor disc that is pushing the load (or fuselage) down in with the middle section of the blade but is trying to lift it with the tips. The effect of twist in forward flight is that more power is required above an advance ratio (μ) of 0.2 due to the increase in profile power on the advancing tip rather than the increase in Mach number resulting in a decrease in L/D. The effect on stall inception is minimal. Stall inception was defined as the C_T/σ where the blade torsion or pitch link loads increase rapidly (C_m). Also, in general, increasing blade twist increases hub and blade vibratory loads. The paper also provides some suggested solutions such as the use of sweep and taper on the blade tip but also the use of active twist, so that there is azimuthal variation in the twist that gives a reduction in twist on the advancing side. Some helicopters, such as the UH60-A, reverse the twist at the tip so that the loss is reduced in forward flight, while maintaining the performance in hover (Leishman 2005).

These parameters, which have relatively large influence on the design, have become the focus of design and optimisation and are modified to suit the requirements of the rotor. Figure 9.2 shows some of the blade planforms that have been developed.

9.2.1 High-fidelity and Low-fidelity Models

The development of helicopter rotor optimisation has been slower than for other rotors such as turbines and compressor blades. This is due to the complexity and unsteady nature of the flow around a helicopter rotor blade (Celi 1999). Also, the problem is multi-disciplinary, involving aerodynamics, aeroelastics, dynamics and flight control, with all of these aspects coupled to each other. This has lead to the development of comprehensive codes that are able to take into consideration all these disciplines (Celi 1999) by combining modules of codes for each specific discipline into a single unified design tool.

A recent example is the work by Rajmohan et al. (2011). Here, GT-Hybrid, a CFD solver, was coupled with a dynamics code, DYMORE, to assess a manoeuvre for a UH60-A rotor. The effects of the blade geometry and wake model fidelity were explored in steady flight, comparing loose and tight coupling of the codes. GT-Hybrid was used as a high-fidelity CFD code, employing the Navier–Stokes equations near the blade and wake models further away. DYMORE uses lifting line theory and an auto-pilot algorithm to perform a fully-trimmed aeroelastic simulation. Differences in the predicted aerodynamics, such as stall on the retreating side, were found to exist, even due to the options of loosely and tightly coupling the two codes, suggesting that the coupling of disciplines will play a key part in developing good optimisation tools in the future, when computational power increases and CPU cost is reduced. Biedron and Lee-Rausch (2012) have used an unstructured flow solver (FUN3D) loosely coupled with with CAMRAD II in comparing experimental and CFD data and how they are integrated for the UH60-A rotor. Celi also suggests that the optimisation method should be decoupled from the analysis so that it is usable with different analysis tools, hence allowing the optimisation technique to use the best version of the analysis code at any time.

Figure 9.2 Various blade planform profiles. Courtesy of J.F. Perry.

The development of optimisation techniques has progressed quite quickly in the structural aspect of rotor design for objectives such as vibration reduction (Glaz et al. 2007, 2009). This was not the case on the aerodynamic front, due to the lack of modelling techniques (Celi 1999) that are both efficient and accurate, although there have been significant recent improvements. Nevertheless, several authors (Allen et al. 2010; Celi 1999; Collins et al. 2008; Walsh 1991) have devised a variety of more or less successful optimisation techniques. Each attempt is typically limited either by the efficiency of the method or the accuracy of the results. The reason for this is that high-fidelity CFD simulations are necessary to accurately capture the effects of design changes, especially for rotor aerodynamics, but a number of these CFD solutions are required for the process and each calculation can take a long time and incurring a significant computational cost. For example, Walsh (1991), who initially optimised a blade without including a wake model for the analysis, found that including one made significant contributions to the design of the final optimal rotor blade. Even in the case of fixed-wing aircraft for a single-point optimisation problem, the use of higher-fidelity solvers is desirable. An example is the work by Rallabhandi and Mavris (2007) for sonic boom optimisation, albeit for fixed-wing problems.

To overcome the computational cost problem, a lower-fidelity model or metamodel can be used. A metamodel is a model of a model. In the example in Section 10.3.4, a lower-fidelity model of the high-fidelity CFD model is used to obtain the performance of a number of designs. The metamodel is able to predict the performance of unknown design points faster, but with the same accuracy as, a higher-fidelity model. It is able to do this because it is given a reduced number of parameters relative to the CFD method and uses only these parameters to find patterns in the change of the output required. It does this accurately because the data it is provided with are based on high-fidelity aerodynamics obtained from CFD.

Some work has also been done in using both techniques dynamically. An example is the work by Kolencherry and Crossley (2012), where a kriging metamodel was constantly updated with promising high-fidelity data as the optimisation was carried out. Collins et al. (2008) also used a combination of low- and high-fidelity models to optimise a scaled BO-105 rotor in hover and forward flight for noise and power. An initial set of designs was obtained and used to obtain Pareto-optimal designs, a few of which were selected for validation using the high-fidelity solver, with the results used to update the lower-order model.

CFD Methods

In the first attempts at optimisation of rotors, the CFD methods employed were simplified. Walsh (1991), for example, attempted to optimise a rotor in hover while maintaining the performance in forward flight. HOVT[1] (a strip theory momentum analysis with no wake model in the initial analysis) was used to obtain the hover horsepower and CAMRAD[2], a comprehensive code, for the forward flight horsepower and manoeuvre and trim conditions. Even at this early stage, the idea of using CFD methods for optimisation was attractive. CAMRAD was used because it could be coupled with CFD methods that could be used to obtain better models of effects, such as transonic effects

1 Rotor hover analysis code developed by Alfred Gessow.
2 Comprehensive Analytical Model of Rotorcraft Aerodynamics and Dynamics.

(Adelman and Mantay 1989). Only some change in the geometry of the blade was captured and hence only a slight improvement was obtained. The conclusions were that further validation and testing would be required.

In the work done by Bohorquez et al. (2010) for optimisation of small rotors where one of the key objectives was cost-effectiveness, the computational cost of the solutions was further reduced by using incompressible steady calculations with one-equation models of turbulence.

The optimisation method by Le Pape and Beaumier (2005) employed elsA,[3] a package developed by ONERA.[4] RANS[5] was used for the hover calculations. Computations were performed on a coarse CFD mesh during optimisation and then a fine mesh was used once the minimum was found. In addition, the HOST[6] was coupled with elsA, to include aeroelastic and dynamic disciplines in the analysis (in other words, deformable blades), since blade torsion has a significant influence on the aerodynamics of the rotor. So at each optimisation step, the HOST code computed the blade deformation (flap, bending and torsion) and the rotor angles (flap and lead-lag) and a new grid was built around the deformed blade to be computed with CFD. However, no recomputation of the total trim was done, making the coupling slightly inaccurate, especially if forward flight was being considered.

Le Pape (2005) later extended his work to forward flight. elsA was only used for hover optimisation since forward-flight calculations would take a lot more time and a trimming model would be required as well. HOST was used as the aerodynamic model for the forward-flight cases as well as in hover, where it was coupled with elsA. It used a lifting-line model with a prescribed wake. However, these types of solvers are not very good at determining the performance of complex blade shapes in forward flight.

Imiela (2009) optimised the aerodynamic properties of rotor blades considering both hover and forward flight using CFD, whilst constraining structural and aeroelastic deformation as calculated by CSD. The solutions were based on high-fidelity CFD/CSM analyses that had weak fluid–structure coupling (in other words, each discipline was calculated alternately) combined with an optimisation algorithm. FLOWer[7] was used as the aerodynamic simulation package and HOST for blade dynamics and elasticity, although the optimisation itself was focused on aerodynamics. The structural model was not modified during the optimisation.

It can be seen that using a high-fidelity model directly for optimisation of rotor blades is either limited to simpler cases such as hover, or lower fidelity must be accepted for more complex cases such as forward flying rotors. To be able to model forward flight efficiently, a lower-order model is required, but its accuracy still needs to be maintained, especially because of its complexity. Metamodels have this ability.

Metamodels

Metamodels have been used extensively in the optimisation of turbomachinery blades, as reported in the works of Mengistu and Ghaly (2008) and Samad and Kim (2008).

---

3 (French acronym meaning 'Software Package for Aerodynamics Simulation'.
4 Office National d'Études et de Recherches Aérospatiales.
5 Reynolds-averaged Navier–Stokes equations.
6 Helicopter Overall Simulation Tool.
7 Open CFD software.

Mengistu and Ghaly (2008) used an artificial neural network (ANN) as a metamodel, along with a genetic algorithm (GA), to optimise turbomachinery blades. They used high-fidelity CFD data to train the ANN to predict the performance of the turbomachinery blades using interpolation between pre-computed designs. The prediction weights were updated using an optimisation algorithm to minimise the error between the network output and the training data (obtained using CFD). The training strategy was enhanced using a GA and gradient-based optimisation approach. There was no rule for determining the number of nodes in the layer of neurons, but a good initial guess was the average of the input and output nodes. This number of nodes was increased to create an optimal trained network. If the number of nodes was too small, underfitting occurred, high error values were obtained and too much generalisation occurred, and vice versa. This is explained further and clearly demonstrated in Section 10.3.4.

In terms of helicopter rotors, optimisation using metamodels has been performed mostly with the aim of vibration and noise reduction. An example is the work done by Glaz et al. (2007, 2009). Here, it is suggested that for the particular case of vibration reduction, kriging is the best surrogate method. However, it depends on several factors that affect the choice of approximation method, such as the sampling density, the scheme used to select the points and the parameters of the metamodel. It is suggested that since no single metamodel is generically the best, it may be useful to use a combination of metamodels instead. Three metamodels were used: polynomial response surface, kriging and radial-basis neural networks. These were combined by a weighted-average surrogate algorithm. The conclusions drawn were that the most accurate method depended on sample size and the error evaluation method and that the most accurate metamodel did not necessarily produce the best design. Also, in the work done by Liu et al. (2012), it was shown that for more complicated cases, kriging outperforms radial basis functions (RBFs), whereas for simpler cases RBFs are more accurate and efficient. Celi (1999) also mentioned that the 'connection between predictions and accuracy of the optimisation may be more subtle than appears at first glance'.

Imiela (2009) used DACE (Lophaven et al. 2002), a kriging metamodel in MATLAB, for the optimisation of a rotor in forward flight. Kriging is similar to RBF interpolation. It is a method of interpolation but the weights involved are driven by the data rather than arbitrary functions. Usually, the weights are functions of the distances of points from the required data point. This means that an estimate of the error is also available. The aim is to determine the weights such that the variance of the estimator is low (Bohling 2005).

One reason kriging is popular is that it has few assumptions; in fact, the only real assumption is that the design space or equation is continuous (Forrester and Keane 2009). For example, Glaz et al. (2010a,b) used kriging to predict the unsteady lift, drag and moment under dynamic stall for rotor sections on a helicopter blade. The Navier–Stokes equations were used to compute the aerodynamics and the dynamic stall was modelled as a function of time and time history. This non-linear mapping function that was numerically found, linking the input to the output, was modelled using kriging at each time step. In effect, the surrogate model had two dimensions: the kriging for each time step and the SBRF (surrogate-based recurrence framework) to complete the prediction including the time histories. This worked well, with error values less than 3%. There are a number of types of kriging apart from ordinary kriging, including universal, co-kriging and blind kriging (Forrester and Keane 2009). Kriging, however, has been found to be less accurate when the data is sparse (Forrester and Keane 2009).

Peter and Marcelet (2008) compared the performance of four different metamodels. They found that kriging and ANN were two of the most accurate methods, and that kriging was more efficient. Ahmed and Qin (2010) compared RBFs and a number of kriging methods as metamodels to optimise hypersonic spiked blunt bodies. They explain how kriging works and that it can be as good a metamodel if not better than an RBF. Rendall and Allen (2010) did some work on improving the efficiency of RBFs. They used 'greedy' RBFs, in which the prediction of a point is dependent on a smaller selection of points from the entire design space, thereby avoiding the calculation of each point's prediction in relation to the required output. They suggest that the greedy RBF will be useful in compressing data and improving efficiency.

In the work of Sun et al. (2003), response surface models (RSMs) were used to obtain the performance of interpolated design variables for rotor sections in pitching motion. However, they also included a regression model to give it more flexibility in fitting the surface, similar to kriging. In addition, they used t-statistics (Sun et al. 2003) to determine the significance of each variable to the objective function and hence determine the accuracy of the response surface.

Hajela (1999) also discussed a number of metamodels used in rotor optimisation, such as immune network models, where the design and performance parameters are simulated as antibodies and their corresponding antigens. Their relation is quantified by a matching function, which works similarly to the fitness value in a GA (a non-gradient based optimiser).

Other methods include the use of a Fourier transform to predict the flow-solver data, as in the work done by Nadarajah and Tatossian (2008). Their goal was to introduce a Mach number variation into an existing non-linear frequency domain (NLFD) framework and then introduce a time-varying cost function through an adjoint boundary condition to redesign a NASA rectangular supercritical wing. They state that there is a need to develop cost-effective optimal control techniques for unsteady aerodynamics and that an efficient method may be the use of periodic methods such as the linearised frequency domain. However, these methods become highly inaccurate when there are strong non-linearities although further validations with high fidelity software show that this problem can be overcome (Tatossian et al. 2008).

The efficiency of the NLFD approach is dependent on the periodicity of the flow. In the case of rotor aerodynamics, representing the variations of loading in the frequency domain can require a larger number of modes to represent the data than other methods, such as the proper orthogonal decomposition method (POD). The POD method, as an interpolation technique, was used by Oyama et al. (2010) to optimise transonic aerofoils for lift/drag ratio. The aerofoils were represented as splines with the design parameters being the control points, and a POD technique was used to obtain the lift-to-drag ratio for new aerofoils that existed on the Pareto front.[8]

A modified version of the POD technique, called the gappy POD method, was developed by Bui-Thanh et al. (2004) and Willcox (2006) for aerodynamic applications. This

8 The Pareto front is the subset of design points with the best compromise in all the objectives for optimisation. It forms a boundary or 'front' in the design space.

method is used to interpolate for incomplete or 'gappy' data by solving a linear system of equations to give a set of POD modes (Bui-Thanh et al. 2004). Such a method could be used to fill in for incomplete experimental data, or for unknown conditions and even inverse design. In the work done by Gunes et al. (2006), a comparison is made of kriging and POD-based interpolation to reconstruct the unsteady flowfield around a cylinder. It was found that the POD method is more accurate when there are many snapshots and smaller gaps in the field. When this was not the case, kriging provides a more accurate solution.

Metamodels can be very useful in reducing computational cost and expanding databases as long as they can give accurate predictions that the optimiser can rely on. The usual way of testing the accuracy is to validate an unknown point in the database with its actual value obtained from the high-fidelity solver. In the work by Tahara et al. (2008), two optimisers were compared. An error value was introduced, which was the difference between the measured and expected improvement by the optimiser. In a similar way, this approach can be applied to metamodels, where the difference between the predictions and actual values are used to validate the model.

If the cost of running an additional set of data is high, another method of testing the prediction accuracy of a metamodel is via cross-validation (Forrester and Keane 2009). Here, the design space is divided into subsets, then each one is removed and the metamodel is trained without it. The sets are then reintroduced and compared with the prediction. The average of these is the error. The problem with metamodels is that they fit the existing data but the accuracy of their predictions outside of this trained area is questionable.

The gain in computational cost through the use of metamodels increases with the number of objectives but the accuracy decreases. The reasons for this are stated by Karakasis and Giannakoglou (2004) who suggest that even in the comparison of the case of single- and multi-objective optimisation, within the first generation, the single-objective optimsation individuals cluster around the optimum region and the metamodel has sufficient points to reliably evaluate them, whereas in the case of a Pareto optimisation technique, the available information is spread over a front. To counter this, a number of these can be combined with weighted formulae.

In Karakasis and Giannakoglou (2004), a metamodel was built for each new member of the population and was limited to a local vicinity of approximately 100 to 150 individuals. It was found that this worked better than building an overall global metamodel in terms of better approximations.

9.2.2 Sampling Methods

Before a metamodel can be used, the sampling of points in the design space must be determined. This is what is known as 'design of experiments' (DOE). The DOE depends on many factors such as the metamodel to be used, the cost of running the high-fidelity software, the dimensions of the problem and the shape of the design space. Usually, Latin hypercube sampling (LHS), and its variants, such as minimum bias design and orthogonal arrays, are preferred (Mack et al. 2007). Crowley et al. (2012) used a nested LHS method, where there were LHS hypercubes of increasing resolution within each other.

Mackman et al. (2011) compared kriging and RBF methods for various cases including aerodynamic coefficients for aircraft manoeuvres. They compared a number of sampling methods and using error estimations, determining which were the most accurate and the quickest to converge by dedicating a portion to exploration and some to refinement of non-linearities.

Other common sampling methods are the factorial methods, including full and fractional factorial (Cameron et al. 2011). Some less common ones are also available, such as the Delauney sampling method (Petrone et al. 2011). This method treats the design space like an unstructured grid, dissecting it using the Delauney triangulation method into simplexes that are repeatedly refined to the optimal refinement for the objective. The Morris–Mitchell criterion is another method, using a criterion that best satisfies an exchange technique iteratively (Forrester and Keane 2009).

9.2.3 Optimisation and its Challenges

There are several difficulties involved in engineering optimisation, some of which are (Hirsh 1998):

- *Unevaluable and infeasible points/regions*: Not all points in the space are legitimate designs – inevaluable points cause the simulation to crash and infeasible points are not physically realisable designs. This problem is addressed via constraints.
- *Expensive evaluation*: Simulators are usually designed for accuracy more than efficiency and therefore they can take a non-negligible amount of time to evaluate a point. A variety of methods, efficient and more accurate, should be used to counteract this. For example the use of both low- and high-order models or time-marching and harmonic balance techniques.
- *Badly structured space*: The structure of the space of good designs may be difficult to search. Therefore, the use of global optimisation methods should be used.
- *Multiple local optima*: The search space may have a large number of local optima and this can trap gradient-based local optimisers and even global optimisers. They may be true local optima or apparent local optima, such as those caused by noise (due to round-off errors, approximations in the model or in the grid, inexact solvers and so on). Engineering intuition and the use of methods such as GAs can counter this problem.
- *Non-smoothness*: This often results from table look-up or numerical problems that usually appear as discontinuities in objective functions. Metamodels can be used to smooth surfaces, leading to better results.

The second point is particularly true for rotor optimisation, so work on rotor optimisation has tended to lean towards gradient-based optimisation techniques; the number of solutions required for non-gradient based optimisation is high, which increases the total computational cost of optimising helicopter rotors. For example, Le Pape and Beaumier (2005) attempted to maximise the FM of a rotor by modifying the geometry of the blade in an automated routine using CONMIN, a gradient-based optimiser developed by Vanderplaats (1973).

However, non-gradient based methods, especially evolutionary type methods, have been used more for other types of rotors such as compressor and turbine blades, and usually for a single point condition. For example, in Zhao et al. (2008), the optimisation of gas turbine blades is performed using a neural network trained model and a GA. The original data was obtained using a Navier–Stokes solver. Bézier curves were used to define the geometry and their control points were used as design parameters. In Samad and Kim (2008), a Reynolds-averaged Navier–Stokes (RANS) code was coupled with a multi-objective GA to find the optimum blade shape for a transonic axial compressor rotor to improve adiabatic efficiency and total pressure ratio. Again, Bézier curves were used for the definition of the blade geometry. In this case, the GA was used to find a global optimum region and then a refined search was made using Newton's method, which is gradient-based. In addition, clustering was used to find the Pareto-optimal solution or front. Mengistu and Ghaly (2008) also used a GA to optimise turbomachinery blades. In their case, the performance is measured by the adiabatic efficiency and the total pressure coefficient. The design variables include the back pressure of a cascade and its shape parameters. There are two terms that make up the optimisation function. One attempts to maximise the efficiency or minimise the total pressure-loss coefficient, while the other term eliminates large differences in the above between the design and off-design points. The last term is a penalty term, including geometric and aerodynamic constraints such as the exit flow angle, spacing to chord ratio and stall margin. The weightings of the penalty coefficient are set by the user. They were successful in obtaining an optimised blade.

Evolutionary methods have also become attractive for fixed-wing aircraft. For example, Watanabe et al. (2008) uses Euler simulations with a GA to optimise a passenger plane wing for shock waves; that is, the lift-to-drag ratio. Constraints were applied, such as the thickness at the root, the reference area, the minimum volume of the wing or fuel cells and the range of angle of attack. Steady-state CFD calculations of rotors also attract evolutionary optimisation methods. For example, Xiong et al. (1997) used an extended compact GA to optimise horizontal-axis wind turbines (HAWT) rotor blades for maximising power by modifying twist and chord.

In terms of helicopter rotors, evolutionary-type optimisation was applied by Glaz et al. (2007, 2009) in the optimisation of forward-flying rotors for vibration and noise reduction. Sun et al. (2003) used GAs to optimise rotor sections in pitch-translation motions for forward flight simulation.

Allen et al. (2010) started off with aerofoils and fixed wings, but applied their methods to optimise rotor twist in hover. Their method directly modifies the design shape and its meshing, and one of the key features is the parameterisation technique used, which is linked to RBF interpolation and allows local and global control of the grid coordinates. The optimisation itself was based on a gradient-based parallel optimiser. The differences between having local and global control over the mesh and shape deformation is highlighted. The process, even for hover of a relatively coarse grid, was lengthy. Application of such a method to forward flight might render it impractical.

Bohorquez et al. (2010) tried to reduce the computational load by using a combination of blade element momentum theory coupled with lookup tables and experimental data to optimise a small rotor in hover for better power loading.

Typically, the rotor-blade design problem is also complicated by conflicting performance requirements between hover and forward flight (Le Pape and Beaumier 2005). The method by Le Pape (2005) is very versatile, optimising for anhedral, sweep, twist, chord and aerofoil distribution of rotor blades. These variables could be optimised for the whole blade or for only part of it. In hover, collective pitch was an added variable so that thrust could be varied and maximum FM could be achieved. In the multi-objective optimisation process, one flight condition was optimised for, whilst the other acted as a constraint. For example, for the 7A ONERA rotor, twist was optimised in hover, whilst constraining performance in forward flight.

Imiela (2009) used a GA to optimise the rotor and the MATLAB DACE kriging toolbox as a metamodel. Hover and forward flight were both considered, whilst constraining structural integrity within boundaries. A number of optimisation algorithms were tested

- CONGRA, a conjugate-gradient based method (which works by using gradients as well as finding new search directions based on former iterations)
- SUBPLEX, a non-gradient based method
- EGO, which uses a metamodel in a GA.

The conclusions were that CONGRA does not reach the optimum fast enough, and that it was dependent on the step size. SUBPLEX had difficulties with the tip optimisation, almost producing a rectangular blade. This may be due to issues with accuracy. EGO was the most efficient and the least-error prone, which suggests that metamodels are promising for rotor optimisation.

What to Optimise For?

The design objectives are usually specified as an objective function containing the relevant performance parameters as components that are weighted. In some cases, the best compromise between all the performance parameters is found and a design is selected from this subset. One example is the Pareto front method. Another is goal programming (Deb 2001), where instead of finding a point that is the maximum or minimum of an objective function, a solution is found that satisfies a goal or reaches it with minimum distance and satisfying constraints. The paper by Deb (2001) describes this approach in more detail, including a method independent of weight factors.

Where objective functions are used, they should have the following properties (Tan et al. 2002):

- Complete, so that all pertinent aspects of the decision problem are presented (all the objectives are captured for example, stall on the retreating side, compressibility effects on the advancing side, pitching moments and so on).
- Operational, in that they can be used in a meaningful manner (the optimiser must be able to access the performance and make a decision based on that).
- Decomposable, if disaggregation of objective functions is required or desirable. For example, the rotor with the best advancing side performance should be obtainable by adding or removing components from the objective function.
- Non-redundant, so that no aspect of the decision problem is considered twice. Sometimes it is possible that two components of an objective function capture the same performance metric in addition to others. It is important to make sure that this does not happen as it would result in one performance parameter being weighted more than intended.
- Minimal, such that there is no other set of objective functions capable of representing the problem with a smaller number of elements. This avoids the above problem as well. For example, moment curves on a blade over a revolution can capture pitching moment performances as well as stall and shock effects. Therefore, the use of, for example, the drag curve to capture the stall effects is unnecessary and would result in redundancies and added unnecessary workload.

Hall and Hall (2010) attempted to find the optimum radial and azimuthal distribution of blade element thrust that would give minimum rotor-induced drag, similar to reducing downwash for fixed-wing aircraft, where elliptic loading is the most ideal. With increasing forward speed, the induced power (in the form of induced drag or torque) first decreases and then increases sharply because the rotor must remain trimmed (especially in roll) even though the region of low speed and flow reversal on the retreating side increases. Their aim was to find the optimal distribution for low induced power by optimising the circulation distribution independent of the rotor geometry and also reducing the gaps in the wake on the retreating side at moderate advance ratios (μ). In their paper, they present the theory and a computational model for finding minimum power and the circulation distribution for that minimum power.

In the work done by Le Pape (2005), maximising the hover FM was the main objective in hover. In forward flight, torque coefficient (C_Q) was optimised, while constraining the thrust coefficient (C_T). The same objectives were used in Imiela's work (Imiela 2009), with the addition of constraining the propulsive force in forward flight.

In Bohorquez et al. (2010), a small rotor was optimised in hover for power loading rather than FM. This parameter was actually the maximum power-loading envelope obtained at various disk loadings so that the results were independent of disk loadings.

Lee and Hajelat (1996) performed multidisciplinary optimisation, and for forward flight the objective was to avoid stall while maintaining trim conditions and power used; in other words, torque. In addition, they had limitations on blade frequencies, pitching and rolling moments, hub shears, aeroelastic stability margins, autorotation index and other material constraints.

In the Blue-Edge design (Rauch et al. 2011), the objective was to actively reduce blade–vortex interaction (BVI) noise while constraining forward flight performance. This also happened to be one of the first cases where high-fidelity CFD was used in the optimisation loop to obtain better hover performance. This was achieved by modifying the anhedral and twist of the blade.

The objective function should encompass all the objectives as well as the constraints to obtain the required performance. For example, the objective in Walsh's paper (Walsh 1991) was to reduce the required power in hover without compromising forward flight; in other words, to lower the total power curve of Figure 9.3. The resulting optimised blade had very different characteristics from that obtained by Le Pape and Beaumier (2005), for which the objective was the increase of FM and which only considered hover, as shown in Figure 9.4.

Gradient-based Optimisation

Gradient-based methods use the gradient of a performance function in order to find the direction that leads to the optimum solution. One of the more commonly used pieces of software is CONMIN, developed by Vanderplaats (1973) at NASA. It has been used by a number of authors (Le Pape and Beaumier 2005). The memorandum by Walsh (1991) states that the CONMIN optimiser was used to optimise the aerodynamic performance of rotor blades by selecting the point of taper initiation, root chord, taper ratio and maximum twist, which minimise hover horsepower while not degrading forward flight performance. CONMIN was also used by Celi (1999) for aerodynamic and vibration reduction optimisation.

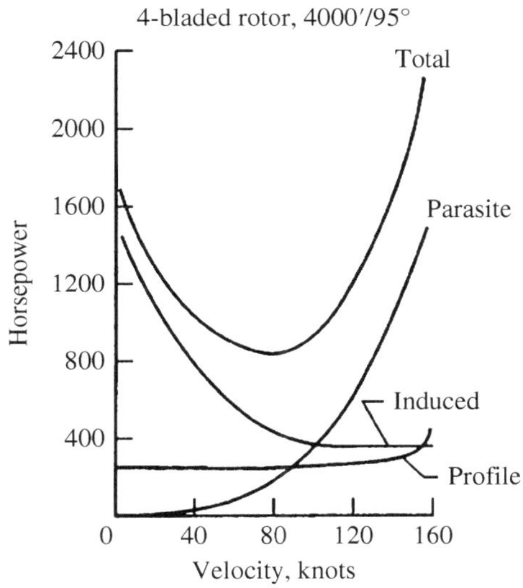

Figure 9.3 Figure showing power as a function of velocity of forward flight. Source: Walsh 1991.

Blade by Walsh (1991)

Blade by Le Pape and Beaumier (2005)

Figure 9.4 Rotor blades optimised for different objectives.

CONMIN can optimise constrained linear and non-linear problems. It uses finite difference methods to find the gradient and can also use the conjugate direction methods by Fletcher and Reeves (1964) for unconstrained minimisation problems. The optimisation can also be performed without any conditions of constraint. Also, in order to test other designs that may not otherwise be found, constraints can be added after an optimum is found and continued from that point. Such an option may be desirable when minimising functions for which one or more constraints are difficult to evaluate.

CONMIN works by iteratively updating the design so that at iteration i, $\overline{x}^i = \overline{x}^{i-1} + \alpha \overline{s}^i$ where $\overline{s}$ is a vector direction and α is the distance in that direction. $\overline{s}$ is determined so that for an arbitrary small α, the optimisation function is reduced (since one is trying to find minimum) the most; in other words the steepest descent is found for which no constraints are violated. If the function violates a constraint, a direction $\overline{s}$ is found to overcome the constraint with minimal increase in the optimisation function (Hicks et al. 1974). Figure 9.5 is a summary of how the CONMIN method works for three points (A, B and C) under various constraint scenarios.

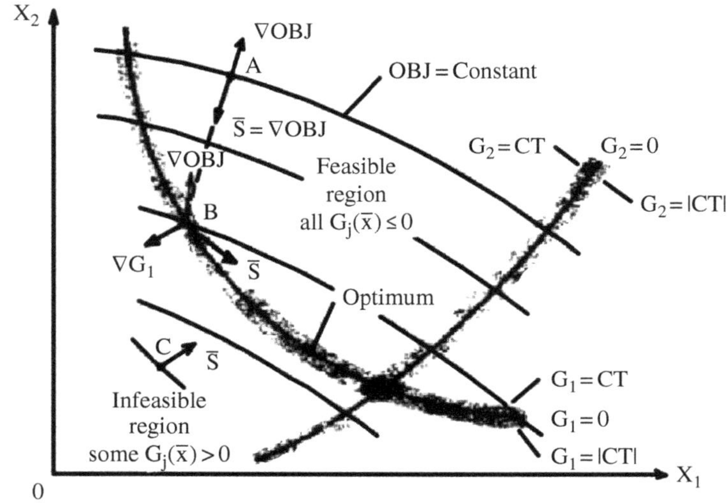

Figure 9.5 Operation of the gradient-based optimiser, CONMIN. Three points, A,B and C are shown in various locations with regard to the constraints. At A, the steepest descent is used to reach the minimum; at B, a constraint is encountered and the point attempts to move along the constraint contour; and at C, the point attempts to overcome the constraint. From Hicks et al. 1974.

CONMIN works efficiently if there are few constraints and the data is relatively smoothly varying. In the case of total aerodynamic rotor optimisation, where there are many more constraints and the data is uneven, it may be less efficient and may fail to reach the global optimum. There is a high probability of the solution getting trapped in local optima.

The tendency of gradient-based methods to stagnate near local optimum regions of the design space is a major problem. This is why in the work carried out by Walsh (1991), it was found that the starting design point had a significant influence on the final geometry obtained. Therefore, it was necessary to start the optimisations from various points in the search space so that a number of optima are obtained, allowing the best of these to be selected.

Schwabacher et al. (1998) worked to overcome this problem. They used gradient-based methods to optimise various shapes, such as aircraft and yachts, using a learning algorithm based on previous optimisation procedures to determine where the best initial design point should be. Similarly, Chen and Lee (2002) tried to improve gradient based methods by using a 'gradient-forecasting search method' to dynamically adjust prediction steps so that it could overcome local optima.

Mohammadi and Pironneau (2004) used a gradient-based optimisation method to optimise supersonic aircraft for sonic boom. They made the suggestion that in the case of multi-criteria optimisation, shape optimisation is conflicting. Therefore, the use of the Pareto optimality method attempts to improve all aspects of the body by minimising a convex combination of all criteria. However, they suggest that the argument in such linear combinations leads to problems such as the existence of many sub-optima, requiring global optimisation tools such as GAs, which are simple but very slow. They therefore propose that the solution may well involve a combination of gradient and evolutionary methods.

Figure 1.1 General Atomics RQ-1A Predator. (Photo: USAF Museum).

Figure 1.11 Northrop Grumman X-47B UCAS-D. Photo: Rob Densmore.

Advanced UAV Aerodynamics, Flight Stability and Control: Novel Concepts, Theory and Applications,
First Edition. Edited by Pascual Marqués and Andrea Da Ronch.

Companion website: http://www.wiley.com/go/marques/advanced_UAV_aerodynamics

Figure 2.3 Airframe of the MA THOR Heli VTOL-UAS. Image: Marques Aviation Ltd.

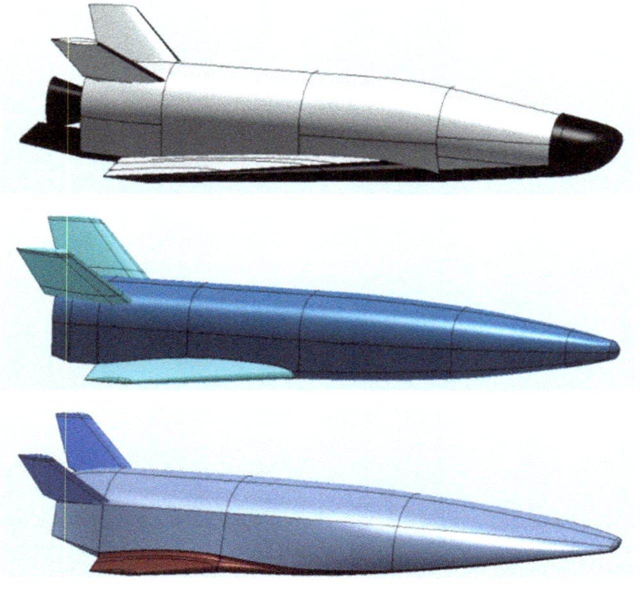

Figure 3.3 ORV-WSB body configurations: top, rather blunt; middle, sharp; bottom, spatulared.

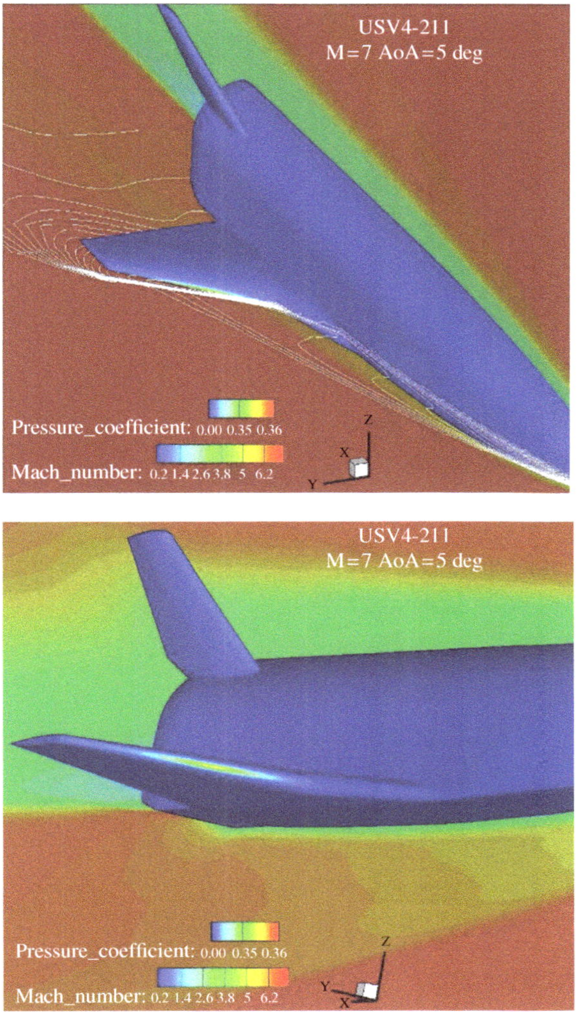

Figure 3.62 Contours of pressure coefficient over the FTB_4-211 surface. Mach number contours on vehicle symmetry plane and wing plane.

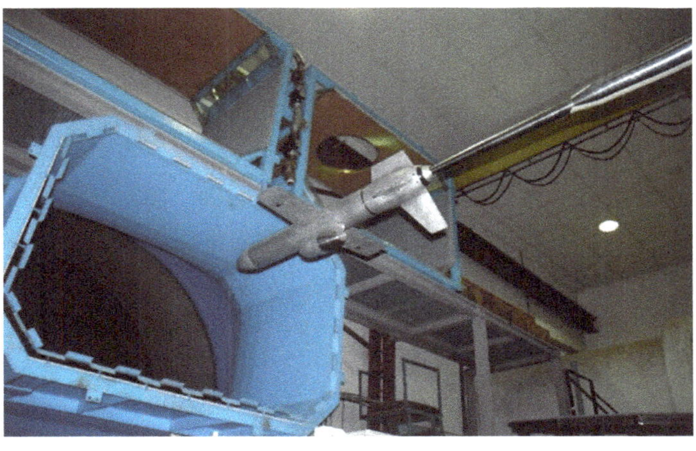

Figure 5.3 Diana aircraft model supported by sting.

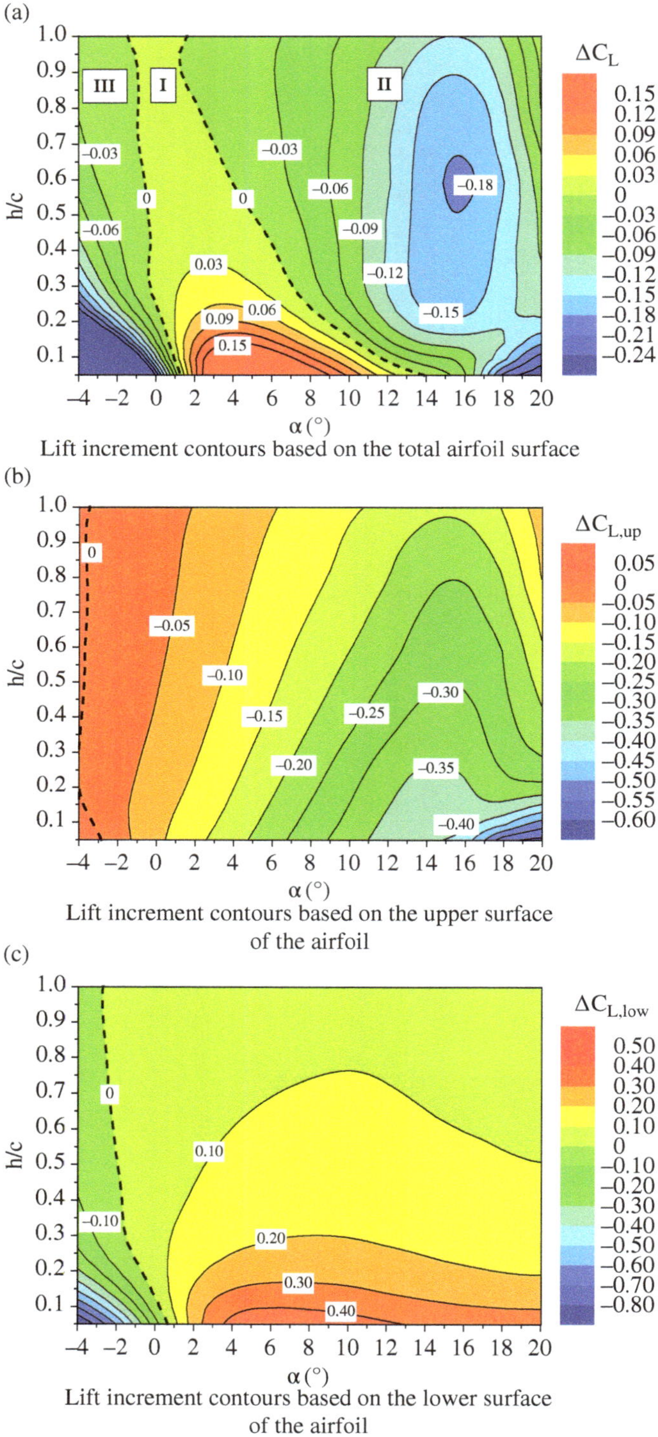

Figure 6.3 Lift increment contours of the NACA4412 airfoil in the AOA–ride-height plane.

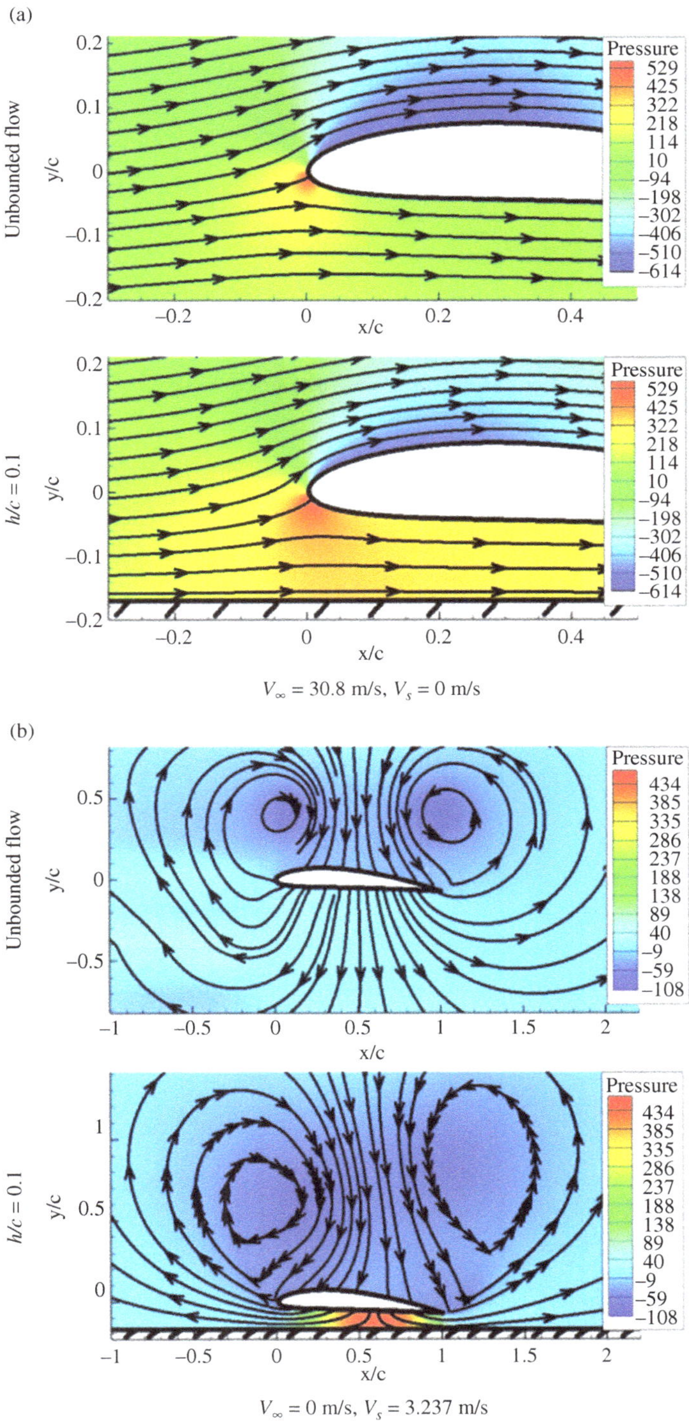

Figure 6.33 Streamlines and pressure contours around NACA 4412 at $\theta = 4°$.

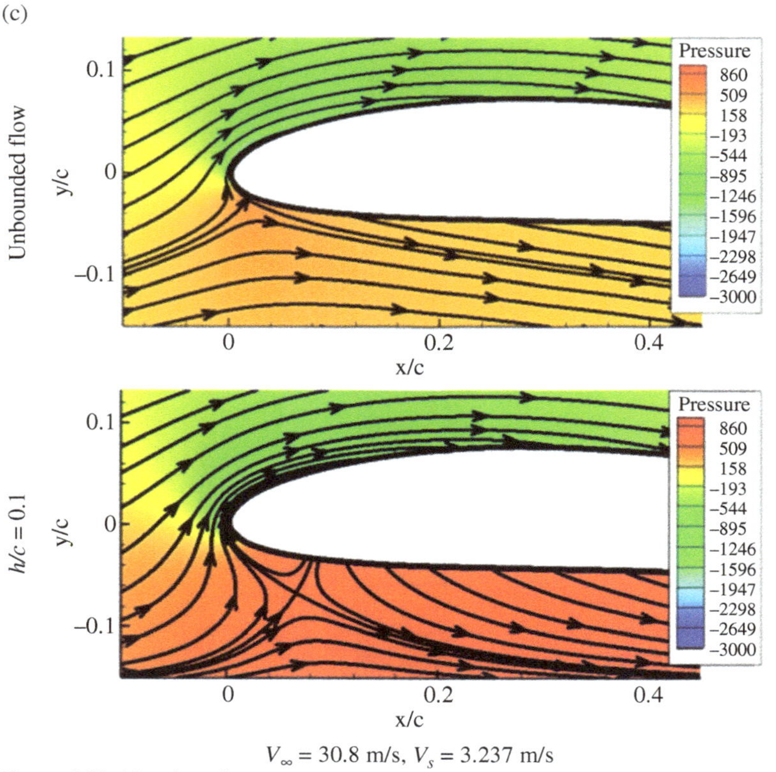

Figure 6.33 (*Continued*)

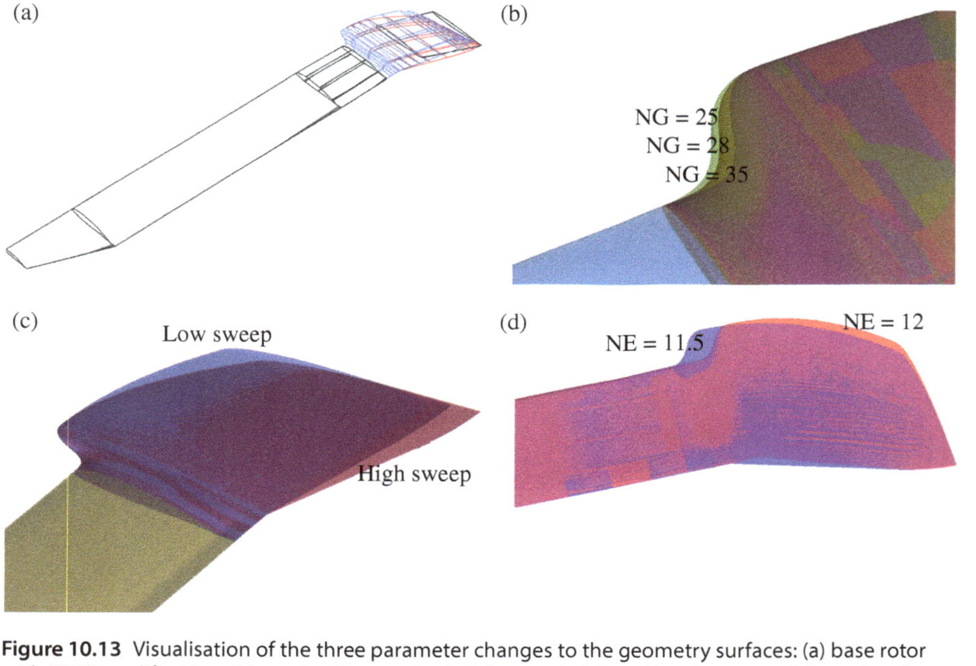

Figure 10.13 Visualisation of the three parameter changes to the geometry surfaces: (a) base rotor with BERP modification; (b) varying the gradient of the notch; (c) varying the sweep; (d) varying the initiation of the notch.

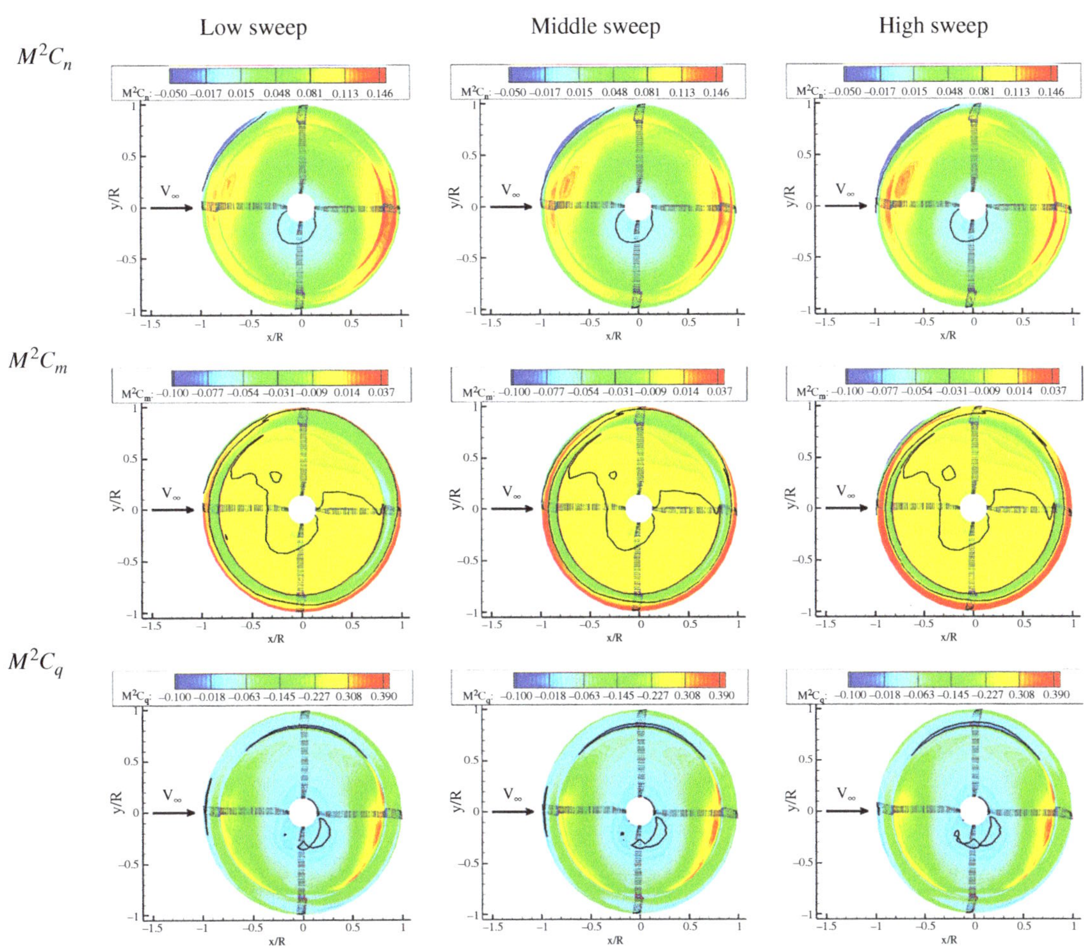

Figure 10.19 M^2C_n, M^2C_m and M^2C_q for NE = 11.5, NG = 35 and variable sweep. The black line indicates the 0 value.

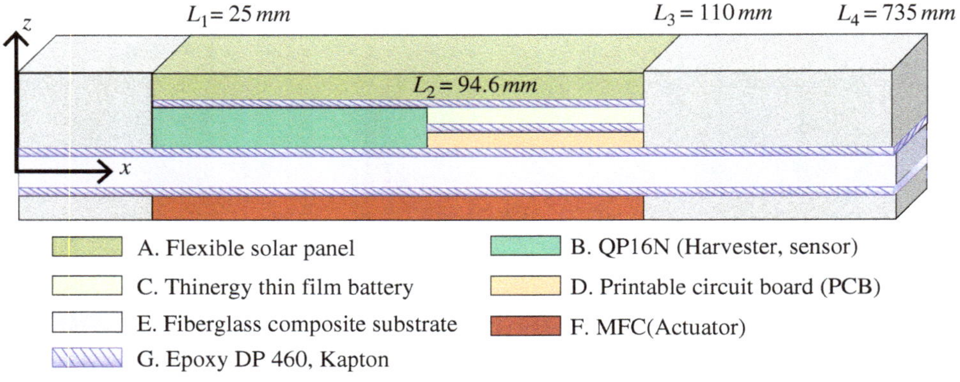

Figure 14.1 Multifunctional wing spar incorporating self-sensing, self-harvesting, self-storage and self-control.

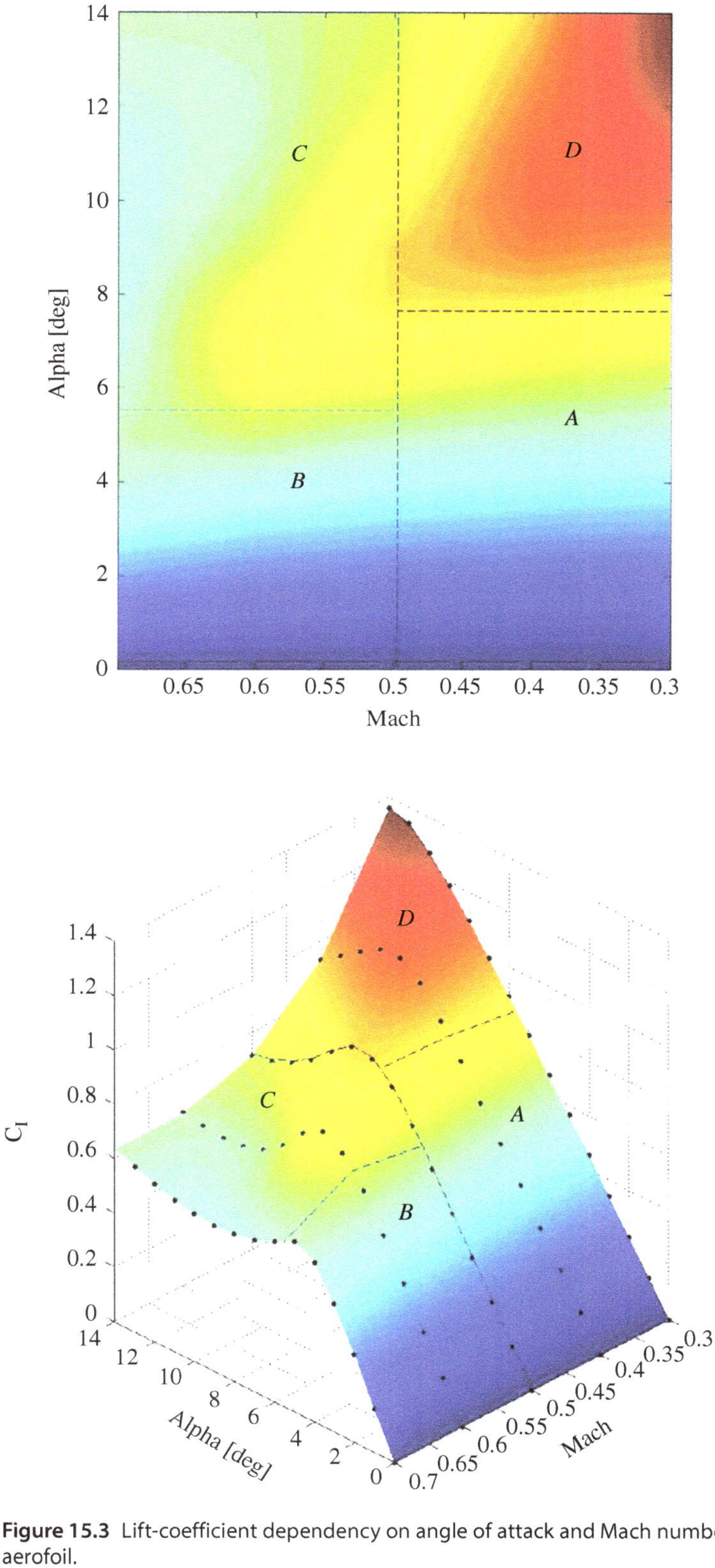

Figure 15.3 Lift-coefficient dependency on angle of attack and Mach number for a NACA 0012 aerofoil.

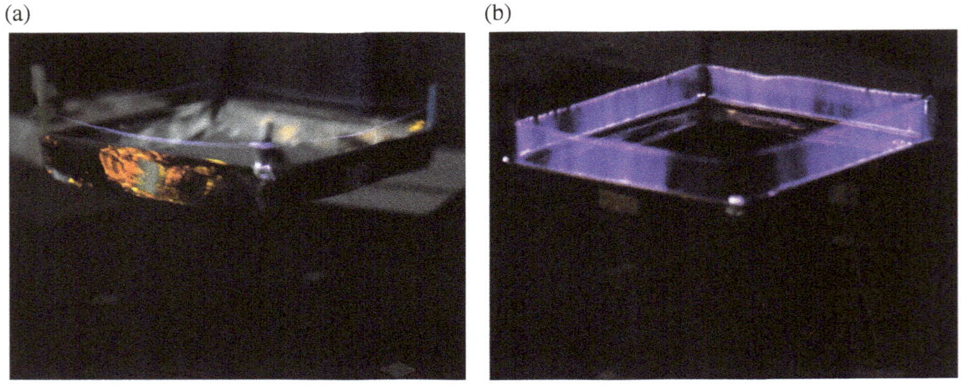

Figure 16.3 DC discharge on a lifter: (a) glow regime; (b) streamer regime.

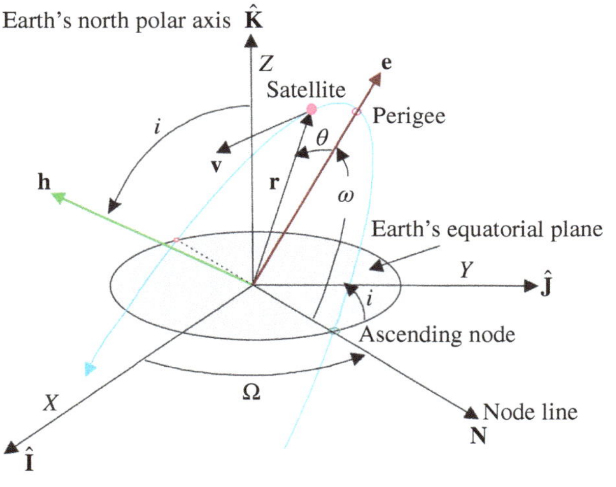

Figure 18.1 Classical orbital elements.

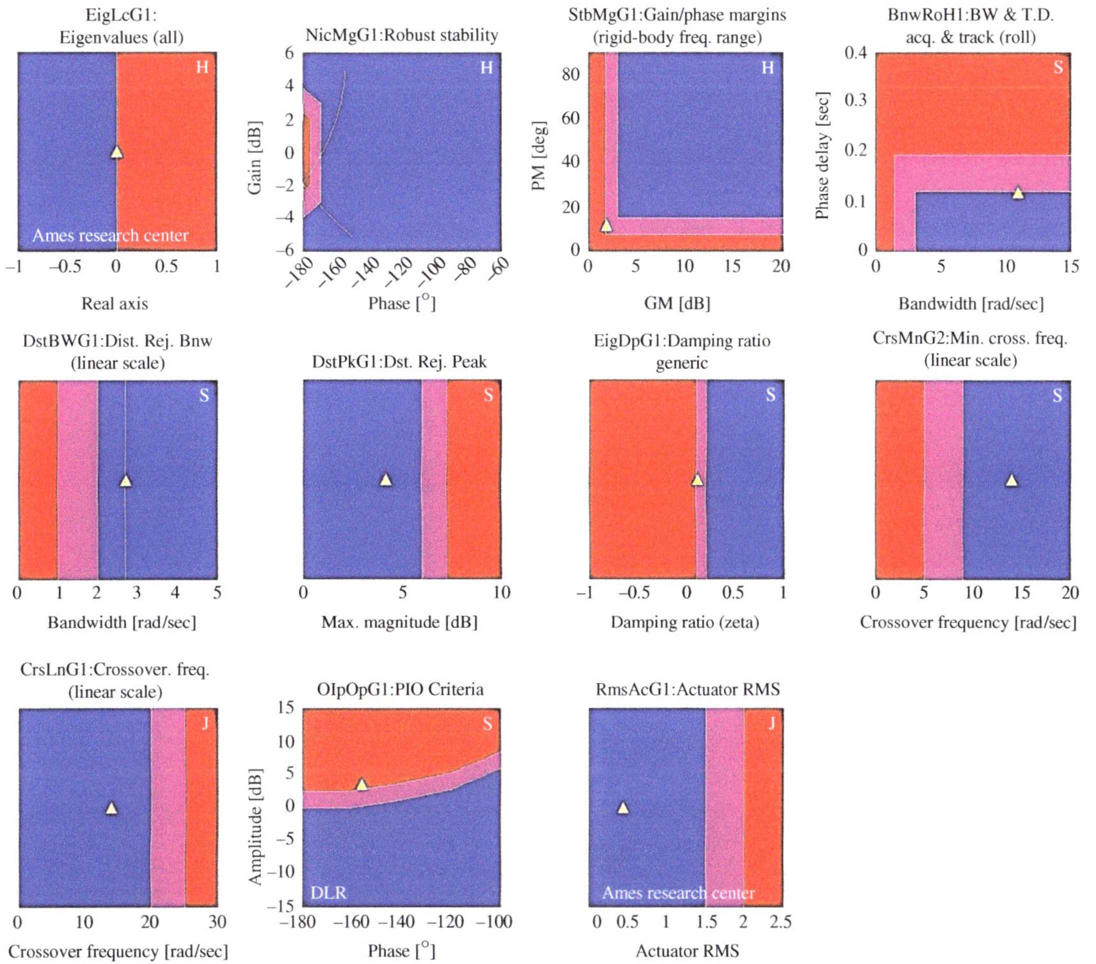

Figure 21.24 CONDUIT handling quality window for initial design.

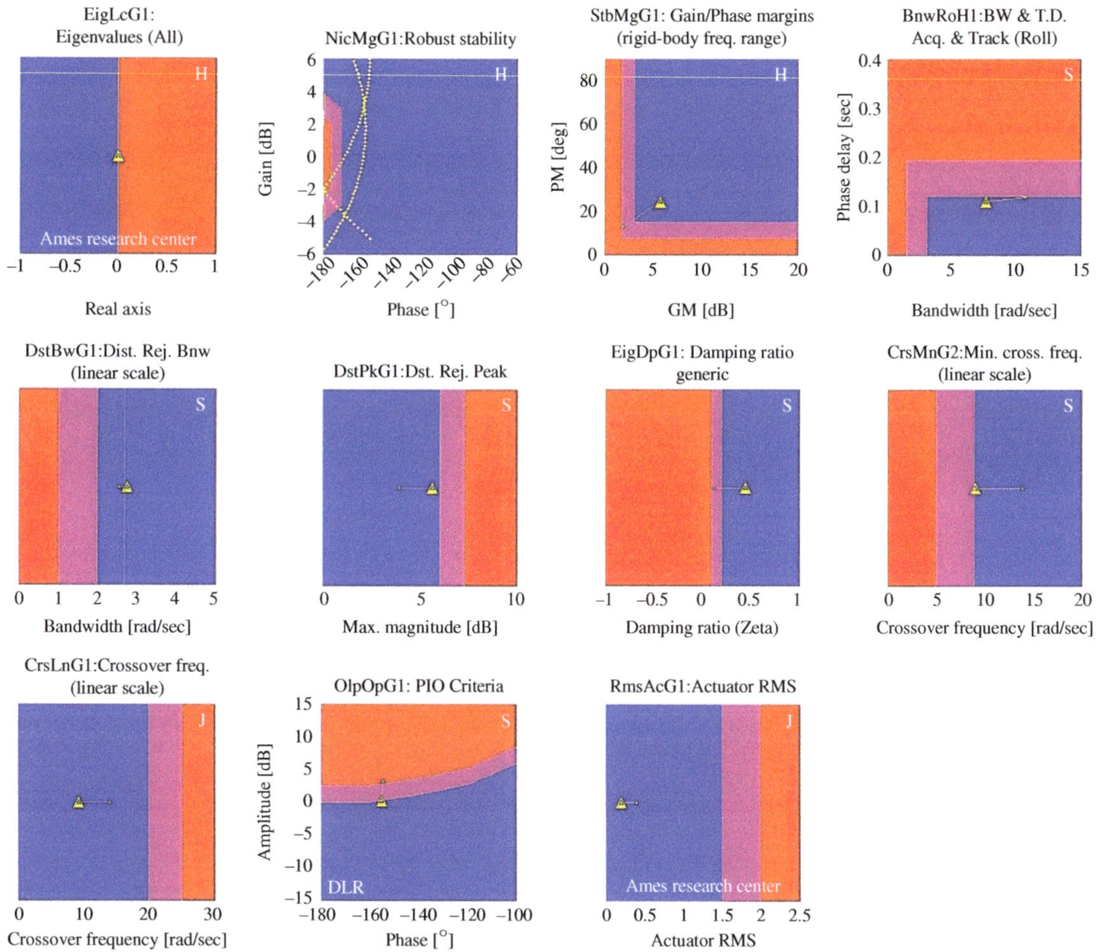

Figure 21.25 CONDUIT handling quality window for final design.

Lehner et al. (2010) also propose a hybrid-optimisation technique for multi-disciplinary optimisation of a subsonic aircraft. The optimiser was a non-gradient-based GA one, but at smoother points in the design space and where the data is continuous, a gradient-based method was used.

Non-gradient-based optimisation is expensive, especially for cases where each solution is costly, such as forward flight rotor optimisation. This is why Le Pape (2005), for example, used a GA for hover optimisation but used CONMIN for the forward flight optimisation. This again re-iterates the importance of using lower-order metamodels to improve non-gradient-based optimisation efficiency.

Another option is to use codes, such as ADIFOR, that can find derivatives analytically and semi-analytically and hence considerably reduce computational time (Celi 1999).

Non-gradient-based Optimisation

Non-gradient-based optimisation methods do not rely on the computation of gradients of a function and hence are not trapped in local optima. This is their greatest advantage and they therefore belong to a group termed 'global optimisation' techniques. Although the convergence to a global optimum is not guaranteed, the probability of it occurring is very high. Also, new designs in the search space are easily introduced, increasing the chances of a truly global optimum design for the required objective.

Typically, these methods consist of a performance metric and a selection phase based on it. The advantage is that sample points can be created and combined with other points in the population so that new designs are easier to create than with gradient-based methods. Some examples are statistical methods, GAs, evolutionary programming, simulated annealing (although it does involve some discretised determination of gradients), and branch and bound.

GAs simulate the evolutionary process involved in natural selection, with an initial population creating newer generations, with each new generation increasing in 'fitness' by the processes of crossover and mutation within the individuals.

Another method similar to GAs is the use of simulated annealing algorithms (SAs) (Hirsh 1998). SAs are based on an analogy with the way metals cool and anneal. They start from an initial point and move from one point to another until they reach an optimum. At each iteration, a new point replaces the current point dependent on a probability, P. If the new point is better than the current point, $P = 1$; if not, it has a lower value. This still gives it the chance of trying new paths and subsequently escaping local optima. At the start, the probability of moving to a worse point is high, but decreases as the optimisation progresses. SAs differ from GAs in that they follow a path, whereas GAs combine points from various regions in the search space to obtain new solutions. GAs also have a high inertia in switching attention to a new area compared to SAs. The best method is dependent on the problem. GAs maintain a population of points rather than one, and they hence are beneficial in that way. SAs are useful where the optimisation problem is large and has many local optima around the global optimum (Dulikravich et al. 1999).

Badyrka et al. (2010) carried out a study comparing three optimisers: GAs, repulsive particle swarm optimiser (RPSO) and a direct geometric search algorithm (VTDirect). The RPSO method is relatively new and was developed by Mishra (2006).

In the PSO, the data points simulate the individuals (or particles) of a swarm. Each has a position and velocity in the swarm and keeps in memory the overall best position it has been in. The individuals communicate the global or local best position to all members. Generally, it is known that PSO methods suffer from premature convergence

(Azaba and Ollivier-Gooch 2012). Therefore, modified versions of this method tend to be used. In the repulsive version, the particles are allowed to have a wider local search ability and this prevents the algorithm from getting stuck in local optima. Particles are allowed to learn from a predetermined number of random individuals (Mishra 2006). Furthermore, particles are not allowed to move into areas outside the constraints specified. In the regrouping version, once premature convergence is detected, the particles are automatically regrouped around the optimum found but with a smaller design space. The optimisation process is then repeated (Azaba and Ollivier-Gooch 2012).

The VTDirect method works by systematically subdividing the design space into quadrants. The division is such that the objective function value (OFV) is kept at the centre and the decision of how to divide is based on the size of the quadrant and the OFV. A quadrant is most likely to be divided if it is large and has a small OFV.

Badyrka et al. (2010) came to the conclusion that the RPSO method has a lot of promise and can have similar performance levels to GAs. It has fewer control parameters than a GA. The authors also suggest that for more complex optimisations, it may be good to couple the RPSO with a GA.

Vytla et al. (2010) employed a hybrid GA–PSO optimiser to optimise a train nose for reduced drag. They found the combined method to be more efficient and robust than either optimiser on its own. Similarly, Poloni et al. (2000) used a hybrid GA and gradient-based method along with an ANN to optimise the keel of a yacht for high lift and low drag, parameterised by Bézier curves. They describe a number of Pareto-based selection methods. One of the novel methods described is a version of a multi-directional crossover. They also suggest that replacing individuals rather than whole generations is more efficient. This makes it easier to parallelise the whole procedure. Dulikravich et al. (1999) also used a hybrid method, combining a number of optimisation techniques including GAs, SAs, gradient methods and the Nelder–Mead (NM) simplex method to create a more robust optimiser. Their method switches between these techniques where they work best. For example, the gradient method is employed when the variance from the GA's output is small and the NM method is used when the random behaviour of the GA is prevalent. This is controlled by a set of rules which can be found in further detail in the paper. The rules determine which method is the most efficient and accurate based on aspects such as the diversity of the selection and so on.

Other non-gradient methods include population-based incremental learning and ant-colony optimisation.

Régnier et al. (2005) state that there are three classifications of non-gradient based optimisation:

- *Apriori – Decide and Search:* The decision maker decides what is required and the solution is found, for example through the use of a weighting system. Only one Pareto solution is found.
- *Progressive and Sequential – Decide and Search:* Optimisation and decision-maker are intertwined. The preferences are sequentially updated. Multiple runs are required to obtain a Pareto solution.
- *Aposteriori – Search and Decide:* A single optimisation run provides a set of solutions that the decider can choose from.

The last of these is of particular interest since no target performance is required. Rather, the best design within the constraints on performance is required. Both the weighted sum and the Pareto method can be used. The advantage of using a weighted sum is that it selects a small region in the database where the optimum can be found. The design variables do not change much in this area. However, this reduces the flexibility of finding another such region unless the weights are changed. Therefore, the Pareto method is used to find the best compromise of the performance parameters and, typically, the results of the weighted sum method should lie somewhere on the Pareto front (Carrese et al. 2011). The advantage of using a Pareto method is it allows the designer to see the best compromise before making the decision. However, this can still be a difficult task if the number of design variables is high and if the properties of the Pareto subset are not properly analysed. More information can be found in the paper by Daskilewicz and German (2012).

Some other search methods such as Tabu, the NM and Powell's method are given in the paper by Hajela (1999). Also, the paper by Sobieszczanski-Sobieski and Haftka (1997), although relatively dated, has some useful insights into the various methods used in multi-disciplinary optimisation.

The cost of running high-fidelity software in order to find optimum solutions usually prevents the use of global optimisation techniques. Therefore, rotor optimisers usually consist of an approximate model or metamodel, as discussed in Section 9.1.2 (an approach known as surrogate-based optimisation). These are used to predict the behaviour of flow without the use of the high-fidelity software, such as a highly accurate flow solver. Once the optimal solution has been reached, the high-fidelity solver is used to validate the optimum solutions.

Methods for Genetic Algorithms

The performance metric (mentioned in Section 9.3.3) for a GA is called a *fitness* value. The success of a GA depends on its ability to perform a balanced amount of exploration as well as exploitation of any promising regions. A good GA usually favours exploration in the beginning of the search and gradually shifts to exploitation (Hirsh 1998). In general, the GA has a number of stages. It begins with a population of individuals from which a selection of two parents are made based on their fitness value. These parents are crossed over to create offspring that have a new combination of the design parameters from their parents. Mutation might occur, which alters the offspring. Based on the fitness value of the offspring, it survives into the next generation.

The fitness value is determined by the objective of the optimisation and can be obtained using a number of different methods, some of which are given in Tan et al. (2002). These authors performed a survey of various evolutionary methods for multi-objective (MO) optimisation and compared them quantitatively and qualitatively. A list of the methods was given in their Table 1.

Figure 9.6, reproduced from their paper, is a comparison of the trends in the usage of each of these methods. As can be seen, the most popular schemes over years up to the publication of Tan et al. (2002) are Pareto, ranking, and weights. Overall, the paper states that methods that used Pareto, ranking, goals and preference, preserved important non-dominant individuals via the method of switching preserved strategy. Also, the distribution along the Pareto front and the robustness in a noisy environment was more or less the same. It was also observed that elitism was found to be an important strategy for obtaining good performance. Elitism allows high-performing individuals to enter the next generation without the selection, crossover or mutation operations.

These methods are still the most popular in GAs methods today. A few newer methods that are as robust and have good performance have been developed. An example is the conservation technique called 'crowding'. This is described in the work by Hirsh (1998). Crowding takes into consideration values other than fitness, such as diversity preservation for example.

A number of methods exist for selecting the parents from a population. The aim of the selection process should be to select individuals who are generally fitter so that the pressure is directed towards better-performing individuals and hence generations. One of the simplest and most straightforward methods is the roulette-wheel method, in which the probability of selection is proportional to the fitness of the individual. Another method is the rank-based method, which is more useful when there is a wide range of fitness values. Here, the individuals are ranked and their probability of selection is proportional to their rank rather than their fitness value.

There are a number of crossover methods as well. Usually the design parameters are converted to binary format (the gene) and then a random point is selected where the gene is split and part of it is swapped with the other parent. In floating-point methods any one design parameter is swapped with that of the other parent. Other methods include heuristic crossover, linear crossover, arithmetic crossover and so on (Hirsh 1998). In the case of the optimisation of turbo-machinery blades, Mengistu and Ghaly (2008) used two methods of crossover: arithmetical, where two parents are linearly combined, and heuristic, which uses the fitness of the parents to determine the search direction and then creates the new offspring.

Similarly, there are numerous ways that mutation can be applied. The simplest is to randomly switch a random bit of the binary format of the design parameter. Other methods are more progressive, with the probability of mutation dropping as the

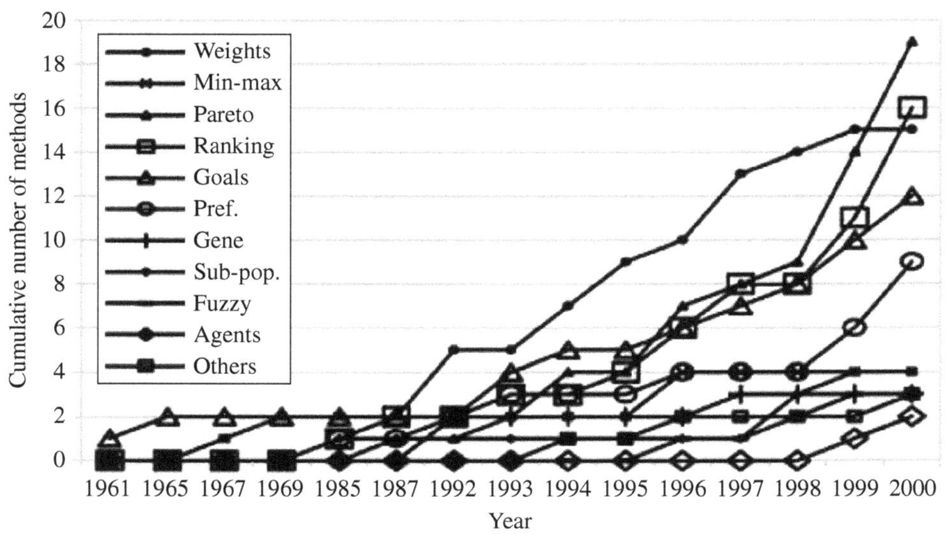

Figure 9.6 Comparison of the methods used in optimisation techniques developed in the last 50 years. It can be seen that ranking, Pareto and weights are the techniques that have become most popular. Reproduced from Tan et al. (2002).

optimisation matures (Hirsh 1998). Mutation can create new designs or reintroduce old designs in the gene pool and improve the diversity of the design search space or gene pool. This is important because after many generations, selection could drive all the bits into a 1 position (using the example of binary coding, to either a 1 or a 0). Mutation allows the designer to reintroduce the missing bit. Also, it allows retesting. However, the number of mutations must be kept to a minimum to reduce randomness (Whitley 1994). Sometimes, binary representation can mean that there are big changes in the binary form for small changes in parameter. Therefore, in these types of cases, the *Gray code* or *reflected binary code* (Gray 1953) is sometimes used. Here, two successive parameter values differ in a single bit in their binary form; in other words, instead of counting up from 7 to 8 for example, which results in a binary change of 4 bits (0111 to 1000), this code allows a conversion of only one bit (0111 to 0101), which translates to counting from 7 to 5.

Constraints can be applied to the resulting offspring. One approach is to treat the gene as faulty and to try to repair it (Hajela 1999). A simpler approach is the penalty method, in which the fitness value is diminished based on the constraint that has been violated and by how much it is violated. If the constraint is a hard one, the offspring is removed from the new population. For a soft constraint, although its fitness value is compromised,it is allowed to join the new population based on its new fitness value (Hajela 1999).

Hirsh created a design optimiser called GADO (Genetic Algorithm Design Optimiser) (Hirsh 1998) used for engineering design, and set out how it could be used to improve the search for the optimum function in a GA. These are some of its features:

- *Screening module (SM)*: The idea of the SM is to predict the merit of a proposed design before the simulator is used to evaluate it. The SM selects a point that is promising (using extrapolation and not simulation). Before a point is selected, the SM finds the K nearest neighbours (usually two, but this is dependent on the sharpness of the optimum) and checks if it is above a certain threshold value before adding it to the new sample population.
- *Diversity maintenance module*: This ensures that the selected points are not very similar to each other, so that the the search space is fully represented. It prevents the algorithm from being trapped in local optima. If at a certain point it is found that the points are quite similar, the module rebuilds the population using previous points.
- *Guided crossover*: This forms search directions without computing gradients, done by joining pairs of points, ranking these directions and taking a step in the best direction. Directions are ranked by the ratio of the difference in the fitness to the distance between them. It works as follows:
  - One point is selected, c1.
  - For every other point in the population, a mutual fitness is calculated as below:

$$\text{Mutualfitness}(A, B) = \frac{(\text{fitness}(A) - \text{fitness}(B))^2}{\text{Euclidean distance}(A, B)^2} \tag{9.2}$$

  A choice for the second point, c2 is made by maximising the mutual fitness.
  - c1 and c2 are swapped if necessary so that c1 has the highest fitness.

- The result of the crossover is a point along the line joining c1 and c2, selected at random from the small region around c1:

$$\text{Result} = L \times c1 + (1 - L) \times c2 \tag{9.3}$$

where L is a uniformly distributed random number in the interval $[1-0.2x, 1+x]$ and $x = 0.75 \times (1-n) + 0.25$ where n is number of iterations. Nevertheless, this should not be used as the only crossover operator (Hirsh 1998).
There are many other methods such as the heuristic method: If the parents are $\overline{X}$ and $\overline{Y}$ such that $\overline{Y}$ is fitter than $\overline{X}$, then the new born is $\overline{Y} + r \cdot (\overline{Y} - \overline{X})$ where r is a random value selector between 0 and 1. Other methods include point, random (which produces a lot of diversity), arithmetic and linear.

- *Dynamic Penalty*: The GA uses an 'adaptive penalty' approach for handling constraints. When the GA focuses too much on feasibility it reduces the penalty coefficient and vice versa.

A number of selection strategies are described in [?], two of which are:

- *Fitness proportional (roulette wheel) selection: The probability of an individual being selected is dependent on its fitness.*
- *Rank-based selection: Each individual's probability of being selected depends on its fitness rank rather than its actual fitness. The most commonly used methods of carrying out this selection are:*
  - tournament selection, in which two candidates selected and the fitter of the two is selected for reproduction.
  - weight series selection, in which each individual is assigned a weight dependent on fitness rank (weights usually taken to be a decreasing arithmetic or geometric series) and then proportional selection is performed using the weights rather than the actual fitness.

Rank-based selection is considered to be more appropriate for use in domains where the fitness range is extremely wide.

In steady state GAs, each new individual replaces an old individual, so the population remains constant. In generational GAs, the entire population or a portion of it is replaced simultaneously. Steady-state GAs converge faster than generational ones.

Two mutation operators are described:

- uniform mutation, in which each component of a solution vector is replaced with a random value uniformly selected from the component range
- non-uniform mutation, which acts, a the beginning of the optimisation, like uniform mutation, but then becomes more and more conservative about the amount of change it makes to a vector component as the optimisation progresses.

Toffolo and Benini (2003) did some work on improving MO optimisation by enhancing the diversity of the gene pools created by GAs. In their algorithm (GeDEM), genetic diversity was a real objective in itself, measured as the variable or optimisation function distance to neighbours. Benini also used GeDEM to optimise calibration of centrifugal pumps (Benini and Cenzon 2009). One of the interesting aspects of the GA used was in the selection of parents. A reproduction pressure was included that selected fitter individuals to crossover, but this pressure varied with the generation so that at the start of the optimisation, the pressure allowed for more diversity, but towards the end, more exploitation of the best genes rather than exploration of the design space took place.

Another advantage of GAs is that they can be easily parallelised to reduce the time required (Janiga 2008). González et al. (2006) created a parallel computing modified GA called the 'hierarchical asynchronous parallel evolutionary algorithm', which proved to use slave machines more efficiently. It also created a number of populations at each generation instead of a single population (González et al. 2006).

Other Optimisation Methods

Hassan and Charles (1997) used a reverse-design method to design aerofoil blades for a hovering helicopter. Here, the pressure distribution was specified by the user and the aerofoil geometry that best produced this distribution was created for each radial station along the blade span. CAMRAD/JA was used to compute the rotor trim state as well as the far-wake-induced incidence. The actual modification and analysis of the geometry was carried out by the solver RFS2, which had a built-in ability to generate new grids of the geometry created at each iteration. This concept can work well, but a three dimensional approach must be taken rather than aggregating many 2D sections to form a blade.

Another method of optimisation or design that has generated a lot of interest is adjoint-based optimisation. This method computes the sensitivity of the shape to the objective. Jameson et al. (1998) use an adjoint approach in which a target pressure distribution is specified for a fixed wing. The Navier–Stokes equations are used to iteratively correct the shape function to obtain the required distribution by reducing a constrained cost function by gradient-based optimisation. Mani et al. (2012) developed an adjoint based method to optimise aerofoils for dynamic stall. The objective was to reduce the peak pitching moment while maintaining lift. This was successfully carried out with considerably less computational effort than performing a time-dependent simulation for the same amount of flexibility in the design. Nielsen et al. (2009) applied an adjoint based method for the design of rotors in hover using Navier-Stokes equations with the aim of creating a general time-dependent adjoint-based optimisation method for rotorcraft. The results were comparable with those of independent approaches and showed improvement in FM.

Liu et al. (2007) also used a reverse engineering method to design rotor blades for HAWTs. They modified the twist, chord and aerofoil in the form of Bézier curves for each section along the blade and merged these 2D designs smoothly along the span. The strip theory was used, taking into account hub and tip losses, cascade effects and so on. A blade design was obtained that had an overall increase in the output power. However, for a more accurate prediction and hence a more validated result, it is important to optimise while taking into account the effects of the flow field over the entire 3D region. Inverse design methods have been explored for a long time. Lekoudis and Sankar (1984) used an inverse potential flow method to prescribe the skin friction coefficient and obtain a pressure distribution from it for a wing. A similar method was used to design aerofoils and nacelle inlets for a target pressure distribution.

Drayna et al. (2011) created a direct non-linear sensitivity solver to optimise a scramjet inlet. This could be incorporated into an existing solver without having to rewrite any of the original solver code, which is what would be required in the case of an adjoint sensitivity solver. It is more accurate than a linearised solver and also requires no new boundary conditions.

9.2.4 Parameterisation Techniques

Parameterisation is essentially numerical coding of the design variables to obtain a general form to represent a particular shape. It is a key part of the optimisation process. Samareh (n.d.) states that a good parameterisation technique for multidisciplinary shape optimisation should be:

- automated
- provide consistent geometry changes across all disciplines
- provide sensitivity derivatives
- fit into product development time
- have a direct connection to the CAD system for design
- produce a compact and effective set of design variables.

However, the initial main focus in this case will be the last point made. This is because the optimisation method desired is one that will not require grid regeneration until the end of the process. The coupled method has been used by a number of authors, such as Le Pape (2005), in which, for the elsA solver, an in-house analytic grid-generator was used to create a grid for each evaluation. However, in this project, the optimisation technique was decoupled from the high-fidelity CFD model and linked to it through a lower fidelity model.

Hàjek describes a number of techniques used to parameterise aerofoils in the light of aerodynamic optimisation (Hájek 2007). This paper states that the choice of technique strongly influences the optimisation. Similarly, Castonguay and Nadarajah (2007) performed a study on four different parameterisation methods for the optimisation of aerofoils using an inverse method. Some of the methods described in these two papers are outlined in the following subsections.

Mesh Point Approach

Here each mesh point can be independently moved. This is one of the easiest methods to implement and allows a lot of flexibility in the design but can lead to discontinuous geometries and possibly too many control variables (Castonguay and Nadarajah 2007).

Joukowski Transformation

The Joukowski transformation consists of transforming a circle in the complex plane via the transformation

$$z = \varepsilon + \frac{1}{\varepsilon} \tag{9.4}$$

The circle should pass through the point $\varepsilon = 1$. The aerofoil shape is controlled by varying the centre of the circle. This method was especially advantageous in the past because it enabled the plane potential flow to be analytically solved.

Splines

Splines use piecewise polynomial approximations of curves. B-splines use the same idea but are slightly more complex in that they are built on linear combinations of base functions. Non-uniform rational B-splines (NURBS) are a further development. Ghaly and Mengistu (2003) show that Bézier curves are used because of the simplicity of their implementation. However, they are global and hence a change in a single control point changes the shape of the entire blade and so the designer has less control over local regions. To accommodate this, B-splines are used. These work just like Bézier curves but have more complex interpolation functions. These allow more local control, but are still not accurate in describing conic sections such as rounded leading edges. Therefore, NURBS are used since these allow local control and can also reproduce conic sections as accurately as required by industrial standards.

Hick–Henne 'Bump' Functions

This involves modelling small or moderate perturbations of 'baseline' aerofoil shapes, especially in inverse design (Sun et al. 2003). The perturbation is expressed as a linear superposition of 'bump' functions of the form

$$y = \sin^t\left(x^\beta\right) \tag{9.5}$$

where β is used to control the maximum of the bump function, located at $x = \pi / 2^{1/\beta}$ and t controls the width of the function (typically $t = 3$). This allows specific regions to be refined, thereby reducing the number of variables; the bump functions also ensure continuous gradients of the shapes. However, they are not orthogonal and hence are incapable of representing the full set of continuous functions (Castonguay and Nadarajah 2007).

PARSEC

PARSEC (Castonguay and Nadarajah 2007; Hájek 2007) is used mainly for subsonic and transonic aerofoils. The aerofoil shape is expressed as an unknown linear combination of suitable base functions. Twelve important geometric characteristics are selected as the control variables in such a way that the shape can be determined by solving a linear system with them. The variables are shown in Figure 9.7. The advantage of PARSEC is that no baseline shape is needed, a wide range of aerofoil shapes can be generated, constraints are easy to impose and the impact of individual PARSEC design parameters on aerodynamic properties can be easily predicted. The disadvantage though, is that it cannot cover as wide a range as spline curves.

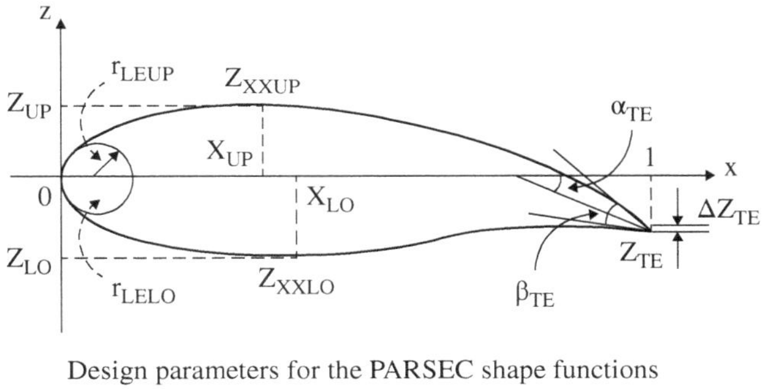

Design parameters for the PARSEC shape functions

Figure 9.7 PARSEC variables of an aerofoil for parameterisation (Hájek 2007).

B-splines, NURBS

For complex geometries, the use of Bézier curves, splines and NURBS is quite popular, especially in the optimisation of compressor blades (Samad and Kim 2008; Zhao et al. 2008) mainly because they allow a greater degree of freedom, with the use of fewer variables, and because they can produce smooth, discontinuity-free curves. However, the implementation of constraints is not as straightforward as in numerical representations of physical measurements. Philip Schneider gives a detailed explanation of how NURB curves work and his article (Schneider n.d.). Joh (2003) used NURBS with 36 points to fully describe the M6 wing. This shows the versatility of the technique, but also the high number of parameters used to describe a simple body. Ghaly and Mengistu (2003) describe a method for finding the minimum number of control points to represent turbomachinery blades using NURBS within a specified tolerance. An objective function to be minimised is a representation of the error, and the weights and control point coordinates are the parameters being optimised for. In this way, coordinate-point data curves can be converted to NURB curves and vice versa. In the case of Ghaly and Mengistu (2003), simulated annealing was used to optimise for the minimum number of points.

In Samad and Kim (2008), cubic spline interpolation was used to define the blade. In other words a separate cubic spline was created for each interval; these are popular because they produce smooth curves and are easy to implement. In total, 20 variables were reduced to 6.

Le Pape (2005) used Bézier curves to parameterise all the distributions of the variables for rotor optimisation, except for the aerofoils.

The use of splines. Bézier curves and NURBS allow localised control and a reduced number of variables and brings the ability to represent continuous curves as well as discontinuities in the geometry. However, the link between the values and the design variables are not intuitive to the user.

Kulfan and Busoletti (2006) also carried out some work on parameterisation. They divided the parameterisation into two classes, one for the shape function – aerofoils and so on – and the other for body cross-sections such as fuselages and nacelles. For an aerofoil, the leading edge and trailing edge are specified and the curve between these extremes is what is modified. Berstein polynomials were used to do this. This technique was adapted to parameterise aerofoils and wings (Young 2011).

Vanderplaat also describes a parameterisation method in the appendix of his paper (Hicks et al. 1974) where the shape of an aerofoil is defined with an equation:

$$t = \delta\left(A\sqrt{x} + a_1 x + a_2 x^2 + \ldots a_n x^n\right) \tag{9.6}$$

where t is the thickness of the aerofoil and δ is the thickness-to-chord ratio for the initial aerofoil. The square root term yields a parabolic leading-edge term. The coefficients A, $a_1, a_2, \ldots, a_n$ are the variables perturbed by the optimisation function.

Other parameterisation methods include Fourier series, piecewise polynomial and orthogonal polynomial. NACA has in effect a parameterisation system of 4, 5 and 6-digit sections. In general, most helicopter optimisation problems rarely exceed 10–30 design variables (Celi 1999). The aim in terms of parameterisation is to maintain a low number of design parameters to optimise, while giving the designer as much flexibility in the specific design variables as possible.

9.3 Summary

From the literature survey, it seems that of non-gradient and gradient based methods, the former seems to be gaining popularity for more complex cases such as rotor-blade design. The limiting factor seems to be the computational effort required to apply this method with high-fidelity simulation data. However, several lower-order models can be used to accurately predict interpolated data given a set of high-fidelity data. They are able to maintain accuracy and increase efficiency at the same time by orders of magnitude, because they are given only a limited set of inputs that are actually the objectives of the designer. Based on the patterns in changing these input parameters, they make the link between the input and output. This seems to be more in line with the objective of fine-tuning the aerodynamic design of a blade.

The adjoint-based method offers a lot of flexibility to the design and is more suitable for designing from scratch or from highly unoptimised designs, where a big design change is expected. Therefore it is not explained in much detail here.

The genetic algorithm is a popular non-gradient-based method that provides a simple efficient implementation. Other non-gradient-based methods can be used as well. However, as a starting point, this method is selected. A number of metamodels are of interest. This part of the optimisation process is quite important, as it is the part that overcomes the problem of efficiency with non-gradient-based methods. The kriging and neural-network methods particularly show a lot of potential for accurate interpolation and ease of coupling with the optimiser.

The design parameters and objectives are also key components of the optimisation process. Our literature survey has provided important pointers in selecting the best techniques to parameterise designs and the factors to consider when creating an objective function. These two parts are highly dependent on the user and their experience. This method, however, is aimed at assisting, rather than replacing the designer. Therefore, some of it is reliant on user-experience.

In the next chapter, the aim is to show how to initiate the optimisation process from an existing blade that is already somewhat optimised for a specific objective. Metamodels will be employed with Navier–Stokes CFD as the high-fidelity flow model. The novelty of our approach is the combination of simple methods that result in efficient and accurate optimisation of well-defined parameters of rotor blades, which are already designed with specific objectives in mind, in order to obtain the best performance out of the design. This is obtained by decoupling the optimisation process directly from the high-fidelity solver and linking it through a lower-order model that relies on the CFD model for its accuracy. The developments made in helicopter rotor blades due to optimisation can be first tested using unmanned air vehicles (UAVs) to minimise cost and increase safety. In addition, this method can be applied to UAVs themselves in order to obtain higher performance in terms of endurance, agility and so on.

Bibliography

Acree, Jr. CW 2010 Integration of rotor aerodynamic optimization with the conceptual design of a large civil tiltrotor *American Helicopter Society Aeromechanics Specialist Conference*, San Francisco, California, USA.

Adelman HM and Mantay WR 1989 Integrated multidisciplinary optimization of rotorcraft: A plan for development.

Ahmed MYM and Qin N 2010 Metamodels for aerothermodynamic design optimization of hypersonic spiked blunt bodies *48th AIAA Aerospace Sciences Meeting Including the New Horizons Forum and Aerospace Exposition*, Orlando, Florida, USA. AIAA 2010-1318.

Allen CB, Rendall TCS and Morris AM 2010 CFD-based twist optimization of hovering rotors. *Journal of Aircraft* **47**(6), 2075–2085.

Anderson, Jr. JD 1995 *Computational Fluid Dynamics, The Basics with Applications.* McGraw-Hill International Editions.

Azaba MB and Ollivier-Gooch C 2012 High order aerodynamic optimization using new hybrid sequential quadratic programing-particle swarm intelligence technique *50th AIAA Aerospace Sciences Meeting including the New Horizons Forum and Aerospace Exposition*, Nashville, Tennessee, USA. AIAA 2012-0730.

Badyrka JM, Jenkins RM and Hartfield RJ 2010 Aerospace design: A comparitive study of optimizers *48th AIAA Aerospace Sciences Meeting Including the New Horizons Forum and Aerospace Exposition*, Orlando, Florida, USA. AIAA 2010-1311.

Benini E and Cenzon M 2009 Calibration of a meanline centrifugal pump model using evolutionary algorithms. *Proceedings of the Institution of Mechanical Engineers Part A: Journal of Power and Energy* **223**, 835–847.

Biedron RT and Lee-Rausch EM 2012 An examination of unsteady airloads on a uh-60a rotor: Computation versus measurement *American Helicopter Society 68th Forum*, Fort Worth, Texas.

Bohling G 2005 Kriging. *Kansas Geological Survey.*

Bohorquez F, Pines D and Samuel PD 2010 Small rotor design optimization using blade element momentum theory and hover tests. *Journal of Aircraft* **47**(1), 268–283.

Bui-Thanh T, Damodaran M and Willcox K 2004 Aerodynamic data reconstruction and inverse design using proper orthogonal decomposition. *AIAA Journal* **42**, 8.

Cameron L, Early JM and McRoberts R 2011 Metamodel assisted multi-objective global optimisation of natural laminar flow aerofoils *29th AIAA Applied Aerodynamics Conference*, Honolulu, Hawaii.

Caradonna FX 1992 The application of CFD to rotary wing problems. Technical report, NASA.

Caradonna FX and Tung C 1981 Experimental and analytical studies of a model helicopter rotor in hover. Technical report, NASA.

Carrese R, Winarto H and Li X 2011 Integrating user-preference swarm algorithm and surrogate modeling for airfoil design *49th AIAA Aerospace Sciences Meeting including the New Horizons Forum and Aerospace Exposition*, Orlando Florida. AIAA-244265-371.

Castonguay P and Nadarajah SK 2007 Effect of shape parameterization on aerodynamic shape optimisation *45th AIAA Aerospace Sciences Meeting and Exhibit*, Reno, Nevada, USA.

Celi R 1999 Recent applications of design optimization to rotorcraft – a survey *55th Annual Forum of the American Helicopter Society.*

Chen CM and Lee HM 2002 An efficient gradient forecasting search method utilizing the discrete difference equation prediction model. *Applied Intelligence Journal* **16**, 43–58.

Collins K, Bain J, Rajmohan N, Sankar L, Egolf TA, Janakiram RD, Brentner K and Lopes L 2008 Toward a high-fidelity helicopter rotor redesign framework *AHS 64th Annual Forum*, Montréal, Canada.

Cook CV 1972 The structure of rotor blade tip vortex. Technical Report CP-111, AGARD.

Crowley D, Robertson B, Douglas R and Mavris DN 2012 Aerodynamic surrogate modeling of variable geometry *50th AIAA Aerospace Sciences Meeting including the New Horizons Forum and Aerospace Exposition*, Nashville, Tennessee, USA. AIAA 2012-0268.
Daskilewicz MJ and German BJ 2012 Observations on the topology of Pareto frontiers with implications for design decision making *50th AIAA Aerospace Sciences Meeting including the New Horizons Forum and Aerospace Exposition*, Nashville, Tennessee, USA. AIAA 2012-0148.
Deb K 2001 Nonlinear goal programming using multi-objective genetic algorithms. *Journal of the Operational Research Society* **52**, 291–302.
Drayna TW, Candler GV and Johnson HB 2011 A nonlinear sensitivity solver for aerodynamic optimization and rapid design space exploration *49th AIAA Aerospace Sciences Meeting including the New Horizons Forum and Aerospace Exposition*, Orlando Florida. AIAA 2011-26.
Dulikravich GS, Martin TJ, Dennis BH and Foster NF 1999 Multidisciplinary hybrid constrained GA optimization *EUROGENâ€™99 – Evolutionary Algorithms in Engineering and Computer Science: Recent Advances and Industrial Applications*, Jyvaskyla, Finland.
Fletcher R and Reeves C 1964 Function minimization by conjugate gradients. *British Computer Journal* **7**(2), 149–154.
Forrester AIJ and Keane AJ 2009 Recent advances in surrogate-based optimization. *Progress in Aerospace Sciences* **45**, 50–79.
Ghaly WS and Mengistu TT 2003 Optimal geometric representation of turbomachinery cascades using NURBS. *Inverse Problems in Science and Engineering* **11**(5), 359–373.
Glaz B, Friedmann PP and Liu L 2007 Surrogate based optimization of helicopter rotor blades for vibration reduction in forward flight. *Structural Multidisciplinary Optimisation Journal* **35**, 341–363.
Glaz B, Goel T, Liu L, Edmann P and Haftka R 2009 Multiple-surrogate approach to helicopter rotor blade vibration reduction. *AIAA Journal* **47**(1), 271–282.
Glaz B, Liu L, Friedmann P, Bain L and Sankar L 2010a A surrogate based approach to reduced order dynamic stall modeling *6th AIAA Multidisciplinary Design Optimization Specialist Conference*, Orlando, Florida, USA. AIAA 2010-3042.
Glaz B, Liu L, Friedmann PP, Bain J and Sankar LN 2010b A surrogate based approach to reduced-order dynamic stall modelling *51st AIAA/ASME/ASCE/AHS/ASC Structures, Structural Dynamics, and Materials Conference*, Orlando, Florida, USA.
González L, Whitney E, Srinivas K and Périaux J 2006 Optimum multidisciplinary and multi-objective wing design in CFD using evolutionary techniques *Proceedings of the 3rd International Conference on Computational Fluid Dynamics, ICCFD3*, Toronto, Canada.
Gray F 1953 Pulse code communication US Patent 2632058, filed Nov. 1947.
Gunes H, Sirisup S and Karniadakis G 2006 Gappy data: to krig or not to krig?. *Journal of Computational Physics* **212**, 358–382.
Hájek J 2007 Parameterization of airfoils and its application in aerodynamic optimization *Proceedings of the 16th Annual Conference of Doctoral Students*, Prague, pp. 233–240.
Hajela P 1999 Nongradient methods in multidisciplinary design optimization – status and potential. *Journal of Aircraft* **36**(1), 255–265.
Hall KC and Hall SR 2010 A variational method for computing the optimal aerodynamic performance of conventional and compound helicopters. *Journal of the American Helicopter Society* **55**(4), 42006.

Harrison R, Stacey S and Hansford B 2008 BERP IV the design, development and testing of an advanced rotor blade *American Helicopter Society 64th Annual Forum*, Montreal, Canada.

Hassan AA and Charles BD 1997 Airfoil design for helicopter rotor blades – a three-dimensional approach. *Journal of Aircraft* **34**(2), 197–205.

Hicks RM, Murman EM and Vanderplaats GN 1974 An assessment of airfoil design by numerical optimization. Technical report, NASA.

Hirsh R 1998 *GADO: A Genetic Algorithm for Continuous Design Optimization* PhD thesis State University of New Jersey.

Imiela M 2009 High-fidelity optimization framework for helicopter rotors *35th European Rotorcraft Forum*, Hamburg, Germany.

Jameson A, Martinelli L and Pierce NA 1998 Optimum aerodynamic using the Navier-Stokes equations. *Theoretical and Computational Fluid Dynamics* **10**, 213–237.

Janiga G 2008 A few illustrative examples of CFD-based optimization – heat exchanger, laminar burner and turbulence modeling In *Optimisation and Computational Fluid Dynamics* (ed. Thévenin D and Janiga G) Springer.

Joh CY 2003 Development of a NURBS-based wing design optimization system *Proceedings KORUS 2003. The 7th Korea-Russia International Symposium on Science and Technology*, 28 June–6 July 2003, University of Ulsan, Republic of Korea.

Johnson W 2010a NDARC – NASA design and analysis of rotorcraft: Theoretical basis and architecture *American Helicopter Society Aeromechanics Specialist Conference*, San Francisco, California, USA.

Johnson W 2010b NDARC – NASA design and analysis of rotorcraft validation and demonstration *American Helicopter Society Aeromechanics Specialist Conference*, San Francisco, California, USA.

Karakasis MK and Giannakoglou KC 2004 On the use of surrogate evaluation models in multi-objective evolutionary algorithms *European Congress on Computational Methods in Applied Sciences and Engineering (ECCOMAS).*

Keys C, Tarzanin F and McHugh F 1987 Effect of twist on helicopter performance and vibratory loads *European Rotorcraft Forum*, Arles, France.

Kolencherry NJ and Crossley WA 2012 Multi-fidelity optimization strategies using genetic algorithms and sequential kriging surrogates *50th AIAA Aerospace Sciences Meeting including the New Horizons Forum and Aerospace Exposition*, Nashville, Tennessee, USA. AIAA 2012-0152.

Kulfan BM and Busoletti JE 2006 Fundamental parametric geometry shape representations for aircraft component shapes *11th AIAA/ISSMO Multidisciplinary Analysis and Optimization Conference*, Portsmouth, Virginia. AIAA 2006-6948.

Le Pape A 2005 Numerical aerodynamic optimization of helicopter rotors: Multi-objective optimization in hover and forward flight conditions *31st European Rotorcraft Forum*, ONERA, France.

Le Pape A and Beaumier P 2005 Numerical optimization of helicopter rotor aerodynamics performance in hover. *Aerospace Science and Technology* **9**, 191–201.

Lee J and Hajelat P 1996 Parallel genetic algorithm implementation in multidisciplinary rotor blade design. *Journal of Aircraft* **33**(5), 962–969.

Lehner SG, Lurati LB, Bower GC, Cramer EJ, Crossley WA, Engelsen F, Kroo IM, Smith SC and Willcox KE 2010 Advanced multidisciplinary optimization techniques for efficient subsonic aircraft design *48th AIAA Meeting Including the New Horizons Forum and Aerospace Exposition*, Orlando, Florida, USA. AIAA 2010-1321.

Leishman JG 2005 *Principles of Helicopter Aerodynamics* 2nd edn. Cambridge University Press.

Lekoudis SG and Sankar LN 1984 A method for designing three-dimensional configurations with prescribed skin friction. *Journal of Aircraft* **21**(11), 924–926.

Liu J, Han Z and Song W 2012 Efficient kriging-based aerodynamic design of transonic airfoils: Some key issues *50th AIAA Aerospace Sciences Meeting including the New Horizons Forum and Aerospace Exposition*, Nashville, Tennessee, USA. AIAA 2012-0967.

Liu X, Chen Y and Ye Z 2007 Optimization model for rotor blades of horizontal axis wind turbines. *Frontiers of Mechanical Engineering in China Journal* **2**(4), 483–488.

Lophaven SN, Nielson HB and Søndergaard J 2002 DACE, A MATLAB kriging toolbox, version 2.0. Technical Report IMM-TR-2002012, Technical University of Denmark.

Mack Y, Goel T, Shyy W and Haftka R 2007 Surrogate model-based optimization framework: A case study in aerospace design. *Studies in Computational Intelligence Journal* **51**, 323–342.

Mackman TJ, Allen CB, Ghoreyshi M and Badcock KJ 2011 Comparison of adaptive sampling methods for generation of surrogate aerodynamic models *49th AIAA Aerospace Sciences Meeting including the New Horizons Forum and Aerospace Exposition*, Orlando Florida. AIAA 2011-1171.

Mani K, Lockwood BA and Mavriplis DJ 2012 Adjoint-based unsteady airfoil design optimization with application to dynamic stall *American Helicopter Society 68th Annual Forum*, Fort Worth, Texas, USA.

Martin PB and Leishman JG 2002 Trailing vortex measurements in the wake of a hovering rotor blade with various tip shapes *58th Annual Forum of the AHS International*, 11–13 June 2002, Montreal, Canada.

Mengistu T and Ghaly W 2008 Aerodynamic optimization of turbomachinery blades using evolutionary methods and ANN-based surrogate models. *Journal of Optimization Engineering* **9**, 239–255.

Mishra SK 2006 Global optimization by particle swarm method: A Fortran program North-Eastern Hill University.

Mohammadi B and Pironneau O 2004 Shape optimization in fluid mechanics. *Annual Review of Fluid Mechanics* **36**, 255–279.

Nadarajah SK and Tatossian C 2008 Multi-objective aerodynamic shape optimization for unsteady viscous flows. *Journal of Optimization Engineering* **11**, 67–106.

Nielsen EJ, Lee-Rausch EM and Jones WT 2009 Adjoint-based design of rotors using the Navier-Stokes equations in a noninertial reference frame *American Helicopter Society 65th Forum*, Grapevine, Texas, USA.

Orr SA and Narducci RP 2009 Framework for multidisciplinary analysis, design, and optimization with high-fidelity analysis tools. Technical Report CR-2009-215563, NASA.

Oyama A, Nonomura T and Fujii K 2010 Data mining of Pareto-optimal transonic airfoil shapes using proper orthogonal decomposition. *Journal of Aircraft* **47**(5), 1756–1762.

Pereira J, Chopra DI, Bohorquez F, Chehab M, Couch R, DuVall T, Park J and Singh B 2002 406-UM TerpRanger light helicopter upgrade. Technical report, University of Maryland.

Peter J and Marcelet M 2008 Comparison of surrogate models for turbomachinery design. *WSEAS Transactions on Fluid Mechanics.*

Petrone G, de Nicola C, Quagliarella D, Witteveen J, Axerio-Cilies J and Iaccarino G 2011 Wind turbine optimization under uncertainty with high performance computing *29th AIAA Applied Aerodynamics Conference*, Honolulu, Hawaii. AIAA 2011-3806.

Poloni C, Giurgevich A, Onesti L and Pediroda V 2000 Hybridization of a multi-objective genetic algorithm, a neural network and a classical optimizer for a complex design problem in fluid dynamics. *Computer Methods in Applied in Applied Mechanics and Engineering* **186**, 403–420.

Rajmohan N, Marpu RP, Sankar LN, Baeder JD and Egolf TA 2011 Improved prediction of rotor maneuvering loads using a hybrid methodology *American Helicopter Society 67th Forum*, Virginia, USA.

Rallabhandi SK and Mavris DN 2007 Aircraft geometry design and optimisation for sonic boom reduction. *Journal of Aircraft* **44**(1), 35–47.

Rauch P, Gervais M, Cranga P, Baud A, Hirsch JF, Walter A and Beaumier P 2011 Blue Edge: The design, development and testing of a new blade concept *American Helicopter Society 67th Forum*, Virginia, USA.

Régnier J, Sareni B and Roboam X 2005 System optimization by multi-objective genetic algorithms and analysis of the coupling between variables, constraints and objectives. *International Journal for Computation and Mathematics in Electrical and Electronics Engineering* **24**(3), 805–820.

Rendall TCS and Allen CB 2010 Evaluation of radial basis functions for CFD volume data interpolation *48th AIAA Aerospace Sciences Meeting Including the New Horizons Forum and Aerospace Exposition*, Orlando, Florida, USA. AIAA 2010-1414.

Samad A and Kim KY 2008 Shape optimization of an axial compressor blade by multiobjective genetic algorithm. *Proceedings of the Institution of Mechanical Engineers Part A: Journal of Aerospace Engineering* **222**, 599–611.

Samareh JA n.d. Survey of shape parameterization techniques for high-fidelity multidisciplinary shape optimization. *AIAA Journal* **39**(5), 877–884.

Schneider PJ n.d. NURB curves: A guide for the uninitiated *Develop, the Apple Technical Journal* No. 25. http://www.mactech.com/articles/develop/issue_25/schneider.html.

Schwabacher M, Ellman T and Hirsh H 1998 Learning to set up numerical optimizations of engineering designs. *Artificial Intelligence for Engineering Design, Analysis and Manufacturing* **12**, 173–192.

Sobieszczanski-Sobieski J and Haftka RT 1997 Multidisciplinary aerospace design optimization: Survey of recent developments. *Structural Optimization* **14**, 1–23.

Sun H, Kim Y, Lee S and Lee D 2003 Technical note: Aerodynamic design of helicopter rotor blade in forward flight using response surface methodology. *Journal of the American Helicopter Society* **48**(4), 300–304.

Tahara Y, Peri D, Campana EF and Stern F 2008 Computational fluid dynamics-based multiobjective optimization of a surface combatant using a global optimization method. *Journal of Marine Science Technology* **13**, 95–116.

Tan KC, Lee TH and Khor EF 2002 Evolutionary algorithms for multi-objective optimization: Performance assessments and comparisons. *Journal of Artificial Intelligence* **19**, 253–290.

Tatossian C, Nadarajah SK and Castonguay P 2008 Aerodynamic shape optimization of hovering rotor blades using a non-linear frequency domain approach *46th AIAA Aerospace Sciences Meeting and Exhibit* number AIAA 2008-322.

Toffolo A and Benini E 2003 Genetic diversity as an objective in multi-objective evolutionary algorithms. *Evolutionary Computation Journal* **11**(2), 151–167.

Vanderplaats GN 1973 CONMIN – A FORTRAN program for constrained function minimization, user's manual. Technical report, NASA.

Vytla VV, Huang PG and Penmetsa RC 2010 Response surface based aerodynamic shape optimization of high speed train nose *48th AIAA Aerospace Sciences Meeting Including the New Horizons Forum and Aerospace Exposition*, Orlando, Florida, USA. AIAA 2010-1509.

Walsh JL 1991 Performance optimization of helicopter rotor blades. Technical Report 104054, NASA.

Watanabe T, Matsushima K and Nakahashi K 2008 Aerodynamic shape optimization of a near-sonic passenger plane using computational fluid dynamics. *Proceedings of the Institution of Mechanical Engineers Part G: Journal of Aerospace Engineering* **222**, 1025–1035.

Whitley D 1994 A genetic algorithm tutorial. *Statistics and Computing* **4**, 65–85.

Willcox K 2006 Unsteady flow sensing and estimation via the gappy proper orthogonal decomposition. *Computers and Fluids* **35**, 208–226.

Xiong L, Yan C and Zhiquan Y 1997 Optimization model for rotor blades of horizontal axis wind turbines. *Frontiers of Mechanical Engineering in China* **2**(4), 483–488.

Young DP 2011 Technical notes: Note on parameterization of airfoils. *AIAA Journal* **49**(1), 254–255.

Zhao H, Wang S, Han W and Feng G 2008 Aerodynamic design by jointly applying s2 flow surface calculations and modern optimization methods on multistage axial turbines. *Frontiers of Energy and Power Engineering in China Journal* **2**(1), 93–98.

10

Framework for the Optimisation of a Helicopter Rotor Blade with an Approximate BERP Tip

Numerical Methods and Application

Catherine S. Johnson, Mark Woodgate and George N. Barakos

School of Engineering, University of Liverpool, UK
School of Engineering, University of Glasgow, UK

10.1 Introduction

10.1.1 BERP Blade

The BERP blade has a 'paddle-shaped' tip due to a forward displacement of the planform (creating a notch), followed by sweep, as shown in Figure 10.1. It is a benchmark design because of its improved performance in forward flight, winning it a world speed record for helicopter flight (Perry et al. 1998). Some CFD analysis was carried out for this blade in order to find the specific reasons for its improved performance, and to analyse what happens in the flowfield around it (Brocklehurst and Duque 1990). However, at the time, computational power was limited and today, higher fidelity CFD methods are available. The aim of this chapter is to apply CFD and optimisation methods to investigate if further performance enhancements can be obtained by fine-tuning specific features of the BERP tip. At the same time, the methods outlined in Chapter 9 are applied. A brief history of the BERP blade development is presented below.

The BERP tip was designed for high-speed forward flight without compromising hover performance (Harrison et al. 2008). The problem associated with the fast forward flight regime, is that the effects of compressibility, such as transonic flow and shock-waves, become significant, especially on the advancing blade. Typically, thin aerofoils are used but these tend to stall more easily at the high angles of attack that occur on the retreating side. The first step in the design of the BERP was the aerofoil selection. The aerofoils were selected such that thinner sections could be used to enable higher forward flight speeds. Camber was introduced to improve the stall capability of the blade on the retreating side, and the increased pitching moments were alleviated by having a reflexed aerofoil inboards. The resulting blade is reported to behave well in terms of control requirements and twist loads (Robinson and Brocklehurst 2008).

The planform was then optimised to reduce high-Mach-number effects by first sweeping the tip of the blade back. This moved the aerodynamic centre of the swept part backwards, causing control problems in the pitch axis. To counteract this, the

Advanced UAV Aerodynamics, Flight Stability and Control: Novel Concepts, Theory and Applications,
First Edition. Edited by Pascual Marqués and Andrea Da Ronch.

Companion website: http://www.wiley.com/go/marques/advanced_UAV_aerodynamics

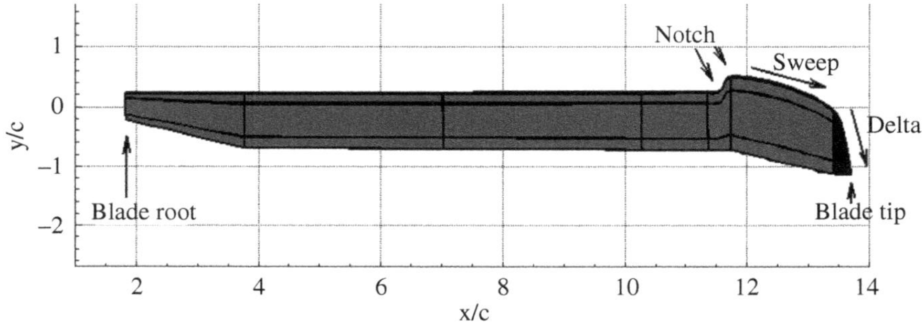

Figure 10.1 Definition of the parameters of the BERP-like blade.

swept part was translated forward, which introduced a notch on the leading edge of the blade. The notch corners were smoothed to avoid flow separation. A 'delta' tip was also incorporated, so that a stable vortex formed at higher angles of attack on the retreating side to delay stall (Brocklehurst and Duque 1990).

One of the characteristics of the blade is that blade stall occurs first inboards of the notch and does not spread outwards. This is because at high angles of attack, such as at the retreating side, the vortex formed travels around the leading edge and the flow over the swept part remains attached. The BERP blade shows similar performance to a standard rotor blade in low-speed flight, but superior performance in forward flight due to the absence of drag rise and flow separation (Brocklehurst and Duque 1990). In hover, the figure of merit (FM) was improved due to the minimisation of blade area and, overall, there were no penalties in hover performance. At high speeds, blade vibration was also reduced, as were control loads for manoeuvres (Brocklehurst and Duque 1990).

The next sections describe how a BERP-like rotor blade tip can be optimised for improved performance given certain constraints. The same method can be applied to UAVs to improve any aspect of their performance within any given constraints.

10.2 Numerical Methods

Fully analysing a rotor using time-marching CFD methods, even when parallel computing is used, can take hours of clock time. Another technique that can be used to obtain the performance of the rotors to the same accuracy (provided a sufficient number of modes is used), is the harmonic balance (HB) method (Woodgate and Barakos 2009).

With HB, the time taken to perform the same calculation can be an order of magnitude less than the time taken using time-marching. This greatly improves the efficiency of the optimisation process, making it a more usable technique for rotor design. The number of snapshots (N_S) obtained is given as

$$N_S = 2N_H + 1 \tag{10.1}$$

where N_H is the number of modes. Here, four modes were used, resulting in nine snapshots; in other words, every 40°. Since this is a four-bladed rotor, it results in a blade snapshot at every 10° of azimuth. This is because blade 1 of the rotor will be captured at 40, 80, 120, 160, 200, 240, 280, 320 and 360° of azimuth and since blade 2 is 90° away from blade 1, in the same snapshots, blade 2 will be at 130, 170, 210, 250, 290, 330, 370 (which is 10), 410 (which is 50) and 450 (which is 90) degrees. Similarly for blades 3 and 4, there will be a shift of 10° from the previous blade, resulting in a snapshot at every 10° of azimuth. The method has been shown to give results of similar accuracy to time-marching methods in Woodgate and Barakos (2009). A brief summary of the method is given here.

The HB method uses the governing (Navier–Stokes) equations in the frequency domain. Equation (10.2) represents the governing equations, where $Q(t)$ is the matrix of each solution variable such as pressure, density and velocity components, and $R(t)$ is the residual for each of these variables; these are assumed to be periodic.

$$F(t)=\frac{dQ(t)}{dt}+R(t)=0 \tag{10.2}$$

Then, expressing the solution as a Fourier series with a fixed number of modes, N_H:

$$Q(t)=\hat{Q}_o+\sum_{n=1}^{N_H}\left(\hat{Q}_c\cos(\omega nt)+\hat{Q}_s\sin(\omega nt)\right) \tag{10.3}$$

$$R(t)=\hat{R}_o+\sum_{n=1}^{N_H}\left(\hat{R}_c\cos(\omega nt)+\hat{R}_s\sin(\omega nt)\right) \tag{10.4}$$

$$F(t)=\hat{F}_o+\sum_{n=1}^{N_H}\left(\hat{F}_c\cos(\omega nt)+\hat{F}_s\sin(\omega nt)\right) \tag{10.5}$$

where ω is the rotational speed of the rotor. A Fourier transform of Eq. (10.5) gives

$$\hat{F}_o=\frac{\omega}{2\pi}\int_o^{2\pi/\omega}F(t)dt=\hat{R}_o \tag{10.6}$$

$$\hat{F}_c=\frac{\omega}{\pi}\int_o^{2\pi/\omega}F(t)\cos(wnt)dt=\omega n\hat{Q}_s+\hat{R}_c \tag{10.7}$$

$$\hat{F}_s=\frac{\omega}{\pi}\int_o^{2\pi/\omega}F(t)\sin(wnt)dt=-\omega n\hat{Q}_c+\hat{R}_s \tag{10.8}$$

This gives a system of N_T equations, where $N_T=2N_H+1$, for the Fourier series coefficients:

$$\hat{R}_o=0 \tag{10.9}$$

$$\omega n\hat{Q}_s+\hat{R}_c=0 \tag{10.10}$$

$$-\omega n\hat{Q}_c+\hat{R}_s=0 \tag{10.11}$$

This is expressed in matrix form as:

$$\omega M\hat{Q}+\hat{R}=0 \tag{10.12}$$

where M is an $N_T\times N_T$ matrix.

$R(t)$ is a function of $Q(t)$ and it is non-linear. Therefore each coefficient $\hat{R}_k$ depends on all the coefficients $\hat{Q}_k$ and hence must be solved iteratively. There are a number of ways that this can be carried out.

In the pseudo-spectral method, the equations are transformed back to the time domain and the period is split into N_T equal discrete time intervals or snapshots, represented as Q_{hb} and R_{hb}. Then Eq. (10.12) is decomposed to form,

$$\omega D Q_{hb} = R_{hb} \tag{10.13}$$

where $D = E^{-1}ME$, and E is a transformation matrix transforming $\hat{Q}$ and $\hat{R}$ to Q_{hb} and R_{hb}. The diagonal of D is zero and pseudo time-marching can then be applied to the HB equation

$$\frac{dQ_{hb}}{dt} + \omega D Q_{hb} + R_{hb} = 0 \tag{10.14}$$

However, other methods exist that use less memory, which can be a limiting factor in the use of the HB method. More details are given in Woodgate and Barakos (2009) and Jang et al. (2012).

The helicopter multi-block method is a parallel CFD code that is used routinely for the analysis of rotors (Steijl et al. 2005). The method was validated for popular rotor cases like the Caradonna and Tung (1981) and the ONERA 7A and 7 AD rotors (Steijl and Barakos 2008). These efforts are described in the literature (Steijl et al. 2006).

In this study, the turbulence model used was the $\kappa - \omega$ two-equation model. The calculations were run with a third-order spatial scheme, and the unsteady Reynolds averaged Navier–Stokes (URANS) equations. No wake model was used. Multiblock structured grids were used, the details of which are given in Section 10.5. The computations were all trimmed to give the same thrust-to-rotor solidity values. Geometric solidity was used and calculated as follows:

$$\sigma = \frac{N_b \times A_b}{\pi R^2} \tag{10.15}$$

where N_b is the number of blades and A_b is the area of the blade, calculated by integrating the chord along the radius, R. No aeroelastic or structural model was used and the blades were treated as rigid blades. Once an initial solution was obtained, it was possible to use this solution as a starting point for another variation of the BERP-like tip.

10.3 Optimisation Method

As seen in Chapter 9, there has been substantial research and interest in the use of genetic algorithms (GAs) in the optimisation of rotors. However, application and development in the past has mostly occurred in the design of compressors, turbines or electric rotors, where the ratio of the blade size to the actual rotor is quite low (Mengistu and Ghaly 2008; Samad and Kim 2008; Zhao et al. 2008). In the case of helicopter and other rotorcraft rotors, gradient-based methods have been popular, because they have a lower demand for computational effort, time and cost. Non-gradient based methods should be able to identify global optima, but the high computational effort involved has rendered these methods impractical for rotor optimisation.

Nevertheless, there are a number of methods that could reduce the computational effort in terms of CPU time as well as storage and, with the advances made in computing power, the use of these methods in rotor optimisation is becoming more feasible. The use of metamodels or surrogate models can reduce the calculation time and is expected to counteract the extra optimisation time required to apply non-gradient based optimisation techniques whilst still maintaining global optimisation. Figure 10.2 is a map of the stages involved in the optimisation. It is an aposteriori-type method according to Régnier et al. (2005). The work described here aims to demonstrate a framework, allowing different aspects of rotors and fuselages to be aerodynamically optimised given an existing design as a starting point. For real helicopter rotors, the initial designs would already be near optimum and the aim is to capture the aerodynamic effects of any design changes and adjust the variables of the design to find an optimum that will lead to better rotor performance.

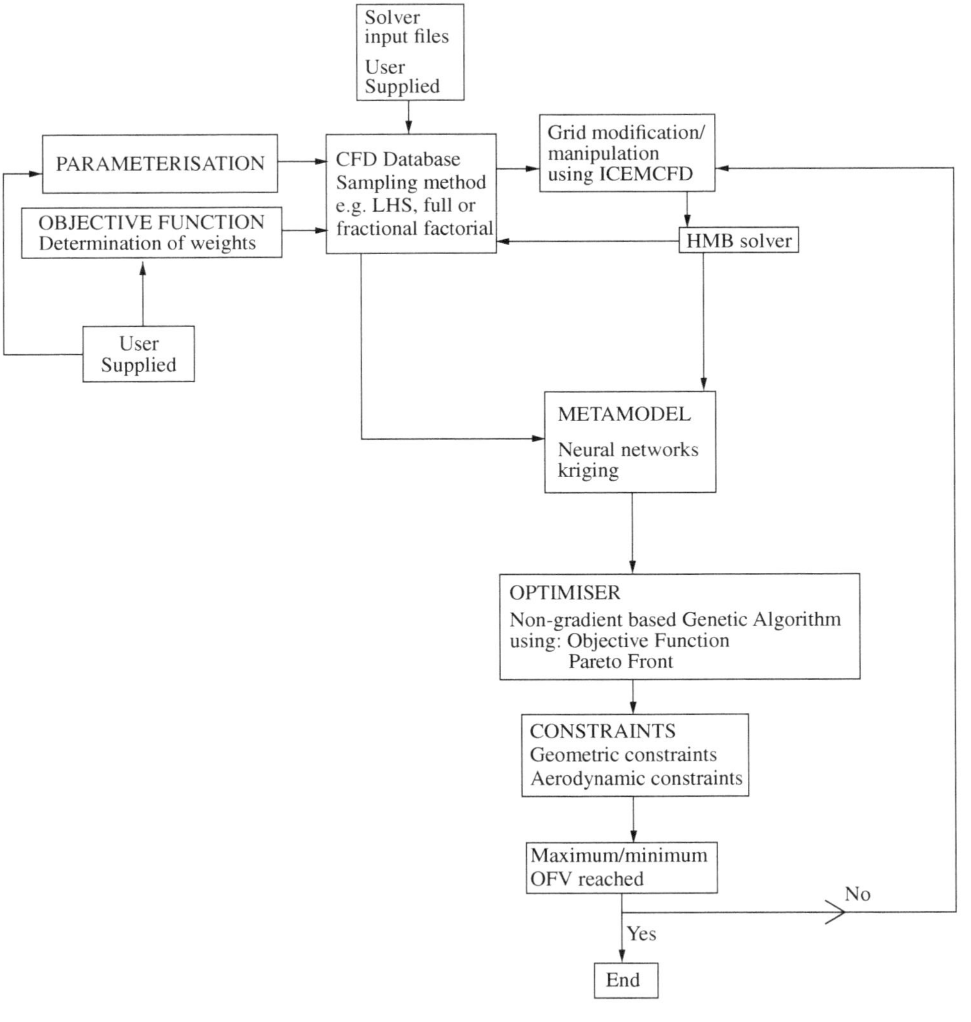

Figure 10.2 Map of the processes involved in the optimisation of rotors.

10.3.1 Optimisation Framework

To improve the efficiency a non-gradient-based method, the high-fidelity solver is decoupled from the optimisation loop. It is, however, available to the optimiser through a database. Therefore, the first step in the process is to create a database or population of designs. It is important for the design parameters to be clearly defined, along with the boundaries of the database and the constraints. These must be user-defined and are based on experience and the requirements of the optimisation case. A parameterisation technique must be created if it does not already exist for the case. The boundaries of these parameters are then set, based on engineering expertise of the problem and the case in question. Once these have been determined, a sampling method is used to generate a number of design points to create the database or design space using the high-fidelity solver. For each case, the geometry is modified and run using the high-fidelity solver.

The next step is to determine the performance parameters that capture the objectives of the optimisation and to then combine these into a single value that the optimiser can use to determine the overall performance of an individual design. This again is dependent on the experience of the user, but is also guided by the performance of the designs in the database and further supported by other optimisation criteria such as the Pareto front method (Daskilewicz and German 2012; Oyama et al. 2010).

The optimiser now uses the database (which represents the design space) to determine the performance of a design point. However, it may need to evaluate the performance of a design point that may not exist in the database and therefore the CFD solver will have to be used. This would incur a large computational cost and time, because such requirements tend to occur often. To overcome this, a metamodel is used to predict the performance of unknown design points based on the performance of the existing points in the database by some interpolation technique. The accuracy or the tolerance value of the error in these predictions is what determines the resolution of the design parameter values in the database. The metamodel can be validated using a small number of additional design points obtained using the high-fidelity solver. A selection of different metamodels were tested in our work, including artificial neural networks (ANNs), kriging, polynomial fits and proper orthogonal decomposition (POD). Of these, the first two were the most successful in terms of accuracy and efficiency. The optimiser can then employ the predictions of the metamodel to obtain the performance of new designs quickly. After a number of iterations, the designs are expected to converge to an optimum or a cluster of optimal designs. The optimiser is able to exercise constraints specified by the user. The final new design is then validated using the high-fidelity solver.

The sampling method used was a full factorial method due to the small number of points in the database. However, for a larger-scale optimisation project with more design parameters, such as the optimisation of a multi-segment fixed wing planform, other sampling methods may provide better efficiency without loss of accuracy. Such methods include the fractional-factorial method and Latin hypercube sampling (LHS), amongst others.

The rest of this section describes each stage of the optimisation in more detail and the following sections present an application example and results.

10.3.2 Sampling the Design Space

There are a number of ways that the database can be populated (Simpson et al. 2001). In the example used here, the full-factorial method was used. Short descriptions of the fractional factorial and the LHS methods are also given here.

The full factorial method analyses all other variables for each design variable. Therefore it has the highest number of design points. The fractional factorial method uses every other point of the full factorial method; in other words, it is a sparser version of the full factorial database. If the area of the design space where the optimum is expected to be found is known, then the points selected using this method can be such that there is a higher concentration of design points in that area; that is, with those parameter values. This can be found by repeating the optimisation a number of times with additional points in the database each time, a process also known as 'adaptive sampling' or 'updating' (Forrester and Keane 2009).

The LHS method attempts to use a design parameter value only once or as few times as possible in populating the database. So, for example, if there are two parameters with equal number of design parameters, the LHS will select the design points along the diagonal line (or on a hyperplane for larger numbers of parameters) of the database. If there are more variables for one design parameter than another, then a random additional point is created for the component that has fewer variables. Sometimes it is necessary to also include the parameter values of the boundary of the database for enhanced accuracy.

The more points in the database, the more accurate the metamodel predictions will be. This means that the full factorial method will always produce the most accurate results. However, for cases where the computation time and cost of obtaining the performance of each design point is high, it is important to reduce the number of such calculations by selecting as few points as possible, as in the case of fractional factorial and LHS methods. These methods can produce good results if the points are selected carefully and if some additional points on the boundary of the database are included. Cameron et al. (2011) used an automated adaptive sampling method to optimise laminar flow aerofoils. In his method, the Pareto front was updated after a set number of generations and the mean square error of the kriging metamodel was used to calculate the probability that a new sample added at any given point would dominate members of the existing Pareto front. This process was automated in the genetic algorithm used, NSGAII (Cameron et al. 2011), and repeated until a set sample size was obtained. Forrester et al. (2008) also describe an optimal LHS method based on adaptive sampling.

10.3.3 Objective Function

The objective function determines the outcome of the optimisation. An ideal objective function should be, according to Tan et al. (2002):

- complete, so that all aspects of the decision problem are presented
- operational – it can be used in a meaningful manner
- decomposable, if disaggregation of the function is required
- non-redundant, so that no aspect of the decision is considered twice
- minimal, so that the function considers the minimum required for a decision.

In line with these properties, the objective function is presented as the summation of weighted components that quantify the objectives. The actual values of each component that makes up the objective function may vary by orders of magnitude, meaning that the objective function would be biased based on the range of values, rather than the sensitivity of the value to performance. Therefore, each performance parameter is scaled with the corresponding value of a reference design, which is usually the original design. So a generic objective function can be presented as:

$$\mathrm{OFV} = \sum_{i=0}^{n} w_i \frac{P_i}{P_i^r} + C \tag{10.16}$$

where OFV is the objective function value, w_i is the weight assigned to the ith performance parameter, P_i is the ith performance parameter value for the design being evaluated, P_i^r is the performance parameter value for the reference design and C is a constant value added so that if the reference design was the design being analysed, the OFV would be zero; that is: $C = -\sum_{i=0}^{n} w_i$.

The weights for the function are guided by the initial CFD data. For example, if there are two performance parameters, P_1 and P_2, for an optimisation problem and the objective is primarily to improve P_1 and secondarily to improve P_2, then for each design point a ratio can be found between P_1 and P_2. Assume the average of all these ratios is 1:1.5 for P_1:P_2. Then the coefficient weight to weigh P_1 more than P_2 in the objective function is $1.5/(1+1.5) = 0.6$ or $w_1 = P_2/(P_1 + P_2)$. In other words, on average, the weight of P_1 must be ≥ 0.6. However, this should only be used as a guide value, because individually the deviation from this ratio can be large. This method is a simple way to obtain guidance in determining the weights, but it can be further refined with the use of standard deviation and other attributes.

The objective function method differs from what is known as the Pareto method of optimisation. The Pareto method tries to find the best compromise in performance for the designs; in the Pareto subset, an increase in one performance parameter will result in a decrease in another. Therefore it creates a boundary or front of design points. In contrast, the objective function method tends to concentrate the optimum designs to a cluster in the design space as opposed to spreading the optimum design along a front. In other words, it selects designs in a region of the Pareto front. In the example shown in this chapter, the Pareto front was found using a GA in which the elite members of the population were the ones that fulfilled the Pareto conditions, as opposed to a selection of the fittest individuals in the weighted method. Both methods were used and the results show that the selected optima using the weights method were also members of the Pareto front.

10.3.4 Metamodels

The cost of running the high-fidelity solver to analyse every new design created by the optimiser can be quite high in terms of computational effort. To avoid this, a lower-fidelity approximation of the performance parameters based on the high-fidelity data in the database can be used until the final result is obtained. This allows the optimiser to access the accuracy that comes with the high-fidelity model with the efficiency of the low-fidelity model or metamodel (model of a model). Four metamodels are described here.

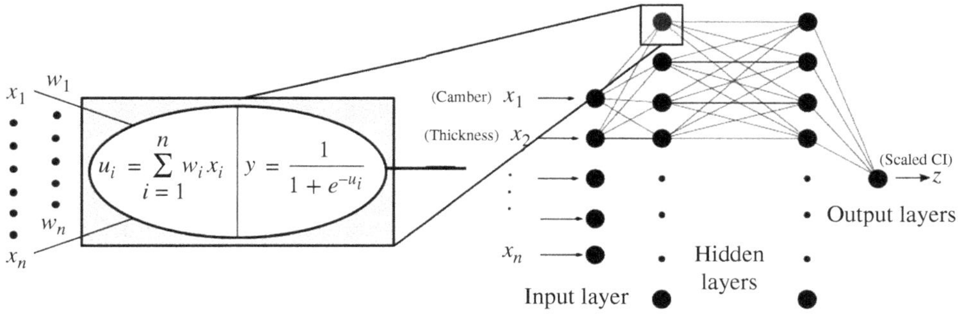

Figure 10.3 An example of a neural network trained to receive inputs such as camber and thickness to obtain the lift coefficient. Adapted from Spentzos (2005).

Artificial Neural Networks

An ANN interpolates based on patterns obtained from a set of data, similar to how the biological brain learns (Samarasinghe 2006). Figure 10.3 is a schematic of the structure of a multilayer feed-forward ANN. It consists of a number of neurons connected to every other neuron in the next layer, from input to output (Spentzos 2005). The layers between the input and output are known as 'hidden layers' and make up what is known as the perceptron. It is here that 'learning' takes place. Each neuron is associated with a weight and an activation function. The weight determines how much influence a neuron has on the output, and the activation function keeps the values within bounds and gives the ANN the ability to be differentiable so that error corrections can be made using, for example as in this case, a gradient descent method. The activation case used most commonly and in all the cases presented here is the sigmoidal function (Samarasinghe 2006), generically shown in Eq. (10.17),

$$y = 1/\left(1+e^{-x}\right) \tag{10.17}$$

There are two phases for ANNs: training and predicting. In the training phase, a data set is available to the ANN; that is, both input and output. The weights of the neurons are randomly chosen and the ANN makes a prediction. The error between this predicted output and the target output is then fed back through the layers and the weights are adjusted accordingly. The full set of data, known as an epoch, is fed in repeatedly and the error back-propagated until it converges to a preset value. In this work, this is done via a conjugate gradient descent method.

For example, let us assume that there are n initial inputs of the ANN (i=1,2,…,n). If the hidden layers have J nodes each, then for the first layer, these are all connected to the n inputs through neurons with weights w_{ij} (i=1,2,…,n, j=1,2,…,J). Each input is first weighted with the weight for that input and that neuron:

$$u_j = \sum_{i=0}^{n} x_i \times w_{ij},\ i = 1,2,\ldots,n \text{ and } j = 1,2,\ldots,J \tag{10.18}$$

This is then passed through the activation function to get the output from that neuron:

$$y_j = \frac{1}{1+e^{-u_j}},\ j = 1,2,\ldots,J \tag{10.19}$$

This is repeated at each node at every hidden layer until the final output layer is reached for each pattern. The optimum number of hidden layers depends on the complexity of the problem. If there are more dependencies between input parameters, then more layers are required to deal with these inter-dependencies; the ANN must be 'smarter'. In general, increasing the number of layers makes an ANN smarter and increasing the number of neurons per layer makes an ANN more accurate (Spentzos 2005). Also, it is critical that during every epoch, the patterns are introduced in a random order. This ensures faster learning, avoids 'memorising' and increases the capability of the ANN to tackle situations it has not been trained for (Spentzos 2005).

In the training phase, the difference between the target output and the predicted output (the error value), is used to determine the changes in the weights of the neurons through a back-propagation technique based on chain differentiation. For the ANN used in this project, the error was corrected after all patterns had been through the system: one epoch rather than after each pattern. The mean squared error can be found as

$$E=\sum_{i=1}^{K}\left|z_i-t_i\right| \text{ or as} \frac{1}{2}\sum_{i=1}^{K}\left(z_i-t_i\right)^2 \tag{10.20}$$

where K is the total number of outputs, t_i is the ith target output, and z_i is the ith output obtained from the final output layer. The aim now is to find the output that minimises E. The method used is the gradient-descent with momentum method (Spentzos 2005), which works as follows.

The error change due to a change in the output is found by differentiating Eq. (10.20) and is given by

$$\frac{\partial E}{\partial z_i}=z_i-t_i \tag{10.21}$$

Therefore, the change in the error due to the change in the input to the node that created z_i is

$$\frac{\partial E}{\partial u_j}=\frac{\partial E}{\partial z_i}\frac{\partial z_i}{\partial u_j}=\left(z_i-t_i\right)\left(z_i\left(1-z_i\right)\right) \tag{10.22}$$

where $\partial z_i/\partial u_j$ is found by differentiating Eq. (10.19), and where z_i is the equivalent of y_j for the output layer. Similarly, the change in E due to a change in the hidden-layer node input (which is similar to x in Eq. (10.19) is

$$\frac{\partial E}{\partial x_j}=\frac{\partial E}{\partial z_i}\frac{\partial z_i}{\partial u_j}\frac{\partial u_j}{\partial x_j}=w_{ij}\left(z_i-t_i\right)\left(z_i\left(1-z_i\right)\right) \tag{10.23}$$

This is repeated all the way to the first input and the total error is calculated. Then the weights are changed by a small amount, in proportion to the error calculated and the input to each neuron. So w_{ij} can be updated as:

$$\Delta w_{ij}=\eta\frac{\partial E}{\partial x_j}+\alpha\Delta w_{ij}, w_{ij}=w_{ij}+\Delta w_{ij} \tag{10.24}$$

In Eq. (10.24), α is termed the 'learning momentum', because it scales up the change in the weight at each iteration (it is multiplied by the previous weight change), and η is termed the 'learning rate', because it scales up the movement along the gradient. The higher these values are, the lower the training time required. However, values that are too high can cause instability and a failure to converge. To further improve the speed of training the ANN, it is allowed to skip a number of steps periodically before the weights are modified as a ratio of the last written weights. A further improvement is the use of an adaptive learning rate. This means that the learning rate changes according to the change or gradient in error for each epoch or set of epochs, which allows the training to occur faster when the ANN is learning quickly and slower if it is learning slowly. The learning rate, η, is multiplied by a fraction less than one if the ratio of the present error to the previous error falls above a certain value and greater than one if it falls below a certain value. In this way, η is constantly adapting to the ANN's learning ability.

Another important parameter to consider in training the ANN is the error convergence. This value must not be smaller than the error resolution of the data points used for training. For example, if the CFD data is accurate to within 1% of its value, then the ANN training must not force the difference between the predicted and target value to be less than 0.01. If so, the ANN prediction trends bend and flex to try and pass through each point inside that tolerance, resulting in over-fitting and large training times. Figure 10.4 shows a comparison of ANN predictions when trained to different convergence values for a test case, which is described in full in Johnson and Barakos (2011). For the purpose of demonstrating the effects of the ANN parameters, this test case is explained briefly here. The aim was to optimise the camber and thickness values of section of a NACA 5-digit rotor for a blade in forward flight and using the moment, lift and drag parameters. These loads are compared against the camber parameter. As can be seen in the figure, the best compromise is a convergence value of about 1% in that the trend is captured. A smaller convergence value overfits the data, because it falls within the error tolerance of the training data itself. The length of time for training required increased exponentially with smaller convergence values. The smallest convergence took about 25 times longer than the intermediate, which took approximately six times longer than the biggest convergence value. Also, using a fixed η took the ANN about 1.25 times longer to train. Using the adaptive η results in a different weights file, but the differences in the overall predictions are negligible, as shown in Figure 10.4.

The number of hidden layers and neurons required to get accurate predictions is determined by the physics of the test case, and the number of inputs and outputs (Spentzos 2005). It was found that if more inputs and output variables are used, then a 'smarter' ANN is required: one that allows more degrees of freedom. Therefore an increased number of layers and neurons are required. Having too many layers for a small number of inputs and outputs can cause overfitting of the data. Figure 10.5 shows the predictions with different numbers of layers and neurons for a number of outputs belonging to the rotor section test case. The C_l and C_d curves show that when the output variables are increased, an increase in the number of layers or neurons has approximately the same accuracy as a single-output prediction with fewer layers or neurons. The green curve is the most inaccurate because it has a large number of outputs and the least number of neurons and hidden layers. Increasing the number of neurons increases the accuracy of the predictions. This is seen more clearly in Figure 10.5c for the average C_m. For a single output, the ANN with 2 layers and 21 neurons has a more

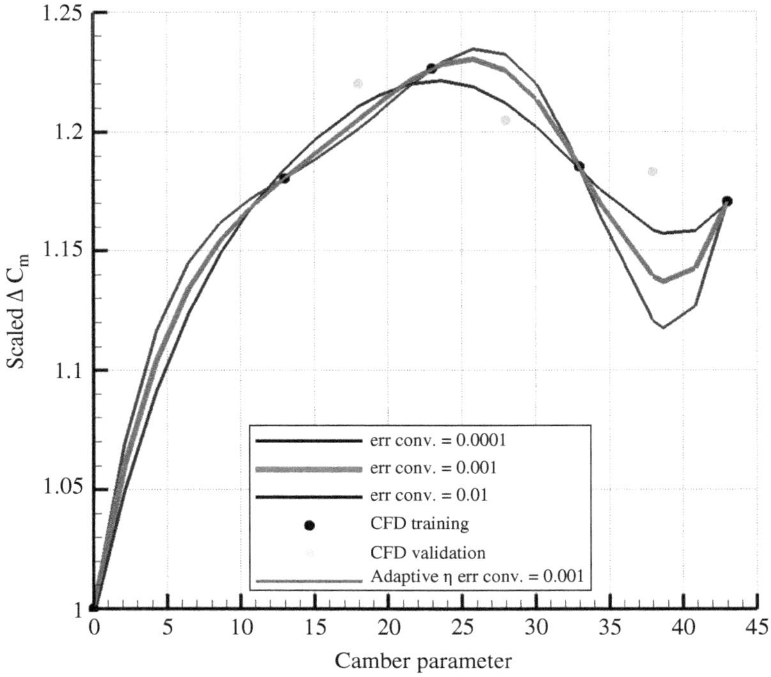

Figure 10.4 Error convergence comparison for an example ANN. This is for the optimisation of a NACA 5-digit rotor section on a blade in forward flight. The *x*-axis is the camber value of the section and the *y*-axis is the peak-to-peak moment of the section over a full revolution. The various curves and the legend correspond to the training of the ANN to converge at different error values, as well as a comparison with the ANN that had an adaptive learning rate. The black dots represent the data used for training and the pink dots are the accurate CFD data not used in the training set; that is, validation data. The full test case analysis can be found in Johnson and Barakos (2011).

accurate curve fit than the ANN with 2 layers and 15 neurons. For the 7 output case for the same output in Figure 10.5d, a higher degree of freedom is required and hence an increase in the number of layers is required to get a smooth curve fit through the data, similar to the single-output prediction. Increasing the number of neurons does not improve the prediction accuracy.

This shows the importance in validating the metamodel before using it in the optimiser. In this work, the metamodel is always validated with additional high-fidelity data unknown to the metamodel. Once the ANN has been trained, it can then be used to accurately predict the output for any input within the limits used for training.

For large values along the x-axis of the activation function – in other words, for large input values – the resolution of the output is reduced, as suggested by Eq. (10.17). Hence, for more accurate results, the inputs are normalised to between 0 and 1. Input data that is not within the range of the original data can be predicted as well (extrapolated from the data). These inputs must be included in the data before being normalised, so that *all* data is always between 0 and 1. These inputs for which the outputs are unknown can be deleted from the data file in the training stage and re-introduced in the prediction stage. However, it has been found that the ANN is highly inaccurate in

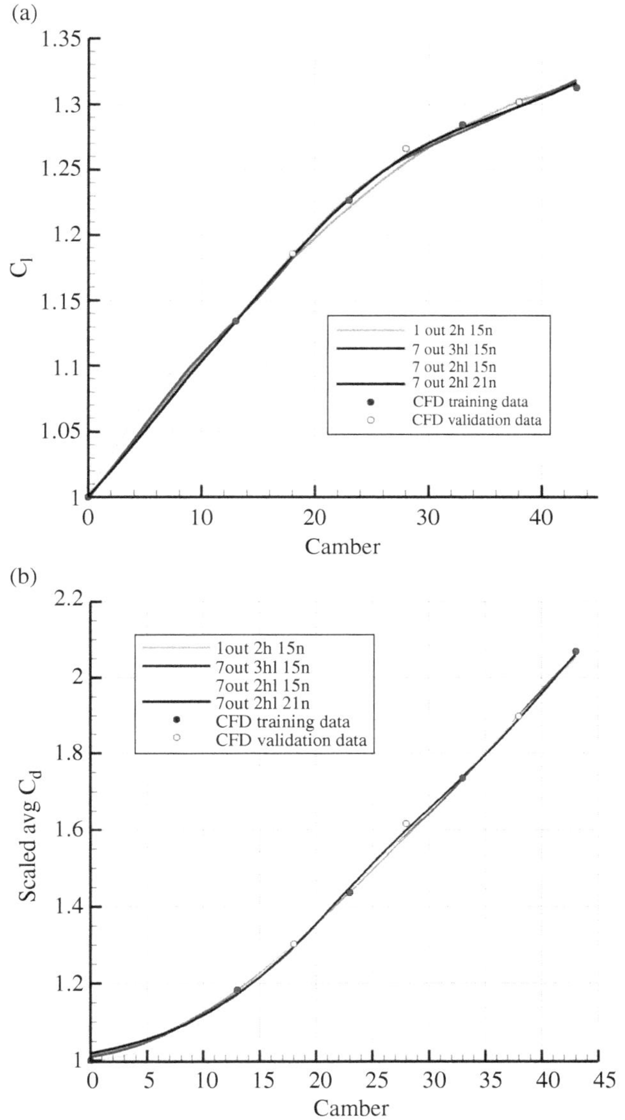

Figure 10.5 Comparison of ANN prediction for scaled average C_l, C_d and C_m with different numbers of outputs (out), hidden layers (hl) and neurons (n). The number of inputs is constant at two. This is for the optimisation of a NACA 5-digit rotor section on a blade in forward flight. The *x*-axis is the camber value of the section and the *y*-axis is the average moment, lift and drag of the section over a full revolution. The various curves and the legend correspond to the training of the ANN with different numbers of neurons and layers. The black dots represent the data used for training and the cyan dots are the accurate CFD data not used in the training set: the validation data. The full test case analysis can be found in Johnson and Barakos (2011).

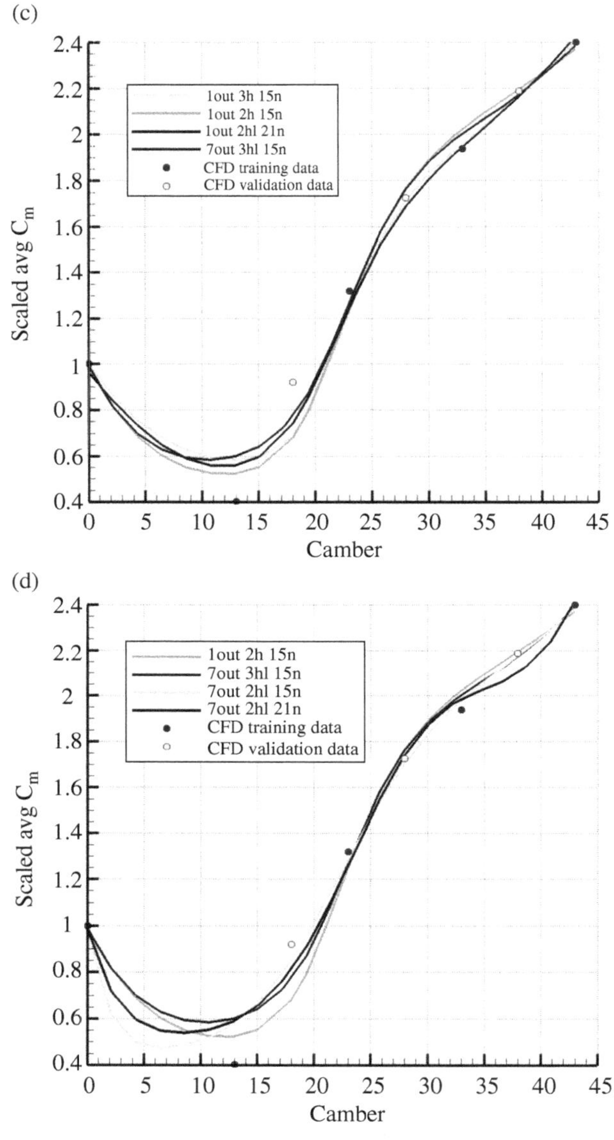

Figure 10.5 *(Continued)*

predicting extrapolated data, so it is suggested that the outputs of the input parameters on the boundaries of the design space should always be known values.

Once the inputs have been normalised, the normalised data are used to train the ANN. Training can continue until a set epoch number is reached or until the values converge to a pre-defined error value between the predicted and target results. In this case, training is always carried out until convergence to a specified error margin.

After training the ANN, the final weights are written to a file. This file is used to predict the output for a normalised set of inputs and then the predictions are denormalised to obtain the predicted results.

Polynomial Fit

Given a number of points, the aim is to fit a polynomial curve through them or create a polynomial equation that satisfies all the points given. Hence, the more known points there are, the more reliable the predictions are and the greater the degree of the polynomial can be. For example, if there are only two points, only a straight line can be used to predict data that falls anywhere on the plane. If three points are known, then a quadratic curve can be fitted through the points and hence the predictions are more accurate, and so on. The maximum degree of the polynomial is dependent on the number of points available, the limit being a function of the number of unknowns in a set of simultaneous equations. If (x_1,y_1) (x_2,y_2) and (x_3,y_3) are three points on a plane and a quadratic polynomial is to be fitted through these points, then,

$$Ax_1^2 + Bx_1 + C = y_1 \tag{10.25}$$

$$Ax_2^2 + Bx_2 + C = y_2 \tag{10.26}$$

$$Ax_3^2 + Bx_3 + C = y_3 \tag{10.27}$$

Using Gaussian elimination or lower–upper (LU) decomposition or any other method for solving simultaneous equations, the coefficients A, B and C can be found.

It is possible to use a polynomial to fit non-linear data. However, if the data do not follow a polynomial trend, having too high a degree of polynomial can result in overfitting the curve. Therefore, normally, the degree of polynomial is limited to six or less (Mathews and Fink 2004).

Figure 10.6 shows a comparison of polynomial fits to the data used to analyse the ANN in Section 10.4.1. With a polynomial of order four, the data are not as accurate as the ANN, but with an order of ten, the polynomial fits closer, although it is less smooth and still not quite as accurate. The advantage of the polynomial technique is that it does not require training time as the ANN does. However, as it is not likely that more than four or five known points will exist for a variable, the order of the polynomial is limited. The ANN will therefore be a better selection for the BERP-tip example described later in this chapter. The method can be very effective if a law exists for the variation of the output. For example if C_d versus C_l follows a parabola, then a polynomial can be used.

Kriging

The kriging approximation method is a more complex version of the polynomial fit method. It uses sample data points to build a model that can be used to predict the output or performance of interpolated design points by fitting a low-order polynomial through the data points but allowing the predictions along these polynomials to deviate based on a correlation model, such as a Gaussian distribution of all the existing points. The Gaussian distribution's characteristics are based on the correlation between the sample points; in other words, on their proximity to each other. This allows the kriging parameters to change with the prediction point, giving the approach more flexibility while still maintaining accuracy (Lophaven et al. 2002). Assume the output or performance, P, is represented as:

$$P(x,y) = f(x,y) + Z(x,y) \tag{10.28}$$

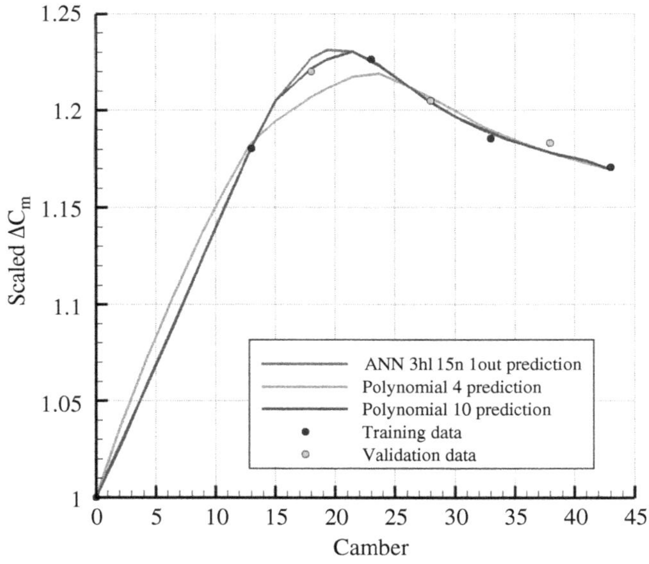

Figure 10.6 Comparison of ANN prediction accuracy with polynomial of order 4 and 10 for ΔC_m. Three hidden layers (hl) and 15 neurons (n) were used for the ANN. Number of inputs is kept constant at two. The solver training and prediction comparison is included. From Johnson and Barakos 2011.

where $f(x, y)$ is the polynomial and $Z(x, y)$ is the correlation model; the latter is based on the Gaussian distribution for all the cases in this chapter. The correlation between all the sample points with each other, and the correlation between the required point and all the points is found as the Gaussian distribution of the distances between all these points with a 'roughness' parameter θ for each design parameter. So, Z is actually a function of θ as well as x and y; that is, $Z(\theta, x, y)$. All the parameters and values are normalised and then denormalised, at the end, so that the mean of $Z(x, y)$ is 0. Also, the data are normalised in a scalar way over each parameter between 0 and 1. So the correlation between P and F can be described as:

$$F\beta \approx P \tag{10.29}$$

where β can be approximated as:

$$\beta = \left(F^T Z^{-1} F\right)^{-1} F^T Z^{-1} P \tag{10.30}$$

and the variance can be estimated as:

$$\sigma^2 = \frac{1}{m}(P - F\beta)^T Z^{-1}(P - F\beta) \tag{10.31}$$

where m is the number of initial data points. The matrix Z, and β and σ^2 depend on θ, where θ is effectively a width parameter that determines how far the influence of a data point extends (Forrester and Keane 2009). It is similar to a weight put on each input depending on its influence on the output. A low value means there is a high correlation and the influence of each point affects many other points. A large value means there is less correlation and hence it has less influence on another point. In this way, θ can also be used to find the design parameters that have the most influence on the performance parameters.

The optimum value of θ is defined as the maximum likelihood estimator, the maximiser of

$$-\frac{1}{2}\left(m\ln\sigma^2 + \ln|Z|\right) \tag{10.32}$$

where $|Z|$ is the determinant of Z (Lophaven et al. 2002). This value is found iteratively, although for small cases with few inputs, a good estimate can be made by comparing predictions with the original data.

The Gaussian correlation function, $Z(x, y)$ can be found as the covariance matrix:

$$\mathrm{Cov}(i,j) = \exp\left(-\Sigma\left[\theta_x\left(x_i - x_j\right)^2 + \theta_y\left(y_i - y_j\right)^2\right]\right) \tag{10.33}$$

$$\text{or } \exp\left(-\Sigma\,[d^2]\right) \tag{10.34}$$

$$\mathrm{Cov}(i,r) = \exp\left(-\Sigma\left[\theta_x\left(x_i - x_r\right)^2 + \theta_y\left(y_i - y_r\right)^2\right]\right) \tag{10.35}$$

where i and j are the sample points and r represents the required points, x and y are the design parameters and d is the distance between the points, which in this case is squared; the value to which it is raised can be between 0 and 2 and represents the differentiability of the response function with respect to the design parameters. Values close to 0 indicate that the function is not differentiable or smooth and a value closer to 2 is for differentiable functions (Lophaven et al. 2002).

The weight matrix, λ, which contains all the weights to obtain the required point from each point in the design space, can then be found as

$$\lambda = \mathrm{Cov}(i,j)^{-1}\,\mathrm{Cov}(i,r) \tag{10.36}$$

Then, the prediction can be made as the summation of each appropriately weighted objective function value

$$\hat{P}_r = \sum_{i=0}^{n} \lambda_i \times P_i(x,y) \tag{10.37}$$

More details can be found Mantovan and Secchi (2010) and Lophaven et al. (2002). For the polynomial fits, up to order-two polynomials are common. In some cases, constant values are sufficient, such as the mean value of all the data.

Figure 10.7 compares the data in Figure 10.6 as predicted by the polynomial, ANN and kriging metamodels. The kriging method tends to smooth out the surface more than the ANN, although this can be dealt with by optimising its parameters. Nevertheless, the ANN and the kriging both seem to perform consistently well across all the cases with little change in parameters, slightly more so for the ANN than the kriging method. The advantage of the kriging is that training time is not required. Both methods are capable of making reasonably accurate predictions.

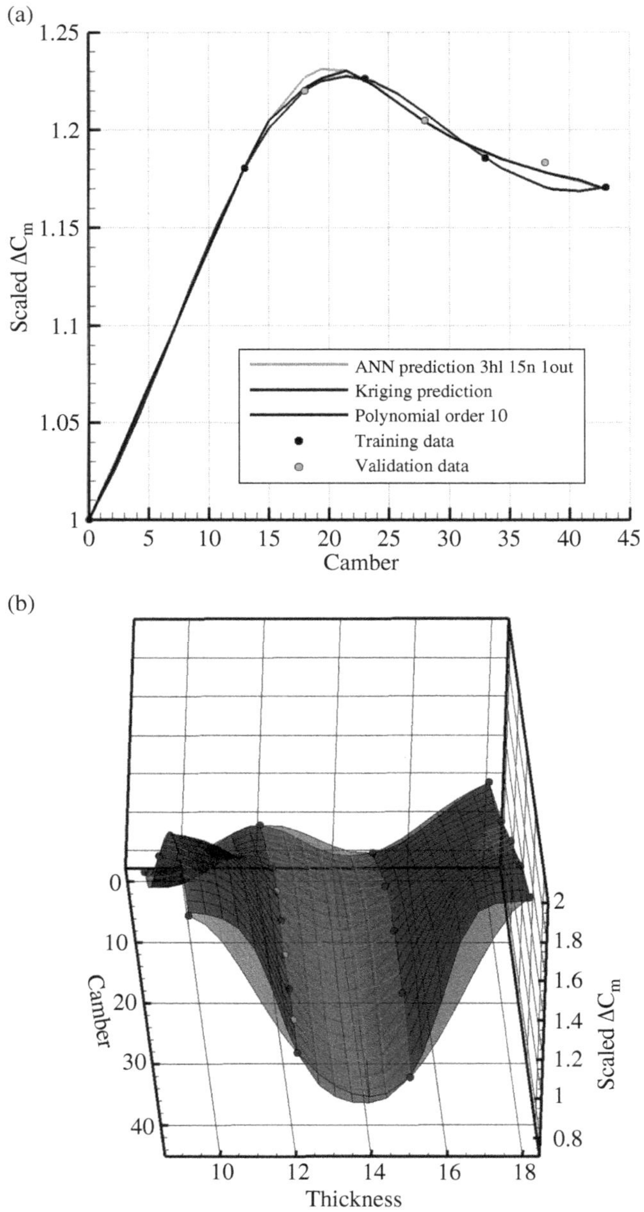

Figure 10.7 Comparison of ANN prediction accuracy with polynomial of order 10 and kriging for ΔC_m. Three hidden layers (hl) and fifteen neurons (n) were used for the ANN. Number of inputs was kept constant at two. The solver training and prediction comparison is included. On the right, the blue surface is the ANN prediction and the red is the kriging prediction.

Proper Orthogonal Decomposition

Proper orthogonal decomposition (POD) is a mathematical technique that is used in many applications, including data compression (Lawson 2007). An analogy can be made with the Taylor series or the Fourier transform, where the part of the infinite equation that does not add significantly to the data is neglected. In the case of fluid dynamics, the flow is decomposed into modes. The modes that make major changes to the flow are retained, and the rest are ignored.

The principle behind POD is that any function can be written as a linear combination of a finite set of functions, called basic functions. There are a number of different ways of performing this decomposition, two of which are the Karhunen–Loève decomposition (KLD) and singular value decomposition (SVD) methods. The SVD method can be described as follows.

Let A be the data matrix, an $m \times n$ matrix where $m > n$. The SVD equation is:

$$A = USV^T \tag{10.38}$$

where U (whose columns are called left singular vectors – output basis vectors for A) is an $m \times \mathrm{mo}$ matrix, where mo is the number of modes, S (whose diagonal elements are called singular values ordered in decreasing value order) is a $\mathrm{mo} \times \mathrm{mo}$ matrix and V^T (whose rows are called the right singular vectors – input basis vectors for A) is $\mathrm{mo} \times n$ matrix. More details can be found in Lawson (2007).

The Karhunen–Loève expansion is usually used to represent stochastic processes (in other words, a family of random variables with respect to time) as a combination of random deterministic time functions. The method of snapshots for the KLD method was introduced by Sirovich (Everson and Sirovich 1995). A snapshot is a set of data at certain spatial points at one particular time. So, the data matrix consists of two dimensions, the first being the spatial grid point data and the second being time. The snapshots are taken at regular intervals over the period of flow. In the KLD method, any snapshot can be expanded using eigen functions. Taking velocity as an example,

$$u(x,t) = u_m(x) + \sum_{i=1}^{N_t} a_i(t)\Phi_i(x) \tag{10.39}$$

where $u_m(x)$ is the mean velocity, $\Phi_i(x)$ is the spatial KLD mode and $a_i(t)$ is the temporal KLD eigen function. To generate this decomposition, the first step is to calculate the covariance of the input data matrix,

$$B = (1/N)(A^T A) \tag{10.40}$$

The resulting matrix is symmetric and has dimensions of $N \times N$. The eigenvalues and eigen vectors of the covariance matrix are then computed. The matrix containing the eigenvectors are temporal KLD eigen functions $a_i(t)$. The original data matrix is then multiplied by the eigenvector to produce $\Phi_i(x)$. The eigenvalues represent the amount of energy stored in the mode; in other words, its contribution to the overall flow.

Typically enough modes are stored to capture 99% of the energy:

$$\sum_{n=1}^{m} \lambda_n / \sum_{n=1}^{\infty} \lambda_n > 0.99 \tag{10.41}$$

where λ_n is the nth eigen value. Generally, the KLD mode is faster and requires fewer modes to reconstruct the flow. Also, it is less sensitive to load imbalance and hence larger block sizes can be used. Even with increased number of snapshots – larger matrices – the KLD execution time increases slower than the SVD method. The number of snapshots has a larger effect on execution time than the number of spatial point data. In addition, a method has been found for compressing data as it is obtained for each time step from the solver, using the SVD method. No such method has been found for the KLD (Lawson 2007).

The POD is not only used as a compression tool to reduce the size of data stored, but can also be used to predict missing data for required inputs, for what has come to be commonly known as 'gappy' data (Willcox 2006). From the theory shown, the model is reconstructed using the eigen functions that carry the highest energy. In this case, the aim is to be able to predict the loads of aerofoils that were not solved for using CFD. There are two methods of doing this: the POD/Galerkin method and the POD/interpolation method. The POD/interpolation method is more suitable with data that does not have much correlation (Joyner 2001).

The field data is reproduced as follows:

$$U = \sum_{i=1}^{p} \alpha_i \Phi^i \tag{10.42}$$

where U is the data that exists, p is the number of modes used for reconstruction, α_i is the ith temporal POD coefficient and Φ^i is the ith spatial POD basis vector.

The file containing the original data set (gappy file), U, and the required file are usually organised as the value at each spatial point down; in other words, each row represents the value of U at a different point in space, and for each snapshot or time step across; in other words, each column represents a time step, so that you have a matrix of $n_{\text{space}} \times n_{\text{snapshots}}$. Using, for example, the ΔC_l values for the aerofoils, the thickness values could replace the spatial data points and the camber values as the snapshots. So, for example, the case described in Section 10.3.4 is used where ΔC_l is the data matrix for A and it would look like this:

	0	**13**	**23**	**33**	**43**
9	0.977276	1.033416	3.000609	5.007123	7.083982
12	1.000000	0.642496	2.269264	3.946593	5.717450
15	1.137035	0.074969	1.296992	2.585237	4.025577
18	1.407559	0.693965	0.077982	0.971107	2.044992

A mask vector, n^k describing where the data is missing must first be created as follows:

$$n_i^k = 0 \text{ if } U_i^k \text{ is missing or incorrect} \tag{10.43}$$

$$n_i^k = 1 \text{ if } U_i^k \text{ is known and available} \tag{10.44}$$

where U_i^k is the ith element of the vector U_k. Let g be the solution vector that has some elements missing. Then,

$$\tilde{g} = \sum_{i=1}^{p} b_i \Phi^i \tag{10.45}$$

where $\tilde{g}$ is the repaired vector. The error between the data that exists in both the solution with the missing data and the repaired vector is given by E,

$$E = g - \tilde{g} \tag{10.46}$$

Only the existing data are compared, using the mask vector to mask out the new inputs. The coefficient b_i is varied so as to reduce the error. b can be found by differentiating Eq. (10.46) with respect to b such that

$$Mb = f \tag{10.47}$$

where $M_{i,j} = (\Phi^i, \Phi^j)$ and $f = (g, \Phi^i)$. Solving this for b gives g and the missing data can be inserted into the original data matrix (Bui-Thanh et al. 2004).

Now, if a number of snapshots are missing the first step is to fill in the missing data with random values or averaged values for the required points. This data matrix is used to obtain the POD basic vectors. As described above, these matrices are used to obtain a corrected data matrix from Eq. (10.45). The values from these intermediate repaired data are now used to reconstruct the missing data for the next iteration. This is continued until the algorithm converges or the maximum number of iterations is reached. For a more in-depth explanation of the KLD method, see the work done by Ly and Tran (1998).

This method, however, is reliant on the availability of data in as many positions as possible and requires at least two elements of data in each snapshot. It also does not perform well with few data points. With smaller files, within each snapshot, if in addition to the required point another point is missing, the accuracy of the prediction is highly compromised. For example, take the load, average C_m at a thickness of 12% and camber of 23 for the rotor optimisation case described in Section 10.3.4. In the data shown in Table 10.1, the difference is more than double the actual value. This large effect is because the initial matrix is small and so every point missing represents a large portion of the initial matrix.

For the matrix shown earlier, predictions were made for the aerofoil of camber 33, thickness 12 and 18 using an ANN and the gappy POD method and compared. The results are shown in Figure 10.8 and, as can be seen, the POD does not perform well

Table 10.1 Comparison between CFD and ANN predictions for the average pitching moment coefficient.

Prediction method	Avg C_m
CFD	*1.296992*
1 point missing	1.28567
2 points missing	3.47769

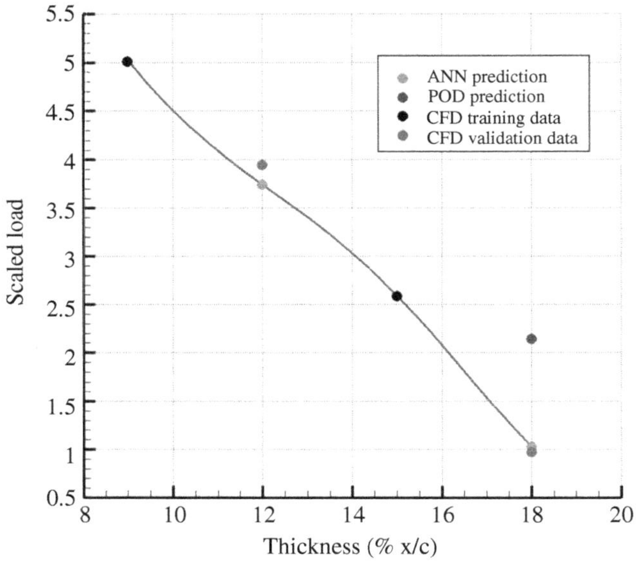

Figure 10.8 Comparison of ANN and POD predictions trained with the original database of 20 CFD points, less the two that are predicted.

with sparse data. Even with a greater number of points, such as 20 variables for each design parameter, which is far above what is practically viable with high-fidelity CFD software, the ANN has superior performance, as demonstrated here. The ANN predictions of 21 points of average C_m were used to show the POD's dependency on the amount of data available to it, as shown in Figure 10.9. With one point missing, the error is small, but as the number of missing points increases, the error grows. Also, the position of the missing data plays a role. If there are many points missing in one location, as opposed to points spread over the full domain, the error can be larger, as shown in Figure 10.9. This is one of the main advantages that the ANN has over the POD interpolation method, in that even though ANNs require more time for training, the POD would require more points for higher accuracy, which reduces its efficiency.

The gappy POD method was originally used to reconstruct images that were 'damaged' or unclear, such as face recognition, satellite maps and CFD flow field data. In these cases, more data is available and at a comparatively smaller resolution. Therefore, the POD interpolation technique works well. However, in the case where very little data is available (which is expected when high-fidelity CFD data is required) the trends are not predicted as accurately as when other methods are used.

10.3.5 Genetic Algorithm Optimisation Method

For the optimisation, a non-gradient method in the form of a genetic algorithm (GA) was implemented and combined with the metamodels. Figure 10.10 shows the analogy and terminology used as applied to the optimisation of an aerofoil case: selection, crossover, mutation, competition, survival. Two parents are first selected based on a roulette-wheel technique. The roulette wheel is a file containing the full population of

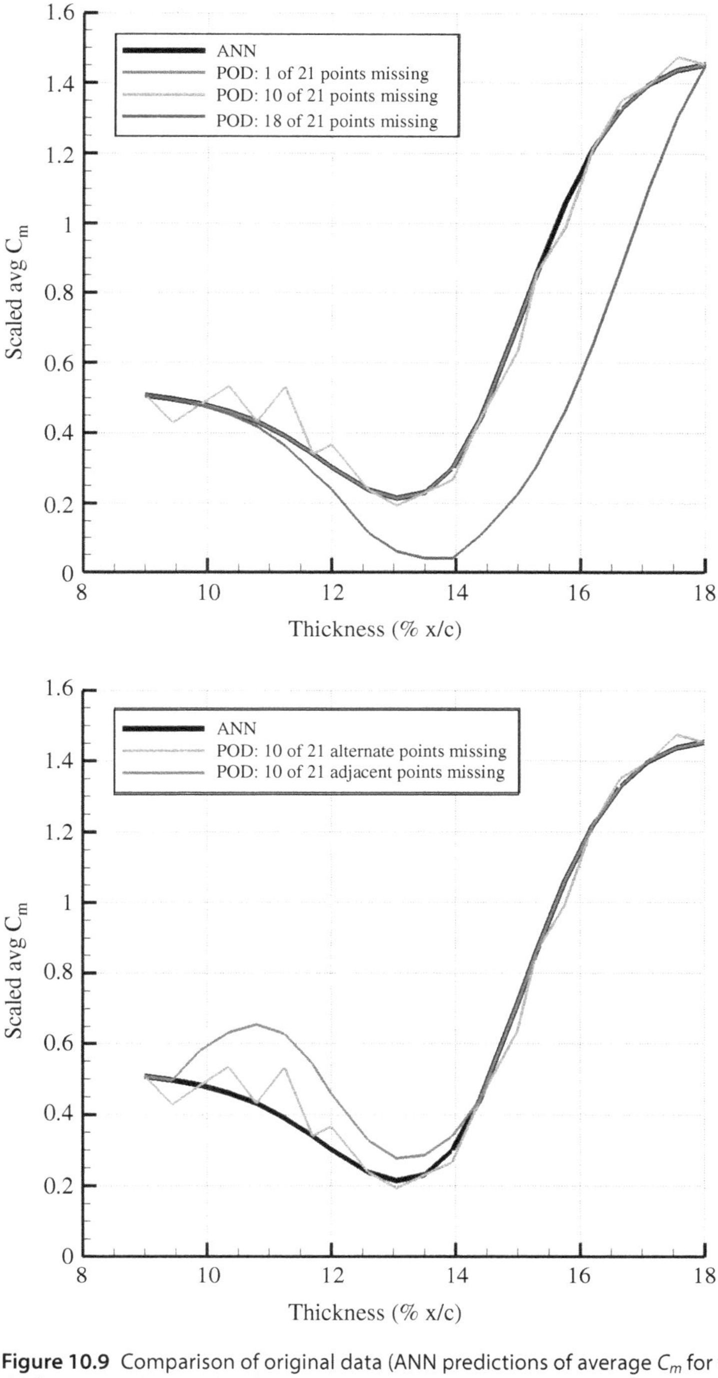

Figure 10.9 Comparison of original data (ANN predictions of average C_m for varying NACA aerofoil thicknesses at $R = 50\%$) with POD interpolation predictions, comparing number and position of training data points.

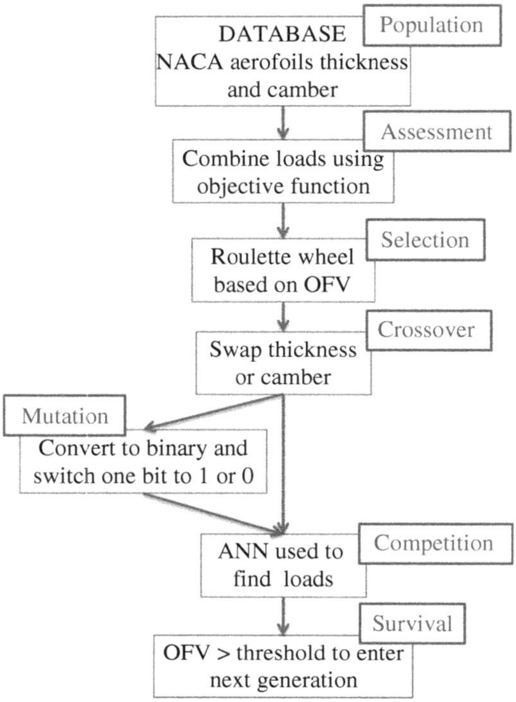

Figure 10.10 Outline of the genetic algorithm employed for an aerofoil selection case and the analogy with genetics. OFV, objective function value.

design points. However, each design point takes up as much space (in other words, it is repeated) in the file as is proportional to its fitness or performance by duplication; in other words, the wheel is biased towards fitter individuals. A random selection is made from it, but since the wheel is biased, the evolution leads to better designs being created. The proportionality function for space on the roulette wheel is user-defined from linear to exponential and is necessary for convergence and stability. This is because if the number of individuals in the database is very high, the percentage of space taken up by fitter individuals on the roulette wheel reduces and the selections become less biased and more random. Therefore a better fitness assignment rule would be an exponential one rather than a linear one, for example.

Once the parents have been selected, their 'genes' are swapped or crossed over. A number of crossover methods have been developed. The most commonly used is random crossover and it is used in this chapter. Here, a gene or more are randomly selected and swapped, producing two new offspring. In the illustration in Figure 10.10, either the thickness or camber is selected randomly and swapped between the two aerofoils.

For the mutation stage, the offspring parameters are converted to binaries of 10 bits, analogous to genes. This representation is simple and effective. In the natural world, the genome is the most basic form of discretisation. Similarly, in the computing world, binary is the basic form of discretisation. A random bit is then chosen and changed to either 1 or 0. Mutation is necessary since it has been found that after a number of generations, some characteristics of the genes get 'lost' (Hirsh 1998). Mutation allows for these characteristics to be re-introduced into the gene pool and it also increases diversity, which allows the global optimum to be found. Its probability is kept low by allowing only one point to be changed and there is a 50% chance that it will be changed to a different value. So the probability of 1 bit changing is 1 in 20 for 10 bits. This prevents the change in the phenotype from happening too often, which would cause the process to lose its evolutionary driving force. The probability of mutation can be increased by selecting more than one point for mutation or by switching the bit in the gene rather than assigning it a value it may already have or by swapping two bits in the code.

The resulting offspring is then assessed by employing the trained ANNs (along with its normalisation and denormalisation functions) and combining their output using the user-defined objective function. This objective function is then modified to stay within the user-defined constraints. There are two types of constraint exercised. One type limits the boundaries of the design space so that the predictions of the metamodel are accurate. The other includes physical constraints:

- geometric, such as. thickness, minimum tip chord length
- aerodynamic, such as angle of attack, stall margin, drag-divergence Mach number.

There are two ways to deal with these constraints: either as hard constraints or as soft constraints (Hajela 1999). Soft constraints are beneficial initially as they increase the diversity of the population, preventing the GA from terminating prematurely before the global optimum is found.

After a number of such iterations, a pool of the offspring characteristics is created and a threshold value is set so that only the majority of the fitter individuals survive and pass on into the next generation pool. The fittest individuals are always carried through into the next generation. This is termed 'elitism'. While the GA can converge without the help of elitism, the convergence takes longer and has a lower probability of being the global maximum since only mutation is capable of re-introducing new design characteristics or 'alleles' back into the pool, and these may not be the best designs. Using elitism ensures that the best genes still exist in the gene pool (or 'live longer') and hence there is a higher probability of reaching the maximum value (Hirsh 1998). Cloning is avoided as it can change the selection process unfairly.

10.3.6 Pareto Front Optimisation

The optimiser used here employs an objective function to determine the optimum and so the selection pressure is towards a small area in the design space. However, another way of finding the optimum is to find the designs that provide the best compromise between all the performance parameters. This is known as the 'Pareto front'. The advantage of using a Pareto-front-type optimiser (PFO) is that it provides a range of design points that represent the best combination of performance measure parameters. In essence, the PFO method provides the user with the designs that give the best performance for each parameter and subsequently leaves the weighting of these performance parameters to be decided at the end of the optimisation process.

The advantage of using an objective-function-type optimiser (OFO) is that it allows designs that do not necessarily fall on the Pareto front to be included in the genetic

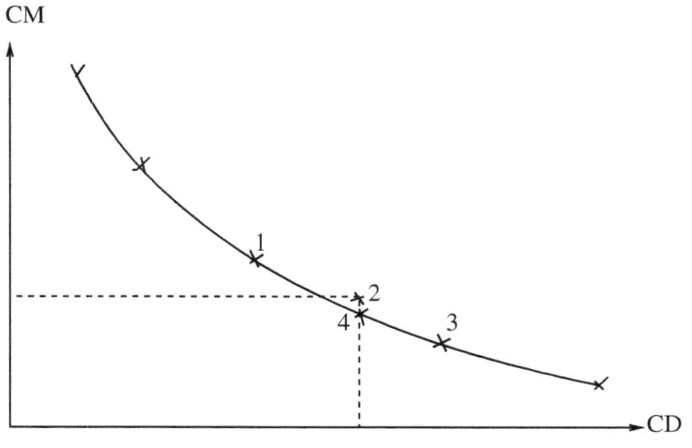

Figure 10.11 Pareto front and objective function optimisers.

algorithm's selection, based on the end-point required. So, rather than spreading out the performance along a front, it clusters it at a specific objective.

For example, in Figure 10.11, if C_M has less value in the objective of the design, point 2 is still a good design even though it may not fall on the Pareto front. However, having the Pareto front is a good indicator that there may be a slightly better design, possibly at point 4.

10.4 Parameterisation Technique

Parameterisation plays a key part in increasing the efficiency of the optimisation process. The fewer the design variables used to define a shape (for a fixed amount of flexibility), the smaller the initial population of high-fidelity data (since the number of dimensions of the database is reduced), the less training time required for the metamodels, the more accurate their predictions are given a fixed number of samples, and the faster the output of the optimiser.

Some techniques use fewer parameters. These techniques, however, can reduce the intuitiveness of the shape representation or reduce the flexibility of the design change. For example, splines and NURBS can be used to represent the data quite efficiently for the amount of flexibility available, but the movement of control points and weights is not directly intuitive to how the shape is being changed. However, these methods are very useful for the design of compressor blades, for example, as these are typically defined as splines. Also, high flexibility may not be required if the aim is to fine-tune specific design parameters rather than to change a shape completely.

Another method uses equations, where the coefficients involved are varied to change the design and the equations are then matched at their end points. This technique allows a lot of flexibility, since any number of equations can be used to define a shape. The matching can be used to reduce the number of independent variables. With experience, these variables can become somewhat intuitive to the change in the shape. This is useful in planform design and was also used by us for the BERP-like tip and fuselage parameterisation.

Table 10.2 Desirable features of an ideal parameterisation method.

1. Well behaved: should produce smooth and realistic shapes.
2. Mathematically efficient and a numerically stable process that is fast, accurate and consistent.
3. Require relatively few variables to represent a large enough design space to contain optimum aerodynamic shapes for a variety of design conditions and constraints.
4. Allows specification of key design parameters such as leading-edge radius, boat-tail angle, aerofoil closure (examples specific to aerofoils).
5. Provide easy control for designing and editing the shape of a curve.
6. Intuitive – geometry algorithm should have an intuitive and geometric interpretation.
7. Systematic and consistent – the way of representing, creating and editing different types of geometries must be the same.
8. Robust – the represented curve will not change its geometry under geometric transformations, such as translation, rotation and affine transformations.

Overall, the parameterisation method should be tailored to what best works for each case. Kulfan and Busoletti (2006) discuss a number of methods used to parameterise aerofoils and also discuss a new method of parameterisation. They also listed a number of desirable features of an ideal parameterisation method, referring specifically for aerofoils, (Kulfan and Busoletti 2006), although the characteristics could also be applied to other cases too. These characteristics are shown in Table 10.2.

For the BERP-like blade equations, each design variable required was independently defined but allowed a smooth curve to be created. The process was automated as described below. Figure 10.1 shows some defining features for the parameterisation of the BERP-like blade. The parameterisation technique developed for this optimisation process allows for the following design features to vary:

- the sweep angle
- the gradient of the BERP notch
- the spanwise position of the notch.

To do this, both the leading and trailing edges of the BERP tip are modified. Referring to Figure 10.12, the leading edge is defined by three equations and the trailing edge by two. For the leading edge, the first part is defined by a sigmoid curve that represents the notch region. The sigmoid equation is given by:

$$y = \frac{\Delta y}{1 + e^{-g\left(x - x_o + \Delta x/2\right)}} \tag{10.48}$$

where

- Δy is the notch height – the notch length in the chord-wise direction
- Δx is the total width of the notch – the notch length in the spanwise direction
- g is the gradient of the notch
- x_o is where the notch starts from.

The x coordinate of the notch maximum is defined by the user and is kept constant except when the notch position needs to be varied. The g value changes the gradient.

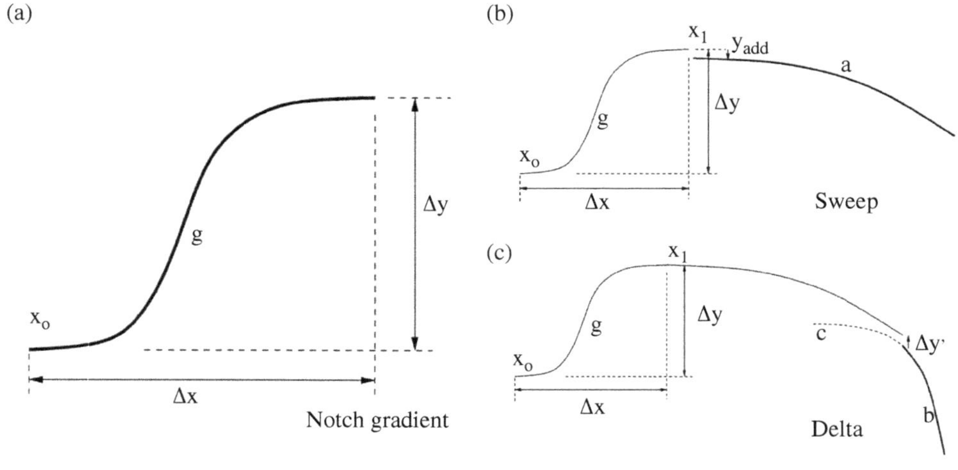

Figure 10.12 Definition of BERP-like design parameters.

The second part is used to define the sweep. It represents the part of the leading edge after the notch as a parabola:

$$y = -a\left(x - x_1\right)^2 + \Delta y + y_{\text{add}} \tag{10.49}$$

where a is the gradient of the parabola used to alter the sweep, x_1 corresponds to the notch end and the beginning of sweep, Δy is the notch height and y_{add} is an additional y offset value to ensure that the y ordinate of the parabola starts at the same position as the notch height. The value of y_{add} is computed automatically, once x_1 and Δy are known.

The third part describes the delta tip, which joins to the trailing edge. It is represented as a polynomial of order 2.5:

$$y = -b\left(x + c\right)^{2.5} - \Delta y' \tag{10.50}$$

where b is the gradient of the delta tip, c is the centre where the gradient of the curve becomes 0 (used to match the gradient of the curve to the previous parabola) and $\Delta y'$ is the additional y displacement required to match the curve to the previous parabolic curve. See Figure 10.12.

The two parameters g and a can be changed independently and the rest of the parameters appearing in the equations are automatically adjusted so that the curves match at the point and have continuity. These are the values of Δx, $\Delta y'$ and c. The initial x co-ordinate, x_0 is modified with the gradient of the parabola so that the tip point occurs at the same place for a required sweep. This is why, for different notch positions, different sweep parameters are used to obtain the same sweep distribution. The gradient b is dependent on the trailing edge curve as well. Therefore, the trailing edge must be determined first. The gradient of the trailing edge curve can also be modified independently of the sweep gradient of the leading edge. This allows the tip point of the blade to move in the y-direction, which inherently modifies the chord distribution as well.

The trailing edge is defined first by a linear curve that has the same gradient as the leading edge sweep parabola or a scaled value of it, if required, and then by a polynomial

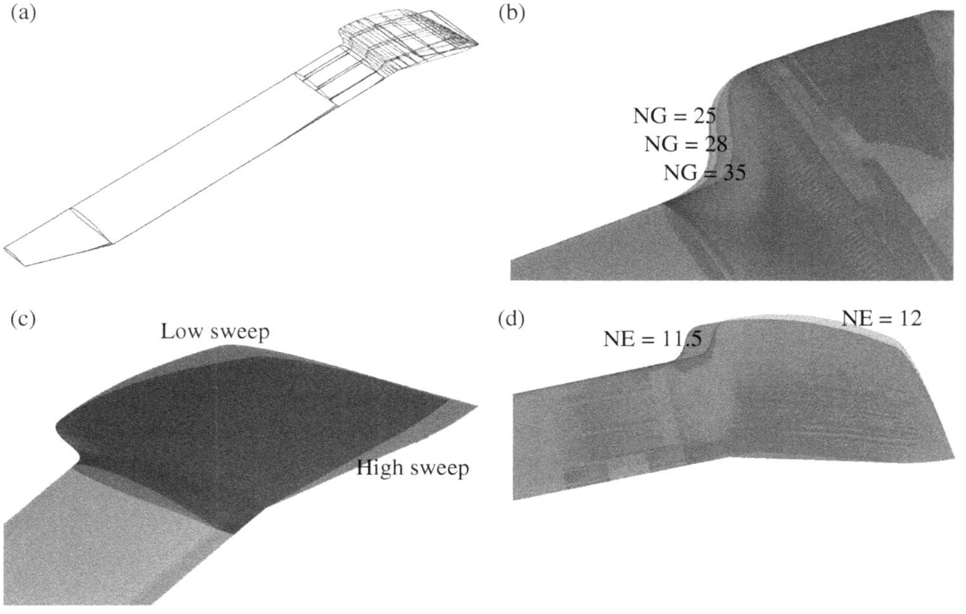

Figure 10.13 Visualisation of the three parameter changes to the geometry surfaces: (a) base rotor with BERP modification; (b) varying the gradient of the notch; (c) varying the sweep; (d) varying the initiation of the notch. (*See insert for color representation of this figure.*)

of order 3.5 that is matched to the point and gradient of the sweep curve that comes before. The trailing edge curve must be specified before hand, as the tip point is required to find the gradient of the delta polynomial so that the leading and trailing edge curves meet at a single point. So the first curve for the trailing edge is given by:

$$y = -a(x - x_1) + \Delta y + y_{\text{add}_{\text{TE}}}, \tag{10.51}$$

and the latter part of the trailing edge is given by:

$$y = -b_{\text{TE}}(x + c_{\text{TE}})^{3.5} - \Delta y_{\text{TE}}. \tag{10.52}$$

All the values and constants correspond to the trailing edge parameters except for the sweep parameter, a which is exactly the same as the leading edge sweep. The gradient of the trailing edge can be increased or decreased relative to the leading edge sweep gradient by scaling it with a factor. Figure 10.13 shows the examples used to build the design space for the optimisation.

The anhedral and twist of the blade were also modified to match the hover performance of the original blade. Once this was done, these parameters were kept constant; they were not optimised any further. Both values were varied linearly during the grid generation process as described in Section 10.5. Overall, three design parameters were varied for the optimisation. These were the sweep, the notch gradient and the notch position.

10.5 Grid and Geometry Generation

10.5.1 Geometry Generation

Figure 10.13 shows how a swept rotor tip can be transformed into a BERP-like tip and the effect of varying the parameters. The base rotor is made up of two sections: the HH-02 inboard (up to $r/R = 0.92$) and the NACA 64A-006 at the tip ($r/R = 1$). The aerofoil is linearly blended towards this latter section. This rotor has a rectangular tip, swept back by 20°, as shown in Figure 10.14. When the BERP planform is applied to this tip, the non-linear variation of the chord means that the thickness changes non-linearly as well. To maintain the thickness so that it decreases linearly, from 9.6% (thickness of the HH-02) to 6% (thickness of the NACA 64A-006), sections are cut from the base rotor such that when they are scaled to the chord length required, they have the thickness value that satisfies the linear variation. The problem with this method is that the point of maximum thickness for each of these sections varies non-linearly, and therefore the blade surface appears to be bumpy. To overcome this, the notch section is blended from the HH-02 to a NACA-64A section of the appropriate thickness and then the rest of the tip is built with NACA-64A sections of the required thickness so that, at the tip, the thickness is 6%.

Another issue that arises is that the HH-02 aerofoil has a tab whilst the NACA 64A-006 does not. So a tab is introduced for the NACA64A and is kept constant until where the tip is rounded off. The tab is introduced by cutting the aerofoil curves at about 20% from the trailing edge and then rotating the latter part of the curves in the longitudinal axis of the blade so that it adds the required thickness for the tab. The curve is then blended by point, tangent and radius to the main part of the curve to remove any kinks.

ICEMCFD™, a geometry and blocking generation software, was used to build the grid. ICEMCFD™ allows automation via 'replay files'. A number of the steps in the grid building exercise were automated in this way, although some manual intervention was required to finalise the grids.

10.5.2 Tip Anhedral

The tip anhedral is implemented as follows. Assuming that 40° of anhedral is to be implemented starting from the station $r/R = 0.918$ to the tip. For each station in between

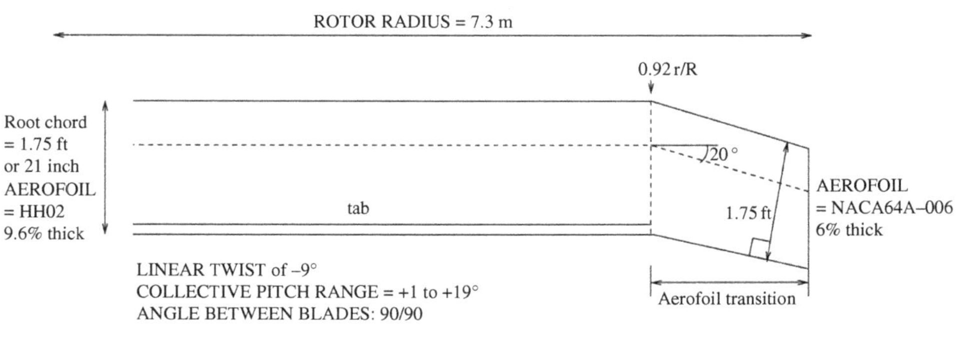

Figure 10.14 Schematic of the baseline rotor blade.

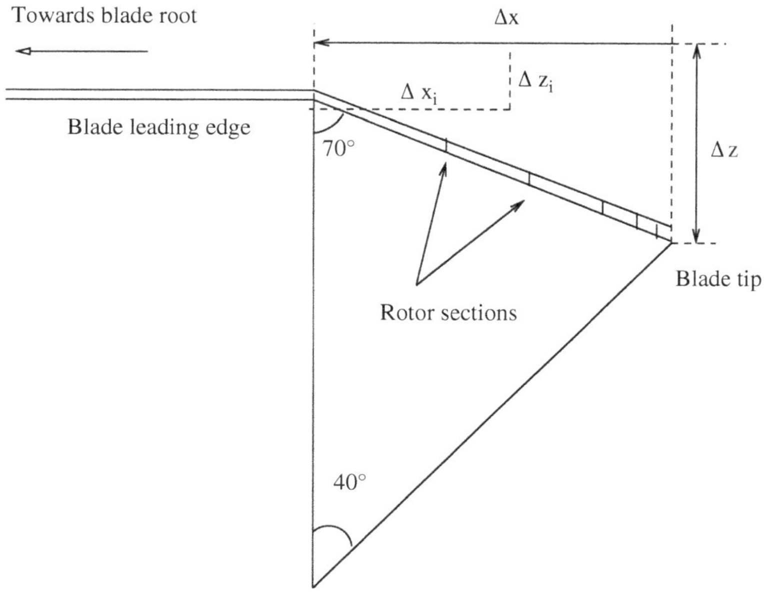

Figure 10.15 Generation of anhedral for the blade tip.

these two stations (r/R = 0.918 and r/R = 1.0), there will be a Δz distance by which that section should be translated downwards (Figure 10.15) which is found as:

$$\Delta z = \Delta x \tan\theta \tag{10.53}$$

Once each station has been translated, then the stations are joined by curves and surfaces are created.

10.5.3 Mesh Generation

The blocking used is of the multiblock type, whichconsists of a C-type topology around the blade set inside an H-type topology containing the full flow domain. Figure 10.16 shows the blocking for a single blade. This was copied and rotated around the azimuth to form a rotor of four blades, which were then used in the computation. Details of the mesh size in the farfield and on the blade are also shown. The total grid size was approximately 11 million cells for all four blades. The spacing perpendicular to solid surfaces was 1×10^{-5} of a chord length, resulting in $y^{+} < 1$. The Reynolds number variation is expected to be between 5 and 10 million for forward flight. Figure 10.16 shows the rounded tip and trailing-edge tab for the blade.

10.6 Flight Conditions

The hover flight conditions for the blade were based on a rotor-tip Mach number of 0.65, which was assumed from a tip speed of 220 m/s at International Standard Atmosphere (ISA) sea-level conditions. The chord of the blade root was 0.534 m and therefore the corresponding Reynolds number was approximately 8 million. The weight of the aircraft was estimated to be 9000 kg; that is, the maximum weight in hover. This results in a C_T of 0.018 based on a rotor radius of 7.32 m.

$$C_T = \frac{T}{1/2\rho A(\Omega R)^2} = 0.018 \tag{10.54}$$

For forward flight, an advance ratio of $\mu = 0.34$ is selected. This represents a relatively high-speed case, where the performance of swept and BERP blades is of interest. The thrust value for forward flight was based on approximately half the maximum hover weight.

These conditions are rough estimates and not exact values for a forward-flying aircraft, because the optimisation process is not being applied for a specific aircraft case.

10.7 Hover Results

First, the hover performance of the original blade was analysed and compared to that of the reference BERP-like blade. For the hovering rotor, the wake is assumed to be steady and the calculations were performed in steady-state mode. Also, it is assumed that the flow is spatially periodic and therefore a $1/N_{\text{blade}}$ segment of the full flow field is used with periodic boundary conditions. The boundary conditions for the farfield boundaries employ a source–sink model (Steijl et al. 2006). These conditions were prescribed at a distance of four rotor radii away from the rotor plane and for the outflow, a potential sink or 'Froude' condition was used (Srinivasan 1994). A quarter segment of the rotor and its surrounding field was created, similar to that shown in Figure 10.16. The mesh size was approximately 9 million cells with 300 cells surrounding the sections on the blade, with a perpendicular spacing of 1×10^{-5} chords.

The objective was to match or better the hover performance in terms of FM over a range of thrust settings using twist and anhedral and not the design parameters to be optimised for in forward flight. This new twist and anhedral blade will then be used to optimise the planform for forward flight. The FM over a range of thrusts was obtained for the original rotor and a BERP variant. Figure 10.17 shows the comparison of the CFD results of these blades for increasing C_T/σ, where σ is the rotor solidity. No experimental data was available for validation. Therefore the results only serve as a relative comparison. The BERP rotor with the same twist and anhedral has a better FM at low thrust. However, at higher thrust values, its performance drops to below that of the original rotor. This means that implementing the BERP-like tip planform reduces the hover performance of the helicopter. However, with a higher twist, the performance of the original rotor is recovered and with an anhedral of 20° implemented, it matches the performance of the original rotor.

The reason for this performance trend is that for the BERP rotor, the loading of the blade increases steeply where the BERP section begins, as can be seen in Figure 10.18. Also the loading inboards is lower than the original rotor. With increasing twist, the inboard loading is increased, which improves the FM. The anhedral reduces the outboard loading and thus the performance of the BERP blade matches the original rotor blade.

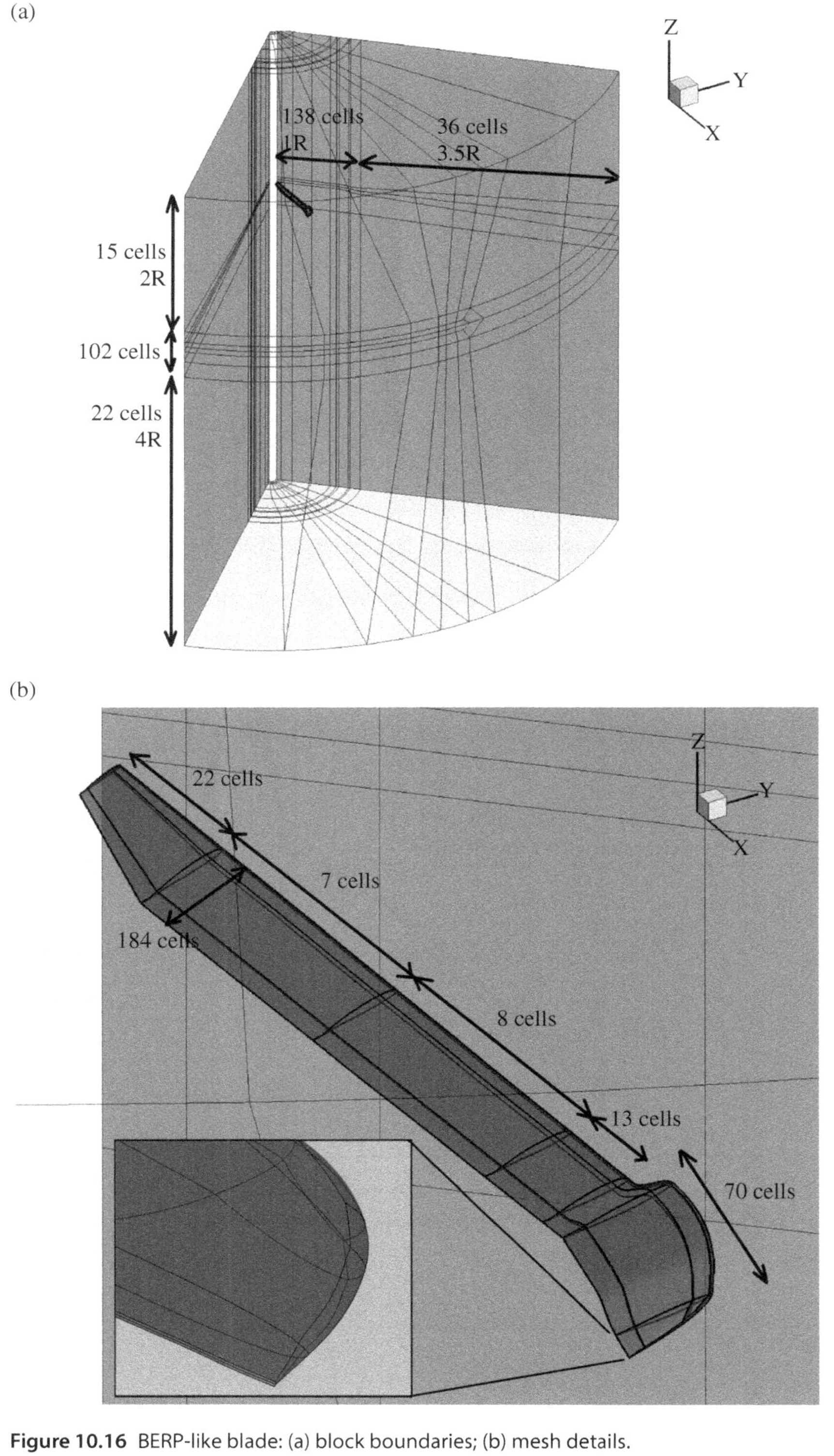

Figure 10.16 BERP-like blade: (a) block boundaries; (b) mesh details.

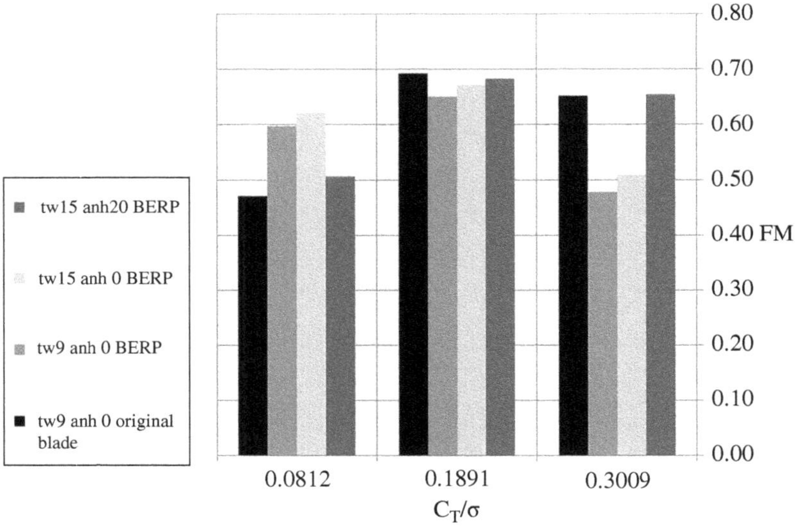

Figure 10.17 Figure of merit vs. thrust coefficient of the original and the BERP variants.

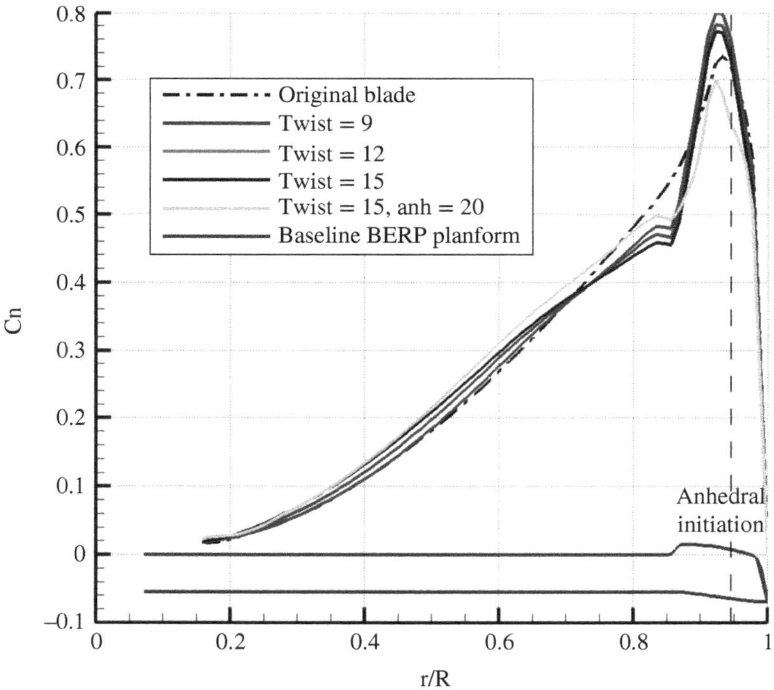

Figure 10.18 Lift distribution along the span with varying twist at 13° of collective.

10.8 Forward Flight Results

In this part, the effect of each of the parameters – tip sweep, notch offset and notch gradient – is analysed before the optimisation is presented. The rotor was trimmed manually to give the same C_T/σ value by adjustment of the collective angle. This corresponds to the thrust value mentioned in Section 10.6. In all cases, the torque includes pressure and viscous components.

Tip-sweep Effects

Table 10.3 shows the effect of sweep on the performance parameters of the BERP rotor. There was considerable loss in thrust with increased sweep. Therefore, for all cases, the rotor was trimmed to give the same thrust of approximately $C_T/\sigma = 0.09$. The obtained loads were very close in terms of C_T/σ (within 5%). The rolling and pitching moments did not change much and were therefore already well trimmed. The results comparing the effect of sweep are shown in Figure 10.19 for the cases with the highest notch gradient and the most inboard notch positions. With more sweep, it can be seen that the lifting load distribution is reduced at the back and outboards on the advancing side, and increased at the front of the disk and more inboards on the advancing side. The pitching moment is mostly negative on the advancing side and mostly positive on the retreating side. With increased sweep, the moment magnitude increases since moment is calculated about the blade pitch axis.

The torque distribution shows a drop in C_Q near the blade notch. C_Q is at its highest at the back of the disk before the notch and lowest at the notch near the advancing side. These extremes increase in magnitude with more sweep. Overall, on the advancing side, the torque reduces with increased sweep and on the retreating side reaches maximum value. The moment distribution has similarities to that of the torque and its magnitude is much higher at the front and back for the more swept tips. The lifting load distributions differ much less than the other performance parameters. So one of the tasks of the optimisation algorithm is to find a good compromise between these two extremes. Figure 10.20 compares the blade loads at four azimuth positions to show the effect of sweep at each. High sweep offloads the tip at the back of the disk and increases it at the front. On the advancing side, lift is maintained till just after the notch, where the highly swept blade loses lift quickly. The torque is low for higher sweep at the tip of the blade. The pitching moment has a more positive value at the tip with more sweep and does not have much of an effect inboards.

Table 10.3 Effect of blade-tip sweep on performance.

NE	NG	Sweep	C_T/σ	C_Q	Avg M^2C_M	ΔM^2C_M
11.75	28	0.09	0.0905	0.000192	0.002527	0.010297
11.75	28	0.13	0.0900	0.000191	0.001640	0.010290
11.75	28	0.21	0.0899	0.000186	0.000197	0.011147

NE, notch position parameter; NG, notch gradient parameter. Average M^2C_M is over one revolution and ΔM^2C_M is the peak-to-peak difference over one revolution.

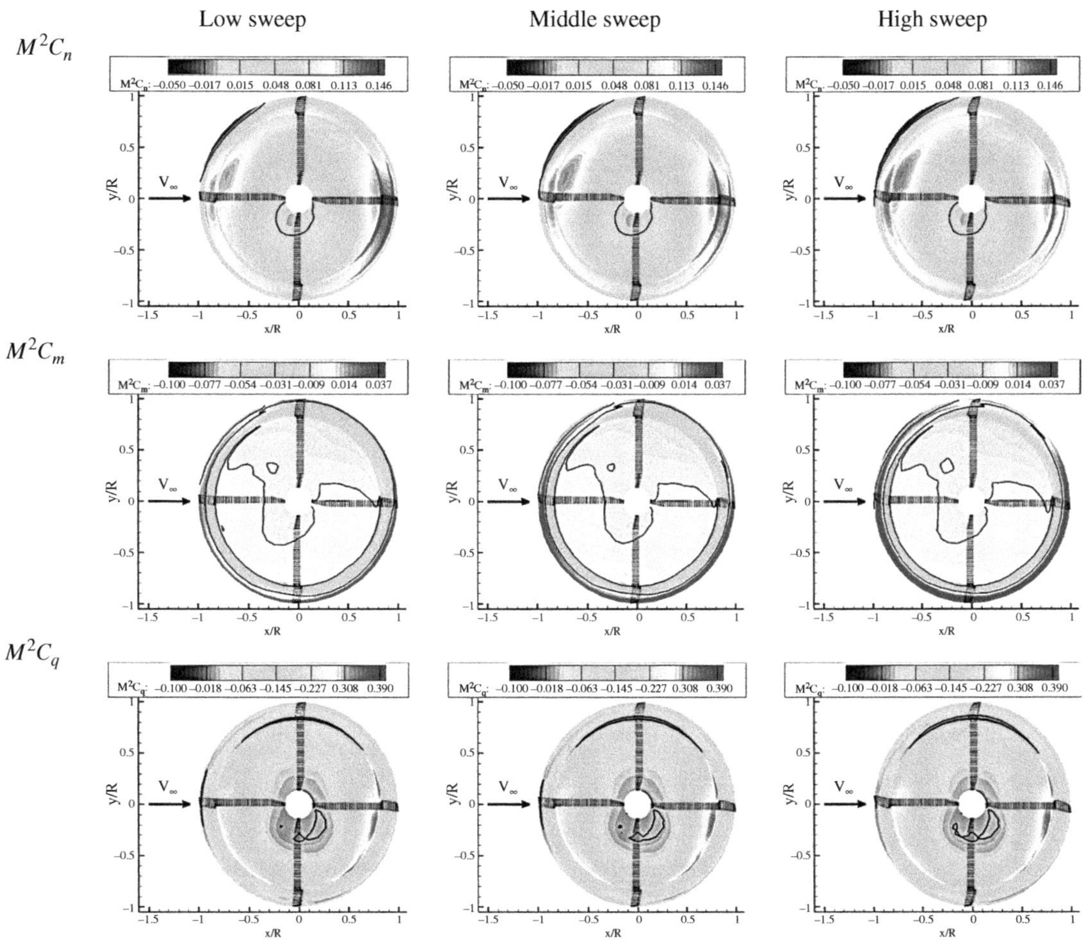

Figure 10.19 M^2C_n, M^2C_m and M^2C_q for NE = 11.5, NG = 35 and variable sweep. The black line indicates the 0 value. (*See insert for color representation of this figure.*)

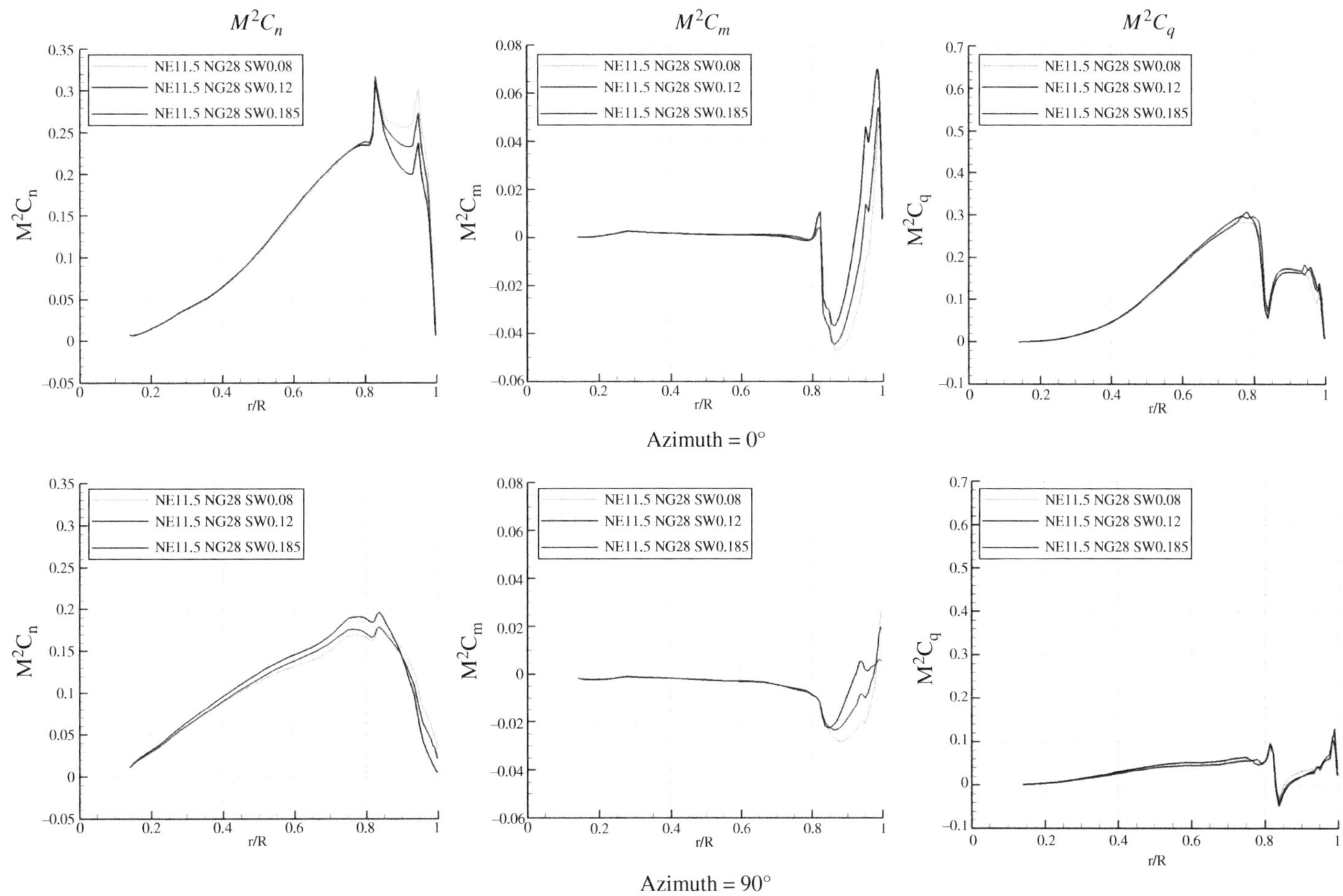

Figure 10.20 Comparisons of the M^2C_n, M^2C_m and M^2C_q with NE = 11.5, NG = 28 and variable sweep.

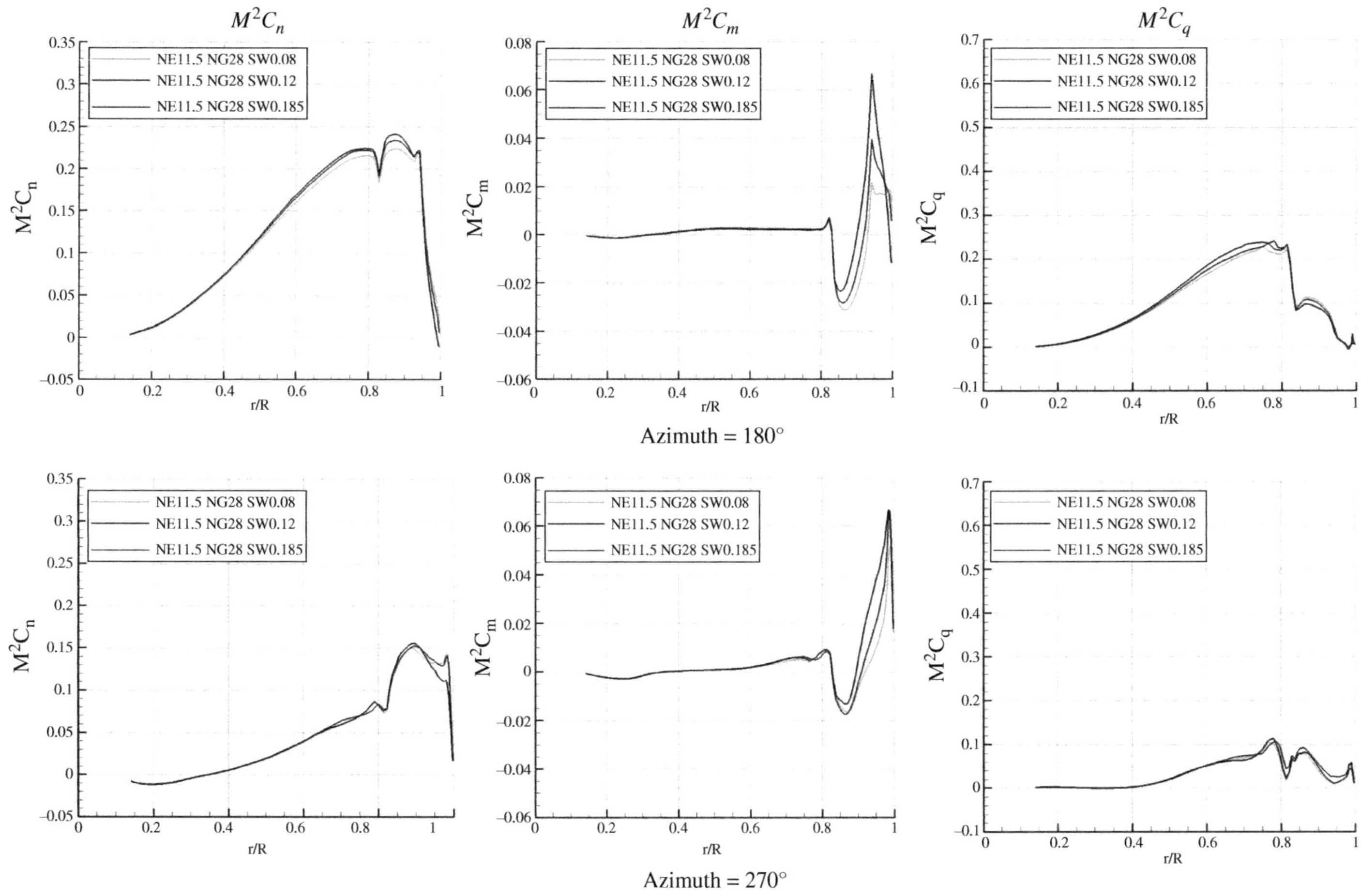

Figure 10.20 *(Continued)*

Effect of Notch Offset

Note that in Table 10.4, the sweep values differ because the gradient of the parabola differs when the position of notch changes to maintain the same sweep. In the Table, the comparison is shown for the highest sweep value for each rotor with a different notch position. The blade tip for these parameters can be seen in Figure 10.21. Figure 10.22 compare the loads when the notch is more inboards. These are trimmed results although thrust does not change much with notch position as shown in Table 10.4. The effect of the notch offset parameter is to amplify the effect of the sweep parameter. For example, the redistribution of lift is reduced at the back and increased at the front of the disk. This is caused by the sweep being larger in magnitude when the notch begins further inboard. The same can be seen for the blade pitching moment in Figure 10.22 where the region near the edge of the disk, where moment is higher, is thinner for the more outboard notch.

For blade torque, the trend is an increase with more outboard position. Where the notch occurs there is a drop in torque and then a continued increase towards the tip. With a more outboard notch, the torque continues to rise for longer prior to reaching the notch, therefore the latter part of the curve is higher. This can be seen in Figure 10.22 where the value of the increased region just before the notch is higher when the notch is more outboard. Also, the torque further out from the notch is higher for the rotor

Table 10.4 Notch position effect on performance.

NE	NG	SWEEP	C_Q	Avg M^2C_M	ΔM^2C_M
11.50	28	0.185	0.000186	0.000296	0.011556
11.75	28	0.21	0.000186	0.000197	0.011147
11.00	28	0.25	0.000184	0.000468	0.010621

NE, notch position parameter; NG, notch gradient parameter. Average M^2C_M is over one revolution and ΔM^2C_M is the peak-to-peak difference over one revolution. $C_T/\sigma = 0.09$.

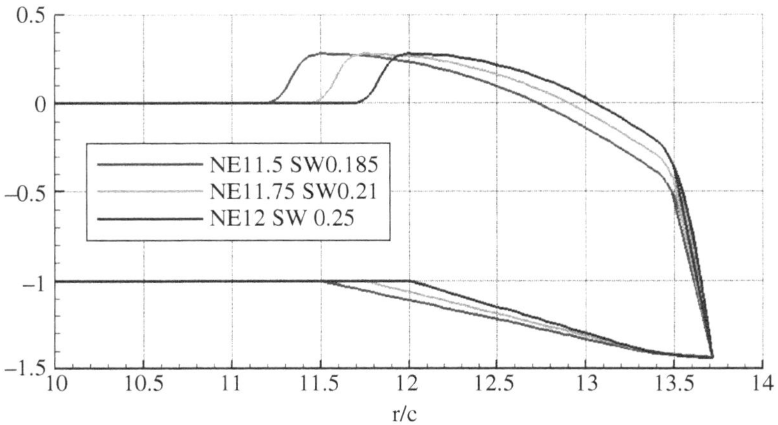

Figure 10.21 Maximum value sweep comparison for each notch offset value.

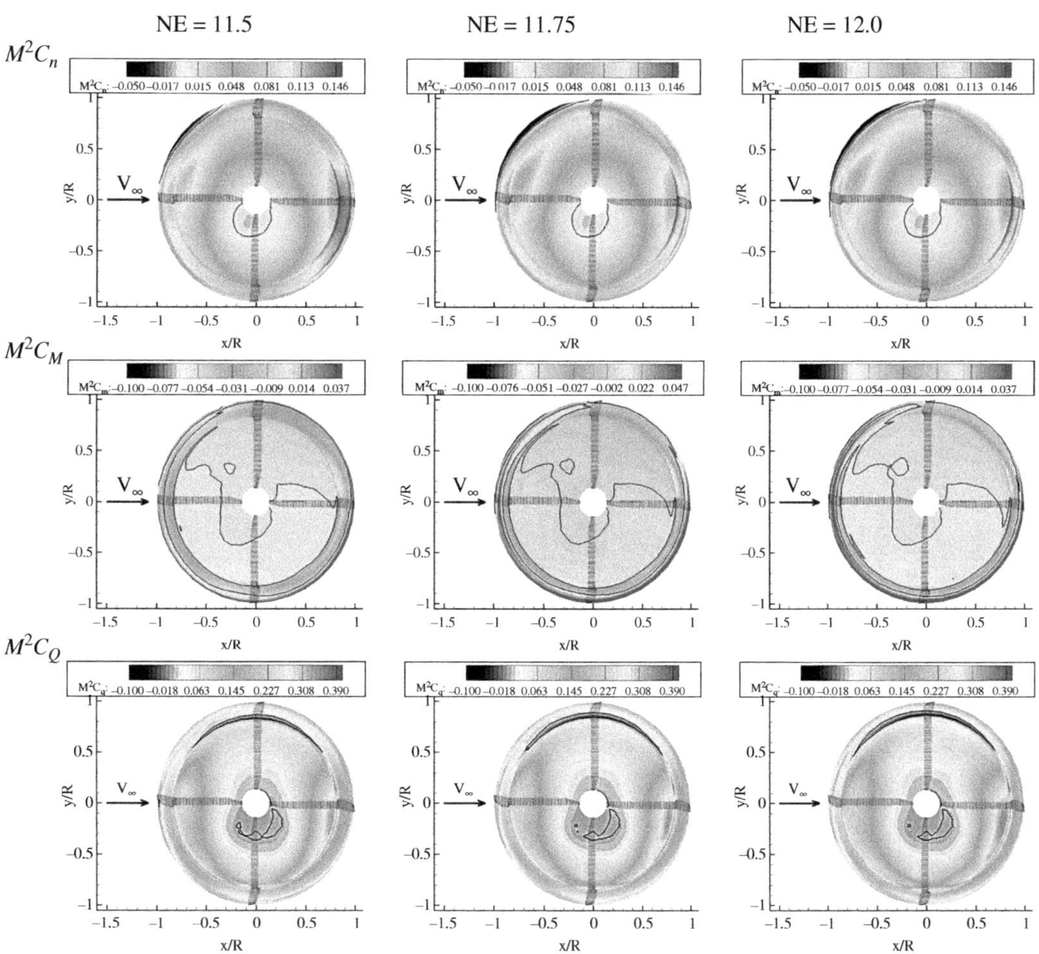

Figure 10.22 M^2C_n, M^2C_M and M^2C_Q for NG = 35 and high sweep. The black line indicates the 0 value.

with the more outboard notch. More inboards, on the advancing side, a decrease in torque is observed over a larger region and this brings the total value of the torque down as shown in Table 10.4.

Table 10.4 also shows that the peak-to-peak moment decreases (also seen in Figure 10.20) but the absolute average moment over a full revolution increases, the further outboard the notch is.

Effect of Notch Gradient

Figure 10.23 compares the performance of different notch gradients. The notch gradient has a smaller influence on the design than the other parameters tested. With a higher notch gradient, there is a slight decrease in negative pitching moment, evident from Table 10.5 where the average moment is slightly higher and the peak-to-peak value is lower, suggesting that a higher notch gradient provides better performance. The torque is not affected much at low sweep, but at higher sweep, the notch gradient has slightly more influence on the torque, as shown by comparing the top and bottom halves of Table 10.5. This suggests that NG has higher significance for designs with more sweep.

Overall Performance Comparison

Figures 10.24 and 10.25 are a summary of the integrated loads of pitching moment and torque over the blade during one revolution for all three design parameters. The peak-to-peak moment value reduces with further outboard notch position and the average moment tends to be more centred around zero when sweep is higher. The torque seems to be mostly affected by sweep on the advancing side. On the retreating side, the differences are more subtle. As a rough initial analysis, this data suggests that a highly swept blade would be optimal. Having a higher notch gradient would also improve the moments and having a notch more inboards would amplify the effects of the sweep. The quantities of the design parameters that make up the optimum design are obtained in the next section.

10.9 Planform Optimisation

The original population obtained from the CFD results contained 27 points. The performance components that make up the objective function selected were the average pitching moment ($C_{m\text{avg}}$), peak-to-peak pitching moment (ΔC_m) and the torque coefficient (C_q). The objective function weights were determined such that on average each of these components has the same influence. This was determined using the data from the original population that was obtained using the high-fidelity CFD solver. First each design was scaled with the baseline design case. The baseline case chosen was a BERP that had similar sweep, the most outboard notch position and the least notch gradient in comparison to the original rotor (the rectangular swept blade) shown in Figure 10.26. The parameters were NE = 12, NG = 25, SW = 0.25. The average ratio of $C_{m\text{avg}}$ to ΔC_m was found to be 2.7893:1 and the average ratio of ΔC_m to C_q was found to be 0.9548:1. Therefore, the ratio of $C_{m\text{avg}}$ to ΔC_m to C_q is obtained as: 2.6634 : 0.9548 : 1.0000. The weight for C_q was then calculated as:

$$w_{C_q} = \frac{2.6634}{2.6634 + 0.9548 + 1} = 0.5767 \tag{10.55}$$

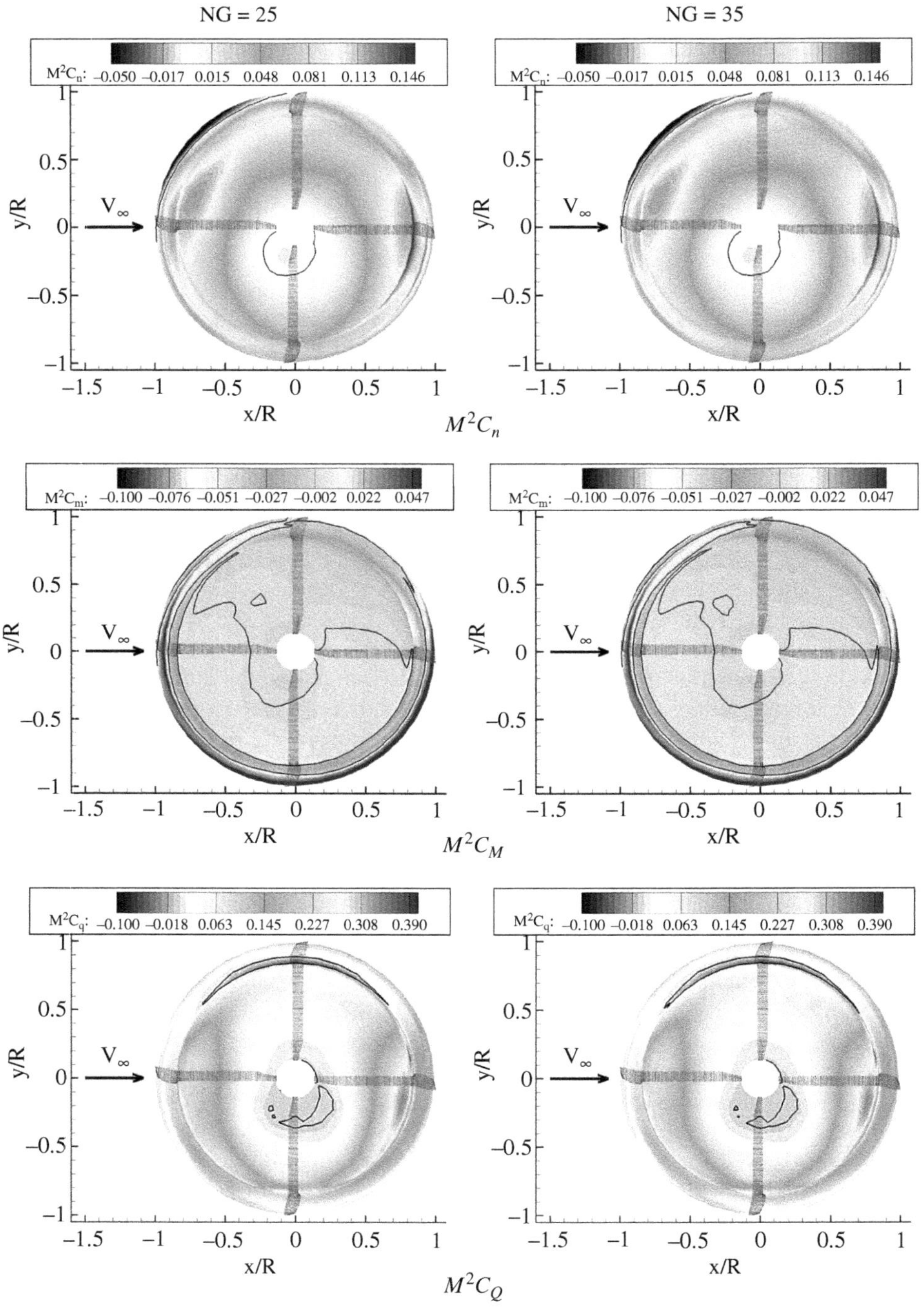

Figure 10.23 M^2C_n, M^2C_M and M^2C_Q for NE = 11.75 and SW = 0.21 with varying NG.

Table 10.5 BERP spanwise notch gradient performance comparison.

NE	NG	SWEEP	C_T/σ	C_Q	Avg M^2C_M	ΔM^2C_M
12.00	25	0.10	0.0909	0.000189	0.002561	0.009611
12.00	28	0.10	0.0907	0.000189	0.002464	0.009566
12.00	35	0.10	0.0911	0.000189	0.002357	0.009489
11.75	25	0.21	0.0898	0.000188	0.000295	0.010960
11.75	28	0.21	0.0898	0.000186	0.000197	0.011147
11.75	35	0.21	0.0898	0.000187	0.000105	0.011019

NE, notch position parameter; NG, notch gradient parameter. Average M^2C_M is over one revolution and ΔM^2C_M is the peak-to-peak difference over one revolution. $C_T/\sigma = 0.09$.

Hence the weights of $C_{m\text{avg}}$ and ΔC_m are given as:

$$w_{C_{m_{avg}}} = 0.5767 / 2.6634 = 0.2165 \quad (10.56)$$

$$w_{\Delta C_m} = 0.5767 / 0.9548 = 0.6040 \quad (10.57)$$

C_q was used as a soft constraint as well. In other words, if it exceeded the reference rotor's C_q, then the objective function value (OFV) was penalised. Since the rotors were trimmed to a $C_T/\sigma = 0.09$, C_T/σ did not need to be constrained. From this data, it was determined that the most influential design parameter was the sweep, followed by the notch position and then the notch gradient. ANNs were trained for each of the performance parameters, as shown in Figure 10.27. The white dots are the GA optimal selection and the black dots are the CFD training data for the ANNs. The dashed line is where the contour level is equal to 1; that is, the value for the baseline design. Each ANN was used to predict a single output with the three available design parameters as inputs. They each consisted of 15 neurons with 2 hidden layers and a convergence tolerance of 1% within the training data. An automatically adjustable learning rate was used to improve the training time required.

The performance parameters were used to find the optimum blade using a GA, and the results were compared with the Pareto front shown in Figures 10.28 and 10.29a. It is clear from Figure 10.28 (the white dots are the GA optimal selection and the black dots are the Pareto front solutions) that selecting an optimum design from the Pareto front when there are several design parameters can still be a difficult task. On the other hand, a weighted function makes it easier to find a particular optimum from the designs that also overlap a part of the Pareto front. The overlap is more clearly seen in Figure 10.29a. The comparison of the optimum with the original baseline blade is shown in Table 10.6 for the trimmed rotors. A much better average M^2C_m was obtained for a slight loss in ΔM^2C_m; at the same time, the torque and thrust coefficient differences are less than 0.5%. The performance of the resulting optimum relative to the baseline design is visualised in Figures 10.29b and 10.30a, which show the difference between the optimum and reference rotor values. The black line indicates the contour line where the difference is 0. Higher values indicate the optimum rotor has a higher value and vice versa. On the moment plot, it can be seen that the optimised blade has higher moments on the

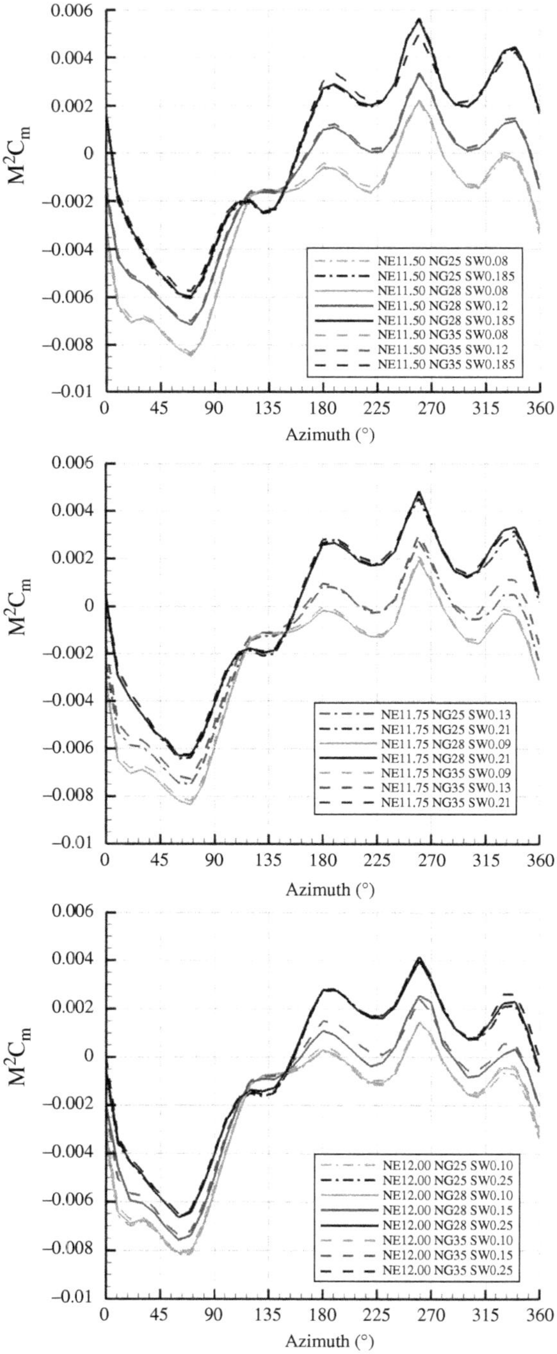

Figure 10.24 M^2C_M integrated over the full blade at each azimuth.

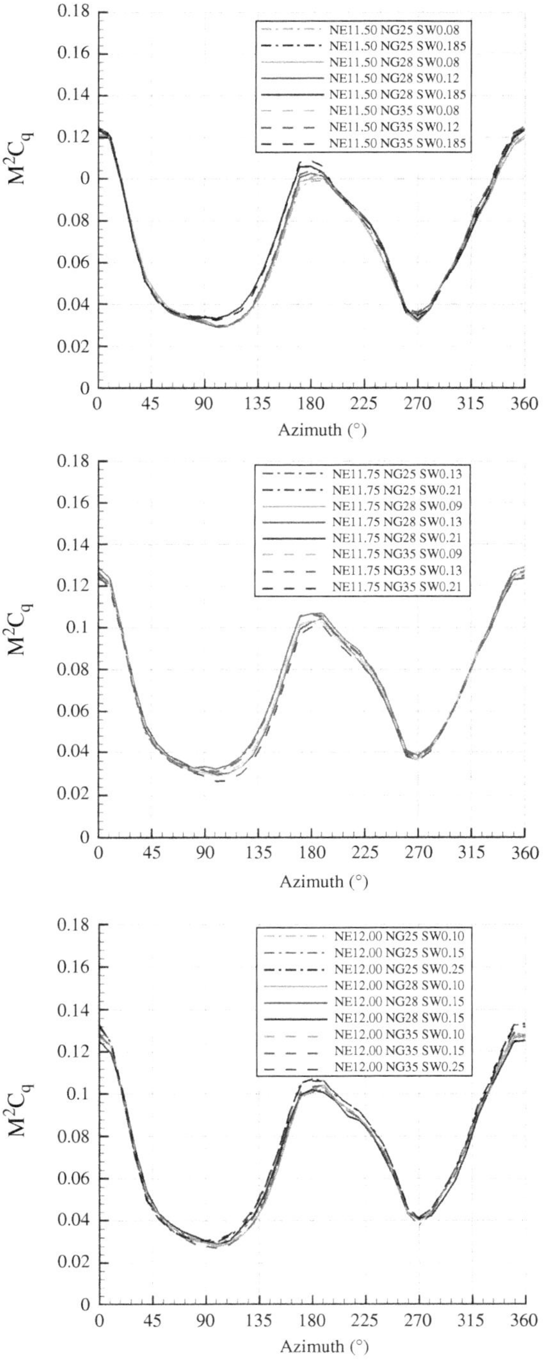

Figure 10.25 M^2C_Q integrated over the full blade at each azimuth. Both pressure and viscous components are included.

Figure 10.26 Baseline BERP-like rotor compared to a swept-tip design. (NE = 12, NG = 25, SW = 0.15).

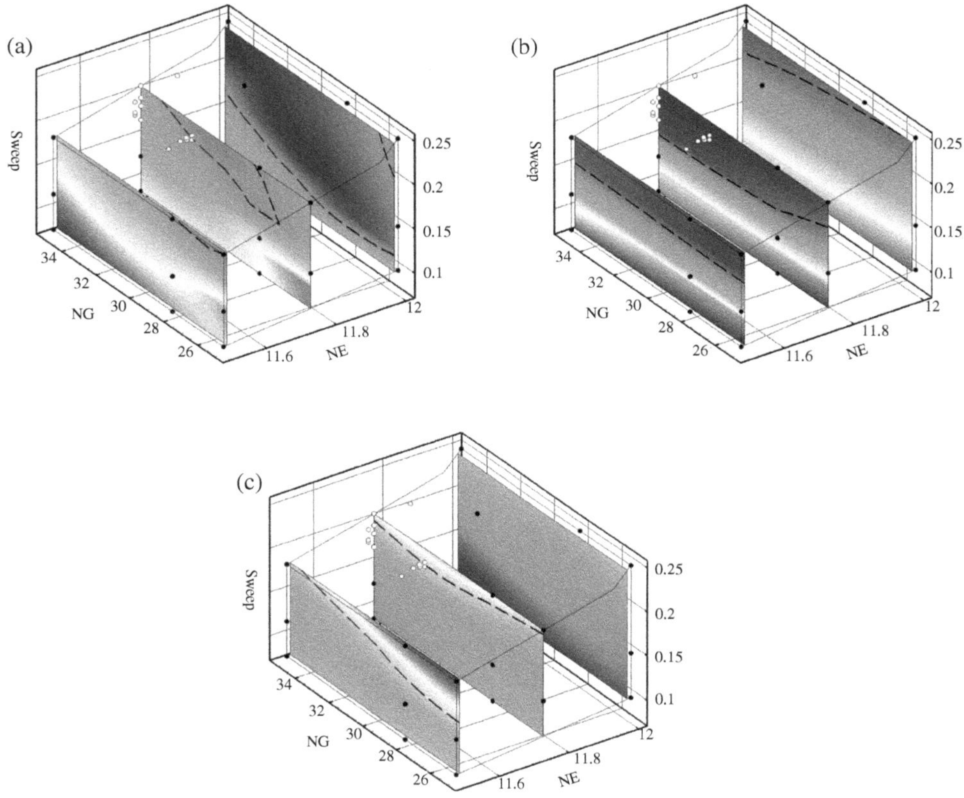

Figure 10.27 ANN predictions with training data and GA selection: (a) C_q, (b) scaled average C_m, (c) scaled ΔC_m.

advancing side. This is to counter the drop in moments that are experienced on the advancing side (see Figure 10.24). The opposite is true on the retreating side but to a smaller scale since the moments are not as far from zero as on the advancing side. For the torque, the decrease in torque for the optimum blade is almost the same as the areas where there is an increase, leading to very little change in the overall torque.

Figure 10.30(b) shows the difference in the objective function components between the optimum and the baseline design. From the scale itself, it is observable that there is an overall increase in the OFV due to the optimisation. The figure also shows the regions where there is an increase in the OFV. This is around the notch region, the sweep region and more inboards on the advancing side and front of the disk.

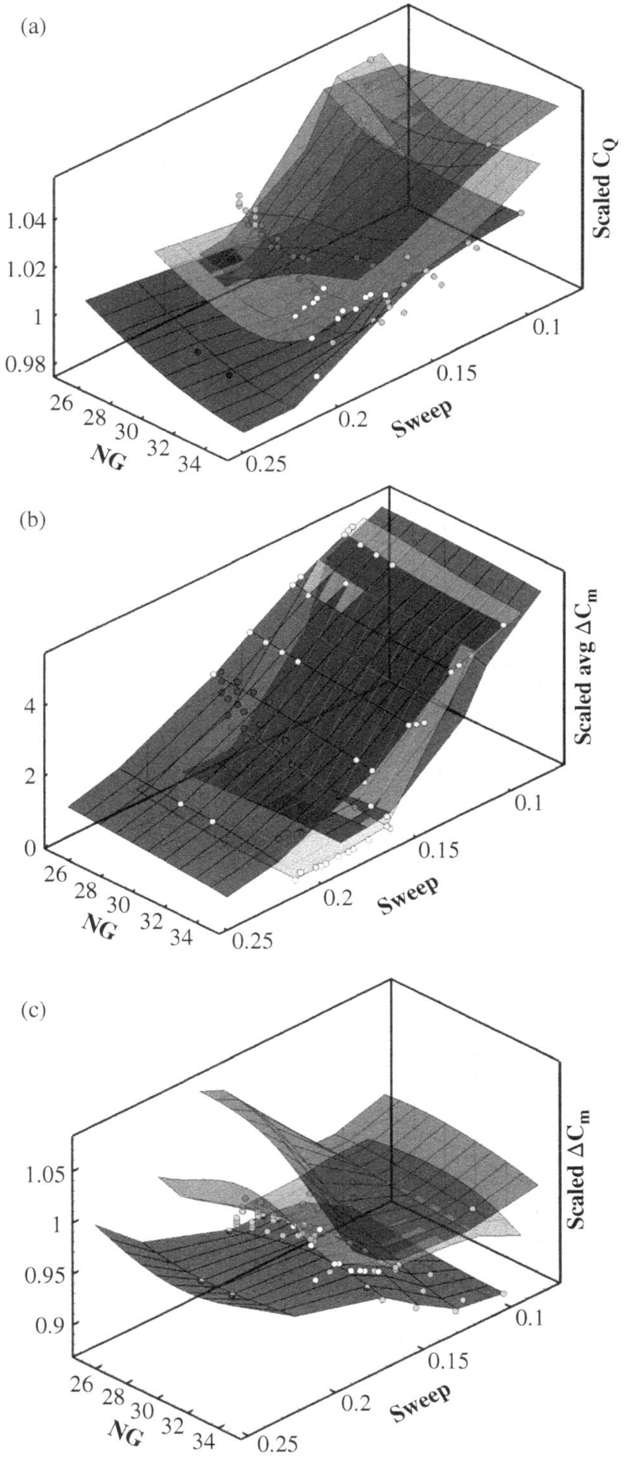

Figure 10.28 Pareto front points compared with GA selection for (a) NE = 11.5, (b) NE = 11.75, (c) NE = 12.

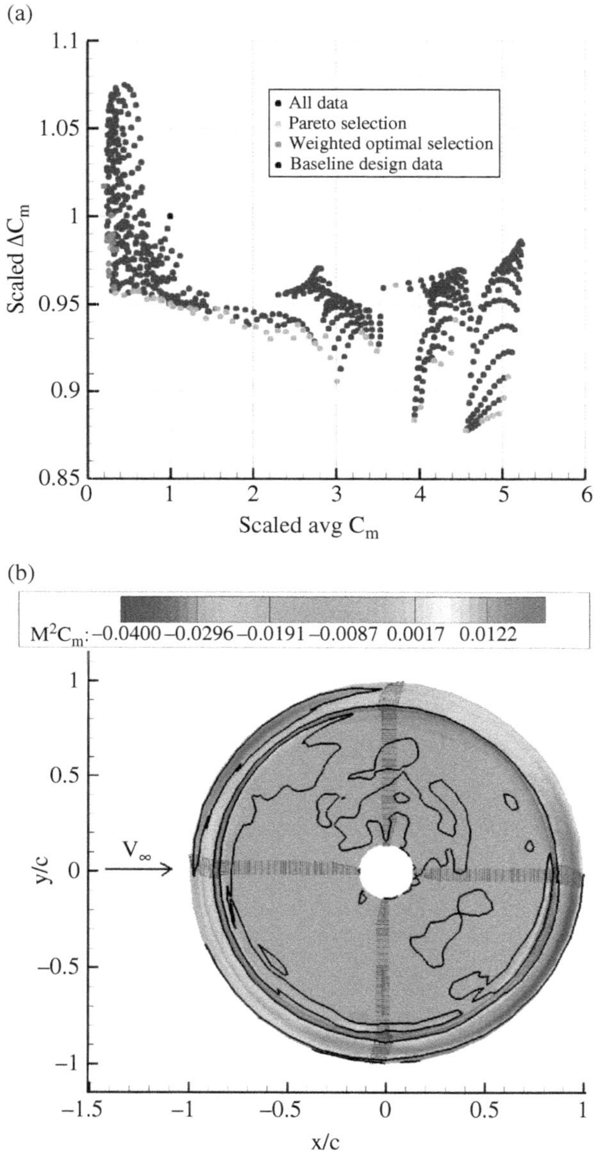

Figure 10.29 (a) Pareto front for the BERP-like design. (b) Difference in M^2C_m for the baseline and the optimum rotor.

The hover performance of the optimum blade was computed and compared to the baseline in Table 10.6. The results were obtained at a collective of 13°. The C_T/σ is slightly less than the baseline design mostly due to the added solidity, but the FM obtained was not substantially changed. Figure 10.31 compares the C_p distribution for the BERP reference and optimised blades at high thrust (the collective was 13°). It can

Table 10.6 BERP and baseline blade (NE = 12, NG = 25, SW = 0.15) performance comparison related to Figure 10.29(b) and 10.30(a).

	NE	NG	SWEEP	C_T/σ	C_q	Avg M^2C_M	ΔM^2C_m
Optimised	11.75	35	0.21	0.0832	0.000171	−0.000373	0.01078
Optimised	After trimming to 0.09			0.0898	0.000187	−0.000105	0.01102
Baseline	12.00	25	0.15	0.0905	0.000186	−0.000517	0.01083
Hover Performance Comparison							
	NE	NG	SWEEP	C_T/σ	FM	Collective (deg)	
Optimised	11.75	35	0.21	0.29	0.687	13	
Baseline	12.00	25	0.15	0.30	0.654	13	

[a] NE is the notch position parameter
[b] NG is the notch gradient parameter
[c] Average M^2C_M is over one revolution
[d] ΔM^2C_M is the peak-to-peak difference over one revolution.

be seen that the optimised planform spreads the loading at the tip over more of the span due to its inboard notch and its higher sweep and reduced tip area. Also from the normal and torque distribution curve, it can be seen that the optimised blade has a reduced load at the tip as well as a reduced torque. Both lead to a very slightly improved Figure of Merit.

Therefore, overall, the optimised blade has better performance than the baseline blade especially in terms of moment where the average pitching moment was reduced to approximately a fifth of the baseline designs.

The ANN predictions were also very accurate. The error in the ANN convergence during training, was set to 1% for each variable and the maximum error in the overall combined objective function between the CFD data and the ANN predictions was 2.7% which was for a design that had a 16% change in design parameters. This suggests that the ANN is a viable metamodel.

10.10 Summary and Conclusions

This chapter has described the optimisation of a BERP-like rotor planform using an optimisation method that employed a genetic algorithm and an artificial neural network that was trained using high-fidelity CFD data from the Helicopter Multiblock Solver CFD solver. The optimisation was carried out for fast forward-flight conditions. The hover performance was first maintained by modifying twist and anhedral. A parameterisation method was then defined to modify the sweep, the notch gradient and the notch position of the BERP-like tip. A full factorial method was used to obtain the sample space, which contained 27 design points. The objective was to reduce the compressibility effects on the advancing side of the rotor and stall in the retreating side. This objective was captured using the average pitching moment and the peak-to-peak pitching moment of the rotor over a full revolution. In addition, torque was

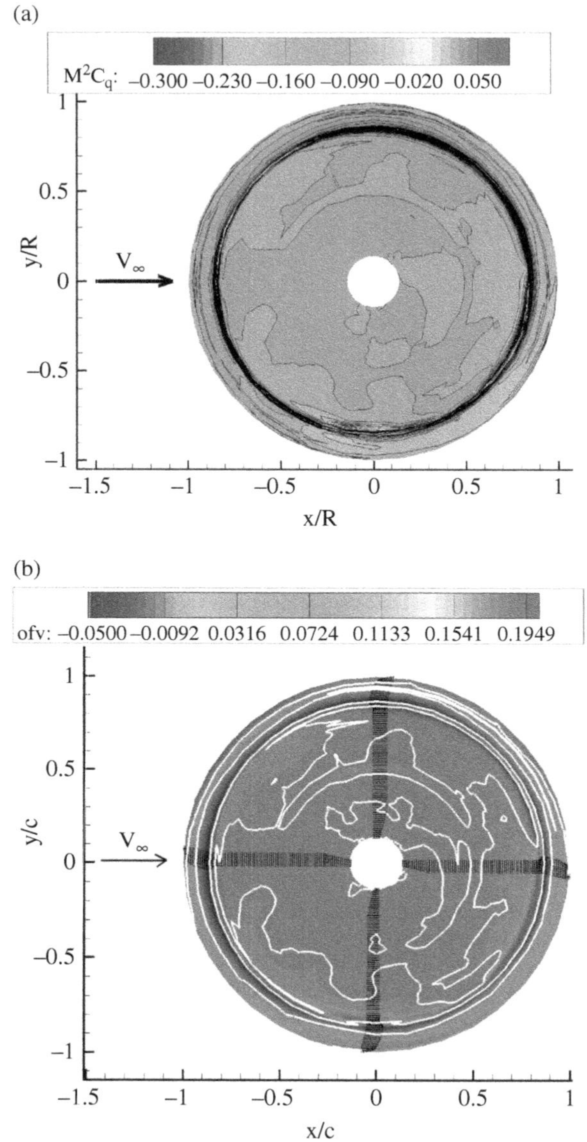

Figure 10.30 (a) Difference in M^2C_q for the baseline and the optimum rotor. (b) Difference in OFV for the baseline and BERP-like rotor.

optimised for and constrained. All cases were trimmed to the same rotor thrust. The outcome was a substantially improved pitching moment performance with slightly reduced torque for the same thrust. This was obtained using high sweep and notch gradients with the notch position at approximately $r/R = 0.86$. Hover performance was not compromised.

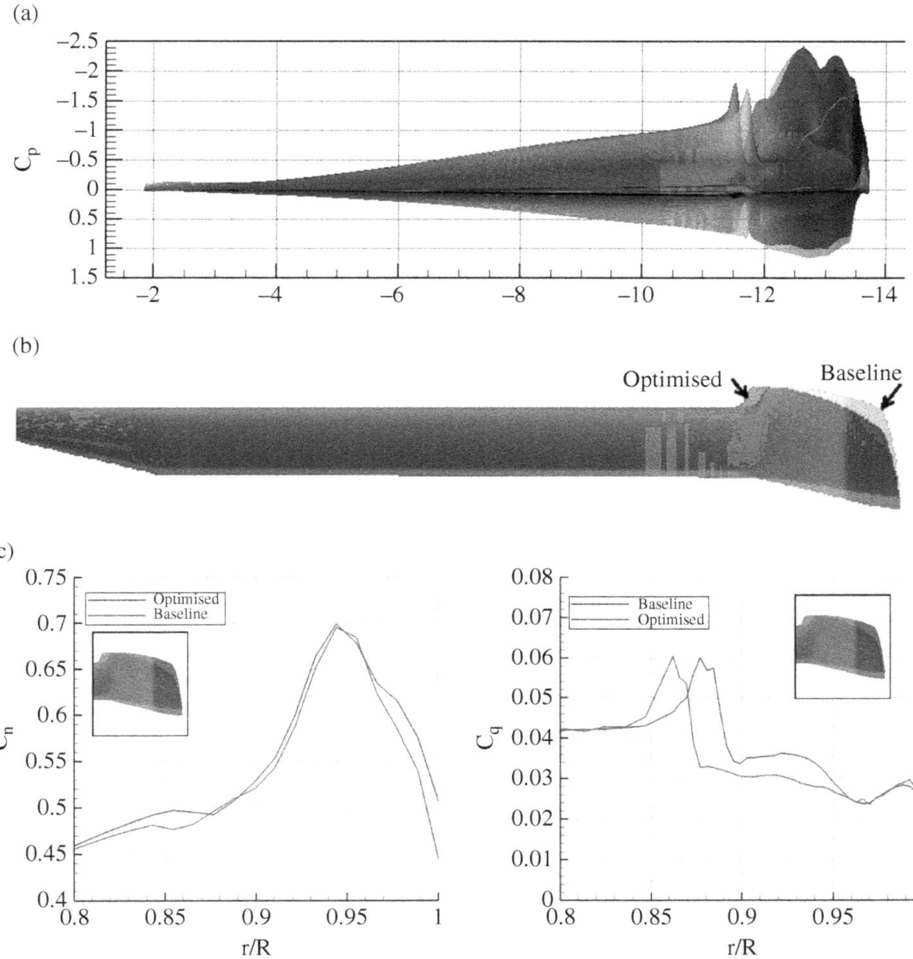

Figure 10.31 Reference and optimised blade at high thrust: (a) C_P distribution comparison; (b) planform comparison; (c) lift and drag distribution along the span of the blade.

The time required to obtain all the data was reduced by using the harmonic balance method and parallel computing. This reduced the the clock time for obtaining the CFD data by an order of magnitude, taking approximately 18 hours when started from a previous solution, compared to 150 hours for time marching over two revolutions (so that the data is periodic). The azimuth resolution for the harmonic balance method was every 10°. This makes the method practical for rotor optimisation in forward flight.

The overall optimisation was limited to the aerodynamic performance of various planform designs and did not include rotor section optimisation. Moderate notch gradient combined with increased sweep resulted in an improvement of the average pitching moment of 80%, although there was an increase of 1.7% in the peak-to-peak moment. The torque did not change much and the hover performance was maintained.

These differences in performance can be just as significant, if not more, for UAVs. They can improve the manoeuvrability and control of UAVS, which increases their agility, as well as their efficiency and endurance.

Bibliography

Brocklehurst A and Duque EPN 1990 Experimental and numerical study of the British Experimental Rotor Programme blade *AIAA 8th Applied Aerodynamics Conference*, Portland, Oregon, USA. AIAA-90-3008.

Bui-Thanh T, Damodaran M and Willcox K 2004 Aerodynamic data reconstruction and inverse design using proper orthogonal decomposition. *AIAA Journal* **42**, 8.

Cameron L, Early JM and McRoberts R 2011 Metamodel assisted multi-objective global optimisation of natural laminar flow aerofoils *29th AIAA Applied Aerodynamics Conference*, Honolulu, Hawaii.

Caradonna FX and Tung C 1981 Experimental and analytical studies of a model helicopter rotor in hover. Technical report, NASA.

Daskilewicz MJ and German BJ 2012 Observations on the topology of Pareto frontiers with implications for design decision making *50th AIAA Aerospace Sciences Meeting including the New Horizons Forum and Aerospace Exposition*, Nashville, Tennessee, USA. AIAA 2012-0148.

Everson R and Sirovich L 1995 The Karhunen-Loève procedure for gappy data. *Journal of the Optical Society of America A* **12**, 1657–1664.

Forrester A, Sóbester A and Keane A 2008 *Engineering Design via Surrogate Modelling – a Practical Guide.* John Wiley.

Forrester AIJ and Keane AJ 2009 Recent advances in surrogate-based optimization. *Progress in Aerospace Sciences* **45**, 50–79.

Hajela P 1999 Nongradient methods in multidisciplinary design optimization – status and potential. *Journal of Aircraft* **36**(1), 255–265.

Harrison R, Stacey S and Hansford B 2008 BERP IV the design, development and testing of an advanced rotor blade *American Helicopter Society 64th Annual Forum*, Montreal, Canada.

Hirsh R 1998 *GADO: A Genetic Algorithm for Continuous Design Optimization* PhD thesis State University of New Jersey.

Jang JS, Choi S, Kwon HI, Im DK, Lee DJ and Kwon JH 2012 A preliminary study of open rotor design using a harmonic balance method *50th AIAA Aerospace Sciences Meeting including the New Horizons Forum and Aerospace Exposition*, Nashville, Tenessee, USA. AIAA 2012-1042.

Johnson C and Barakos G 2011 A framework for optimising aspects of rotor blades. *The Aeronautical Journal* **115**(1165), 147–161.

Joyner M 2001 *An Application of a Reduced Order Computational Methodology for Eddy Current Based Non-destructive Evaluation Techniques* PhD thesis North Carolina State University.

Kulfan BM and Busoletti JE 2006 Fundamental parametric geometry shape representations for aircraft component shapes *11th AIAA/ISSMO Multidisciplinary Analysis and Optimization Conference*, Portsmouth, Virginia. AIAA 2006-6948.

Lawson SJ 2007 *Parallel performance of library algorithms for computational engineering* Master's thesis University of Edinburgh and Liverpool.
Lophaven SN, Nielson HB and Søndergaard J 2002 DACE, A MATLAB kriging toolbox, version 2.0. Technical Report IMM-TR-2002012, Technical University of Denmark.
Ly HV and Tran HT 1998 Proper orthogonal decomposition for flow calculations and optimal control in a horizontal CVD reactor. Technical report, CRSC.
Mantovan P and Secchi P 2010 *Complex Data Modeling and Computationally Intensive Statistical Methods*. Springer-Verlag.
Mathews JH and Fink KD 2004 *Numerical Methods Using MATLAB*. Pearson Education.
Mengistu T and Ghaly W 2008 Aerodynamic optimization of turbomachinery blades using evolutionary methods and ANN-based surrogate models. *Journal of Optimization Engineering* **9**, 239–255.
Oyama A, Nonomura T and Fujii K 2010 Data mining of Pareto-optimal transonic airfoil shapes using proper orthogonal decomposition. *Journal of Aircraft* **47**(5), 1756–1762.
Perry FJ, Wilby PG and Jones AF 1998 The BERP rotor – how does it work, and what has it been doing lately. *Vertiflite* **44**(2), 44–48.
Régnier J, Sareni B and Roboam X 2005 System optimization by multi-objective genetic algorithms and analysis of the coupling between variables, constraints and objectives. *International Journal for Computation and Mathematics in Electrical and Electronics Engineering* **24**(3), 805–820.
Robinson K and Brocklehurst A 2008 BERP IV – aerodynamics, performance and flight envelope *34th European Rotorcraft Forum*, Liverpool, UK.
Samad A and Kim KY 2008 Shape optimization of an axial compressor blade by multiobjective genetic algorithm. *Proceedings of the Institution of Mechanical Engineers Part A: Journal of Aerospace Engineering* **222**, 599–611.
Samarasinghe S 2006 *Neural Networks for Applied Sciences and Engineering – From Fundamentals to Complex Pattern Recognition*. Auerbach Publications.
Simpson TW, Peplinski JD, Koch PN and Allen JK 2001 Metamodels for computer-based engineering design: Survey and recommendations. *Engineering with Computers Journal* **17**, 129–150.
Spentzos A 2005 *CFD Analysis of 3D Dynamic Stall* PhD thesis University of Glasgow.
Srinivasan GR 1994 A free-wake Euler and Navier-Stokes CFD method and its application to helicopter rotors including dynamic stall. Technical Report 93-01: AD-A278 000, JAI Associates.
Steijl R and Barakos G 2008 Sliding mesh algorithm for CFD analysis of helicopter rotorâ€“fuselage aerodynamics. *International Journal for Numerical Methods in Fluids* **58**, 527–549.
Steijl R, Barakos G and Badcock K 2005 A framework for CFD analysis of helicopter rotors in hover and forward flight *31st European Rotorcraft Forum*, vol. **91**.
Steijl R, Barakos G and Badcock K 2006 A framework for CFD analysis of helicopter rotors in hover and forward flight. *International Journal for Numerical Methods in Fluids* **51**, 819–847.
Tan KC, Lee TH and Khor EF 2002 Evolutionary algorithms for multi-objective optimization: Performance assessments and comparisons. *Journal of Artificial Intelligence* **19**, 253–290.

Willcox K 2006 Unsteady flow sensing and estimation via the gappy proper orthogonal decomposition. *Computers and Fluids* **35**, 208–226.

Woodgate M and Barakos GN 2009 Implicit CFD methods for fast analysis of rotor flows *36th European Rotorcraft Forum*, Paris, France.

Zhao H, Wang S, Han W and Feng G 2008 Aerodynamic design by jointly applying s2 flow surface calculations and modern optimization methods on multistage axial turbines. *Frontiers of Energy and Power Engineering in China Journal* **2**(1), 93–98.

11

Active Blade Twist in Rotary UAVs using Smart Actuation

Pascual Marqués

Marques Aviation Ltd, Southport, UK

11.1 Introduction

Helicopter rotors operate in a complex unsteady aerodynamic environment characterised by cyclic variation of centrifugal and aerodynamic loads on the rotating blades (Figure 11.1). Various aerodynamic phenomena occurring on the helicopter rotor generate vibration and noise, and these are most pronounced in forward flight and during descent. In forward flight, the advancing blade can undergo airflow velocities close to the speed of sound; in this transonic regime shock waves can be generated. The retreating blade can experience dynamic stall and stall flutter (Feszty and Nitzsche 2011; Figure 11.2).

Both shock waves and dynamic stall cause vibration on the rotor, which is transferred through the shaft to the fuselage. Moreover, the helical tip vortex developed by each blade interacts with the other blades in a phenomenon known as BVI, which is an additional source of vibratory loads and noise. The amplitude of vibration recorded inside the fuselage of a helicopter is typically about $0.4g$ and the frequency of peak vibration is between 1/rev and N/rev; where, N signifies the number of blades (Konstanzer et al. 2008). Therefore, typical frequencies are in the range of 2–8 Hz depending on the size of the helicopter.

Helicopter noise caused by the rotor aerodynamics typically reaches 80–120 dB in the far-field, and represents the addition of BVI noise, swishing noise and rotational noise. BVI is the main source of the characteristic 'slapping' noise of helicopters. Swishing noise is a broadband noise associated with the random fluctuation of blade lift caused by vortex shedding from the blade ahead, fluctuations in the airflow velocity or flow separation. Rotational noise is produced by the periodic forces applied to the air around the azimuth of the rotor disk. Rotational noise has been associated with vibration of the fuselage (Feszty and Nitzsche 2011; Pagano et al. 2011).

The sources of noise and vibration are strongly coupled, because they are often caused by the same aerodynamic phenomena (such as BVI). Suppression of rotor vibration on helicopters has several advantages, including enhanced forward flight performance, use of lighter aircraft components because parts are exposed to less fatigue loading,

Advanced UAV Aerodynamics, Flight Stability and Control: Novel Concepts, Theory and Applications, First Edition. Edited by Pascual Marqués and Andrea Da Ronch.

Companion website: http://www.wiley.com/go/marques/advanced_UAV_aerodynamics

Figure 11.1 MA THOR Heli VTOL-UAS. Image: Marques Aviation Ltd.

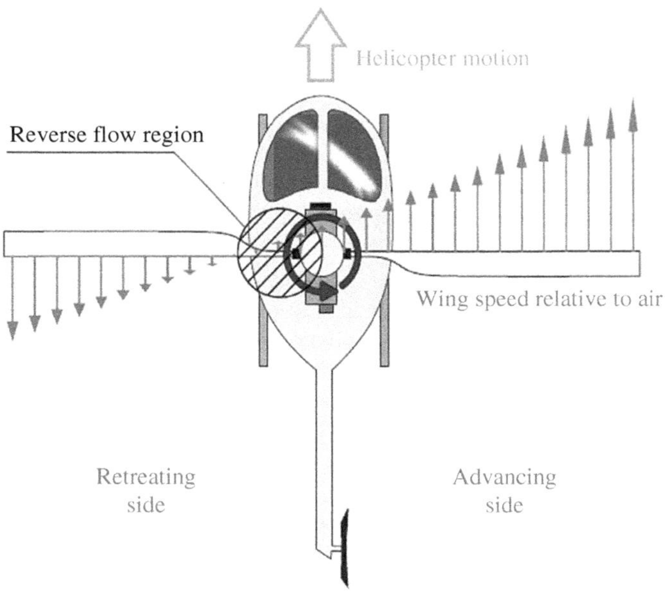

Figure 11.2 Spanwise distribution of relative airflow velocities. Source: Paternoster et al. 2012.

extended component life, and reduced helicopter operational costs (Feszty and Nitzsche 2011). In particular, excessive vibration in the control system limits the maximum forward flight speed before the maximum power of the propulsion unit is reached. Active blade twist attenuates noise and vibration and also improves hover performance because of the better spanwise distribution of aerodynamic loads. There is more uniform downwash in the far wake, leading to lower induced power. However, very low and even negative angles on the tips of rotor blades have the potential to adversely affect forward flight performance and increase vibratory loads, hence the importance of blade twist optimisation (Johnson et al. 2012). Passive devices suppress vibration over a

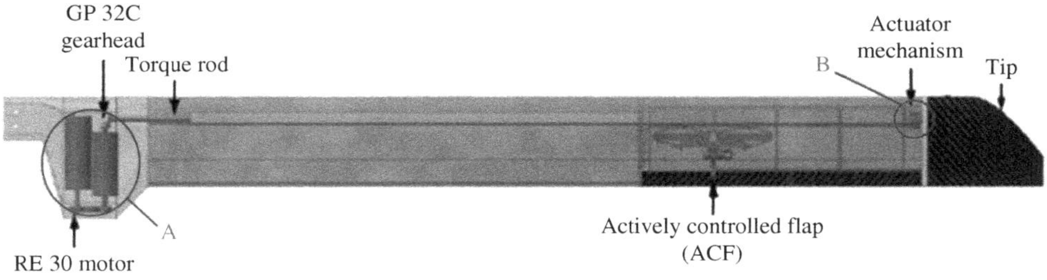

Figure 11.3 Concept of embedding the Maxon A-22 motor and gearbox unit at the root of the SHARCS (smart hybrid active rotor control system) blade. Source: Feszty and Nitzsche 2011.

narrow frequency range, but active blade control systems attain vibration suppression across a broad band of frequencies. In particular, rotor-based active control systems that attenuate vibrations and noise at their source – the rotor blade – are considered superior (Baier and Datashvili 2011; Feszty and Nitzsche 2011; see also Figure 11.3).

There have been considerable efforts in the helicopter community to control the sources of fuselage vibration and rotor noise. Advances in active materials allow the embedding or surface-mounting of multiple lightweight sensors and actuators at selected locations in the rotor blades (Shin et al. 2007). Active materials provide advantages, in terms of weight and power consumption, over the hydraulic systems that have been used previously. Various implementations of embedded actuators have been suggested. The integral twist actuation concept, which uses active materials technology, has shown promising benefits in hover testing with small-scale models (Shin et al. 2007). Advanced active rotor blade technology requires a sensing/actuation system. This can be achieved using piezoelectric transducers integrated within a flexible host structure (Shevtsov et al. 2009). Distributed power actuators capable of modifying the local blade twist enhance a helicopter's flight performance and effectively reduce the noise and vibration levels. Performance benefits include reduction in cruise lift-dependent drag, less rotor power consumption, increased aeromechanical stability, and delay of the onset of stall flutter; thus significantly expanding the flight envelope. Engineering advancements in active rotor blade design include the introduction of high-stroke on-blade piezostack actuators in rotors with trailing-edge flaps. Examples of piezostack actuators are macro-fibre composites (MFCs) and active fibre composites (AFCs), which include a layer of extruded piezoceramic fibres encased in a protective polymer matrix material and interdigitated electrodes that generate electrical fields in the plane of the actuator. However, piezoceramic fibres present technological difficulties in their manufacture, so piezoelectric actuators formed as plate-like patches or layers bonded to the surface of the host structure are used more frequently (Shevtsov et al. 2009). For efficient control of structural vibration, it is important to select the correct type, number and location of piezoelectric patch actuators. This requires a process of optimization. However, the optimization problem in helicopters has a notable constraint in that the actuators must be controlled by a high voltage transmitted from power amplifiers (in other words, piezodrivers), which complicates the design of the hub and rotor.

Design and optimization of adaptive helicopter rotor blades requires an interdisciplinary approach, incorporating structural mechanics, aerodynamics, rotor dynamics, actuator technology, power-electronics and control. The same aerodynamic principles that apply to manned helicopters must also be considered in the design of rotor unmanned aerial vehicles (RUAVs). Often, RUAVs provide an ideal 'low-risk' platform in which to test novel rotor technologies. This chapter presents an evaluation of actuation concepts that underlie active blade twist technology and aspects of integral twist actuation and rotor optimisation. It thus provides a reference in the design of advanced RUAVs. The chapter explores the following themes:

Actuation concepts: the source of vibrations and noise in helicopters, higher harmonic control, individual blade control, the smart spring concept, SMART active flap control, actuation concepts, and applications of active-twist rotor control in RUAVs.

Integral twist actuation: embedded active fibre composite actuators, single crystal piezoelectric fibre composites, the SHARCS project, smart spring systems, active and passive piezoelectric patches, swashplateless helicopter rotors and active servopaddles, progress in rotor-based active control technologies, optimization frameworks for active twist rotor blades, and rotor blade optimization using CFD.

11.2 Actuation Concepts

11.2.1 The Source of Vibration and Noise in Helicopters

Modern helicopters suffer from noise, vibrations, and limited flight envelopes. Improvements in recent years have been due to better rotor aerodynamics, novel planforms and airfoils, advanced tail rotor designs, composite materials, hingeless and bearingless rotors, and passive and active vibration damping (Maucher et al. 2007). Rotor vibrations are more prominent in forward flight. Variations in the spanwise distribution of lift and drag excite the blade's bending modes. This generates alternating rotor hub loads consisting of vertical forces, and lateral and longitudinal mast moments. The vibration frequencies are a multiple of the number of blades and the revolution frequency. Using more rotor blades and a smaller flapping hinge offset reduces vibrations. Deficient blade tracking is also a source of vibrations (Maucher et al. 2007).

The acoustic signal of the helicopter rotor consists of broadband and impulsive noise. The broadband noise originates from boundary-layer effects on the rotor blades (known as self-noise) at high frequencies, and blade–wake interaction (BWI) and BVI at medium frequencies. BWI persists over a large range of flight conditions. The low-frequency impulsive sources are dominant in high-speed forward flight and in descent. In descent flight and low-speed manoeuvres, the rotor blades encounter tip vortices emanating from preceding blades. BVI causes loud slapping noises. In high-speed forward flight, the airflow at the blade tip becomes transonic, producing shock waves and loud noise directed in the flight direction; this is known as high-speed impulsive noise (HSI). Harmonic and higher harmonic loading noise are additional low-frequency sources of noise originating at the main rotor (Maucher et al. 2007).

11.2.2 Higher Harmonic Control

The majority of the research into active control of rotorcraft has attempted to eliminate rotor vibrations and noise problems by addressing the source – the aerodynamic loads – by modifying the periodic aerodynamic loads, such that they no longer produce the detrimental effects (Wilbur and Wilkie 2005). The simplest method is to modify the rotor blade pitch at harmonic frequencies above the rotor rotational frequency. The concept of HHC was introduced in the 1970s and ultimately flight-tested at full scale. Flight tests showed that HHC reduces vibration and noise by changing the pitch at the root of the blades via the swashplate. HHC superimposes the standard blade-pitch variation with a low-amplitude blade pitch at multiples of the rotational frequency (Maucher et al. 2007). In the 1980s, HHC was used in helicopters such as the OH-6A, S-76 and SA 349 Gazelle; the technology was also tested in a wind tunnel (Yu et al. 2002).

11.2.3 Individual Blade Control

Individual blade control (IBC) typically uses hydraulically actuated pitch links to control each rotor blade independently. The advantages of IBC systems over HHC systems include better simultaneous attenuation of vibration and noise, shaft power reduction, and extension of the flight envelope (Maucher et al. 2007). However, the difficulty with IBC arises in providing the necessary hydraulic power in the rotating system (Wilbur and Wilkie 2005). As a result, many active rotor control concepts incorporate on-blade controls, typically trailing-edge flaps or strain-induced blade twist, actuated by piezoelectric systems (Figure 11.4). In the IBC system, actuators are placed on the blade to suppress vibration at the source (Nitzsche et al. 2004). In the past, active IBC has been implemented using discrete and integral actuation. Discrete actuation uses an active actuator embedded in the blade and this activates a trailing-edge servo-flap. However, discrete actuation requires complex displacement amplification devices that must also be robust so that they can operate under the extreme dynamic environment that is characteristic of the helicopter rotor (Nitzsche et al. 2004). Integral actuators are either embedded or bonded to the skin along the blade span, and therefore produce a smooth continuous structural deformation (Nitzsche et al. 2004). Integral actuators require very high voltages to induce sufficient actuation of the AFC elements. Peak-to-peak driving voltages of 3000 V have been employed to generate the required blade twist and suppress rotor vibrations (Kessler et al. 2003).

Successful implementation of IBC systems has been hindered in the past by electromechanical limitations of smart material actuators, in particular piezoceramic actuators. The limited deformation capabilities of piezoceramic actuators necessitates the use of complex displacement amplification mechanisms or the application of very high voltages

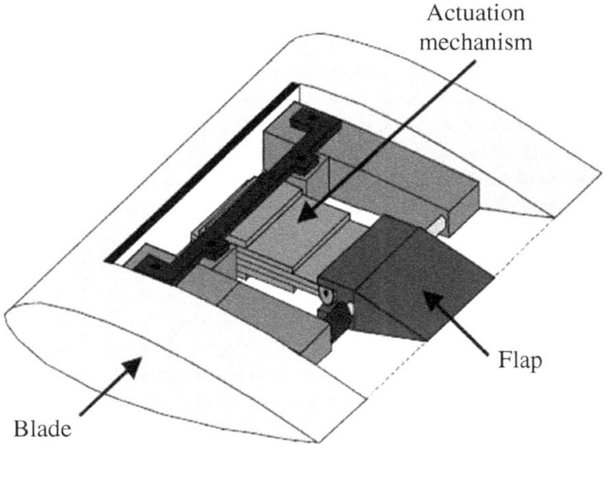

Figure 11.4 Trailing-edge flap and actuation system. Source: Paternoster et al. 2012.

to control rotor vibration. The smart spring concept provides IBC and overcomes the limitations of piezoceramic actuators (Nitzsche et al. 2004).

11.2.4 Smart Spring Concept

Suppression of rotor vibrations in helicopters using IBC can be achieved by altering either the time-varying loads on each blade or the stiffness of the blades according to the variable aerodynamic loads (Nitzsche et al. 2004). The smart spring concept actively alters the blade dynamic stiffness to suppress rotor vibration. The system is particularly effective when positioned near the blade root to alter its boundary conditions. The smart spring concept does not require the transmission of high voltages along the blade span and reduces the likelihood of the electromagnetic interference that was present in previous IBC systems (Nitzsche et al. 2004). Thus, by altering the structural response of the blade, rather than the aerodynamic characteristics, the smart spring embodies a unique approach to helicopter IBC.

11.2.5 SMART Active Flap Control

Straub et al. (2004) developed a full scale SMART system that uses piezoelectric actuated blade flaps. Previous simulations and model-scale wind tunnel tests showed that the SMART system achieves an 80% reduction in vibration, 10-dB noise attenuation for a helicopter passing overhead, and higher aerodynamic performance. Tests in a whirl tower demonstrated the functionality, robustness, and control authority of the active flap system. The effectiveness of the SMART system is due to the success in adapting piezoelectric materials and actuators to rotor blades, and meeting actuator–blade integration requirements. The application of the SMART technology was perceived as a precursor for the elimination of the swashplate (Straub et al. 2004).

11.2.6 Actuation Concepts

The main advantages of active twist are that the aerodynamic profile does not change and there are no moving parts. Grohmann et al. (2006) described different actuation concepts that deform the helicopter rotor blades and control rotor aerodynamics. Their analysis focuses on blade twist and airfoil camber variation that can be implemented using piezoceramic actuation. The direct-twist concept and trailing-edge flaps have also been the focus of attention of previous research (Maucher et al. 2005). Discrete flaps are the most straightforward means to obtain camber variation. Leading-edge flaps increase maximum airfoil lift incurring separation and fluctuations of the pitching moment and are used for dynamic stall alleviation. A method of achieving blade twist involves the so-called servo effect of a trailing edge flap, whereby deflection of a servo flap generates aerodynamic pitching moments that twist the blade. The change in aerodynamic pitching moment twists the rotor blade aeroelastically. The actuation system uses piezoceramic stack actuators. A mechanical amplification of the active stroke is necessary. The servo effect was demonstrated on a BK117 helicopter which was flight tested by Eurocopter in 2005 (Jänker et al. 2006; Roth et al. 2006; see also Figure 11.5). An electronic system comprising a controller unit, power electronics, data acquisition and data and power transmission is required to operate piezoelectrical driven actuators (Maucher et al. 2007).

Figure 11.5 BK117 fitted with servo flaps. Source: Maucher et al. 2007.

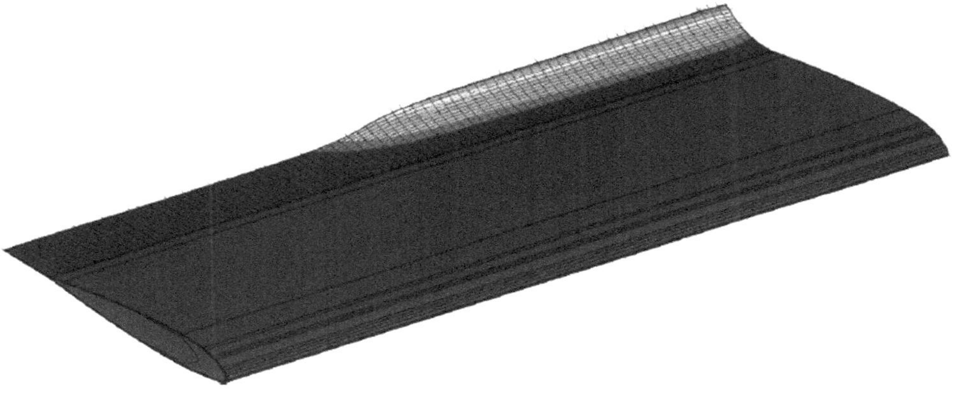

Figure 11.6 ATE technology provides smooth deformation of the helicopter blade (Grohmann et al. 2006)

Blade twist is also achieved by structure-borne twist actuation. This causes a spanwise variation of lift distribution without affecting the aerodynamic pitching moment. Blade twist concepts embed plies of piezoelectric fibres in the rotor blade's skin. The active fibres are arranged in a way that induces strain at ±45° from the blade spanwise axis to generate maximum twisting moment. Control of each individual rotor blade using on-blade actuation by active materials offers advantages in terms of weight, power consumption and bandwidth compared to systems actuating the rotor blade root (Maucher et al. 2007). The active trailing-edge (ATE) concept provides continuous blade deformation by displacing the trailing edge upwards and downwards (Figure 11.6). Similar to trailing-edge flaps, ATE aims to twist the blade aeroelastically using the servo effect; that is, the change in aerodynamic pitching moment twists the rotor blade (Maucher et al. 2007). In an ATE system, an active tab may be fitted to the trailing edge of the helicopter blade. Alternatively, the ATE actuator may be integrated into the airfoil

Figure 11.7 Geometry of the Advanced Technology Rotor. Photo: Marques Aviation Ltd.

(Grohmann et al. 2006). The actuator consists of a multi-morph bender that includes piezoelectric ceramics and glass-fibre reinforced plastics. The ATE concept allows smooth deflection of the airfoil contour in the chordwise direction. An ATE leaves no gaps at the ends of the deflected active surfaces in the spanwise direction, thus preventing parasitic drag and discrete wake vortices.

Optimization of the ATE actuator requires a multidisciplinary approach. Aero-servo-elastic optimization demands the smart aerostructure be flexible for large active deflections, as well as having the minimum mass, low electric power consumption, and substantial aerodynamic effectiveness. The minimum mass of the piezoelectric actuator is important, but there may be a need for an additional balancing weight to keep the centre of mass at 25% of the chord length. Structural interfaces and power amplifiers also add weight to the actuation system. Blade twist is also achieved using an active reflexed skeleton line, which removes variations in pitching moment. The active reflexed skeleton line is realised by deflecting the ATE in one direction at its upstream root while the downstream chord portion is deflected inversely. An active blade tip controls tip vortices, and therefore BVI and transonic effects such as HSI. The ATE actuator is easily integrated into an active blade tip. An ATE does not contain components close to the leading edge and can be fitted to the varying blade chord and small airfoil thickness of parabolic blade tips (Figure 11.7).

Adaptive helicopter rotor concepts such as active blade twist use large assemblies of piezoelectric actuators integrated in composite structures, such as glassfibre-reinforced plastics. Different sets of piezoelectric actuators, such as MFC actuators, are employed to simplify the manufacturing process. MFCs are piezoelectric devices that have structural flexibility and high actuation authority. However, a limitation of piezoelectric actuators is that they require a high voltage input (Bilgen et al. 2011). The modular design of the ATE actuator simplifies maintenance because the actuator is easily detached from the host blade structure for replacement.

Piezoceramics are a suitable actuation technology for morphing helicopter rotor blades, because of their efficient electrical power supply, robustness in hostile environmental conditions, such as when there are high levels of vibration, and high bandwidth. In particular, d33-mode actuation is suitable for rotor blades. Piezoceramics offer potential for applications in the commercial helicopter market because of their low manufacturing cost and easy integration, reliability and maintenance (Grohmann et al. 2006).

For an active rotor blade, the actuation must be integrated into the aerodynamic shape without incurring substantial additional weight. Flap and bending stiffness of an active design should be equivalent to those in a passive rotor blade. Also, the centre of gravity must remain at the 25% chord line. The actuation mechanism of an active blade must withstand the large loads and the strains of the blade structure caused by high centrifugal forces, flapping accelerations, lead-lag bending, flapping, and blade torsions. In addition, modularity of the active devices is desirable, so as to facilitate maintenance and repair in active rotors (Maucher et al. 2007). In their project, Maucher et al. (2007) employed a three-layer (tri-morph) bender made up of piezoceramic actuators and a glassfibre-reinforced plastic core. The ATE had no discrete hinges and the authors presented three ATE options:

Smart tab: The bending actuator is attached to the airfoil's trailing edge. The tab presents a long lever from the blade's neutral axis at 25% chord. Thus, the smart tab achieves a higher aerodynamic moment for a given deflection of the actuator. However, additional mass is necessary to balance the centre of gravity. The smart tab brings a degradation in aerodynamic performance, as the centre of pressure of the profile is no longer at 25% chord. Hence, only small smart tabs may be used (Maucher et al. 2007; see also Figure 11.8).

ATE: The bending actuator is integrated within the airfoil's trailing edge and does not change its aerodynamic shape. A flexible filler material maintains the shape of the airfoil and ensures the continuous transition between the deflected and the passive trailing edge in the spanwise direction. This continuity reduces parasitic drag and discrete wake vortices of the deflected ATE (Maucher et al. 2007). The ATE achieves

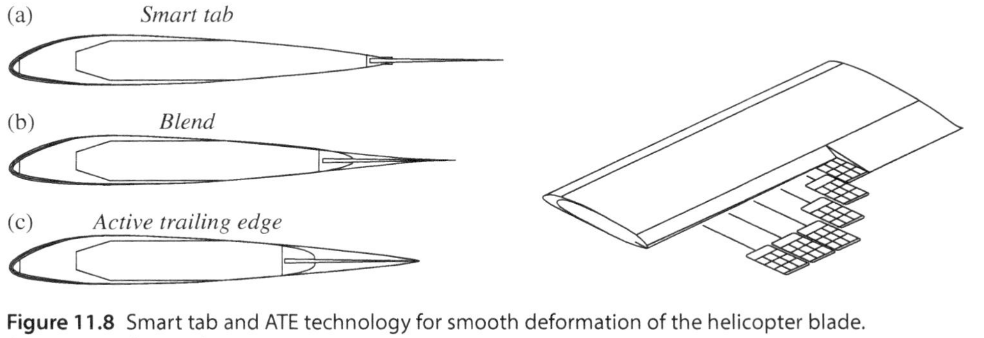

Figure 11.8 Smart tab and ATE technology for smooth deformation of the helicopter blade. Source: Maucher et al. 2007.

high authority. Its smaller leverage to the neutral axis of the blade requires less additional mass to equilibrate the centre of mass. Its stiffer design results in higher resonance frequencies than the smart tab. Furthermore, the ATE does not contain components close to the leading edge and permits the design of an active blade tip. In fact, the ATE can be combined with varying blade chord and small airfoil thickness at the blade tip.

Blend: A blend of the two concepts.

11.2.7 Applications of Active-twist Rotor Control in RUAVs

RUAVs require the application of active-twist rotor control technology (Wilbur and Wilkie 2005). Active-twist rotor control systems use piezoelectric fibre composite actuators and achieve vibration and noise reduction, enhancements in rotor performance, active blade tracking, and stability augmentation. As in piloted helicopters, the rotor system of unmanned helicopters encounters a highly variable aerodynamic environment (Figure 11.9), which induces high vibratory loads, noise, poor performance, instabilities, difficulty in maintaining rotor blade track, and limitations on load capacity and forward flight speed. Therefore, active twist may be applied to RUAVs that incorporate a conventional helicopter main-rotor/tail-rotor configuration. Examples of these vehicles include the A-160 Hummingbird, RQ-8A Fire Scout, Vigilante, Bell HV-911 Eagle Eye, and MA THOR Heli.

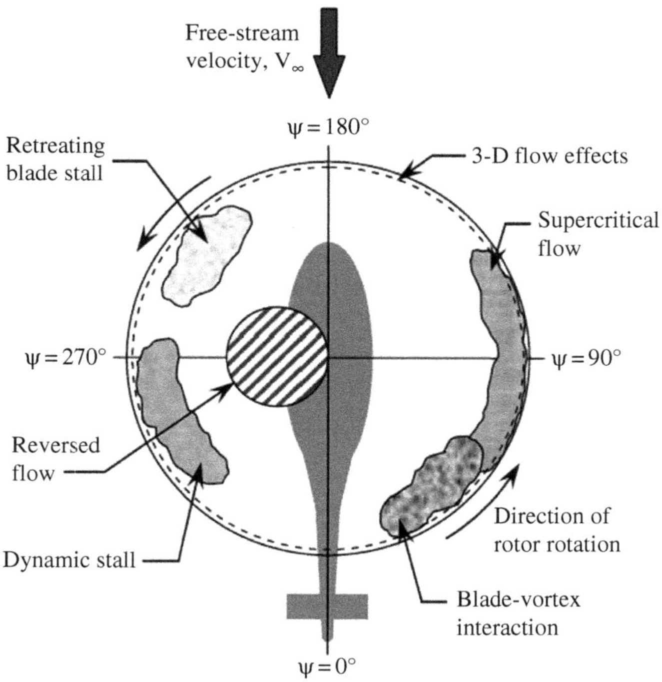

Figure 11.9 The complex aerodynamic environment in forward flight in which the RUAV's rotor operates. Source: Wilbur and Wilkie 2005.

11.3 Integral Twist Actuation

11.3.1 Embedded Active Fibre Composite Actuators

The integral twist actuation concept offers several potential benefits: there is redundancy in operation, it does not increase the profile drag of the blade (something which is characteristic of discrete flap concepts), and the actuators are embedded in the composite construction, therefore becoming part of the load-bearing structure. Therefore, the active blade becomes a truly integrated multifunctional structure, and this facilitates their construction and assembly.

Shin et al. (2007) examined the integral twist deformation of blades and the aeroelastic behaviour of blades equipped with active fibre composite actuators embedded in them. Rotor performance was evaluated in different flight conditions. The aeroelastic model utilised in the analysis consists of a three-dimensional electroelastic beam with geometrical-exactness, coupled with finite-state dynamic inflow aerodynamics (Shin et al. 2007). A system identification method estimates the harmonic transfer function of the rotor. The linear time-periodic (LTP) and linear time-invariant (LTI) components of the transfer functions are extracted. The helicopter rotor becomes an LTP system during forward flight due to a periodicity in aerodynamics, so the LTP components exist and may impact the closed-loop response. Shin et al. (2007) designed a vibration-minimizing controller that implements a classical disturbance rejection algorithm with some modifications. The closed-loop controller was numerically simulated and the capability for hub-vibratory load reduction was demonstrated.

11.3.2 Single Crystal Piezoelectric Fiber Composites

Park and Shin (2007) presented the design of a novel integral twist-actuated blade that uses single-crystal piezoelectric fibre composites. These composites efficiently reduce the vibration and acoustic signal of helicopters. Previously, the active twist rotor (ATR) blade had been investigated by Park and Shin (2007). The ATR blade utilizes piezoelectric fibre composite actuators, such as AFC or MFC, that are embedded within the composite blade structure and cause direct twisting deformation of the blade. The ATR blade could effectively reduce vibration and noise in rotorcraft, but a high input voltage of about 4000 V_{pp} was required. Nonetheless, single-crystal piezoelectric materials can produce strain levels five times larger than those of conventional piezoceramics. Park and Shin (2007) suggested the development of a new ATR blade that incorporates single crystal MFCs. The new ATR concept is known as the Advanced Active Twist Rotor (AATR) blade. The AATR blade can reduce the vibration and acoustic signals of the helicopter more efficiently. However, much lower input voltages and active region lengths are used than in previous ATR blade concepts. Design and analysis of the blade structure consisted of a two-dimensional cross-section analysis and a one-dimensional beam analysis. The cross-section modelled as a two-cell composite beam is shown in Figure 11.10.

The sectional properties obtained in terms of the length of the active region are shown in Figure 11.11. As the active region length is increased, the flapwise-bending stiffness decreases considerably, but the torsional and chordwise-bending stiffness remain unaltered (Park and Shin 2007).

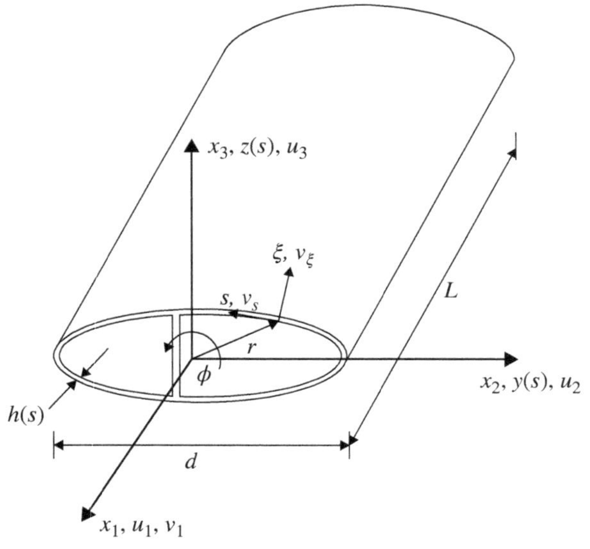

Figure 11.10 Two-cell, thin-walled beam used for the two-dimensional cross-section analysis. Source: Park and Shin 2007.

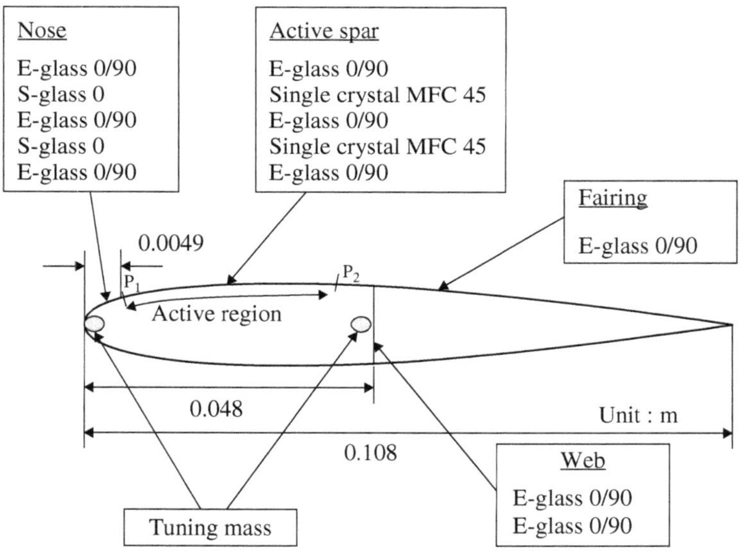

Figure 11.11 Cross-section of the AATR blade (NACA0012 airfoil). Source: Park and Shin 2007.

Figure 11.12 shows the dynamic frequency response of the AATR blade under the aerodynamic load in hover. At 5/rev excitation frequency, the AATR blade achieves a tip twist actuation amplitude of 4° with 75% lower input voltage and 20% smaller active region than an ATR blade. The AATR blade more effectively attenuates helicopter vibration (Park and Shin 2007).

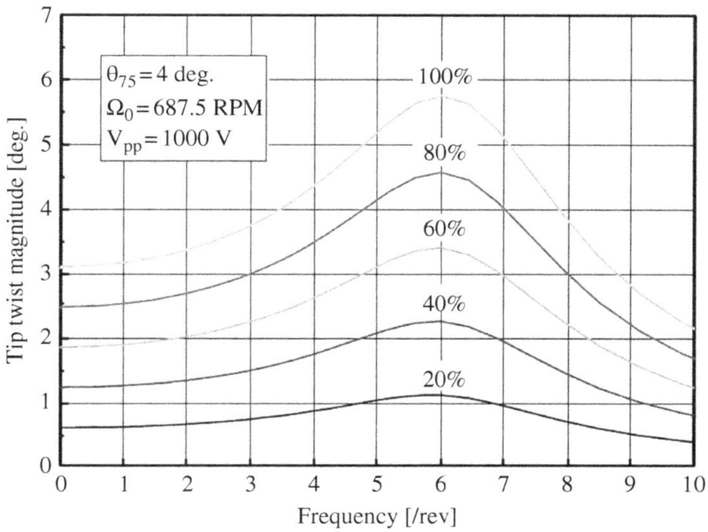

Figure 11.12 Frequency responses in hover. Source: Park and Shin 2007.

Park et al. (2008) later identified various issues regarding the application of AATR technology in full-scale rotors. Single-crystal piezoelectric actuators are difficult to manufacture in large sizes. Also, severe brittleness in the material needs examining. Park et al. (2008) proposed a new AATR, the AATR-II blade, which has an improved twist actuation performance that reduces vibratory loads in forward flight using much lower input voltages. The AATR-II blade requires only 10% the input voltage of the ATR design.

11.3.3 The SHARCS Project

The appearance of vibration and noise is strongly coupled in helicopter; when one is reduced, the other one is amplified. It is therefore desirable to introduce simultaneous reductions of vibration and noise using multiple or hybrid active control devices in each blade. Mander et al. (2008) described the design of the Active Pitch Link (APL) device developed for the SHARCS project. This project demonstrates simultaneous reduction of vibration and noise in helicopters using a hybrid system that integrates three independent active control systems on the helicopter blade: APL, ACF and ACT. The APL and ACF systems primarily reduce vibration, whereas ACT reduces noise.

The project involved the design and construction of a four-bladed scaled rotor of 1.096 m radius, and the testing of the system in a wind tunnel. Static tests showed that the system stiffness could be varied continuously between two extremes to attain vibration and noise suppression. The final design fulfilled the three main design criteria: observability, controllability and failsafe design. The last of these is important because it ensures that when the actuator fails, the APL retains the function of a conventional pitch link. The design is easily scalable to larger sizes, thus providing opportunities for flight testing in full-scale helicopters. Feszty et al. (2009) and Feszty and Nitzsche (2011) have demonstrated the capabilities of the SHARCS hybrid control

Figure 11.13 Active pitch link installed in the Carleton whirl tower facility. Souce: Feszty and Nitzsche 2011.

concept using whirl tower testing (Figure 11.13). The hybrid control combines a flow control device (such as ACF or ATR), and a structural or stiffness system on the helicopter blade. The structural control system uses an APL that dynamically controls the torsional stiffness of the blade.

11.3.4 Smart Spring System

In helicopter rotors composed of identical blades, the rotor hub cancels the majority of the vibratory loads except for the harmonic multiples of the number of blades. Control of the blade aeroelastic response has been a matter of research interest in the helicopter industry. Smart materials have been studied extensively. Smart materials have a limited stroke, but they are able to generate large forces. The limitations of smart materials are overcome with the use of smart springs, which indirectly control vibration of the rotor (Oxley et al. 2009). The smart spring system is particularly effective in altering the boundary conditions of a structure and it is incorporated at the root of the blade. The smart spring functions as a mechanical filter that attenuates the higher harmonic vibratory loads caused by the unsteady aerodynamic loads on the blade. The technology is best suited for integration with an IBC system and another flow-control device, such as a servo flap or an ACF-embedded blade (Nitzsche et al. 2005). It has been shown that small changes in the nominal stiffness of the blade root can attenuate vibratory loads with up to 90% efficiency. This is because modifying the boundary conditions usually significantly influences the vibration characteristics of a mechanical system. The forces transmitted through the pitch link with the smart spring were 44% of the original value at 4/rev and 62% of the original value at 8/rev. These are modest gains compared with other active vibration reduction mechanisms used in helicopters (Oxley et al. 2009).

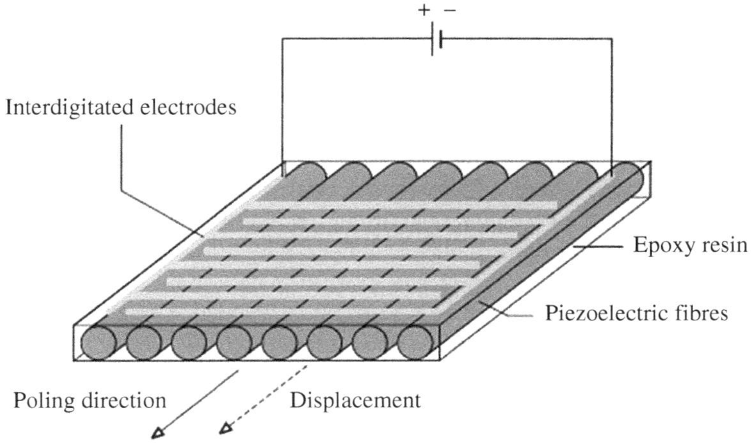

Figure 11.14 Components of an AFC actuator. Source: Paternoster et al. 2012.

11.3.5 Active and Passive Piezoelectric Patches

Shevtsov et al. (2009) compared the efficiency of rotor-blade vibration suppression by active (controlled) and passive (shunted by electrical circuit) piezoelectric patches. For reasonable load capacitance, inductance and resistance, effective suppression of oscillations was achieved at frequencies above 1 kHz only. Active rotor-blade technology requires a sensing–actuation system. Piezoelectric transducers disposed on a controlled flexible host structure can reduce noise and vibration levels, reduce lift-dependent drag, minimize rotor power consumption, improve aeromechanical stability, and delay the onset of stall flutter. MFCs or AFCs consist of a layer of extruded piezoceramic fibres encased in a protective polymer matrix (Figure 11.14). Interdigitated electrodes generate electrical fields in the plane of the actuator. Optimal placement of an optimal number of piezoelectric actuators is essential for the successful structural control of the blade.

Passive piezodamping concepts offer a low-cost solution and do not require complicated electronics. The shunt circuit does not require an external power source, is small, and maintains its integrity under centrifugal loading. Passive shunt damping consists of connecting a passive electrical circuit to the electrode terminals of a piezoceramic patch attached to the host structure. Considering a shunted harmonically excited piezoelectric plate, the passive mode can be effective for vibration suppression in high natural modes only. Shevtsov et al. (2009) concluded that active vibration damping is more efficient for harmonic excitation. In contrast, the passive mode is suitable for the suppression of large-spectrum oscillations caused by impulse loads.

11.3.6 Swashplateless Helicopter Rotors and Active Servopaddles

Leishman (2006) estimated that the swashplate, hub and shaft account for 35% of the entire helicopter parasitic drag. Therefore, eliminating the swashplate and related hydraulic actuation and replacing them with control inputs directly on the blade should yield significant performance improvements. Other advantages of the swashplateless rotor include decreased empty weight of the helicopter, parasitic drag, number of mechanical parts, maintenance requirements, fuel consumption, and acquisition and operating costs (Brindejonc 2009). Stringent volumetric constraints in the rotor blade, and the high centrifugal and vibration loads experienced near the tip of the blade make elusive the practical implementation of on-blade aerodynamic controls. Brindejonc (2009) reported the design and testing of a new actively controlled Hiller-type servopaddle to achieve helicopter rotor control. The rotor is swashplateless and the blades are coupled to a servopaddle fitted with a piezo-electrically actuated aileron positioned behind the trailing edge of the paddle. The aileron is deflected by the actuator, generating lift and a variation of the paddle pitch moment and lift. The system consists of a blade, paddle and aileron assembly. The active servopaddle is capable of generating both cyclic and collective inputs. A system of this type reduces mechanical complexity, parasitic drag and weight. This, in turn, reduces the fuel consumption of the aircraft. The evaluation of this novel concept for a full-scale production helicopter requires substantial analytical development and systematic testing at different scales. An appropriate controller incorporated into the flight control system must be used for actuating the paddles.

11.3.7 Progress in Rotor-based Active Control Technologies

There are four types of rotor-based active control technologies: ACF, ATR, ACT and APL. The ACF, ATR and ACT technologies are classified as blade-based mechanisms and APL is a hub-based device (Feszty and Nitzsche 2011; see also Figure 11.15).

ACF represents advanced technology. It is a blade-based active control technology that has been flight tested in several research programs, such as the Eurocopter in 2005 (Roth et al. 2006). The ACF can usually be optimized to reduce either vibration or noise, at the expense of the other (Straub et al. 2004). ATR also attenuates noise and vibratory

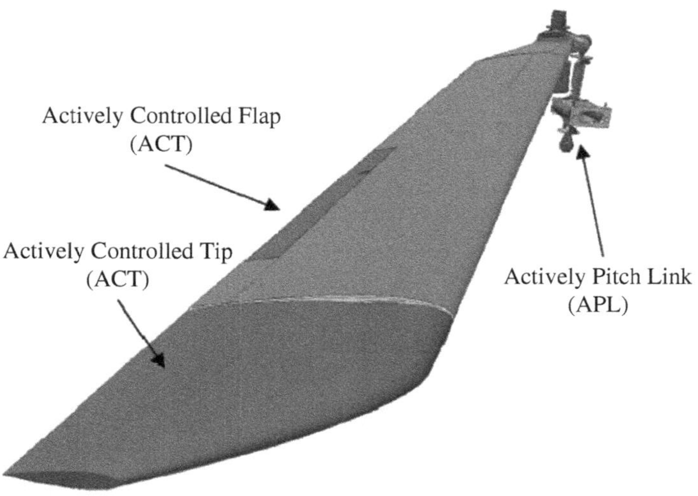

Figure 11.15 SHARCS hybrid concept with four active blade control systems. Image: Feszty and Nitzsche 2011.

loads, but it requires much higher voltages and consumes more power. The absence of moving components in ATR makes it more reliable and suitable for harsh environments, such as those with dust or snow. The ACT system modifies the anhedral of the blade tip and it is activated by the pilot some 30 s before entering a particular flight regime. The principle of operation of ACT consists of reduction of BVI noise and vibration, which is achieved by displacing the helical tip vortex and thus controlling the BVI miss-distance, or the blade-to-tip-vortex distance. The APL system changes the blade pitch angle and hence the aerodynamic loads (Feszty and Nitzsche 2011).

11.3.8 Optimization Framework for Active Twist Rotor Blades

Optimization of ATR blades that use embedded anisotropic piezocomposite actuators was conducted by Mok (2010). Optimization of active twist blades is a complex task that involves a highly nonlinear design space with tightly coupled design variables. The optimization framework for blade cross-sectional design combines different analytical components including cross-sectional analysis (UM/VABS), an automated mesh generator, a beam solver (DYMORE), a three-dimensional local strain recovery module, and a gradient based optimizer within MATLAB. In the mathematical optimization problem, the static twist actuation performance of a blade is maximized within a series of blade constraints: locations of the centre of gravity and elastic axis, blade mass per unit span, fundamental rotating blade frequencies, and the blade strength and strain fields. Resolutions to the limitations of the optimisation process include mesh overlapping, element distortion, trailing-edge tab modelling, electrode modelling, foam implementation of the mesh generator, and sensitivity of the optimization scheme. An example of the effectiveness of this process is its application to a scale model of the AH-64D Apache blade, which demonstrated the capability of this framework to explore the nonlinear design space of complex planforms. The optimization framework is an effective tool to design high-authority active twist blades that reduce helicopter vibrations (Mok 2010).

Active materials, commonly called smart materials, have properties that change by an external input, including stress, temperature, pH (acidity), electric or magnetic fields, and moisture. The deformation of shape-memory alloys is induced and recovered by means of temperature changes. Electrorheological and magnetorheological materials are fluids that undergo a dramatic change in their viscosity in the presence of an electric or magnetic field, respectively. Piezoelectric materials produce a voltage upon application of a mechanical stress, and vice versa. Therefore, a structural property, such as stress, is easily controllable using an electric field. There are two types of piezoelectric materials: piezoceramic (PZT) and piezofilm (PVDF). PZT is used as an actuator due to its high strength. MFCs were developed at NASA Langley. They follow the same principles as AFCs and use piezoelectric fibres and inter-digitated electrodes. Unidirectional piezoceramic fibres are embedded into a thermosetting polymer matrix and sandwiched between Kapton sheets layered with copper inter-digitated electrodes. MFCs have lower fabrication costs than AFCs. Mok (2010) designed an advanced ATR blade based on the AH 64D Apache blade but at 1/6 scale. This had 2.5 times more authority than its predecessor (Figure 11.16). Advanced airfoils, enhanced blade planform and MFC integrated actuator design were introduced to the nonlinear design space using a nested-loop optimization approach.

Figure 11.16 AH-64 Apache helicopter. Image: Petty Officer 3rd Class Shawn Hussong, US Navy.

11.3.9 Rotor Blade Optimization Using CFD

The work of Johnson and Barakos (2011) and Johnson et al. (2012) emphasises the fact that the design of rotor blades is complex and involves several disciplines of engineering, including aerodynamics, structural analysis, dynamics, aeroelasticity and control systems, thus requiring a multidisciplinary approach. Johnson et al. (2012) presented a framework for the optimisation of various design parameters of rotor blades in the forward-flight dynamic condition (unsteady conditions). Optimization methods consisted of CFD combined with meta-models, including artificial neural networks, kriging interpolation, and a non-gradient-based optimiser using genetic algorithms. The optimization approach was also demonstrated in the application of linear twist of rotors in hover (steady case). Rotor blade tip optimisation included the anhedral and sweep of the UH60-A rotor blade in forward flight and a BERP-like rotor in forward flight; constraining hover performance in both cases. The findings of Johnson et al. (2012) suggest optima in agreement with engineering intuition and the metamodels were also validated with high-fidelity CFD data. The CFD database required for the metamodel was extensive. Therefore, the harmonic balance alternative for obtaining the CFD data was used, rather than time-marching, to increase computation efficiency during the optimisation of the BERP-like blade tip. Johnson et al. (2012) successfully combined a metamodel with high-fidelity CFD data to efficiently obtain high-resolution performance using a non-gradient-based method.

11.4 Summary

As in piloted helicopters, the rotor system of RUAVs encounters a highly variable aerodynamic environment. The simplest method of modifying the periodic aerodynamic loads that cause helicopter noise and vibration is to implement HHC, which changes the pitch at the root of the blades via the swashplate. HHC superimposes the standard blade pitch variation with a low amplitude blade pitch at multiples of the rotational frequency. IBC controls each blade independently using hydraulically-actuated pitch links. IBC systems are superior to HHC, however the challenge with IBC arises in providing the necessary hydraulic power. Unlike IBC, the smart spring concept does not require the transmission of high voltage along the blade span. The smart spring alters the structural response of the blade, rather than the aerodynamic characteristics. In the servo effect method, deflection of a servo flap using piezoceramic stack actuators generates aerodynamic pitching moments that twist the blade aeroelastically. The ATE concept provides continuous deformation of the blade trailing edge upwards and downwards, and twists the blade aeroelastically using the servo effect; that is, by exploiting the change in aerodynamic pitching moment of the blade. The ATE actuator incorporates piezoelectric ceramics and glass fibre reinforced plastics and may be integrated into the airfoil. Different sets of piezoelectric actuators are employed to simplify the manufacturing process; such as MFC actuators that offer structural flexibility and high actuation authority. Three methods have been proposed for the integration of actuation into the aerodynamic shape: Smart tab, ATE, and a blend of these two. The helicopter industry has been reluctant to pursue active rotor control systems aggressively due to the complexity, development costs, and inherent risks. However, the risk is reduced by implementing and testing active rotor control technology on RUAVs first. This also allows advanced RUAV concepts to acquire the active blade twist technology.

Advances in active materials allow embedding or surface-mounting multiple lightweight sensors and actuators at selected locations in the rotor blades. Integral twist deformation of blades using active fibre composite actuators embedded in the composite blade offers several benefits, including redundancy in operation, low profile drag, and the actuators becoming part of the load bearing structure. Novel integral twist actuated blades use single crystal piezoelectric fibre composites that efficiently reduce the vibration and acoustic signal of the helicopter. Single crystal piezoelectric materials can produce strain levels five times larger than those of conventional piezoceramics. The AATR blade incorporates single crystal MFC and requires much lower input-voltage and active region length than ATR blade concepts. The SHARCS project demonstrated simultaneous reduction of vibration and noise in helicopters using a hybrid system that integrates three independent active control systems on the helicopter blade: APL, ACF and ACT. APL and ACF primarily reduce vibration, whereas ACT reduces noise. The smart spring acts as a mechanical filter that attenuates the higher harmonic vibratory loads and it is a technology best suited for integration with an IBC system and another flow-control device, such as a servo flap or an ACF-embedded blade. Active vibration damping is efficient for harmonic excitation. In contrast, passive piezoelectric patches are suitable for the suppression of the large spectrum oscillations caused by impulse load. An actively controlled Hiller-type servopaddle achieves helicopter rotor control using a swashplateless configuration in which the blades are coupled to a servopaddle fitted with a piezoelectrically-actuated aileron at the trailing edge of the paddle. There are four types of rotor-based active control technology. The ACF, ATR and ACT technologies are blade-based mechanisms and APL is a hub-based device. Optimization of active twist blades is a complex task that involves a highly non-linear design space with tightly coupled design variables. Rotor blade optimisation is becoming an increasingly important part of the design process as engineers push further the boundaries of helicopter aerodynamic efficiency and performance. Recent optimization work included CFD combined with meta-models, ANNs, kriging interpolation, and a non-gradient based optimiser using GAs. Such technologies and optimisation methods need incorporating for the design of advanced RUAVs.

References

Baier H and Datashvili L (2011). Active and morphing aerospace structures – A synthesis between advanced materials, structures and mechanisms. *International Journal of Aeronautical & Space Science. 12*(3): 225–240.

Bilgen O, Friswell MI, Kochersberger KB and Inman DJ (2011). Surface actuated variable-camber and variable-twist morphing wings using piezocomposites. *52nd AIAA/ASME/ASCE/AHS/ASC Structures, Structural Dynamics and Materials Conference.* 4–7 April. Denver, Colorado.

Brindejonc A (2009). *Study of a swashplateless helicopter rotor with active servopaddles.* Doctoral thesis. L'École Nationale Supérieure d'Arts et Métiers, Paris.

Feszty D and Nitzsche F (2011). Review of active rotor control research in Canada. *International Journal of Aeronautical & Space Science. 12*(2): 93–114.

Feszty D, Nitzsche F, Mander A, Coppotelli G, Vetrano F, Riemenschneider J and Wierach P (2009). Whirl tower demonstrations of the SHARCS hybrid control concept. *65th Annual Forum of the American Helicopter Society.* 27–29th May. Grapevine, Texas.

Grohmann B, Maucher C and Jänker P (2006). Actuation concepts for morphing helicopter rotor blades. *25th International Congress of the Aeronautical Sciences.* 3–8 September. Hamburg, Germany.

Jänker P, Hermle F, Friedl S, Lentner K, Enenkl B and Müller C (2006). Advanced piezoelectric servo flap system for rotor active control. *32nd European Rotorcraft Forum.* Maastricht, The Netherlands.

Johnson CS and Barakos GN (2011) A framework for optimising aspects of rotor blades. *The Aeronautical Journal. 115*(1165): 147–161.

Johnson CS, Woodgate M and Barakos GN (2012). Optimisation of aspects of rotor blades in forward flight. *International Journal of Engineering Systems Modelling and Simulation. 4*(1/2): 79–93.

Kessler C, Fuerst D and Arnold UT (2003). Open loop flight test results and closed loop status of the IBC system on the CH-53G helicopter. *American Helicopter Society 59th Annual Forum.* 6–8 May. Phoenix, Arizona.

Konstanzer P, Enenkl B, Aubourg PA and Cranga P (2008). Recent advances in Eurocopter's passive and active vibration control. *American Helicopter Society 64th International Annual Forum*, Vol. *1*. Montreal, Quebec.

Leishman JG (2006). *Principles of Helicopter Aerodynamics.* Cambridge University Press.

Mander A, Feszty D and Nitzsche F (2008). Active pitch link actuator for impedance control of helicopter vibration. *American Helicopter Society 64th Annual Forum.* 29 April–1 May. Montréal, Canada.

Maucher CK, Grohmann BA and Jänker P (2005). Review of adaptive helicopter rotor blade actuation concepts. *Proceedings of the 9th Adaptronic Congress.* 31 May–1 June. Göttingen, Germany.

Maucher CK, Grohmann BA, Janker P, Altmikus A, Jensen F and Baier H (2007). Actuator design for the active trailing edge of a helicopter rotor blade. *Proceedings of the 33rd European Rotorcraft Forum*. Kazan, Russia.

Mok JW (2010). *Design optimization for active twist rotor blades*. Doctoral thesis. The University of Michigan, Ann Arbor, USA.

Nitzsche F, Zimcik DG, Wickramasinghe VK and Yong C (2004). Control laws for an active tunable vibration absorber designed for rotor blade damping augmentation. *The Aeronautical Journal*. *108*(1079): 35–42.

Nitzsche F, Feszty D, Waechter D, Bianchi E, Voutsinas S, Gennaraetti M, Coppotelli G and Ghiringelli GL (2005). The SHARCS Project: Smart hybrid active rotor control system for noise and vibration attenuation of helicopter rotor blades. *American Helicopter Society*. Paper 52. pp. 1–15.

Oxley G, Nitzsche F and Feszty D (2009). Smart spring control of vibration on helicopter rotor blades. *Journal of Aircraft*. *46*(2): 692–696.

Pagano A, Ameduri S, Cokonaj V, Prachar A, Zachariadis Z and Drikakis D (2011). Helicopter blade morphing strategies aimed at mitigating environmental impact. *Journal of Theoretical and Applied Mechanics*. *49*(4): 1233–1259.

Park JS and Shin SJ (2007). A preliminary design on the second generation integral twist-actuated blade. *International Conference on Computational & Experimental Engineering and Sciences*. *4*(4): 265–269.

Park JS, Shin SJ and Kim DK (2008). Design and vibratory loads reduction analysis of advanced active twist rotor blades incorporating single crystal piezoelectric fiber composites. *International Journal Aeronautical and Space Sciences*. *9*(2): 18–33.

Paternoster A, Loendersloot R, de Boer A and Akkerman R (2012). Smart actuation for helicopter rotor blades. In Berselli G, Vertechy R and Vassura G (eds). *Smart Actuation and Sensing Systems – Recent Advances and Future Challenges*. InTech.

Roth D, Enenkl B and Dieterich O (2006). Active rotor control by flaps for vibration reduction: Full scale demonstrator and first flight test results. *32nd European Rotorcraft Forum*. Maastricht, the Netherlands.

Shevtsov S, Soloviev A, Acopyan V and Samochenko I (2009). Helicopter rotor blade vibration control on the basis of active/passive piezoelectric damping approach. *4th International Conference on Physics and Control*. 1–4 September. Catania, Italy.

Shin S, Cesnik CES and Hall SR (2007). Design and simulation of integral twist control for helicopter vibration reduction. *International Journal of Control, Automation, and Systems*. *5*(1): 24–34.

Straub FK, Kennedy DK, Stemple AD, Anand VR and Birchette TS (2004). Development and whirl tower test of the SMART active flap rotor. *SPIE's International Symposium on Smart Structures and Materials*. 14–18 March. San Diego, California.

Wilbur ML and Wilkie WK (2005). Active-twist rotor control applications for UAVs. *Proceedings of the 24th Army Science Conference*. 29 November–2 December. Orlando, Florida.

Yu YH, Tung C, van der Wall B, Pausder HJ, Burley C, Brooks T, Beaumier P, Delrieux Y, Mercker E and Pengel K (2002). The HART-II test: Rotor wakes and aeroacoustics with higher-harmonic pitch control (HHC) inputs – The Joint German/French/Dutch/US Project. *American Helicopter Society 58th Annual Forum*. 11–13 June. Montréal, Canada.

1.3 Hybrid Aircraft

12

Hybrid Aircraft Aerodynamics and Aerodynamic Design Considerations of Hover-to-Dash Convertible UAVs

Ron Barrett

The University of Kansas, Lawrence, KS, USA

12.1 Why Hover-to-Dash Conversion is Important

The general concept of hovering in place like a helicopter, then dashing at high speeds is not a new one. Although vertical takeoff and landing craft can be traced as far back as the time of da Vinci, truly high performance rotorcraft would not come about until the 1930s, with the record setting flights of the FA-61. Although a plethora of helicopters came into being in the 1940s and early 1950s, forward-flight speeds significantly beyond 100 kts would remain elusive until some unusual aircraft configurations were employed. Because of the drive to takeoff and land vertically and to dash while between these two flight phases, different combinations of powerplants, rotors, propellers, wings and thrusters were tried and tried again and again. Indeed, aircraft like the Fairey Rotodyne was a wonderful example of this chimera-like approach to VTOL-capable, high-speed dashing aircraft [1]. Figure 12.1 shows an artist's rendition of the aircraft in high speed flight. The aircraft was specifically designed for the London city center to Paris city center route without the use of conventional airports. Although the aircraft first flew in 1957, with a maximum speed of nearly 200 mph, and made 350 flights with not a single noise complaint, the primary reason given for program cancellation was indeed related to the noise it generated [2].

12.1.1 Runway Elimination

Clearly, aircraft like the Fairey Rotodyne highlighted the market demand for an aircraft that would operate from heliports but be able to cruise at high speed. The idea of flying from city center to city center appealed and still appeals to many air travelers. Of course, the comparatively high cost, low speed and limited range of helicopters make such flights unattractive to all but the wealthiest of today's travelers. Indeed, the appearance and successful operation of airports like London City and Gatwick and LaGuardia highlight the desire of the traveling public to travel to and from urban cores, but also pose great challenges. Indeed, the real estate occupied by airports like these is among the most valuable in the world.

Advanced UAV Aerodynamics, Flight Stability and Control: Novel Concepts, Theory and Applications, First Edition. Edited by Pascual Marqués and Andrea Da Ronch.

Companion website: http://www.wiley.com/go/marques/advanced_UAV_aerodynamics

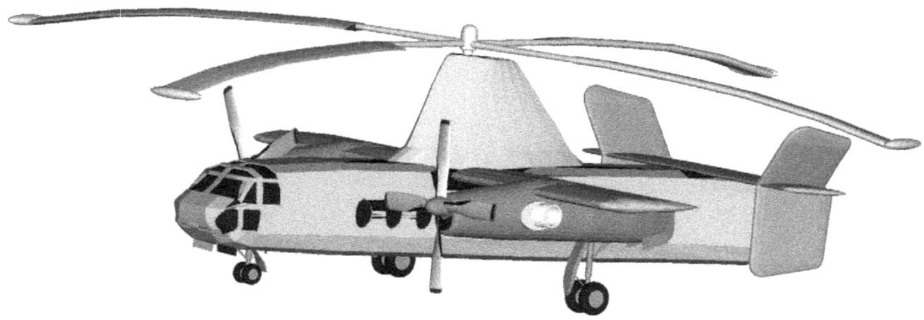

Figure 12.1 Fairey Rotodyne tipjet-turboprop autogyro.

As such, the cost of the land often drives up landing fees. Additionally, the urban core typically presents stringent noise limitations, which leads to curfews and other constraints on flight operations. With all of that said, there are still flights and urban airports continue to operate profitably. Because the cost of the land and noise-restricted airspace are typically two of the greatest burdens borne by such airports, complete elimination of runways and shallow flightpaths would significantly alleviate these burdens. To do this, it is easy to argue that the aircraft operating from such airports must necessarily be capable of vertical or near-vertical flight.

12.1.2 Civil Vertical Flight Operations

The importance of vertical flight operations for many aircraft cannot be overstated. Among the most compelling justifications, as stated above, are simply associated with the cost of real estate. Indeed, in highly congested urban areas, many vertical flight operations are simply conducted from building rooftops. Compounding the challenges of the high cost of urban real estate is the trouble associated with shallow commercial aircraft glideslopes and the associated large swathes of land that encounter high noise levels as aircraft take off and climb. Figure 12.2 highlights the tremendous cost of real-estate for urban heliports and airports. Even with the most advanced "Clean Sky" initiatives envisioned, more than £14.6 billion of real estate is affected by the flight corridor [3].

If one considers a heliport at the same location with the same passenger density, then a comparison of the real estate affected can be made. Figure 12.3 shows that there is a reduction of two orders of magnitude. The area impacted is indeed so small that it lies within the footprint of the terminal of an equivalent urban airport. As the surrounding buildings are not influenced by the noise footprint of a large-scale heliport, noise-based curfews would be eliminated, and flight operations restricted by prevailing winds would be mitigated, if not also eliminated, leading to far more open (and profitable) flight operations. Further, because local communities would be much less significantly impacted, the total number of heliports with similar flight operations could increase dramatically, leading to a large growth in total air passengers.

12.1.3 Military Vertical Flight Operations

While the case for civil vertical flight operations of high-speed aircraft is easy to make given the restrictions associated with the congested urban landscape, military flight operations of VTOL-capable high-speed aircraft have also been shown to be very useful

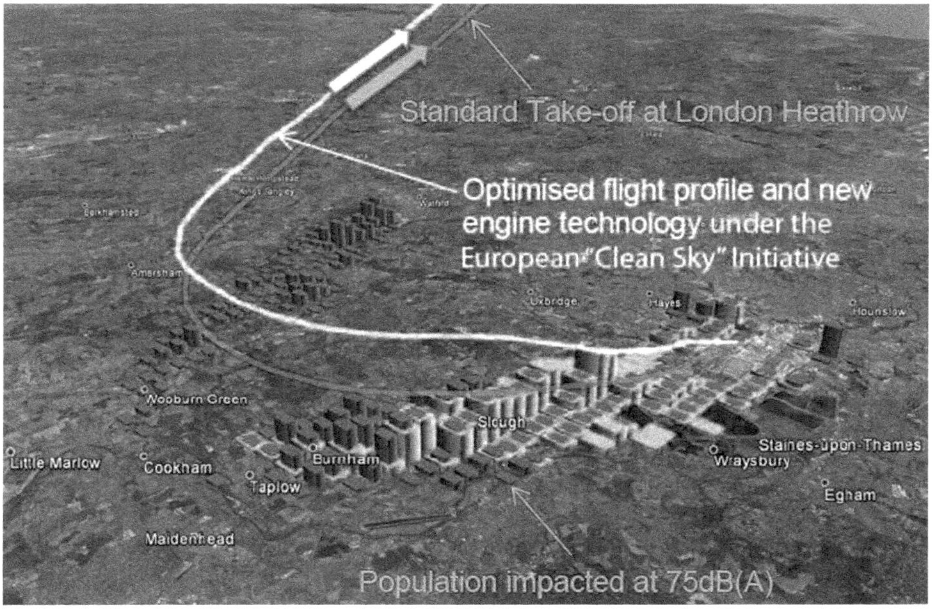

Figure 12.2 Physical noise footprint of real estate impacted at 75 db(a) of representative urban airport (London Heathrow).

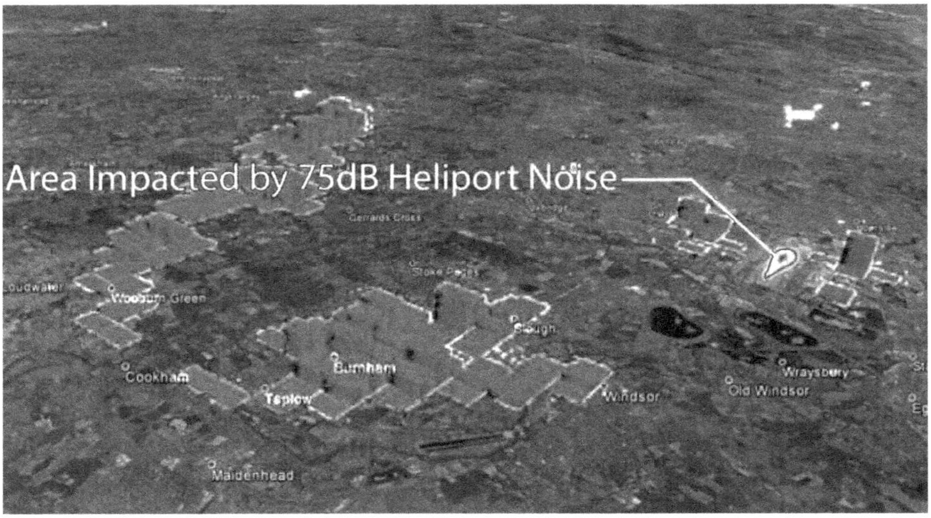

Figure 12.3 Physical noise footprint of real estate impacted at 75 db(A) of representative urban heliport.

and are therefore in high demand. Figure 12.4 shows a hovering missile prototype, the XQ-138 hovering around (at the time), one of the US Army's Future Combat System prototypes. Although capable of more than an hour of hover endurance, the XQ-138 could also pop up and dash at speeds in excess of 350 kts TAS.

Figure 12.4 XQ-138 hovering missile prototype performing battle damage assessment on a US Army Future Combat System prototype; Redstone Arsenal, Alabama, 2002. Image: Barrett.

During these four months of exercises, it was shown that extremely close inspection of damaged armored assets was not only desirable, but absolutely necessary. Figure 12.5 shows XQ-138 flight operations as part of live-fire exercises at Eglin Air Force Base, Florida. Although the T-60 tank endured a direct hit by a Javelin missile, which produced roughly 15 s of smoke from the vehicle, the penetration hole was extremely small and impossible to pick out from any fixed-wing asset. Only upon inspection by an XQ-138 from just 5 m away was it possible to even see the slightly smoking 10-cm hole, which indicated that the vehicle had indeed been killed. From combat experience in theatres such as Iraq and Afghanistan and extensive XQ-138 live-fire testing, it is obvious that overhead observation of targets from hundreds of meters is not sufficient to perform proper battle damage assessment (BDA). Only extremely close inspection of damaged armor can truly assess their ability to engage in further combat. Indeed, the precision approach and landing capabilities of the XQ-138 was so good that one method of post-engagement BDA was simply to land on the vehicle, shut the engine down and listen for signs of life within via microphones in the aircraft feet. Certainly no class of conventional takeoff and landing aircraft could even come close to such definitive BDA capabilities.

Similarly, target identification and close-quarters engagement was shown to be extremely effective with hovering-missile-type aircraft. Figure 12.6 shows the XQ-138 delivering 40-mm submunitions with 30-cm accuracy at the McKenna Military Operations in Urban Terrain (MOUT) range in Fort Benning, Georgia, in 2003. Because the basic flight operations of hovering missiles were concretely proven more than a decade ago, the capabilities of such weapon systems has only increased – from organic weapons with organic reach to collocated weapons with strategic reach.

Figure 12.5 XQ-138 hovering missile prototype performing battle damage assessment on a destroyed soviet-era T-60 tank, Eglin Air Force Base, Hellfire Range, Florida, 2002. Image: Barrett.

Figure 12.6 XQ-138 hovering missile prototype delivering 40-mm submunitions against a target from a standing hover, Ft. Benning McKenna MOUT Range, 2003. Image: Barrett.

12.2 Aircraft Mission Profiles and Sizing Chart Structure

Several of the most important tools that are used by aircraft design engineers are the "Mission Profile" and the "Sizing Chart." These pictorial representations of the way an aircraft is supposed to fly, under what conditions it must fly, and how big its powerplant and lifting systems are have been employed for nearly a century. Foundational texts such as Roskam's *Airplane Design* series [4] employ such tools and reference common industry practice.

Many VTOL aircraft have comparatively nondescript mission profiles, as seen in Figure 12.7 [5]. With vertical ascent and descent, the profiles take the form of a simple box and quite often do not include climb or descent legs with slopes, so as to avoid generation of acoustic footprint sweeps on large swathes of the ground. The mission profile of a common fixed-wing aircraft – a KC-135 – can be seen in Figure 12.8. Clearly, because the aircraft is not capable of VTOL flight, its mission is shaped quite differently to that of a VTOL aircraft. Indeed, many conventional takeoff and landing (CTOL) aircraft have quite similarly shaped mission profiles including (especially) features of pronounced climb and descent stages with an extended range in between. However, if one looks at a convertible aircraft (that is, an aircraft that is capable of "converting" from hover-mode to "dash/airplane" flight modes), then a completely different profile appears.

12.2.1 Challenging Mission Profile: VTOL Launch, Recovery, High-speed Dash

Mission profiles of convertible aircraft often represent some of the most challenging aircraft design problems known in the aerospace industry. Starting with aircraft like the Convair XFY-1 and Lockheed XFV-1 Pogo, vertical takeoff and landing flight phases bracket high-speed dash segments. More precisely, these aircraft hover for takeoff with their noses pointed skyward, then transition by pitching over so that the aircraft deck is mostly horizontal. The aircraft then achieves great speeds during cruise and dash flight modes with this deck level, often greater than a conventional aircraft with the same powerplant, as the wings of convertible aircraft can be much smaller than those of a conventional aircraft. Then as the convertible aircraft come in for landing, the aircraft transitions (or rather, "converts") again from high speed flight to vertical flight by pulling nose up, entering a hover flight state, eventually touching down like a helicopter. Modern convertible aircraft are found most commonly in the UAV world. Among the several UAVs that are capable of full-transitioned flight between VTOL segments and high-speed dash flight phases, the MicroAutonomous Systems (MAS) XQ-138 is one of the best performers on the market today. At the end of their Micro Aerial Vehicle (MAV) program, a flyoff was held at Quantico, Virginia. MAS's engineers produced the Lu-MAV VTOL coleopter MAV. Although the only aircraft that flew without crashing and performed all intended flight functions among the competitors, DARPA chose another team for the following Organic Aerial Vehicle (OAV) program. The MASs team then teamed up with Boeing Future Combat Systems (FCS) supplier ST Aerospace for an RDT&E effort to bring the XQ-138 to life.

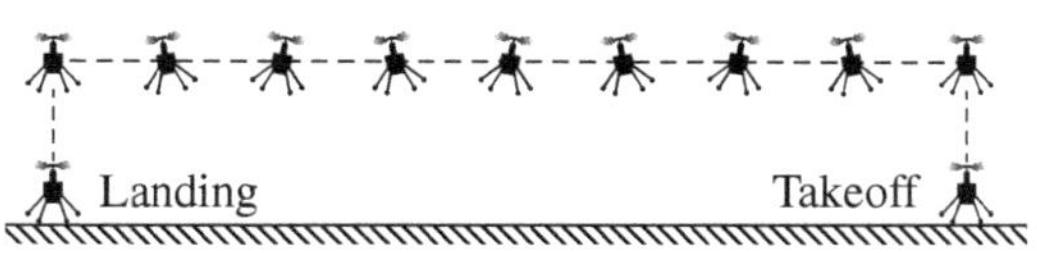

Figure 12.7 Hovering aircraft straight-line range mission profile [5].

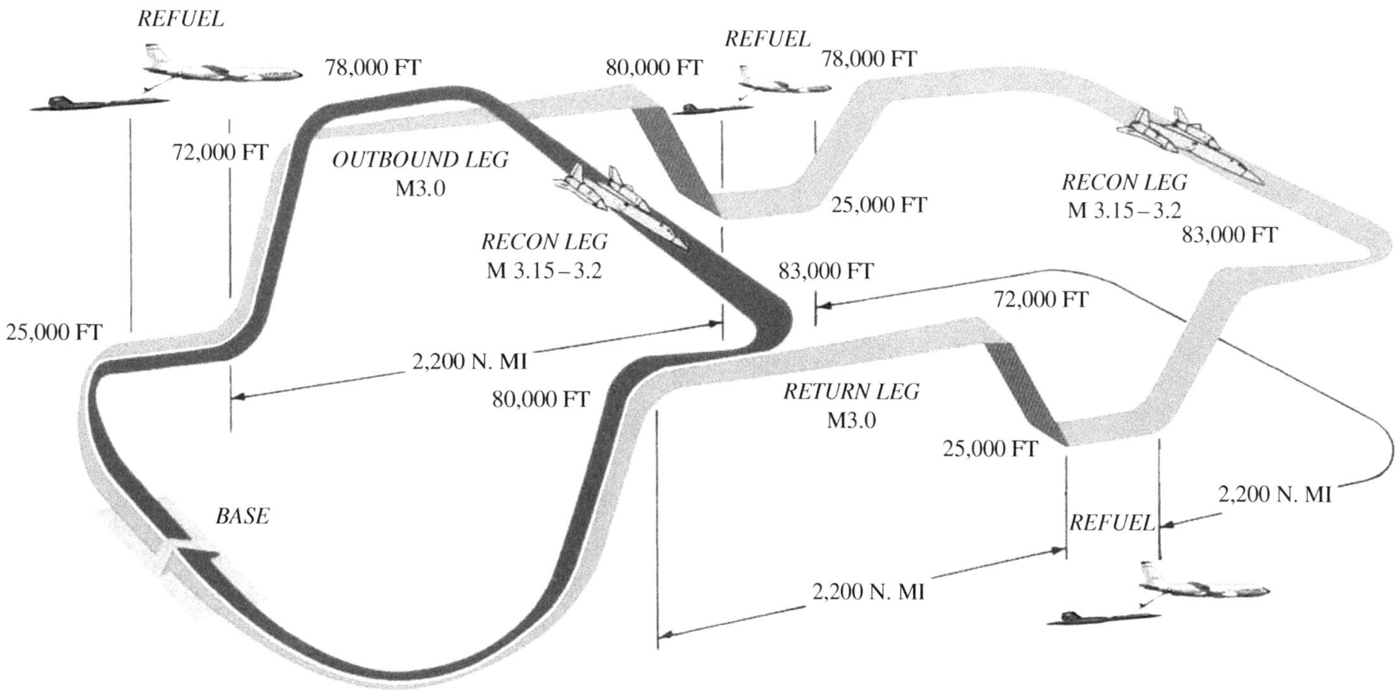

Figure 12.8 Sample military aircraft (KC-135) mission profile.

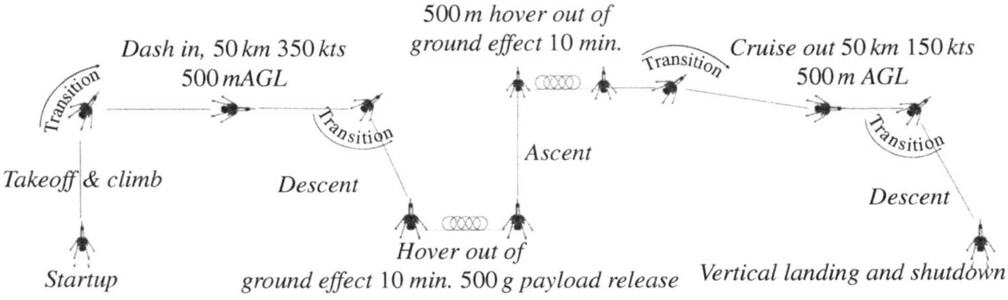

Figure 12.9 Mission profile of the MicroAutonomous Systems XQ-138. Image: Barrett.

The kickoff meeting of the XQ-138 program was held in Singapore in July of 2001 and settled on the extremely challenging mission profile shown in Figure 12.9.

The mission profile above clearly includes performance levels that have never been achieved by subscale aircraft before. In spite of this, the MAS team took the project on and embarked on this most robust program. To achieve this high level of performance, the aircraft was forced to shed total wetted area and profile drag. The problem with eliminating wetted area is that most aircraft depend upon finite-sized wings to provide lift. These wings, while driving up wetted area, allow the aircraft to takeoff and land at low speeds. Because the XQ-138 would not depend upon wing area for generation of lift at low speeds (it would necessarily use the rotor in the ducted fan for lift), the wing could essentially be totally eliminated.

12.2.2 Convertible Aircraft: Elimination of Stall Sizing Lines

A standard tool which is used to lay out aircraft is the so-called "aircraft sizing chart." This chart typically tracks thrust-to-weight ratio (T/W) as a function of wing loading (the amount of weight that can be lofted by a given amount of wing area, W/S). In general, most aircraft are sized by high lift considerations, often associated with takeoff and landing conditions, in addition to cruise, dash, and/or maneuver flight states [6]. The takeoff and landing flight phases necessarily occur at speeds that are so low that flap systems are needed to maintain lift, and yet still large amounts of wing area are often prescribed. Even if one examines a design of a supersonic aircraft [7], then even at relatively moderate wing loadings, the C_{Lmax} considerations of the wing are still critical in laying out the overall size of the wing. Indeed, Figure 12.10 clearly shows that this prize-winning supersonic aircraft design is firmly defined by landing C_{Lmax} considerations. If an aircraft designer could somehow generate lift at low speeds without the use of a wing, then the very shape of the area of possible solutions on the sizing chart would be dramatically expanded.

12.2.3 Changes in the Aircraft Sizing Chart

Because many convertible aircraft like the XQ-138 rely upon lift being directly generated from a propulsion device, all of the traditional stall sizing lines essentially disappear from the aircraft sizing chart. In their stead appears a different set of constraints, which are relatively independent from the traditional constraints [7].

These "transition corridors," rather than being simple functions of wing and power loading, are often mapped out by using wind tunnel and/or flight test data. One of the main reasons why such transition corridors are developed from primarily experimental methods is because the actual aerodynamics of the situation are so complicated that current computational techniques are unable to properly capture the actual behavior of the aircraft. Given the departure of hard and fast stall lines, the design point shown on Figure 12.10 moves dramatically to the right, typically against the stall corners of the transition corridor, which typically increases effective wing loading by factors of three or more. With the considerable increase in wing loading comes a corresponding reduction in wetted area, which in turn allows for much higher speeds throughout all operational altitudes.

12.3 Convertible Coleopter Design, Wind Tunnel and Flight Testing

Coleopters are not new concepts by any stretch of the imagination. Indeed, the "heyday" of inhabited coleopters occurred in the 1950s, with the SNECMA Coléoptère and the various incarnations of the Hiller VZ-1 (Figure 12.11) [8]. The basic definition of a coleopter is an aircraft that uses a ducted propulsor, such as a single-ducted fan, as the primary mechanism for both lift and propulsion in all flight modes. Although there have been a wide range of ducted fan aircraft over the years, comparatively few have successfully flown, with even fewer making full-transitioned flight and only one having ever been patented and placed into high-rate serial production.

12.3.1 The Roots: Heinkel Wespe and Lerchle

Some of the earliest coleopters were designed during World War II as a series of concepts to skirt problems associated with allied bombing of German runways. Indeed, the Heinkel Lerche (Lark) and Wespe (Wasp) were to be powered by counter-rotating ducted rotors, takeoff in vertical flight modes, transition, intercept aircraft at high speeds, then return for a vertical landing. The earliest incarnation of the German coleopters was the Heinkel Wespe. Designed at the Vienna branch of the Heinkel Aircraft Company, the tail-sitting aircraft was touted as being able to takeoff and land without the need for runways [9]. This capability, of course was of great interest to the Luftwaffe, which was dealing with ever-increasing cratering of their runways and destruction of their cities and military complexes. As the Heinkel designers worked in late 1944 to pitch the Wespe designs to the Luftwaffe, objections were raised about the lack of a wing and the inherent disorientation that pilots would experience as they transitioned from conventional aircraft to an axisymmetric aircraft. At the same time, concerns about the dreaded transition flight phase induced a change of design from a pure ducted-fan coleopter to one that sported stub wings. This can be seen in the Lerche design shown in Figure 12.12. Fortunately, neither aircraft made it off the drawing boards and the war in Europe concluded long before either could get off a single shot. However, the line of designs continued with punctuated success. Indeed, the postwar Heinkel Aircraft Company even proposed a tail-sitting aircraft design as late as 1956.

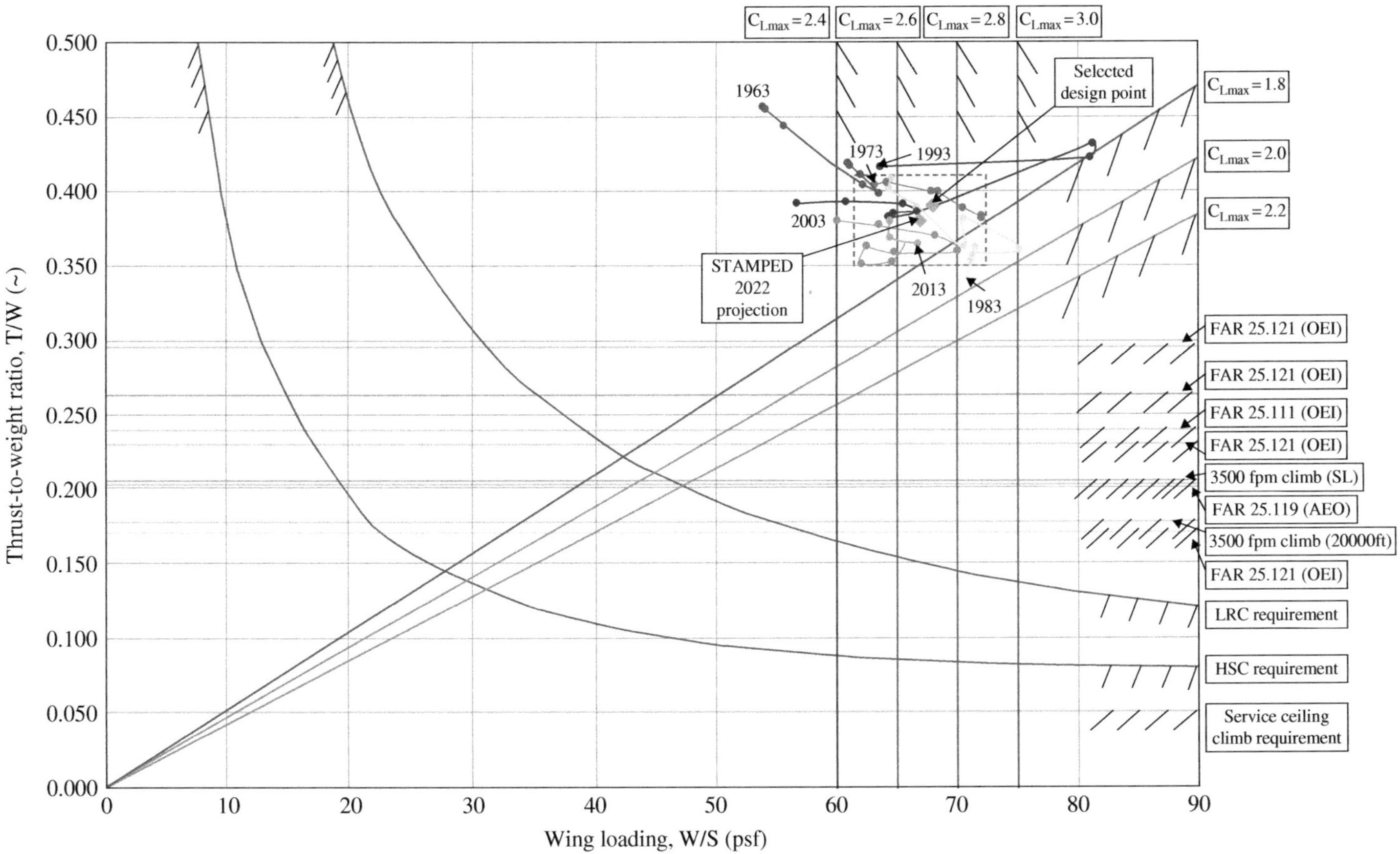

Figure 12.10 Example sizing chart for a high-speed aircraft.

Figure 12.11 Hiller VZ-1 Pawnee.

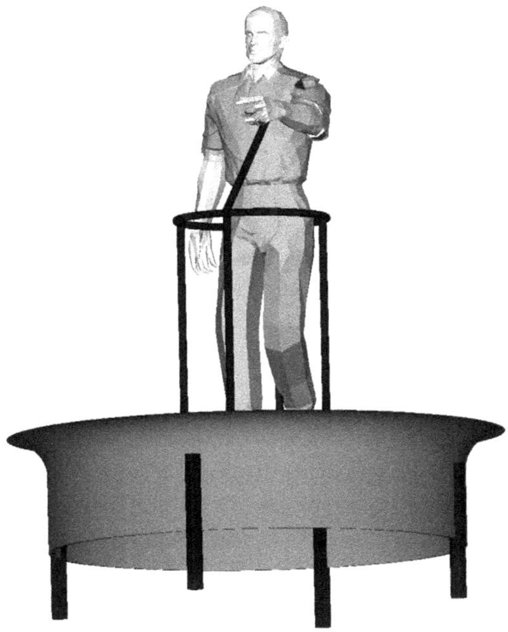

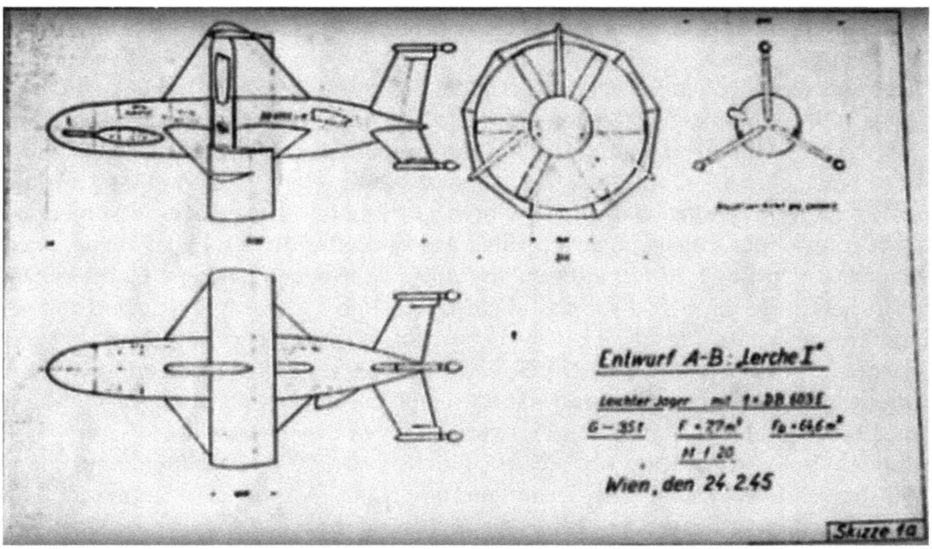

Figure 12.12 Kurt Reiniger's Lerche concept 1945 [9].

12.3.2 Troubles with Coléoptère and Pogos: Pitchback Instabilities

Throughout the 1950s a wide range of VTOL/VSTOL/STOVL aircraft concepts were tried, at a breakneck pace. For the aerospace historian, this rich "Wild West" period of aircraft design had never really been seen in peacetime, before or since. Although exhilarating in terms of pace of innovation, many corners in analysis and logic of the concepts of operations were clearly cut. The SNECMA Coléoptère represented one of the fastest coleopters to make it to production. As with all of the earlier designs, it was designed to takeoff and land from a very confined space. Lockheed, with their XFV-1, and Convair, with their XFY-1, were also exploring the possibility of vertical take-off aircraft that could then transition via a pitch-over maneuver and fly out at extremely high speeds.

While these programs each led to full-scale aircraft prototyping and flight testing, several profound problems surfaced. The first and most obvious of these involved the human factors. As pilots climbed aboard these aircraft, their downward vision was obscured. This obscuration by objects like "the pilot's nose" meant often that only one eye at a time could accurately see the ground approaching. Accordingly, the pilots lost parallax vision, which in turn compromised depth perception. Given that the delicate tail fins were just in front of comparatively light landing gear, the risk of empennage damage during takeoff and landing was pronounced and often led to damaged aircraft components.

Compounding the human factors, troubles associated with near-ground flight operations also included the blanking of the control surfaces due to the presence of a large ground-effect-induced recirculation donut. This aerodynamic structure, which necessarily formed as the aircraft approached the ground plane, would reduce the dynamic pressure across the empennage to nearly zero. As a result, just when the pilots needed the most amount of control authority, they had the least. Of course, given ever-more robust landing gear, designers could simply have pilots "zoom" for a takeoff and "plant it" for a landing, but out-of-ground-effect (OGE) freeflight also posed great challenges.

These nontrivial challenges were observed first in the Hiller VZ-1 Pawnee flight tests and later in the punctuated coleopter designs, which have regularly come around every 20–30 years after those early aircraft. To understand these OGE problems, one simply needs to reach back a bit into time, to a proposal submitted to DARPA in 2001. At the time, two main competing concepts were being pitched to the Federal Government by a pair of contractors. One contractor, Allied Aerospace Industries, had no experience with vertical-flight ducted-fan aircraft; the other contractor, Lutronix Corp. of Del Mar, CA had more than 1800 successful coleopter flights under its belt over the preceding three years. As a result of such extensive knowledge of how coleopters behave, the Lutronix Team explained to DARPA how coleopters suffer tremendously from pitchback instabilities and necessary excessive control force application during these atmospheric conditions. Figure 12.13 is extracted from their proposal.

Although the Lutronix team worked very hard to warn DARPA of impending doom should they choose to fund such a configuration, DARPA sponsored it anyway. The most public demonstration of this catastrophic pitchback instability occurred in September of 2003 as the resulting iSTAR vehicle was hit by light crosswinds and proceeded to catastrophically crash in front of an assembled crowd of military commanders and press at Fort Eustis, Virginia. As Gary Downs, the director of Allied UAV Systems said following the catastrophic crash, "We didn't have the control authority needed for the prevailing conditions." [11]

The assessment of the crash included an examination of the forward-flight capabilities. Downs explained the "newly discovered" pitchback instabilities further to the press: "Unfortunately, this effect limited the forward speed to a mere 26 kph (16 mph)

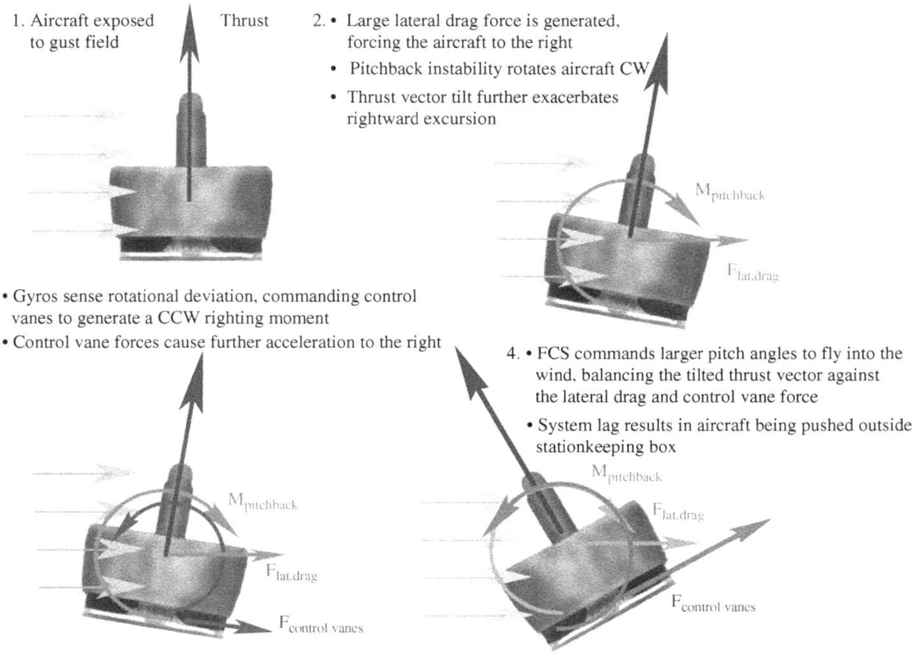

Figure 12.13 Fundamentals of Coleopter pitchback instability [10, 11].

and caused erratic handling in windy conditions." So although the effect was well understood, predicted and reported to funding agencies long before the first flight test, such designs still come around at regular intervals and still face the same issues, which are generated by the fundamental physics of the configuration.

12.3.3 Control Power – The Necessity for Proper Control Effector Coupling

Although conventional coleopters like the Coléoptère, VZ-1 and iSTAR all suffer from pitchback issues and lack of control authority, not all coleopter designs are inhibited by this characteristic. Curiously enough, Reiniger and Schulz of the Heinkel Design bureau both recognized this issue all the way back in 1944 and laid out designs that specifically counter this tendency. This was done by placing the aerodynamic center of the tail of the Heinkel Wespe (Figure 12.14), several mean geometric wing chords downstream from the aircraft center of gravity, and it was seen as a critical design feature. Only is this way was it possible to maintain acceptable pitch and yaw control authority. Given that as of 1944 the Heinkel Design Bureau had extensive experience with real, inhabited, high performance aircraft design, they understood exactly how to lay such things out.

While the Heinkel design engineers were far from clairvoyant, they understood aircraft design and fundamental physics very well. The aircraft designers of Lutronix Corporation of Del Mar, CA called upon these early historical lessons as they laid out the designs of the Lutronix LuMAV family of aircraft, starting in 1994. Although these aircraft would become the DoD's first MAVs, they were based on some very old design

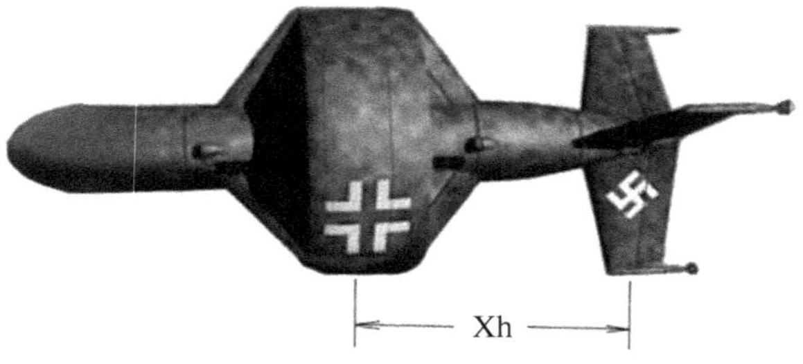

Figure 12.14 Heinkel Lerche with tail control effector distance highlighted.

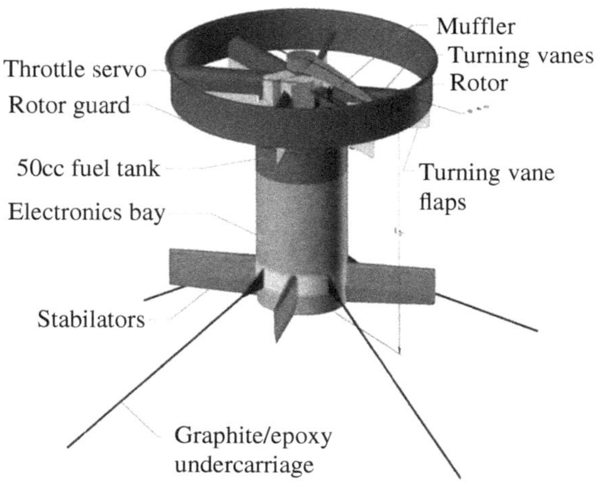

Figure 12.15 Lutronix LuMAV hovering coleopter for the DoD Counterdrug Technology Office.

principles. Figure 12.15 shows the earliest freeflight version of the LuMAV family of hovering coleopters. The reader will notice that unlike the Heinkel Lerche and Wespe, the LuMAV flew without a nose assembly. Because the mission specification and profile called for only low-speed hovering flight, the aircraft was not designed for transition (as was the Lerchle and Wespe). Although not specifically designed for high-speed forward flight, it was shown that the aircraft could still translate at more than twice the speed of the iSTAR and could handle gusts that were so severe that the local Coast Guard HH-65 fleet of helicopters were grounded during flight testing at MacDill Air Force Base, Florida in 1998. The general design consideration that evolved from the experiences with the LuMAV program was that the aerodynamic centers of the empennage control effectors should be approximately one rotor diameter or more away from the aircraft's center of gravity.

As DARPA chose the iSTAR configuration, fortunately, the Lutronix team managed to make contact with the Boeing Future Combat Systems team. This group included a number of players from around the world, including Singapore Technologies Aerospace (ST Aero). Because the Lutronix team, arguably, had at the time the best uninhabited coleopter designs in the world, ST Aero signed an agreement to partner with Lutronix Corp. so as to bring take the coleopter designs out of the hover state and towards full-transitioned flight modes, thus realizing their potential. So began the RDT&E effort of the XQ-138 Convertible Coleopter family of aircraft.

12.3.4 High-performance Coleopter Flight Testing: Control Authority and the Establishment of the Transition Corridor

Starting with handshakes, mission specification and profiles in July of 2001, the Lutronix team was absorbed into the new company, MASs. The team worked at a breakneck pace in the Fall of 2001 with many members of the team logging well over 100 work hours/week. The result of this "wartime pace" of RDT&E effort would be a first flight date just 11 weeks after first start of funding. Although the team was well aware of the "one rotor diameter" separation between the aircraft center of gravity and the aerodynamic center of the empennage effector, the team laid out a very special empennage configuration, which called upon the highest normal force gradient effectors known in the aerospace industry. These "grid fins" were typically used on Soviet and Russian missiles and rockets exactly because they have such tremendous normal force gradients.

The chief designer of the XQ-138, as it happened, was also instrumental in reverse-engineering the grid-fins for the US government and was well aware of how they work, how to size them, where and when to employ them. Indeed, they are critical features distinguishing the XQ-138 family of aircraft from other coleopters. In short, the crux of US Patent 6,502,787 [12] was that the fuselage, powerplant and propulsion core of the Heinkel Wespe and Lerche were fitted with a grid-fin assembly akin to that of the Russian AA-12. as shown in Figure 12.16. In spite of the best efforts and analysis techniques of the XQ-138 design team, the first three flight tests in December of 2001 yielded dangerously low levels of control effectiveness. So although more than a half-century had passed since Reiniger and his team at Heinkel had first laid out the Wespe, his lessons still rang true: More control effectiveness was needed, spacing the empennage roughly 1 rotor diameter from the aircraft center of gravity. Accordingly, the first significant geometric change that was made to the earliest variant of the XQ-138 was to grow the empennage length by one third of a rotor diameter. To this day, nearly all variants of the XQ-138 still fly with the volume coefficient established by the Heinkel team in 1944.

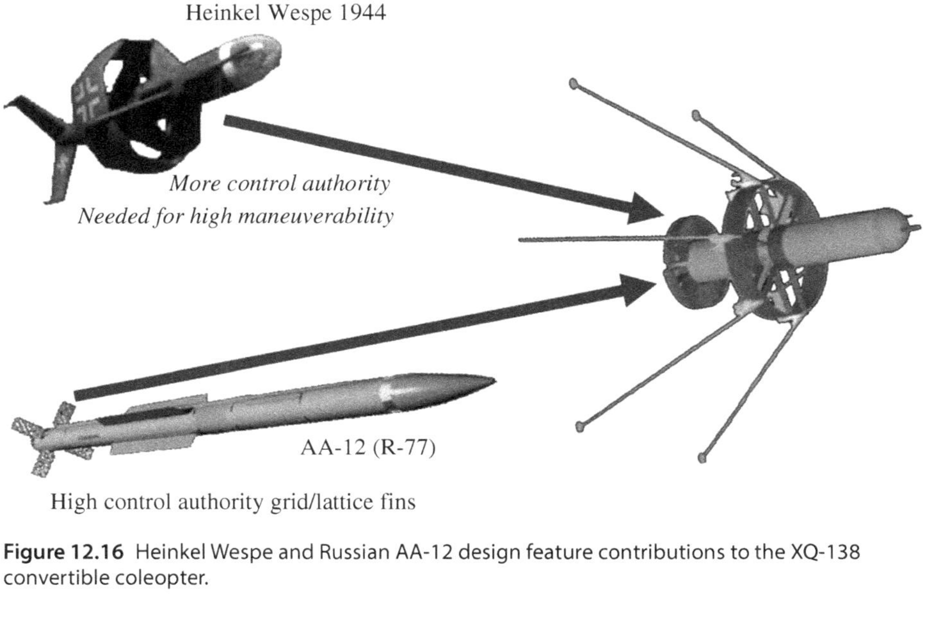

Figure 12.16 Heinkel Wespe and Russian AA-12 design feature contributions to the XQ-138 convertible coleopter.

The result of such high levels of control authority was simply astounding. By January of 2002, the MASs team had achieved full conversions from vertical to horizontal flight modes. As part of the preparations for Asia Aerospace 2002, the design team had several of the aircraft undergo full transitions at 38 °C and 100% humidity while breaking speeds of more than 100 kts. These top speeds would eventually reach more than 350 kts with high-speed variable-pitch rotors and faired boattail speed kits (Figure 12.17). In short, the performance of the XQ-138 was nothing short of stellar: Hover times in excess of an hour, high-speed dash at up to 350 kts for nearly an hour.

Not surprisingly, in 2004, ST Aerospace purchased the exclusive rights to the entire line of aircraft related to the XQ-138 and took it into serial production under the trade name of "The ST Aero Fantail." To date, the ST Aero Fantail is thought by many to be the only fully convertible coleopter ever to have achieved high-rate production on the commercial market.

As part of the RDT&E effort, the MASs-ST Aero team also flew more than a thousand flight tests on the XQ-138 and its follow-on variants. During flight tests, some extremely dangerous flight modes were explored. The most unstable and dangerous of these included high-rate tail slides and high-speed pull-ups. On no less than three occasions, aircraft departed the flight path, tumbling uncontrollably end over end, eventually crashing in the swamps next to the Strait of Johor. A montage of a flight path departure is shown in Figure 12.18.

Figure 12.17 High-speed variant of the XQ-138.

Figure 12.18 Montage of XQ-138 flight path departure at 150 kts, 5.9*g* turning pullup.

Clearly from Figure 12.18, the aircraft departure was violent and lead to a complete hull loss. As the flight test data was analyzed, it was determined that a violent flow-separation pitch-up instability was induced by flow separation off the aircraft forward fuselage, blanking part of the flow into the propulsor. So in spite of the high levels of control authority available from the gridfin assembly, there are clearly limits. Similarly, flight path departures were also experienced in comparatively low-speed vertical descents. These effectively put the empennage right in the middle of the vortex generated in the vortex-ring state (similar to what is experienced by helicopters). With experience, pilots figured out that quite often extremely high power settings can be used to increase the dynamic pressures over the empennage, which in turn restores control power and can overcome these departure dynamics. Accordingly, nearly fifty flights were conducted at the edges of the flight envelope, resulting in full-flight path departures, but with all but six recovered by pilots.

A general rule combining transition and departure concerns evolved from these flight tests. That is, if steady-state flight could be achieved in a trimmed state at a given angle of attack and total flight speed, then any power settings higher than those required for steady state flight would most often lead to recovery from departure. Conversely, it was often observed that lower power settings at high angles of attack almost always led to (often catastrophic) stability and control issues.

Flight testing was conducted between Auburn, Alabama, Eglin AFB, Florida, Ft. Benning, Georgia and Singapore at altitudes ranging from sea level through 500-m AGL. So as to discover the edge of the flight envelope, over a span of three years (2002–2005) the aircraft was regularly flown "to its limits." Figure 12.19 shows the results of these flight tests. A very clearly defined departure boundary was established; above the boundary, the angle of attack was deemed to be so high that in steady-state flight

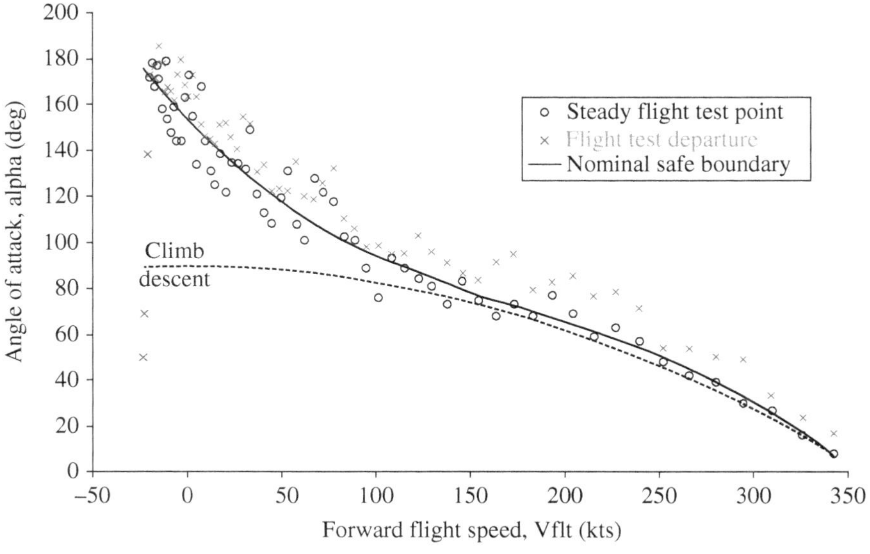

Figure 12.19 XQ-138HS flight test data showing steady state and departed flight boundary.

the aircraft would depart, mostly due to flow separation on the nose being ingested into the rotor guard. Below the line, with positive forward flight speeds, both stability and control was generally assured. It should be noted that as the aircraft would descend in a "tail slide," various helicopter-related instabilities would present themselves. The reader will see three red cross-marks left of the zero forward flight speed line at comparatively low angles of attack. The reader will note that the aircraft in these flight conditions is moving backwards through the sky with slight rearward motions at speeds beyond 20 kts. In these flight states a recirculation zone forms around the grid fins, which nullifies longitudinal control authority.

Clearly, Figure 12.19 forms the kernel of the flight director boundaries, which set the limits of the aircraft during transition and forward flight. The reader will see that departures were even noted at high forward flight speeds. These generally corresponded to extremely high-*g* pullups being executed, again inducing flow separation off the nose, causing rotor blanking and high pitching moments.

12.4 Future: Convertible QuadCopter Design and Flight Testing

Although coleopters are among the leading classes of aircraft which are capable of full body hover-to-dash conversions, there are several other types of aircraft that are coming to market which can also convert and are making inroads into markets such as those for subscale UAVs and toys. Figure 12.20 shows Edissey, LLC's Quadrocket family of convertible UAVs. As this chapter is being penned, the aircraft and its designers are working to set class speed records.

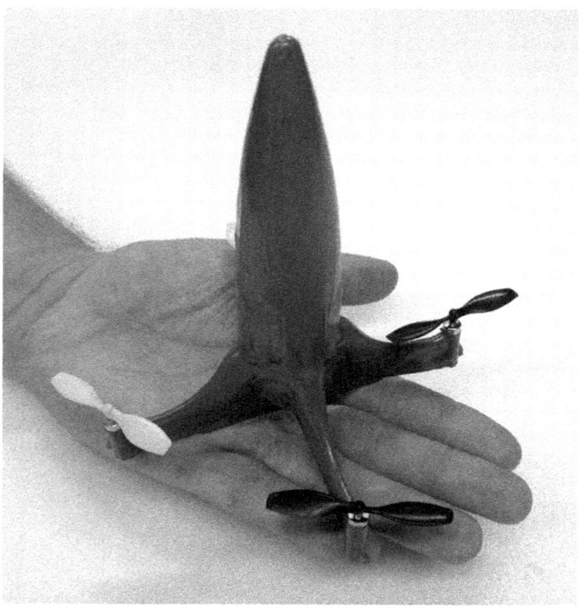

Figure 12.20 Edissey convertible "Quadrocket" UAV.

12.5 The Extreme: Hover-to-Supersonic Dash Aircraft

The major drivers for conversion, as stated earlier, are clearly related to vertical flight operations during takeoff and landing coupled with extremely high dash speeds. If one takes this to the extreme, then the top speeds push well into the supersonic. To achieve this, a ducted fan geometry is not a suitable solution. Rather, relying upon folding rotors that can become wings is a far more suitable option. From extensive flight testing, it is clear that there are currently no practical upper bounds on the size of the aircraft under consideration.

Rather, it is perfectly possible that such coleopters could be stretched to comparatively large sizes to do all kinds of missions. If one examines Figure 12.21, then one can see that aircraft in excess of 10 m of fuselage length are more than feasible. If one looks only at a comparatively small aircraft, just 2 m in length (as shown in the design point on the chart), then many useful missions could be accomplished. If the aircraft were to be refueled in flight (as shown in Figure 12.22), then the aircraft could stay airborne for extremely long periods of time. Because the XQ-138 first demonstrated that aircraft of this class could be quite handily armed with small arms well over a decade ago, it is perfectly possible that the concept of collocated close air support could be realized in the very near term. If one considers the fastest "time in contact" response that the military generally works with when considering close air support, the lower bound is 5 min. If one uses this number in laying out operations of current interest, then it can be seen that just 9 tracks are needed for Iraq and 10 are required for full coverage in Afghanistan. Other critical missions for such aircraft include SOF support as well as counter-piracy. So clearly such aircraft could be of tremendous utility to military planners if they would only realize that such aircraft are just an RDT&E effort away.

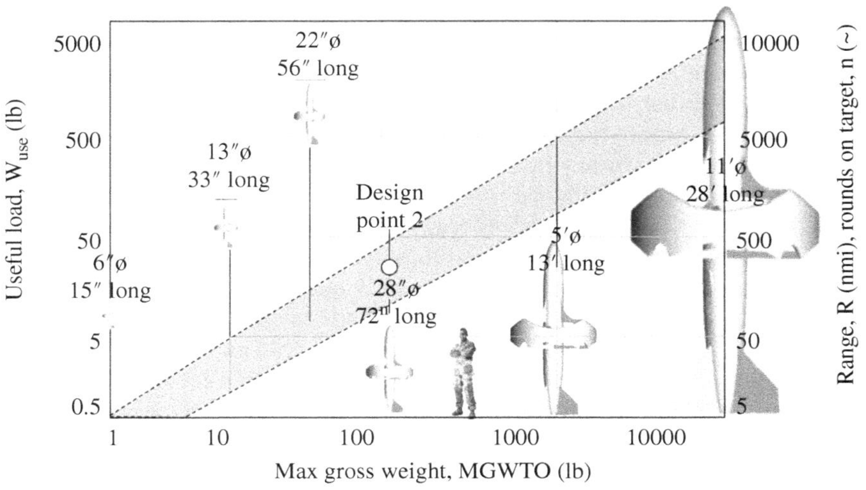

Figure 12.21 Tactical and strategic coleopter designs.

Figure 12.22 2-m coleopter refueling in flight.

If one stretches the capabilities of the convertible aircraft deep into high-performance levels, then it is easy to see that via clever arrangements, a supersonic variant of the aircraft appears. The "Supersonic Hovering Aerial Vehicle" (SHAV) is one way to achieve efficient, long-endurance hover capability while maintaining efficient supersonic dash capability. The aircraft would be built around a variable-cycle turboshaft engine, which is capable of declutching from the rotor system, allowing the propulsion core to augment thrust in the high subsonic through the low supersonic range. A variable engine cycle is more than feasible. Indeed, the J58 propulsion core has been flown for more than a half-century in the SR-71. Figure 12.24 shows one configuration that allows for this combined hover and dash capability. The folding blade-wings have such high fineness ratios that wave drag is minimized throughout the flight envelope. If such an aircraft were to undergo full-scale development then the 5-min TIC response figures of Figure 12.23 collapse from 9 and 11 tracks for full-country coverage to coverage of the entire country from a single base. If counterpiracy missions are examined, then large swathes of the Indian Ocean can be covered from a single surface vessel with the horizontal deck space limitations of a corvette or cruiser, as shown in Figure 12.25. It is obvious that such tremendous performance significantly outstrips the capabilities of regular navies of today. Because the technologies which are fundamentally enabling to aircraft like the SHAV are mature and ready for integration, it is simply a matter of time until aircraft like these are integrated into armed forces.

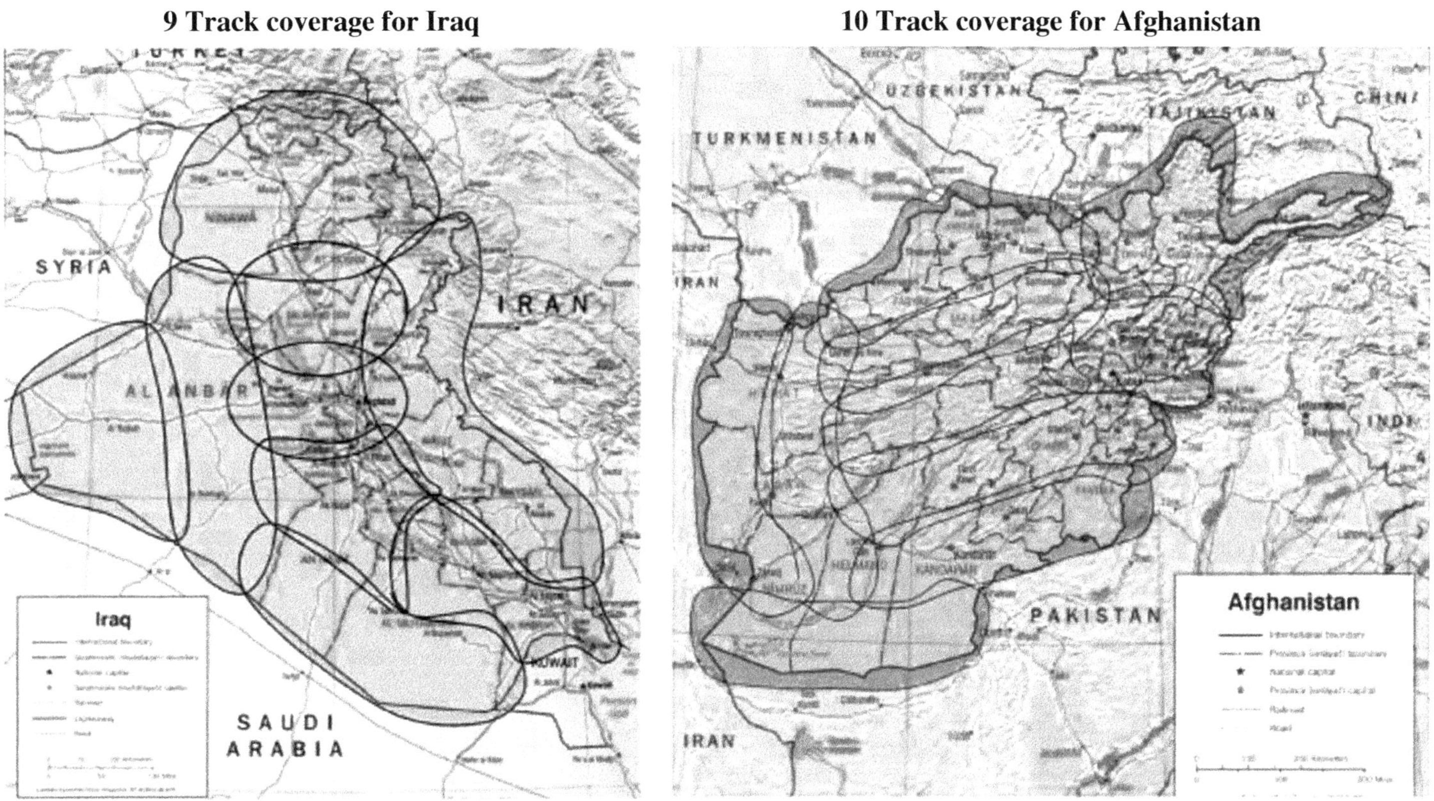

Figure 12.23 2-m coleopter tracks for 5-min time-in-contact reponse collocated close air support.

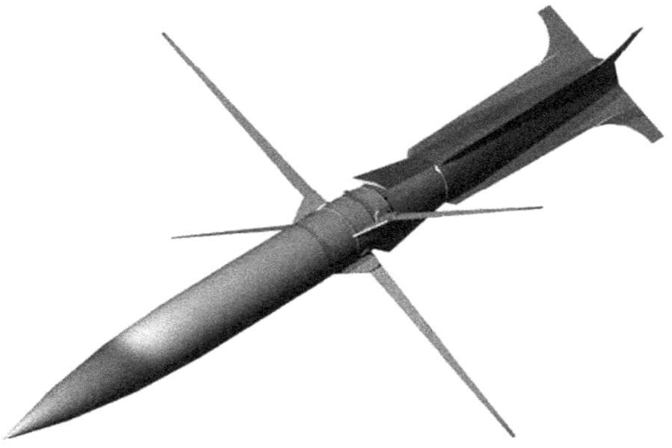

Figure 12.24 Supersonic hovering aerial vehicle [13, 14].

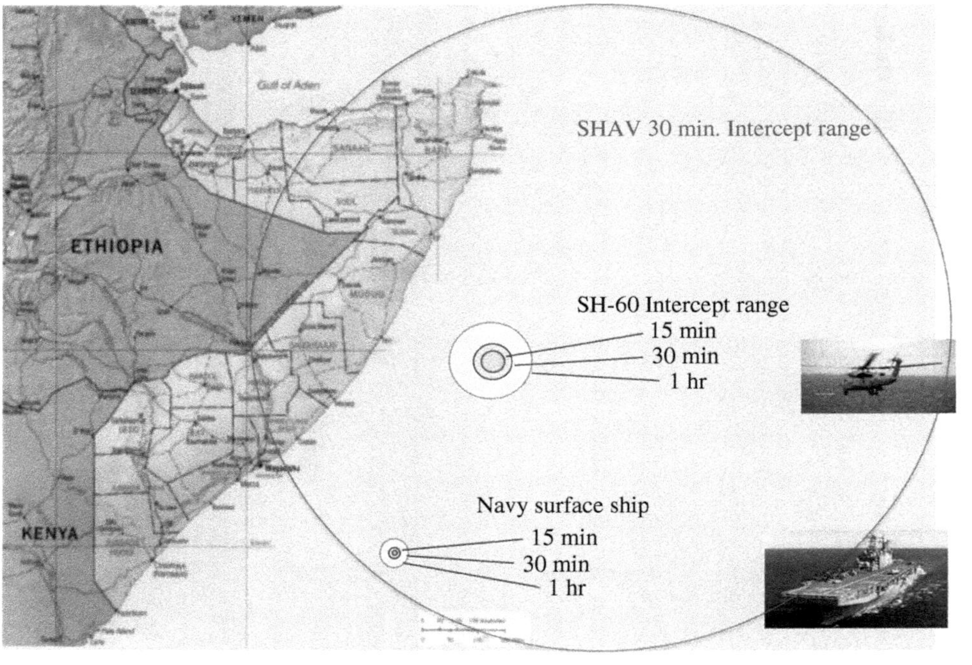

Figure 12.25 Counterpiracy coverage area of a 1000-lb maximum gross weight at takeoff supersonic hovering aerial vehicle.

References

1 Barrett, R., "Convertible aircraft," Short Course offered through Barrett Aerospace Technologies, Lawrence, Kansas, USA, 2009.
2 Charnov, B.H. *From Autogiro to Gyroplane: The Amazing Survival of an Aviation Technology*, Westport, Connecticut: Praeger Publishers, 2003.

3 Albert, M., Cadot-Burillet, D., and Williams, A., "Integration of mission trajectory management functions into clean skytechnology evaluation process," Paper no. 100, *Proceedings of the 5th Challenges in European Airspace (CEAS) Conference*, 7–11 September 2015, Technical University of Delft, Holland.
4 Roskam, J., *Airplane Design*, Volumes I–VIII, published by Design and Research Corporation, Lawrence, Kansas 2014.
5 Barrett, R.M., "A pattern of mishandling of patentable technology at Auburn University: A review and response to Auburn University's Office of the Vice President for Research," BAT Technical Report, Vol. 1, No. 1, Oct. 2003.
6 Barrett, R., "Typical refueling aircraft mission profile," AE 521 Aerospace Design I Lecture Manual, p. 147, University of Kansas, Lawrence, Kansas USA 2016.
7 Brown, M., Van de Kerkhof, K., Klein, R., et al., "The supersonic multi-utility aeroelastic reconfigurable test-bed," Winner of the American Institute of Aeronautics and Astronautics Graduate Aircraft Design Competition, The Technical University of Delft, Delft, Holland, 2014.
8 Anon., "Flight test photo of the Hiller VZ-1 Pawnee," Smithsonian Institution, Washington, DC, July 2015.
9 Reiniger, K., "Heinkel Lerche I design drawing," Heinkel Flugtechnische Werke (VFW), 24 February 1945.
10 Barrett, R., "Typical aircraft sizing chart including STAMPED data overlay," AE 721 Advanced Aerospace Design I Lecture Manual, p. 169, University of Kansas, Lawrence, Kansas USA 2016.
11 Barrett, R., "Coleopter configured aircraft," AE 781 Rotary-Wing Aerodynamics Lecture Manual, p. 118, University of Kansas, Lawrence, Kansas USA 2016.
12 Barrett, R., "Convertible vertical take-off and landing miniature aerial vehicle," 6,502,787, US Utility Patent, February 22, 2002.
13 Barrett, R., Bramlette, R., and Honea, R., "Convertible Aerial Vehicle," US Patent Invention Disclosure Filed 24 August 2014.
14 Barrett, R., "Supersonic hovering air vehicle," US Patent Application US 9004393, October 2010.

Part II

Novel Concepts in Unmanned Aircraft Flight Stability and Control

2.1 Fixed-wing (airplanes)

14

Autonomous Gust Alleviation in UAVs

Ya Wang[1] and Daniel J. Inman[2]

[1] *Department of Mechanical Engineering, SUNY, Stony Brook, NY, USA*
[2] *Department of Aerospace Engineering, University of Michigan, Ann Arbor, MI, USA*

14.1 Introduction

When a gust hits a small, umnanned air vehicle, stability can easily be lost. Various active control schemes have been proposed in the literature to cancel wing vibrations induced by gusts (see for instance Regan and Jutte 2012). The expression 'gust alleviation' often refers to load alleviation. Here, however, we are referring specifically to reducing the vibrations induced in a wing by a gust using harvested energy. The whole control scheme is automatic, or autonomous, in a self-contained way, and does not reduce the power available to the propulsion system. In particular we focus on a system that uses energy absorbed from wing vibrations during normal flight to power a control system that will cancel vibrations induced when the aircraft is subject to a gust. In order to focus on the details of the system design, the air-flow is modelled in the simplest way, but using the Dryden power spectral density (PSD); see for instance McClean (1990). Other simplifying assumptions include linearity of the structural response and simplified modelling of the composite structure using the rule of mixtures. The basic idea put forth here is that normal "clear sky" induced wing vibrations can be harvested. The resulting energy stored is released later when the aircraft experiences a gust; it is used to run an active feedback control system for gust alleviation. The discussion here is limited to looking at wing vibrations only and does not consider rigid-body motions (Richardson et al. 2011). The focus is on how to use a limited amount of energy to control wing vibrations caused by a gust.

Flutter suppression for UAVs is also an issue. In May 2014, Lockheed Martin announced that its X-56A UAV will be fitted with an anti-flutter system as it moves to use lighter, more flexible wings. The move to lighter wings provides the opportunity for a drastic increase in efficiency, but it makes the plane much more susceptible to flutter and gusts. In fact, UAVs in general are more likely to push the flutter envelope than manned aircraft are. Thus flutter suppression, while not specifically addressed here, provides a motivation for the use of harvested energy to perform control. Suppression of flutter by simply adding an energy-harvesting component is discussed by de Marqui et al. (2011).

Advanced UAV Aerodynamics, Flight Stability and Control: Novel Concepts, Theory and Applications, First Edition. Edited by Pascual Marqués and Andrea Da Ronch.

Companion website: http://www.wiley.com/go/marques/advanced_UAV_aerodynamics

The system discussed here has the following components. First an energy-harvesting scheme collects energy from the vibrations of the wing under normal flight conditions. An energy storage system is provided by a thin-film battery. A sensor and actuator are needed for implementing a control law. Since the control law must use as little energy as possible, a controller reduces the energy required to suppress vibrations. The system, along with the controlling electronics, is then integrated into the spar of a small UAV and bench tested to both validate the modelling, hence the predictions, and to illustrate a working system. This approach is the first to attempt gust alleviation without resorting to the use of discrete control surfaces (ailerons).

14.2 The Composite Spar

The heart of the autonomous gust alleviation system is a multifunctional spar. A multifunctional structure is a structure that serves several functions besides its primary function of providing strength and packaging. The idea grew out of work on structural batteries performed by the US Naval Research Labs (Thomas and Qidwa 2005). This showed a weight saving by integrating a battery function into a structural wing. Here we examine a wing spar that performs not only its strength function but also the functions of computing, harvesting, sensing, energy storage and control. The energy-harvesting component uses the piezoelectric effect (Erturk and Inman 2011) to transduce low-level wing vibrations into usable electrical energy. Recall that the piezoelectric effect allows certain materials to emit an electric field when subject to a mechanical strain and experiences a mechanical strain when a voltage is applied across the material. There are a variety of different types of piezoelectric materials, so a question immediately arises regarding which one is best to use in this situation. Furthermore, the piezoelectric effect is also used for sensing and for actuation, the key elements in a closed-loop control system. Again, the question arises as to which material makes the best choice for each function. Ideally, the answer would be the same for all three functions of harvesting, sensing and actuation. Besides these functions, the composite spar must also provide computing circuitry to manage the harvesting and storage electronics as well as the gust-alleviation control law, making the system autonomous.

The multifunctional wing spar of interest here is illustrated schematically in Figure 14.1. For weight reduction and strength purposes, the beam-like wing spar is mounted on a fiberglass composite substrate (E) of mass 17.8 g, with a total length of 593.7 mm, width of 38 mm and thickness of 2.38 mm. The most common piezoelectric material for actuation and sensing is lead zirconate titanate (PZT). The harvesting, sensing and actuating PZT layers are placed at the root of the wing spar, since this section will experience the largest strain during normal wing vibration or wind-gust disturbance. The PZT-based harvester/sensor (B) is layered on the top surface of the fiberglass substrate. The PZT actuator (F) is layered on the bottom surface of the fiberglass substrate. The thin-film battery (C) allows for power storage from harvesting and provides the energy supply for wind-gust alleviation. The electronic module combines conditioning, sensing and control circuitry, on a single layer of printed circuit board (PCB). The harvested energy from wing vibrations appears as an alternating voltage, which needs to be rectified and smoothed in order to charge the battery. These

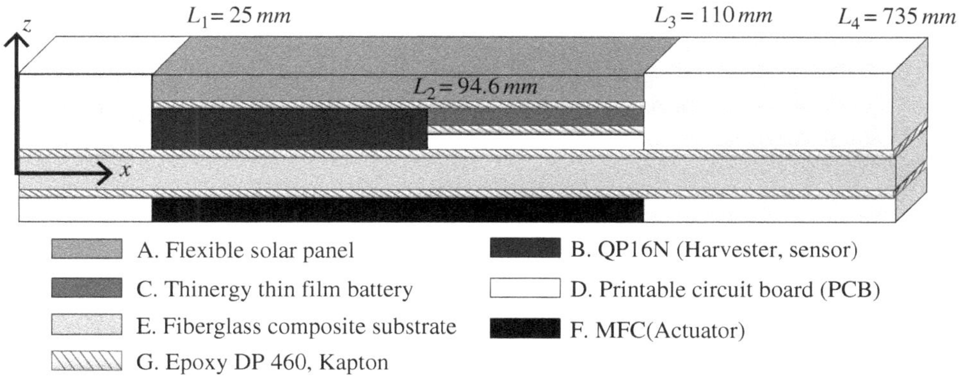

Figure 14.1 Multifunctional wing spar incorporating self-sensing, self-harvesting, self-storage and self-control. (*See insert for color representation of this figure.*)

multifunctional layers, together with the fiberglass composite substrate, form the multiple-layer wing spar. ScotchWeld^{TM} DP460 epoxy from 3 M is bracketed by Kapton to be used in the individual layer layup. Both of them are grouped together as layer G in Figure 14.1.

Next, consider the class of piezoelectric materials and which types would be most suitable for various required functions. Ceramic materials versus polymer-based piezoelectric materials are considered here because of their generally higher coupling coefficients. We further restrict our examination to commercially available piezoceramics. The main piezoceramic configurations are:

- monolithic PZT
- macro fibre composites (MFCs)
- a packaged monolithic PZT called QuickPack, which is encased in Kapton and fitted with electrical leads.

The piezoelectric effect is a three-dimensional effect that causes the material to strain when subject to an electric field and produces an electric field when subject to mechanical strain. Temperature effects are also important but here we will operate within acceptable temperature limits so we may effectively ignore temperature considerations.

Ignoring temperature effects, a simplified version of the constitutive laws covering the piezoelectric effect are (IEEE Standard 1987):

$$
\begin{aligned}
S &= s^E T + dE \\
D &= dT + \varepsilon^T E
\end{aligned}
\tag{14.1}
$$

Here, S is the mechanical strain, T is the mechanical stress, E is the electric field strength, D is the dielectric displacement, s^E is the compliance measured at zero electric field, and ε^T is the dielectric constant measured at zero stress. The piezoelectric constant, d, determines the coupling between mechanical and electrical fields. Since the piezoelectric effect is three-dimensional, Eq. (14.1) is really a full tensor relationship. In most circumstances the piezoelectric material is in the shape of a thin plate with the

Table 14.1 Geometric and material properties of multifunctional wing spar.

Layer	Length (mm)	Width (mm)	Thickness (mm)	Mass (g)	Young's modulus (GPa)
Solar panel	85	28	0.2	0.69	52
QP10N	59.6	25.4	0.38	2.25	67
Battery	25.4	25.4	0.18	0.46	55
PCB	25.4	25.4	0.20	0.23	60
Fiberglass substrate	735	28	10	185.2	71
MFC	85	28	0.18	2.92	63
Foam	650	28	0.76	4.10	60
Fiberglass composite					
3D epoxy	1580.4	28	0.008	0.35	3
Kapton	1580.4	28	0.0075	0.41	3.7

electric field perpendicular to the resulting strain and with the respective coupling being denoted by d_{31}. However, it is possible to arrange the material so that the electric field and the strain are in the same direction (Hagood et al. 1993), in which case the coupling coefficient is d_{33}. Experiments show that d_{33} is about three times larger than d_{31}. This led to the invention of MFCs, which use fibres of PZT laced in a Kapton casing with interdigitated electrodes (Wilkie et al. 2000; Bent and Hagood 1997). The resulting material makes an outstanding actuator, with the added ability of being flexible, unlike other ceramic-based actuators.

Several previous studies have examined the differences in piezoceramic performance as sensors, actuators and energy-harvesting devices. Sodano et al. (2004; 2005) investigated the sensing qualities of MFCs and later considered the energy-harvesting abilities of the three different piezoceramic configurations mentioned above (Sodano 2006). A later work (Bilgen et al. 2011) considered the effect of volume or mass, a critical issue in aircraft, as an additional constraint in evaluating piezoceramics' relative performance as sensors, actuators and harvesting devices. The resulting study shows clearly that MFCs have the best actuation ability and QuickPack has the best harvesting and sensing ability. This is the reason for the arrangement of sensing, harvesting and actuating elements in Figure 14.1. Note that only one piezoceramic is used for both sensing and harvesting. The energy from the solar panel is not considered here but can be easily integrated to provide additional energy.

To model the layered composite, a beam model is used, with the changes in stiffness and modulus modelled by discontinuity functions in the x direction and a smeared modulus used in the z direction, applying the rule of mixtures. Table 14.1 gives the physical parameters of each component. This is followed by an energy formulation of the distributed-parameter electromechanical multifunctional wing spar used to derive the dynamic equations of motion.

Under the plane-stress assumption and indicating the directional properties by subscripts, the linear constitutive equations for piezoelectric transducers excited in the

z-direction and deformed in the x-direction (the tensor version of Eq. (14.1) for the planer geometries used here) become:

$$\begin{aligned} T_1 &= c_{11}^E S_1 - e_{31} E_3 \\ D_3 &= e_{31} S_1 + \varepsilon_{33}^S E_3 \end{aligned} \tag{14.2}$$

This expression is appropriate for a monolithic d_{31} domain PZT harvester and sensor such as the QuickPack-type QP10N used here. On the other hand, the actuator material is chosen to be an MFC, which is deformed and excited in the x-direction so that the larger d_{33} coefficient is used. Then the reduced constitutive equations become:

$$\begin{aligned} T_3 &= c_{33}^E S_3 - e_{33} E_3 \\ D_3 &= e_{33} S_3 + \varepsilon_{33}^{S^*} E_3 \end{aligned} \tag{14.3}$$

The above piezoelectric coefficients are related to the elastic modulus (c, e) and the dielectric permittivity of the piezoelectric material (ε) via the following relations:

$$\left\{ \begin{aligned} c_{11}^E &= \frac{s_{11}^E}{\left(s_{11}^E\right)^2 - \left(s_{12}^E\right)^2}, \\ e_{31} &= \frac{s_{11}^E d_{31} - s_{12}^E d_{31}}{\left(s_{11}^E\right)^2 - \left(s_{12}^E\right)^2}, \\ \varepsilon_{33}^S &= \varepsilon_{33}^T - \frac{2 d_{31}^2 \left(s_{11}^E - s_{12}^E\right)}{\left(s_{11}^E\right)^2 - \left(s_{12}^E\right)^2}. \end{aligned} \right\} \tag{14.4}$$

$$\left\{ \begin{aligned} c_{33}^E &= \frac{s_{11}^E}{s_{11}^E s_{33}^E - \left(s_{13}^E\right)^2}, \\ e_{33} &= \frac{s_{11}^E d_{33} - s_{13}^E d_{31}}{s_{11}^E s_{33}^E - \left(s_{13}^E\right)^2}, \\ \varepsilon_{33}^{S^*} &= \varepsilon_{33}^T - \frac{\left(s_{33}^E d_{31}^2 - 2 s_{13}^E d_{31} d_{33} + s_{11}^E d_{33}^2\right)}{s_{11}^E s_{33}^E - \left(s_{13}^E\right)^2}, \end{aligned} \right\} \tag{14.5}$$

Again S is the mechanical strain, T is the mechanical stress, D is the dielectric charge displacement, and E is the electrical field strength; c is the elastic stiffness constant, s is the elastic compliance constant, d and e are piezoelectric constants, and ε is the permittivity constant. The subscripts 1 and 3 are tensor notations indicating the geometric directions. The superscript E refers to evaluating the parameter at a constant electric field. The superscript S implies the parameter is evaluated at a constant mechanical strain and likewise T denotes that the parameter is evaluated at constant mechanical stress.

The total potential energy in the beam is:

$$U = \frac{1}{2}\int_V S_{xx} T_{xx} dV. \tag{14.6}$$

Here S_{xx} and T_{xx} denote the stress and strain vectors of both the structural and the QuickPack/MFC layers over the wing spar volume V. The structure here includes all the components in Figure 14.1. Let $w_{rel}(x,t)$ denote the transverse displacement relative to the clamped end of the wing spar. It is assumed that the wing-to-fuselage connection forms a clamped boundary condition, decoupling the rigid-body motion from elastic wing vibrations. The non-zero strain component at an arbitrary position in the x-direction along the neutral axis, and at a given time t, is given by:

$$S_{xx}(x,z,t) = -z\frac{\partial^2 w_{rel}(x,t)}{\partial x^2}. \tag{14.7}$$

The multifunctional wing spar is assumed to be linear elastic so according to Hooke's law the structural stress T_{xx} is the product of the global elastic modulus E_s and the strain vector S_{xx}. In this case the structural strain energy becomes:

$$U_s = \frac{1}{2}\int_0^L E_s I_s \left(\frac{\partial^2 w_{rel}(x,t)}{\partial x^2}\right) dx. \tag{14.8}$$

Here, $E_s I_s$ stands for the bending stiffness of the structure, which is taken as constant across each of the four individual longitudinal sections in Figure 14.1. In order to accommodate the spatially dependent ending stiffness in the x direction Heaviside step functions (denoted H) are used to yield:

$$\begin{aligned} E_s I_s(x) = {} & E_1 I_1 [H(L_1 - x)H(x)] \\ & + E_2 I_2 H(L_2 - x)H(x - L_1) + E_3 I_3 H(L_3 - x)H(x - L_2). \end{aligned} \tag{14.9}$$

where the bending stiffness variation in the z direction of each longitudinal section is estimated using the rule of mixtures (ROM) and the cross-section transformation (CST) [13].

Following the constitutive relations given in Eqs. 1.2–1.5, the stress component in the d_{31} harvesting/sensing QuickPack layer and in the d_{33} MFC actuation layer are given by:

$$\begin{aligned} T_1 &= E_{11}^E S_1 - d_{31} E_{31}^E E_3^o; \\ T_3 &= E_{33}^E S_3 - d_{33} E_{33}^E E_3^a. \end{aligned} \tag{14.10}$$

respectively. Here, S_1 and S_3 both represent the strain vectors S_{xx} in the x direction. The superscripts o and a denote 'output' from the PZT harvester/sensor at the top in Figure 14.1 and 'actuation' from the MFC actuator at the bottom in Figure 14.1. Since the output voltage $v^o(t)$ across the QuickPack output layer and the actuation voltage $v^a(t)$ across the MFC actuator is the product of the thickness of each PZT layer with the instantaneous electric field $E_3^o(t)$ and $E_3^a(t)$, respectively, the total strain energy in each transducer layer becomes:

$$U_{po} = \frac{1}{2}\int_{L1}^{L2}\left[E_{31}^E I_p^0 \left(\frac{\partial^2 w_{rel}}{\partial x^2}\right)^2 - v(t) J_p^0 \frac{\partial^2 w_{rel}}{\partial x^2}\right] dx,$$

$$U_{pa} = \frac{1}{2}\int_{L1}^{L2}\left[E_{33}^E I_p^a \left(\frac{\partial^2 w_{rel}}{\partial x^2}\right)^2 - v^a(t) J_p^a \frac{\partial^2 w_{rel}}{\partial x^2}\right] dx, \tag{14.11}$$

Again the subscripts *po* and *pa* denote the harvesting/sensing (output) layers and the actuation layer, respectively. The bending stiffness of each layer is derived using the parallel axis theorem. The coupling terms J_p of the harvesting/sensing QuickPack and actuation MFC are defined as:

$$J_p^O = \iint_{po} z \frac{d_{31} dE_{31}^E}{h_{po}} dydz, J_p^a = \iint_{pa} z \frac{d_{33} dE_{33}^E}{h_{pa}} dydz, \tag{14.12}$$

Here h_{pa} and h_{po} represent the thickness of the harvesting/sensing QuickPack and actuation MFC, respectively. Let $w(x,t)$ denote the transverse displacement of the multifunctional beam in the absolute frame of the reference. The total kinematic energy of the multifunctional wing spar is then given by:

$$T_e = \frac{1}{2}\int_0^L \rho(x) A_s(x) \left(\frac{\partial w(x,t)}{\partial t}\right)^2 dx. \tag{14.13}$$

Here $\rho(x)$ stands for the density and $A_s(x)$ for the cross-sectional area of the wing spar in the x direction. Note the x-dependence as the beam density and area change moving along the longitudinal axis. The internal electric energy of the harvesting/sensing QuickPack layer and the actuation MFC layer is:

$$E_{ie}^{o,a} = \frac{1}{2}\int_{Vpo,a} E_j^{o,a} D_j^{o,a} dV_{po,a}. \tag{14.14}$$

After substituting the expressions for the electric field and the electric displacement from the constitutive Eq. (14.2) into Eq. (14.14), the internal electric energy becomes:

$$E_{ie}^{o,a} = \frac{1}{2}\int_{L1}^{L2} J_p^{o,a} v(t)^{o,a} \frac{\partial^2 w_{rel}}{\partial x^2} dx + \frac{1}{2} C_p^{o,a} v(t)^{o,a2}. \tag{14.15}$$

Here the capacitance term C_p of the harvesting/sensing QuickPack and the actuation MFC is defined as:

$$C_p^o = \frac{\varepsilon_{ji}^s A_p^o}{h_{po}}; C_p^a = \frac{\varepsilon_{ji}^{s*} A_p^a}{h_{pa}}. \tag{14.16}$$

The constant A_p^o is the electrode surface area of the harvesting and sensing QuickPack and A_p^a denotes the electrode surface area of the MFC providing the actuation.

The non-conservative work is due to the harvesting/sensing QuickPack output energy and the MFC actuation energy, given by

$$E_{nc}^{o,a}(t) = Q^{o,a}(t)v^{o,a}(t). \tag{14.17}$$

Here $Q^{o,a}$ denotes the output charge or actuation charge of the corresponding piezoceramic device.

An assumed mode method based on the Rayleigh–Ritz formulation is used to build a dynamic model useful for predicting the behavior of the multifunctional spar. The assumed mode method represents the vibration response $w_{rel}(x,t)$, relative to the base as a finite series expansion of admissible trial functions $\Phi_r(x)$ and unknown modal coordinates $\eta_r(t)$ of the form:

$$w_{rel}(x,t) = \sum_{r=1}^{N} \Phi_r(x)\eta_r(t), \tag{14.18}$$

Here N is the number of modes kept in the series subject to convergence conditions and the bandwidth of interest. The admissible trial function $\Phi_r(x)$ has to satisfy the clamped-free boundary conditions. For cantilever beams a standard admissible trial function for long thin beams is:

$$\Phi_r = 1 - \cos\left(\frac{(2r-1)\pi x}{2L}\right). \tag{14.19}$$

The extended Hamilton's principle satisfies:

$$\int_{t1}^{t2} (\delta T_e - \delta U + \delta E_{ie} + \delta E_{nc})dt = 0. \tag{14.20}$$

Taking the variation suggested in Eq. (14.20) results in the Euler–Lagrange equations for the multifunctional beam:

$$\frac{d}{dt}\frac{\partial T_e}{\partial \dot{q}} - \frac{\partial T_e}{\partial q} + \frac{\partial U}{\partial q} - \frac{\partial E_{ie}}{\partial q} = \frac{\partial E_{nc}}{\partial q}. \tag{14.21}$$

where the over dot indicates ordinary differentiation with respect to the temporal variable t. Here the generalized coordinates are the modal coordinates $\eta(t)$ and the voltage $v(t)$. Substituting the assumed mode solution into the potential energy term yields:

$$U^{o,a} = \frac{1}{2}\sum_{i,l=1}^{N}\left[\eta_i(t)\eta_l(t)k_{il}^{o,a} - \eta_i(t)v^{o,a}(t)\theta_i^{o,a}\right]. \tag{14.22}$$

Likewise, substitution into the kinetic energy term yields:

$$T_e^{o,a} = \frac{1}{2}\sum_{i,l=1}^{N}\left(\dot{\eta}_i(t)\dot{\eta}_l(t)m_{il}^{o,a} + 2\dot{\eta}_i(t)\int_0^L \rho A_s(x)\Phi_i \frac{\partial w_b}{\partial t}dx\right) + \frac{1}{2}\int_0^L \rho A_s(x)\left(\frac{\partial w_b}{\partial t}\right)^2 dx. \tag{14.23}$$

With the assumed mode expansion, the internal energies become:

$$E_{ie}^{o,a} = \frac{1}{2}\sum_{i=1}^{N}\left(\eta_i v^{o,a}\theta_i^{o,a} + C_p^{o,a}v^{o,a}\right) \tag{14.24}$$

where the coupling coefficient is defined as:

$$\theta_{ie}^{o,a} = \int_{L1}^{L2,L3} J_p^{o,a}\Phi_i'' \, dx. \tag{14.25}$$

and the prime denotes ordinary differentiation with respect to the spatial variable x.

These energy expressions, along with Euler–Lagrange equations, define the mass and stiffness matrices. The mass elements are given by:

$$m_{il} = \int_0^L \rho A_s(x)\Phi_i\Phi_l dx, \tag{14.26}$$

and the stiffness elements are given by:

$$k_{il}^{o,a} = \int_0^L E_s I_s \Phi_i'' \Phi_l'' dx + \int_{L1}^{L21L3} E_{31,33}^E I_p^{o,a}\Phi_i''\Phi_l'' dx. \tag{14.27}$$

The equations of motion then become:

$$\begin{aligned} &\sum_{l=1}^{N} m_{il}\ddot{\eta}_l + k_{il}\eta_l - \theta_i^o v_i^o - f_i = 0, \\ &C_p^o \dot{v}^o + \frac{v^o}{Z^o} + \sum_{i=1}^{N}\left(\theta_i^o \dot{\eta}_i^o\right) = 0. \end{aligned} \tag{14.28}$$

Here the Z^o denotes the impedance of the output circuit. In the case considered here, Z^o is modelled as a pure resistance. The term f_i is the force due to the base excitation and is given by:

$$f_i = -\int_0^L \rho A_s(x)\Phi_i(x)\frac{\partial^2 w_b}{\partial t^2} dx. \tag{14.29}$$

The composite also exhibits some natural damping, approximated here by assuming proportional damping (also called Rayleigh damping) of the form

$$\mathbf{C} = \kappa\mathbf{M} + \gamma\mathbf{K}. \tag{14.30}$$

The constants κ and γ are determined by experimental procedures detailed later. Inserting damping into Eq. (14.8) and rewriting the equivalent Euler–Lagrange Eq. (14.28) into matrix form, the dynamic equation of motion becomes the coupled system:

$$\mathbf{M}\ddot{\eta} + \mathbf{C}\dot{\eta} + \mathbf{K}\eta = \mathbf{f} + \theta^o \mathbf{v}^o, \tag{14.31}$$

$$C_p^o \dot{\mathbf{v}}^o + \frac{\mathbf{v}^o}{Z^o} + \theta^o \dot{\eta}^o = 0. \tag{14.32}$$

For the damping matrix, the coefficients κ and γ are determined using the damping ratios associated with the first and second modes: 0.37% and 0.1% respectively, based on experiments and the values given in Table 14.1. With these damping ratios, the matrix **C** must satisfy the following relations in order to uncouple the modal equations:

$$2\omega_{ni}\zeta_i = \Phi_i^T \mathbf{C}\Phi_i = \kappa + \gamma\omega_{ni}^2. \tag{14.33}$$

Here ω_{ni} and ζ_i represent the ith natural frequency and its associated damping ratio as measured. Equation (14.33) is then solved to yield $\kappa = 1.34$ and $\gamma = 1.89 \times 10^{-7}$. The damping matrix becomes:

$$\mathbf{C} = \kappa\mathbf{M} + \gamma\mathbf{K} = 1.34\mathbf{M} + 1.89e^{-7}\mathbf{K}. \tag{14.34}$$

A solution is formulated from the equations of motion by assuming that the ambient normal vibration applied to the wing is harmonic and applied to the clamped end of the wing spar $w_b(t)$. Then the base acceleration $a_b(t)$ and the forcing vector $f_i(t)$ of the wing will have a harmonic form as well, so that the base motion, acceleration, and force become:

$$w_b(t) = w_b e^{j\omega t}, a_b(t) = a_b e^{j\omega t}, f_i(t) = -m_i a_b e^{j\omega t}. \tag{14.35}$$

The harmonic forcing excitation leads to a harmonic solution for the generalized displacement coordinates

$$\eta(t) = \eta e^{j\omega t}, \tag{14.36}$$

and the generalized voltage:

$$v(t) = Ve^{j\omega t}. \tag{14.37}$$

Substitution of Eqs. (14.36) and (14.37) into Eqs. (14.31) and (14.32) yields for the steady-state response

$$\left(-\omega^2\mathbf{M} + j\omega\mathbf{C} + \mathbf{K}\right)\eta^{\mathrm{o}} = \mathbf{f} + \eta^{\mathrm{o}}V^{\mathrm{o}}, \tag{14.38}$$

and:

$$\left(j\omega C_p^o + \frac{1}{Z^o}\right)V^{\mathrm{o}} + j\omega\theta^o\eta^{\mathrm{o}} = 0. \tag{14.39}$$

If the superscript Tr stands for transpose, the steady-state solutions are solved to yield:

$$\eta^{\mathrm{o}} = \left(-\omega^2\mathbf{M} + j\omega\mathbf{C} + \mathbf{K} + \frac{-j\omega\theta^o\theta^{oTr}}{j\omega C_p^o + \frac{1}{Z^o}}\right)^{-1}\mathbf{f}. \tag{14.40}$$

$$V^{\mathrm{o}} = \frac{-j\omega\eta^{\mathrm{o}}\theta^{oTr}}{j\omega C_p^o + \frac{1}{Z^o}}, \tag{14.41}$$

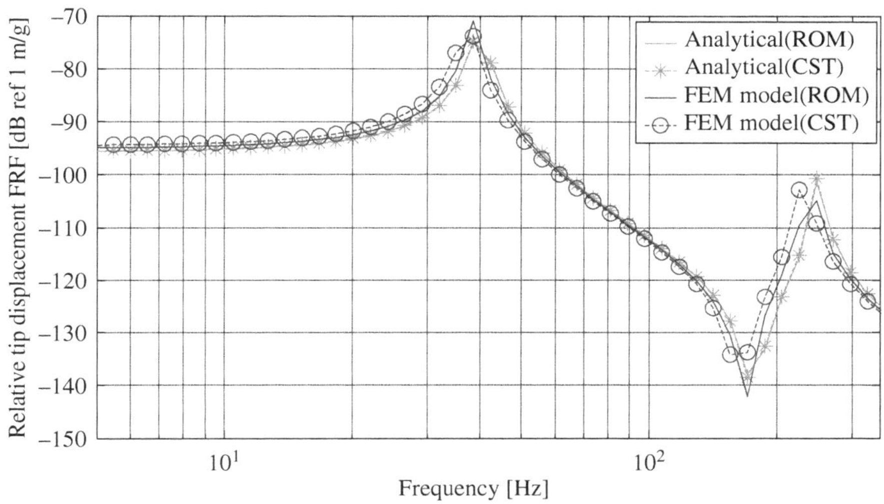

Figure 14.2 Frequency response prediction using both Rayleigh–Ritz and FE modelling. FRF, frequency-response function; ROM, rule of mixtures; CST, cross-section transformation.

With the values given in Table 14.1 and the damping values given in Eq. (14.34), the frequency-response function for the relative displacement can be numerically analyzed using four approaches and two methods. The two methods are an 'analytical' solution, using Rayleigh–Ritz, and a finite element (FE) approximation. In each method both the ROM and CST are implemented for modal frequency predictions. The four modelling approaches based on the above analysis are compared in Figure 14.2 using a displacement to base acceleration frequency-response function. The base acceleration is represented in terms of the acceleration of gravity, $g = 9.81\,\mathrm{m/s^2}$. Note that the ROM and CST approaches agree using both the analytical model solved by Rayleigh–Ritz and the FE model. The approximate analytical models are within 5% root mean square (RMS) error of the FE models.

The analytical model is useful in designing a control law because it is of much lower order then the FE model needed to obtain the same accuracy. The analytical model also keeps much of the physics transparent, unlike the FE approach.

14.3 The Energy-harvesting and Storage Component

The harvesting and storage component consists of the monolithic PZT coupled through electronics to the thin-film battery. The original concept for this part of the multifunctional spar was a self-charging structure (Anton et al. 2010). The structure consisted of a PZT harvesting material, a thin-film battery mounted on a thin-beam substrate, as pictured in Figure 14.3. The circuitry required to charge the battery was left external to the beam in this initial implementation. However, a full strength and dynamic analysis was performed to show the true multifunctional aspects of the design. Charge/discharge testing under dynamic excitation verified that there was transfer of electrical energy

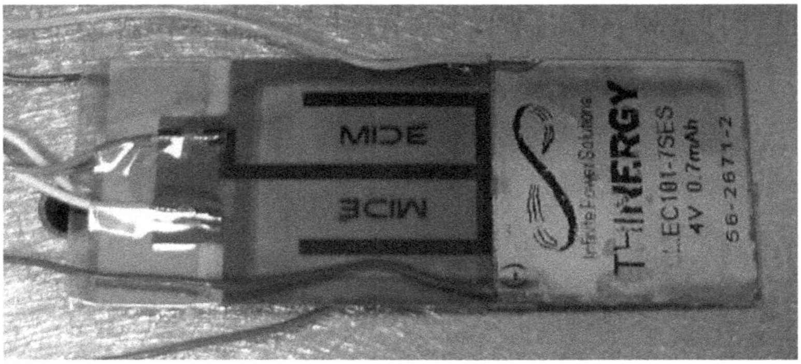

Figure 14.3 Photograph of a self-charging structure.

from the PZT to the battery, demonstrating a self-charging operation. A three-point bend test of the system, illustrated in Figure 14.3, showed that the strength of the multifunctional structure was around 100–150 MPa. As expected, the PZT and thin-film battery formed the critical failure points for bending. However, the dynamic tests show clearly that no electrical or mechanical failure occurs in the device for vibration excitation levels up to 7.0 g, providing a reasonable level of robustness.

The amount of energy required to provide a control action is larger than the energy directly available from harvesting. Thus the scenario is to harvest ambient energy during normal flight conditions, store it and release this larger amount of energy later when called for. The battery requires a DC voltage to charge it but the PZT provides an AC voltage. Thus a rectifying circuit and smoothing capacitor are required. For the discussion here, a simple resistance is used to model the circuitry for the sake of pointing out optimality conditions and keeping the text brief.

Modelling the electronics simply as a resistance, the steady-state solutions given in Eqs. (14.40) and (14.41) reveal the value of the harvested output voltage-to-base input acceleration frequency-response function to be

$$\frac{v^o(\omega)}{a_f e^{j\omega t}} = \frac{j\omega\theta^{oTr}}{j\omega C_p^o + \dfrac{1}{Z^o}} \left(-\omega^2 \mathbf{M} + j\omega \mathbf{C} + \mathbf{K} + \frac{-j\omega\theta^o\theta^{oTr}}{j\omega C_p^o + \dfrac{1}{Z^o}} \right)^{-1} \mathbf{f} \tag{14.42}$$

The value of the resistance experienced by an energy harvesting is well known to have an optimum based on the physical properties of the harvesting transducer (PZT). Figure 14.4 shows a power harvested versus load resistance for a typical PZT harvester showing the clear existence of an optimal resistance value.

In reality the circuit that takes the AC voltage appearing across the PZT and coverts it to smooth DC voltage suitable for charging the battery is fairly complex and is a topic of current research (Dunbar and Popovic 2014). Roundy and Wright (2004) and Liang and Liao (2009) used an equivalent circuit of piezoelectric generator representation for a single degree of freedom system representative of the first vibration mode of a distributed parameter model of the structure (Elvin and Elvin 2009). This last approach used the assumed modes method to create an equivalent electrical circuit using Matlab's

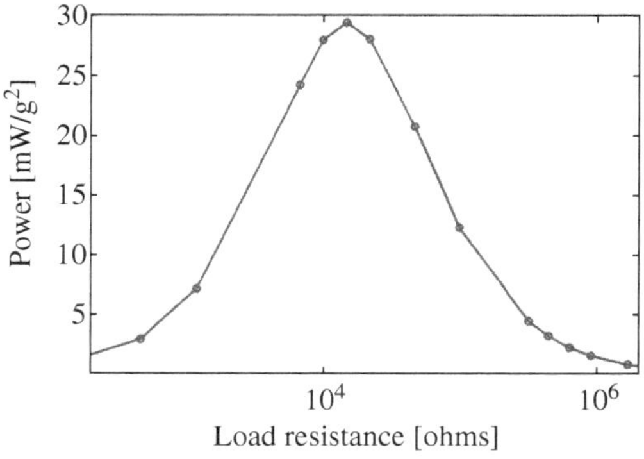

Figure 14.4 Power out versus load resistance for a PZT energy-harvesting transducer, showing the existence of clear optimal value of resistance. The dots are experimental values.

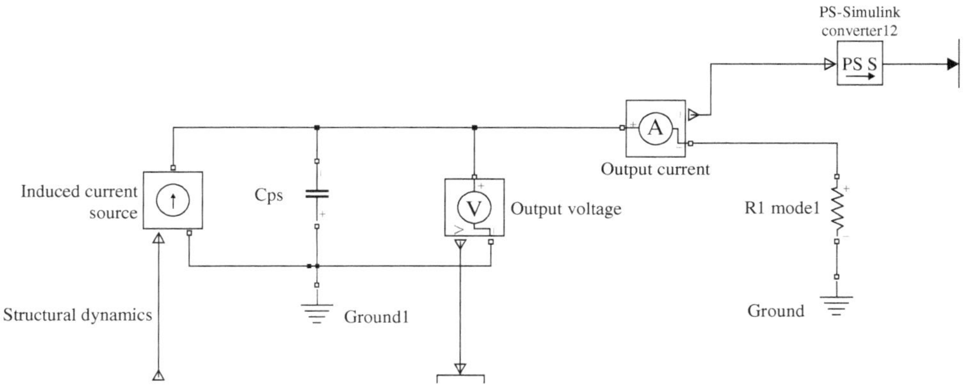

Figure 14.5 The equivalent circuit for first-mode piezoelectric generator with resistive impedance.

Simulink program adapted to multiple modes. Their method allowed an understanding of the dynamics of harvesting from more than one mode of vibration. Figure 14.5 shows the equivalent current sources circuit for the first-mode piezoelectric generator.

It is well known (see Pederson et al. 1966, for example) that a complex conjugate matching load impedance will deliver the maximum power output. Unfortunately, impedance matching for low-frequency energy harvesting usually requires an impractically large inductor. A suitable compromise is to use resistance matching, which provides suboptimal harvested power for a frequency band around resonance (Kong et al. 2010). A 0.1-MΩ resistance was found to maximize the harvested power for the multifunctional wing spar under consideration here. The frequency-response function of the harvested voltage output to the input of base acceleration for the first two vibration modes is given in Figure 14.6 for the 0.1-MΩ load resistance. Fifty assumed modes were used in order to ensure the convergence of the first two modes. The ambient normal wing-vibration acceleration is simulated with 0.2g RMS acceleration over a frequency bandwidth of 5–300 Hz.

Figure 14.7 shows a plot of the frequency spectrum of the instantaneous harvested power for a 0.1-MΩ resistance load.

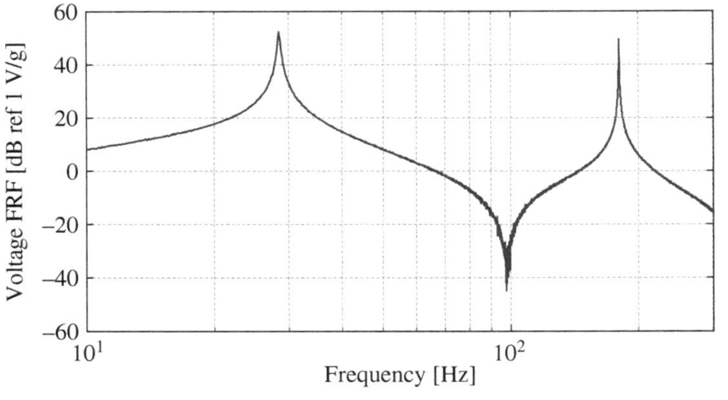

Figure 14.6 The output voltage to base acceleration frequency-response function for a 0.1-MΩ load resistance.

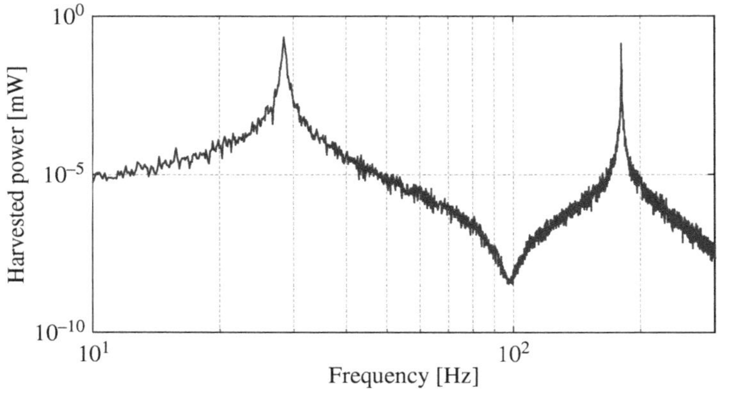

Figure 14.7 The harvested power spectrum for a 0.1-MΩ load resistance.

Defining the self-sensing sensitivity as the derivative of the sensing voltage v^s with respect to the PZT strain S^s_{xx}, the following relations can be derived for the sensitivity:

$$\frac{\partial v^s}{\partial S^s_{xx}} = \frac{h^s_y d_{31} E^E_{31}}{C^o_p} = \frac{d_{31} E^E_{31} h^o_p}{\varepsilon^s_{31} h^s_x}. \tag{14.43}$$

Here, h^s_y and h^s_x denote the width and the length of the piezoelectric sensor, respectively. The current flow into the sensor is neglected because of the large impedance (0.1 MΩ), which is equivalent to open-circuit conditions. More details of the circuits used here can be found in Wang (2012).

14.4 Reduced Energy Control Law

Because in general the amount of energy available from harvesting is limited, the active control law chosen to perform gust alleviation must use the least amount of energy possible to perform the required reduction in vibration. The standard method of minimum energy control is to use optimal control and penalize the control effort. However, it was found by Wang and Inman (2011) that a control algorithm that switches a control input from a constant voltage supply to the feedback signal provided from a more traditional controller greatly reduces the amount of energy required. Several different common control laws were examined, as well as their voltage-limited versions; so-called reduced energy controllers (RECs). They were used to control the free response of a cantilever beam. The laws examined were

- positive position feedback (PPF; Fanson and Caughey 1990)
- linear quadratic regulator (LQR; see, for instance, Levine 1996)
- proportional-integral-derivative (PID; see for instance O'Dwyer 2009).

A non-linear controller consisting of feeding back the product of position and velocity was also examined (Castelazo and Lee 1990). To form a fair comparison, each controller was given the same initial conditions and then designed to give a settling time of 0.85 s. The settling time was chosen to be quicker than any passive damping treatment could possibly achieve. The power consumption for each case was measured and the results are reported in Table 14.2.

It is clear from Table 14.2 that the use of a clipped or 'on-off' controller greatly reduces the power required to produce the same closed-loop response. Therefore, the control law chosen to use in a self-powered gust-alleviation system, employs a REC technique. Note that the alternative to active control is to add viscoelastic damping treatments and/or to make the wing stiffer. Damping treatments are not physically able to achieve this short settling time and making it stiffer would add substantially more weight.

14.5 Gust Modelling

Holbit (1988) provides an excellent introduction to gust loads applied to aircraft. The book covers gust loads on aircraft but especially focuses on continuous turbulence gust loads. Wright and Cooper (2007) provide an updated treatment of gust loading. Basic concepts are defined and explained in both textbooks. Varito et al. (2005) examined use of modal control methods to provide gust alleviation for a sensor craft concept. The SensorCraft, as it was called, was an unmanned 'flying wing' concept aircraft featuring a very flexible wing. As such, it was very vulnerable to gust-induced loads and vibration. Gust-load alleviation was used to reduce the craft's structural strength requirements. The gust response was controlled by inboard control surfaces. It was shown that the structural modes are well intermixed with the rigid-body modes, meaning that they will be highly coupled. This creates a situation where gust-induced vibration of the wings could destabilize the rigid-body modes and hence the aircraft. Varito et al. (2005) constructed a hardware test of a closed-loop gust-alleviation system that controlled the first two wing bending modes in the NASA Langley Transonic Dynamics Tunnel (TDT). The same group (Silva et al. 2006) performed a rigorous time-domain system identification on their hardware in the wind tunnel, a SensorCraft demonstrator called the High Lift-to-Drag Active (HiLDA) wing model. They then used the model to design a controller to eliminate the negative effects of a gust, using sinusoidal gusts created by gust vanes in the tunnel to emulate the real world.

Vartio et al. (2008) presented a linear quadratic Gaussian control law for gust-load alleviation and tested it in the TDT. They illustrated the ability to control the moments

Table 14.2 Different control systems for controlling a cantilever beam.

	Unit	Open-loop	PPF	REC-PPF	PID	REC-PID	Non-linear	REC-non-linear	LQR	REC-LQR
Initial displacement, velocity	mm, mm/s	(4.8,0)	(4.8,0)	(4.8,0)	(4.8,0)	(4.8,0)	(4.8,0)	(4.8,0)	(4.8,0)	(4.8,0)
Settling time T_s	s	10.8	0.85	0.85	0.85	0.85	0.85	0.85	0.85	0.85
Maximum voltage	V	N/A	450	130	450	130	450	130	450	130
Maximum current	mA	N/A	0.5	0.7	0.5	0.7	0.5	0.7	0.5	0.7
Average power	mW	N/A	10.6	5.73	15.5	6.47	16.7	5.46	15.52	6.20

applied at the root by the first few modes. They used strain gauges to measure the first two bending modes and the first torsional mode, as well as accelerometers to measure the acceleration at the wing tip. Shao et al. (2010) developed a fuzzy logic/neural network based control law for gust alleviation and tested its effectiveness in a wind tunnel applied to sinusoidal and random gusts and showed a 20–27% reduction in the response due to a gust. In 2012, NASA performed a state-of-the-art review of gust alleviation (Regan and Jutte 2012). They focused on both modelling and active control of atmospheric disturbances during flight. Their report provides both a historical survey of the topic as well as a state-of-the-art assessment. All of the above mentioned gust-alleviation control laws were implemented using discrete control surfaces (flaps, ailerons, and so on). The work presented here does not use discrete control systems to counter gust vibrations.

Gusts may be modelled in several ways. An excellent and detailed analysis of atmospheric wind models particular to lower flying UAVs can be found in Etele (2006). In simple terms, wind experienced in flight can be thought of as made up of mean wind, waves, and turbulence. Turbulence is either discrete or continuous. Discrete models of turbulence can be represented as deterministic in a form proportional to $[1 - \cos(\theta)]$, where the amplitude is the amplitude of the gust and θ is proportional to the length of the gust and some aircraft parameters. Here the Dryden PSD, Φ_{Dry}, is used to model the gust, which is given by:

$$\Phi_{\mathrm{Dry}}(\omega) = \frac{\sigma_v^2 L_v}{\pi U_0} \frac{1 + 3\left(\dfrac{L_v \omega}{U_0}\right)^2}{\left(1 + \left(\dfrac{L_v \omega}{U_0}\right)^2\right)^2} \tag{14.44}$$

Here U_0 is the trim velocity of the aircraft, L_v is the vertical scale of turbulence, and σ_v is the root mean square of the vertical gust velocity. In developing the Dryden model, atmospheric turbulence is assumed to be a stationary random process. Here the vertical gust velocity σ_v for clear air, a cumulus cloud, and a severe storm are: 0.5 m/s, 2.0 m/s, and 4.0 m/s, respectively. In order to generate a signal corresponding to clear sky (normal wing vibration) and wind gust signals of the required intensity, scale lengths, and PSD functions, a Gaussian white noise source of $n(t) \sim \mathrm{N}(0,1)$ with a PSD function of 1, is amplified by a wind-gust gain K_g and filtered by a wind-gust transfer function $G_v(s)$, for a given flight velocity and height. A schematic representation of wind-gust generation for open-loop and closed-loop tip displacement response of the wing spar is shown in Figure 14.8.

The relationship between the PSD of the output signal $v_g(t)$ and the input signal $n(t)$ is given by:

$$\Phi_v(\omega) = K_g^2 \left|G_v(\omega)\right|^2 \Phi_N(\omega) \tag{14.45}$$

Choosing the noise source to be Gaussian white noise – $\Phi_N(\omega) = 1$ – then the wind-gust transfer function becomes:

$$\left|G_v(\omega)\right|^2 = \frac{\Phi_v(\omega)}{K_g^2} \tag{14.46}$$

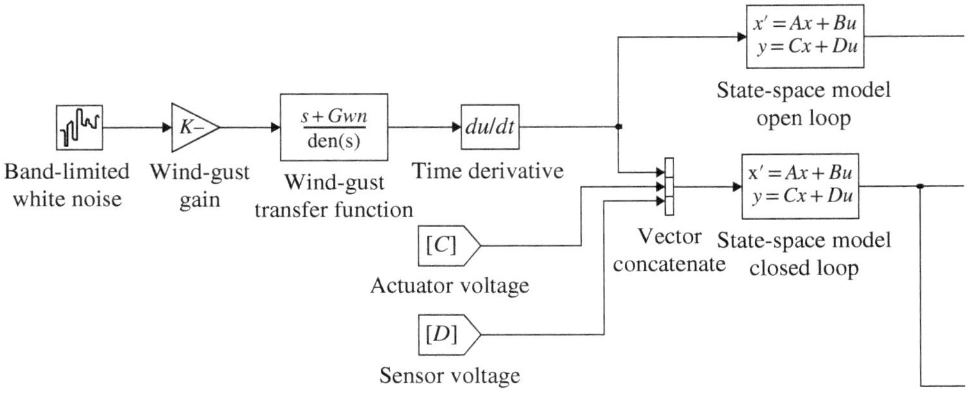

Figure 14.8 Block diagram of wind-gust signal generation for open-loop and close-loop tip displacement response.

Here the values for K_g and G_v are given by (s is the Laplace variable):

$$K_g = \sqrt{\frac{3\sigma_v^2 U_0}{\pi L_v}}, \quad G_v(s) = \frac{s + \dfrac{U_0}{\sqrt{3}L_v}}{\left(s + \dfrac{U_0}{L_v}\right)} \tag{14.47}$$

Figure 14.9 shows an example of simulated clear-sky normal wing vibration (dotted line) and cumulus cloud wind gust (solid line) signals, in acceleration units ($g = 9.81\,\text{m/s}^2$).

These PSD models are used in the next section to determine just how a self-powered gust-control system would work on a small UAV. The idea is to harvest the signal in the dotted line (clear sky) in Figure 14.9 over a long time period in order to control the signal in the solid line – the gust response – over a short time.

14.6 Experimental Validation

The modelling described above was validated with experimental data using a composite 'spar'. The spar consisted of a 504 × 38 × 3.05 mm aluminum substrate within the dimensions of a beam designed to fit in the tapered wing cavity of a specific radio-controlled model airplane. The length-to-thickness ratio of this design satisfies the required spanwise rigidity for normal flight. The rectangular substrate is configured with one layer of QuickPack QP10n patch (forming the PZT harvester/sensor layer suggested in Figure 14.1) and a layer of Smart Material Corp MFC 8528P1 patch (forming the PZT actuator). Both PZT layers were bonded with Kapton and 460 3D epoxy layers under 0.8 atm vacuum for 6 h to minimize variations in epoxy thickness and to eliminate air gaps. The PZT layers were positioned 54 mm away from the root end of the beam, corresponding to the beam length attached to the fuselage when the wing spar is inserted into the wing and fuselage. This location experiences the largest strain during flight, hence the placement of the piezoelectric devices at the root of the beam, as pictured in Figure 14.10.

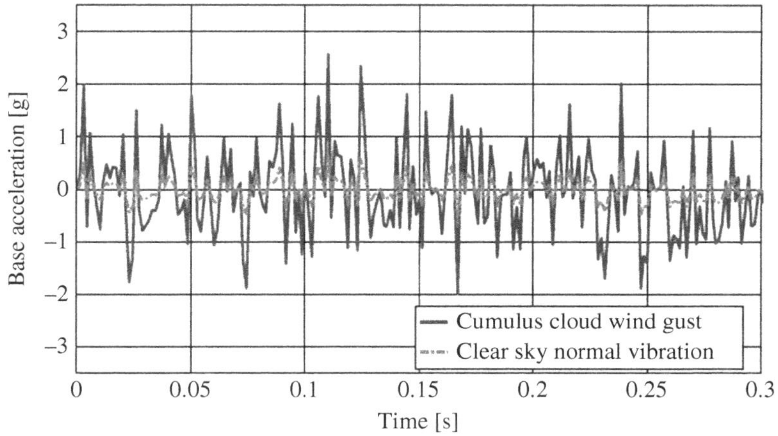

Figure 14.9 Normal wing vibration and wind gust acting on multifunctional wing spar base, $U_0 = 15\,\text{m/s}$, $L_{v=}350\,\text{m}$.

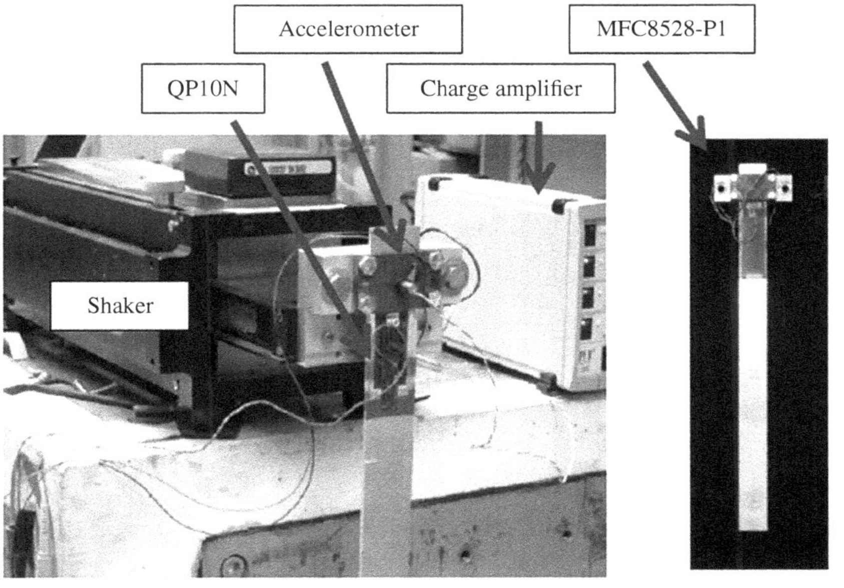

Figure 14.10 Wing spar validation test: left, front view; right, back view.

The experimental apparatus is shown in Figure 14.11, where the representative spar is clamped to an APS Dynamics ELECTRO-SEIS long-stroke shaker with frequency range 0.1–200 Hz to provide the simulated excitation and sine sweeps for model validation. The shaker is driven by an APS dual-mode power amplifier. Normal vibration signals are generated for harvesting purposes and wind-gust signals for control purposes. The white noise and Dryden PSD functions are coded in Matlab/Simulink and implemented via dSPACE. The absolute tip displacement is measured using an MTI LTC-50-20 laser sensor.

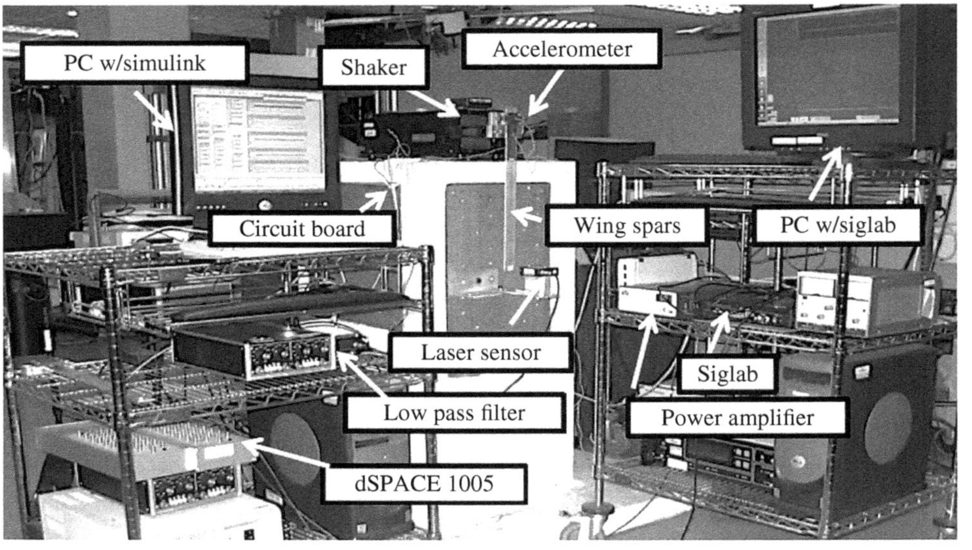

Figure 14.11 Configuration of the test of the gust alleviation system using REC laws from harvested energy.

14.7 Performance

With the validated model in hand, numerical experiments were conducted using the Dryden PSD to simulate aerodynamic loads. The numerical experiments allowed clarification of the conditions under which the proposed self-powered, gust-alleviation system will be practical. In order to compare control performance for a range of control parameters the frequency spectrum of the measured relative tip displacement is calculated. The various control gains and corresponding vibration reduction levels are shown in Figure 14.12 and listed in Table 14.3. A modified version of the positive position feedback control law is used. This is called positive strain feedback (PSF). The PSF control algorithm uses strain feedback from a strain gauge rather than using a position sensor, as is required in standard PPF. The remaining control law is exactly the same as PPF. However, by using strain directly, the need for an external sensing source and power supply is completely eliminated, allowing the system to be self-contained. The Laplace transfer function of PSF for the first mode (driven by a PSF op-amp and a voltage buffer op-amp) is:

$$\frac{v_o}{v_i} = k\frac{\omega_f^2}{s^2 + 2\zeta_f\omega_f s + \omega_f^2} \tag{14.48}$$

The PSF filter parameters: gain k, damping ratio ζ_f, and filter natural frequency ω_f are given by:

$$k = \frac{R_2}{R_1}\left(1 + \frac{R_5}{R_4}\right), \quad \omega_f^2 = \frac{1}{C_1C_2R_2R_3}, \quad \zeta_f = \frac{C_1(R_2 + R_3)}{2\omega_f} \tag{14.49}$$

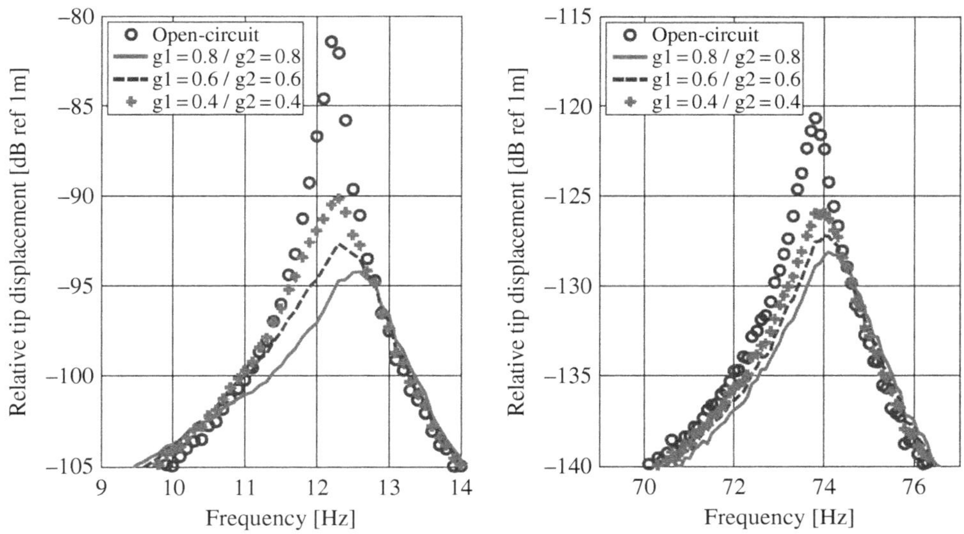

Figure 14.12 Control performance of PSF controllers for different control gains for the first (left) and second (right) modes. Damping ratio of mode 1: $\zeta_1 = 0.15$ and mode 2: $\zeta_2 = 0.35$.

Table 14.3 Control performance vs PSF control gains.

Two-mode control gains	0.8	0.6	0.4
First mode reduction (dB)	13	11	9
Second mode reduction (dB)	8	7	6

In addition, experimental measurement shows that when the control gain is fixed $(0.1 < g_1, g_2 < 1.2)$, changing the controller's damping ratio $(0.10 < \zeta_1 < 0.55, 0.30 < \zeta_2 < 0.75)$ has little effect on the control performance.

A Spectral Dynamics SigLab 20–42 data acquisition system was used for recording the absolute-tip-displacement-to-base-acceleration frequency-response function, where a PCB Piezotronics U352C67 accelerometer was employed for input acceleration measurements. Similar to Eq. (14.42), the absolute-tip-displacement-to-base-acceleration frequency-response function becomes:

$$\frac{w_{\text{abs}}(L,\omega)}{a_f} = \frac{1}{-\omega^2} + \Phi^T(L)\left(-\omega^2\mathbf{M} + j\omega\mathbf{C} + \mathbf{K} + \frac{-j\omega\theta^o\theta^{oT}}{j\omega C_p^o + \dfrac{1}{Z^o}}\right)^{-1}\mathbf{M} \tag{14.50}$$

These are in agreement with time-domain measurements, which are not shown here for space considerations (see Wang 2012).

The purpose of the frequency-response plots in Figure 14.13 is two-fold. First the figure shows that the same control performance (11-dB reduction for Mode 1, 7-dB reduction for Mode 2 from open-circuit control (dashed line)) is reached using the

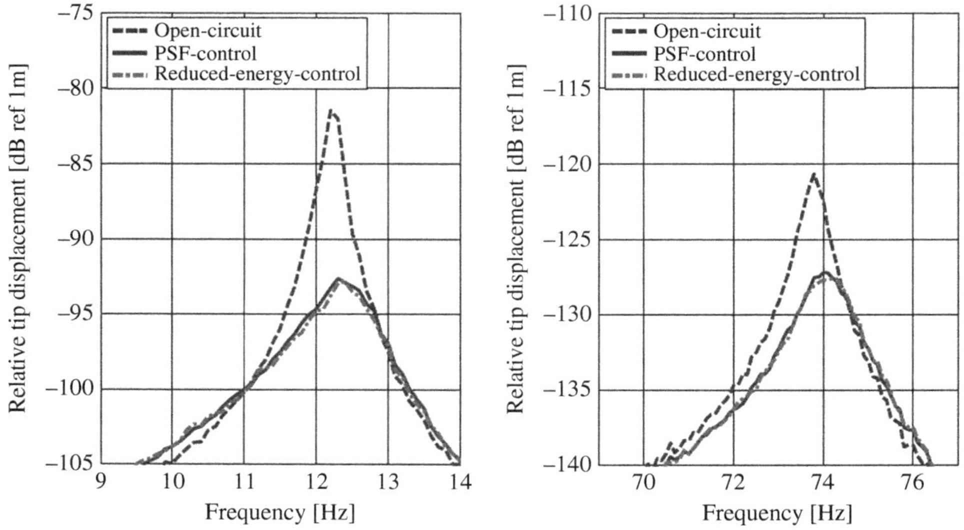

Figure 14.13 Vibration suppression performance using the PSF and REC laws for (left) mode 1 and (right) mode 2.

PSF controller (solid line) with gains $g_1 = 0.6$, $g_2 = 0.6$, damping ratios $\zeta_1 = 0.15$, $\zeta_2 = 0.35$; and using the REC controller (dotted line) with gains $g_1 = 0.8$, $g_2 = 0.8$, damping ratios $\zeta_1 = 0.15$, $\zeta_2 = 0.35$, saturation voltage of 120 V.

Secondly, Figure 14.13 also illustrates that the time-domain steady-state response under the PSF and REC controllers is much better than the open-loop response. Both the PSF controller (solid line) and the REC controller (dashed line) are designed to reduce the RMS value of the relative tip displacement from 3.0 mm for the open-circuit case (dotted line) down to 1.2 mm when using active control. Of course the difference is that the REC requires an order of magnitude less energy to achieve the same response.

Figure 14.14 repeats the same information presented in Figure 14.13 except in the time domain. The plot shows three curves of wing tip displacement versus elapsed time. The large-amplitude time response is the wing tip oscillating as the result of an applied gust with no control applied. The other two lower-amplitude responses are the wing-tip response with the two different control laws, illustrating the reduction in maximum deflection achieved by the gust-control law. Again, the REC law achieves the same result as the traditional controller but uses an order of magnitude less energy.

The actuation voltage was amplified (×200 gain) using a TREK 2220 power amplifier, and applied across the MFC actuator. Both the REC and the PSF control laws were tuned to an identical closed-loop performance. The actuation voltage and current required by each law was measured and the power calculated for each. These values are listed in Table 14.4. Note that in practice, this same result is achieved with a small printed circuit board having a mass of a few grams powered by a small battery using very little power. The average harvested power P^{hav} was 0.3 mW. All values in Table 4 correspond to an 11 dB reduction in the 1st Mode and 7 dB reduction in the 2nd Mode compared to the open-circuit gust response. The energy E_{tr} is the energy consumed for transient vibration control and the power required to achieve steady state vibration

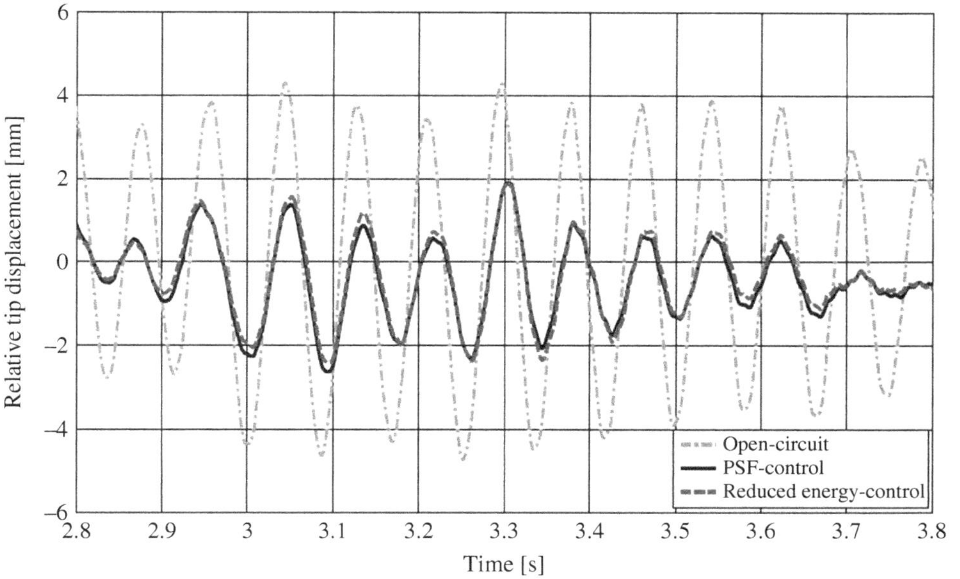

Figure 14.14 Control performance using PSF and REC laws compared to the open-loop response (time history of relative tip displacement response).

Table 14.4 Power and energy elements associated with PSF and REC laws.

Element		PSF control	REC	Ratio of reduction
Voltage RMS	V	100	82.3	18%
Current RMS	mA	0.065	0.053	18%
Required energy E_{tr}	mJ	6.6	1.6	76%
Average power P_{st}	mW	1.2	1.0	17%

suppression is P_{st}. The settling time t_s is defined as the time needed for the controlled response to reach 40% of the amplitude of the open loop response, and is 0.8 s for this case (see Wang and Inman 2013a and 2013b for more details and examples).

The main question is does this work? To provide an answer, some simple calculations are made to show that the answer is, 'it depends.' While no universal answer can be given for all situations, the following formulations can guide the answer to this question. For any wind-gust duration t_g, the total energy required to perform control E_{tot} can be calculated by:

$$E_{\text{tot}} = E_{\text{tr}} + P_{\text{st}} * \left(t_g - t_s\right) \tag{14.51}$$

The total energy required for the REC control law becomes:

$$E_{\text{tot}}^{\text{REC}} = 1.6 + 1.0 * \left(t_g - 0.8\right). \tag{14.52}$$

The time t_h required to harvest enough energy to control a wind gust of duration t_g is given by:

$$t_h = \frac{E_{\text{tot}}}{P^{\text{harv}}} \tag{14.53}$$

With these formulas in mind, whether or not the self-powered gust-control system works depends on a number of factors. First, the clear-sky condition must charge up the battery until it contains enough energy to control the gust. The time to do this depends on the values of the clear-sky vibration, the dynamics of the wing and the efficiency of the harvesting system. The energy required of the battery for the control system to work depends on the physical properties of the gust: duration, the trim velocity of the aircraft, the vertical scale of turbulence, and the root mean square of the vertical gust velocity.

Numerical simulations were used to obtain some nominal values of how long it takes to harvest energy before control can be performed. The power required for a 28-dB reduction of the first mode and a 37-dB reduction of the second mode in tip displacement is 6.42 mW, which is 40 times higher than the harvesting power of 0.16 mW. Thus in order to control a 1-s wind gust, 30 s of clear-sky harvesting is required.

14.8 Other Considerations

This chapter only looked at harvesting effects under flutter speed. However, several other aspects of the proposed system are worth discussing. These are the effect of harvesting on the aircraft's flutter speed, the use of solar energy, and associated electronics. Harvesting energy involves extracting energy from a structure and, as such, provides some vibration suppression or additional damping to the structure. Erturk et al. (2010) built an piezoaeroelastic model on different aeroelastic effects at the flutter boundary of a piezoaeroelastic airfoil. Based on a two degree of freedom pitch-plunge model typically used to study flutter, they investigated the energy-harvesting ability of a flexible wing in a flow. They added piezoelectric coupling to the plunge degree of freedom and termed the system 'piezoaeroelastic' to indicate the interplay between structure, piezoelectricity and aerodynamics. While their goal was to see how much energy could be extracted from a flow (as used in the previous sections) they also discovered that the 'shunt' damping phenomenon had a noticeable effect on the flutter speed, or the speed at which flutter is initiated. Based on a lumped-parameter state-space model with aerodynamic loads obtained using the lumped-vortex model they investigated both linear and non-linear energy-harvesting scenarios. Furthermore, their linear piezoaeroelastic predictions were verified against wind-tunnel tests, adding credibility to their results. Their results clearly showed that the shunt effect of the piezoaeroelastic harvester increased the flutter speed by 5.5%. In an experiment defined to measure this, a 5.1% increase in flutter speed was measured.

Next consider including the energy-harvesting capability of the solar panel indicated in Figure 14.1. There are numerous articles, laboratory experiments and flight tests involving harvesting energy from the sun using a variety of different types of solar panels embedded in or layered onto wings (see Anton et al. 2010 for a review). The most interesting studies are those of Thomas and Qidwai (2005) and Qidwai et al. (2009), who examined replacing the skin of a small UAV with a structural battery, effectively launching the field of multifunctional structures. Obviously, solar panels can harvest much more energy than is available from wing vibrations. However, solar power is not always available, so creating a system that uses both solar and vibration sources is a worthwhile endeavor.

The circuits for harvesting each type of energy are shown in Figure 14.15. The main difference between the two is that piezoelectric harvesting produces an alternating current (AC) and solar panels produce a direct current (DC). The oscillatory voltage from the piezoelectric material requires rectification and an efficient power-conditioning circuit if the voltage is to be used to charge the battery (a thin-film battery in this case). A full-wave rectifier followed by a linear voltage regulator is used to convert the AC voltage to DC. Because of the importance of matching impedance to encourage the maximum transfer of energy, a non-linear switching circuit, the parameters of which can be tuned to allow for impedance matching, is used. The first stage is a diode bridge, used to rectify the AC output of the piezoelectric element to DC. This is followed by a second stage consisting of a buck-boost converter designed to run in discontinuous conduction mode.

The switches' switching frequencies are designed to be considerably larger than the vibration frequencies. During each switching period, the input voltage to the buck-boost converter is virtually a DC voltage. The three intervals of the switching cycle are:

1) In the interval $D_1 T_S$, the switch turns on and the piezoceramic element charges the inductor and the inductor current increases with a slope of *v*rect/*L*.
2) During the second interval, $D_2 T_S$, the switch turns off and the inductor is discharged through the load and the inductor current decreases with a slope of *vo*/*L* until it reaches zero.
3) During the remaining switching cycle, the switch remains off and the inductor current remains at zero.

The advantage of this circuit is that the input impedance of the DC-DC converter is inherently a resistance (Kong, et al. 2010) with a value given by

$$R_{\text{in}} = \frac{v_{\text{rect}}}{\frac{1}{T_s}\int_0^{D_1 T_S} i_L \, dt} = \frac{2L}{D_1^2 T_s} \tag{14.54}$$

where *v*rect is the rectified voltage, L is the inductor value, I_L is the current through the inductor, T_S is the switching period, and D_1 denotes the duty cycle. By adjusting the duty cycle of the gate signal for the switch, the optimal resistive load can be emulated by the buck-boost converter given by

$$D_{1,opt} = \sqrt{\frac{2L}{R_{in,opt} T_S}} \tag{14.55}$$

The output power is not a function of the load. More details about the resistive impedance matching circuit can be found in Kong et al. (2010).

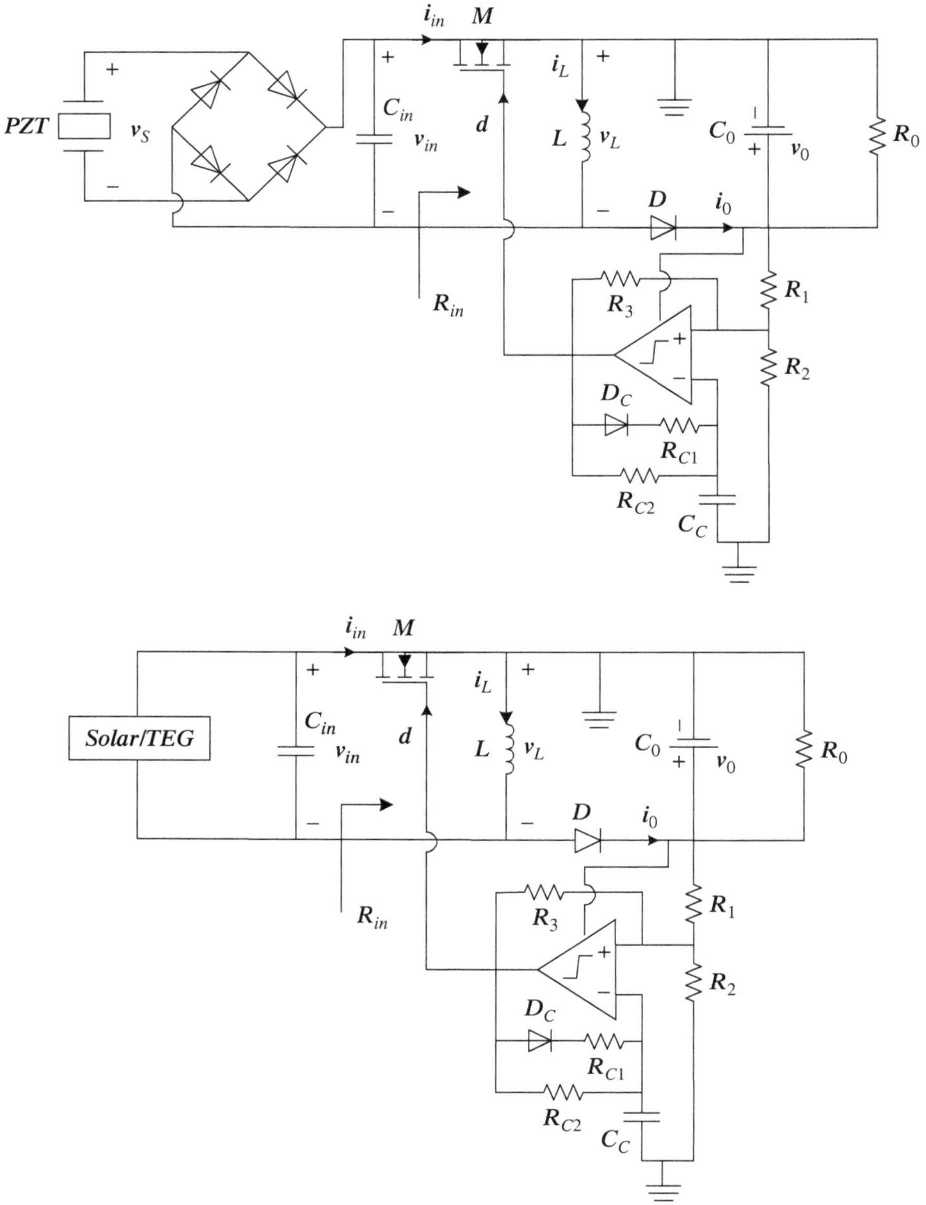

Figure 14.15 Sample circuitry needed to convert electrical energy from transducers into useable voltage for chargin: left, for piezoelectric transduction; right, for solar panel transduction.

A similar circuit for converting the voltage from the solar panel to a form capable of charging the battery is shown on the right of Figure 14.15. Since the solar-generated voltage is DC, no rectifier is needed. However, the buck-boost converter is used, generating an equivalent resistance as given in Eq. (14.54). Each harvesting technique needs

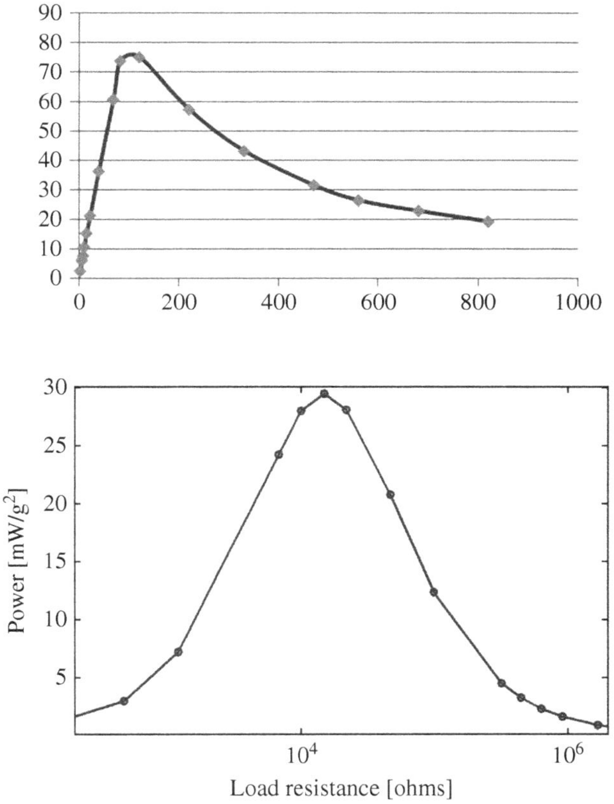

Figure 14.16 Optimal load resistance for: (left) solar harvesting and (right) piezoelectric harvesting.

to see an optimal resistance, as shown in Figure 14.16. Note that the two optimal resistances are orders of magnitude apart, so the duty cycle in each circuit must be chosen to provide the optimal resistance in order to charge a single battery simultaneously using both solar and vibration energy.

The circuit side of energy harvesting is still a highly researched area. Some are using adaptive circuits to maximize the amount of power extracted from the transducer.

14.9 Summary and Discussion

In summary, a novel concept of simultaneous energy harvesting and gust alleviation for a small UAV was presented. Vibration energy from clear-sky wing oscillations is harvested and shunted to a storage battery. A sensor system determines if the wing is experiencing a gust load and turns on an active control system to cancel the vibratory effects of the gust. The concept of a multifunctional wing spar is used to provide all the required functions: sensing, harvesting, energy storage, and control. Sensing, harvesting, and actuation use the piezoelectric effect. Because the amount of energy available from harvesting clear-sky wing conditions is very low, a recently developed reduced energy control law is implemented to reduce control power while preserving control performance. Numerical simulations and experimental results show that the tip displacement due to a wind-gust disturbance can be reduced by 11 dB and 7 dB for the first and second modes, respectively.

Although not yet implemented, the autonomous version of the self-powered gust-control system would work as follows. The PZT as sensor would constantly measure strain and simultaneously charge the battery. When the strain exceeded a preset value, indicating a gust, the control system would turn on and reduce the strain to desired values, effectively cancelling the gust. When the strain dropped below this desired value the controller would shut down and the system would continue to recharge the battery with the vibration induced by the clear-sky conditions. As pointed out, this only works for one gust. If a second gust occurs before the battery is charged sufficiently then the controller will be less effective. However, if a second gust occurs after a sufficient delay, then the system will again cancel the effects of the gust.

Acknowledgement

The authors gratefully acknowledge the support from the U.S. Air Force Office of Scientific Research under the grants F9550-06-1-0326 'Energy Harvesting and Storage Systems for Future Air Force Vehicles' and F9550-09-1-0625 'Simultaneous Vibration Suppression and Energy Harvesting' monitored by Dr B.L. Lee.

References

Anton, S.R., Erturk, A., and Inman, D.J., 2010, 'Multifunctional self-charging structures using piezoceramics and thin-film batteries,' *Smart Materials and Structures*, **19** (11): 115021–115035. doi: 10.1088/0964-1726/19/11/115021

Bent, A.A., and Hagood, N.W., 1997, 'Piezoelectric fiber composites with interdigitated electrodes,' *Journal of Intelligent Material Systems and Structures*, **8** (11): 903–919.

Bilgen, O., Wang, Y., and Inman, D.J., 2011, 'Electromechanical comparison of cantilevered beams with multifunctional piezoceramic devices,' *Mechanical Systems and Signal Processing*, **27**: 763–777. doi: 10.1016/j.ymssp.2011.09.002.

Castelazo, I.A., and Lee, H., 1990, 'Nonlinear compensation for flexible manipulators,' *Journal of Dynamic Systems, Measurement and Control, Transactions of the ASME*, **112**: 62–68.

De Marqui, Jr., C., Vieira, W.G.R, Erturk, A., and Inman, D.J., 2011, 'Modeling and analysis of piezoelectric energy harvesting from aeroelastic vibrations using the doublet-lattice method,' *ASME Journal of Vibration and Acoustics*, **133** (1):011003.

Dunbar, S., and Popović, Z., 2014, 'Low-power electronics for energy harvesting sensors,' *Wireless Power Transfer*, **1**, 35–43.

Elvin, N.G., and Elvin, A.A., 2009, 'A general equivalent circuit model for piezoelectric generators,' *Journal of Intelligent Material Systems and Structures* **20**: 7.

Erturk, A., Vieira, W.G.R, De Marqui, Jr., C., and Inman, D.J., 2010. 'On the energy harvesting potential of piezoaeroelastic systems,' *Applied Physics Letters*, **96**: 184103.

Erturk, E., and Inman, D.J., 2011. *Piezoelectric Energy Harvesting*, John Wiley & Sons.

Etele, J., 2006, *Overview of Wind Gust Modelling with Application to Autonomous Low-Lever UAV Control*, Technical report CR 2006-221, Defence R&D Canada.

Fanson, J.L. and Caughey, T.K., 1990, 'Positive position feedback control for large space structures,' *AIAA Journal*, **28**: 717–24.

Hagood, N.W., Kindel, R., Ghandi, K., and Gaudenzi, P., 1993. 'Improving transverse actuation using interdigitated surface electrodes,' in: *North American Conference on Smart Structures and Materials*, N. Albuquerque, SPIE Paper No. 1917-25: pp. 341–352.

Hoblit, F.M. (1988). *Gust Loads on Aircraft: Concepts and Applications*. American Institute of Aeronautics and Astronautics.

IEEE Standard, 1987, '*Institute of Electrical and Electronics Engineers Standard on Piezoelectricity*,' ANSI/IEEE Standard 176–1987.

Kong, N., Ha, D.S., Erturk, A., and Inman, D.J., 2010, 'Resistive impedance matching circuit for piezoelectric energy harvesting,' *Journal of Intelligent Material Systems and Structures*, **21** (13): 1293–1302.

Levine, W.S., 1996, *The Control Handbook*, CRC Press; IEEE Press.

Liang, J.R., and Liao, W.H., 2009, 'Piezoelectric energy harvesting and dissipation on structural damping,' *Journal of Intelligent Material Systems and Structures* **20**: 27.

McLean, D., 1990, *Automatic Flight Control System*, Prentice Hall.

O'Dwyer, A., 2009, *Handbook of PI and PID Controller Tuning Rules*, Imperial College Press.

Pederson, D.O., Studer, J.J., and Whinnery, J.R., 1966, *Introduction to Electronic Systems, Circuits, and Devices*, McGraw-Hill.

Qidwai, M.A.S., Thomas, J.P., and Pogue III, W.R., 2009, 'Structure-battery composites for UUVs: multifunctional interaction effects,' in *Proceedings of the 50th AIAA/ASME/ASCE/AHS/ASC Structures, Structural Dynamics, and Materials Conference*.

Regan, C.D., and Jutte, C.V, 2012, *Survey of Applications of Active Control Technology for Gust Alleviation and New Challenges for Lighter-weight Aircraft*, Techical report, TM-2012-216008, NASA.

Richardson, J.R., Atkins, E.M., Kabamba, P.T., and Girard, A.R, 2011, 'Envelopes for flight through stochastic gusts,' *Proceedings AIAA Atmospheric Flight Mechanics Conference*, Portland Oregon. AIAA paper 2011–6213.

Roundy, S., and Wright, P.K., 2004, 'A piezoelectric vibration based generator for wireless electronics,' *Smart Materials and Structures* **13**: 12.

Shao, K., Wu, Z., Yang, C., Chen, L., and Lv, B., 2010, 'Design of an adaptive gust response alleviation control system: simulations and experiments,' *Journal of Aircraft*, **47** (3): 1022–1029.

Silva, W.A., Vartio, E., Shimko, A., Kvaternik, R.G., Eure, K.W., and Scott, R.C., 2006, 'Development of aeroservoelastic analytical models and gust load alleviation control laws of a SensorCraft wind-tunnel model using measured data.' In *47th AIAA/ASME/ASCE/AHS/ASC Structures, Structural Dynamics, and Materials Conference*, 1–4 May 2006, Newport, Rhode Island. AIAA 2006–1935.

Sodano, H.A., Park, G., and Inman, D.J., 2004, 'An investigation into the performance of macro-fiber composites for sensing and structural vibration applications,' *Mechanical Systems and Signal Processing*, **18** (3): 683–697.

Sodano, H.A., Park, G., and Inman, D.J., 2005, 'Comparison of piezoelectric energy harvesting devices for recharging batteries,' *Journal of Intelligent Material Systems and Structures*, **16** (10), 799–808.

Sodano, H.A., Lloyd, J., and Inman, D.J., 2006, 'An experimental comparison between several active composite actuators for power generation,' *Smart Materials and Structures*, **1**: 1211–1216.

Thomas, J.P., and Qidwai, M.A., 2005, 'The design and application of multifunctional structure-battery materials systems,' Journal of The Minerals, Metals & Materials Society, **57** (3): 18–24.

Vable, M., 2002, *Mechanics of Materials*, Oxford University Press.

Vartio, E.J., Shaw, E.E., and Vetter, T., 2008, 'Gust load alleviation flight control system design for a SensorCraft vehicle,' in *26th AIAA Applied Aerodynamics Conference* 18–21 August 2008, Honolulu, Hawaii. AIAA 2008–7192.

Vartio, E., Shimko, A., Tilmann, C.P., and Flick, P.M., 2005, 'Structural modal control and gust load alleviation for a Sensorcraft concept,' in *46th AIAA/ASME/ASCE/AHS/ASC Structures, Structural Dynamics and Materials Conference* 18-21 April, Austin, Texas. AIAA 2005–1946.

Wang, Y., 2012, *Simultaneous Harvesting and Control*, PhD Dissertation, Virginia Tech.

Wang, Y., and Inman, D.J., 2011, 'Comparison of control laws for vibration suppression based on energy consumption,' *Journal of Intelligent Material Systems and Structures*, **22** (8): 795–809.

Wang, Y., and Inman, D.J., 2013a, Simultaneous energy harvesting and gust alleviation for a multifunctional wing spar using reduced energy control via piezoceramics, *Journal of Composite Materials*, **47** (1): 125–146.

Wang, Y., and Inman, D.J., 2013b, Experimental validation for a multifunctional wing spar design with sensing, harvesting and gust alleviation capabilities, *IEEE/ASME Transaction on Mechatronics*, **18** (4): 1289–1299.

Wilkie, W.K., Bryant, G.R., and High, J.W., 2000, 'Low-cost piezocomposite actuator for structural control applications,' in: *Proceedings of the SPIE 7th Annual International Symposium on Smart Structures and Materials*, Newport Beach, CA.

Wright, J.R., and Cooper, J.E., 2007, *Introduction to Aircraft Aeroelasticity and Loads*, John Wiley and Sons.

16

Flow Structure Modification Using Plasma Actuation for Enhanced UAV Flight Control

Antonio J. Conesa Torres

Experimental Aerodynamics Branch, INTA, Madrid, Spain

16.1 Introduction

In recent years, plasma actuators have been intensively studied and developed, representing a fascinating and promising technology for active aerodynamic flow control [1]. There is worldwide interest in increasing the roles and capabilities of unmanned aerial vehicles (UAVs) for both military and civil use [2]. If a vehicle has to be optimized for diverse roles and sometimes conflicting conditions and objectives, active aerodynamic flow control becomes a powerful tool for adapting the vehicle from one condition to another. So active flow control emerges as a promising strategy to enhance the aerodynamic performance and manoeuvre capability of UAVs.

Plasma actuators, specifically those based on dielectric barrier discharge (DBD) actuation, have a number of characteristics that make them ideal for active flow control, namely they are all electric, lightweight, and have fast responses among other properties. The following paragraphs describe recent investigations on plasma actuators based on DBD actuation for aerodynamic uses. The list is not exhaustive and many references are not given; the goal is to present the initial experiments and some of the most relevant research related to flight control of UAVs, showing the ability of these devices to control the airflow around lifting surfaces.

Most of the research conducted has focused on separation control over airfoils or wing sections. There are numerous experiments that have shown successful reattachment of separated flows. Roth [3] studied the effect of a One Atmosphhere Uniform Glow Discharge (OAUGD®) panel on a NACA 0015 airfoil by flow visualization. The chord was 12.7 cm, the freestream velocity 2.85 m/s and the Reynolds number around 10^4. The results show that flow separation is delayed by using the plasma discharge to produce an acceleration of the flow in the boundary layer.

Corke and his group at The University of Notre-Dame have investigated extensively the lift enhancement on airfoils [4–6]. In these papers, diverse aspects concerning separation control in stationary and oscillating airfoils at high angles of attack have been

Advanced UAV Aerodynamics, Flight Stability and Control: Novel Concepts, Theory and Applications, First Edition. Edited by Pascual Marqués and Andrea Da Ronch.

Companion website: http://www.wiley.com/go/marques/advanced_UAV_aerodynamics

studied along with steady and unsteady actuations and position of the actuator in order to simulate slats and flaps.

A plasma jet vectoring is used by Benard et al. [7] to control the flow separation of a NACA 0015 airfoil at a freestream velocity of 15 m/s. Wall-normal jet, co-flow and counter-flow were tested in a steady mode of actuation. The better results were obtained with a co-flow jet, delaying by 3° the stall angle of attack.

A comparison between two types of actuators based on DBD actuation (AC and ns) was performed by Kelley et al. [8]. Both actuator types successfully reattached the separated flow up to Ma = 0.4.

Several methods have been employed to create control forces. Patel et al. [2] used DBD actuators to manipulate the leading-edge vortex and the vortex breakdown structure to modify the lift over the 1303 unmanned combat air vehicle wing at high angles of attack. The tests were conducted at a freestream velocity of 15 m/s and the Reynolds number was 4.12×10^5, based on the root chord of 0.19 m. Several DBD plasma actuators were located at the leading edge of the inboard and outboard halves of the wing to provide longitudinal control without the hinged control surfaces. The results showed an increase in the lift characteristics of the wing for plasma-on cases. In the 18–35° angle of attack range, the control authority was significantly better than a 5° deployment of a traditional trailing-edge device (flap), thereby extending the operational flight envelope of the wing.

Lopera et al. [9] used a 47°-sweep scaled 1303 unmanned combat air vehicle model to assess a new design for flight control at low angles of attack. The wing was modified to produce an aft separation ramp on the windward side near the trailing edge. Plasma actuators were mounted at the onset of the windward-surface separation ramp. Experiments were conducted at a chord Reynolds number of 4.33×10^5, based on the mean chord of 0.2 m and freestream velocity of 15 m/s. The results showed that plasma actuation causes a reduction in the lift coefficient compared with the baseline case. Results also showed that unsteady (pulsed) plasma actuation has a better response than steady actuation. No moment measurements were performed, but from the reduction of lift it could be inferred that rolling and pitching moments could be generated by the placement of actuators at different parts of UAV.

Vorobiev et al. [10] have used plasma actuators to control the roll of a wing. A plasma actuator was placed into the suction side of both half wings, at the three-quarter-chord location ($x/c = 0.75$). The wing had a 0.2-m chord and a span of 0.91 m with a NACA 0009 profile. The driven waveform was a positive saw-tooth signal with a voltage of 16 kV at a frequency of 2.9 kHz. The results show that plasma actuation produced lift enhancements equivalent to a 3° deflection at 2 m/s, falling to a fraction of a degree at a freestream of 10 m/s.

Flight control using wing-tip plasma actuation was investigated by Boesch et al. [11]. DBD plasma actuation around the wing tip of a NACA 4418 with a rounded tip was proposed. The actuator adds momentum in the direction opposite to that of the flow forming the tip vortex. A variation of 1% of the lift coefficient was obtained at a freestream of 15 m/s.

An experimental investigation of airfoil circulation control by DBD actuation has been performed on a symmetric NACA64-2-A015 airfoil [12]. The curved airfoil trailing edge allows circulation control. The tests were conducted at three freestream velocities (10 m/s, 15 m/s and 20 m/s). Preliminary results indicate an increase in lift coefficient of approximately 0.1 for low Reynolds number.

The DBD plasma actuator has been used as a virtual Gurney flap [13]. The tests were carried out over a NACA 0012 airfoil and the actuator was placed over the pressure side, near the trailing edge, giving rise to tangential counter-flow. The results showed an increase in the lift and indicated that the virtual Gurney flap is equivalent to a mechanical Gurney flap of 1% chord length.

The literature reviewed in this section paragraphs shows the investigation of 2D bodies. This chapter will mainly focus on DBD plasma actuators to manipulate and control the airflow around a complete UAV: a 3D model. Two aspects will be investigated:

- enhancement of the lift by delaying the flow separation and increasing the angle of attack at which stall occurs, achieved by plasma actuators located at the wing leading-edge
- generation of aerodynamic control moments, by plasma actuators located at the trailing edge (wing and V-tail).

16.2 Aerodynamic Flow Control

In today's society, concerned with the economic and environmental impact of human activity, any device or method that makes better use of resources becomes the solution of choice. Flow control appears to be a means to follow this trend, due to the large number of activities in which fluids are involved and the improvements that can be obtained through their efficient control. The concept of flow control is not new; in 1904 a little-known physicist revolutionized fluid dynamics in the Third International Mathematics Congress held in Heidelberg (Germany). Ludwig Prandtl was a 29-year-old university professor, his presentation was ten minutes long and the congress proceedings paper ran to only eight pages. Entitled *Über Flüssigkeitsbewegung bei sehr kleiner Reibung* (On fluid motion with very small friction), it introduced the concept of boundary-layer theory and demonstrated how active flow control by suction can control the boundary layer and delay separation.

Flow control is a multidisciplinary research area which is making rapid advances. It aims to manipulate a flow field in order to get a better desired flow condition [14, 15]. Two main types of flow control can be distinguished: passive control and active control. Passive aerodynamic flow control usually consists of a surface geometrical modification, such as vortex generators on a wing, which extracts energy from the flow to realize the control process. Passive control devices act all the time and in all conditions and regimes. This actuation is clearly inefficient and usually increases induced drag. On the other hand, active flow control acts on demand and only if there is a need. When a need arises the active control system responds by adding momentum or energy to the flow. The active flow control can be activated, regulated and shut off at will, but involves power consumption and an additional complication of the system. An active control device must meet the technical challenge but also all other non-technical issues. The benefit must exceed the total cost of the operation: maintenance and others. It must be effective and efficient, giving a significant effect with just a little effort or energy input.

A review of actuators for active flow control can be seen in the paper by Cattafesta and Sheplak [16], which includes an evaluation of the strengths and drawbacks of each.

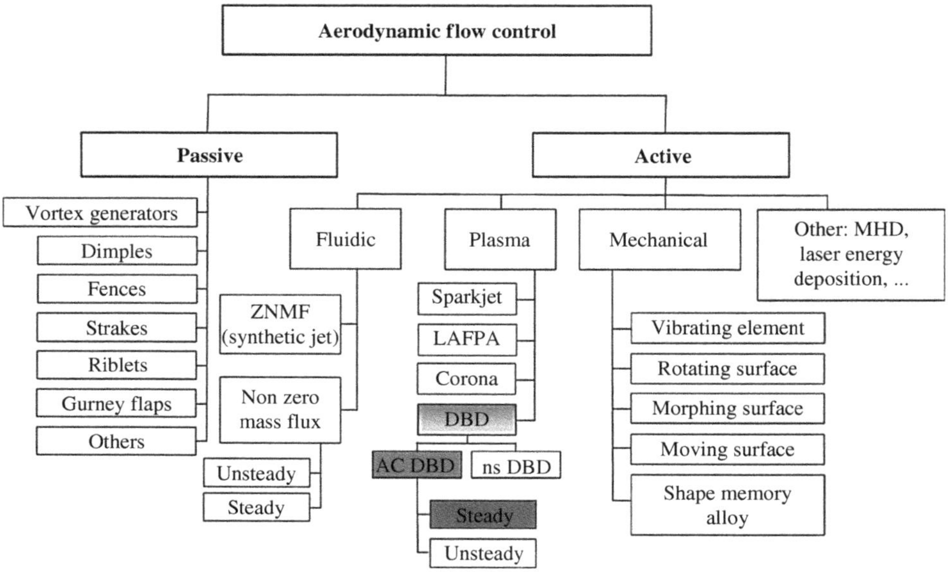

Figure 16.1 Aerodynamic flow control classification.

Figure 16.1 shows the types of devices (a non-exhaustive list) that can be found under each of the two flow control categories. Some devices may belong to both categories, but are classified in the strategy under which they were initially developed. For example, the Gurney flap was initially developed as a small flat plate that was attached to the trailing edge of the rear wing in racing cars, but has evolved into a device that can be fixed or actuated mechanically or can incorporate a jet or a plasma actuator to improve its performance when fired. The shaded devices shown in the figure will be used in this investigation.

Another classification employed by Wood [17] in his paper on flow control on UAVs could also be followed. Initially plasma actuators were not included in his study. His method for classifying actuators is based on a five-step labelling scheme:

- type of control (active or passive)
- interaction with the flow (geometric, fluidic or plasma in this case)
- the mechanism that produces an aerodynamic change (attached or separated flow)
- how the actuator operates in time (steady or unsteady)
- the aerodynamic force vector that is modified (lift, drag or lift/drag).

A technical maturity evaluation of the different control-effector concepts is presented in Wood's paper, along with a correlation with four UAV classes (micro, meso, macro and mega UAVs).

16.3 Plasma Actuators

A plasma actuator is a novel method of active aerodynamic flow control. The plasma actuator is a generic term that is used to refer to a wide range of devices based on the use of electric discharges at atmospheric pressure. The plasma actuation can modify the structure of a flow field, allowing the control of a flow around bluff bodies, airfoils, aircraft and low-pressure turbines.

An actuator is a device that converts an electrical input into a desired physical quantity [16]. Plasma actuators may produce different actuation mechanisms or effects on the flow field to be controlled. The effect of momentum has been the first and most used so far, inducing near-surface flow velocities up to 8 m/s. Another mechanism of actuation is the shock effect, as produced by nanosecond pulses, which induce a rise of gas pressure or temperature above the electrode, inducing flow velocities around 1 m/s or less. The chemical effect, which introduces new elements into the flow field as ions (positive and negative), electrons, free radicals, metastable elements and other species, is another aspect to be considered in the actuation of plasma devices.

The plasma concept was first used by Irving Langmuir [18], Nobel Prize winner in chemistry 1932, to describe partially ionized gases. Simply stated, plasma is a set of quasi-neutral elements with free electric charge carriers, which presents a collective behaviour. This ionized gas reacts to externally applied electric and magnetic fields.

Plasma is created by providing sufficient energy, usually thermal energy or a high-intensity electromagnetic field, to the atoms or molecules. The result is that a gas with electrons, positive and negative ions, free radicals and other species is obtained. The chemistry can be very rich, and depends on the gas involved. Two of the most important characteristics of a plasma are the electron temperature (which can go from 10^{-2}–10^{5} eV) and the degree of ionization; that is, the ratio of ionized particles to the total number of particles, given by Eq. (16.1).

$$\alpha_i = \frac{N_e}{N_e + N_n} \tag{16.1}$$

where N_e and N_n are the electron density and density of neutral species respectively, and α_i the degree of ionization. A gas is considered to be weakly ionized if the degree of ionization is less than 10^{-4} [19]. In a weakly ionized gas, the collision of charged particles with neutrals is more important than the interaction with other charged particles. As the degree of ionization increases, Coulomb interactions become more important. A weakly ionized gas is commonly called a cold plasma. If the plasma is in thermodynamic equilibrium is a thermal plasma, the temperatures of electrons, ions and neutral particles are equal. A plasma out of thermodynamic equilibrium is a non-equilibrium plasma, and the electron temperature is higher than the temperature of the other particles.

As can be seen in Figure 16.1, plasma actuation is divided into different discharge types. The first type studied in aeronautical applications was the corona discharge. A DC corona discharge is a very simple plasma actuator that can have several different configurations: two wires, point-to-plane, wire-to-plate and others. These discharges are based on electron avalanches (the Townsend mechanism), produced by the multiplication of some initial electrons.

In aerodynamic flow control the configuration is formed with two wires. One of the wires has a smaller diameter than the other, as shown in Figure 16.2a. When a high voltage is applied to the anode (the smallest wire), an intense electric field is created, ionizing the air around the wire. This deployment is called a positive corona. Figure 16.2b shows a photograph of wire-to-wire discharge in a flat plate; the plate is painted white for better visualization.

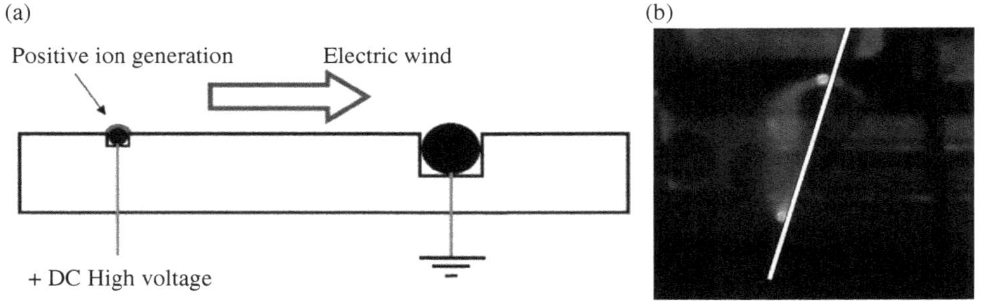

Figure 16.2 Discharge types: (a) Schematic of a positive corona discharge; (b) Photography of a DC discharge on a flat plate.

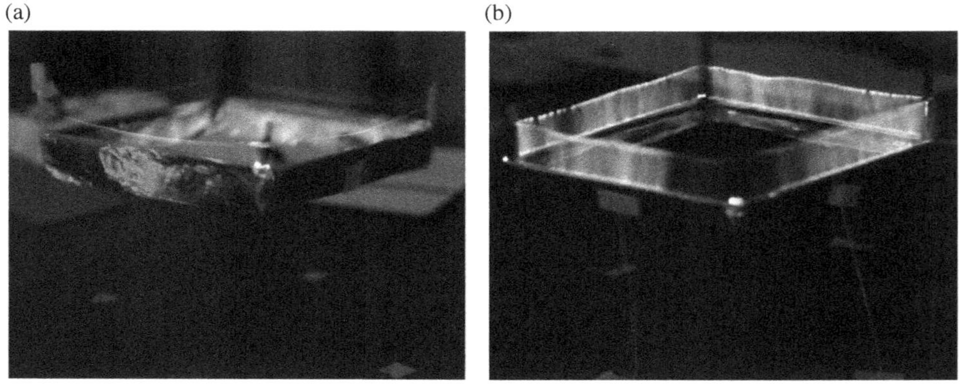

Figure 16.3 DC discharge on a lifter: (a) glow regime; (b) streamer regime. (*See insert for color representation of this figure.*)

The main disadvantages of the DC corona discharge are the glow-to-arc transition, in which efficacy is lost, and the sensitivity to several parameters, such as the humidity. However, the DC corona discharge setup is very simple and can be efficient in the generation of electric wind. They can therefore be used to develop thrusters powered by ionic wind for use in long-duration near-space vehicles [20, 21]. Figure 16.3 shows two lifters using this type of electric propulsion.

The lifters are raised about 10 cm from the table and are held with threads. The white adhesive tape can be seen on the table. A thrust-to power ratio of more than 120 N/kW was obtained in the High Speed Laboratory at INTA.

Although DBD, which will be explained in the next section, is more often used than DC corona discharge in airflow control applications, the latter is still a viable option. More details can be found in a very comprehensive review by Moreau [22].

Corona discharge and dielectric barrier discharge plasma actuators are both based on the electrohydrodynamic (EHD) interactions that are obtained when an electric field acts on a plasma. Due to the quasi-neutral character of the plasma, the applied electric field induces forces on the charged particles. Collisions between charged particles and neutrals produce a momentum transfer mechanism.

These actuators give excellent authority control at low velocities and Reynolds numbers of a few hundreds of thousands. When the flow velocity is higher, the momentum mechanism loses efficiency and actuators based on shock effects are more appropriate. The following methods have been proposed to control flows at high speed, including transonic, supersonic and hypersonic regimes.

Sparkjets (also called a pulsed plasma jet actuator or a plasma synthetic jet actuator) is a hybrid between a synthetic jet and a plasma actuator. It was developed at the Johns Hopkins University Applied Physics Laboratory by Cybyk et al. [23]. The device consists of electrodes housed in a small cavity and a discharge orifice or exit. An electric discharge (spark) across the electrodes produces a rapid heating in the cavity, which generates a high pressure and an injection of high-momentum fluid into the flow through the discharge orifice. The actuation is unsteady and each cycle is composed of three stages: energy deposition, discharge and recovery. This actuator can generate flows up to 300 m/s and repeatable pulses are obtained for frequencies up to 5 kHz [24]. Some improvements in the design have been performed by Popkin et al. [25]. Control of high-speed flows can be achieved with these devices.

Localized arc filament plasma actuators (LAFPAs), developed at The Ohio State University (OSU), also operate through gas heating. An electric discharge across pin electrodes produces strong localized, short-duration perturbations into transonic and supersonic flows. These rapidly heated regions allow flow instabilities to be excited. LAFPAs have been successfully used in the control of supersonic jets and produce wave–boundary-layer interactions [26, 27].

Finally, overviews of different plasma technologies under study for aeronautical and astronautical applications can be found in the literature [28, 29]. These include plasma aerodynamic flow control, on-board power generation and plasma-assisted combustion among others.

16.4 Dielectric Barrier Discharge

Volume dielectric barrier discharges (DBDs) or silent discharges were first proposed, for ozone production, by Werner von Siemens in 1857. Developments since then have been significant. Current applications include plasma display panels, surface cleaning and treatment, pollution control, CO_2 lasers, water purification, excimer lamps, polymer and textile treatments and ozone synthesis [30]. The most recent developments have been plasma actuators for active aerodynamic flow control and for sterilization and biomedical applications.

The publication by Roth et al. [31] in 1998 was the beginning of an exponential growth in research into aerodynamic applications of plasma discharges. Roth's discharge, called OAUGDP® (One Atmosphere Uniform Discharge Plasma), was initially dedicated to decontamination purposes, but when the authors realised the presence of a secondary airflow of several metres per second, the plasma flow control era began.

Since then, DBD actuators have become one of the hottest topics in fluid dynamics flow control due to their numerous advantages:

- no moving parts (mechanical reliability)
- low weight implying fewer inertial problems
- simplicity

- compactness
- robustness
- low power consumption
- fast response times
- the possibility of both steady and unsteady operation
- easy installation and easy application on the surface allowing retrofitting
- the ability for real-time control at high frequencies.

There are several disadvantages of DBD actuators though: the use of high voltages, problems associated with electromagnetic interference and the low efficiency of energy conversion.

A SDBD (surface dielectric barrier discharge) plasma actuator in its simplest configuration consists of two metallic electrodes placed on either sides of a dielectric medium and disposed in an asymmetric arrangement. When a high-voltage AC signal, exceeding the breakdown threshold, is applied to the electrodes, the air on both sides of the dielectric material will be ionized in the regions with the largest electric field potential. One of the electrodes is coated with an insulating material in order to prevent plasma formation. Usually, the grounded electrode is isolated, preventing plasma formation there and it is called a grounded, buried or encapsulated electrode.

A schematic diagram of the asymmetric surface dielectric barrier discharge plasma actuator can be seen in Figure 16.4. In this scheme, the lower electrode is encapsulated with the aerodynamic surface.

The weakly ionized air above the dielectric, which cover the encapsulated electrode, is subjected to the electric field gradient produced by the electrode configuration, which results in a body force in the direction of decreasing electric field potential acting on the neutral air. The geometry presented in Figure 16.4 causes a tangential jet flow that moves to the right in the scheme. Further insights into the mechanisms of electrohydrodynamic interactions may be found in the literature [32, 33].

The velocity of the tangential jet flow, or electric wind, produced by a plasma actuator increases with the applied voltage and frequency. The maximum velocity is near the wall, about 0.5–1.0 mm above the surface. Thus, the electric wind induced by a DBD can modify the boundary-layer properties and hence the flow-field structure. Several laboratory experiments have been carried out to characterize plasma actuators used in

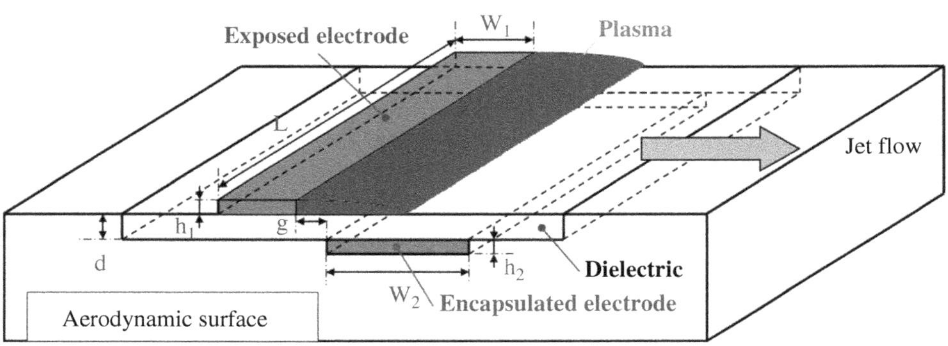

Figure 16.4 Schematic SDBD and geometric characterization.

wind tunnel tests. All the data presented in the following figures have been taken in a quiescent environment with $U_\infty = 0$ (no flow).

Figure 16.5 shows two examples of the velocity profile induced by a single actuator with a Kapton® dielectric layer of 0.2 mm and another actuator with a glass of 1 mm as the dielectric layer, as measured using a Pitot tube.

In Figure 16.6, the graphs of induced body force versus input voltage are shown for different frequencies. The dielectric medium is 2 mm thick glass and the electrodes are made of adhesive copper with a zero gap.

For a given voltage, the maximum thrust is achieved by increasing the frequency, on condition that discharge mode saturation does not occur. However, due to the effect of the actuator capacitance, the actuator is rapidly saturated with the frequency. In this situation, it is better to increase the voltage, while keeping the frequency at a lower value. Figure 16.7a shows a saturated plasma actuator that corresponds to the tipping

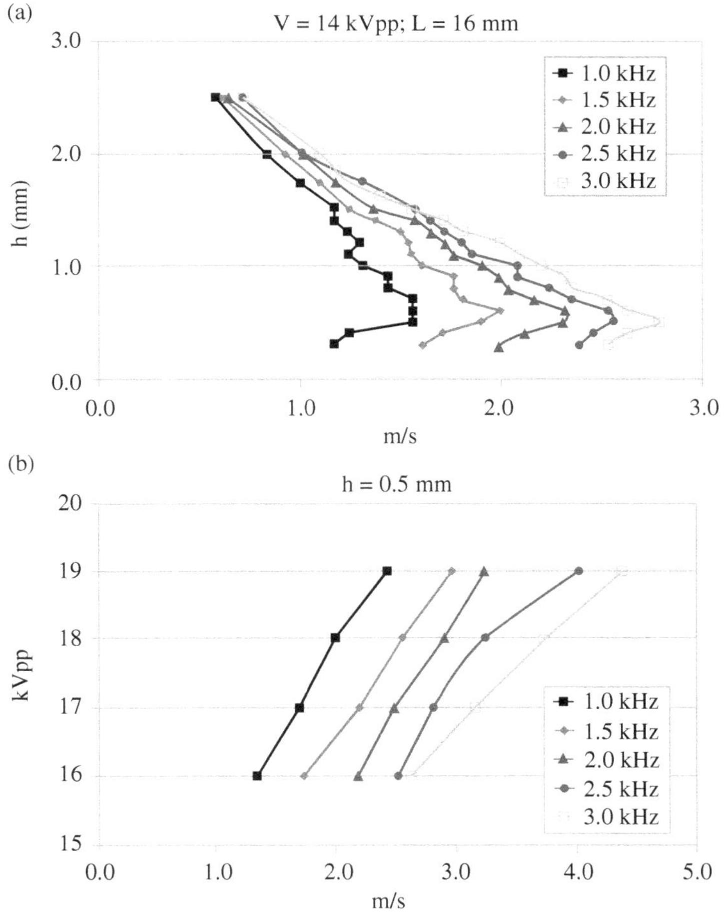

Figure 16.5 Voltage and frequency influence on velocity profiles: (a) dielectric layer Kapton®; Pitot located at 16 mm from trailing-edge exposed electrode; (b) dielectric layer glass; Pitot located at 0.5 mm above the actuator and at 16 mm from trailing-edge exposed electrode.

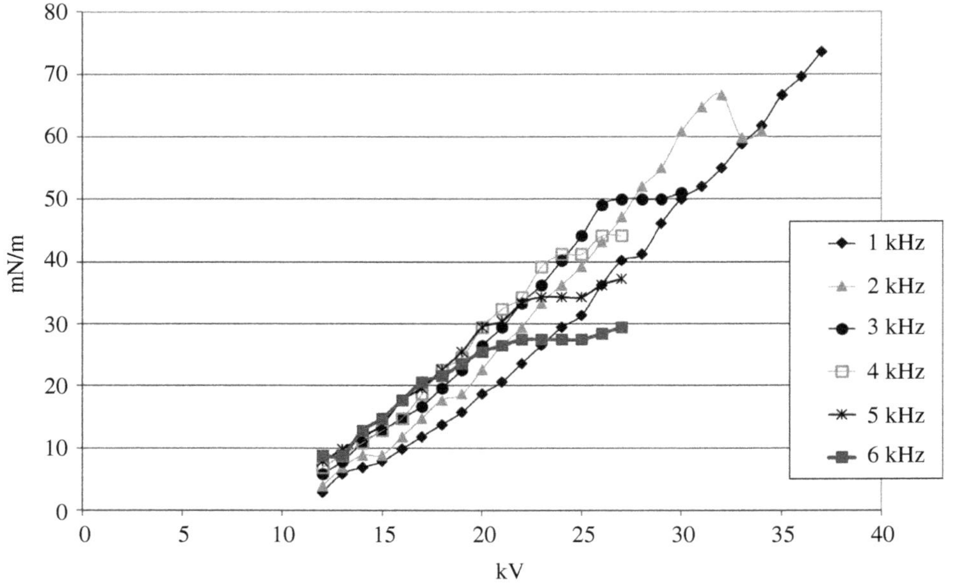

Figure 16.6 Induced thrust from a single dielectric barrier discharge plasma actuator with a 2 mm thick glass dielectric for different AC frequencies of the applied voltage.

point in Figure 16.6. The actuator does not induce more thrust and there is a risk of dielectric failure. The dielectric is perforated in Figure 16.7b. Figure 16.7a also shows a problem that is frequent in multiple DBD actuators: parasitic cross talk. There is an induced jet in the opposite direction to the main jet, which reduces efficiency.

In order to optimize the DBD plasma actuator performance, several parametric studies have been performed [34–36]. The main results are as follows. The height of the exposed electrode has a strong influence on the electric field, whereas its width and material seems to have a null or a weak influence on the performance. A low h_1, as depicted in Figure 16.4, produces better results. Normally, the exposed electrode consists of metallic strips or metallic wires. If the encapsulated electrode width increases (w_2), the length of plasma increases, giving an improvement in performance, but there is a value above which no further improvement is obtained. The actuator may be gapless, or have an overlap or a gap g between the electrodes, normally ranging from 5 to −5 mm. It seems that a slight gap ($g = 1$ or 2 mm) gives the best results. In this investigation, the electrodes will have a slight overlap of less than 0.5 mm, in order to obtain a uniform discharge. The material of the dielectric layer is important, because the dielectric constant ε affects the maximum thrust generated by the actuator before the discharge mode saturates. If the voltage is fixed, a high dielectric constant is preferred, although the electrical consumption is higher. However, if the power is fixed, a low dielectric constant material gives better results, allowing greater applied voltages [36]. Typical media used as the dielectric include Teflon®, glass, Kapton®, Macor®, Bakelite, aluminium oxide, quartz, Delrin® and poly(methyl methacrylate) (PMMA). There is no a perfect dielectric medium; the most appropriate depends on the particular application case.

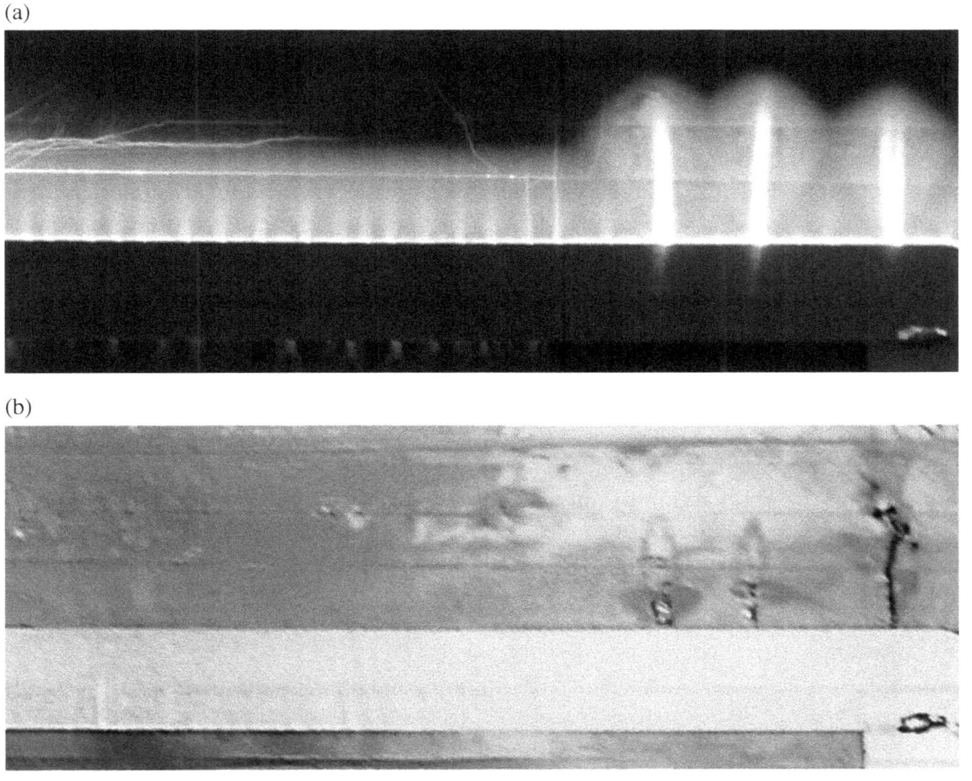

Figure 16.7 Photograph of a DBD: (a) The actuator is on saturated regime and the peak efficiency is lost; (b) The same actuator damaged by an excessive current.

Of great importance is the possibility of a DBD plasma actuator to tailor the body force and the induced jet through the design of the electrode arrangement. Several configurations are visualized in the next examples using a Schlieren system, in order to explore this capability. The Schlieren technique is used to study density fields in transparent media [37, 38], taking advantage of the relation between the fluid density and the index of refraction in the medium. The Gladstone–Dale equation (16.2) is a simplified relation, which holds quite well in gases.

$$\frac{n-1}{\rho}=C \tag{16.2}$$

where n is the refraction index, ρ the density, and C the Gladstone–Dale constant, which is a function of the gas and depends slightly on the wavelength. When the plasma is formed, a light density variation is produced, and this can be detected with the Schlieren system, enabling visualization of the starting vortex and the steady wall jet. The typical Z-configuration for Schlieren systems with mirrors has been employed. The system has two parabolic mirrors (diameter 200 mm and focal length 2500 mm), a light source, a condensing lens, a pinhole, flat mirrors, a knife, and a high-speed digital video camera (IDT, model Y3M).

For the coloured electrode configuration at the bottom of Figure 16.8, the two vortices and their evolution to a tangential wall jet induced by the steady plasma actuation in quiescent ambient air is shown. Initially, the double DBD actuator generates two vortices that go downstream (to the left). The top photograph shows the suction of the second vortex by the first; ultimately a tangential jet flow is established (not shown). The photograph also shows cross talk in the first electrode.

The example shown in Figure 16.9 is known as plasma synthetic jet actuator, which was developed by Santhanakrishnan [39]. This actuator mimics a synthetic jet, but it may operate in either a steady and unsteady manner. Two opposite jet are formed initially, and when these jets collide a normal wall jet is formed. In this example, the two DBDs are equal and with the same input waveform, but by changing these parameters a directional jet may be obtained.

The next example, Figure 16.10, shows an actuator that produces two tangential wall jets. It sucks air from above the electrode and throws it away through both edges. The initial two vortices develop to two wall jets in a quiescent environment. This configuration is used in UAV wing leading-edge tests in a wind tunnel. When the actuator is subjected to the wind action produced by the wind tunnel, the wall jet at the actuator trailing edge is maintained, but the jet at the actuator leading edge will develop into a fixed vortex. That vortex is supposed to add high momentum into the boundary layer and delay the flow separation.

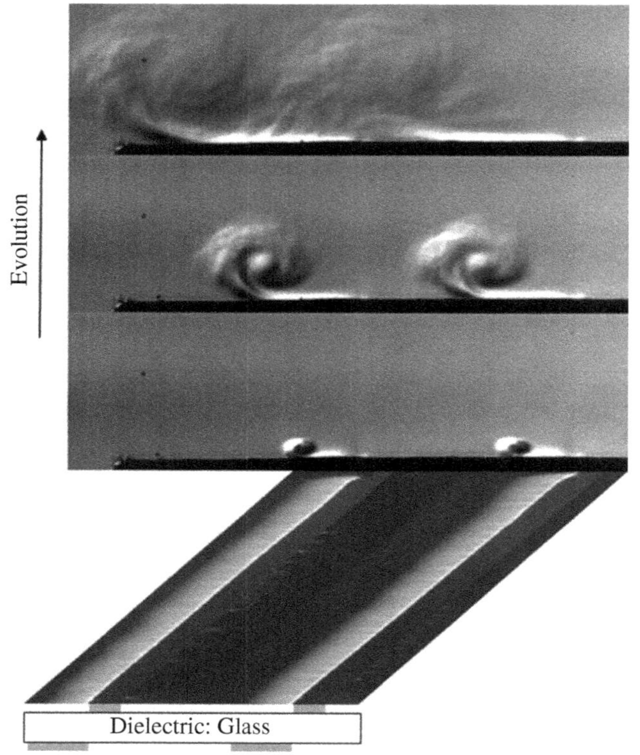

Figure 16.8 Double DBD with electrode and dielectric configuration, and the flow evolution.

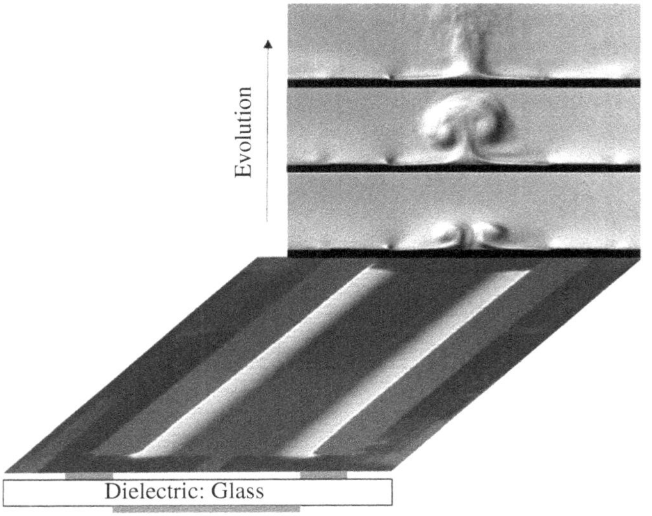

Figure 16.9 Wall normal jet; electrode and dielectric deployment, and flow evolution.

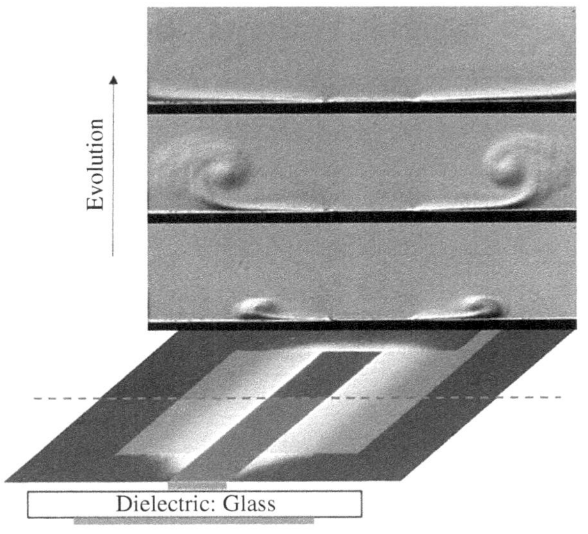

Figure 16.10 Double tangential wall jet; electrode and dielectric configuration, and flow evolution.

Finally, in Figure 16.11 an unsteady actuation may be seen. The actuator generates a tangential wall jet in steady operation. By changing the waveform signal, the actuator goes from a steady to an unsteady device. In this case, the input signal is AM modulated, giving a train of vortices, as depicted in the evolution images. Normally, the unsteady actuation is realized by the modulation of the voltage waveform by a pulse-burst mode. The burst mode consists of switching an AC cycle on and off at a given frequency. The ratio of the actuation time and the period of excitation gives the duty cycle; a steady

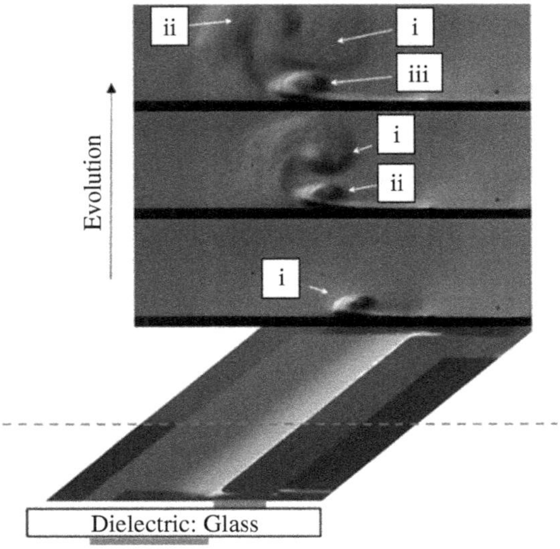

Figure 16.11 Tangential wall jet with AM modulation; electrode and dielectric arrangement, and vortex evolution.

actuation has a 100% duty cycle. In separated flows, the Strouhal number plays a fundamental role that is defined as:

$$St = \frac{fc}{U_\infty} \tag{16.3}$$

where f, c and U_∞, are respectively the excitation frequency, the chord length and the freestream velocity.

When the Strouhal number $St \approx 1$, an optimum separation control can be achieved, and this is better than that obtained with a steady signal. Moreover, by reducing the duty cycle, the electrical consumption is reduced. Control improvement using the burst mode has been demonstrated in other active flow control devices, such as synthetic jets. Both methods excite the instabilities in the separated shear layer, allowing flow attachment.

Recent investigations have shown that nanosecond pulse discharges are more efficient than microsecond discharges. They are effective over a wide range of Mach numbers, up to $Ma = 0.85$ [40]. It seems that the shock effect has a greater influence on separation control than the momentum effect. The configuration of a nanosecond-pulse DBD (ns-DBD) plasma actuator is equal to that of an AC-DBD actuator, while differing in its input waveform; that is, a short pulse (tens of nanoseconds). The Joule effects are important in this type of actuation and the rapid heating of the air around the actuator produces compression waves. The ns-DBD mode is one of the most promising concepts in the DBD area.

To end this section, two new DBD designs will be discussed. The first is the use of the sliding discharge from pumped gas lasers to generate the excited species. The design uses three electrodes, sharing elements of DC corona discharge and DBD arrangements [41]. Two electrodes are exposed and one electrode is encapsulated, and between them there is a dielectric barrier, as shown in Figure 16.12.

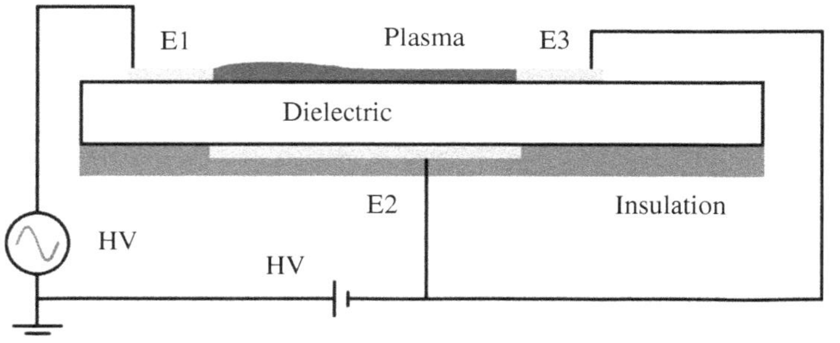

Figure 16.12 Sliding discharge.

Several electrode connections are possible: if the electrode E1 is connected to an HV AC signal and the electrodes E2 and E3 are grounded, the discharge is similar to a normal DBD. This arrangement may have some advantages because the electrode E3 allows charge neutralization, preventing its accumulation on the dielectric. If connections as depicted in Figure 16.12 – with E2 and E3 connected to a negative high-voltage DC – the discharge will be like a normal DBD between E1 and E2, but streamers will propagate from E1 to E3 as in a corona discharge, similar to what happens in Figure 16.3b. Thus large regions of plasma are achieved with a reduced voltage.

The last type of actuator treated in this chapter is the plasma vortex generator. Unlike conventional DBD actuators that are located spanwise, these actuators are yawing to the incoming flow. The plasma vortex generator is oriented along the freestream direction to produce an induced jet in the spanwise direction [42]. The interaction of this spanwise jet with the oncoming flow gives a streamwise vortex. The vorticity added by the actuator increases the circulation along the actuator length. If a collection of actuators is arranged in the wing parallel to the incident airflow and with the induced jets facing each other, multiple streamwise vortices can be induced in the flow. This actuation mimics the effect of fixed vortex generators (passive control devices). Control separation has been demonstrated to be effective with this configuration.

From the above comments it can be seen how the growing research on dielectric optimization, electrode configurations, high-voltage equipment, driving waveforms and other issues are improving DBD actuation and expanding its applications.

More details on issues related to the SDBD plasma flow control actuators and their performance can be found in several excellent review papers [22, 36, 43, 44].

16.5 Experimental Setup

The following subsections describe the facility and equipment used in the experimental investigation.

16.5.1 Wind Tunnel Description

The measurements were conducted in the Wind Tunnel No. 2, in the Experimental Aerodynamics Branch of INTA. The facility is a closed-circuit wind tunnel with an open test-section measuring $2.8 \times 1.9\,m^2$ by 4.9 m long. The test section is octagonal.

Figure 16.13 UAV model in the wind tunnel test section with reference axis.

Models are supported by a sting and variations of angle of attack and sideslip are allowed. The maximum wind speed is up to 50 m/s, with Reynolds numbers up to 3.8×10^6/m [45].

The quality of the flow is always an important question in wind tunnel testing, but in flow control applications it has become crucial, because small disturbances in the boundary layer can greatly affect the external flow. The turbulence intensity in the wind tunnel has a not negligible effect on the aerodynamic performance of the tested model [32]. An increased level of freestream turbulence injects a significant amount of high-momentum fluid and the flow can better resist the adverse pressure gradient. In order to obtain a good flow quality, the wind tunnel settling chamber has an 18-cm thick aluminium honeycomb grid and three turbulence-reducing screens, giving a turbulence intensity of 0.27%.

Figure 16.13 shows the wind tunnel with the model in the test section.

16.5.2 Model

The model used in this investigation, as a proof-of-concept of plasma flow control experiments, is an UAV developed at INTA. The DIANA UAV project is a versatile aerial target system to simulate the real threats of current and future weapons. In addition, this system is a platform for research on high-speed UAV technologies. The wind tunnel model used in the current experimental work was made of aluminium and is 52% scale replica of the real UAV. In Figure 16.14, a schematic of the whole model can be seen and some geometric specifications are given in Table 16.1.

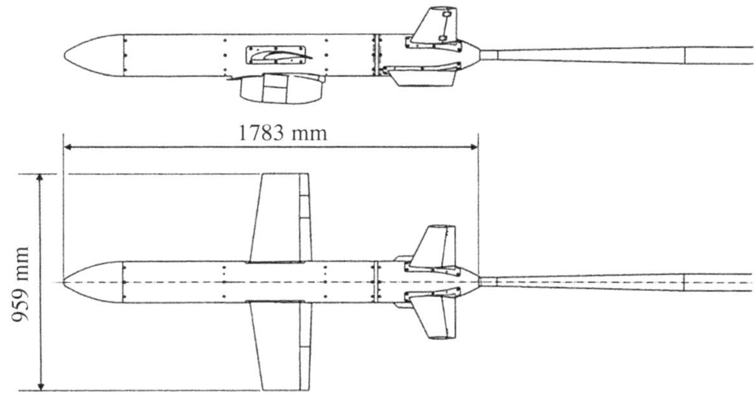

Figure 16.14 Wind tunnel model.

Table 16.1 Specifications of Diana UAV model.

Parameter	Type/value
Vehicle span	959 mm
Wing surface area	0.23 m^2
Vehicle length	1783 mm
Mean aerodynamic chord	242 mm
Sweep at wing leading edge	8°
Sweep at V-tail leading edge	20°
Wing taper ratio	0.70
V-tail taper ratio	0.49

The total wing span is 0.959 m and the fuselage length is 1.783 m. An internal balance is mounted inside the UAV model, and then the balance is attached to the sting, which joins with the sting support, allowing changes of angle of attack and sideslip. The measured forces and moments are given in balance axes, so the coordinate system is moving with the model. In Figure 16.13 the balance axes are depicted on the model.

This model is not the most suitable for wind tunnels tests, where electromagnetic interference is expected. Usually, tests are conducted with non-metallic models, avoiding these problems. A further complication of the current tests (model and set-up) is that the balance must be attached to the model made of aluminum with no isolation between them. The internal balance is composed of six Wheatstone bridges, with four strain gauges for each one. These bridges may be subject to electromagnetic interference and parasitic currents. In short, the model will have good stiffness, but there will also be difficulties with the measurements. However, it may still be a good test bench to verify some of the claims made earlier regarding easy installation and application on existing surfaces with no changes or with only light modifications and also about overcoming some of the disadvantages of EMI problems and use of high voltage. Another interesting aspect is that a reduced power will be applied in order to avoid interference, and with this limitation the actuator must be efficient.

16.5.3 DBD Actuators

A single-surface DBD is the basis for the plasma actuators employed in this investigation. DBD actuators used in these tests were handcrafted. This fact may introduce asymmetries in the results due to differences between actuators on larboard and starboard. For the same reason, the actuator may be non-uniform spanwise.

The actuator assembly is described in the section on the analysis of the results since several arrangements have been used, with slight differences between them. The dielectric material used was Kapton®, perhaps not the best material, but flexible, of low thickness and allowing easy retrofitting.

16.5.4 Measurements

The UAV model has a six-component internal strain gauge force balance fabricated by ROLLAB, which allows measurement of the overall aerodynamic forces and moments. Table 16.2 shows the nominal forces and moments.

The balance was calibrated before conducting the tests in the wind tunnel; AIAA recommendations were followed [46]. Force and moments data were obtained from the internal balance through a data acquisition system comprising an HP power supply, NEC amplifiers and a PXI-based data acquisition system. These forces and moments were reduced to engineering units and coefficients form using in-house software. The aerodynamic coefficients are commonly defined as:

$$C_F = \frac{F}{\frac{1}{2}\rho U_\infty^2 S} \tag{16.4}$$

$$C_M = \frac{M}{\frac{1}{2}\rho U_\infty^2 S l_{ref}} \tag{16.5}$$

where F and M are forces and moments determined from time-averaged balance measurements, ρ is the air density, U_∞ is the free stream velocity, S is the wing surface area and l_{ref} is the reference length (the mean aerodynamic chord for M_y and the wing span for M_x and M_z).

Table 16.2 Forces and moments of I6B128 balance.

Normal force	3000 N
Side force	2000 N
Axial force	800 N
Pitch moment	150 Nm
Yaw moment	80 Nm
Roll moment	100 Nm

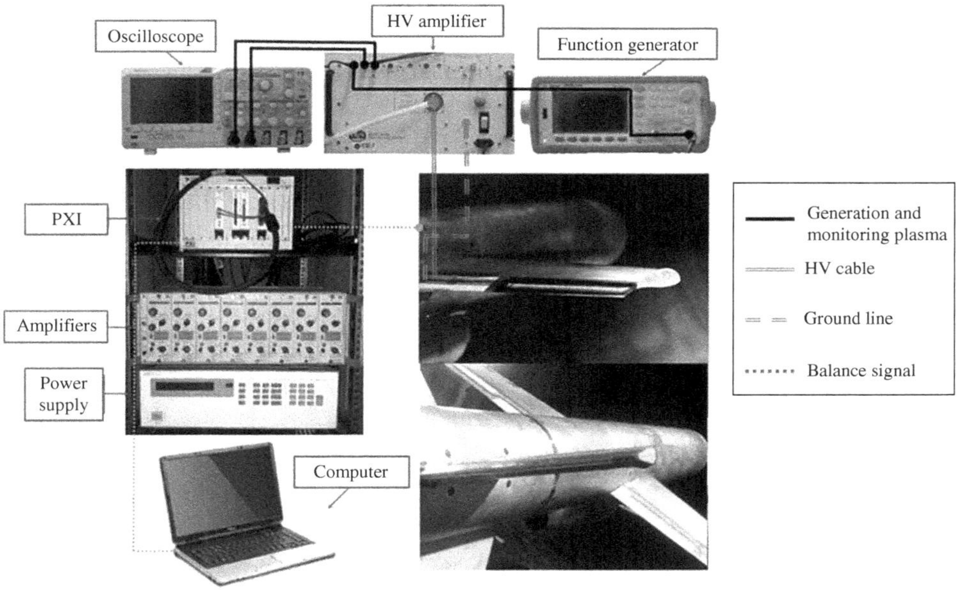

Figure 16.15 Equipment and connections.

This equipment may be seen in Figure 16.15. The same equipment was used in the calibration process. The analogue-to-digital converter used in calibration and tests was 16-bit. For every angle of attack, 10 000 samples were taken and averaged, at a sampling rate of 1000 samples per second. No wind tunnel corrections were applied to the data obtained, since only the difference or changes in the aerodynamic coefficients due to plasma flow control are of interest for this investigation.

Prior to each test, a tare calibration was made in order to discount the model weight from the aerodynamic forces. The deflections of the sting and balance was accounted for to obtain the real angle of attack of the model under the action of aerodynamic forces and moments [47].

The plasma actuation produces high-intensity radio-frequency electromagnetic noise. The force balance and instrumentation may be influenced by the electric and magnetic fields. In order to ensure that they were not materially affected, verification and validation tests were performed with no wind and plasma actuation.During the development of the tests, multiple verifications were performed in order to keep the influence of disturbances on measurements under control. If interference was observed, the results were discarded and the driven voltage was reduced until the disturbance disappeared or remained constant and known. The results presented in this chapter have no differences between plasma-off and plasma-on output with no wind, or any difference has been removed, so electromagnetic fields do not influence these data.

16.5.5 High-voltage Equipment

The generation of weakly ionized air between the two electrodes was produced by a high voltage amplifier TREK model 20/20C (±20 kV, ±20 mA, slew rate greater than

450 V/μs). The output voltage and the frequency could be varied by means a function generator (Agilent model 33509B) which output is amplified by the Trek equipment. The maximum attainable frequency without distortion depends on the load and slew rate of the waveform used. The high voltage amplifier Trek has two BNC to measure its voltage and current output. In order to monitor these parameters a four channel Tektronix DPO2004-B was used. The accuracies are voltage monitor 0.1% of full scale and current monitor 1% of full scale.

Figure 16.15 shows the equipment used in the experimental work and their connections. The ionized region can be observed on the upper surface of the aileron and on the lower surface of the ruddervator.

16.6 Results and Analysis

The objective of the current investigation is to demonstrate that plasma actuation may manipulate the flow over the lifting surfaces (wing and V-tail planes) of an unmanned aerial vehicle in order to enhance performance and to control its flight. Specifically, the extension of the angle of attack envelope by delaying flow separation, and the possibility of hingeless control through trailing-edge manipulation.

16.6.1 Actuation on Leading Edge

The lift-to-drag ratio and stall characteristics of the wing are of vital importance to the take-off distance and climbing speed, among others key performances of aircraft [48]. Flow separation is provoked when the flow faces an adverse pressure gradient near the lifting surface. Separation induces an unstable shear layer, leading to large-scale vortices and an unsteady and large wake.

Prior to the wind tunnel tests, an effectiveness proof of the actuator design was performed on a NACA 4418 airfoil. The DBD plasma actuator comprised one exposed electrode made of adhesive copper foil tape, a dielectric layer and the body of the model as a virtual electrode, which was grounded. The exposed electrode was 5 mm in width, and the dielectric layer was made of three layers of Kapton® insulating tape with an approximate depth of 0.2 mm (polyimide and adhesive). The upper exposed electrode was connected to the TREK amplifier's high-voltage output, and the encapsulated electrode was grounded to the TREK device. A plasma sheet was formed on the leading and trailing edges of the exposed electrode, similar to the example shown in Figure 16.10. The electrode leading edge was located at chord location $x/c = 0.1$.

Flow-field velocity was investigated by means of particle image velocimetry (PIV). The PIV system used during the experiment was a model made by TSI Inc. The equipment consists of two Nd:YAG lasers that are capable of running at a 10–30 Hz pulse repetition rate, emitting low power pulses (10–100 mJ) at 532 nm wavelength (frequency doubled). Images of the flow field during experiments were captured by a charge coupled device (CCD) camera (TSI PowerView Plus 4MP) with 2048 × 2048 pixels resolution. The camera operated in a double-frame mode. The seeding particles used in the experiment were 1-μm-diameter droplets of olive oil. Velocity vectors were calculated using conventional cross-correlation algorithms. Interrogation windows 32 × 32 pixels in size with 50% overlap in both horizontal (streamwise) and vertical directions.

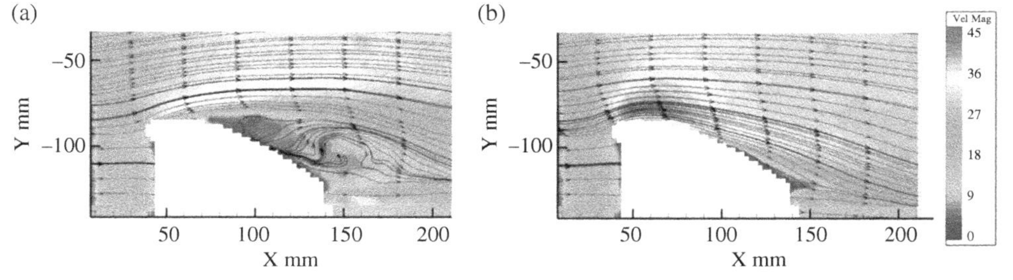

Figure 16.16 Time-averaged contour velocity field and streamlines of NACA 4418 airfoil at 29 m/s and angle of incidence 21°: (a) no plasma actuation; (b) plasma actuation with sinusoidal waveform at 15 kV_{pp} and 2 kHz.

Erroneous vectors were replaced by interpolation of the neighbouring velocity vectors. The laser beam illuminated from upper surface of the airfoil, so there was a shadow region on the pressure surface where the particles could not be seen by the CCD camera and no data were obtained [49, 50].

Several waveforms, driven voltages and frequencies were applied to the actuators at free streams from 6–29 m/s, with sweeps of angle of attack. Figure 16.16 shows one of these tests at 29 m/s. Time-averaged PIV streamlines confirmed that the airflow followed the airfoil surface, showing that plasma actuation is able to reattach the flow at $\alpha = 21°$.

Once satisfied that the actuator was efficient on a airfoil, a similar arrangement was placed on the UAV model to measure the modification in forces and moments due to plasma actuation.

The actuators ran the span of each half wing. The encapsulated electrode length was 350 mm, with a width of 10 mm, while the exposed electrode was 350 mm long and 5 mm wide. The dielectric between the electrodes was composed of three layers of Kapton® with an overall thickness of 200 μm. The electrodes, made of copper adhesive foil tape, had a slight overlap and the encapsulated electrode was downstream of the exposed electrode. The intersection line was located at chord location $x/c = 0.05$ on the suction surface. The ends of the electrodes were rounded to avoid discharge points and the actuators were operated in a steady manner, with a sine waveform, characterised by its carrier frequency and peak-to peak voltage, which was used to drive the TREK amplifier.

This set was isolated from the UAV model with two layers of Kapton® tape, and due to the low isolation a plasma sheet was formed on the leading edge of the exposed electrode, as shown in Figure 16.17.

Figure 16.18 shows a small but important part of the total set of data taken to evaluate the effect of leading-edge actuators on the UAV lift. The solid lines correspond to no-plasma actuation and the dashed lines to plasma actuation with 9 kV_{pp} voltage at a frequency of 3 kHz. The aerodynamic characteristics were tested three times and then averaged.

The two graphs correspond to tests performed a month apart and show the repeatability of the measured data. The actuator of Figure 16.18a was dismounted and one month later remounted by hand in the same manner as the first actuator; Figure 16.18b

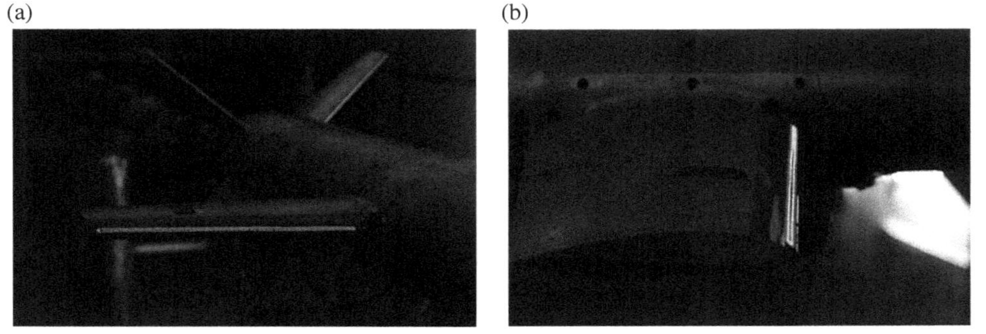

Figure 16.17 Plasma actuation on starboard wing leading-edge: (a) front view; (b) side view.

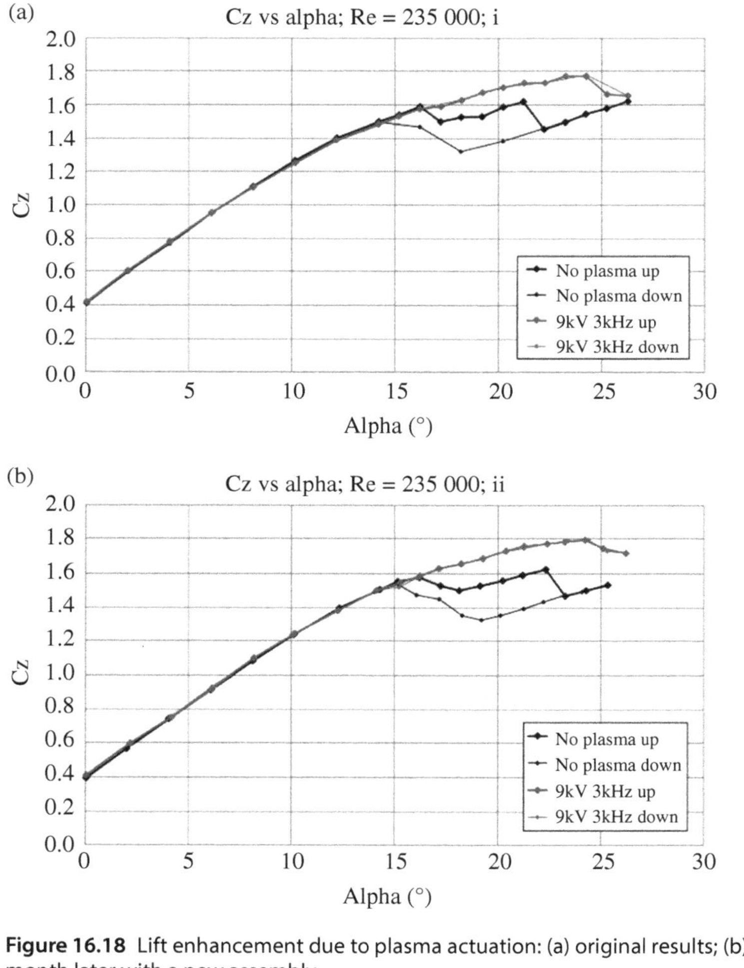

Figure 16.18 Lift enhancement due to plasma actuation: (a) original results; (b) repeated test one month later with a new assembly.

shows the data obtained. The important thing about this figure is that a hysteresis loop can be seen; this has previously been reported at low Reynolds numbers. This post-stall hysteresis is formed when flow separation occurs, producing a loss of lift; when the angle of attack decreases the flow reattaches at a lower angle of attack, creating the hysteresis loop [51]. In both examples the hysteresis loop is suppressed by the plasma actuation.

The graphs also show that up to an angle of 16°, plasma actuation has no effect or is not observable because the flow is not separated. When flow separation occurs in the baseline case, plasma actuation is able to maintain the flow attached and to delay the angle of attack from 16° to 24°. An increase in the lift coefficient is also observed. The c_z (balance axes) increases from nearly 1.6 to 1.8.

Force balance test results have shown important changes in the lift characteristics of the UAV model with plasma actuation compared with no-plasma actuation. However, further research is needed to understand the effects of the mechanisms. The flow re-attachment could be due to the momentum injection into the boundary layer, or due to the vorticity, or due to a forced transition, or due to something else.

As mentioned in Section 16.5.2 less power is used than in PIV tests, for which peak-to-peak voltage was 15 kV_{pp}. This voltage introduced strong electromagnetic interference in the balance signals and was discarded.

In any case, steady DBD plasma actuation has demonstrated the ability to recover the stall condition at a Reynolds number of 2.35×10^5, based on the mean chord of 0.242 m and freestream velocity of 15 m/s, and to suppress the hysteresis loop that occurs on the aerodynamic forces.

16.6.2 Actuation on Trailing Edge

The generation of moments to control the UAV without moving surfaces will be addressed in this subsection. Specifically, this investigation focus on the development of DBD actuators as alternatives to conventional control surfaces that employ mechanical actuators and linkages, which add weight, are complicated and need maintenance, and produce noise and vibration. The plasma actuators were mounted on the trailing edge of the wing and ruddervator.

Aileron

Flight control implies that the movement of an aircraft about different axes is controlled. Several control surfaces are used to achieve this objective. The aileron controls the rolling moment. In this section, a plasma actuator will be developed in order to manipulate the flow and generate moment control on a UAV.

The signal voltage needed to generate a useful control moment with a metallic aileron is too high and introduces electromagnetic interference in the balance signals. To solve this problem, similar flaps and ailerons were fabricated with additive manufacturing, giving non-conductive surfaces that mimicked the originals.

A DBD actuator was assembled in the larboard wing. The encapsulated electrode was made of aluminium foil tape, its leading edge was located at chord location $x/c = 0.96$ in the suction surface and continued beyond the trailing edge in the pressure surface. The dielectric medium was five layers of Kapton® and the exposed electrode was made of copper foil tape with a width of 3 mm. The actuator length was 160 mm and began at

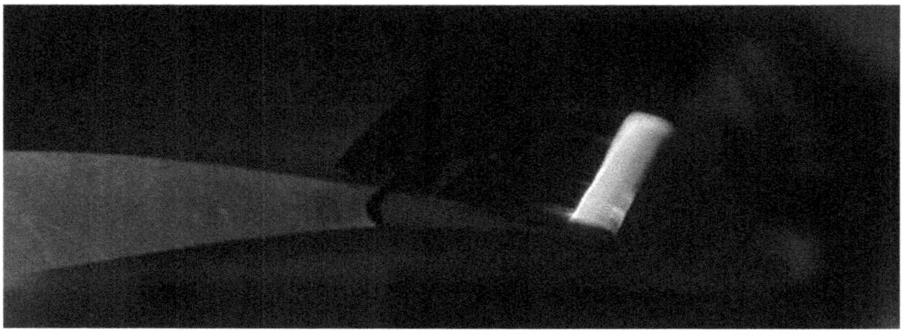

Figure 16.19 Plasma view on the larboard wing aileron.

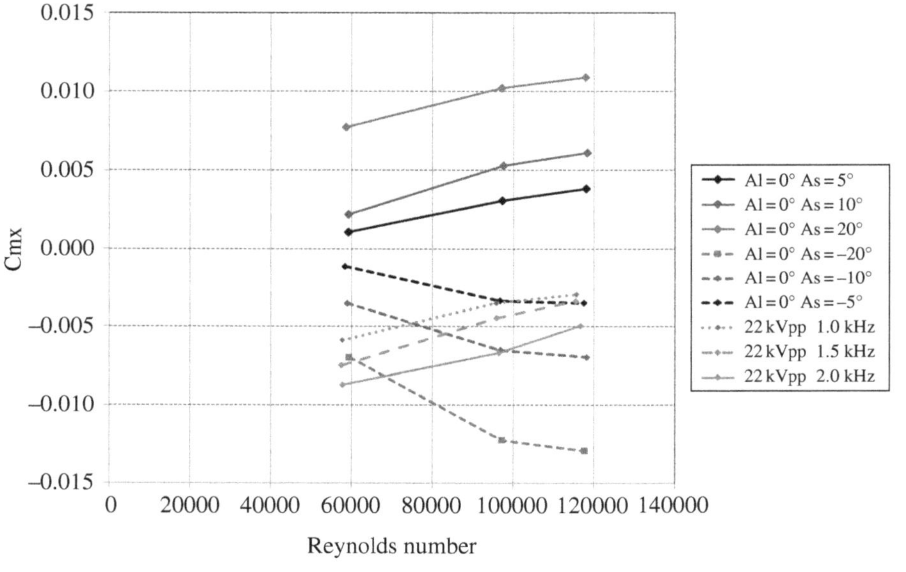

Figure 16.20 Comparison of mechanical effector and plasma actuator with steady sine waveform at 22 kV peak-to-peak voltage. Al, aileron larboard; As, aileron starboard.

10 mm from wing tip. Figure 16.19 shows the actuator driven with a sine waveform at 22 kV_{pp} at a frequency of 2 kHz.

In order to evaluate the plasma actuator as a rolling moment device, a sweep of deployment of the original starboard wing aileron was conducted. The δ values were 0°; ±5°; ±10°, and ±20°, and the rolling moment coefficients were obtained, C_{mx}. These values were compared with the results acquired with plasma actuation on the larboard wing. All data correspond to a zero angle of attack.

A positive deflection of the starboard wing aileron generates a positive rolling moment, whereas the plasma actuator located on the larboard wing produces a lift increase that generates a negative rolling moment. The comparison between results obtained with plasma actuation and with a mechanical moving surface are shown in Figure 16.20. The results correspond to free stream values of about 3.5 m/s, 6 m/s and

7.5 m/s. It is noted that at low Reynolds numbers the plasma actuator achieves the same moment coefficient as an aileron deflection of −20°.

Ruddervator

A ruddervator is the movable airfoil at the trailing edge of a V-tail that performs the functions of both rudder and elevator. In this latter application, a plasma actuator that allows circulation control was mounted in a modified ruddervator. The rounded trailing-edge airfoil was made by additive manufacturing for the same reasons as in the previous case. This rounded edge means that the Kutta condition is not locked, as in a sharp-edged airfoil. The circulation around the ruddervator and hence the lift can be changed with the displacement of the stagnation point.

The V-tail profile, an Althaus symmetric airfoil, was trailing-edge rounded at chord location $x/c = 0.92$. The plasma actuator was as follows: the encapsulated electrode was made of aluminium foil tape and located in the suction surface, its trailing edge was located at the intersection of the trailing-edge airfoil and the ruddervator plane of symmetry. Its length was 200 mm and the width was 10 mm. The dielectric medium was four layers of Kapton® and the exposed electrode was made of copper foil tape with a width of 4 mm. It was located on the pressure surface and there was no gap between it and the encapsulated electrode.

Measurements were performed at zero static angle of attack, for a range of speed from 3–5 m/s. In these conditions, voltage time series proportional to the aerodynamic moments were recorded using the PXI digital data acquisition system. As in the previous tests, the time series were sampled at 1 kHz for 10 s for a total of 10 000 samples. These samples were averaged to give the time-averaged moments. The following results are the average of five of these acquisitions.

Actuators were located on both ruddervators and tested. The device on starboard ruddervator can be seen in Figure 16.21, which shows a sweep of voltage with a fixed frequency of 1 kHz. Figure 16.22 shows the comparison of plasma actuation and mechanical deflection ($\delta_{r\,=}$ −5°, −10° and −20°) on the starboard ruddervator. The three moment coefficients weare obtained from the balance signals.

The plasma actuator produces a loss of lift in the modified and fixed ruddervator that mimics the behaviour of a negative deflection of the mechanical effectors. The three generated moments are equivalent to a deployment of 7° or 8° of the original ruddervator at a Reynolds number equal to 9×10^4. Thus, plasma actuation as used in this application is useful at flows at low velocity or with a reduced density.

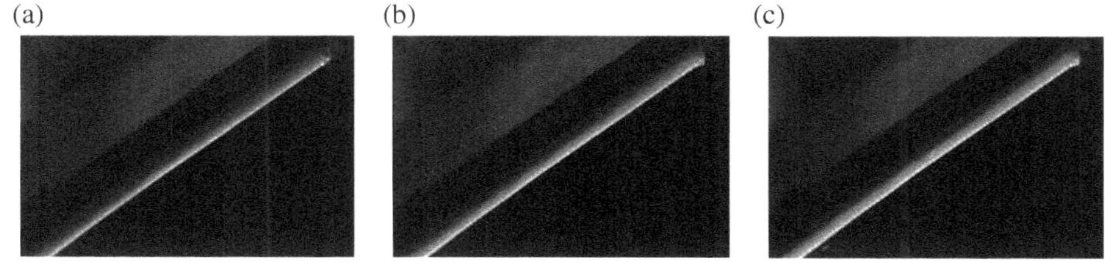

Figure 16.21 Plasma actuation on starboard ruddervator: (a) sine waveform 16 kV_{pp} and 1 kHz; (b) sine waveform at 18 kV_{pp} and 1 kHz; (c) sine waveform 20 kV_{pp} and 1 kHz.

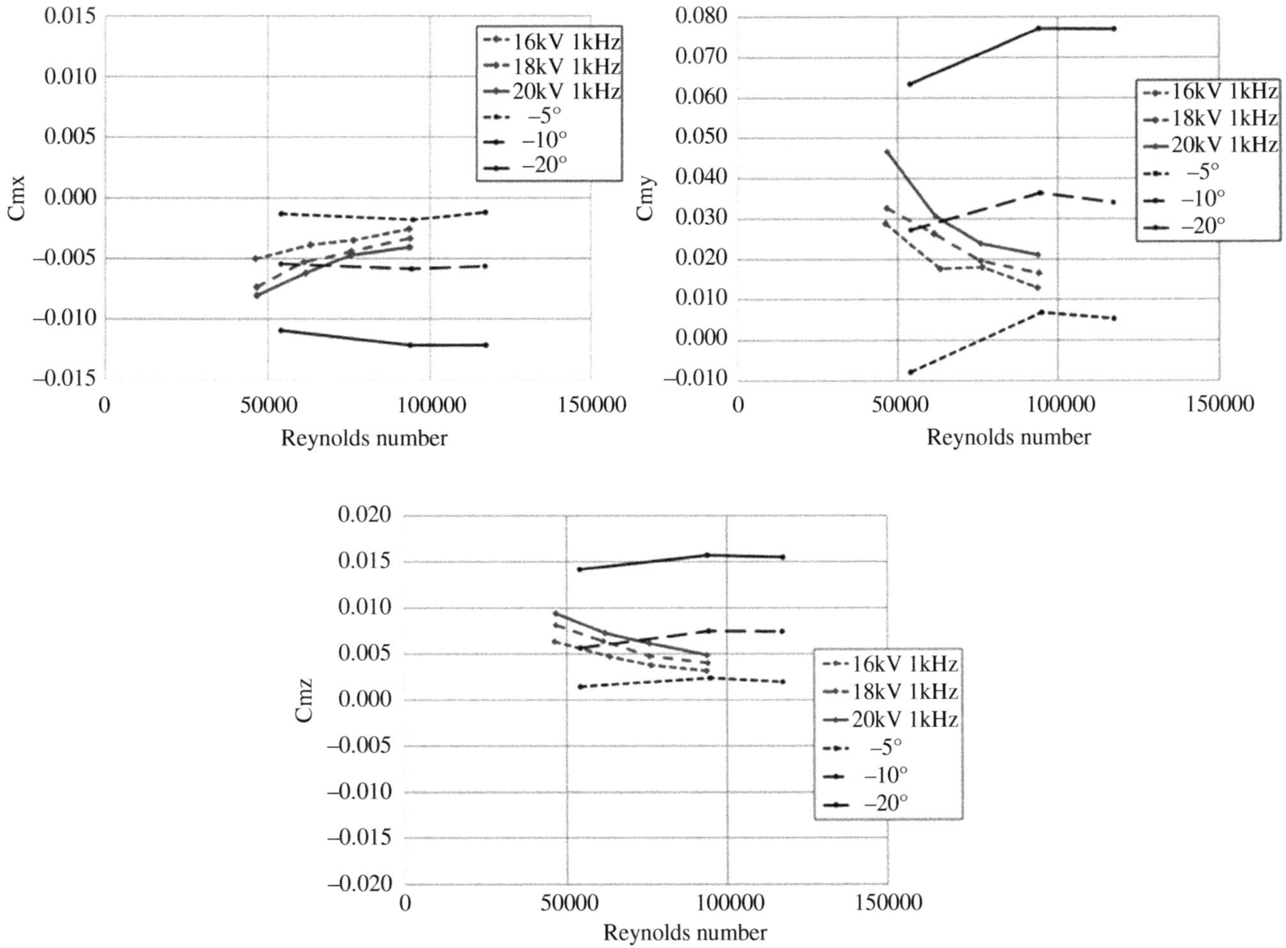

Figure 16.22 Effect of negative deflections of mechanical ruddervator and plasma actuator: cm_x, cm_y and cm_z.

This investigation is still in progress, but the small sample of preliminary results shows the potential of this technology. Among the various issues that are being investigated it is worth mentioning:

- the influence of waveform (sine, square, triangle, saw-tooth)
- unsteady actuation with variable duty cycle in order to optimize the consumption
- the electrode shape (all data given in this chapter are with straight-edged electrodes, but electrodes with a serrated or undulating trailing edge to induce a 3D component could improve the performance)
- plasma actuation on a Gurney flap
- use of multiple actuators.

16.7 Conclusions

There is a growing interest in plasma actuators for aeronautical applications, particularly in DBD-based actuation. This chapter has tried to give an overview of these devices and some practical applications in UAVs. From the results, DBD plasma actuators seem to have the potential to replace mechanical control surfaces at low Reynolds numbers. This would lead to a weight reduction, a fabrication cost reduction, higher reliability and faster response times, among other benefits.

The results have shown that plasma actuation on the leading edge of an UAV can delay flow separation, suppress the hysteresis loop and enhance the lift of the model. The tests performed with actuators located on the trailing edge have proven their ability to generate moments of control in the three axes with enough authority to control the attitude of a UAV.

All tests were evaluated only for steady plasma actuation, and better results can be expected with unstable operations, an approach that also implies less energy consumption. A new generation of UAVs with plasma actuators replacing or enhancing traditional aerodynamic movable control surfaces seems to be just around the corner.

Acknowledgments

These lines are to thank everyone who has contributed to a greater or lesser extent to the achievement of this research, especially Juan Carlos Martín for his help and company in the long days of testing, Santiago Martín for the additive manufacturing models and finally to the DIANA project staff, in particular Belén Gutiérrez.

References

1 Touchard G. (2008) Plasma actuators for aeronautical applications – State of art review. *International Journal of Plasma Environmental Science and Technology*, **2**: pp. 1–25.
2 Patel, M.P., Ng. T.T., Vasudevan, S., Corke, T.C., and He, C. (2007) Plasma actuators for hingeless aerodynamics control of an unmanned air vehicle. *Journal of Aircraft*, **44**, 4: pp. 1264–1274.

3 Roth, J.R. (2003) Aerodynamic flow acceleration using paraelectric and peristaltic electrohydrodynamic effects of a one atmosphere uniforms glow discharge. *Physics of Plasmas*, **10**, 5: pp. 2117–2126.
4 Corke, T.C., Jumper, E.J., Post, M.L., Orlov, D., and McLaughlin, T.E. (2002) Applications of weakly-ionized plasmas as wing flow-control devices. *40th AIAA Aerospace Sciences Meeting and Exhibit*. AIAA Paper 2002-0350.
5 Post, M.L., and Corke, T.C. (2004) Separation control using plasma actuators – dynamic stall control on an oscillating airfoils. *2nd AIAA Flow Control Conference*. AIAA Paper 2004-2517.
6 Corke, T.C., He, C., and Patel, M.P. (2004) Plasma flaps and slats: an application of weakly-ionized plasma actuators. *2nd AIAA Flow Control Conference*. AIAA Paper 2004-2127.
7 Benard, N., Jolibois, J., Touchard, G., and Moreau, E. (2008) A directional plasma-jet device generated by double DBD actuators – an active vortex generator for aerodynamic flow control. *4th Flow Control Conference*. AIAA Paper 2008–3763.
8 Kelley, C.L., Bowles, P.O., Cooney, J., He, C., Corke, T.C., Osborne, B.A., Silkey, J.S., and Zehule, J. (2014) Leading-edge separation control using alternating-current and nanosecond-pulse plasma actuators. *AIAA Journal*, **52**, 9: pp. 1871–1884.
9 Lopera, J., Ng, T.T., Patel, M.P., Vasudevan, S., Santavicca, E., and Corke, T.C. (2007) Aerodynamic control using windward-surface plasma actuators on a separation ramp. *Journal of Aircraft*, **44**, 6: pp. 1889–1895.
10 Vorobiev, A.N., Rennie, R.M., Jumper, E.J., and McLaughlin, T.E. (2008) Experimental investigation of lift enhancement and roll control using plasma actuators. *Journal of Aircraft*, **45**, 4: pp. 1315–1321.
11 Boesch, G., Vo, H.D., Savard, B., Wanko-Tchatchouang, C., and Mureithi, N.W. (2010) Flight control using wint-tip plasma actuation. *Journal of Aircraft*, **47**, 6: pp. 1836–1846.
12 Kotsonis, M., Pul, R., and Veldhuis, L. (2013) Experimental study on airfoil circulation control using plasma actuators. *31st AIAA Applied Aerodynamics Conference*. AIAA Paper 2013-3164.
13 Feng, L.H., Choi, K.S., and Wang, J.J. (2015) Flow control over an airfoil using virtual Gurney flaps. *Journal of Fluid Mechanics*, **767**: pp. 595–626.
14 Gad-El Hak, M. (1996) Modern developments in flow control. *Applied Mechanics Reviews*, **49**: pp. 365–379.
15 Gad-El Hak, M. (2000) *Flow Control. Passive, Active, and Reactive Flow Management*. Cambridge University Press.
16 Cattafesta III, L.N., and Sheplak, M. (2011) Actuators for active flow control. *Annual Review of Fluid Mechanics*, **43**: pp. 247–272.
17 Wood, R.M. (2002) A discussion of aerodynamic control effectors (ACEs) for unmanned aerial vehicles (UAVs). *1st AIAA Technical Conference and Workshop on Unmanned Aerospace Vehicles*. AIAA Paper 2002-3494.
18 Langmuir, I. (1928) Oscillations in ionized gases. *Proceedings of the National Academy of Sciences*, **14**, 8: pp. 627–637.
19 Inan, U.S., and Golkowski, M. (2011) *Principles of Plasma Physics for Engineers and Scientists*. Cambridge University Press.
20 Masuyama, K., and Barrett, S.R.H. (2013) On the performance of electrohydrodynamic propulsion. *Proceedings of The Royal Society*, **469**: 20120623.

21 Moreau, E., Benard, N., Alicalapa, F., and Douyère, A. (2015) Electrohydrodynamics force produced by a corona discharge between a wire active electrode and several cylinder electrodes – Application to electric propulsion. *Journal of Electrostatics*, **76**: pp. 194–200.

22 Moreau, E. (2007) Airflow control by non-thermal plasma actuators. *Journal of Physics D: Applied Physics*, **40**: pp 605–636.

23 Cybyk, B.Z., Wilkerson, J.T., and Grossman, K.R. (2004) Performance characteristics of the SparkJet flow control actuator. *2nd AIAA Flow Control Conference*. AIAA Paper 2004-2131.

24 Narayanaswamy, V., Raja, L.L., and Clemens, N.T. (2010) Characterization of a high-frequency pulsed-plasma jet actuator for supersonic flow control. *AIAA Journal*, **48**, pp. 297–305.

25 Popkin, S.H., Cybyk, B.Z., Land III, H.B., Emerick II, T.M., Foster, C.H., and Alvi. F.S. (2013) Recent performance-based advances in SparkJet actuator design for supersonic flow applications. *51st AIAA Aerospace Sciences Meeting*. AIAA Paper 2013-0322.

26 Samimy, M., Adamovich, I., Webb, B., Kastner, J., Hileman, J., Keshav, S., and Palm, P. (2004) Development and application of localized arc filament plasma actuators for jet flow and noise control. *42nd AIAA Aerospace Sciences Meeting and Exhibit*. AIAA Paper 2004-0184.

27 Webb, N., Clifford, C., and Samimy, M. (2013) An investigation of the control mechanism of plasma actuators in a shock wave-boundary layer interaction. *51st AIAA Aerospace Sciences Meeting*. AIAA Paper 2013-0402.

28 Suchomel, C.F., Wie, D.V., and Risha, D. (2003) Perspectives on cataloging plasma technologies applied to aeronautical sciences. *37th AIAA Plasmadynamics and Lasers Conference*. AIAA Paper 2003-3852.

29 Plasma aerodynamics: current status and future directions. Plasma Aerodynamics Discussion Group. AIAA Plasmadynamics and Lasers Technical Committee. June 2014.

30 Kogelschatz, U. (2003) Dielectric-barrier discharges: their history, discharge physics, and industrial applications. *Plasma Chemistry and Plasma Processing*, **23**, 1: pp 1–46.

31 Roth, J.R., Sherman, D., and Wilkinson, S.P. (1998) Boundary layer flow control with one atmosphere uniform glow discharge surface plasma. *36th AIAA Aerospace Sciences Meeting*. AIAA Paper 98-0328.

32 Enloe, C.L., McLaughlin, T.E., Van Dyken, R.D., Kachner, K.D., Jumper, E.J., and Corke, T.C. (2004) Mechanisms and response of a single-dielectric barrier plasma actuator: plasma morphology. *AIAA Journal*, **42**: pp. 589–594.

33 Enloe, C.L., McLaughlin, T.E., Van Dyken, R.D., Kachner, K.D., Jumper, E.J., and Corke, T.C. (2004) Mechanisms and response of a single-dielectric barrier plasma actuator: geometric effects. *AIAA Journal*, **42**: pp. 595–604.

34 Roth, J.R., and Dai, X. (2006) Optimization of the aerodynamic plasma actuator as an electrohydrodynamic (EHD) electrical device. *44th AIAA Aerospace Sciences Meeting and Exhibit*. AIAA Paper 2006-1203.

35 Thomas, F.O., Corke, T.C., Iqbal, M., Kozlov, A., and Shatzman, D. (2009) Optimization of dielectric barrier discharge plasma actuators for active aerodynamic flow control. *AIAA Journal*, **47**: pp 2169–2178.

36 Benard, N., and Moreau, E. (2014) Electrical and mechanical characteristics of surface AC dielectric barrier discharge plasma actuators applied to airflow control. *Experiments in Fluids*, **55**: 1846.

37 Settles, G.S. (2001) *Schlieren and Shadowgraph Techniques. Visualizing Phenomena in Transparent Media*. Springer-Verlag.
38 Holder, D.W., and North, R.J. (1963) Schlieren methods, NPL notes on Applied Science No. 31. Her Majesty's Stationery Office, London.
39 Santhanakrishnan, A., and Jacob, J.D. (2006) Flow control using plasma actuators and linear/annular synthetic jet actuators. *3rd AIAA Flow Control Conference*. AIAA Paper 2006-3033.
40 Roupassov, D.V., Nikipelov, A.A., Nudnova, M.M., and Starikovskii, A.Y. (2008) Boundary layer separation control by nanosecond plasma actuator. *44th AIAA/ASME/SAE/ASEE Joint Propulsion Conference*. AIAA Paper 2008-5067.
41 Louste, C., Artana, G., Moreau, E., and Touchard G. (2005) Sliding discharge in air at atmospheric pressure: electrical properties. *Journal of Electrostatics*, **63**: pp. 615–620.
42 Jukes, T., and Choi, K. -S. (2013) On the formation of streamwise vortices by plasma vortex generators. *Journal of Fluid Mechanics*, **733**: pp 370–393.
43 Corke, T.C., Enloe, C.L., and Wilkinson, S.P. (2010) Dielectric barrier discharge plasma actuators for flow control. *Annual Review of Fluid Mechanics*, **42**: pp 505–529.
44 Wang, J.J., Choi, K.-S., Feng, L.H., Jukes, T.N., and Whalley, R.D. (2013) Recent developments in DBD plasma flow control. *Progress in Aerospace Sciences*, **62**: pp 52–78.
45 Shyy, W., Lian, Y., Tang, J., Viieru, D., and Liu, H. (2008) *Aerodynamics of Low Reynolds Numbers Flyers*, Cambridge Aerospace Series
46 Calibration and use of internal strain-gage balances with applications to wind tunnel testing, Technical report R-091-2003, AIAA, 2003.
47 Barlow, J.B., Rae, W.H.Jr., and Pope A. (1999) *Low-Speed Wind Tunnel Testing*, 3rd ed. John Wiley & Sons.
48 Post, M.L., and Corke, T.C. (2004) Separation control on high angle of attack airfoil using plasma actuators. *AIAA Journal*, **42**, 11: pp. 2177–2184.
49 Raffel, M., Willert, C.E., Wereley, S.T., and Kompenhans J. (2007) *Particle Image Velocimetry – A practical Guide*, 2nd edn, Springer-Verlag.
50 Adrian, R.J., and Westerweel, J. (2011) *Particle Image Velocimetry*, Cambridge University Press.
51 Ananda, G.K., Sukuman, P.P., and Selig, M.S. (2012) Low-to-moderate aspect ratio wings tested at low Reynolds numbers. *30th AIAA Applied Aerodynamics Conference*. AIAA Paper 2012-3026.

17

Constrained Motion Planning and Trajectory Optimization for Unmanned Aerial Vehicles

Seid H. Pourtakdoust[1] *and Jalal Karimi*[2]

[1] *Center for Research and Development in Space Science and Technology, Sharif University of Technology, Tehran, Iran*
[2] *Space Research Institute, MUT, Tehran, Iran*

17.1 Introduction

Increasing complexity and advances in hardware and software technologies within the realms of control, robotics, electronics and artificial intelligence have led to the enhanced design, development and utility of unmanned air vehicles (UAVs) in a wide range of applications. Improvements in the level of autonomy, effective decision-making with reduced workload, reduction of human user/operator errors, as well as increases in the operating range of UAVs are considered highly desirable. This chapter is devoted to one of the main topics required to enhance the autonomy of UAVs, namely constrained motion planning (CMP).

This chapter is arranged as follows: motion planning is described in Section 17.1. Section 17.2 is devoted to UAV dynamics and its corresponding internal constraints. Environmental constraints are introduced in Section 17.3. Sections 17.4–17.6 cover three approaches to CMP: a modified particle swarm optimization algorithm for offline motion planning is described in Section 17.4, Section 17.5 develops a dynamic hybrid particle swarm optimization algorithm for real-time motion planning and the process of multi-variable-objective dynamic CMP optimization is discussed in Section 17.6.

17.1.1 Motion Planning

Motion planning (MP) in general refers to the derivation of a feasible plan to steer the vehicle from an initial point to a target destination while observing a multitude of system-related and/or environmental constrains. It is, however, desirable to generate such an optimal plan using existing resources while simultaneously minimizing certain cost functions. In practice, many MP problems require real-time solutions in a dynamic non-definite environment. Thus for proper mission planning, one needs to realistically observe and utilize the full dynamic capabilities of the UAV. In this sense, the terminology of 'constrained motion planning', in contrast to other expressions used,

Advanced UAV Aerodynamics, Flight Stability and Control: Novel Concepts, Theory and Applications,
First Edition. Edited by Pascual Marqués and Andrea Da Ronch.

Companion website: http://www.wiley.com/go/marques/advanced_UAV_aerodynamics

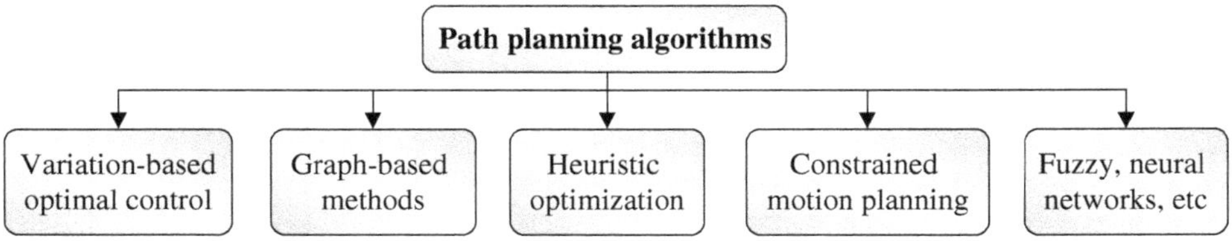

Figure 17.1 Categorizing path planning algorithms.

such as 'path planning' and/or 'trajectory optimization', is used to emphasize the fact that the MP problem and trajectory optimization will be performed while considering the vehicle's full nonlinear dynamics plus all physical and environmental constraints.

17.1.2 Categorizing Path Planning Algorithms

As shown in Figure 17.1, there are a few algorithms specifically developed to handle the basic problem of MP. Variation-based optimal control schemes [1, 2], graph-based methods [3, 4], heuristic optimization methods [5–7] as well as the CMP approach [8–14] are among the more important existing algorithms. However, most of the algorithms in the first three categories have only used simplified dynamic models of the vehicle or the kinematics equations. On the other hand, UAV system dynamic constraints and performance limitations need to be considered in MP to ensure a practical executable flight path. In this regard CMP is an approach that inherently handles all of these constraints.

Frazzoli [8–10] was the first to propose the concept of manoeuvre-based motion planning – CMP – using the idea of manoeuvre automation. In his approach, distinctive portions of the vehicle's overall motion, called motion primitives, are utilized to generate the total trajectory. The motion primitives include a set of constant trim states (TSs) and manoeuvre paths that are used to transfer the vehicle from one TS to another. Thus, by generating a library of countable trim states (TSs) and corresponding transition manoeuvres, the MP problem is transformed into a hybrid optimization problem with discrete and continuous design variables.

17.1.3 Motion Planning Problem Complexity Level

Motion planning algorithms are typically evaluated against two criteria: completeness and complexity. In a complete algorithm, feasible solutions are found and if there are no feasible solutions, failure will be declared. On the other hand, the computational complexity of the motion planning algorithm is classified as a PSPACE-hard problem in which the complexity grows exponentially along with the dimension of the configuration space [15].

17.1.4 Motion Planning Problem Characteristics

The behavioural model of a vehicle in flight can be described by a series of coupled, nonlinear differential equations. In this case, the only possible solution for this set of nonlinear equations is found through numerical integration. Adding the constraints of the motion planning to the motion equations will further complicate this set of equations and make reaching a feasible solution more difficult. This level of complexity of

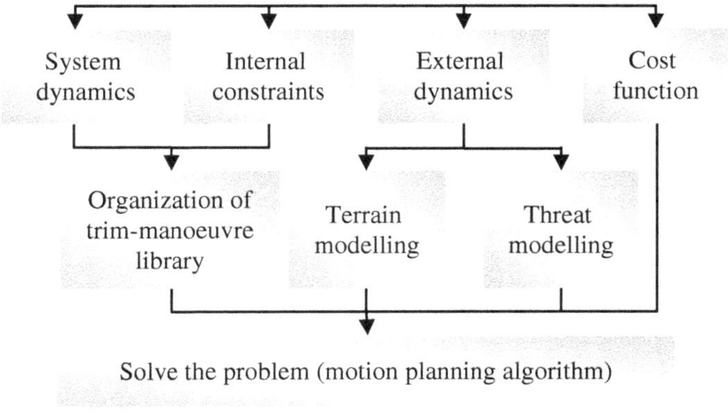

Figure 17.2 CMP problem flow chart.

the equations cannot be easily overcome by numerical optimization algorithms. Even if one is able to optimally solve this set of complex equations, the corresponding solution cannot be used in real time.

17.1.5 Constrained Motion Planning Problem

The basic CMP problem can be broken down into four main parts: the system dynamics, internal constraints, external constraints and the cost function. Pertinent definitions of these main parts along with their classification, modelling aspects, subparts and their role and place within the CMP problem will be covered in following subsections. Figure 17.2 shows the core of the CMP flowchart.

17.2 UAV Dynamics and Internal Constraints

17.2.1 UAV Missions and Classification

Unmanned aerial vehicles (UAVs) have been increasingly utilized for aerial missions in the past decade. Remote sensing, surveillance, delivery, search and rescue, broadcast, communication relay and combat attack of air and ground targets are among the many types of missions flown by UAVs. Due to this variety of possible missions, there are several classes if UAV, as shown in Table 17.1 [16]. According to Table 17.1, the most suitable class of UAV for motion planning is tactical, because of the suitable operating altitude, although other classes can also benefit from MP over parts of their mission profiles as well. Mission planning of tactical UAVs should be performed while observing terrain and threats. However, regardless of the type or classification of UAV, their general arrangements and configurations in terms of modelling and simulation are usually similar to manned aircraft.

17.2.2 UAV Simulation and Modelling

As it is our intention to produce some physical results, a typical tactical fixed-wing UAV is selected for implementation purposes. The general arrangement of the selected

Table 17.1 UAV classifications [12].

Designation	Mass (kg)	Operating altitude (km)	Endurance (h)
Micro UAV	<1	<0.15	2
Mini UAV	1–14	0.030–3	15
Tactical UAV	14–450	0.450–5.5	20
MALE UAV	450–13500	5.5–15	50
HALE UAV	450–13500	15–24	50

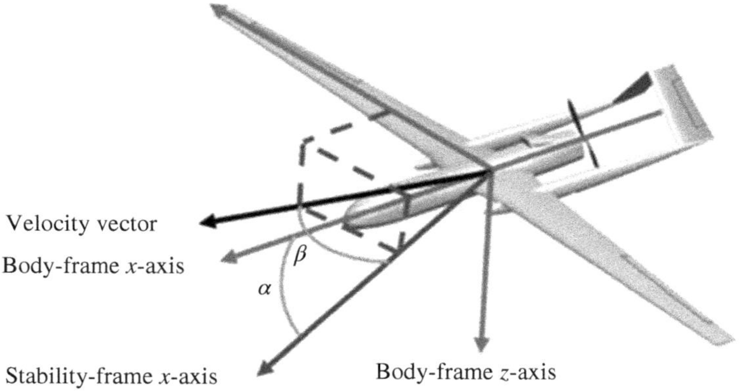

Figure 17.3 Typical shape of a UAV in body frame.

UAV is shown in Figure 17.3; it has four conventional controls: elevator, aileron, rudder and throttle setting. Some additional characteristics and limitations of the selected UAV are listed in Table 17.2.

For the purposes of MP and verification, one needs to develop an accurate simulation model for the UAV as well as its corresponding environmental constraints; the latter are discussed in Section 17.3. The most general form of the nonlinear 6DOF governing equations of motion, derived from basic physical laws, in the body-coordinate system assuming the flat Earth model [17] is utilized for the CMP simulation.

$$\dot{u} = vr - wq + \frac{-\mathrm{mg}\sin(\theta) + \mathrm{F_{Ax}} + \mathrm{F_{Tx}}}{\mathrm{m}} \tag{17.1}$$

$$\dot{v} = -ur + wp + \frac{\mathrm{mg}\sin(\phi)\cos(\theta) + \mathrm{F_{Ay}} + \mathrm{F_{Ty}}}{\mathrm{m}} \tag{17.2}$$

$$\dot{w} = uq - vp + \frac{\mathrm{mg}\cos(\phi)\cos(\theta) + \mathrm{F_{Az}} + \mathrm{F_{Tz}}}{\mathrm{m}} \tag{17.3}$$

$$\dot{p} = -\frac{\left(-\mathrm{I_y I_z} + \mathrm{I_{xz}}^2 + \mathrm{I_z}^2\right)qr + \left(\mathrm{I_{xz} I_y} - \mathrm{I_{xz} I_x} - \mathrm{I_{xz} I_z}\right)pq - \mathrm{I_z}\bar{L} - \mathrm{I_{xz}}\bar{N}}{\mathrm{I_x I_z} - \mathrm{I_{xz}}^2} \tag{17.4}$$

Table 17.2 Characteristics of the UAV.

Mass, m (kg)	80
Wing area, S (m2)	0.9
Wing span, b (m)	2.12
Velocity range, V (m/s)	75, 115
Sideslip angle range, β (deg)	−10, 10
Angle-of-attack range, α (deg)	−8, 14
Body-axis angular rates range, p, q, r (deg/s)	−45, 45
Elevator-angle deflection range, δ_E (deg)	−20, 10
Aileron-angle deflection range, δ_A (deg)	−15, 15
Rudder-angle deflection range, δ_R (deg)	−20, 20
Throttle-setting allowable range, δ_T	0, 1

$$\dot{q} = \frac{(I_z - I_x)pr - I_{xz}(p^2 - r^2) + \bar{M}}{I_y} \tag{17.5}$$

$$\dot{r} = -\frac{(I_x I_y - I_x{}^2 - I_{xz}{}^2)pq + (I_x I_{xz} + I_{xz} I_z - I_{xz} I_y)qr - I_{xz}\bar{L} - I_x \bar{N}}{I_x I_z - I_{xz}{}^2} \tag{17.6}$$

$$\dot{\phi} = p + q\sin(\phi)\tan(\theta) + r\cos(\phi)\tan(\theta) \tag{17.7}$$

$$\dot{\theta} = q\cos(\phi) - r\sin(\phi) \tag{17.8}$$

$$\dot{\psi} = \frac{q\sin(\phi) + r\cos(\phi)}{\cos(\theta)} \tag{17.9}$$

$$\begin{aligned}\dot{x} &= u\cos\theta\ Cos\psi + v(\sin\phi\sin\theta\cos\psi - \cos\phi\sin\psi)\\ &\quad + w(\cos\phi\sin\theta\cos\psi + \sin\phi\sin\psi)\end{aligned} \tag{17.10}$$

$$\begin{aligned}\dot{y} &= u\cos\theta\sin\psi + v(\sin\phi\sin\theta\sin\psi + \cos\phi\cos\psi)\\ &\quad + w(\cos\phi\sin\theta\sin\psi - \sin\phi\cos\psi)\end{aligned} \tag{17.11}$$

$$\dot{h} = u\sin\theta + v\sin\phi\cos\theta + w\cos\phi\cos\theta \tag{17.12}$$

where u, v and w are the UAV velocity vector components in the body frame, respectively. Similarly, the body-frame components of the vehicle angular velocity vector are denoted by p, q and r, respectively. Finally, the vehicle attitude orientation with respect to the local Erath tangent plane are define through the well-known Euler angles, denoted by the variables ϕ, θ and ψ.

It should be noted that the aerodynamic and thrust forces and moments acting on the UAV are also present in the equations as the forcing functions, whose values need to be known a priori. They can be specified via various theoretical or experimental/empirical methods, depending on the desired level of accuracy. One of the most common form of aerodynamic modelling is, however, through the use of what is called the 'aircraft stability and control derivatives', which that can in turn be functions of the UAV instantaneous flight condition, namely the Mach number and altitude. Existing engineering and numerical software, such as Fluent [18] and Digital Datcom [19], are available for computation of these derivatives. The engine thrust can also be modelled as a function of the static thrust, throttle setting, flight altitude and the Mach number.

17.2.3 Organizing the Trim-manoeuvre Library

One of the basic ingredients of the manoeuvre-based approaches to CMP is offline generation of a library of manoeuvres representing the motion primitives. A set of trim and transition manoeuvre paths are initially determined based on the performance and dynamic capabilities of the UAV under study. These offline paths are selected within the operational flight envelope of the UAV for MP and will produce the so-called trim-manoeuvre library.The trim path characteristics include the velocity, altitude, turn rate and rate-of-climb. The transition manoeuvre characteristics include their 3D trajectory along with their corresponding flight timespans. The trim path timespans are not specified in the library and are determined by the CMP algorithm, as described in Section 17.4.1.

UAV Trim Problem

The concept of a singular or equilibrium point of a time-invariant system is introduced in the theory of nonlinear control systems [20]. For a nonlinear system having the general form $\mathbf{F}(\mathbf{x},\dot{\mathbf{x}},\mathbf{U})=\mathbf{0}$, a singular point is a point that satisfies $\mathbf{F}(\mathbf{x},\dot{\mathbf{x}},\mathbf{U})=\mathbf{0}$, with $\dot{\mathbf{x}}=\mathbf{0}$ and $\mathbf{U}=$ constant. This concept has a strong intuitive appeal for aerospace applications, as in steady-state flight conditions the vehicle flies with constant motion and control parameters. In these conditions the general nonlinear equations of motion (Eqs. (17.1)–(17.6)) turn into a set of algebraic equations that result in the TSs. For this purpose, three typical TSs that could occur in the flight of a UAV are investigated:

- steady wings-level flight ($V=\text{constant}, \dot{h}=\dot{\psi}=0$)
- steady level turning flight ($V=\text{constant}, \dot{\psi}=\text{constant}, \dot{h}=0$)
- steady climbing flight ($V=\text{constant}, \dot{h}=\text{constant}, \dot{\psi}=0$).

In general, the conditions corresponding to $\text{V}=\text{constant}$, $\dot{h}=\text{constant}$ and $\dot{\psi}=\text{constant}$ can gives all possible TSs. The steady-state flight can be calculated using Eqs. (17.1) and (17.12) while imposing $\dot{V}, \dot{\alpha}, \dot{\beta}, \dot{p}, \dot{q}, \dot{r}=0$ and $\mathbf{U}=\text{constant}$. Therefore, a set of nonlinear algebraic equations are obtained and these can be solved to get the TSs and controls. However, another viable approach to solving this set of nonlinear algebraic equations is through optimization. In this respect, an optimization problem with a design vector of $\mathbf{W}=[\alpha, \beta, \delta_E, \delta_A, \delta_R, \delta_T]^T$ can be solved by minimizing the following cost function [21]:

$$F=\min\left(\dot{V}^2+10\dot{\alpha}^2+\dot{\beta}^2+\dot{p}^2+100\dot{q}^2+\dot{r}^2\right) \tag{17.13}$$

where δ_E, δ_A, δ_R are the elevator, aileron and rudder control deflections respectively, and δ_T denotes to the engine throttle setting. If the cost function converges to zero in the optimization process, with a predefined small tolerance, one can conclude that the vehicle has been trimmed and $\dot{V}$, $\dot{\alpha}$, $\dot{\beta}$, $\dot{p}$, $\dot{q}$ and $\dot{r}$ have converged to zero. In this sense, a flight condition is called trimable if the cost function goes below 10^{-5}. The optimization approach used for the trim problem is a modified version of the particle swarm optimization (PSO) algorithm [17] developed by the authors, which is explained in the next section.

The Modified PSO Algorithm

Particle swarm optimization (PSO) is a population-based optimization algorithm initially introduced by Kennedy and Eberhart in 1995 [22]. The core idea was based on a simulation of simplified animal social behaviours such as fish schooling or bird flocking. One of the inherent characteristics of the PSO algorithm is that it works efficiently in continuous environments.

In basic PSO [23], vectors $\mathbf{x}_i(t)$ and $\mathbf{v}_i(t)$ denote the position and the velocity of particle i in the search space at time or iteration t. The particle position at iteration $(t+1)$ is updated in a current movement by:

$$\mathbf{x}_i(t+1) = \mathbf{x}_i(t) + \mathbf{v}_i(t) \tag{17.14}$$

The velocity vector reflects both the experienced knowledge of a particle as well as socially exchanged information from the neighbouring particles. The velocity vector of particle i at iteration $t+1$ is updated as:

$$\mathbf{v}_i(t+1) = \omega\mathbf{v}_i(t) + \mathrm{c}_1\mathrm{r}_1\big(\mathbf{y}_i(t) - \mathbf{x}_i(t)\big) + \mathrm{c}_2\mathrm{r}_2\big(\hat{\mathbf{y}}(t) - \mathbf{x}_i(t)\big) \tag{17.15}$$

Where $\mathrm{c}_1, \mathrm{c}_2$ are constant parameters, $\mathrm{r}_1, \mathrm{r}_2$ are random numbers between $[0,1]$, $\mathbf{y}_i(t)$ is the personal best position of particle i from the initialization at iteration t and $\hat{\mathbf{y}}(t)$ represents the best position found by neighbours at iteration t. ω is an inertia weight that controls the exploration and exploitation properties of the swarm. The modified PSO is enhanced with the following strategies in order to promote the algorithm performance:

The inertia weight parameter The inertia weight is selected randomly as proposed in [24]:

$$\omega = \mathrm{rnd} \tag{17.16}$$

Where rnd is a uniform random number in the range [0, 1].

Velocity vector saturation The velocity vector of the saturated particles is randomized:

$$\mathbf{v}_i(t+1) = \begin{cases} \mathbf{v}_i(t+1) & \text{if } |\mathbf{v}_i(t+1)| \le \mathrm{V}_{\max} \\ \mathrm{rnd}_{\mathrm{Vmax}} & \text{if } |\mathbf{v}_i(t+1)| > \mathrm{V}_{\max} \end{cases} \tag{17.17}$$

where $\mathrm{V}_{\max}$ is the maximum value, considered for each element of the velocity vector and $\mathrm{rnd}_{\mathrm{Vmax}}$ generates a uniformly distributed random number between $-\mathrm{V}_{\max}$ and $\mathrm{V}_{\max}$. The initial velocity vector components are also randomly selected within $[-\mathrm{V}_{\max}, \mathrm{V}_{\max}]$.

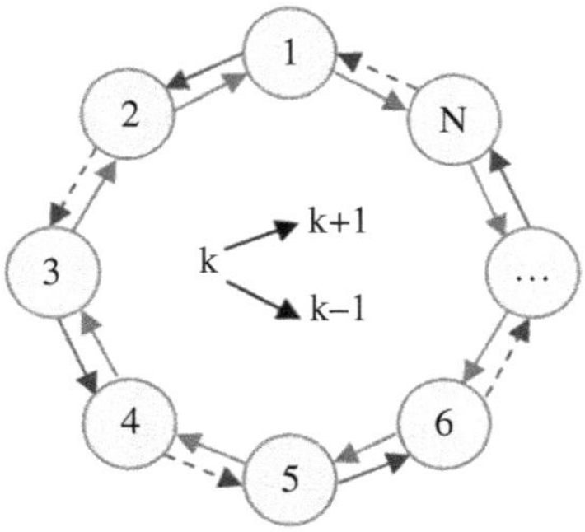

Figure 17.4 Ring-neighborhood structure.

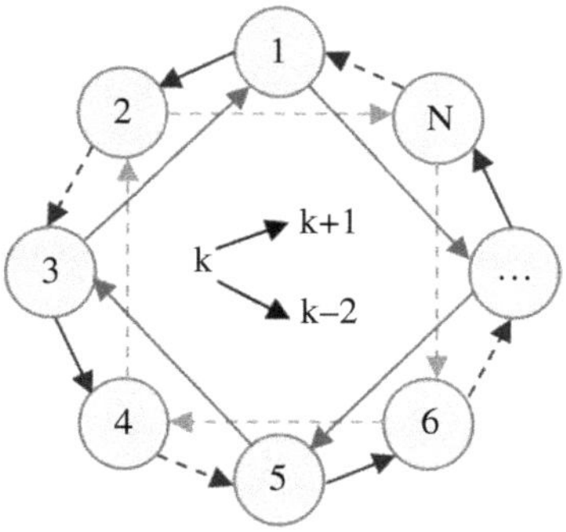

Figure 17.5 Singly-linked ring structure.

Social network structure In PSO, each particle moves, following a leader; this feature is realized through the social term of Eq. (17.15). A leader can be global to all the swarm, or local to a particle's neighbourhood. In the latter case, there are as many local leaders as neighbourhoods. Having more than one leader in the swarm translates into more attractors or good spots within the search space. Therefore, the use of neighbourhoods is a natural approach to fight premature convergence [23]. Particles in the same neighbourhood communicate with one another by exchanging information about moving towards a better position. The flow of information through the swarm depends on the neighbourhood structure. In a highly connected neighbourhood structure, the information about the best particle in the swarm is quickly transmitted through the whole swarm. This means faster convergence, which implies a higher risk of converging to a local extremum. There are various neighbourhood structures, such as star, ring, wheel and von Neumann [23]. In a ring neighbourhood structure, as in Figure 17.4, a particle k has two neighbours, particles $k-1$ and $k+1$. In turn, particles $k-1$ and $k+1$ have particle k as one of their neighbours. In this way, there is a mutual attraction between consecutive particles, forming overlapped clusters. The slow convergence of the ring structure has been empirically shown [25]. So a topology of singly-linked rings has been proposed [26, 27], as presented in Figure 17.5. This topology keeps two neighbours for each particle, but breaks the mutual attraction between neighbours. In this way, the information is transmitted faster through the whole swarm than in the original ring topology. Therefore, the singly-linked ring keeps the exploration at the search space, and increases the exploitation of the best solutions. In the present work, the singly-linked ring structure is used.

The pseudo-code for the modified PSO is presented in Figure 17.6. An important advantage of using the optimization problem to solve for the TSs is its potential for augmentation with constraints to satisfy all of the vehicle's dynamic, motion-parameter and control limitations. By solving the trim problem of the selected UAV within its original flight envelope, several interesting results are realized, as discussed in the next sections.

The Trimability Envelope

The trimability of UAVs in steady wings-level flight throughout the flight envelope is investigated using the modified PSO algorithm. The conventional flight envelope, based on the UAV 3DOF model and its aerodynamic and thrust forces, called the h–V graph,

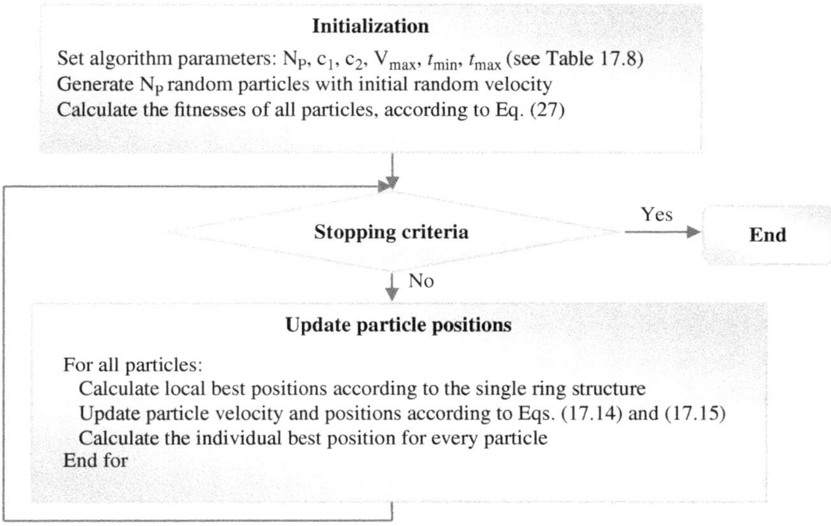

Figure 17.6 Pseudo-code for modified PSO.

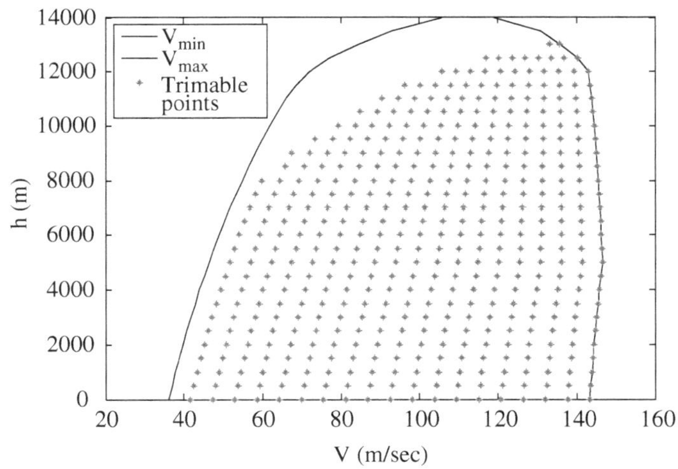

Figure 17.7 The general trimability envelope for steady wings level flight.

is usually established by considering the minimum (stall) and maximum velocity for different flight altitudes [28]. The flight envelope can be subsequently used to show the trimability range for various steady-state flights, including the steady wings-level flight, steady climb and steady turns (around some mean operational altitude), as shown in Figures 17.7–17.9.

Forming the Trim States

Based on the data provided in Table 17.2 and the flight envelopes of Figures 17.6–17.8 of the UAV under study, 45 different TSs around a mean operational altitude are selected

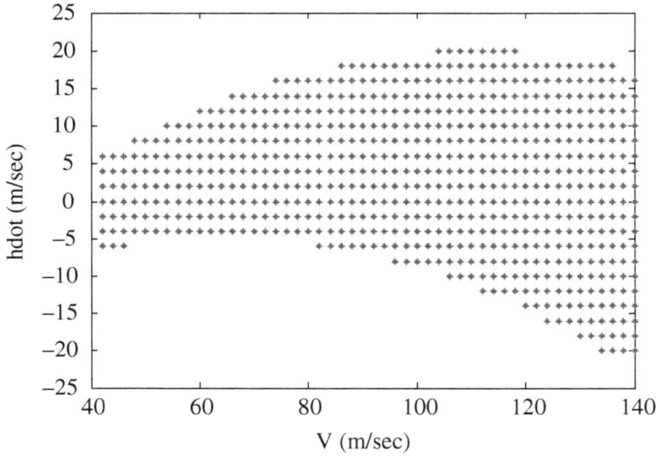

Figure 17.8 The steady climb trim envelope at a mean operational altitude of 2000 m.

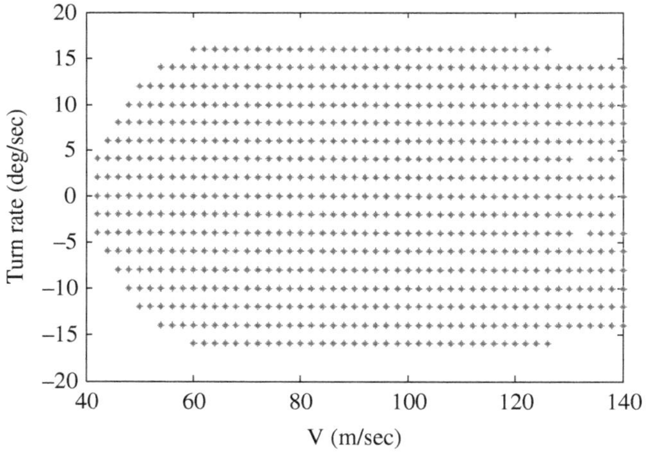

Figure 17.9 The steady turn trim envelop at a mean operational altitude of 2000 m.

Table 17.3 Selected flight conditions for the 45 trim states.

Mean operational altitude, h (m)	2000
Velocity, V (m/sec)	75, 85, 95, 105, 115
Altitude rate, $\dot{h}$ (m/ sec)	−6, −3, 0, 5, 10
Heading angle rate, $\dot{\psi}$ (deg/ sec)	−10, −5, 0, 5, 10

for the purpose of CMP implementation. These states are compatible with the possible flight conditions of Table 17.3. As mentioned before, the solution set of the required TS parameters is determined via a new heuristic optimization approach developed by the authors [27]. Subsequently, an index number is assigned to each TS in the library.

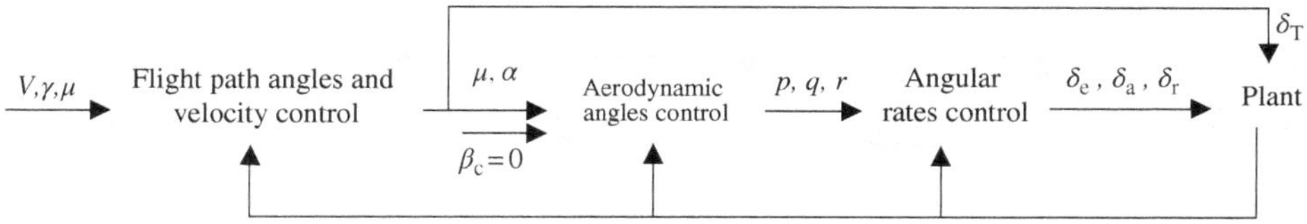

Figure 17.10 The block diagram of the backstepping method for manoeuvre design [19].

For example, TS-1 is characterized by $V = 75\,(\text{m/sec})$, $\dot{h} = 0$ and $\dot{\psi} = 0$. The TSs are derived for a mean operating altitude of 2000 m in accordance to its low altitude missions.

Generation the Manoeuvre Trajectories

The manoeuvre paths are trajectories that connect two different TSs. In the current study, to establish the communication network between the existing 45 TSs, a total of 398 manoeuvre trajectories are determined and stored in the library. The strategy behind extraction of these manoeuvre trajectories is to allow the following transitional flights:

- Manoeuvre between different steady wings level TSs (to increase or decrease the UAV velocity in steady flight).
- Manoeuvre between different constant rate of climb TSs (with identical flight velocities).
- Manoeuvre from a higher velocity TS to a lower velocity TS with a larger rate of climb.
- Manoeuvre from a lower velocity TS to a higher velocity TS with a smaller rate of climb.
- Manoeuvre between different steady turns (with identical flight velocities).

Note that different strategies can also be applied, depending on the mission at hand. Once the strategy is determined and the number of transitional manoeuvres is established, a nonlinear constrained backstepping method is used for their design as a part of the library of manoeuvres. This approach not only generates smooth translational manoeuvres, but also considers all dynamic and performance limitations of the UAV; see Figure 17.10. More details can be found in the literature [11].

As an example of the suggested backstepping approach for manoeuvre design, the manoeuvre between two TSs is presented [19]. In this manoeuvre, the UAV transitions from a steady wings-level TS of $V = 75\,(\text{m/sec})$, $\dot{h} = 0$, $\dot{\psi} = 0$ to a steady turning flight TS of $V = 75\,(\text{m/sec})$, $\dot{h} = 0, \dot{\psi} = 10\,(\text{deg/sec})$. The commanded variables V, χ and γ are assumed to change linearly during the manoeuvre. The behaviour of the UAV in this transitional manoeuvre is illustrated in Figure 17.11 through Figure 17.14. Note that there are slight negligible differences between the final states reached and the desired values, but these are acceptable for our purposes and the vehicle limitations are satisfied. A 3D view of the resultant transfer trajectory is shown in Figure 17.14.

Utilization of the Symmetry Property

In addition to the relative equilibrium, the manoeuvre automaton (MA) concept additionally benefits from the symmetry property (SP), another important characteristic of mechanical control systems [29]. SP stipulates that the dynamics of most human-built vehicles such as aircraft and ships are invariant with respect to translation and rotation

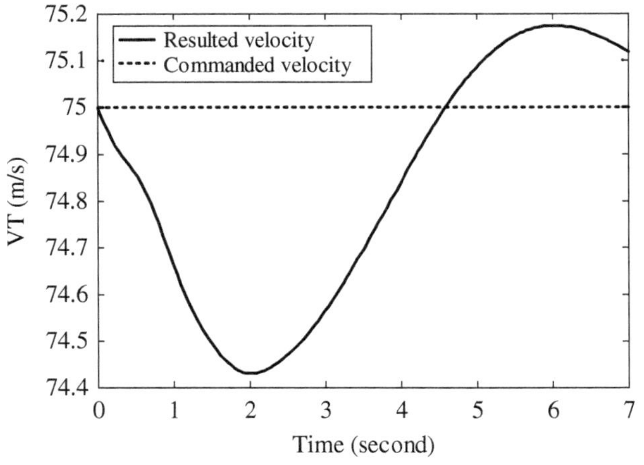

Figure 17.11 Resulting and commanded velocity in a transitional maneuvering flight.

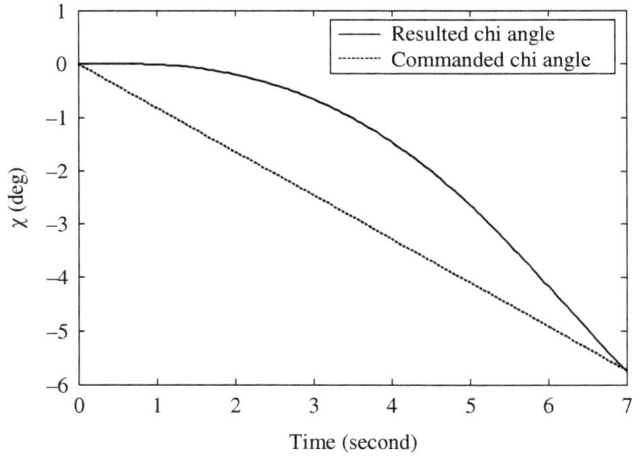

Figure 17.12 Resulting and commanded heading angle in a transitional maneuver flight.

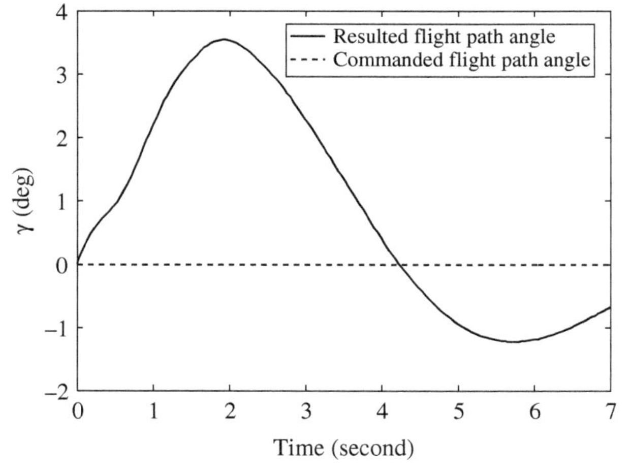

Figure 17.13 Resulting and commanded flight path during the transitional maneuver.

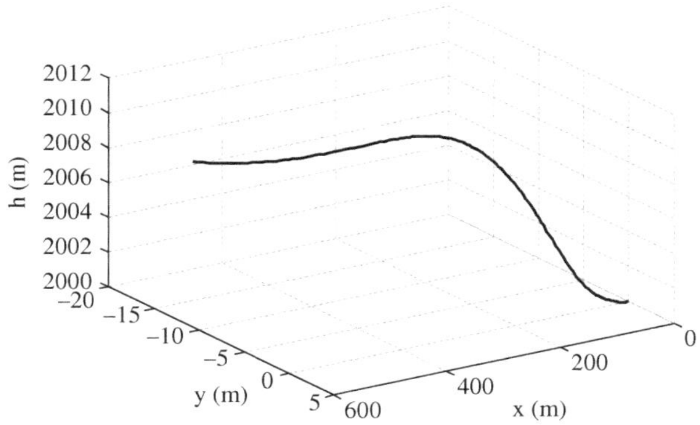

Figure 17.14 The resulting 3D trajectory for a typical transitional maneuver.

about a vertical axis under fairly reasonable assumptions. For the problem at hand, namely CMP, the vertical axis can be any axis parallel to the direction of the local gravitational field. Related valid assumptions are the homogeneous and isotropic nature of the standard atmosphere as well as the constant gravitational acceleration for any nominal operating altitude. As a result, the SP property makes the aircraft heading immaterial for manoeuvre design. In other words, the SP property allows one to utilize the trim and manoeuvre trajectories, determined at a pre-selected heading angle such as zero, to develop a new trajectory on any new heading angle using a simple transformation matrix:

$$\begin{bmatrix} x_{\text{new}}(t) \\ y_{\text{new}}(t) \\ h_{\text{new}}(t) \end{bmatrix} = \begin{bmatrix} \cos(\psi) & -\sin(\psi) & 0 \\ \sin(\psi) & \cos(\psi) & 0 \\ 0 & 0 & 1 \end{bmatrix} \begin{bmatrix} x(t) \\ y(t) \\ h(t) \end{bmatrix} \tag{17.18}$$

where $[x(t), y(t), h(t)]^{\text{T}}$ are the coordinates of an existing trajectory within the library and $[x_{\text{new}}(t), y_{\text{new}}(t), h_{\text{new}}(t)]^{\text{T}}$ correspond to the new trajectory coordinates for a different heading angle.

17.3 Environmental Constraints

In general, the external constraints include the set of limitations that are dictated by the flight environment, which for the CMP problem includes terrain obstacles and the threat areas. The processes for modelling the terrain and threats are described in the following subsections.

17.3.1 Terrain Modelling

In most practical situations, accurate and updatable terrain maps exist. Therefore, it is assumed that the terrain constraint is known and that digital elevation maps (DEM) are available for terrain modelling. Even with this assumption, one of the common difficulties of MP is the burden of massive DEM data handling, especially for online applications. The problem gets more serious when one needs to interpolate through this database repeatedly, in order to obtain the elevation at unknown geographic locations.

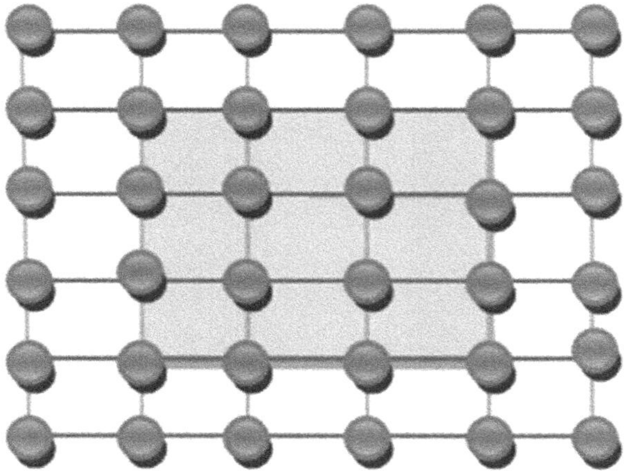

Figure 17.15 Points for a Catmull–Rom surface patch.

Fast and smooth interpolation schemes usually use smaller patches of DEM data in order to satisfy the continuity property. For the purpose of CMP, a special type of spline, referred to as 'cardinal splines' [30] is used. Cardinal splines overcome the main disadvantage of cubic splines, namely the lack of local control. They also eliminate the need to solve a set of linear algebraic equations for terrain modelling, which could be cumbersome by itself. Another desired feature of cardinal splines is that the specified curve passes through all of the control points, a nice characteristic that does not happen for the other types of spline. The idea of cardinal splines or the Catmull–Rom curve can be easily extended to a surface that is fully defined by a rectangular grid of data points. Analogous to the Catmull–Rom curve segment involving four points, a single Catmull–Rom surface patch is specified by 16 control points. Theses control points are used for derivation of coordinates belonging to any other points within the area delimited by them (see Figure 17.15). The expression for the elevation height interpolated by a Catmull–Rom surface can be presented as:

$$elev(x,y) = \begin{bmatrix} x^3 & x^2 & x & 1 \end{bmatrix} \mathbf{B}\mathbf{P}_z\,\mathbf{B}^{\mathrm{T}} \begin{bmatrix} y^3 \\ y^2 \\ y \\ 1 \end{bmatrix} \tag{17.19}$$

where

$$\mathbf{B} = \begin{bmatrix} -0.5 & 1.5 & -1.5 & 0.5 \\ 1 & -2.5 & 2 & -0.5 \\ -0.5 & 0 & 0.5 & 0 \\ 0 & 1 & 0 & 0 \end{bmatrix} \tag{17.20}$$

where $\mathbf{P}_z$ is a matrix containing the elevation of the 4×4 points in the patch, and $[x, y]$ specifies the coordinates of the point whose elevation height needs to be determined via interpolation.

Figure 17.16 The terrain penalty function.

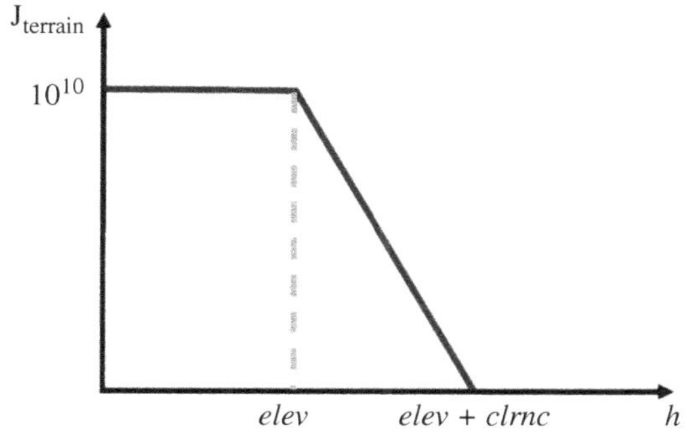

17.3.2 Terrain-related Penalty Function

Hitting the terrain will result in a failed mission. In order to avoid the terrain, a penalty function ($J_{terrain}$) is formulated below and depicted as a function of altitude in Figure 17.16.

$$J_{terrain} = \begin{cases} 10^{10} & h \leq elev \\ \dfrac{10^{10}\left(elev + clrnc - h\right)}{clrnc} & elev < h \leq elev + clrnc \\ 0 & h > elev + clrnc \end{cases} \tag{17.21}$$

where h denotes to the local vehicle flying altitude and $elev$ is the terrain elevation. The clearance ($clrnc$) term is defined as a measure of the vehicle's minimum safety height above the ground, as its value will usually be mission dependent. As can be easily deduced from Eq. (17.21), the cost function heavily penalizes an MP if its trajectory hits the terrain. In fact, a high assigned penalty is to ensure that terrain collision will not occur. On the other hand, if the UAV flies above the clearance altitude as desired, it will not be penalized at all. And finally, if it passes below the clearance altitude, but does not hit the terrain, it will be penalized by a linear cost relationship.

17.3.3 Threats Modelling

One of the key threats to low flying UAVs is from anti-aircraft artillery and manned portable air defence systems that may be either randomly or deterministically arranged in a battle field. Each of these hostile systems has an effective coverage range and a delimiting zone that potentially increases the threat to UAVs and even could terminate its mission.

The areas of threat can be categorized as deterministic or stochastic. The position of stochastic threats is not known a priori, and they are the most difficult threats to handle in a real-time MP algorithm. For the purpose of stochastic threat modelling, both the position and number of threats are taken as random. In this chapter, deterministic and stochastic threats are both considered. The response to deterministic threats is calculated in offline CMP, while the response to stochastic threats is a real-time CMP problem.

To model the stochastic threats to a UAV flight, a predetermined probability (pr_{thrt}) function is used. This strategy will establish a random number of threats during the successive runs of the CMP algorithm. The significant threat position and height is also randomly arranged ahead of the UAV flight area:

$$\begin{bmatrix} x_{\mathrm{threat}} \\ y_{\mathrm{threat}} \\ h_{\mathrm{threat}} \end{bmatrix} = \begin{bmatrix} x_{\mathrm{P1}} + \mathrm{R_r}\left(1+\mathrm{rnd}\right)\cos(\eta_{\mathrm{t}}) \\ y_{\mathrm{P1}} + \mathrm{R_r}\left(1+\mathrm{rnd}\right)\sin(\eta_{\mathrm{t}}) \\ elev\left(x_{\mathrm{threat}}, y_{\mathrm{threat}}\right) \end{bmatrix} \tag{17.22}$$

where

$$\eta_{\mathrm{t}} = \eta_{\mathrm{c}}\,\mathrm{rnd} + \tan^{-1}\left(\frac{y_{\mathrm{F}} - y_{\mathrm{P1}}}{x_{\mathrm{F}} - x_{\mathrm{P1}}}\right) \tag{17.23}$$

The constant $\mathrm{R_r}$ is the minimum range required for the UAV to identify the threat centre location. η_{c} is the UAV camera field-of-view angle and rnd is a uniformly distributed random number in the range [0, 1]. $[x_{\mathrm{P1}}, y_{\mathrm{P1}}]$ denotes the starting coordinates of each trajectory segment and $[x_{\mathrm{F}}, y_{\mathrm{F}}]$ is position of the UAV target point.

17.3.4 Threat Zone Entrance Penalty Function

As entering a threat zone itself could be associated with high risk, a penalty function is defined to ensure its avoidance. Assuming $\mathbf{P}_{\mathrm{threat}} = [x_{\mathrm{threat}}, y_{\mathrm{threat}}, h_{\mathrm{threat}}]^{\mathrm{T}}$ to be the centre of each threat and considering an effective destruction range ($\mathrm{R_e}$) for the threats, the threat zone penalty function is defined as follows:

$$\mathrm{J}_{\mathrm{threat}} = \begin{cases} 10^9 & \text{If } PX < \mathrm{R_e} \\ 0 & \text{If } PX \geq \mathrm{R_e} \end{cases} \tag{17.24}$$

where PX is the distance of the vehicle's current position $[x_1, y_1]^{\mathrm{T}}$ to the threat centre:

$$PX = \sqrt{\left(x_{\mathrm{threat}} - x_1\right)^2 + \left(y_{\mathrm{threat}} - y_1\right)^2} \tag{17.25}$$

According to Eq. (17.24), the penalty function will be zero if the vehicle is out of the threat zone. On the other hand, if the vehicle enters the threat zone, it will be penalized by a large constant value of 10^9, which would ensure that the CMP path does not enter a threat zone in practice.

17.4 Off-line CMP

As mentioned in the previous sections, the UAV trim-manoeuvre library should be constructed offline. Subsequently, there are two alternatives for MP: an offline approach and an online or real-time strategy. The latter will be discussed in Section 17.5. In this section, the off-line CMP problem solution scheme is introduced. In this regard, it is important to note the following points:

First, in line with the MA strategy, the vehicle starts its mission from a starting point with a predefined TS of known timespan. Subsequently, the proposed CMP optimally determines a sequence of TSs and their corresponding timespans from the trim-manoeuvre library. Once the TSs are determined, their connecting transfer or manoeuvre trajectories would be automatically specified from the trim-manoeuvre library as well.

Second, as described above, the MP problem is now regarded as a hybrid optimization problem, as the design variables include both the discrete (index number of TSs) variables as well as the continuous variables; that is, the flight timespan of the selected trim trajectories.

17.4.1 The Off-line Motion Planning Algorithm

In this respect, the modified PSO algorithm described in Section 17.2.3.2 needs to be hybridized to handle the off-line CMP problem. The design vector of the CMP problem can be presented as:

$$\mathbf{W} = \left[T_1, T_2, \ldots, T_{l/2}, t_1, t_2, \ldots, t_{l/2} \right]^{\mathrm{T}} \tag{17.26}$$

where T_i is the index number of ith TS and t_i is its corresponding timespan. The design vector total length is assumed to be l.

The initial UAV trim equilibrium (TS) is chosen by the user/operator along with its known timespan, t_{max}, as the initial UAV flight condition. This initial condition usually depends on the launching system and/or the UAV classification. Note that the design vector does not need this initial TS, and only uses its information to proceed with the optimization scheme. Next, the initial PSO position vector needs to be set. In this way, a feasible TS is randomly picked from the library as its first component T_i, in accordance to the selection strategy described in Section 17.2.3. The other components are chosen in a similar fashion. The continuous components of the position vector $t_i, t_{i+1}, \ldots$ will be initialized using uniformly distributed random numbers in the interval $[t_{min}, t_{max}]$.

As mentioned above, half of the optimization design vector is discrete parameters or indices (T_1, T_2, ..., $T_{l/2}$), while the other half is continuous variables (t_1, t_2, ..., $t_{l/2}$) representing continuous timespans of the TSs. As the modified PSO is a continuous optimization approach, it needs to be empowered (hybridized) for discrete variables using a simple strategy [31]. According to this strategy, when the position update process is completed, the discrete elements of the total design vector are rounded to the nearest valid integers. Therefore, for discrete state variables, the representation of position and velocity vectors remains the same as the continuous ones, as do the velocity and position update equations. Finally, the stopping criterion of the hybridized PSO algorithm is set as a predetermined maximum number of iterations.

17.4.2 Cumulative Objective Functions

The cumulative cost function considered for optimality comprises four terms, some of which have already been introduced:

- terrain avoidance or $J_{terrain}$, as presented in Eq. (17.21),
- threat or J_{threat}, as presented in Eq. (17.24),
- penalty to ensure the target state is reached: $J_1 = \sqrt{(x_F - x_T)^2 + (y_F - y_T)^2}$,
- penalty to minimize the mean altitude: $J_2 = \dfrac{\sum_{i=1}^{N} h(i)}{N}$.

Thus we will have a cumulative cost function, presented below:

$$J = J_{terrain} + J_{threat} + J_1 + J_2 \quad (17.27)$$

where N is the total number of the segments in each trajectory.

17.4.3 Off-line CMP Simulations and Results

The modified PSO routine is developed in MATLAB and runs on a PC with a 2.0 GHZ CPU and 1.0 GB of RAM. A model of real terrain is used for all case studies. The real-terrain data are interpolated by cardinal splines, as described in Section 17.3.1. In the simulated scenarios, the trajectory starts from $[x_0, y_0, h_0] = [0, 0, elev(x_0, y_0) + clrnc]$ and ends at the target point $[x_F, y_F, h_F] = [20000, 25000, elev(x_F, y_F) + clrnc]$. The maximum allowable design vector length is set to $l = 80$, but the algorithm determines the optimal value by itself. Two scenarios with two and three deterministic threats are included in the scenario; see Table 17.4 for the threat coordinates. The parameter sets selected for the modified hybrid PSO algorithm are presented in Table 17.8. Figures 17.17 and 17.18 show the 3D trajectories in the presence of two and three threat zones, respectively. It is important to note that the maximum number of iterations for off-line CMP is set to 200 to have a reasonable run time. Obviously, more iterations can result in better trajectories at the expense of more computational time. The next section evaluates the performance of the current off-line CMP algorithm.

Table 17.4 Position of threats for simulating the off-line CMP.

x_T (m)	y_T (m)
8000	10000
5000	18000
14000	17500

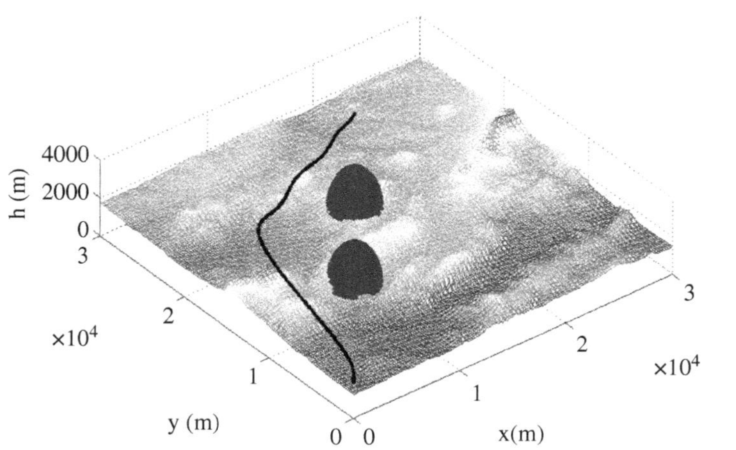

Figure 17.17 Resulting optimal trajectory for offline CMP in the presence of two threats.

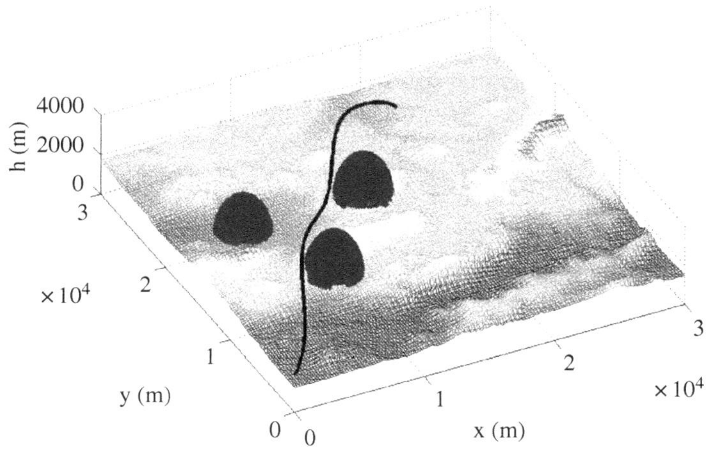

Figure 17.18 Resulting optimal trajectory for offline CMP in the presence of three threats.

Table 17.5 Modified PSO performance evaluation.

Number of deterministic threats	Terrain avoidance (%)	Threat avoidance (%)
24	90	79.16

17.4.4 Modified Hybrid PSO Performance Evaluation

In order to evaluate the performance of the modified hybrid PSO for off-line CMP, ten successive runs are performed using the initial conditions outlined in Section 17.4.3. During these runs, the UAV has impacted the terrain only once. Therefore 90% of the time, terrain obstacles are avoided. In addition, 79.16% of deterministic threat zones have been avoided. This represents a relatively good performance by the off-line CMP. The performance results are presented in Table 17.5. This off-line CMP algorithm is significantly better for real-time applications, as discussed in the next section.

17.5 Real-time CMP

There are several approaches to real-time path planning in the literature. Frazzoli et al. [30] have utilized the MA idea with four varieties of rapidly exploring random trees algorithm (RRT), varying in complexity and completeness for the MP problem. Saunders et al. [31] have combined the RRT, as an a-priori planner of waypoint paths, with a reactive planner that dynamically avoids obstacles. Kothari [32] has utilized the RRT for waypoint generation along with what is called an *anytime* algorithm for dynamic obstacle avoidance. The RRT algorithm and cubic Bezier spiral curves have also been used for 3D smooth path planning of a UAV in cluttered natural environments [33]. An improved heuristic A* algorithm has also been implemented for UAV path planning in real time [34].

More recently, different dynamic heuristic optimization algorithms have been developed and tested against complex dynamic optimization benchmark problems. The dynamic hybrid particle swarm optimization (DHPSO) algorithm [12, 13], also developed by the authors, is one such recent scheme that is potentially appropriate for real time CMP. The key benefits of this approach will be in using the complete dynamic model of the UAV to assess the vehicle's full capabilities while observing the system and environmental constraints for real-time applications. Real-time CMP still utilizes the trim-manoeuvre library, allows for stochastic threats and uses DHPSO algorithm to solve the MP problem. Most importantly, a segmented approach combined with a look-ahead strategy is devised in the new definition of the design vector. This enables the proposed DHPSO algorithm to perform real-time CMP in presence of terrain and random threats.

17.5.1 Solving the Real-time CMP Problem

In the real-time CMP problem, it is important to note the several points.

Dynamic response to threats: Threats may occur in practice at random times and positions. In other words, the location and timing of a threat is usually unknown. The MP therefore requires a dynamic optimization strategy. A required feature of a dynamic optimization algorithm is its convergence speed, as it needs to be fast enough to detect dynamic changes (for example, random threats at any time), track them and find a suitable solution before other dynamic changes occur. To enable the proposed algorithm to function in real time, a segmented approach is proposed in the optimization process. The dynamic changes are allowed to take place just before the algorithm's run time reaches the flight time of the previous optimal segment of the trajectory. In other words, the algorithm must provide the next segment of the optimal trajectory while flying over the previously generated segment of the optimal trajectory.

Shortened design vector: Inspired by the predictive control concept, a finite running number (sequence) of TSs and their required timespans is selected as the optimization design vector. In other words, in this case the design vector does not include all trajectory segments from the start to the target point. In this regard, the online CMP algorithm is only allowed to find an optimal value of the shortened design vector, before the next occurrence of any dynamic change. For example, the shortened design vector can be presented as $\mathbf{W} = [T_i \; T_{i+1} \; t_i \; t_{i+1}]^{\mathrm{T}}$, where, T_i, T_{i+1} are index numbers of two TSs from the library and t_i, t_{i+1} are their corresponding timespan.

Non-optimality: Moreover, in order to provide a look-ahead property for the algorithm, when a dynamic change occurs, the above optimal design vector will not be completely optimal. In other words, when the optimal design vector is $\mathbf{W}^* = [T_i \; T_{i+1} \; t_i \; t_{i+1}]^{\mathrm{T}}$, the online CMP algorithm only utilizes the first segment of $\mathbf{W}^*$, namely $[T_i \; t_i]^{\mathrm{T}}$ as the optimal trajectory to be utilized. Additionally, when a dynamic change occurs, the final condition of the current optimal trajectory will be used as the initial condition for the next iterations.

17.5.2 Illustration of the Real-time CMP Algorithm

The pseudo-code of the DHPSO algorithm is presented in Figure 17.19 and further explained in the following subsections.

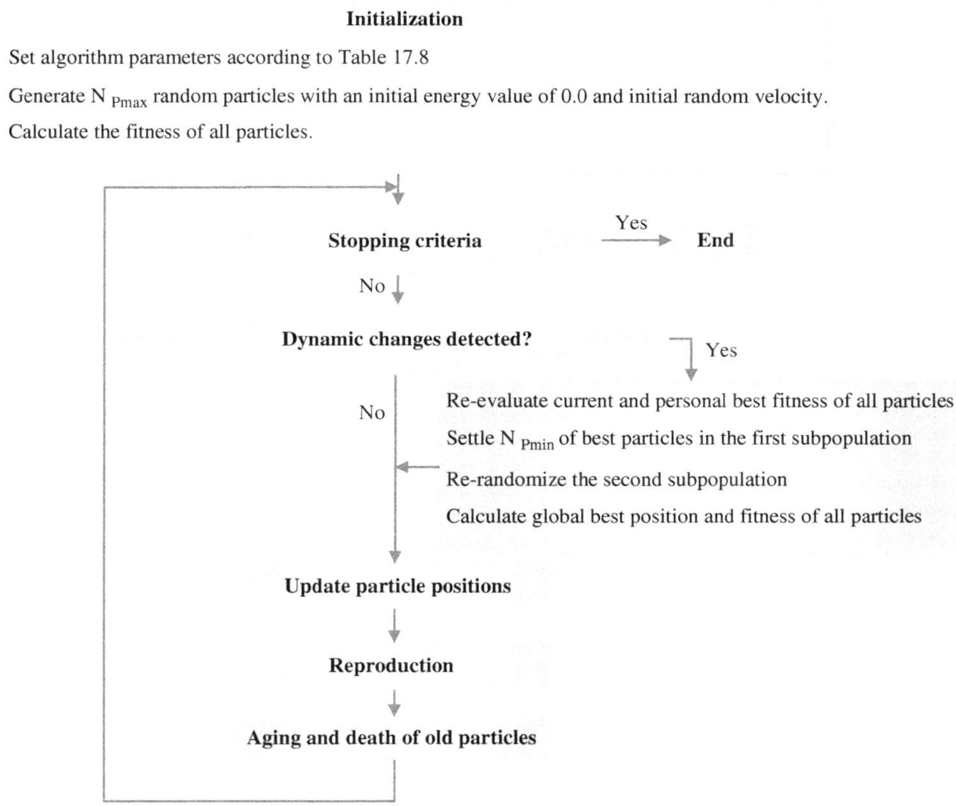

Figure 17.19 Pseudo-code for the real-time CMP

Initialization

In contrast to the basic PSO, the number of particles in DHPSO is allowed to change. This is in order to improve the tracking and converging aspects of the algorithm. Therefore, the algorithm is initialized with the maximum number of particles (N_{Pmax}) that are not allowed to fall below a minimum threshold number (N_{Pmin}). However, the particle number can change in the interval (N_{Pmin}, N_{Pmax}), during the reproduction and death steps of the optimization process. In addition, the swarm is divided into two subpopulations, where N_{Pmin} particles belong to the first subpopulation and the remaining particles belong to the second subpopulation.

Detection of Dynamic Changes

When a dynamic change occurs, the fitness of all particles and their best experiences are re-evaluated. The first subpopulation is filled by the N_{Pmin} best particles. All other particles remaining in the second subpopulation are reinitialized to new random positions. The global best position is then calculated among the individual best positions of the current particles in the swarm.

Updating Particle Positions

In DHPSO, position updating for both subpopulations is determined according to Eq. (17.14). The velocity vector of the first subpopulation is updated according to Eq. (17.15). For the second subpopulation, a strategy similar to differential mutation [35] is used for the velocity update:

$$\mathbf{v}_i(t+1)=\begin{cases}-\left(\omega\mathbf{v}_i(t)+\mathrm{c}_1\mathrm{r}_1\left(\mathbf{y}_i(t)-\mathbf{x}_i(t)\right)+\mathrm{c}_2\mathrm{r}_2\left(\overline{\mathbf{y}}_k(t)-\mathbf{x}_i(t)\right)\right) & \text{if rnd}>0.5\\ \left(\omega\mathbf{v}_i(t)+\mathrm{c}_1\mathrm{r}_1\left(\mathbf{y}_i(t)-\mathbf{x}_i(t)\right)+\mathrm{c}_2\mathrm{r}_2\left(\overline{\mathbf{y}}_k(t)-\mathbf{x}_i(t)\right)\right) & \text{if rnd}\le 0.5\end{cases} \tag{17.28}$$

where $\mathrm{c}_1, \mathrm{c}_2$ are constant parameters, $\mathrm{r}_1, \mathrm{r}_2$ are random numbers in the range [0, 1], ω is the inertia weight that controls the exploration/exploitation abilities of the swarm, $\mathbf{y}_i(t)$ is the personal best position of particle i from initialization to iteration t, $\hat{\mathbf{y}}(t)$ represents the best position found by the swarm at iteration t and $\overline{\mathbf{y}}_k(t)$ is the best position of the randomly chosen kth particle in the first subpopulation. The inertia weight is selected randomly, similar to the offline CMP problem. DHPSO is also empowered to handle discrete variables, as explained in Section 17.4.1 for the offline CMP problem.

Reproduction

Using the quadratic interpolation method, a reproduction procedure is introduced. In this procedure, at every iteration, the two nearest neighbours of the ith particle are identified: n1 and n2. The distances are calculated via Euclidian norm. A new infant can probably be generated by three parents, i, n_1, n_2. The probability of reproducing a new particle in a minimization problem is determined by:

$$P_i(t)=\frac{J_{\text{worst}}(t)-J_i(t)}{J_{\text{worst}}(t)-J_{\text{best}}(t)} \tag{17.29}$$

where $J_i(t)$ denotes the fitness of particle i in iteration t, $J_{\text{worst}}(t)$ is the worst fitness of the iteration t and $J_{\text{best}}(t)$ is the global best fitness found up to iteration t.

If $P_i(t)>\text{rnd}$, then a new particle is generated and placed according to the quadratic interpolation method. The new particle will be placed randomly in the search space if the resulting point is not a minimum for a minimization problem. In this strategy, the position of the new particle in a minimization problem is calculated as:

$$x_{\text{new}}(t,j)=\begin{cases}\dfrac{a}{2b}, & \text{If } b>0\\ \text{rnd}, & \text{If } b\le 0\end{cases} \tag{17.30}$$

where

$$\begin{aligned}a&=\left(x_{n_1}(t,j)^2-x_{n_2}(t,j)^2\right)\mathrm{J}_i(t)+\left(x_{n_2}(t,j)^2-x_i(t,j)^2\right)\mathrm{J}_{n_1}(t)+\left(x_i(t,j)^2-x_{n_1}(t,j)^2\right)\mathrm{J}_{n_2}(t)\\ b&=\left(x_{n_1}(t,j)-x_{n_2}(t,j)\right)\mathrm{J}_i(t)+\left(x_{n_2}(t,j)-x_i(t,j)\right)\mathrm{J}_{n_1}(t)+\left(x_i(t,j)-x_{n_1}(t,j)\right)\mathrm{J}_{n_2}(t)\end{aligned} \tag{17.31}$$

Here, $x_i(t,j)$, $x_{n_1}(t,j)$, $x_{n_2}(t,j)$ and $x_{\text{new}}(t,j)$ are the jth components of the parents and the new particle position vectors at iteration t, respectively.

The velocity vector assigned to the new particle is the average velocity of its parents. Reproduction is allowed for the members of both subpopulations, but the infants are not allowed to reproduce in the iteration in which they were just born. They can take part in the reproduction process in later iterations.

An advantage of this reproduction strategy is that it improves the exploration and exploitation abilities; the exploitation improves by generating more particles at locations with better fitness. On the other hand, the probability of reproducing new random particles results in acceptable exploration ability.

Aging and the Death of Old Particles

Inspired by natural life [12, 36], the initial age assigned to every particle increases linearly with iterations as:

$$\mathbf{e}_i(t+1) = \mathbf{e}_i(t) + \Delta\text{a} \tag{17.32}$$

where Δa is a constant value. The initial age for a particle is chosen as zero, which is also taken for the infants. When a particle's age value reaches one and the swarm size is not less than N_{Pmin}, the corresponding particle dies.

The death strategy effectively adjusts the population size and applies a self-regulating property to the swarm. It should also be noted that the size of the first subpopulation is fixed to N_{Pmin}, but the size of the second subpopulation is not fixed and varies between $[0, \text{N}_{\text{Pmax}} - \text{N}_{\text{Pmin}}]$ due to the reproduction and death of particles.

Stopping Criteria

The algorithm stops if the trajectory has reached/passed the target point. The algorithm stops if the following conditions are met:

$$\mathbf{x}_1.\mathbf{x}_2 < 0 \text{ and } \|\mathbf{x}_2\| < \text{d} \tag{17.33}$$

where $\mathbf{x}_1, \mathbf{x}_2$ are defined as:

$$\mathbf{x}_1 = \left[x_\text{F} - x_0, y_\text{F} - y_0\right]^\text{T}, \mathbf{x}_2 = \left[x_\text{F} - x_1, y_\text{F} - y_1\right]^\text{T} \tag{17.34}$$

where $[x_0, y_0]^\text{T}$, $[x_1, y_1]^\text{T}$ and $[x_\text{F}, y_\text{F}]^\text{T}$ denote the starting point, current position and final location, respectively. The parameter d is a user-defined fixed value that determines when the algorithm must be stopped after having passed the target. The DHPSO algorithm's parameter settings are as presented in Table 17.A.

17.5.3 Optimality Objective Function

The DHPSO cost function consists of three main parts.

- a function to handle terrain avoidance or $\text{J}_{\text{terrain}}$ presented by Eq. (17.21).
- a function to handle stochastic threats are considered; the penalty function of Eq. (17.24) is still used, but random threats are modeled as described in Section 17.3.3

- a function that combines the minimum altitude objective (to increase the surveillance capability, while remaining undetected by threats by encouraging flight within valleys) with the minimum distance from the target objective to yield:

$$J_1 = 0.05\, i \sqrt{\left(x_F - x_{P2}\right)^2 + \left(y_F - y_{P2}\right)^2} + h_{P2} \tag{17.35}$$

where h_{P2} corresponds to the altitude of the end point in every segment of the selected trajectory and the parameter i indicates the current dynamic change index. Thus, as the number of dynamic changes increases and the vehicle converges toward the target point, this term dominates and steers the UAV towards the target destination.

Therefore, the cumulative cost function for the DHPSO to be minimized will be:

$$J = J_1 + J_{\text{terrain}} + J_{\text{threat}} \tag{17.36}$$

17.5.4 Real-time CMP Simulations and Results

In the simulated scenarios, the trajectory starting point and the target location are the same as those utilized for the offline CMP problem; see Section 17.4.3. However, the design vector length is now set to only $l = 4$.

The cumulative objective function is now represented by Eq. (17.36). As previously mentioned, the dynamic changes occur in real time. In other words, the motion planner is allowed to derive the next segment of the trajectory while the UAV flies over the current segment of the optimal trajectory. This provides more realistic operational conditions. Here, randomly generated threats are also allowed to be included. However, while multiple stochastic threats can happen, they are assumed to be separated by 10 km.

One other noteworthy issue that particularly helps real-time implementation capability is the use of Catmull–Rom surfaces for rapid terrain modelling and interpolation. With a massive volume of DEM data, Catmull–Rom surfaces have enabled the DHPSO algorithm to be applied in real-time applications. In Figure 17.20, two threats have emerged in front of the vehicle during the flight. As can be seen, the first threat

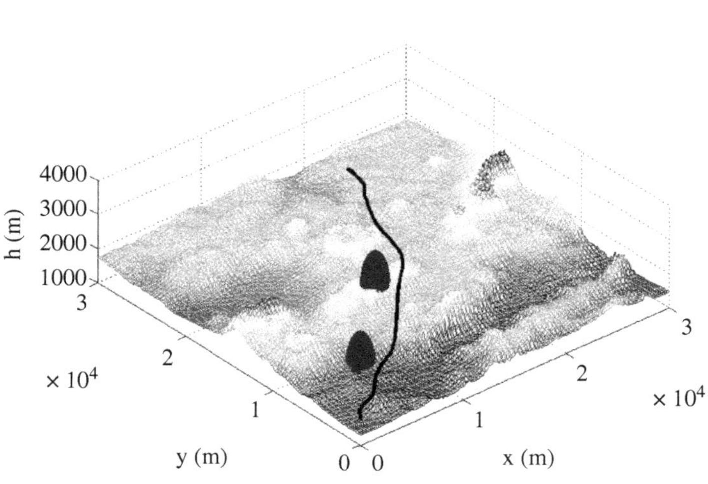

Figure 17.20 Resulting trajectory in the presence of two threat zones.

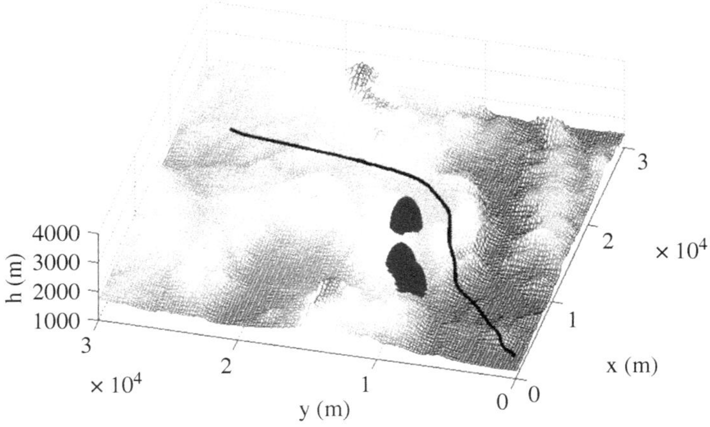

Figure 17.21 Resulting trajectory in the presence of three threat zones.

Table 17.6 DHPSO performance through terrain/threat obstacles for different successive runs.

Case	l	Number of random threats	Terrain avoidance (%)	Threat avoidance (%)
1	4	197	100	90
2	6	185	100	100

occupied a valley with lower altitude, so the vehicle had to change its trajectory and thus avoided both threats successfully. The total flight time of this mission was about 433.49 s, close to the required computer run time. Figure 17.21 shows the resulting CMP trajectory for a case with three stochastic threats, where all threats have been successfully avoided. This mission flight time lasted 459.83 s.

17.5.5 DHPSO Performance Evaluation

Avoiding terrain obstacles is considered the most important issue in MP. In other words, if the resulting trajectory does not avoid the terrain and collides with it, both the mission and the algorithm are said to have failed. Another important constraint is to avoid stochastic threats. Arriving in threat zones does not constitute a failure of the algorithm, but it is undesirable.

For performance evaluation and analysis, 100 successive runs are performed using the DHPSO algorithm. The results are presented in Table 17.6, which shows that the UAV has not impacted the terrain and has reached the target destination in all of the runs. In other words, the mission has never failed. In addition, within these 100 runs, a total of 197 random threats were produced and 90% of the resulting trajectories avoided the threat zones.

To assess the impact of the look-ahead strategy on the performance of the proposed algorithm, Table 17.6 contains another row of results corresponding to still another change in the design vector length (l), from 4 to 6. But again, similar to the case of $l = 4$ when a dynamic change occurs, only the first two elements of the design vector will be selected as optimal. This change in the length of the design vector provides a two-step look-ahead property. A hundred successive runs of this case indicate that not only do the trajectories not impact the terrain but also they avoid all the threat zones that have randomly popped-up in the flight scenery; see Table 17.6.

The above summary is indicative of good performance characteristics of the real-time CMP algorithm.

17.6 Variable Objective CMP

Motion planning can be formulated as a multi-objective optimization problem as well. Some researchers have treated multi-objective CMP problem as single-objective [36–38]. However, this is unfavourable, as in some cases it may require prior knowledge of the biases or sensitivity level of the optimizer with respect to each of the objectives. Pareto-front-based multi-objective evolutionary algorithms [23] have also been used [39–41].

Additionally, the vehicle may need to perform its mission with varying objectives. For instance, part of the flight may be executed with a minimum time objective and the remaining part with a minimum altitude objective; the two cannot be handled at the same time. Therefore, dynamic multi-objective optimization approaches are also of growing interest and are explored in the remaining part of this chapter, following a brief review of some related works.

Peng et al. [42] have proposed a dynamic multi-objective path planning algorithm in the sense of model predictive control. They have also developed Pareto-optimal solution sets using only a kinematic model of the UAV. To select the best path emanating from their online path planning approach, Bayesian networks and fuzzy logic were used to quantify the bias of each of the optimization objectives.

The complete and desirable features of DHPSO for real-time CMP have been enhanced to solve variable-objective dynamic MP problems as well. Multi-objective dynamic particle swarm optimization (MODPSO) [14] was used to do this. MODPSO solves the multi-objective optimal trajectory planning in a piecewise fashion for each segment of the mission, while a variable-objective strategy (VOS) selects an optimal trajectory from the resulting Pareto-optimal set.

17.6.1 Multi-objective Functions

One of the key issues in the multi-objective optimization problem is often the existence of conflicting objectives, where improving one objective may cause deterioration in the others. Therefore, the main task of multi-objective optimization schemes is to produce a solution set of optimum compromises. This set of optimal design variables is referred to as the non-dominated set or the Pareto-optimal set and the corresponding design objective functions values are referred to as the Pareto front [23]. To introduce and develop variable-objective CMP, the three following objectives (J_1, J_2 and J_3) are defined and used in this current study. These distinct objectives are formulated as follows:

$$J_1 = J_{terrain} + J_{threat} \quad (17.37)$$

The second objective (J_2), to be considered separately, combines two objectives within itself, namely the distance to the target and the flight altitude; see Eq. (17.38), which is the same as Eq. (17.35).

$$J_2 = 0.05\, i\sqrt{\left(x_{\mathrm{F}} - x_{\mathrm{P2}}\right)^2 + \left(y_{\mathrm{F}} - y_{\mathrm{P2}}\right)^2} + h_{\mathrm{P2}} \tag{17.38}$$

The third objective denoted by (J_3), includes the flight time of the current segment of the trajectory plus the estimated time of arrival to the target point:

$$J_3 = \left(\sum_{j=1}^{l/2-1} \left(t_{T_j} + t_{\mathrm{M}_{j,j+1}} \right) + t_{\mathrm{T}_{l/2}} \right) + t_{\mathrm{go}} \tag{17.39}$$

Here, $t_{\mathrm{M}_{j,j+1}}$ is the flight time of the manoeuvre from the TS j (T_j) to TS j + 1 (T_{j+1}), l is the design vector length and t_{go} is the estimated time of arrival to the target point, calculated by:

$$t_{\mathrm{go}} = \frac{\sqrt{\left(x_{\mathrm{F}} - x_{\mathrm{P2}}\right)^2 + \left(y_{\mathrm{F}} - y_{\mathrm{P2}}\right)^2}}{V_{\mathrm{mean}}} \tag{17.40}$$

where V_{mean} denotes the UAV's mean velocity, as given in Table 17.7.

Table 17.7 Parameter settings for the three proposed algorithms.

Parameter description	Symbol	Modified PSO	DHPSO	MODPSO
Safe altitude at which the vehicle flies above the ground surface	$clrnc$ (m)	500	500	500
Probability of random threats occurrence in each piece of trajectory	pr_{thrt}	–	50%	50%
UAV camera field of view	η_{c} (deg)	–	20	20
UAV mean velocity	V_{mean} (m/sec)	–	–	95
Threat effective radius	$\mathrm{R_e}$ (m)	2000	2000	2000
Minimum range required for the UAV to identify the threat location	$\mathrm{R_r}$ (m)	–	5000	5000
Design vector length	l	Algorithm determines	4, 6	6
Parameters of Eqs. (17.15) and (17.28)	$\mathrm{c_1, c_2}$	2	2	2
Number of particles	$\mathrm{N_p}$	50	–	–
Minimum and maximum allowable particles	$\mathrm{N_{P_{min}}}$	–	10, 30	10, 30
Minimum and maximum allowable time of TSs	$t_{\mathrm{min}}, t_{\mathrm{max}}$ (sec)	2, 10	2, 10	2, 10
Estimated total number of trajectory pieces	m_{f}	–		25
Particles age increment value	$\Delta\mathrm{a}$	–	0.1	0.1
Maximum value of particles velocity vector, Eq. (17.17).	$\mathrm{V_{max}}$	12	12	12

17.6.2 The Variable-objective Strategy

As previously mentioned, the MODPSO algorithm enables the vehicle to perform its mission with varying objectives in flight. In this regard, a VOS is introduced. According to this strategy, the UAV follows a minimum-time objective at the beginning of flight. The objective varies to a linear function of minimum time and minimum altitude in the middle phase and finally converts to a minimum altitude objective in the terminal phase of the mission. Therefore, a multi-variable objective cost function is defined as:

$$J_{opt} = \min\left(\begin{cases} y_1, & k \leq m_f/3 \\ \dfrac{y_2 - y_1}{m_f/2 - m_f/3}(k - m_f/3) + y_1, & m_f/3 \leq k < m_f/2 \\ y_2, & i \geq m_f/2 \end{cases}\right) \tag{17.41}$$

where

$$y_1 = J_1 + J_3 \tag{17.42}$$

$$y_2 = J_1 + J_2 \tag{17.43}$$

Here, m_f is the estimated total number of trajectory segments that is selected by the user (see Table 17.7), and k corresponds to the current piece of the trajectory.

17.6.3 Illustration of the MODPSO Algorithm

As previously indicated, the CMP is now categorized as a hybrid dynamic multi-variable objective optimization problem, since in addition to all the features of a dynamic optimization problem (Section 17.5.1), a set of three distinct objectives is now defined (Section 17.6.1). After the generation of multiple trajectories (for each piece of the mission), the VOS is used to pick the final optimal trajectory for that segment. The pertinent pseudo-code of the MODPSO algorithm is presented in Figure 17.22. The features of the MODPSO algorithm that are different to the DHPSO are explained in the following subsections.

Filling the Archive

An extra feature of the MODPSO over the DHPSO is that it uses an archive to store the non-dominated particles. The main objective of an archive in a multi-objective algorithm is to keep track of all non-dominated solutions found up to that point. Here, the number of non-dominated solutions or archive members is not restricted. As the next requirement, a non-dominated particle of the archive must avoid entering the terrain or threat zones. The archive is initialized after the initialization step and is reinitialized upon detection of a dynamic change step.

Detection of Dynamic Changes

When a dynamic change takes place, the final condition of the current optimal trajectory will be used as the initial condition for the next iteration. Then, all particles are reinitialized to new random particles. Utilizing the Pareto optimality, the archive is filled by the newly generated non-dominated particles. Then the global best position is calculated among the archive members by implementing the VOS.

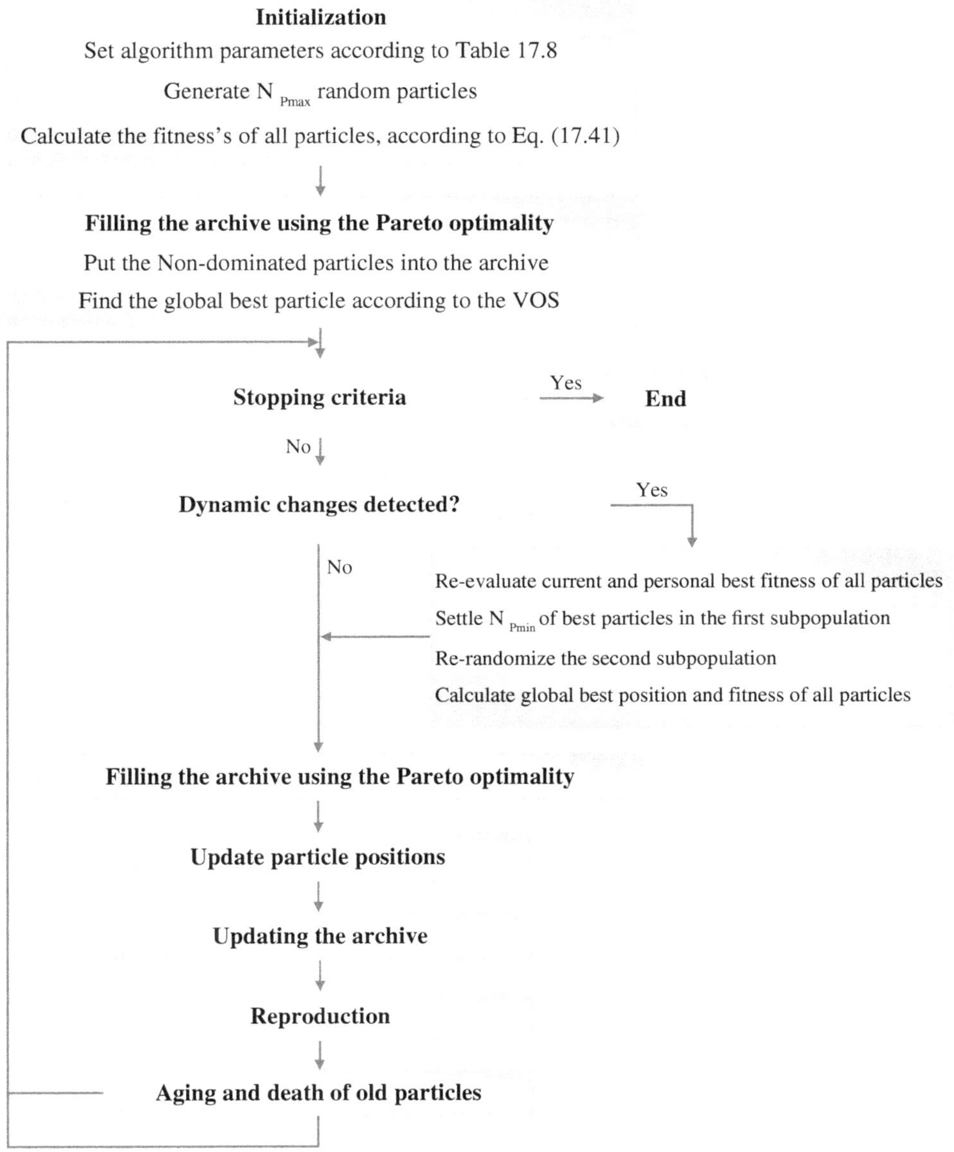

Figure 17.22 Pseudo-code for the MODPSO algorithm.

Updating the Archive

In this step, all of the particles in the swarm are compared with the archive members; dominated members of the archive are deleted and the non-dominated ones are replaced in the archive. Then, the global best particle and its fitness ($J_{\text{best}}(t)$) are chosen using the VOS.

17.6.4 Variable-objective CMP Simulations and Results

The trajectory starting and target points are again similar to those for the algorithms for offline and real-time CMP. Similar to the DHPSO, the algorithm run time for each segment of the trajectory is less than or equal to the UAV total flight time. In addition to the terrain obstacles, randomly generated threats are also included in the scenario. In Figure 17.23, two threats are randomly generated and placed in the scenario. In this case, the resulting UAV path optimizer initially proposes to circle around the first threat to the right. When the second threat is randomly placed in a valley, it forces the remaining trajectory to curve around it to the left.

A scenario with three random threats is also shown in Figure 17.24, where the resulting trajectory moves around the first two threats to the left and manoeuvres to the right of the third threat to optimally reach the target.

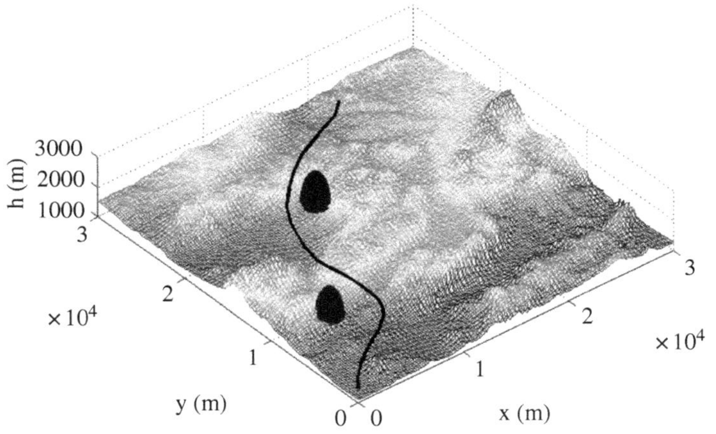

Figure 17.23 Scenario 1: 3D trajectory with two threats.

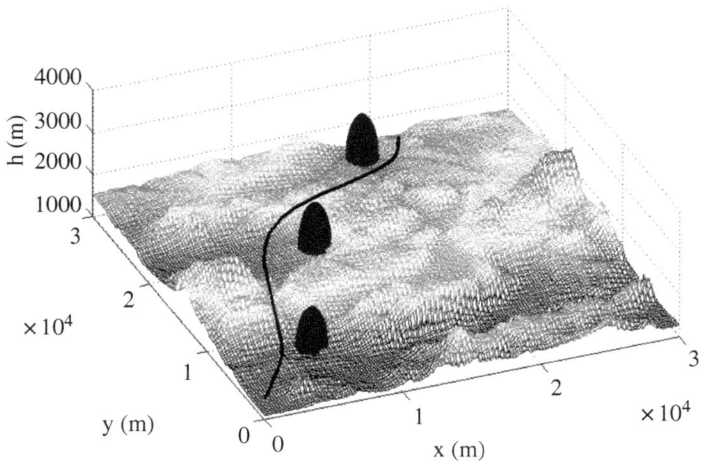

Figure 17.24 Scenario 2: 3D trajectory with three threats.

Table 17.8 Some performance characteristics for Scenarios 1 and 2.

Scenario	Flight time (sec.)	Mean flight altitude (m)	Number of trajectory pieces	Miss distance (m)
0 threat	380	1899.8	26	112
1 threat	353.23	1941.3	27	103
2 threats	408.22	2013.4	29	141
3 threats	363.64	1928.4	28	33
4 threats	470.38	2148.8	30	59

Table 17.9 MODPSO performance through terrain/threat obstacles for different successive runs.

Total number of random threats	Terrain avoidance (%)	Threat avoidance (%)
342	100	100

Specific performance characteristics of these scenarios are shown in Table 17.8. Comparing two- and three-threat scenarios reveals that the algorithm performs better for the three-threat scenario for flight time and mean altitude. Nevertheless, in both scenarios the target point is reached. The number of trajectory pieces for each scenario is also noted in Table 17.8.

17.6.5 MODPSO Performance Evaluation

For this analysis, 150 successive runs are performed. The results are presented in Table 17.9. As seen in this table, the UAV has not impacted the terrain and has reached the target point for all 150 cases. In other words, the mission has never failed and a total of 342 random threats have been produced. Therefore, the average number of pop-up threats is around 2.28 for each scenario. A key reason for this satisfactory performance is the selection of the design vector length equal to six ($l = 6$), thus activating a two-step-forward look-ahead property, as previously discussed.

17.6.6 Pareto Fronts for a Selected Segment of Trajectory

As mentioned previously, for the VOS, each piece of the optimal trajectory will be chosen from the archive members. The resulting Pareto front and the optimal objective found by the VOS (J_{opt}; see Eq. (17.41)) for the 12th segment of the trajectory is shown in Figure 17.25.

17.7 Summary and Conclusions

Three different versions of the CMP problem – including offline CMP, real-time CMP and variable-objective CMP – are investigated. To simulate more complicated and realistic cases, it is assumed that terrain DEM data are available, but the threat zones are considered to be deterministic (for offline CMP) and random (for real-time and

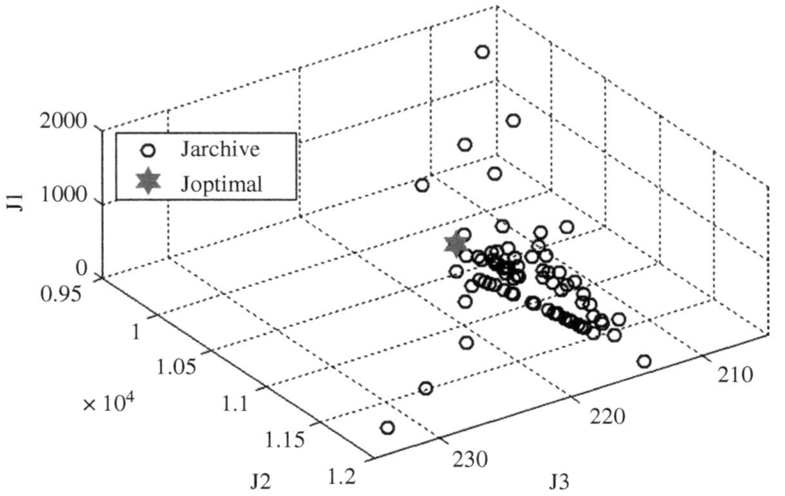

Figure 17.25 Pareto front and optimal fitness of the 12th piece of the trajectory.

variable-objective CMP). The large number of DEM terrain data points is handled by utilizing cardinal spline surfaces. Application of Catmull–Rom surface patches notably decreases the computational run times of the algorithms. In line with the MA concept, the full nonlinear 6DOF vehicle dynamics and constraints are considered in generating a feasible optimal trim-manoeuvre library of trajectories. The TSs are established using a heuristic optimization approach, while a nonlinear backstepping control system design method is used for the manoeuvre trajectories. Through this development, the complicated CMP problem is converted to a hybrid optimization problem.

A modified PSO is initially developed to solve the offline CMP problem, resulting in reasonable performance. The MP problem and its solution was subsequently enhanced using the more innovative DHPSO and MODPSO strategies.

DHPSO is a hybrid approach for dynamic optimization problems with the capability of handling hybrid search spaces. It combines the efficient hybridized features of the PSO in tracking dynamic changes with those of an evolutionary procedure for the reproduction and death of old particles. Utilizing a moving segmented design vector in the optimization process with a look-ahead strategy has enabled DHPSO to work in real-time. Simulation results show that the proposed scheme performs well, making it potentially viable for real-time MP.

Finally, the more complex problem of multi-objective MP in the presence of terrain and random threats is also investigated and solved via a MODPSO algorithm. MODPSO is a hybrid approach for dynamic MO optimization problems in which a VOS technique allows for the selection of the most desirable, objective-based, optimal trajectory in each segment via a three dimensional Pareto set. Again, utilizing a moving segmented design vector in the optimization algorithm with a look-ahead strategy has enabled this approach to work in real-time, even with variable objectives. The simulation results of this case over various scenarios indicate that the proposed MODPSO scheme performs well, thus making it potentially useful for real-time CMP with variable objectives.

In summary, this chapter has reviewed the modelling aspects of aerodynamics, stability and control that play important roles in UAV flight via its full set of nonlinear equations of motion. Subsequently, the UAV model was utilized and integrated with the problem of autonomous UAV CMP using different optimization schemes. The proposed optimization schemes have been progressively enhanced to pave the way for feasible real-time path planning of tactical UAVs. Verification and performance evaluation of the proposed schemes was demonstrated via high-fidelity simulations using realistic mission scenarios, terrain and potential threats.

References

1 S. Twigg, A.J. Calise and E.N. Johnson, On-line trajectory optimization including moving threats and targets, Proceedings of the AIAA guidance, navigation and control conference, Providence, RI, 16–19 August 2004, pp. 807–813.
2 A.R. Babaie and J. Karimi, Optimal trajectory planning for a UAV in presence of terrain and threats. Aerospace and Mechanics Journal of Imam Hossein University, 7(2), (2011) 57–69.
3 J.O. Royset, W.M. Carlyle and R.K. Wood, Routing military aircraft with a constrained shortest–path algorithm, Military Operations Research, 14(3), (2009) 31–52.
4 J.J. Kuffner and S.M. LaValle, RRT-connect: an efficient approach to single-query path planning, Proceedings of IEEE International Conference on Robotics and Automation, 2000, pp. 995–1001.
5 D. Jian and J. Vagners, Parallel evolutionary algorithms for UAV path planning, Proceedings of AIAA 1st Intelligent System Technology Conference, 2004, pp. 1499–1506.
6 M. Guanjun, D. Haibin, L. Senqi, Improved ant colony algorithm for global trajectory planning of UAV under complex environment, International Journal of Computer Science & Application, 4(3) (2007) 57–68.
7 M. Hao, T. Yantao, Z. Linan, A hybrid ant colony optimization algorithm for path planning of robot in dynamic environment, International Journal of Information Technology, 12(3), (2006) 78–88.
8 E. Frazzoli, F. Bullo, On quantization and optimal control of dynamical systems with symmetries, Proceedings of IEEE Conference on Decision and Control, Las Vegas, 10–13 December 2002, vol. 1, pp. 817–823.
9 E. Frazzoli, M.A. Dahleh and E. Feron, Maneuver-based motion planning for nonlinear systems with symmetries. IEEE Transactions on Robotics, 21(6), (2005) 1077–1091.
10 E. Frazzoli, Robust hybrid control for autonomous vehicle motion planning, Ph.D. thesis, Massachusetts Institute of Technology (2001).
11 J. Karimi and S.H. Pourtakdoust, Integrated motion planning and trajectory control system for unmanned air vehicles, Proceedings of the Institution of Mechanical Engineers, Part G: Journal of Aerospace Engineering, 227 (13–18) (2013).
12 J. Karimi, H. Nobahari and S.H. Pourtakdoust, A new hybrid approach for dynamic continuous optimization problems, Applied Soft computing, 12(3), (2012) 1158–1167.
13 J. Karimi, S.H. Pourtakdoust, Optimal maneuver-based motion planning over terrain and threats using a dynamic hybrid PSO algorithm, Aerospace Science and Technology Journal, 2012.

14 J. Karimi, S.H. Pourtakdoust, A real-time algorithm for variable-objective motion planning over terrain and threats, Proceedings of the Institution of Mechanical Engineers, Part G: Journal of Aerospace Engineering, (2014).
15 S.M. LaValle, *Planning Algorithms*, Cambridge University Press, 2006.
16 R.E. Weibel, R.J. Hansman, Safety considerations for operation of unmanned aerial vehicles in the national airspace system, Master of science thesis, Masachusetts Institute of Technology, 2005.
17 B.L. Stevens, F.L. Lewis, *Aircraft Control and Simulation*, John Wiley & Sons, 1992.
18 http://ansys.com/Products
19 http://openae.org/digital-datcom-download
20 J.J. Slotine, W. Li, *Applied Nonlinear Control*, Prentice-Hall, 1991.
21 J. Karimi, S.H. Pourtakdoust and H. Nobahari, Trim and maneuverability analysis of a UAV using s new constrained PSO approach. Journal of Aerospace Science and Technology (JAST), Iranian Aerospace Society, 8(1), (2011) 1–12.
22 J. Kennedy, R.C. Eberhart, Particle swarm optimization, Proceedings of IEEE International Conference on Neural Networks, Perth, Australia, 1995, pp. 1942–1948.
23 A.P. Engelbrecht, *Fundamentals of Computational Swarm Intelligence*, John Willey & Sons, 2005.
24 R.C. Eberhart, Y. Shi, Tracking and optimizing dynamic systems with particle swarms, Proceedings of IEEE Congress on Evolutionary Computation, May 2001, vol. 1, pp. 91–100.
25 J. Kennedy, R. Mendes, Population Structure and Particle Swarm Performance, Proceedings of the 2002 Congress on Evolutionary Computation, CEC02, 2002, pp. 1671–1676
26 A.M. Zavala, A.H. Aguirre, E.V. Diharce, Constrained optimization via particle evolutionary swarm optimization algorithm (PESO), Proceedings of the 2005 conference on Genetic and Evolutionary Computation, 2005, pp. 209–216.
27 A.M. Zavala, A.H. Aguirre, E.V. Diharce, robust PSO-based constrained optimization by perturbing the particle's memory, in *Swarm Intelligence. Focus on Ant and Particle Swarm Optimization*, I-Tech Education and Publishing, 2007, pp. 57–76.
28 J. Roskam, C.T. Edward Lan, *Airplane Aerodynamics and Performance*, DAR Corporation, 1997.
29 E.C. Laskari, K.E. Parapoulos, M.N. Vrahatis, Particle swarm optimization for integer programming, Proceedings of the IEEE 2002 Congress on Evolutionary Computation, Hawaii, HI, USA, 12–15 May 2002, pp. 1576–1581.
30 E. Frazzoli, M.A., Dahleh and E. Feron, Real-time motion planning for agile autonomous vehicles. AIAA Journal of Guidance, Control, and Dynamics, 25(1), (2002) 116–129.
31 J. Saunders, B. Call, A. Curtis, R. Beard and T. McLain, Static and dynamic obstacle avoidance in miniature air vehicles, Proceedings of Infotech Aerospace, Arlington, VA, September 2005. Paper AIAA-2005-6950.
32 M. Kothari, I. Postlethwaite and G. Da-Wei, Multi-UAV path planning in obstacle rich environments using rapidly-exploring random trees, Joint 48th IEEE Conference on Decision and Control and 28th Chinese Control Conference Shanghai, PR China, 16–18 December, 2009, pp. 3069–3074.
33 K. Yang and S. Sukkarieh, 3D smooth path planning for a UAV in cluttered natural environments, IEEE/RSJ International Conference on Intelligent Robots and Systems Acropolis Convention Center Nice, France, 22–26 September 2008, pp. 794–800.

34 X. LiXia, C. Manyi and W.Z. XieMing, Path planning for UAV based on improved heuristic A* algorithm, The Ninth International Conference on Electronic Measurement and Instruments, ICEMI'2009, 2009, pp. 488–493.
35 W. Du, B. Li, Multi-strategy ensemble particle swarm optimization for dynamic optimization, Journal of Information Sciences, 178(15), (2008) 3096–3109.
36 D. Rathbun, B. Capozzi, Evolutionary approaches to path planning through uncertain environments, Proceedings of AIAA Unmanned Aerospace Vehicles, Systems, Technologies and Operations Conference and Workshop, 2002.
37 C. Zheng, L. Li, F. Xu, F. Sun, M. Ding, Evolutionary route planner for unmanned air vehicles, IEEE Transactions on Robotics, 21(4), (2005) 609–620.
38 W. Jiabo, L. Li, L. Teng, W. Zhu, Three dimensional constrained UAV path planning using modified particle swarm optimization with digital pheromones, 3rd International Conference on Engineering Optimization, 2012.
39 S. Mittal, K. Deb, Three dimensional off-line path planning for UAVs using multi-objective evolutionary algorithms, IEEE Congress on Evolutionary Computation, *CEC* 2007, 2007, pp. 3195–3202.
40 G.B. Lamont, J.N. Slear, K. Melendez, UAV Swarm mission planning and routing using multi-objective evolutionary algorithms, IEEE Symposium on Computational Intelligence in Multi-criteria Decision Making, 2007, pp. 10–20.
41 P. J. Vesselin, X.L. Rong, On fusion of multiple objectives for UAV search and track path optimization, Journal of Advances in Information Fusion, 4(1), (2009) 27–39.
42 X. Peng, D. Xu, Intelligent online path planning for UAVs in adversarial environments, International Journal of Advanced Robotic Systems, vol. 9(3), pp. 1–12, (2012).

18

Autonomous Space Navigation Using Nonlinear Filters with MEMS Technology

Seid H. Pourtakdoust and Maryam Kiani

Center for Research and Development in Space Science and Technology, Sharif University of Technology, Tehran, Iran

18.1 Introduction and Problem Statement

Rapid growth of space traffic and low Earth orbit (LEO) satellite systems for remote sensing and communication applications have produced a need for improved autonomous performance for the navigation subsystems that are a vital part of any operational space system. Accordingly, due to the wide variety of space missions and applications, attitude and orbit determination schemes that are independent of Earth-based observation systems are getting more attention for space applications. The current chapter is dedicated to the development and enhancement of autonomous space navigation. In this regard, concurrent orbit and attitude determination (COAD) is investigated in Section 18.2. COAD is especially useful because it greatly reduces the cost and weight of the navigation and control subsystems for small satellites.

Sensor calibration and a good knowledge of dynamic system parameters affect the estimation accuracy, and the problems of simultaneous attitude determination (AD), parameter identification (PI) and measurement system calibration (MSC) are discussed in Section 18.3.

Undoubtedly, the rapid development and growth of microelectromechanical systems (MEMS) technology has played an important role in its use in space projects. MEMS fabrication is similar to what takes place in the chip industry in terms of patterning and surface-processing technologies. Their small size, low cost, acceptable reliability, and ease of interfacing with control subsystems make MEMS-based sensors and actuators ideal for space applications. MEMS can be considered a logical extension of silicon techniques within the realm of mechanical devices. MEMS' lower power consumption leads to lower launch expenses, and this, together with their satisfactory resistance to radiation and vibration have made MEMS-based integrated sensor/actuator packs ideal candidates for autonomous AD and COAD systems.

To meet the desired accuracy requirements of autonomous space navigation subsystems using low-cost MEMS technology, advanced algorithms with a higher computational burden must be developed and exploited. In this sense, the current chapter

Advanced UAV Aerodynamics, Flight Stability and Control: Novel Concepts, Theory and Applications,
First Edition. Edited by Pascual Marqués and Andrea Da Ronch.

Companion website: http://www.wiley.com/go/marques/advanced_UAV_aerodynamics

is devoted to advanced orbit determination (OD) and AD algorithms that can effectively compensate for the effect of low-cost hardware and MEMS sensor packs used in small satellites.

18.2 Concurrent Orbit and Attitude Determination

The problem of orbit and attitude motion of LEO satellites is usually coupled and nonlinear due to the multiple perturbations affecting spacecraft (SC). In this regard, concurrent orbit and attitude determination (COAD) of satellites requires investigation of the correlation of roto-translational dynamics, the results and analysis of which are helpful in reducing satellite mass and power budget requirements.

There are many research studies that consider the problems of OD and AD separately [1–7]. Separate AD or OD has been investigated with different degrees of accuracy using techniques from simple engineering methods to sophisticated advanced filtering schemes. However, using separate OD and AD subsystems imposes higher weight and costs on the space system. In this sense, COAD algorithms are gaining popularity for small low-budget satellite missions that require inexpensive lightweight autonomous navigation subsystems.

18.2.1 Spacecraft Dynamics

A Bayesian framework is normally used for state estimation [8]. However, all Bayesian-based filtering schemes require a mathematical model that properly describes the system's dynamic behaviour. Accordingly, this subsection is a brief introduction of SC kinematics and dynamics for both rotational as well as translational motions.

Attitude Kinematics

Attitude kinematics describes how the attitude of a satellite changes under the influence of its angular velocity. There are various methods to describe the attitude of a satellite, such as Euler angles, quaternion parameters, Gibbs vector, and the direction cosine matrix [9]. The quaternion parameters are the most popular and widespread means to represent the attitude due to their linear propagation equations and their non-singular characteristics for any arbitrary rotation angle. The constraint of unit norm is the only disadvantage that must be observed. The quaternion parameters are described as:

$$\{\mathbf{q}\} = \left\{q_1 \quad q_2 \quad q_3 \quad q_4\right\}^T = \left\{\mathbf{e}\sin(\phi/2) \quad \cos(\phi/2)\right\}^T \tag{18.1}$$

where $\mathbf{e}$ is the unit vector of rotation, and ϕ is the corresponding rotation angle. Quaternion parameters are propagated in time as [10]:

$$\frac{d}{dt}\{\mathbf{q}\} = \frac{1}{2}\begin{bmatrix} k\varepsilon & \omega_z & -\omega_y & \omega_x \\ -\omega_z & k\varepsilon & \omega_x & \omega_y \\ \omega_y & -\omega_x & k\varepsilon & \omega_z \\ -\omega_x & -\omega_y & -\omega_z & k\varepsilon \end{bmatrix} \begin{Bmatrix} q_1 \\ q_2 \\ q_3 \\ q_4 \end{Bmatrix} \tag{18.2}$$

where $\varepsilon = 1-\{\mathbf{q}\}^T\{\mathbf{q}\}$, $\boldsymbol{\omega}_{BI} = [\omega_x \quad \omega_y \quad \omega_z]^T$, and q_4 is considered the scalar part of $\{\mathbf{q}\}$. $\boldsymbol{\omega}_{BI}$ represents the angular velocity vector of the satellite body frame relative to the inertial frame. The diagonal elements of Eq. (18.2) are specially chosen as shown to guarantee the unit norm constraint, even in the presence of rounding errors. In addition, k is a constant to be selected such that $k\Delta t < 1$ (where Δt is the integration time step).

Attitude Dynamics

The nonlinear attitude dynamics of a rigid satellite can be described via Euler's law, according to which, the time rate of the angular momentum of an SC body with respect to the inertial frame equals the total exerted external moments on that body. Expressing this law in the satellite-body coordinate system results in [11]:

$$\begin{aligned}
\left[\dot{\boldsymbol{\omega}}_{BI}\right]^B &= I_B^{B-1}\left(\left[\boldsymbol{\tau}\right]^B - \left[\boldsymbol{\omega}^{BI}\right]^B \times I_B^B\left[\boldsymbol{\omega}_{BI}\right]^B\right), \\
\left[\boldsymbol{\tau}\right]^B &= \left[\boldsymbol{\tau}_c\right]^B + \left[\boldsymbol{\tau}_d\right]^B
\end{aligned} \tag{18.3}$$

where $[.]^B$ denotes expressed quantity in the body-coordinate system, I_B^B is the satellite moment of inertia tensor (matrix) and $\boldsymbol{\tau}$ is the total exerted torque on the SC. $\boldsymbol{\tau}_c$ is the control input torque and $\boldsymbol{\tau}_d$ is taken as the sum of all disturbing torques. Disturbances on an Earth satellite originate from various internal and/or external sources, such as aerodynamic drag, solar radiation pressure, gravity gradient, electromagnetic torque, and fuel sloshing. The level of severity of most disturbing forces/moments depends on the altitude of space vehicle. Aerodynamic drag and the gravity gradient are the most effective sources of disturbance torques for an LEO satellite.

The aerodynamic drag force produces a perturbing torque about the satellite centre of mass that can be expressed in the body coordinate system as [11]:

$$\left[\boldsymbol{\tau}_a\right]^B = \frac{1}{2}\rho V^2 A C_d \left(\left[\mathbf{u}_v\right]^B \times \left[\mathbf{s}_{cp}\right]^B\right) \tag{18.4}$$

in which ρ is the atmosphere density, C_D is the drag coefficient, $[\mathbf{u}_v]^B$ is the unit vector of the satellite velocity in the body-coordinate system, $[\mathbf{s}_{cp}]^B$ is the position vector of the satellite pressure centre with respect to satellite centre of mass, A is the satellite cross-sectional area, and V is the magnitude of the satellite velocity.

The gravity gradient torque can also be modelled in the body-coordinate system as [11]:

$$\left[\boldsymbol{\tau}_{gg}\right]^B = \frac{3\mu}{r^3}\left(\mathbf{c}_3^B \times I_B^B \mathbf{c}_3^B\right) \tag{18.5}$$

where μ is the Earth gravitational parameter, r is the satellite distance from the Earth's centre, and $\mathbf{c}_3^B$ is the third column of the inertial-to-body transformation matrix (T^{BI}) that when defined in terms of the quaternion parameters will be as follows [10]:

$$T^{BI} = \begin{bmatrix} q_1^2 - q_2^2 - q_3^2 + q_4^2 & 2(q_1q_2 + q_3q_4) & 2(q_1q_3 - q_2q_4) \\ 2(q_1q_2 - q_3q_4) & -q_1^2 + q_2^2 - q_3^2 + q_4^2 & 2(q_2q_3 + q_1q_4) \\ 2(q_1q_3 + q_2q_4) & 2(q_2q_3 - q_1q_4) & -q_1^2 - q_2^2 + q_3^2 + q_4^2 \end{bmatrix} \tag{18.6}$$

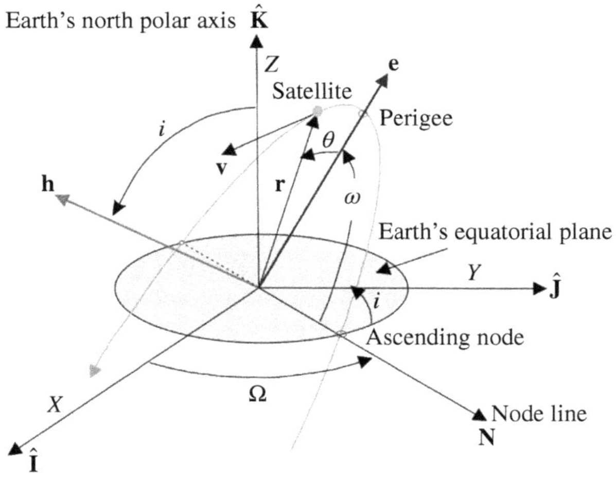

Figure 18.1 Classical orbital elements. (*See insert for color representation of this figure.*)

Orbital Kinematics

Orbit kinematics describes the satellite position as a continuous trajectory with discrete observations at specific times. There are various parameters and coordinate systems one could use to describe an orbit, such as Cartesian inertial coordinates, classical orbital elements, and modified equinoctial elements [12]. Since osculating classical orbital elements seem to provide a good visualization of orbits in three dimensions; they are selected for the orbit descriptions in this chapter. The osculating orbital elements are shown in Figure 18.1 and consist of six parameters that completely define an orbit and the instantaneous satellite position on it. These parameters are the orbit semi-major axis (a), eccentricity (e), inclination (i), longitude of the ascending node (Ω), argument of perigee (ω), and the mean anomaly (M), which is related to the instantaneous position of the SC in its orbit.

Orbital Dynamics

Orbital dynamics describes the motion of a point mass in a central force field. The pertinent governing differential equation for each of the orbital elements following the Gauss approach [11] that includes the effect of any type of perturbing forces on the SC, are presented below:

$$\frac{da}{dt} = \frac{2}{n\sqrt{1-e^2}}\left[e\sin(\vartheta)R + \left[1 + e\cos(\vartheta)\right]S\right] \tag{18.7}$$

$$\frac{de}{dt} = \frac{\sqrt{1-e^2}}{na}\left[\sin(\vartheta)R + \left[\cos(\Psi) + \cos(\vartheta)\right]S\right] \tag{18.8}$$

$$\frac{di}{dt} = \frac{1}{na\sqrt{1-e^2}}\frac{r}{a}\cos(\vartheta + \omega)W \tag{18.9}$$

$$\frac{d\Omega}{dt}=\frac{1}{na\sqrt{1-e^2}}\frac{r}{a}\frac{\sin(\vartheta+\omega)W}{\sin(i)} \tag{18.10}$$

$$\frac{d\omega}{dt}=\frac{\sqrt{1-e^2}}{nae}\left\{-R\cos(\vartheta)+\left[1+\frac{1}{1+e\cos(\vartheta)}\right]\sin(\vartheta)S-\frac{d\Omega}{dt}\cos(i)\right\} \tag{18.11}$$

$$\frac{dM}{dt}=n+\frac{1-e^2}{nae}\left\{\left[\frac{-2e}{1+e\cos(\vartheta)}+\cos(\vartheta)\right]R-\left[1+\frac{1}{1+e\cos(\vartheta)}\right]\sin(\vartheta)S\right\} \tag{18.12}$$

where Ψ denotes the eccentric anomaly and ϑ stands for the true anomaly. R, S and W are the components of the perturbing forces expressed in a moving Cartesian orbital frame. The unit vectors in this moving frame are defined such that **R** is along the radius vector, **S** lies in the local osculating plane, perpendicular to **R**, and follows the direction of the satellite motion, and **W** is perpendicular to both **R** and **S** in the direction of the momentum vector $\mathbf{R}\times\mathbf{S}$. With this formulation, any perturbing acceleration (specific force) can be expressed as $\gamma_p = R\mathbf{R}+S\mathbf{S}+W\mathbf{W}$.

As the aerodynamic drag force as well as the Earth oblation (J2) effects are among the most dominant perturbing forces for LEO satellites, they are modelled using the above notation. The aerodynamic drag force modelled in the moving orbital frame can be represented as [11]:

$$\gamma_p=-D\begin{bmatrix}\dfrac{e\sin(\vartheta)}{\sqrt{1+e^2+2e\cos(\vartheta)}} & \dfrac{1+e\cos(\vartheta)}{\sqrt{1+e^2+2e\cos(\vartheta)}} & 0\end{bmatrix}^T \tag{18.13}$$

where $D=(1/2)\rho V^2 AC_D$, and the satellite velocity can be calculated using the orbital elements [13].

Components of the disturbing force due to Earth oblations in the moving orbital frame are given as [14]:

$$\gamma_p=-\frac{3\mu J_2 R_E^2}{2r^4}\begin{bmatrix}1-3\sin^2(i)\sin(\omega+\vartheta) & \sin^2(i)\sin(2\omega+2\vartheta) & \sin(2i)\sin(\omega+\vartheta)\end{bmatrix}^T \tag{18.14}$$

where R_E is the Earth radius and $J_2=1082.6\times10^{-6}$.

18.2.2 Measurement Package

The measurement package is usually selected based on the mission accuracy requirements and the project budget. To produce a low-cost satellite system, the cost of key components must be reduced while preserving the subsystem performance requirements. In this respect, efficient used of low-cost MEMS-based sensors with proper filtering schemes becomes an important priority in order to satisfy the performance and budget requirements of modern small satellites. Various types of sensor, such as gyros, Sun sensors, star trackers, Earth horizon sensors, and magnetometers, can be used to provide the satellite attitude or the position and/or their corresponding rates autonomously. However, the low cost and weight of the three-axis magnetometer (TAM) makes it an ideal candidate for satellite state estimation. Although TAM measurements in the satellite-body coordinate system contain both the translational and rotational data, its greatest disadvantage is low accuracy in state estimation when used in isolation. To remedy this limitation, sensor fusion techniques can be applied to incorporate other types of assisting sensors. Use of the more accurate Sun sensors next to TAMs is a viable choice, but obviously the AD estimation accuracy may degrade during eclipses. Fusing data from TAMs and Sun sensors using novel filtering techniques therefore provides a low-cost and low-weight measurement package with appropriate performance and a centralized fusion of TAM and Sun sensor is considered in the rest of this chapter. In centralized measurement fusion, multi-sensor data is merged via an augmented measurement vector with all measured data centrally processed to minimize information loss. The advantage of this fusion scheme is its ability to globally obtain the optimal state estimator with a higher computational burden than is easily achievable with the onboard computers or microprocessor technology available today [15].

In almost COAD studies [16–19], direct or pseudo gyro measurements are included in the measurement package, while gyro-less schemes are more common for small satellites. The state estimation for gyro-less SC is motivated by several factors. The failure or degradation of gyros during flight can result in the loss of an SC or impair its mission effectiveness. Gyros may also be too expensive for low-cost, low-weight satellite designs. Consequently, the gyro-less roto-translation dynamics approach [20] has recently received more attention.

TAMs measure the Earth's geomagnetic field in the SC body-coordinate system; the output can be modelled as:

$$\left[\mathbf{B}_{meas}\right]^{B} = T^{BI}\left[\mathbf{B}_{\text{model}}\right]^{I} + \mathbf{v}_{B} \tag{18.15}$$

where $[\mathbf{B}_{\text{model}}]^{I}$ is a function of the satellite orbital position that is derivable from the International Geomagnetic Reference Field (IGRF) model [21]. $\mathbf{v}_B$ represents the measurement noise associated with TAM sensors and is assumed to be zero-mean Gaussian with variance σ_B^2 along each axis.

The Sun sensor is the other reference sensor used; it provides the sunlight direction with respect to the satellite's onboard sensor frame, which is here assumed to be the same as the body frame. Its output in the satellite body coordinate system is modelled as:

$$\left[\mathbf{s}_{meas}\right]^{B} = T^{BI}\left[\mathbf{s}_{ref}\right]^{I} + \mathbf{v}_{sun} \tag{18.16}$$

where $[\mathbf{s}_{meas}]^{B}$ and $[\mathbf{s}_{ref}]^{I}$ are the Sun-direction vector in the body and inertial coordinate systems, respectively. For simulation and filter design purposes, $[\mathbf{s}_{ref}]^{I}$ can be obtained as the difference between the Sun's position in the Earth-centred inertial frame and the inertial instantaneous position of the satellite. $\mathbf{v}_{sun}$ is the Sun-sensor measurement noise, modelled as zero-mean Gaussian white noise with variance σ_s^2 along each axis. In addition, no correlation is assumed regarding the noise associated with the models used for the two sensors.

As indicated earlier, the Sun sensors provide no useful information during eclipses, except for the noise indicated in their model. In turn, the eclipse duration can be recognized in the simulation process via assumption of a shadow pattern. Assuming a cylindrical shadow pattern, the satellite moves in the Sun's shadow when $\beta_{shad.} < \beta_{\lim.}$, defined as follows [14]:

$$\beta_{shad.} = \cos^{-1}\left(\mathbf{s}_{ref}.\mathbf{r}/r\right) \tag{18.17A}$$

$$\beta_{\lim.} = \sin^{-1}\left(R_E/r\right) \tag{18.17B}$$

18.2.3 Minimum Sigma Point Kalman Filter

The extended Kalman filter (EKF) is the simplest nonlinear Bayesian filter most commonly used for state estimation of nonlinear dynamic systems in many fields. EKF is mathematically based on minimum mean-square error estimation performed over the first-order linearization of nonlinear dynamic and measurement systems. However, application of EKF to complex nonlinear problems is often faced with two potential threats, namely filter divergence and estimation performance degradation.

In contrast to EKF, the unscented Kalman filter (UKF) does not rely on a linear approximation of the governing dynamic and the measurement equations. The key idea behind UKF is that with a fixed number of parameters, it should be easier to approximate a Gaussian distribution than an arbitrary nonlinear function [22]. In this sense, UKF is known as a Gaussian-approximation sample-based nonlinear filter in which the posterior probability density function is assumed Gaussian. UKF uses $2n+1$ samples, called sigma points, which are deterministically chosen for accurate estimation of an n-dimensional state vector of a nonlinear system. In fact, state estimation via UKF is provided from a weighted average of sigma points. It is shown that UKF performs much better than EKF, but with an increased computational cost [8]. In fact, the run time of all sample-based algorithms, such as UKF, strongly depends on the number of sample points needing evaluation. Accordingly, to reduce the computational cost of sample-based algorithms, strategies known as reduced sigma point filters (RSPF) have been developed. These lower the number of sigma points, using, for example, simplex point selection strategies that utilize only $n+2$ points [23]. These strategies contain a zero central point. In contrast to central point strategies, other schemes have also evolved that use only $n+1$ equally weighted sigma points, without the need for any central point. It has also been proved that equally non-negative-weighted sigma point sets are numerically more stable and accurate while being more efficient [24].

The modified UKF (MUKF) [24] was developed as a simplex unscented transform (UT) to estimate an n-dimensional state vector using only $n+1$ sigma points. MUKF is based on the Schmidt orthogonal algorithm. The computational intensity and run time of MUKF is demonstrated to be less than for UKF, simplex UKF (SUKF), and EKF, with similar levels of performance. MUKF uses the minimum number of equally weighted sigma points to provide an efficient unbiased estimation for nonlinear dynamic systems within Gaussian filters.

Application of MUKF for COAD as a highly nonlinear estimation problem with a high-dimensional state space has been shown to be advantageous in terms of the required run time and the degree of required accuracy. The following paragraph as well as the Table 18.1 are dedicated to an introduction to and implementation details of the MUKF algorithm.

The MUKF implementation is best described via the following nonlinear discrete time system:

$$\mathbf{x}_{k+1} = f(\mathbf{x}_k, k) + \mathbf{w}_k \tag{18.18A}$$

$$\mathbf{z}_k = h(\mathbf{x}_k, k) + \mathbf{v}_k \tag{18.18B}$$

where $\mathbf{x}_k \in \Re^{n\times1}$ is an n-dimensional state vector, $\mathbf{z}_k \in \Re^{m\times1}$ is an m-dimensional observation vector, and f and h are nonlinear functions representing system and measurement dynamics. In addition, $\mathbf{w}_k$ and $\mathbf{v}_k$ are n-dimensional and m-dimensional process and measurement noise vectors, respectively. It is assumed that the noise vectors $\mathbf{w}_k$ and $\mathbf{v}_k$ are uncorrelated zero-mean Gaussian white noise with covariances denoted by $\mathbf{Q}_k$ and $\mathbf{R}_k$, respectively.

At time k, the estimate of the state vector is denoted by $\hat{\mathbf{x}}_k$ and its covariance matrix is represented by $\mathbf{P}_k$. Inasmuch as $\mathbf{P}_k$ is a positive definite matrix, its Cholesky decomposition is given by $\mathbf{P}_k = \mathbf{S}_k\mathbf{S}_k^T$, where $\mathbf{S}_k$ is a positive lower triangular matrix. The $n+1$ sigma points are selected as follows [24]:

1) Choose the equal weights as $\varpi_i = 1/(n+1); i = 1,\ldots,n+1$.
2) Construct the scale vectors as

$$\Lambda_i = \sqrt{n+1}\left[\underbrace{\left[i(i+1)\right]^{-0.5},\ldots,\left[i(i+1)\right]^{-0.5}}_{i}, -\left(\frac{i+1}{i}\right)^{-0.5}, \underbrace{0,\ldots,0}_{n-i}\right]$$

3) Arrange the sigma points in an $n\times(n+1)$ matrix; $\chi_k = \mathbf{1}\otimes\hat{\mathbf{x}}_k + \mathbf{S}_k\Lambda$, where $\otimes$ represents the Kronecker product.

After the sigma points are chosen, they are propagated and updated according to the pseudo code in Table 18.1.

18.2.4 Results and Discussion

Numerical simulations performed in this subsection are meant to verify the performance of MUKF in the problem of gyro-less COAD. As a demonstration of the capabilities of MUKF, a small LEO satellite is taken in its normal mode of operation. The states of the COAD problem are grouped in a vector $\mathbf{x}_k = \left[\boldsymbol{\omega}_{BI}^T \ \{\mathbf{q}\} \ a \ e \ i \ \Omega \ \omega \ M\right]^T$. The measurement pack consists of TAM and Sun sensors the noise of which are modelled as zero mean Gaussian white standard deviations of 50 nT and 1.8°, respectively. The geomagnetic field is also modelled using the 13th order IGRF 11 [21]. The estimation results are considered for seven orbital periods. Tables 18.2 and 18.3 show the initial conditions as well as the system data required for the simulation.

Applying UKF to the prescribed COAD problem shows that mean average run time over ten Monte Carlo simulations is about 1740 s, while it only takes only 900 s for the MUKF to achieve a similar accuracy level. This comparison indicates that MUKF has made a 48% reduction in run time, and is therefore definitely more efficient and advantageous for online applications. The problem was analyzed on a PC computer with 4 Gb RAM and a CPU of 2.53 GHz. In the next sections, additional details of the results and their analyses are presented.

Table 18.1 Modified unscented Kalman filter (MUKF) algorithm.

Prediction step:	$\chi_{k+1}^{i} = f\left(\chi_{k}^{i}, k\right)$	(18.19)
	$\hat{\mathbf{x}}_{k+1\mid k} = \sum_{i=1}^{n+1} \varpi_i^{(m)} \chi_{k+1}^{i}$	(18.20)
	$\mathbf{S}_{k+1\mid k} = qr\left(\left[\sqrt{\varpi_2}\left(\chi_{k+1}^{2:n+1} - \hat{\mathbf{x}}_{k+1}\right) \quad \sqrt{\mathbf{Q}_{k+1}}\right]\right)$	(18.21)
	$\mathbf{S}_{k+1\mid k} = cholupdate\left(\mathbf{S}_{k+1\mid k}, \chi_{k+1\mid k}^{1} - \hat{\mathbf{x}}_{k+1\mid k}, \sqrt{\varpi_1}\right)$	(18.22)
	$\chi_{k+1} = \mathbf{1} \otimes \hat{\mathbf{x}}_{k+1\mid k} + \mathbf{S}_{k+1\mid k} \Lambda$	(18.23)
Measurement update:	$\hat{\mathbf{z}}_{k+1} = \sum_{i=1}^{n+1} \varpi_i^{(m)} h\left(\chi_{k+1}^{i}, k+1\right)$	(18.24)
	$\mathbf{S}_{k+1}^{zz} = qr\left[\sqrt{\varpi_2^{(c)}}\left(h\left(\chi_{\mid k+1}^{2:n+1}, k+1\right) - \hat{\mathbf{z}}_{k+1}\right) \quad \sqrt{\mathbf{R}_{k+1}}\right]$	(18.25)
	$\mathbf{S}_{k+1}^{zz} = cholupdate\left\{\mathbf{S}_{k+1}^{zz}, h\left(\chi_{\mid k+1}^{1}, k+1\right) - \hat{\mathbf{z}}_{k+1}, \varpi_1^{c}\right\}$	(18.26)
	$\mathbf{P}_{k+1}^{xz} = \sum_{i=1}^{n+1} \varpi_i^{(c)} \left(\chi_{k+1}^{i} - \hat{\mathbf{x}}_{k+1\mid k}\right)\left(h\left(\chi_{k+1}^{i}, k+1\right) - \hat{\mathbf{z}}_{k+1}\right)^{T}$	(18.27)
	$\mathbf{K}_{k+1} = \left(\mathbf{P}_{k+1}^{xz} / \mathbf{S}_{k+1}^{zz\,T}\right) / \mathbf{S}_{k+1}^{zz}$	(18.28)
	$\hat{\mathbf{x}}_{k+1\mid k+1} = \hat{\mathbf{x}}_{k+1\mid k} + \mathbf{K}_{k+1}\left(\mathbf{z}_{k+1} - \hat{\mathbf{z}}_{k+1}\right)$	(18.29)
	$\mathbf{U} = \mathbf{K}_{k+1} \mathbf{S}_{k+1}^{zz}$	(18.30)
	$\mathbf{S}_{k+1\mid k+1} = cholupdate\left\{\mathbf{S}_{k+1\mid k}, \mathbf{U}, -1\right\}$	(18.31)

Table 18.2 Initial simulation conditions for COAD.

State	True values	Estimation values
$\boldsymbol{\omega}_{BI}$ (rad/s)	$[0\ -0.0011\ 0]^{\mathrm{T}}$	$[1e-3\ -0.002\ 1e-3]^{\mathrm{T}}$
$\{\mathbf{q}\}$	$[-0.0739\ -0.7032\ -0.5495\ -0.4450]^{\mathrm{T}}$	$[0.5359\ -0.4794\ 0.3806\ 0.5814]^{\mathrm{T}}$
a (km)	7078.145	6778.145
E	0.07	0.1
i (°)	70	50
Ω (°)	57	45
ω (°)	0	17.18
M (°)	0	17.18

Table 18.3 System and simulation data for COAD.

Mass (kg)	40
$\mathbf{I}_B^B\left(\text{kg.m}^2\right)$	*diag*([1.2 1 1.1])
$\mathbf{Q}_\omega$	$\left(1e-5\right)^2 E_{3\times3}$
$\mathbf{Q}_\alpha$	$\left(1e-6\right)^2 E_{6\times6}$
Sampling interval (s)	10

Estimation Results and Analysis

Figures 18.2–18.4 show true and estimated states in the coupled roto-translational dynamics for up to seven orbital periods for a sample simulation. As shown in these figures, initial fluctuations have diminished over less than two orbital periods and the estimation of the whole system states that include the orbital parameters have converged to their true values after two orbital periods.

To study the performance of the MUKF algorithm, a Monte Carlo simulation is also performed on the noise characteristics of the measurement package. Root mean-square error (RMSE) values of the error magnitudes for attitude, angular velocity magnitude and the orbital elements over ten Monte Carlo simulations are shown in Figures 18.5 and 18.6. These figures show that the large initial errors for the state vector have gradually decreased to zero as the observation data is fed in to the algorithm, indicating good robustness of the MUKF. Magnification of the results illustrates that the AD achieves accuracy levels of better than 0.3° in attitude estimation and better than 0.003°/s in body angular rates. In addition, the OD navigation accuracy in terms of the satellite position and velocity are about 3.85 km and 0.5 m/s respectively along each axis, which is very good considering the use of MEMS sensors in the measurement package.

A summary of this MEMS-based COAD research in comparison with the recent Gyroless work is presented in Table 18.4. As this table shows, despite inclusion of disturbing forces and moments, appropriate selection of a measurement package (sensors) combined with proper implementation of the MUKF filter has improved the attitude and position estimation accuracy without the need for any true/pseudo gyro measurement.

Covariance Analysis and Observability

As the state-estimation problem reconstructs the system states using available observations, system observability demonstrates a uniqueness recovery of the system states from the observed data. In this sense, observability is a prerequisite for successful

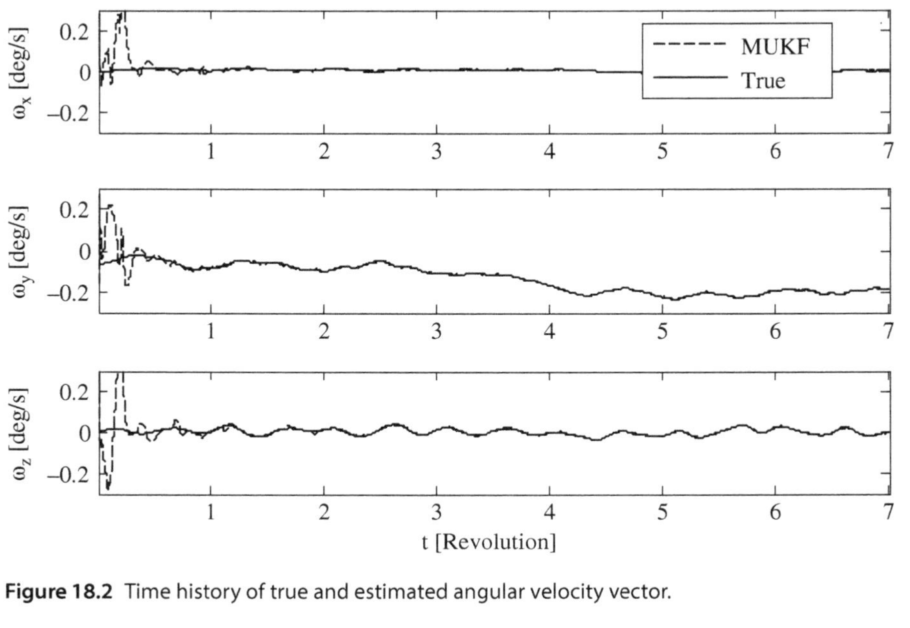

Figure 18.2 Time history of true and estimated angular velocity vector.

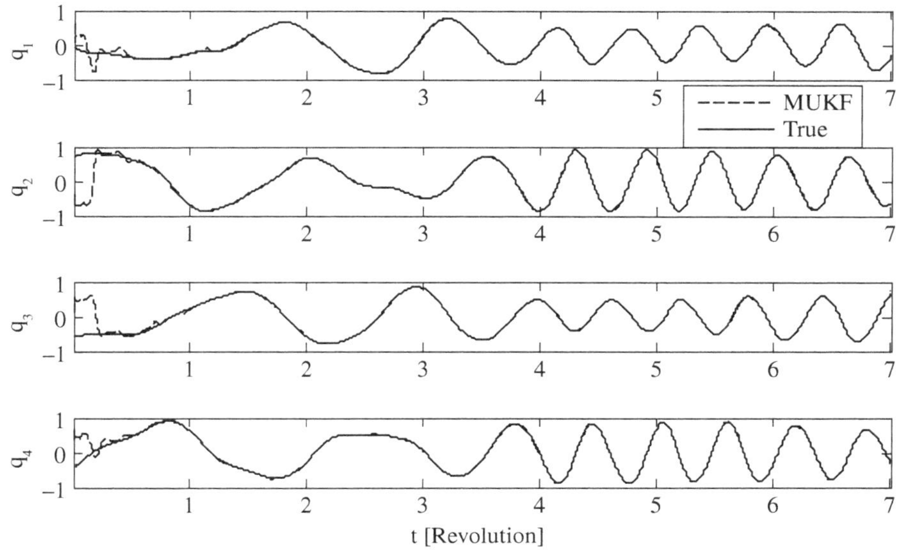

Figure 18.3 Time history of true and estimated quaternion parameters.

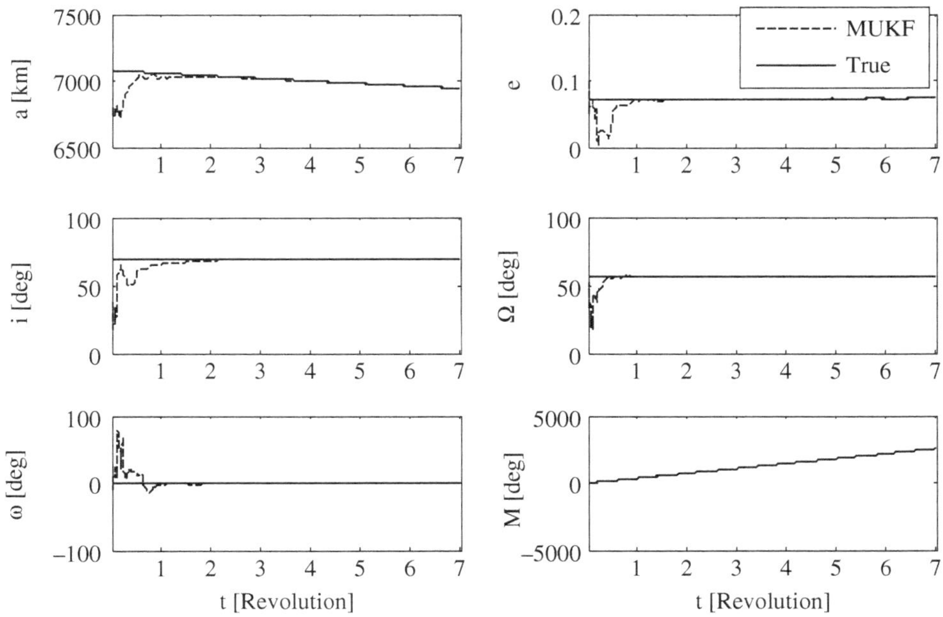

Figure 18.4 Time history of true and estimated orbital elements.

estimation. A system is theoretically observable if the so-called observability matrix (OM) is of full rank [25]. By investigating the roto-translation dynamic system with the observation model, one can guarantee local observability via assessment of the OM rank. The OM rank over the whole sampling time equals 12, indicating full observability of the COAD problem.

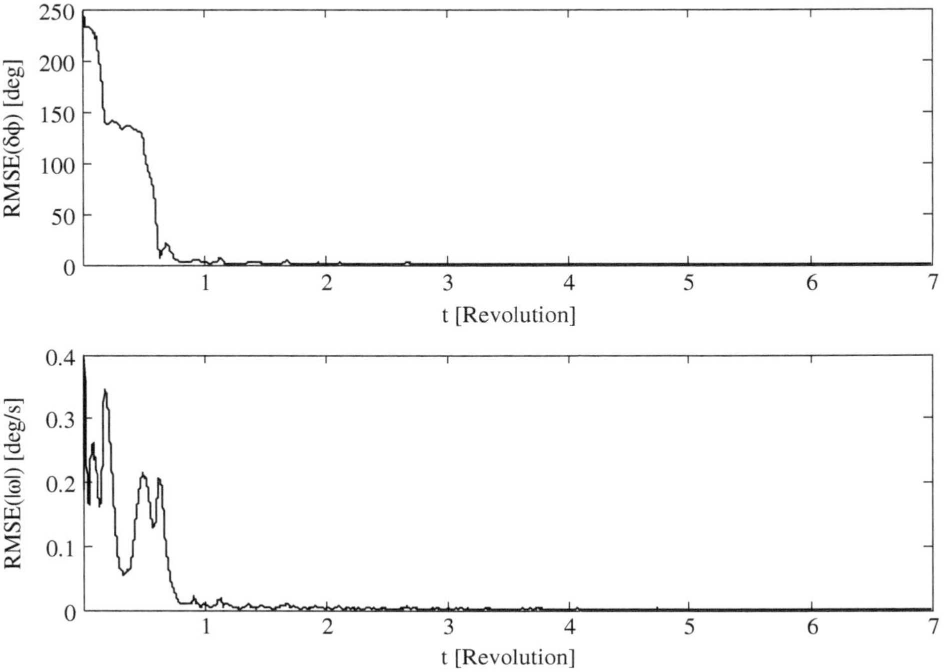

Figure 18.5 RMSE of attitude and norm of angular velocity.

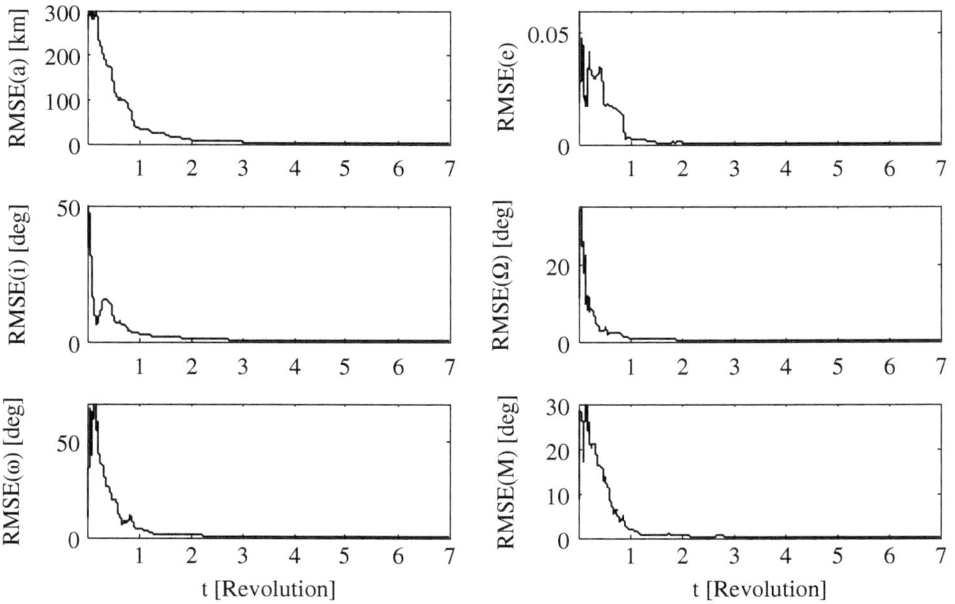

Figure 18.6 RMSE of orbital elements.

Table 18.4 A comprehensive summary of COAD studies in terms of estimation accuracy [20].

Ref	Sensor(s)	Orbit	Filter	Accuracy			
				Position (km)	Velocity (m/s)	Attitude (°)	Att. rate (°/s)
[16]	Gyro, TAM	Non-Keplerian.	EKF	30	30	0.7–1.4	—
[17]	Gyro, TAM	Keplerian.	EKF				—
[18]	Gyro, TAM	Keplerian.	UKF	5	20	0.2	—
[19]	TAM, TAM rate	Non-Keplerian.	EKF	8	5	5	0.03
Current work	TAM, Sun sensor	Non-Keplerian.	MUKF	3.85	0.5	0.3	0.003

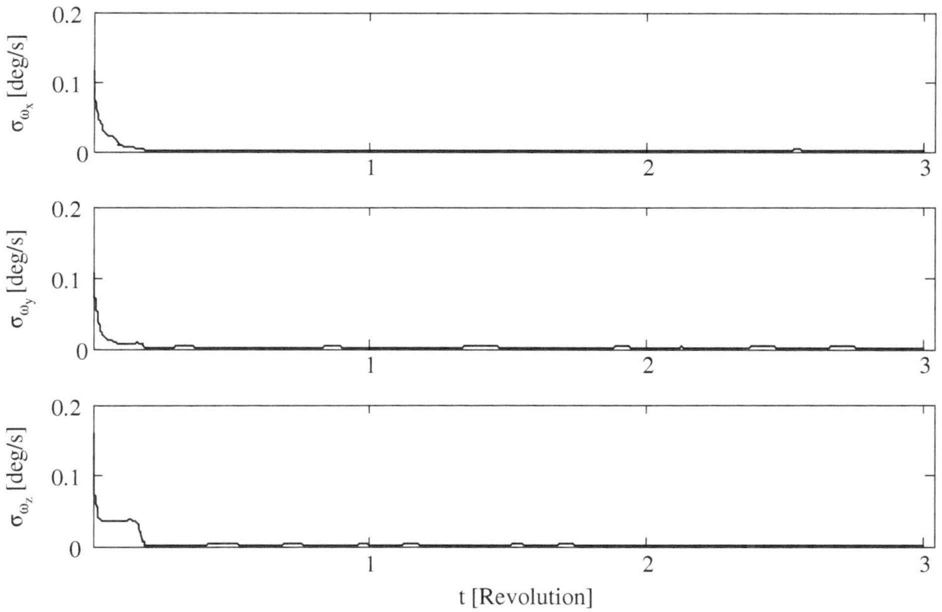

Figure 18.7 Square root of the diagonal elements of matrix **P** appropriated to the angular velocity components.

The standard deviations of the estimated state vector elements are also a suitable measure to assess the system observability [1]. In other words, if the square roots of the diagonal elements of $\mathbf{P} = \mathbf{S}\mathbf{S}^T$ are small in a problem-dependent sense, the system is observable in practice. Since the covariance of the estimated vector is equal to the inverse of the observability Gramian of the linearized system, small values of these quantities imply that the state error covariance matrix is nonsingular, which in turn implies system observability.

Figures 18.7–18.9 illustrate a decreasing trend of the standard deviations, as described above, to acceptable small levels, reaffirming the observability of the proposed system. The time histories of the state standard deviations are plotted for three orbital periods.

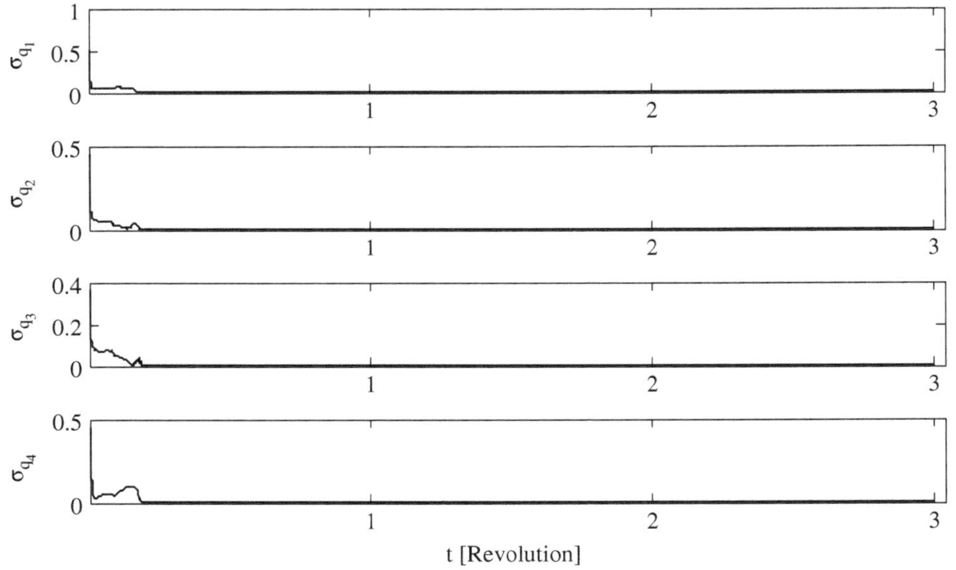

Figure 18.8 Square root of the diagonal elements of matrix **P** appropriated to the quaternion parameters.

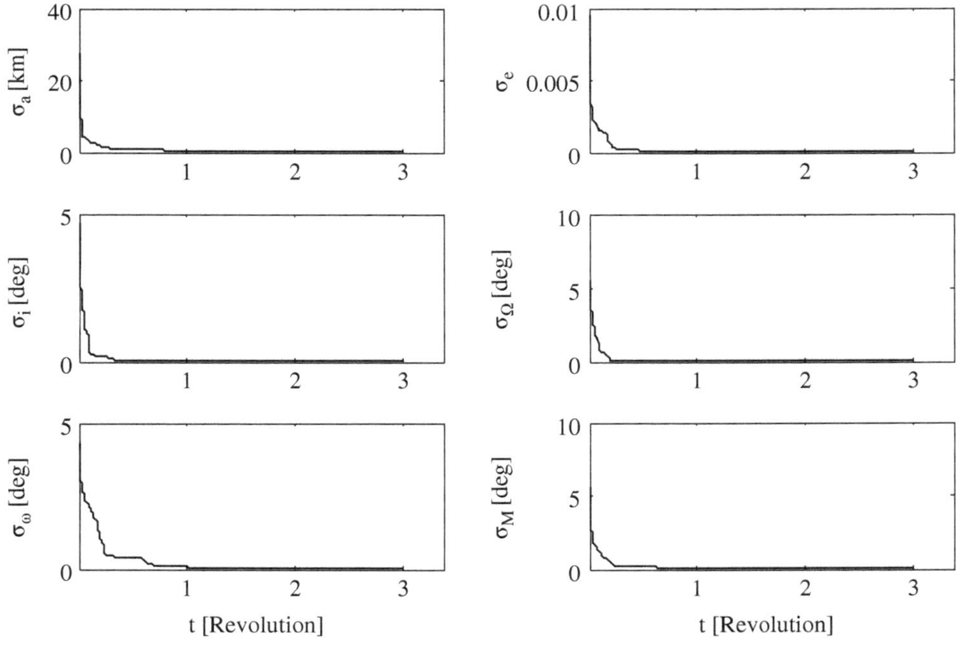

Figure 18.9 Square root of the diagonal elements of matrix **P** appropriated to the orbital elements.

In this regard, it is important to note that in contrast with previous research [19], the proposed COAD problem has no need to include the time rate of the magnetic field as part of the measurement system to achieve complete observability; this is an additional advantage of the MUKF sensor package.

Sensitivity Analysis

The effect of various system and sensor parameters on the observability and estimation accuracy is analysed in this subsection. It is worth mentioning that the filter tuning is of significance in estimation accuracy, which means that good filter tuning can provide a more precise state estimation in a wide range of acceptable system and mission scenarios. It is also notable that to calculate the attitude estimation error, the quaternion error can be calculated as:

$$\{\delta\mathbf{q}\} = \mathbf{q} \otimes \hat{\mathbf{q}}^{-1} \tag{18.32}$$

Since the fourth element of the quaternion is the scalar part related to the rotation angle, its estimation error can be computed as follows:

$$\delta\phi = 2\cos^{-1}\left(\{\delta\mathbf{q}\}_4\right) \tag{18.33}$$

Semi-major axis effect: The effect of the orbital semi-major axis on the absolute error of the estimated process is shown in Table 18.5, indicating that increasing the orbital altitude slightly reduces the estimation accuracy. The key reason for this effect is the altitude dependency of the Earth's magnetic field, which reveals itself via the magnetometers and the IGRF model. As the geomagnetic field weakens with altitude, the precision of the state estimation decreases. Moreover, the attitudinal and orbital parameters are correlated. Attitude and orbital parameters are coupled through the perturbation forces, so a change in the orbital elements affects the attitudinal parameters. However, based on these observations, it is seen that even for severe cases, the estimation results are of good accuracy. Thus one could expect to use the tested algorithm for autonomous navigation and control in a wide range of small satellites in LEO missions.

Orbit inclination effect: Orbit inclination is the most significant orbital parameter affecting observability. It is shown that increasing the orbit inclination improves the accuracy of state estimation. As the satellites moving in high-inclined orbits pass over more asymmetrical regions of the Earth geomagnetic field, the observability is strengthened; thus, better estimation results are achieved. Expectedly, as Table 18.6 indicates, the semi-major axis is the most affected parameter in this sensitivity.

Table 18.5 Effect of the initial semi-major axis on the estimation accuracy [20].

h_a[km]	Attitude error [°]	$\|\Delta a\|$[km]	$\|\Delta e\|$	$\|\Delta i\|$[°]	$\|\Delta\Omega\|$[°]	$\|\Delta\omega\|$[°]	$\|\Delta M\|$[°]
600	0.2	0.3	1e−4	0.08	0.09	0.1	0.08
700	0.3	0.4	1e−4	0.1	0.1	0.2	0.1
800	0.45	0.7	1e−4	0.2	0.17	0.22	0.1

Table 18.6 Effect of the orbit inclination on the estimation accuracy [20].

i[°]	Attitude error [°]	$\|\Delta a\|$[km]	$\|\Delta e\|$	$\|\Delta i\|$[°]	$\|\Delta\Omega\|$[°]	$\|\Delta\omega\|$[°]	$\|\Delta M\|$[°]
50	0.35	0.6	3e−4	0.13	0.12	0.22	0.12
70	0.3	0.4	1e−4	0.1	0.1	0.2	0.1

Table 18.7 Effect of the TAM standard deviation on the estimation accuracy [20].

σ_{TAM}[nT]	Attitude error [°]	$\|\Delta a\|$[km]	$\|\Delta e\|$	$\|\Delta i\|$[°]	$\|\Delta\Omega\|$[°]	$\|\Delta\omega\|$[°]	$\|\Delta M\|$[°]
30	0.2	0.3	1e−4	0.04	0.07	0.15	0.03
50	0.3	0.4	1e−4	0.1	0.1	0.2	0.1
70	0.4	0.7	1e−2	0.2	0.5	0.34	0.12

Table 18.8 Effect of the initial angular velocity on estimation accuracy [20].

ω_0[rad/s]	Attitude error [°]	$\|\Delta a\|$[km]	$\|\Delta e\|$	$\|\Delta i\|$[°]	$\|\Delta\Omega\|$[°]	$\|\Delta\omega\|$[°]	$\|\Delta M\|$[°]
$\omega_0 + 0.01$	0.35	0.5	2e−4	0.2	0.18	0.3	0.12
ω_0	0.3	0.4	1e−4	0.1	0.1	0.2	0.1

Table 18.9 Effect of sample rate on the estimation accuracy [20].

Data sampling rate [Hz]	Attitude error [°]	$\|\Delta a\|$[km]	$\|\Delta e\|$	$\|\Delta i\|$[°]	$\|\Delta\Omega\|$[°]	$\|\Delta\omega\|$[°]	$\|\Delta M\|$[°]
1	0.2	0.3	5e−5	0.02	0.02	0.015	5e−3
0.1	0.3	0.4	1e−4	0.1	0.1	0.2	0.1

Measurement accuracy effect: It is intuitively obvious that more expensive accurate sensors improve state estimation accuracy. This fact is verified in Table 18.7. However, as shown in this table, good nonlinear filters such as the proposed MUKF can compensate for less accurate MEMS-type sensors to provide an acceptable navigation subsystem.

Initial angular velocity effect: Satellites can experience different angular velocities due to perturbations. The effect of initial angular velocity on the precision of the estimation results is presented in Table 18.8. As anticipated, MUKF has low sensitivity with respect to angular velocity forcing. This also shows that MUKF results can be reliable even during manoeuvring phases.

Sampling rate effect: Data-sampling rate directly influences the estimation accuracy. Table 18.9 clearly shows that increasing the sample rate prevents covariance growth, so a more precise estimation is achieved.

Eclipse effect: It is obviously impossible to get any reliable measurements from the Sun sensor during eclipses. The effect of solar eclipses on the state estimation of roto-translational dynamics is illustrated in Figures 18.10 and 18.11 over the first three orbital periods. The vertical dashed bars indicate the occurrence of eclipses.

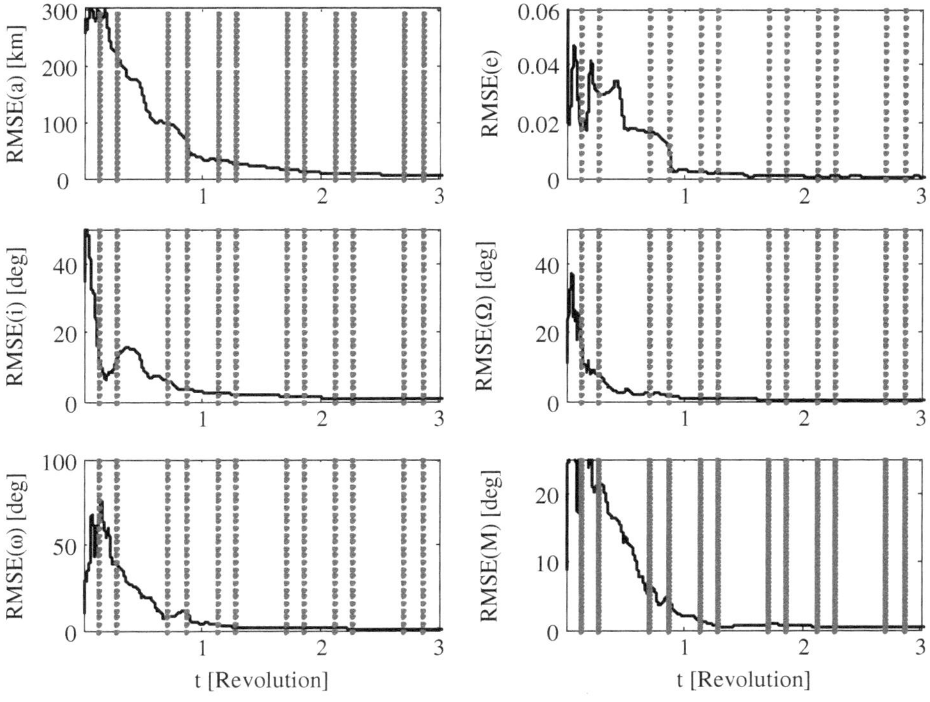

Figure 18.10 RMSE of orbital elements considering eclipse durations.

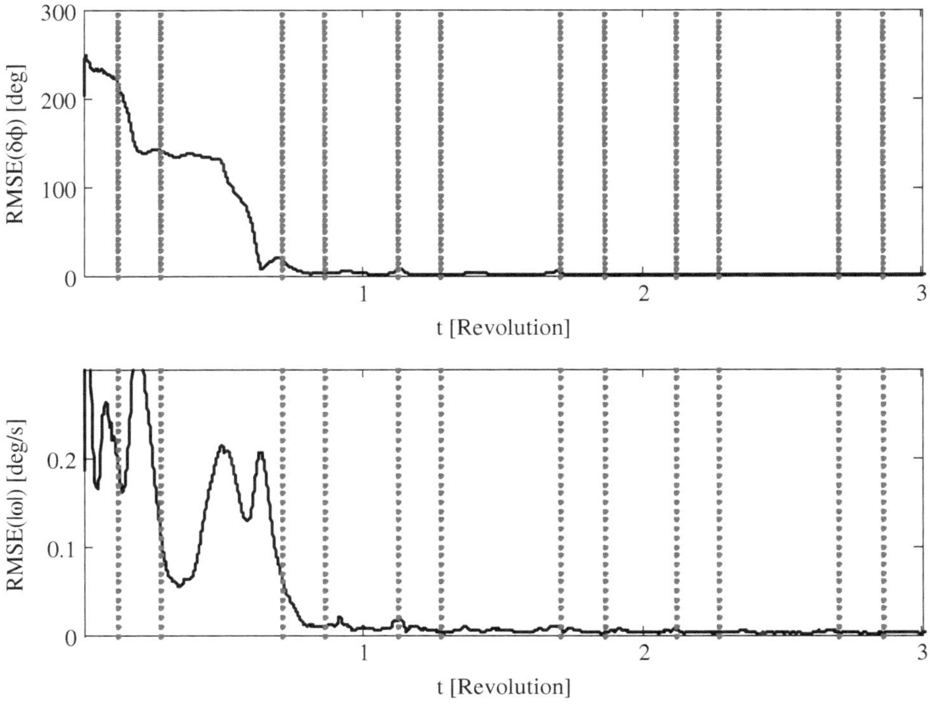

Figure 18.11 RMSE of attitude and angular velocity considering eclipse durations.

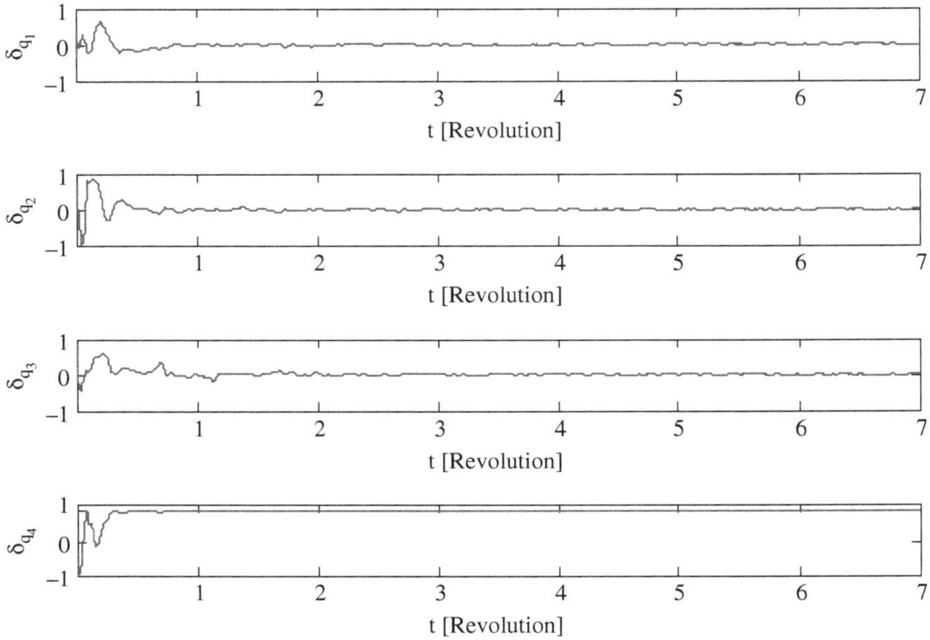

Figure 18.12 Time history of estimation error in quaternion parameters

From these figures, the resulting accuracy is degraded in the first orbital period, but soon after, with little elapsed time and data processing, the MUKF has taken over and the state estimation process has recovered. Therefore, the proposed dynamics and the measurement systems are appropriate even during eclipses or even if the Sun sensor fails.

Noise correlation effect: Correlation between the process and measurement noise seldom occurs in practice but is likely. To investigate the effect of noise correlation on the COAD performance, MUKF is improved using an idea inspired by the work of Xu et al. [26]. Figures 18.12–18.14 demonstrate the acceptable quality of estimation in this case as well.

Moment of inertia effect: The moment of inertia (MOI) matrix is the most important parameter that affects the rotational dynamics of an orbiting satellite. In the current simulation, MOI has been so far considered as *diag*([1.2 1 1.1]), but in order to investigate its effect on COAD performance, it is changed to:

$$\mathbf{I}_B^B = \begin{bmatrix} 200 & 50 & -30 \\ 50 & 240 & 10 \\ -30 & 10 & 100 \end{bmatrix}$$

The time history of the estimation error is shown in Figures 18.15–18.17. These figures clearly show that not only has the estimation error not increased but also, because of more regular changes of angular velocity, the convergence time is improved too.

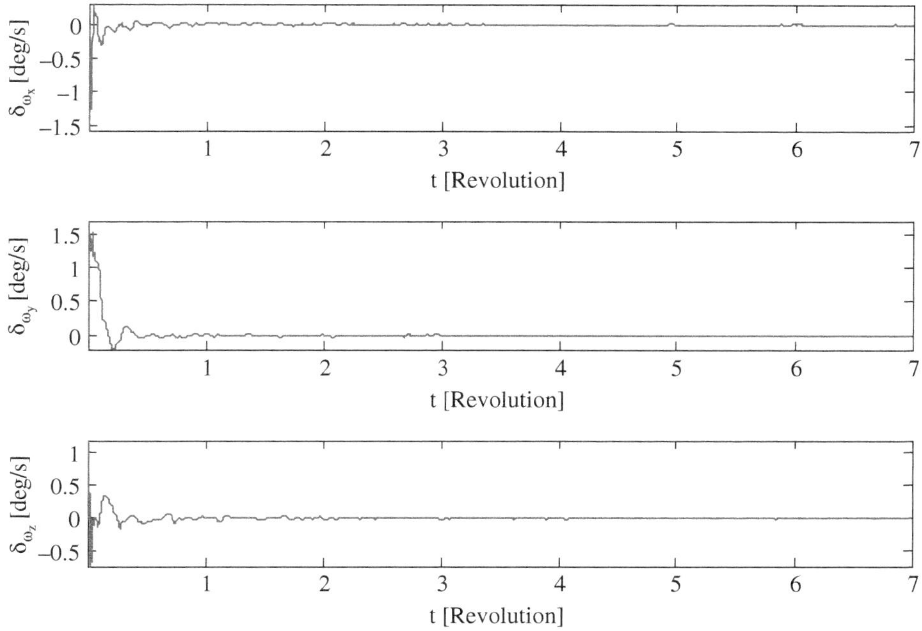

Figure 18.13 Time history of estimation error in angular velocity components.

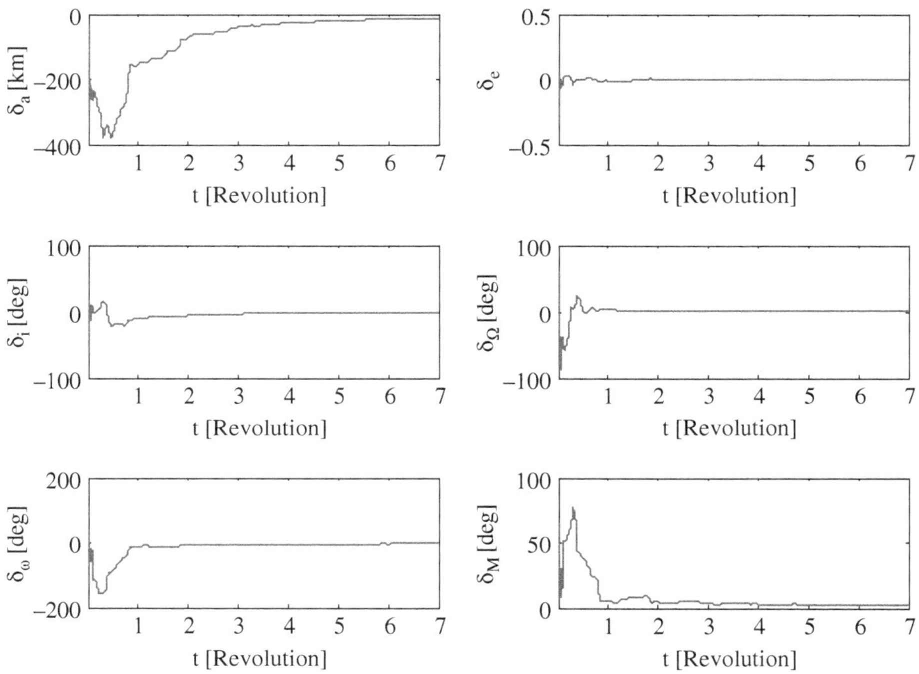

Figure 18.14 Time history of estimation error in orbital elements.

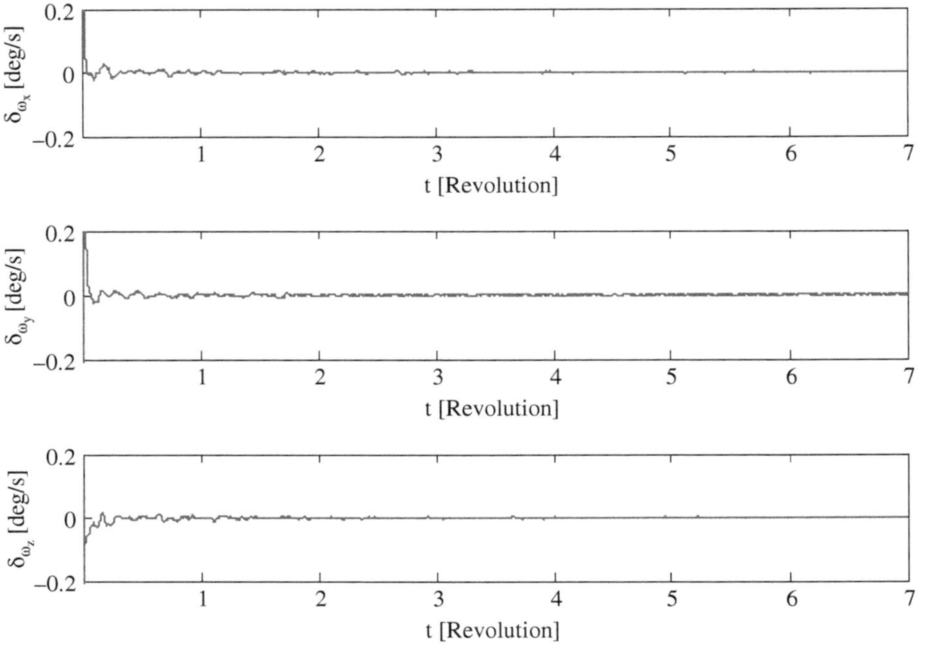

Figure 18.15 Time history of estimation error in angular velocity components.

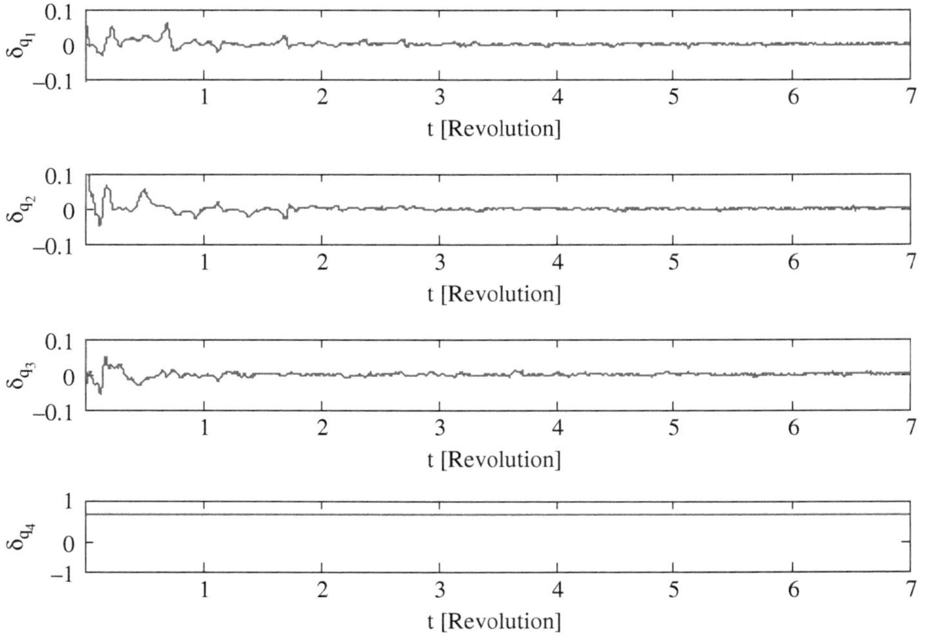

Figure 18.16 Time history of estimation error in quaternion parameters.

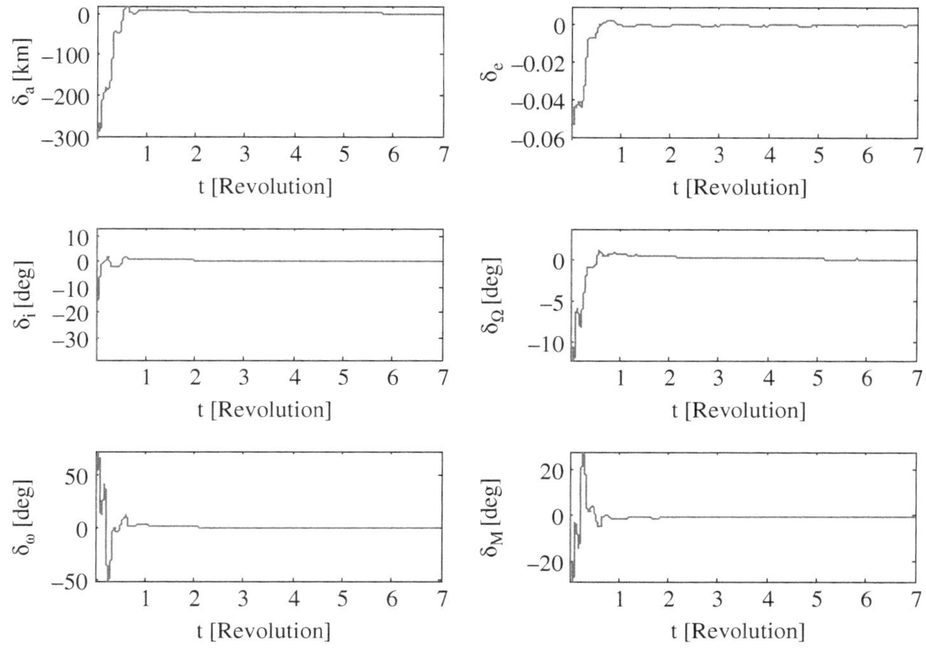

Figure 18.17 Time history of estimation error in orbital elements.

18.3 Concurrent Attitude and System Identification

Accurate knowledge of SC attitude, attitude rate and system parameters is vital for the performance of space tasks such as remote sensing, imaging, pointing, and data transmission. It is important to note that the SC attitude and attitude rate are required to implement closed-loop attitude control for keeping SCs on station or manoeuvring them. But for many space missions there is a direct relationship between accurate identification of the SC system parameters and the desired level of mission success and/or achievements. In this respect, the MOI tensor is one of the key system characteristics, and significantly affects the satellite rotational dynamics and the control subsystem design. In short, not only the SC states, but also its system parameters and the error parameters involved in the measurement package, need to be accurately identified to achieve accuracy and reliable performance. As operational and environmental conditions of orbiting SCs may vary considerably from those assumed in the design and simulation phases, online estimation of dynamics and the measurement system's parameters are of the utmost importance. This section is devoted to concurrent estimation of satellite's attitude, MOI, and the sensor error parameters.

18.3.1 Dynamic System with Sensor Parameter Estimation

A Bayesian framework is used to simultaneously estimate the SC states and parameters. To this end, the nonlinear attitude dynamics of the satellite (Table 18.3) are described by Eq. (18.3). The model assumes that the aerodynamic drag and gravity gradient forces are the most effective forces that produce torques for the selected LEO satellite (described by Eqs (18.4) and (18.5)). The effect of the other perturbing torques can be modelled as process noise.

Again, a centralized data fusion of TAM and Sun sensors provides the measurement data. However, in this case the TAM measurements are assumed to be contaminated with various sources of error – such as the scale factor, misalignment, bias, and random white noise [27] – the accurate online estimation of which is part of the problem:

$$\left[\mathbf{B}_{meas}\right]^{B} = (E+M)T^{BI}\left[\mathbf{B}_{Model}\right]^{I} + \mathbf{b} + \mathbf{v}_{B} \tag{18.34}$$

in which

$$M = \begin{bmatrix} \lambda_1 & \delta_{12} & \delta_{13} \\ \delta_{21} & \lambda_2 & \delta_{23} \\ \delta_{31} & \delta_{32} & \lambda_3 \end{bmatrix} \tag{18.35}$$

whose diagonal elements represent the scale factors, with the misalignment parameters in the off-diagonal elements. $\mathbf{b}$ represents the bias vector and $\mathbf{v}_B$ is the measurement white noise with a predefined covariance. Finally, E represents a 3×3 unit matrix.

The Sun sensor is again as modelled as in Eq. (18.16). The Sun sensor errors are usually negligible in comparison with those of the TAMs, which is a logical assumption. Further, it is assumed that the SC navigational data (position and velocity) is available via an orbital determination subsystem. It needs to be mentioned that are parameters needing estimation are modelled as random constants.

18.3.2 Persistence of Excitation

Complete observability requirement of the complex problem introduced in this section demands satellite manoeuvring for identification purposes. In other words, in order to achieve simultaneous estimation of the SC states, MOI and the TAM parameters, persistent excitation must be guaranteed. As one of the reference manoeuvre trajectories that satisfies the required condition of persistent excitation, the following rate trajectory is used [28]:

$$\begin{aligned} &\vartheta = at; \\ &\boldsymbol{\Omega} = \begin{bmatrix} \sin(\omega_1 t)\sin(\omega_2 t) \\ \cos(\omega_1 t)\sin(\omega_2 t) \\ \cos(\omega_2 t) \end{bmatrix} \\ &\boldsymbol{\omega}_{comm} = \dot{\vartheta}\boldsymbol{\Omega} - (1-\cos(\vartheta))\boldsymbol{\Omega}\times\dot{\boldsymbol{\Omega}} + \dot{\boldsymbol{\Omega}}\sin(\vartheta) \end{aligned} \tag{18.36}$$

where $a = 0.4\pi$ [rad/s], $\omega_1 = 0.01$ [rad/s], and $\omega_2 = 0.004$ [rad/s]. Figure 18.18 shows the proposed reference manoeuvre trajectory. The overall manoeuvre time is taken to be 15 min.

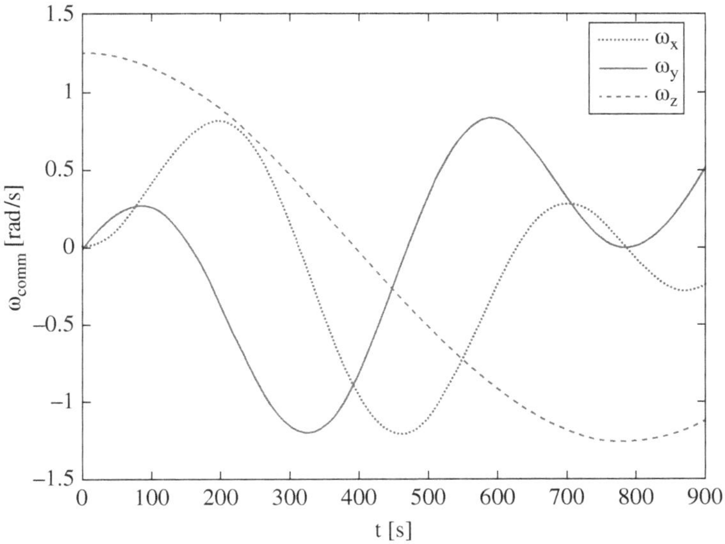

Figure 18.18 Body angular rate command trajectory.

18.3.3 Marginal Minimum Sigma Point Kalman Filter

As pointed out in Section 18.2, the nonlinear problem of SC AD has been extensively investigated over the past couple of decades using ideas and algorithms the details of which are beyond the scope of this chapter. Although EKF has been widely studied and utilized for AD in many previous papers, sample-based algorithms such as MUKF, discussed in Subsection 18.2.3, are emerging rapidly, and are more suitable for online applications in nonlinear systems.

Although MUKF is by itself an efficient algorithm for practical applications, marginalization of partially linear systems, such as that of MSC, can effectively enhance its efficiency via reduction of its computational complexity. The basic idea behind marginalization is to partition the state vector into two parts: a linear part $\mathbf{x}^l$ and a nonlinear part $\mathbf{x}^n$. In the framework of Bayes' rule, the linear state variables can then be marginalized out and estimated using the basic optimal linear Kalman filter (KF). The nonlinear state variables are to be estimated using MUKF. In this way, marginal MUKF (MMUKF) only needs one set of sigma points that adequately describes the statistical properties of the nonlinear part of the state vector. This technique is also referred to as Rao–Blackwellization. While the basic idea of marginal filters and Rao-Blackwellization is also applied for PF, its application to the UKF is new and useful [29].

In our case study, the dimension of the state vector is high ($n = 25$), while at the same time, only the kinematics and dynamics of the attitude system are nonlinear. Accordingly, MMUKF can be effectively utilized in this case. In this regard, TAM error parameters are estimated using the basic KF, while the MOI that appears nonlinearly in Euler equation of motion can be estimated via MUKF. Therefore, in conclusion, the MOI accompanied by the quaternion parameters and the body angular velocities are considered in the nonlinear part of the states, and are to be estimated using the MMUKF algorithm described in Table 18.10.

Table 18.10 The marginal modified unscented Kalman filter algorithm.

1) Initialize for i = 1,…,N = n1 + 1

$$\hat{\mathbf{x}}^n(0)=\mathbf{x}_0^n,\ \hat{\mathbf{x}}^l(0)=\hat{\mathbf{x}}_0^l$$

$$\mathbf{P}(0)=\begin{bmatrix}\mathbf{P}_0^n & 0\\ 0 & \mathbf{P}_0^l\end{bmatrix}=\begin{bmatrix}\mathbf{S}_0^n\mathbf{S}_0^{nT} & 0\\ 0 & \mathbf{P}_0^l\end{bmatrix};\mathbf{S}_0^n=chol\left(\mathbf{P}_0^n,'lower'\right)$$

2) Introduce weights and sigma points matrix
3) Time update
   a) Nonlinear part: Eq. (18.19)–(18.22)
   b) Linear part:

$$\hat{\mathbf{x}}_{k+1|k}^l=A_k^l\hat{\mathbf{x}}_{k|k}^l$$

$$\mathbf{P}_{k+1|k}=A_k^l\mathbf{P}_{k|k}\left(A_k^l\right)^T+\mathbf{Q}_k^l$$

4) Measurement update
   a) Nonlinear part: Eq. (18.24)–(18.31)
   b) Linear part:

$$\mathbf{x}_{k|k}^l=\hat{\mathbf{x}}_{k|k-1}^l+K_k\left(\mathbf{z}_k-\mathbf{h}_k-C_k\hat{\mathbf{x}}_{k|k-1}^l\right)$$

$$\mathbf{P}_{k|k}=\mathbf{P}_{k|k-1}-K_kC_k\mathbf{P}_{k|k-1}$$

$$K_k=\mathbf{P}_{k|k-1}C_k^T\left(C_k\mathbf{P}_{k|k-1}C_k^T+\mathbf{R}_k\right)^{-1}$$

To introduce MMUKF, the dynamic and measurement systems are rewritten in discrete time form as:

$$\mathbf{x}_{k+1}^n=f^n\left(\mathbf{x}_k^n,\ k\right)+\mathbf{w}_k^n \tag{18.37-A}$$

$$\mathbf{x}_{k+1}^l=A^l\left(\mathbf{x}_k^n,\ k\right)\mathbf{x}_k^l+\mathbf{w}_k^l \tag{18.37-B}$$

$$\mathbf{z}_{k+1}=h\left(\mathbf{x}_k^n,\ k\right)+C\left(\mathbf{x}_k^n\right)\mathbf{x}_k^l+v_k \tag{18.37-C}$$

where

$$\mathbf{w}_k=\begin{bmatrix}\mathbf{w}_k^n\\ \mathbf{w}_k^l\end{bmatrix}\sim N\left(\mathbf{0},\mathbf{Q}_k\right),\quad \mathbf{Q}_k=\begin{bmatrix}\mathbf{Q}_k^n & 0\\ 0 & \mathbf{Q}_k^l\end{bmatrix},\quad \mathbf{v}_k\sim N\left(\mathbf{0},\mathbf{R}_k\right) \tag{18.38}$$

$\mathbf{x}^n$ is the n_1-dimensional nonlinear part of the state vector, $\mathbf{x}^l$ is n_2-dimensional linear part of state vector ($n_1+n_2=n$) and $\mathbf{z}_k$ is the m-dimensional observation vector. In addition, f, h, A^l, and C are some nonlinear functions. The process and the measurement noise vectors are $\mathbf{w}_k$ and $\mathbf{v}_k$, respectively. Again, the noise vectors $\mathbf{w}_k$ and $\mathbf{v}_k$ are taken as zero-mean Gaussian (white) with co-variances given by $\mathbf{Q}_k$ and $\mathbf{R}_k$, respectively.

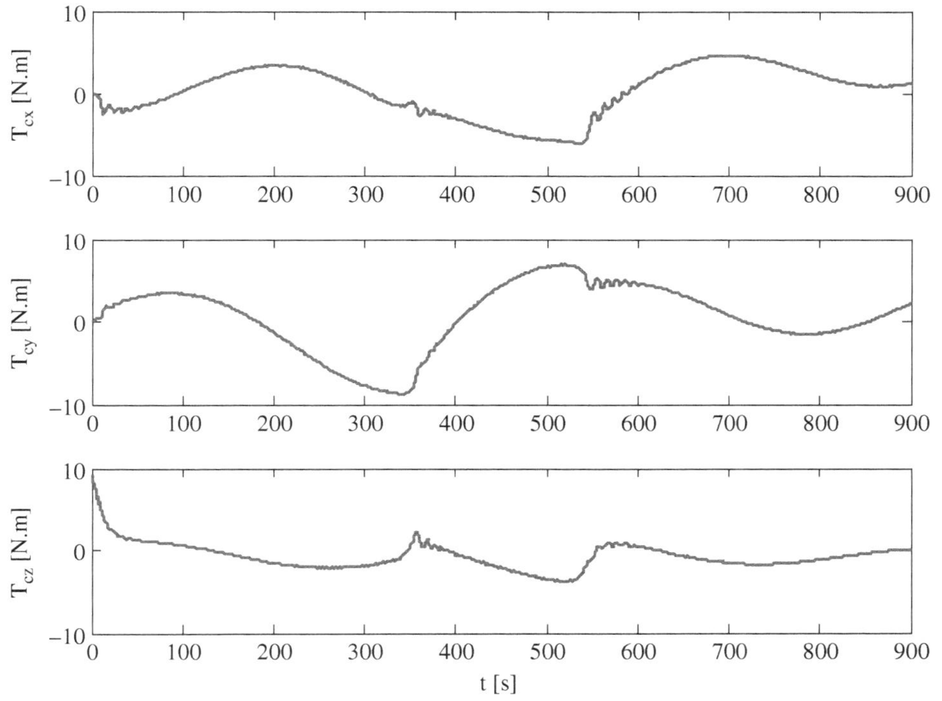

Figure 18.19 Time history of control torque components.

18.3.4 Results and Discussions

This subsection is devoted to the simulation results, which includes estimation of the system states, system parameters (MOI) and the parameters related to the measurement system. Given the discussions of Section 18.3.3, the nonlinear part of the state vector is $\mathbf{x}^n = [\{\mathbf{q}\}^T \ \boldsymbol{\omega}^{BI^T} \ MOIV]^T$ and the linear part contains the measuring-system-related parameters $\mathbf{x}^l = [\boldsymbol{\lambda}\ \boldsymbol{\delta}\ \mathbf{b}]$, where $MOIV = [I_{xx}\ I_{yy}\ I_{zz}\ I_{xy}\ I_{xz}\ I_{yz}]$, $\boldsymbol{\lambda} = [\lambda_1\ \lambda_2\ \lambda_3]$, $\boldsymbol{\delta} = [\delta_{12}\ \delta_{13}\ \delta_{21}\ \delta_{23}\ \delta_{31}\ \delta_{32}]$ and $\mathbf{b} = [b_1\ b_2\ b_3]$.

The control torque required to implement the commanded manoeuvre is computed using a fuzzy self-tuning PID controller [25]. Figure 18.19 depicts the time history of the input control torque components.

It needs to be mentioned that TAM and Sun sensor random noises are modelled as zero mean Gaussian white with standard deviations of 50 nT and 1.8° respectively. The geomagnetic field is modelled using the 13th order IGRF 11 [21]. Initial conditions for the simulation and orbit simulation are presented in Tables 18.11 and 18.12, respectively.

The measurement sampling interval is taken as 0.2 s. $\mathbf{Q}_k^n$ is considered as a diagonal matrix, with elements equal to $(1e-3)^2$ for body angular velocities and zero for the other nonlinear states. Since all the parameters needing estimation are modelled as random constants, $\mathbf{Q}_k^l$ is a 12×12 zero matrix as well.

Application of the three algorithms UKF, MUKF, and MMUKF to the current problem shows that the mean average run time of the MMUKF over ten simulations takes

Table 18.11 Initial simulation conditions.

State	Initial true values	Initial estimated values
Angular velocity (rad/s)	[0 −0.0011 0]	[1e − 3 −0.002 1e − 3]
Quaternion parameters	[−0.0739 −0.7032 −0.5495 −0.4450]	[0.4063 0.7592 −0.3573 −0.3617]
MOI (kg.m^2)	[200 50 −30 50 240 10 −30 10 100]	[160 20 −20 20 160 −20 −20 −20 160]
Scale factors (ppm)	[5 −1 −2]e + 4	[0 0 0]
Misalignments (arcs)	[648 1296 972 648 −648 1296]	[0 0 0 0 0 0]
Bias (nT)	[1500 1000 500]	[0 0 0]

Table 18.12 Initial orbit conditions.

Element	Value	Element	Value
Semi-major axis (km)	7078.145	Eccentricity	0.07
Inclination (°)	70	Longitude of ascending node (°)	57
Argument of perigee (°)	0	Mean anomaly (°)	0

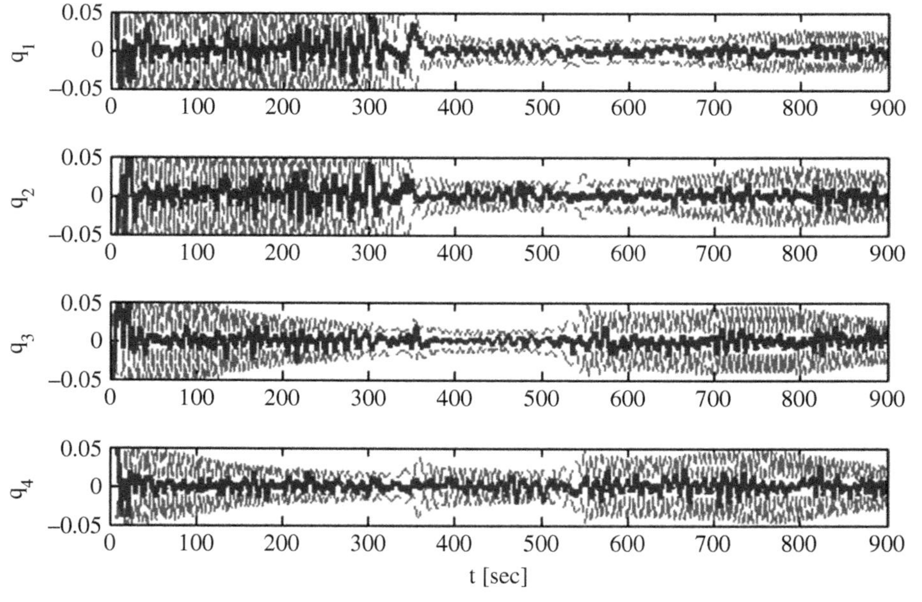

Figure 18.20 Estimation error of quaternion parameters in MMUKF with ±3σ error bound.

only about 1615 s, as opposed to 3588 s for the MUKF and 4957 s for the standard UKF. This comparison indicates that the MMUKF has caused a 68% reduction in run time relative to the standard UKF, and 55% relative to MUKF. Therefore, MMUKF is definitely more efficient for online applications. It also needs to be mentioned that the current problem is implemented and solved on a PC computer with 4G RAM and a CPU of 2.53 GHz.

To further demonstrate MMUKF efficiency, the results of its Monte Carlo simulations are now presented. Figures 18.20–18.25 show state estimation errors (solid lines) along with ±3σ error bounds (dashed lines) taken from the covariance matrix at every time step. This convergence pattern verifies the good performance of the algorithm and indicates that each variable has a different convergence time constant. Figures 18.20 and 18.21 show that the attitude and body angular velocity estimation errors converge within their respective 3σ value after only 40 s, indicative of near-optimal performance of the MMUKF algorithm. Similarly, Figures 18.22–18.24 show convergence to 3σ bounds in 60 sec for the MOI estimation, 30 sec for the scale factors and 20 sec for the misalignment parameters. In contrast to fast convergence of the states and system parameters, it takes about 300 sec to have a converged bias estimation error as denoted in Figure 18.25. Table 18.13 summarizes the RMSE of the state vector.

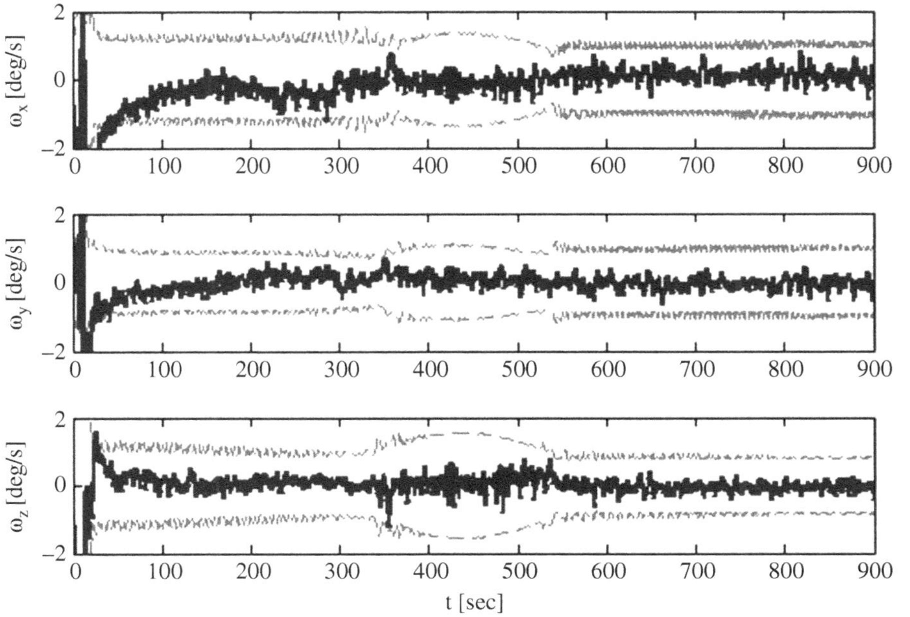

Figure 18.21 Estimation error of body angular velocity in MMUKF with ±3σ error bound.

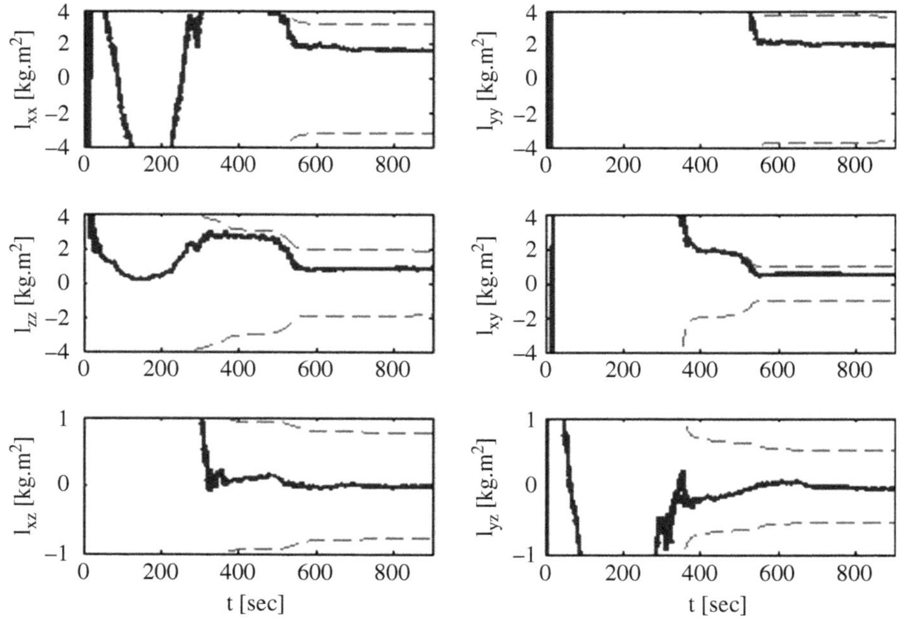

Figure 18.22 Estimation error of moments of inertia in MMUKF with ±3σ error bound.

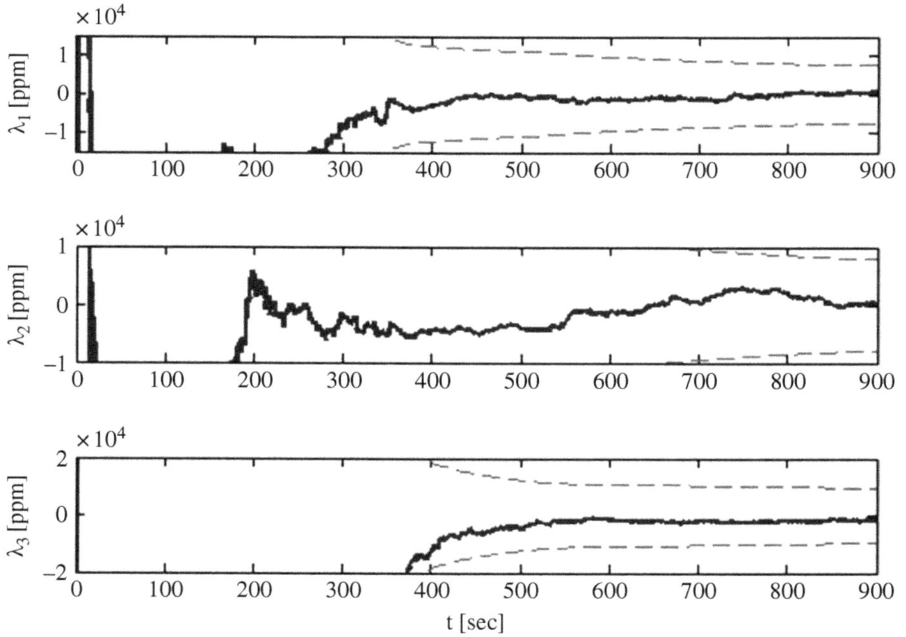

Figure 18.23 Estimation error of scale factors in MMUKF with ±3σ error bound.

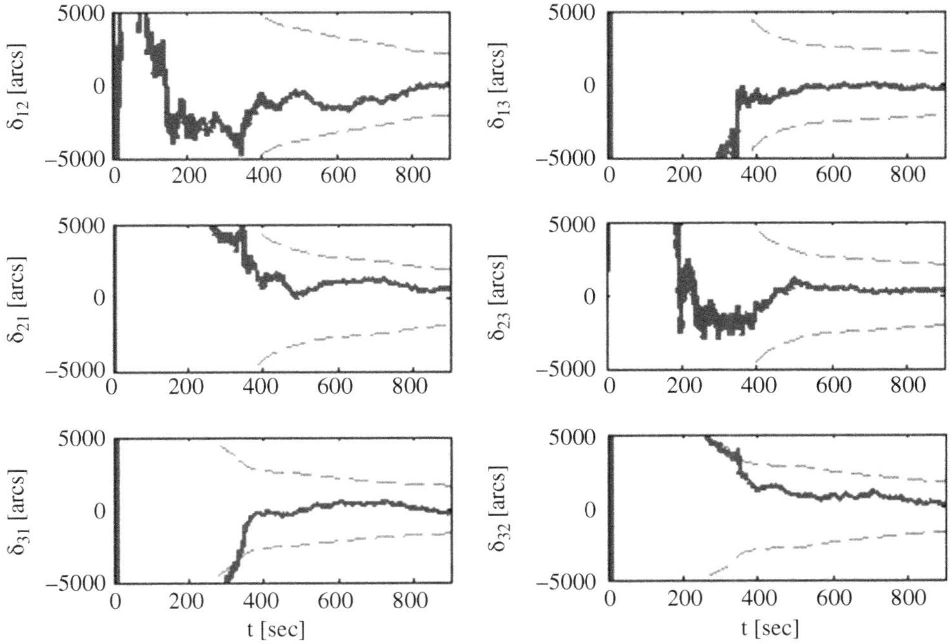

Figure 18.24 Estimation error of misalignments in MMUKF with ±3σ error bound.

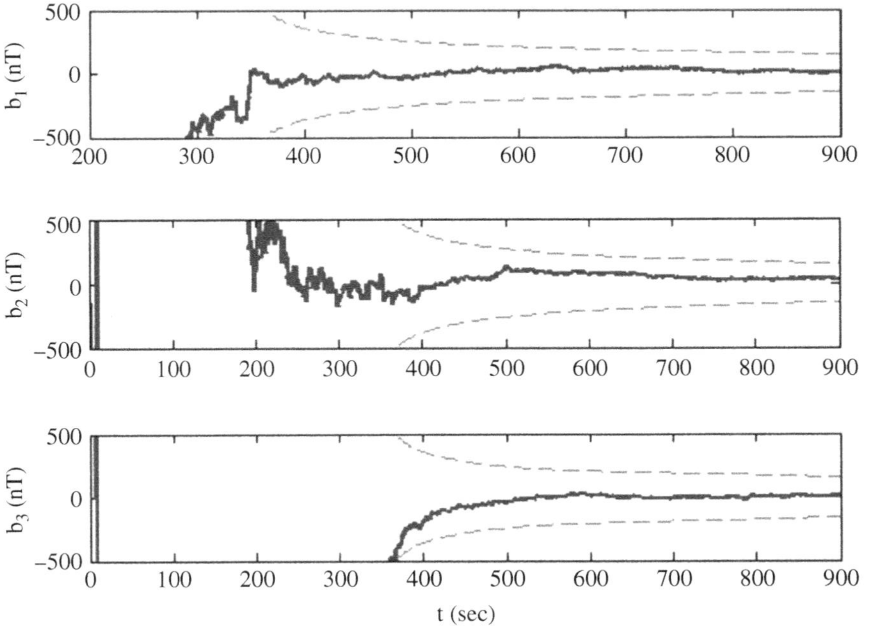

Figure 18.25 Estimation error of biases in MMUKF with ±3σ error bound.

Table 18.13 RMSE of state vector.

State	Root Mean Square Error	State	Root Mean Square Error
Angular velocity (rad/s)	4e − 3	Scale factors (ppm)	[250 200 2000]
Rotation angle (deg)	1.4	Misalignments (arcs)	[100 200 800 270 200 200]
MOI (kg.m^2)	[1.6 0.5 0.01 0.5 1.9 0.02 0.01 0.02 0.83]	Bias (nT)	[1 25 4]

Source: Kiani et al. (2014) [27].

18.4 Summary and Conclusions

Two important topics of spacecraft navigation related to concurrent orbit and attitude determination as well as concurrent system and parameter identification plus sensor calibration are described.

The first part described the feasibility of concurrent gyro-less orbit and attitude estimation for small LEO satellites using a centralized data fusion scheme with a TAM and a Sun sensor. The MUKF was implemented as a robust tool that results in acceptable accurate estimations; its major advantage is the use of a minimum number of sigma points, within the family of sigma point Kalman filters, causing a 48% run time reduction in comparison with standard UKF while preserving the same level of accuracy.

The stability and performance of MUKF were demonstrated via Monte Carlo simulations for a typical LEO satellite. A comprehensive sensitivity analysis over different effective parameters, such as orbital elements, TAM standard deviations and eclipse durations, was performed as part of the numerical simulation in Subsection 18.2.4. The results are promising and indicative of improved and enhanced state-of-the-art capabilities in the COAD estimation problem. The results also indicate that the achieved levels of accuracy satisfy the navigation and control subsystem qualification requirements of small satellites.

In the second part, the complex high-dimensional problem of concurrent attitude determination (AD), parameter identification (PI) and measurement sensor calibration (MSC) was developed and investigated. Accordingly, the SC attitude, attitude rates, moments of inertia tensor (MOI) plus the measuring system parameters, including scale factors, misalignments and biases of the TAMs, were recursively and simultaneously estimated based on the central fusion of the TAM and Sun sensor data. As the pertinent governing system model of the considered problem is a mixture of linear and nonlinear equations, an advanced marginal modified unscented Kalman filter – the marginal modified UKF (MMUKF) – consisting of a Kalman filter (KF) and MUKF was introduced. MMUKF was implemented as a robust tool which gives acceptably accurate estimations for the complex problem posed. MMUKF uses only 14 sigma points to estimate a full 25-dimensional state vector in the case study considered, while standard UKF needs 52 sigma points and MUKF requires 26 sigma points for the same problem. Application of MMUKF has reduced the run time by about 68% as compared with the standard UKF, while preserving the same accuracy level. Stability and performance of the mixed problem of AD, PI and MSC using the proposed MMUKF algorithm were demonstrated via Monte Carlo analysis for a LEO satellite. The results are promising and indicative of a good state-of-the-art capability for simultaneous AD, PI and MSC state/system estimation using a low-cost MEMS TAM plus a Sun sensor measurement package.

References

1 L. Psiaki M., 'Autonomous LEO orbit determination from magnetometer and sun sensor data', *AIAA Paper* **98-4308**, 1998, pp. 1–12.
2 Y.T. Chiang, L.S. Wang, F.R. Chang, and H.M. Peng, 'Constrained filtering method for attitude determination using GPS and gyro', *IEEE Proceedings-Radar Sonar and Navigation*, **149** (5), 2002, pp. 258–264.
3 Ma Zi, A. Ng, 'Spacecraft attitude determination by adaptive Kalman filtering', *AIAA Modelling and Simulation Techniques Conference and Exhibit*, California, 2002. AIAA Paper 2002-5042.
4 J.L. Crassidis and F.L. Markley, 'Unscented filtering for spacecraft attitude estimation', *Journal of Guidance, Control and Dynamics*, **26** (4), 2003, pp. 536–542.
5 D. Choukroun, I.Y. Bar-Itzhack, and Y. Oshman, 'Optimal-REQUEST algorithm for attitude determination', *Journal of Guidance, Control and Dynamics*, **27** (3), 2004, pp. 418–425.
6 A. Carmi and Y. Oshman, 'Robust spacecraft angular rate estimation from vector observations Using interlaced particle filtering', *Journal of Guidance, Control and Dynamics*, **30** (6), 2007, pp. 1729–1741
7 T. Xu, G. Gu, and X. Shen, 'A maneuvered GEO satellite orbit determination using robustly Adaptive Kalman filter', *Proceeding of International Conference on Intelligent System Design and Engineering Application*, Vol. **1**, 2010, pp. 55–59.
8 B. Ristic, S. Arulampalam, N. Gordon, *Beyond the Kalman Filter*, Particle Filters for Tracking Applications, Artech House, 2004.
9 D.M. Shuster, 'A survey of attitude representations', *Journal of the Astronautical Sciences*, **41** (3), 1993, pp. 439–517.
10 P.H. Zipfel, *Modeling and Simulation of Aerospace Vehicle Dynamics*. AIAA, New York, 2000, pp. 122–125 and 181–185.
11 M.J. Sidi, *Spacecraft Dynamics and Control, A Practical Engineering Approach*. 1st edn, Cambridge University Press, New York, 1997, pp. 30–33 and 95–96.
12 G.R. Hintz, 'Survey of orbit element sets', *Journal of Guidance, Control, and Dynamics*, **31** (3), 2008, pp. 785–790.
13 H.D. Curtis, *Orbital Mechanics for Engineering Students*, 2nd edn, Elsevier, 2010, pp. 208–212.
14 D.A. Vallado, W.D. McClain, *Fundamentals of Astrodynamics and Applications*, 4th edn, Microcosm Press and Springer, (2013), pp. 279–280, 303, 565, and 637.
15 J. Bae and K. Youan, 'Attitude estimation for satellite fault tolerant system using federated unscented Kalman filter', *International Journal of Aeronautical and Space Sciences*, **11** (2), 2010, pp. 80–86.
16 J.K. Deutschmann, I.Y. Bar-Itzhack, 'Comprehensive evaluation of attitude and orbit estimation using actual earth magnetic field data', *Journal of Guidance, Control, and Dynamics*, **24** (3), 2001, pp. 616–635.

17 W.Jianqi, C. Xibin, and S. Zhaowei, 'Attitude and orbit determination for small satellite using magnetometer measurement', *Aircraft Engineering and Aerospace Technology*, **75** (3), 2003, pp. 241–246.

18 Y. Xing, X. Cao, S. Zhang, 'Modified UKF for integrated orbit and attitude determination based on gyro and magnetometer', 2nd International Symposium on Systems and Control in Aerospace and Astronautics, 2008, pp. 1–4.

19 M. Abdelrahman, S.-Y. Park, 'Simultaneous spacecraft attitude and orbit estimation using magnetic field vector measurements', *Aerospace Science and Technology*, **15** (8), 2011, pp. 653–669.

20 M. Kiani, Seid H. Pourtakdoust, 'Concurrent orbit and attitude estimation of LEO satellites using minimum sigma points UKF', *Proceedings of the Institute of Mechanical Engineers Part G: Journal of Aerospace Engineering*, **228** (6), 2013, pp. 810–819.

21 C.C. Finlay, S. Maus, C.D. Beggan et. al., 'International Geomagnetic Reference field: the eleven generation', *Geophysical Journal International*, **183**, 2010, pp. 1216–1230.

22 S.J. Julier and J.K. Uhlmann, 'A new extension of the Kalman filter to nonlinear systems', in *Proceedings Of Aero Sense: The 11th Int. Symposium on Aerospace/Defence Sensing, Simulation and Controls*, Orlando, 1997, pp. 54–65.

23 S.J. Julier, 'The spherical simplex unscented transformation', in *Proceedings of the American Control Conferenc*e, Denver, Colorado, 2003, pp. 2430–2434.

24 W.-.Ch. Li, P. Wei, and X.-C. Xiao, 'Novel simplex unscented transform and filter', *Journal of Electronic and Technology of China*, **6** (1), 2008, pp. 61–65

25 L.X. Wang, *A Course in Fuzzy systems and Control*, Prentice Hall, 2002, pp. 257–258.

26 J. Xu, G.M. Dimirovski, Y. Jing, and C. Shen, 'UKF design and stability for nonlinear stochastic systems with correlated noises', *Proceedings of the 46th IEEE Conference on Decision and Control*, New Orleans, LA, USA, 12–14 December 2007.

27 M. Kiani, Seid H. Pourtakdoust, 'Spacecraft attitude and system identification via marginal reduced UKF utilizing the Sun and calibrated TAM sensors', *Scientia Iranica*, **21** (4), 2014, pp. 1451–1460.

28 H. Myung, K. -L. Yong and H. Bang, 'Hybrid estimation of spacecraft attitude dynamics and rate sensor alignment parameters', *International Conference on Control, Automation and Systems*, 2007.

29 M. Briers, S.R. Maskell and R. Wright R., 'A Rao-Blackwellised unscented Kalman filter', in *Proceedings of 6th International Conference of Information Fusion*, Queensland, Australia, 2003.

19

Adaptive Fault-tolerant Attitude Control for Spacecraft Under Loss of Actuator Effectiveness

Qinglei Hu[1], Bing Xiao[1], Bo Li[1] and Youmin Zhang[2]

[1] *Department of Control Science and Engineering, Harbin Institute of Technology, Harbin, China*
[2] *Department of Mechanical and Industrial Engineering, Concordia University, Montreal, Quebec, Canada*

19.1 Introduction

Accurate and reliable control law design for orbital vehicles is a major challenge for designers. Considerable research has been undertaken into designing spacecraft attitude controllers that will function in the presence of the uncertainties and external disturbances that the system will encounter in operation, to guarantee high performance, such as optimal control [1, 2], sliding mode control [3], adaptive and robust control [4, 5] and so on. However, during operations, it is possible that the system becomes abnormal, for instance due to the ageing of components, or actuator and sensor failures. This may result in substantial performance deterioration and even system instability. Fortunately, fault-tolerant control is an effective control strategy is applicable to a large class of subsystem and component faults or failures, giving good performance and with reliability guaranteed for fault-free systems as well as for faulty systems. Researchers in the system control community have proposed a number of methods of fault-tolerant control [6, 7 and references therein]. Specific fault-tolerant control schemes are also covered in the literature too: adaptive control [8, 9], feedback linearization control [10], multiple-model control [11, 12], dynamic inversion control [13] and others [14 and references therein].

The problem of fault-tolerant attitude control design for spacecraft was considered by Cai et al. [15], in which the objective of attitude tracking was achieved with a simple controller structure employing an indirect adaptive method. An alternative fault-tolerant control design for compensation of reaction wheels faults was discussed by Ji et al. [16], who achieved the desired attitude performance by using a time-delay control method. Jiang et al [17] presented an adaptive backstepping sliding mode control scheme for a flexible spacecraft attitude tracking system in the presence of bounded disturbances, unknown inertia parameter uncertainties and even actuator faults, but low boundedness of the actuator fault was required in advance for the designers. Godard et al. [18] discussed the attitude control of a satellite using coordinated movement of the tether attachment points, and derived a nominal sliding mode control law and an adaptive

Advanced UAV Aerodynamics, Flight Stability and Control: Novel Concepts, Theory and Applications, First Edition. Edited by Pascual Marqués and Andrea Da Ronch.

Companion website: http://www.wiley.com/go/marques/advanced_UAV_aerodynamics

fault-tolerant control law for cases when tether deployment suddenly stopped and tether breakage occurred. For the previous research results, it is assumed that the actuator fault occurs instantaneously; that is, that faults are piecewise constant functions of time.

In this chapter, the attitude stabilization of spacecraft during partial loss of actuator effectiveness is modelled by a multiplicative factor. Due to the time-varying feature of the fault, the whole attitude control plant is a multi-input, multi-output (MIMO) system with time-varying gain. For a system with time-varying gain, in Zhang and Ge [19] presented an adaptive neural controller using the principle of sliding mode control and a Nussbaum-type function to handle the unknown high gains and dead zones. However, the system should be written into a triangular control structure. Marino and Tomei [20] derived an adaptive output feedback control algorithm for linear time-varying systems, but the approach required that full knowledge of the sign of high-frequency gain be known in advance. A backstepping control approach combined with online parameter estimator was also discussed by Zhang et al. [21] for linear time-varying single-input, single-output systems. Under the designed controller, all the closed-loop signals are guaranteed to be globally uniformly bounded and the tracking error remains small even in the presence of unknown time-varying parameters. However, the unstructured plant-parameter variations are required to be slow. A new control scheme incorporating with adaptive backstepping technique was considered by Zhou et al. [22]; the control objectives are achieved by introducing an estimator for the bound of the variation rate of parameters with a Nussbaum-type function. However, these methods cannot be directly applied to attitude control of a spacecraft with a time-varying fault and, to the best of our knowledge, there are few papers dealing with the MIMO time-varying issue at present.

The contribution of this chapter is to provide an adaptive fault-tolerant strategy for spacecraft attitude control when there is partial loss of actuator effectiveness. Specifically, by applying an adaptive backstepping control technique, a normal attitude controller is first derived for the rigid spacecraft system in the presence of external disturbances, in which all the actuators are fault-free and operating normally. Then the situation of partial loss of actuator effectiveness is considered, and by using the appropriate transformation of the auxiliary system states, the time-varying MIMO attitude control system can be decoupled into three auxiliary systems. The output of the auxiliary system is considered as the faulty actuator output. To this end, the three new adaptive controllers are developed for the auxiliary systems, to guarantee that outputs of the auxiliary system can follow the normal attitude-control command signals and that the tracking error can remain small enough. A key feature of the proposed strategy is that the design of the fault-tolerant control does not require a fault detection-and-identification (FDI) mechanism to obtain information about the fault. This means large savings in computing power and reductions in response times.

In addition to detailed derivations of the new controllers and a rigorous outlining of all the associated stability and attitude convergence proofs, extensive simulation studies have been conducted to validate the design, and the results are presented to highlight closed-loop performance benefits when compared with conventional control schemes, even under partial loss of actuator effectiveness. The chapter is organized as follows. Spacecraft-attitude mathematical model and control problems are presented in Section 19.2. In Section 19.3, the adaptive backstepping attitude controller is derived in the presence of partial loss of actuator effectiveness. The results of the numerical simulations in Section 19.4 demonstrate the good performance of the proposed scheme. Finally, the chapter is completed with some conclusions.

19.2 Mathematical Model of Flexible Spacecraft and Problem Formulation

This section briefly reviews the Euler parameters description of the attitude motion of a rigid spacecraft. The nonlinear equations of motion, in terms of components along the body fixed control axes, are given by the attitude kinematics and dynamics [23]:

Attitude kinematics:

$$\dot{q} = \frac{1}{2}\left(q^{\times} + q_0 I_3\right)\omega \tag{19.1a}$$

$$\dot{q}_0 = -\frac{1}{2}q^T\omega \tag{19.1b}$$

where $\omega = \left(\omega_1\ \omega_2\ \omega_3\right)^T$ is the spacecraft angular velocity with respect to an inertial frame I and expressed in body-fixed frame B, and the unit quaternion $Q = (q_0, q^T) \in R \times R^3$ describes the attitude orientation of the spacecraft in B with respect to I satisfying $q^T q + q_0^2 = 1$. Note that I_3 denotes the 3×3 identity matrix, and for $\forall q = \left(q_1\ q_2\ q_3\right)^T$, the notation $q^{\times}$ denotes the following skew-symmetric matrix:

$$q^{\times} = \begin{pmatrix} 0 & -q_3 & q_2 \\ q_3 & 0 & -q_1 \\ -q_2 & q_1 & 0 \end{pmatrix} \tag{19.2}$$

Spacecraft dynamics:

When all the actuators run normally, the attitude control system is called a 'normal' system. Then, the dynamics of a rigid spacecraft can be described by:

$$J\dot{\omega} = -\omega^{\times}J\omega + u + d \tag{19.3}$$

where $J \in R^{3\times3}$ represents the positive-definite moment inertia of the rigid spacecraft, $u = [u_1\ \ u_2\ \ u_3]^T$ is the control torque input generated by thrusters, and $d = [d_1\ \ d_2\ \ d_3]^T$ denotes the external disturbance torque.

In case of partial loss of actuator effectiveness, the system in Eq. (19.3) becomes a faulty one. In this chapter, the partial loss of actuator effectiveness is modelled by a multiplicative factor, and then the faulty spacecraft dynamic system can be rewritten as:

$$J\dot{\omega} = -\omega^{\times}J\omega + E\left(t\right)u + d \tag{19.4}$$

where $E(t) = diag(e_1(t), e_2(t), e_3(t))$ denotes the effectiveness factor of spacecraft actuators, such that $0 < e_i(t) \le 1$ $(i = 1,\ 2,\ 3)$. The case when $e_i(t) = 1$ means that the ith actuator works normally; for $0 < e_i(t) < 1$, it means that the ith actuator loses its effectiveness partially, but still works at all times.

For the development of control laws, the following assumptions are imposed:

Assumption 1 For the effectiveness factor $e_i(t)$, it is assumed that its first-order derivative exists, and there exists an unknown bounded positive constant λ such that the condition $|\dot{e}_i(t)| \leq \lambda$ is satisfied. Moreover, λ is inside a closed interval $[\lambda_{\min}, \lambda_{\max}]$.

Assumption 2 The disturbances d considered in Eqs. (19.3) or (19.4) include environmental torques, such as gravitational torque, and are assumed to be bounded. Therefore, it is reasonable to assume that there always exists a constant T_d such that;

$$\|d\| \leq T_d \tag{19.5}$$

The objective in this chapter is to design a fault-tolerant control law for the system in Eqs. (19.1) and (19.4) such that the following targets are achieved in the presence of external disturbances and partial loss of actuator effectiveness:

- The closed-loop system is globally stable, and in which all the signals are bounded and continuous.
- The attitude and angular velocity converge to an arbitrary small set containing the origin, that is, $\|q(t)\| \leq \varepsilon_1$ and $\|\omega(t)\| \leq \varepsilon_2$.

In what follows, we shall develop such a control law for attitude stabilization of spacecraft.

19.3 Adaptive Backstepping Fault-Tolerant Controller Design

In this section, for the normal system without an actuator fault, a baseline attitude controller based on the adaptive backstepping technique is developed. Then, using the baseline controller, the fault-tolerant controller is derived. This guarantees that the actual outputs of the faulty actuators can still follow the normal command inputs, and the fault can be compensated online.

19.3.1 Baseline Attitude Controller Design

For the normal system in Eq. (19.3), let us introduce the following state variables transformation:

$$z_1 = q \tag{19.6a}$$

$$z_2 = \omega - \alpha_1 \tag{19.6b}$$

where α_1 is a virtual control law, which will be given latter. Then, the standard backstepping design procedure is elaborated as follows:

Step1: Choose a Lyapunouv function candidate as

$$V_1 = \frac{1}{2} z_1^T z_1 + (1 - q_0)^2 = 1 - q_0 \tag{19.7}$$

and select an appropriate virtual control α_1 as

$$\alpha_1 = -c_1 z_1 \tag{19.8}$$

where c_1 is a design parameter. Then, the derivative V_1 is

$$\dot{V}_1 = -\dot{q}_0 = \frac{1}{2} z_1^T \left(z_2 + \alpha_1 \right) = -\frac{c_1}{2} z_1^T z_1 + \frac{1}{2} z_1^T z_2 \tag{19.9}$$

Clearly if $z_2 = 0$, then $\dot{V}_1 = -c_1 \|z_1\|^2$ and z_1 is guaranteed to converge to zero asymptotically.

Step 2. For the second error dynamic Eq. (19.6b) for z_2, one obtains

$$J\dot{z}_2 = J\dot{\omega} - J\dot{\alpha}_1 = -\left(z_2 + \alpha_1 \right)^{\times} J\left(z_2 + \alpha_1 \right) - J\dot{\alpha}_1 + u + d \tag{19.10}$$

Now design the control input u as

$$u(t) = \left(z_2 + \alpha_1 \right)^{\times} J\left(z_2 + \alpha_1 \right) + J\dot{\alpha}_1 - \frac{1 + c_2}{2} z_2 - \frac{z_2 T_d^2}{\|z_2\| T_d + \varepsilon e^{-\beta t}} \tag{19.11}$$

By employing Eqs. (19.1) and (19.6), the derivative of α_1 can be calculated as

$$\dot{\alpha}_1 = -c_1 \dot{z}_1 = -\frac{c_1}{2} \left(z_1^{\times} + q_0 I_3 \right) \left(z_2 + \alpha_1 \right) \tag{19.12}$$

Choose a new Lyapunov function $V_2 = V_1 + \frac{1}{2} z_2^T J z_2$, and, in view of Eqs. (19.10)–(19.11), the time derivative of V_2 yields

$$\begin{aligned} \dot{V}_2 &= -\frac{c_1}{2} z_1^T z_1 + \frac{1}{2} z_1^T z_2 + z_2^T \left[-\left(z_2 + \alpha_1 \right)^{\times} J\left(z_2 + \alpha_1 \right) - J\dot{\alpha}_1 + u + d \right] \\ &\le -\frac{c_1}{2} z_1^T z_1 - \frac{c_2}{2} z_2^T z_2 + \|z_2\| \|T_d\| - \frac{\|z_2\|^2 T_d^2}{\|z_2\| T_d + \varepsilon e^{-\beta t}} \\ &\le -\frac{c_1}{2} \|z_1\|^2 - \frac{c_2}{2} \|z_2\|^2 + \varepsilon e^{-\beta t} \\ &\le -c_0 \left(\|z_1\|^2 + \|z_2\|^2 \right) + \varepsilon \end{aligned} \tag{19.13}$$

where ε is a sufficiently small positive scalar control gain, β is an adequate positive constant, c_2 is a positive constant, and $c_0 = \min\left\{ \frac{c_1}{2}, \frac{c_2}{2} \right\}$.

Clearly, if $\|z_1(t)\| > \sqrt{\varepsilon / c_0}$ and $\|z_2(t)\| > \sqrt{\varepsilon / c_0}$, then $\dot{V} < 0$, which implies that $V(t)$ decreases monotonically. Therefore, the state signals are bounded ultimately as

$$\lim_{t \to \infty} \left[z_1(t) \quad z_2(t) \right] \in \left(\|z_1\| \le \sqrt{\varepsilon / c_0}, \|z_2\| \le \sqrt{\varepsilon / c_0} \right) \tag{19.14}$$

which is a small set containing the origin $[z_1 \quad z_2] = 0$; moreover, the larger the values of c_0 and ε selected, the better control performance will be yielded. Using Eqs. (19.6) and (19.14), we have

$$\lim_{t \to \infty} \left[q(t) \quad \omega(t) \right] \in \left(\|q\| \le \sqrt{\varepsilon / c_0}, \|\omega\| \le \left(c_1 + 1 \right) \sqrt{\varepsilon / c_0} \right) \tag{19.15}$$

Thus, from Eqs. (19.13) and (19.15), it can be concluded that the closed-loop system is globally stable. Then the following statements can be presented.

Theorem 1: Consider the normal system given by Eqs. (19.1) and (19.3) and with the normal controller Eq. (19.11), the closed-loop system is globally stable in the sense that all the closed-loop signals are bounded. Moreover, the attitude q and angular velocity ω will ultimately converge to an arbitrary set containing zero; that is, the objectives (a) and (b) as stated in Section 19.2 can be achieved.

Proof: This can easily be obtained from above analysis and therefore the proof is omitted here.

19.3.2 Adaptive Backstepping Fault-Tolerant Attitude Controller Design

In Section 19.3.1, we used adaptive backstepping to designed a stable system for control of a spacecraft system with unknown disturbance torques. However, no actuator fault was considered. In this subsection, a partial loss of actuator effectiveness will be considered. To handle such a fault, for the ith actuator, the following auxiliary system is added:

$$\dot{u}_{act}^{i} = v_i \tag{19.16a}$$

$$\dot{v}_i = e_i(t)u_{com}^{i} \tag{19.16b}$$

$$y_i = u_{act}^{i} \tag{19.16c}$$

where u_{com}^{i} and y_i are the input and output of the auxiliary system in Eqs. (19.16a)–(19.16c), respectively. Note that the output y_i is viewed as the actual output of the ith actuator, and the baseline controller in Eq. (19.11) is denoted u_{nor}. Then, the control objective in this sub-part can be stated as: design control input u_{com}^{i} for auxiliary system Eq. (19.16) satisfying Assumption 1 such that the output y_i can follow the specified desired signal u_{nor}^{i}, which represents the ith element of u_{nor}.

Before going into the details of the design of u_{com}^{i}, the following definition and Lemma are presented:

Definition 1 [24]: A function $N(\chi)$ is called a Nussbaum-type function if the following two properties are satisfied:

$$\text{(P1)} \qquad \lim_{s\to+\infty} \sup \frac{1}{s}\int_0^s N(\chi)d\chi = +\infty \tag{19.17a}$$

$$\text{(P2)} \qquad \lim_{s\to-\infty} \inf \frac{1}{s}\int_0^s N(\chi)d\chi = -\infty \tag{19.17b}$$

Throughout this paper, the even Nussbaum function $N(\chi) = \exp(\chi^2)\cos\left(\frac{\pi}{2}\chi\right)$ is employed.

Lemma 1 [24]: Let $V(t)$ and $\chi(t)$ be smooth functions defined on $[0, t_f)$ with $V(t) \geq 0$, $\forall t \in [0, t_f)$ and let $N(\chi)$ be an even smooth Nussbaum-type function. If the following inequality holds:

$$V(t) \leq f_0 + e^{-f_1 t}\int_0^t g(\tau)N(\chi)\dot{\chi}e^{-f_1\tau}d\tau + e^{-f_1 t}\int_0^t \dot{\chi}e^{-f_1\tau}d\tau \tag{19.18}$$

f_0 represents some suitable constant, f_1 is a positive constant, and $g(\tau)$ is a time-varying parameter that takes values in the unknown closed intervals $I_1 = [g^-, g^+]$ with $0 \notin I_1$, then $V(t)$, $\chi(t)$ and $\int_0^t g(\tau)N(\chi)\dot{\chi}d\tau$ must be bounded on $[0, t_f)$.

Similar to the work of Zhang et al. [21], the following two filters are designed

$$\dot{\xi}_i = A_i\xi_i + k_i y_i \tag{19.19a}$$

$$\dot{\Omega}_i = A_i\Omega_i + \begin{bmatrix} 0 \\ u_{com}^i \end{bmatrix} \tag{19.19b}$$

where $\xi_i = [\xi_{i,1} \quad \xi_{i,2}]^T \in R^2$, $\Omega_i = [\Omega_{i,1} \quad \Omega_{i,2}]^T \in R^2$, $k_i = [k_{i,1} \quad k_{i,2}]^T$ and matrix $A_i = \begin{bmatrix} -k_{i,1} & 1 \\ -k_{i,2} & 0 \end{bmatrix}$ is strictly stable. Then, with the above filters, a state estimate for the system in Eq. (19.16) can be given by:

$$\hat{x}_i = \xi_i + e_i(t)\Omega_i \tag{19.20}$$

We define the estimation error $\tilde{x}_i$ as $\tilde{x}_i = x_i - \hat{x}_i$ with $x_i = [u_{act}^i \quad v_i]^T \in R^2$. In view of Eqs. (19.16) and (19.20), one has:

$$\dot{\tilde{x}}_i = \dot{x}_i - \dot{\hat{x}}_i = A_i\tilde{x}_i - \dot{e}_i(t)\Omega_i \tag{19.21}$$

From Eq. (19.21), the estimation error $\tilde{x}_i$ can be divided into two parts; that is, $\tilde{x}_i = \tilde{x}_i^a + \tilde{x}_i^b$. Then:

$$\dot{\tilde{x}}_i^a = A_i\tilde{x}_i^a \tag{19.22a}$$

$$\tilde{x}_i^b = \int_0^t e^{A_i(t-\tau)}\left(-\dot{e}_i(t)\Omega_i\right)d\tau \tag{19.22b}$$

From Eq. (19.22) and the definition of matrix A_i, one has

$$\left\|\tilde{x}_i^b\right\| \le \int_0^t \left\|e^{A_i(t-\tau)}\right\|\left\|\dot{e}_i(t)\right\|\left\|\Omega_i\right\|d\tau \le \lambda\int_0^t \pi_2 e^{-\pi_1(t-\tau)}\left\|\Omega_i\right\|d\tau \tag{19.23}$$

where π_1 and π_2 are chosen positive constants such that

$$\left\|e^{A_i(t-\tau)}\right\| \le \pi_2 e^{-\pi_1(t-\tau)} \tag{19.24}$$

Thus, we have

$$\left\|\tilde{x}_i^b\right\| \le \lambda\zeta(t) \tag{19.25}$$

where the variable $\zeta(t)$ is the solution of the following equation

$$\dot{\zeta}(t) = -\pi_2\zeta(t) + \pi_1\left(\left\|\Omega_i\right\|^2 + \frac{1}{4}\right) \tag{19.26}$$

Due to the requirement of the strict stability of matrix A_i, there exists a positive definite matrix $P_i \in R^{2\times2}$ satisfying $P_iA_i + A_i^T P_i \le -2I_3$. Then, from Eqs. (19.16b) and (19.20), it follows that:

$$v_i = x_{i,2} = \hat{x}_{i,2} + \tilde{x}_{i,2} = \xi_{i,2} + e_i(t)\Omega_{i,2} + \tilde{x}_{i,2} \tag{19.27}$$

where $x_{i,2}$, $\hat{x}_{i,2}, \tilde{x}_{i,2}$ are the second elements of the variables x_i, $\hat{x}_i, \tilde{x}_i$, respectively. Then, the auxiliary system in Eq. (19.16) can be rewritten in the following form based upon the designed filters in Eq. (19.19):

$$\dot{y}_i = \dot{u}_{act}^i = \dot{v}_i = \xi_{i,2} + e_i(t)\Omega_{i,2} + \tilde{x}_{i,2} \tag{19.28a}$$

$$\dot{\Omega}_{i,2} = -k_{i,2}\Omega_{i,1} + u_{com}^i \tag{19.28b}$$

It can clearly be seen that Eq. (19.28) is written in a triangular nonlinear form. Hence, the standard backstepping controller design step can be implemented into it.

Take the state transformation as:

$$z_{i,1} = y_i - u_{nor}^i \tag{19.29a}$$

$$z_{i,2} = \Omega_{i,2} - \beta_i \tag{19.29b}$$

Where β_i is a virtual control law and will be designed later on. Now, the standard backstepping design procedures are adopted as:

Step 1: From Eqs. (19.28a) and (19.19a), it follows that

$$\dot{z}_{i,1} = \dot{y}_i - \dot{u}_{nor}^i = \xi_{i,2} + e_i(t)\Omega_{i,2} + \tilde{x}_{i,2} - \dot{u}_{nor}^i \tag{19.30}$$

Using the method of Zhou et al. [22], the virtual control law β_i is selected as

$$\beta_i = N(\chi)\bar{\beta}_i e^{-\delta_i t} \tag{19.31}$$

in which $\delta_i \in R$ is a positive constant and χ satisfies the following equation:

$$\dot{\chi} = z_{i,1}\bar{\beta}_i \tag{19.32}$$

and $\bar{\beta}_i$ is selected as:

$$\bar{\beta}_i = \left[m_{i,1} + l_{i,1} + \hat{e}_i(t)\right]z_{i,1} + \xi_{i,2} + z_{i,1}\zeta^2(t)\hat{\lambda} - \dot{u}_{nor}^i \tag{19.33}$$

where $m_{i,1}$ and $l_{i,1}$ are two positive constants, and $\hat{e}_i(t)$ and $\hat{\lambda}$ are the estimation of $e_i(t)$ and λ, respectively. Then, with Eqs. (19.29), (19.30) and (19.33), one has

$$\begin{aligned}\dot{z}_{i,1} - \bar{\beta}_i &= -\left[m_{i,1} + l_{i,1} + \hat{e}_i^2(t)\right]z_{i,1} - z_{i,1}\zeta^2(t)\hat{\lambda} + e_i(t)\Omega_{i,2} + \tilde{x}_{i,2} \\ &= -\left[m_{i,1} + l_{i,1} + \hat{e}_i^2(t)\right]z_{i,1} - z_{i,1}\zeta^2(t)\hat{\lambda} + \hat{e}_i(t)z_{i,2} + e_i\beta_i + \tilde{e}_i(t)z_{i,2} + \tilde{x}_{i,2}\end{aligned} \tag{19.34}$$

where $\tilde{e}_i(t) = e_i - \hat{e}(t)$.

Then choose a new Lyapunov function as

$$V_{i,1} = \frac{1}{2} z_{i,1}^2 + \frac{1}{4l_{i,1}} \left(\tilde{x}_i^a\right)^T P_i \tilde{x}_i^a + \frac{1}{2l_{i,2}} \tilde{\lambda}^2 + \frac{1}{2l_{i,3}} \tilde{e}_i^2 \tag{19.35}$$

where $l_{i,j}(j=1,2,3)>0$ is a design parameter and $\tilde{\lambda} = \lambda - \hat{\lambda}$.

Using Eqs. (19.31), (19.32) and (19.34), the time derivative of $V_{i,1}$ can be calculated as

$$\begin{aligned}
\dot{V}_{i,1} &= z_{i,1}\left(\dot{z}_{i,1} - \bar{\beta}_i\right) + z_{i,1}\bar{\beta}_i + \frac{1}{4l_{i,1}}\left(\tilde{x}_i^a\right)^T \left(P_i A_i + A_i^T P_i\right)\tilde{x}_i^a - \frac{1}{l_{i,2}}\tilde{\lambda}\dot{\hat{\lambda}} + \frac{1}{l_{i,3}}\tilde{e}_i\left(\dot{e}_i - \dot{\hat{e}}_i(t)\right) \\
&\le -\left[m_{i,1} + \hat{e}_i^2(t)\right] z_{i,1}^2 + \hat{e}_i(t) z_{i,1} z_{i,2} + \frac{1}{l_{i,3}}\tilde{e}_i\left(l_{i,3} z_{i,1} z_{i,2} - \dot{\hat{e}}_i(t)\right) + e_i \beta_i z_{i,1} - z_{i,1}^2 \zeta^2(t)\hat{\lambda} - \frac{1}{l_{i,2}}\tilde{\lambda}\dot{\hat{\lambda}} \\
&\quad + \frac{1}{l_{i,3}}\tilde{e}_i \dot{e}_i - \frac{1}{2l_{i,1}}\left\|\tilde{x}_i^a\right\|^2 - l_{i,1} z_{i,1}^2 + z_{i,1}\bar{\beta}_i + z_{i,1}\left(\tilde{x}_{i,2}^a + \tilde{x}_{i,2}^b\right) \\
&\le -m_{i,1} z_{i,1}^2 + \frac{1}{l_{i,3}}\tilde{e}_i\left(l_{i,3} z_{i,1} z_{i,2} - \dot{\hat{e}}_i(t)\right) + \frac{1}{l_{i,3}}\tilde{e}_i \dot{e}_i + \left[e_i N(\chi)\exp(-\delta_i t) + 1\right]\dot{\chi} - \frac{1}{4l_{i,1}}\left\|\tilde{x}_i^a\right\|^2 + \frac{1}{4} z_{i,2}^2 \\
&\quad - z_{i,1}^2 \zeta^2(t)\hat{\lambda} - \frac{1}{l_{i,2}}\tilde{\lambda}\dot{\hat{\lambda}} + z_{i,1}\tilde{x}_{i,2}^b
\end{aligned} \tag{19.36}$$

in which the Young's inequality

$$\hat{e}_i(t) z_{i,1} z_{i,2} \le \hat{e}_i^2(t) z_{i,1}^2 + \frac{1}{4} z_{i,2}^2,\ z_{i,1}\tilde{x}_{i,2}^a \le l_{i,1} z_{i,1}^2 + \frac{1}{4l_{i,1}}\left|\tilde{x}_{i,2}^a\right|^2 \le l_{i,1} z_{i,1}^2 + \frac{1}{4l_{i,1}}\left\|\tilde{x}_i^a\right\|^2 \tag{19.37}$$

is applied.

From Eq. (19.25), it can be easily known that

$$z_{i,1}\tilde{x}_{i,2}^b - z_{i,1}^2\zeta^2(t)\hat{\lambda} \le \lambda\zeta(t)\left|z_{i,1}\right| - z_{i,1}^2\zeta^2(t)\hat{\lambda} \le \lambda\left(\zeta^2(t) z_{i,1}^2 + \frac{1}{4}\right) - z_{i,1}^2\zeta^2(t)\hat{\lambda} = z_{i,1}^2\zeta^2(t)\tilde{\lambda} + \frac{\lambda}{4} \tag{19.38}$$

Now design the update laws as

$$\dot{\hat{e}}_i(t) = proj_{[0,1]}\left(l_{i,3} z_{i,1} z_{i,2}\right) \tag{19.39a}$$

$$\dot{\hat{\lambda}} = proj_{[\lambda_{\min},\lambda_{\max}]}\left(l_{i,2} z_{i,1}^2 \zeta^2(t)\right) \tag{19.39b}$$

where $proj(\cdot)$ denotes the projection operator [25].

Using Eqs. (19.37) and (19.38) and projection operator property [25] $-\tilde{\lambda} proj(l_{i,2} z_{i,1}^2 \zeta^2(t)) \le -\tilde{\lambda} l_{i,2} z_{i,1}^2 \zeta^2(t)$, $-\tilde{e}_i(t) proj(l_{i,3} z_{i,1} z_{i,2}) \le -l_{i,3} z_{i,1} z_{i,2} \tilde{e}_i(t)$, Eq. (19.36) can be rewritten as

$$\dot{V}_{i,1} \le -m_{i,1} z_{i,1}^2 + \frac{1}{l_{i,3}}\tilde{e}_i \dot{e}_i + \left[e_i N(\chi)\exp(-\delta_i t) + 1\right]\dot{\chi} + \frac{\lambda}{4} - \frac{1}{4l_{i,1}}\left\|\tilde{x}_i^a\right\|^2 + \frac{1}{4} z_{i,2}^2 \tag{19.40}$$

Step 2. Now differentiating Eq. (19.29b) along with Eq. (19.28b), one has

$$\dot{z}_{i,2} = \dot{\Omega}_{i,2} - \dot{\beta}_i = -k_2\Omega_{i,1} - \dot{\beta}_i + u^i_{com} \tag{19.41}$$

where β_i is a function of $y_i, u^i_{nor}, \hat{e}_i, \hat{\lambda}, \chi$, and can be calculated as

$$\dot{\beta}_i = \frac{\partial \beta_i}{\partial y_i}\dot{y}_i + \frac{\partial \beta_i}{\partial u^i_{nor}}\dot{u}^i_{nor} + \frac{\partial \beta_i}{\partial \hat{e}_i}\dot{\hat{e}}_i + \frac{\partial \beta_i}{\partial \hat{\lambda}}\dot{\hat{\lambda}} + \frac{\partial \beta_i}{\partial \chi}\dot{\chi} \tag{19.42a}$$

Defining the Lyapunov function $V_{i,2} = V_{i,1} + \frac{1}{2}z^2_{i,2}$, the time derivative of $V_{i,2}$ yields

$$\begin{aligned} \dot{V}_{i,2} \le & -m_{i,1}z^2_{i,1} + \frac{1}{l_{i,3}}\tilde{e}_i\dot{e}_i + \left[e_i N(\chi)\exp(-\delta_i t) + 1\right]\dot{\chi} + \frac{\lambda}{4} - \frac{1}{4l_{i,1}}\left\|\tilde{x}^a_i\right\|^2 + \frac{1}{4}z^2_{i,2} \\ & + z_{i,2}\left(-k_{i,2}\Omega_{i,1} - \dot{\beta}_i + u^i_{com}\right) \end{aligned} \tag{19.42b}$$

Then the control input u^i_{com} is designed as

$$u^i_{com} = -\left(\frac{1}{4} + m_{i,2}\right)z_{i,2} + k_{i,2}\Omega_{i,i} + \dot{\beta}_i \tag{19.43}$$

Submitting the control in Eq.(19.43) into Eq. (19.42) yields

$$\dot{V}_{i,2} \le -m_{i,1}z^2_{i,1} - m_{i,2}z^2_{i,2} + \frac{1}{l_{i,3}}\tilde{e}_i\dot{e}_i + \left[e_i N(\chi)\exp(-\delta_i t) + 1\right]\dot{\chi} + \frac{\lambda}{4} - \frac{1}{4l_{i,1}}\left\|\tilde{x}^a_i\right\|^2 \tag{19.44}$$

With the definition of $V_{i,2}$, it has

$$V_{i,2} \le \frac{1}{2}z^2_{i,1} + \frac{1}{2}z^2_{i,2} + \frac{1}{4l_{i,1}}\|P\|\left\|\tilde{x}^a_i\right\|^2 + \frac{1}{2l_{i,2}}\tilde{\lambda}^2 + \frac{1}{2l_{i,3}}\tilde{e}^2_i \tag{19.45}$$

Integrating both side of Eq. (19.44) from zero to t yields

$$\begin{aligned} \int_0^t \dot{V}_{i,2}e^{\delta_i\tau}d\tau \le & -\int_0^t (m_{i,1}z^2_{i,1} + m_{i,2}z^2_{i,2})e^{\delta_i\tau}d\tau + \int_0^t \dot{\chi}e^{\delta_i\tau}d\tau + \int_0^t e_i N(\chi)\dot{\chi}d\tau \\ & + \int_0^t \left(\frac{1}{l_{i,3}}\tilde{e}_i\dot{e}_i + \frac{\lambda}{4}\right)e^{\delta_i\tau}d\tau - \frac{1}{4l_{i,1}}\int_0^t \left\|\tilde{x}^a_i\right\|^2 e^{\delta_i\tau}d\tau \end{aligned} \tag{19.46}$$

Then the following inequality can be achieved

$$\begin{aligned} 0 \le V_{i,2}(t) \le & V_{i,2}(0) + e^{\delta_i\tau}\int_0^t e_i N(\chi)\dot{\chi}d\tau + \int_0^t \dot{\chi}e^{-\delta_i(t-\tau)}d\tau \\ & + \frac{\delta_i}{2}\int_0^t \left(\tilde{e}^2_i + \tilde{\lambda}^2\right)e^{-\delta_i(t-\tau)}d\tau + \int_0^t \left(\frac{1}{l_{i,3}}\tilde{e}_i\dot{e}_i + \frac{\lambda}{4}\right)e^{-\delta_i(t-\tau)}d\tau \end{aligned} \tag{19.47}$$

where $\delta_i = \min\left\{2m_{i,1}, 2m_{i,2}, \frac{1}{\|P\|^2}\right\}$.

Note that because of the utilization of projection operator for $\hat{e}_i(t)$ and $\hat{\lambda}$, the boundness of $\tilde{e}_i(t)$ and $\tilde{\lambda}$ can be obtained. Then, incorporating with the boundness of $\dot{e}_i$ and λ, it can be seen that $\frac{\delta_i}{2}\int_0^t\left(\tilde{e}_i^2+\tilde{\lambda}^2\right)e^{-\delta_i(t-\tau)}d\tau+\int_0^t\left(\frac{1}{l_{i,3}}\tilde{e}_i\dot{e}_i+\frac{\lambda}{4}\right)e^{-\delta_i(t-\tau)}d\tau$ is bounded. Thus, according to Lemma 1, $V_{i,2}(t)$, $\dot{\chi}$ and $\int_0^t e_iN(\chi)\dot{\chi}d\tau$ are bounded on $[0,t_f)$. Therefore, $z_{i,1}, z_{i,2}, e_i, \lambda$ and $\tilde{x}_i^a$ are bounded on $[0,t_f)$ for all $t_f>0$; with Eqs. (19.16) and (19.29), it can be shown that all the signals in the closed-loop system are bounded on $[0,t_f)$ for all $t_f>0$. Thus, according to Ryan [26], it can be shown that this conclusion is true for $t_f=+\infty$. Moreover, according to Eq. (19.47), one has

$$\left|z_{i,1}\right|=\left|y_i-u_{nor}^i\right|\le\mu_i \tag{19.48}$$

where

$$\begin{aligned}\mu_i=&V_{i,2}(0)+e^{\delta_i\tau}\int_0^t e_iN(\chi)\dot{\chi}d\tau+\int_0^t\dot{\chi}e^{-\delta_i(t-\tau)}d\tau+\frac{\delta_i}{2}\int_0^t\left(\tilde{e}_i^2+\tilde{\lambda}^2\right)e^{-\delta_i(t-\tau)}d\tau\\&+\int_0^t\left(\frac{1}{l_{i,3}}\tilde{e}_i\dot{e}_i+\frac{\lambda}{4}\right)e^{-\delta_i(t-\tau)}d\tau\end{aligned} \tag{19.49}$$

and the value of μ can be adjusted by appropriately choosing the design parameters. Based upon the above analysis, the following theorem can be concluded.

Theorem 2 Consider the auxiliary system Eq. (19.16) under Assumption 1, with the application of controller in Eq. (19.43) and the update laws in Eq. (19.39), The closed-loop system is bounded input and bounded output stable, and the output y_i can follow the specified signal u_{nor}^i as closely as possible by appropriately choosing the design parameters.

From Eqs. (19.48) and (19.49), it can be noted that the tracking error μ_i can be adjusted to a value as small as possible by appropriately choosing the parameters δ_i, $m_{i,1}$, $m_{i,2}$ and $l_{i,j}\left(j=1,2,3\right)$.

19.3.3 Fault Tolerance Analysis

From Eq. (19.16):

$$u_{act}^i(t)=\int_0^t\int_0^s e_i(\tau)u_{com}^i d\tau ds \tag{19.50}$$

According to Theorem 2 and Eq. (19.48), it is reasonable to suppose that

$$u_{act}^i-u_{nor}^i=\Delta u_i \tag{19.51}$$

where Δu_i satisfies the following inequality

$$\left|\Delta u_i\right|=\left|y_i-u_{nor}^i\right|\le\mu_i \tag{19.52}$$

Then under the effect of actual output u_{act}^{i} of the ith actuator, the spacecraft dynamic can be rewritten as

$$J\dot{\omega} = -\omega^{\times} J\omega + u_{act} + d = -\omega^{\times} J\omega + u_{nor} + \Delta u + d \tag{19.53}$$

where $u_{nor} = [u_{nor}^{1} \quad u_{nor}^{2} \quad u_{nor}^{3}]^{T}$ and $\Delta u = [\Delta u_1 \quad \Delta u_2 \quad \Delta u_3]^{T}$ satisfying $\|\Delta u\| \leq \sqrt{\sum_{i=1}^{3} \mu_i^2}$.

From Eq. (19.50), Eq. (19.53) and the controller developed in Eq. (19.43), the whole attitude control system in the presence of a partial loss of actuator effectiveness can be depicted as in Figure 19.1.

In Eq. (19.53), the item $\Delta u + d$ can be viewed as the lumped disturbances. If the constant T_d is chosen as

$$\|\Delta u\| + \|d\| \leq \sqrt{\sum_{i=1}^{3} \mu_i^2} + \|d\| \leq T_d \tag{19.54}$$

then with the normal controller in Eqs. (19.11) and (19.53), the inequality in Eq. (19.13) can be guaranteed.

Therefore, based on Theorem 1, the following theorem can be obtained.

Theorem 3 Consider the system in Eqs. (19.1) and (19.4) with partial loss of actuator effectiveness. If the design parameter T_d is appropriately chosen to satisfy Eq. (19.54), then with the fault-tolerant controller in Eq. (19.43), the attitude system is globally stable in the sense that all the closed-loop signals are bounded. Moreover, the attitude q and the angular velocity ω will ultimately converge to an arbitrary set containing zero; that is, the objectives (a) and (b) as stated in Section 19.2 can be achieved.

Proof: This can easily be obtained from above analysis and therefore the proof is omitted here.

From the above analysis, because the matrix E is not used in the control scheme, there is no need to include a health monitoring unit to identify or estimate which actuator is unhealthy. The fault accommodation/compensation is done automatically and adaptively by the control algorithm. This feature is necessary to build affordable and effective fault-tolerant spacecraft control schemes.

19.4 Numerical Simulations

In this section, the simulation results obtained using the control law in Eq. (19.43) are compared with conventional proportional-derivative (PD) control, first for normal conditions and then then in the presence of the fault. The model parameters for the rigid spacecraft are chosen [15] as $J = \begin{pmatrix} 20 & 0 & 0.9 \\ 0 & 17 & 0 \\ 0.9 & 0 & 15 \end{pmatrix}$ kgm^2 and the initial conditions Are set as $Q(0) = [0.9631 \quad -0.1 \quad 0.15 \quad -0.2]^{T}$ and $\omega(0) = [0 \quad 0 \quad 0]^{T}$ rad/s.

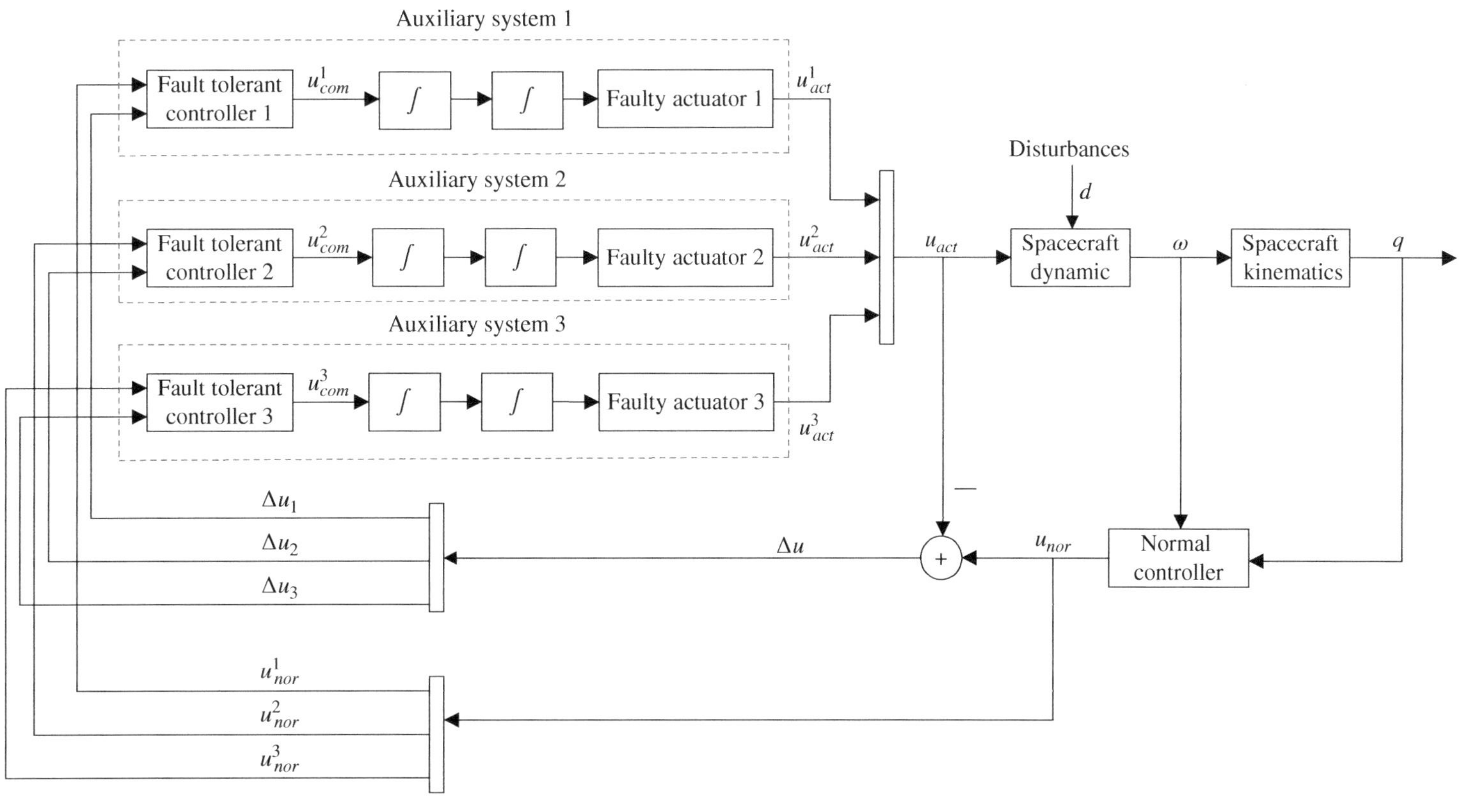

Figure 19.1 Fault-tolerant attitude control system under partial loss of actuator effectiveness.

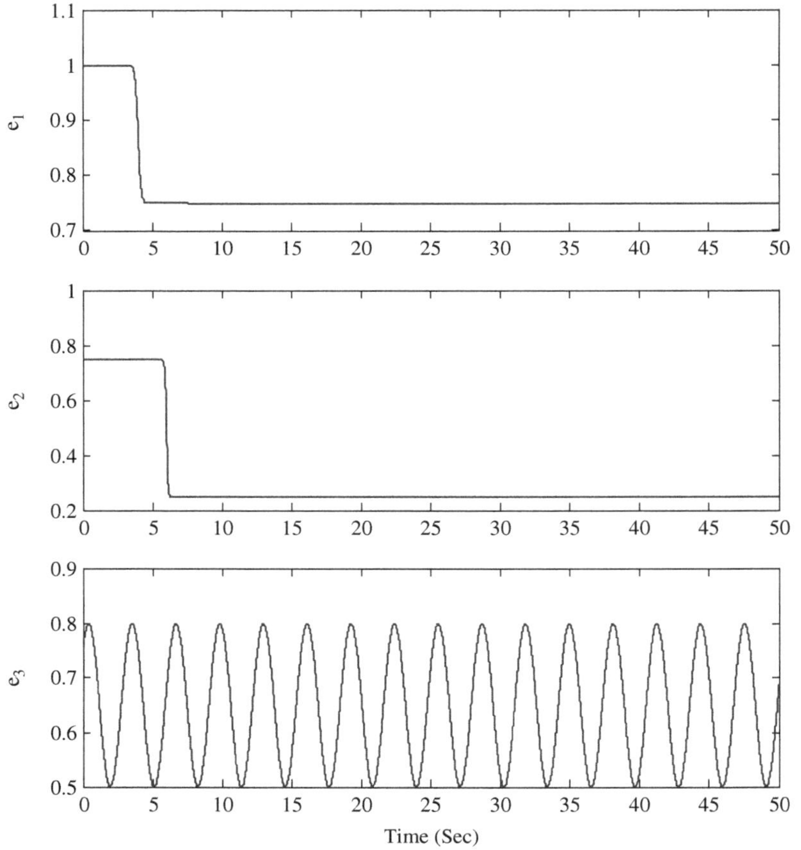

Figure 19.2 Time history of fault leading to partial loss of actuator effectiveness *E*(*t*).

In the simulation, the external disturbance torque is assumed to be $d = (|\omega|^2 + 0.05)[\sin 0.8t \ \ \cos 0.5t \ \ \cos 0.3t]^T$ (Nm). In the actuator fault situation, the fault $E(t) = diag\left(e_1(t), e_2(t), e_3(t)\right)$ is assumed to be as in Eq. (19.55):

$$\begin{cases} e_1(t) = 0.875 + 0.125\tanh\left(-5(t-4)\right) \\ e_2(t) = 0.5 + 0.25\tanh\left(-10(t-6)\right) \\ e_3(t) = 0.65 + 0.15\sin\left(2t + \pi/2\right) \end{cases} \tag{19.55}$$

The time history of the fault is shown in Figure 19.2.

19.4.1 Simulations of PD Control under Normal and Fault Cases

In this simulation, a PD controller is implemented as the attitude stabilization control system, and it is of the form:

$$u = \omega^{\times} J\omega - K_d\omega - K_p q \tag{19.56}$$

where design parameters $K_p = 45$ and $K_d = 20$ are selected in the simulation. The time histories of attitude quaternion, angular velocity and control input are shown in Figures 19.3a, 19.4a and 19.5a, respectively. We can clearly see that although the PD controller can stabilize the normal attitude within 8 s, it fails to stabilize the spacecraft when a fault leads to a partial loss of actuator effectiveness. Despite the fact that there is still some room for improvement with different design control parameter sets, there is not much improvement in the attitude and velocity responses.

19.4.2 Simulations of Proposed Control under Normal and Fault Cases

To show the effect of the proposed fault-tolerant controller in Eq. (19.43), a simulation was performed under the same initial condition and fault scenario. In the simulation, the control parameters for the normal controller Eq. (19.11) were chosen as $c_1 = 5$,

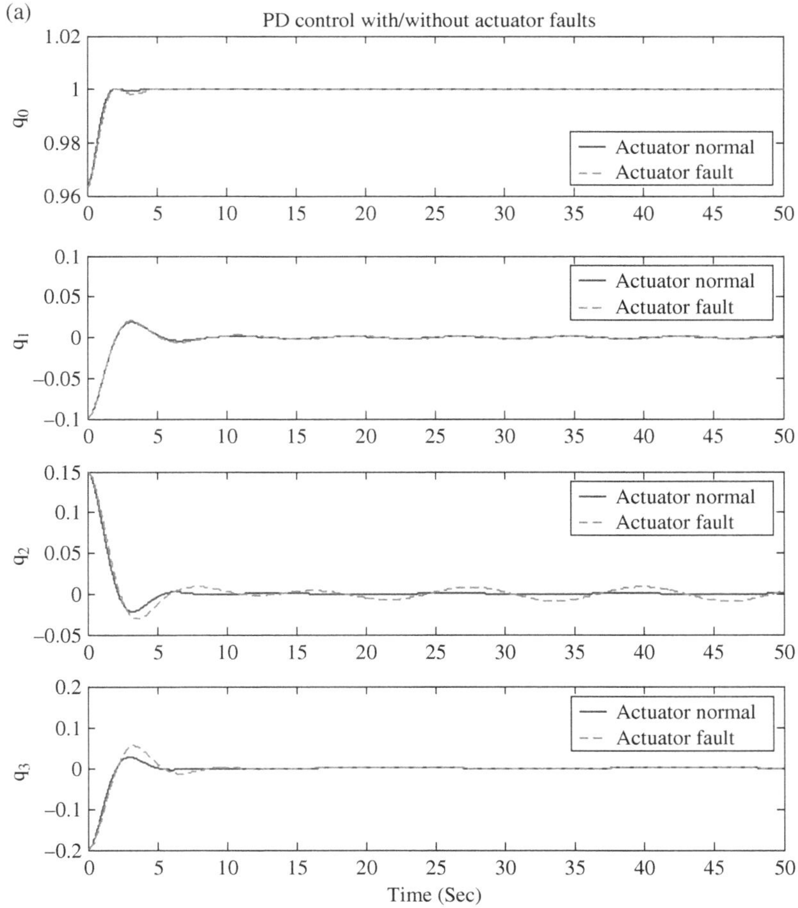

Figure 19.3 Time histories of attitude quaternion.

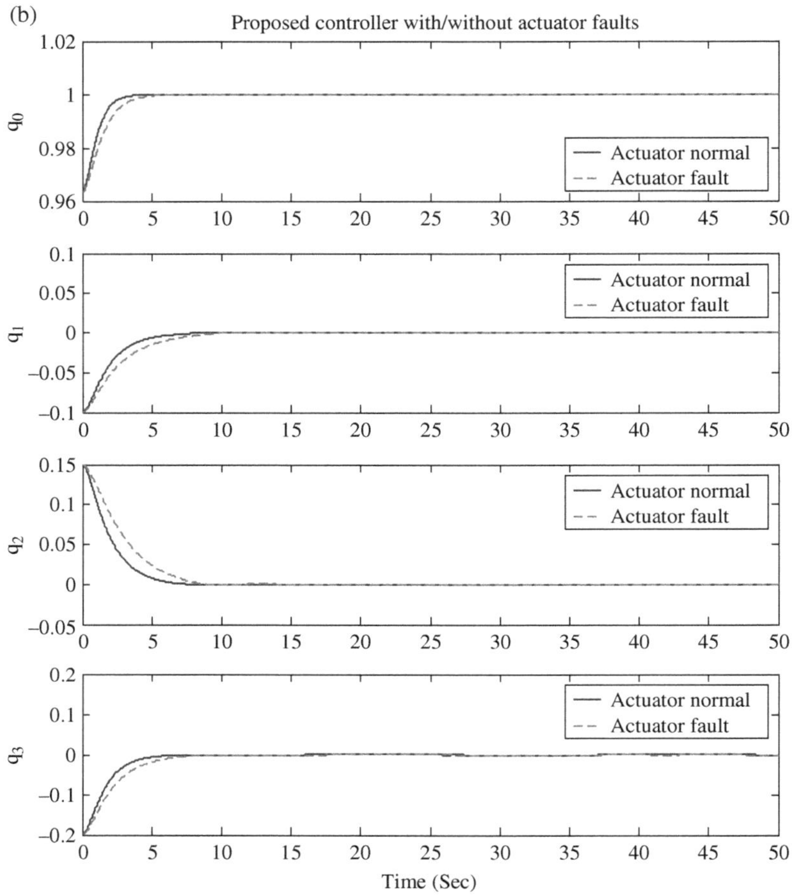

Figure 19.3 (*Continued*)

$c_2 = 20$, $T_d = 1$, $\varepsilon = 0.5$ and $\beta = 2$. For the fault-tolerant controller, Eq. (19.43), the matrix P_i and the design parameters were set as:

$$P_i = \begin{bmatrix} \frac{1}{k_{i,1}} + \frac{k_{i,2}}{k_{i,1}} & -1 \\ -1 & \frac{1}{k_{i,1}} + \frac{k_{i,2}}{k_{i,1}} + \frac{1}{k_{i,1}k_{i,2}} \end{bmatrix}, m_{i,1} = m_{i,2} = 5, k_{i,1} = 4, k_{i,2} = 2.5, l_{i,1} = 1, l_{i,2} = 4 \ and \ l_{i,3} = 2.$$

The time responses of the closed-loop system under the fault-tolerant controller in Eq. (19.43) are shown in Figures 19.3b, 19.4b and 19.5b. Attitude stabilization can be achieved even with a fault leading to an unknown partial loss of actuator effectiveness. As expected, the attitude and the angular velocity will converge to an arbitrary small set containing the origin, as shown in Figures 19.3b and 19.4b. respectively.

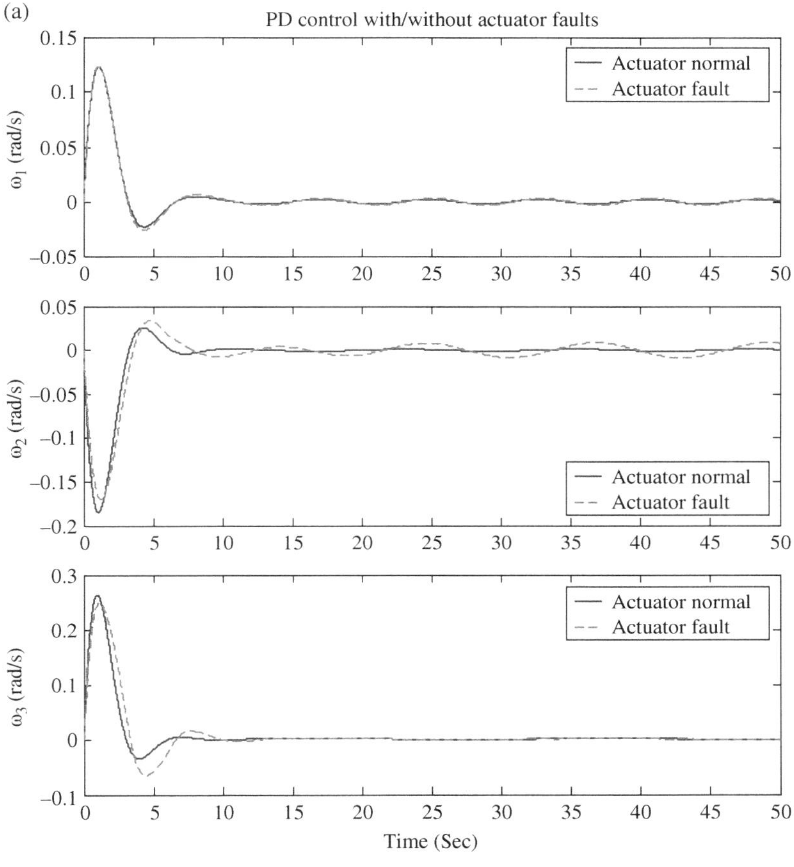

Figure 19.4 Time histories of angular velocity.

The control input error Δu between the actual input u_{act} and the normal control signals u_{nor} is shown in Figure 19.6. The maximum value of Δu is smaller than 0.7. This together with selection of external disturbance d, leads to $\|\Delta u\| + \|d\| < 1$. Thus, $T_d = 1$ is sufficiently large to deal with lumped disturbances, and so faults can be successfully managed with our proposed controller Eq. (19.43).

Summarizing both normal and fault cases, it should be noted that the proposed controllers have significantly better performance than the PD method in both theory and simulations. Also, in the presence of faults, the proposed methods are better than conventional controllers. In addition, extensive simulations were also done using different control parameters, disturbance inputs and even combinations of actuator faults. These results show that in the closed-loop system attitude stabilization is accomplished in spite of these undesired effects. Moreover, the flexibility in the choice of control parameters can be utilized to obtain any desirable performance. These control approaches provide the theoretical basis for the practical application of advanced control theory to flexible spacecraft attitude control system.

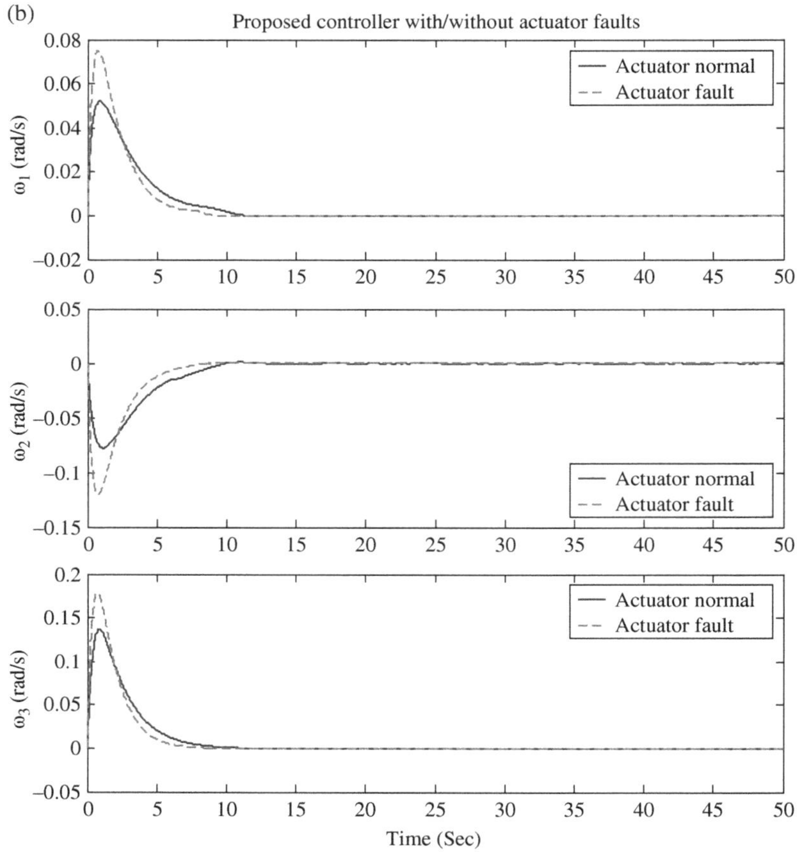

Figure 19.4 *(Continued)*

19.5 Conclusion

A fault-tolerant adaptive backstepping control scheme has been developed for spacecraft attitude stabilization in the presence of both an external disturbance and an unknown partial loss of actuator effectiveness. The proposed scheme does not require the system identification process to identify the faults. The control formulation is based upon Lyapunov's direct stability theorem, incorporating the auxiliary system of the actual control output into the controller synthesis to compensate the actuator faults. The uniform ultimate bounded stability of the system is ensured and its robustness to both disturbance and actuator faults is also guaranteed.

The control designs are evaluated using numerical simulation and comparisons between the developed approach and other referred schemes, with the expected performance of the two being shown. In the simulation, different types of actuator failure scenarios were investigated. Based upon the results presented, it is concluded that the fault-tolerant adaptive backstepping control scheme successfully handles failures if the efficiency of one or several actuators decreases. While the simulation results presented

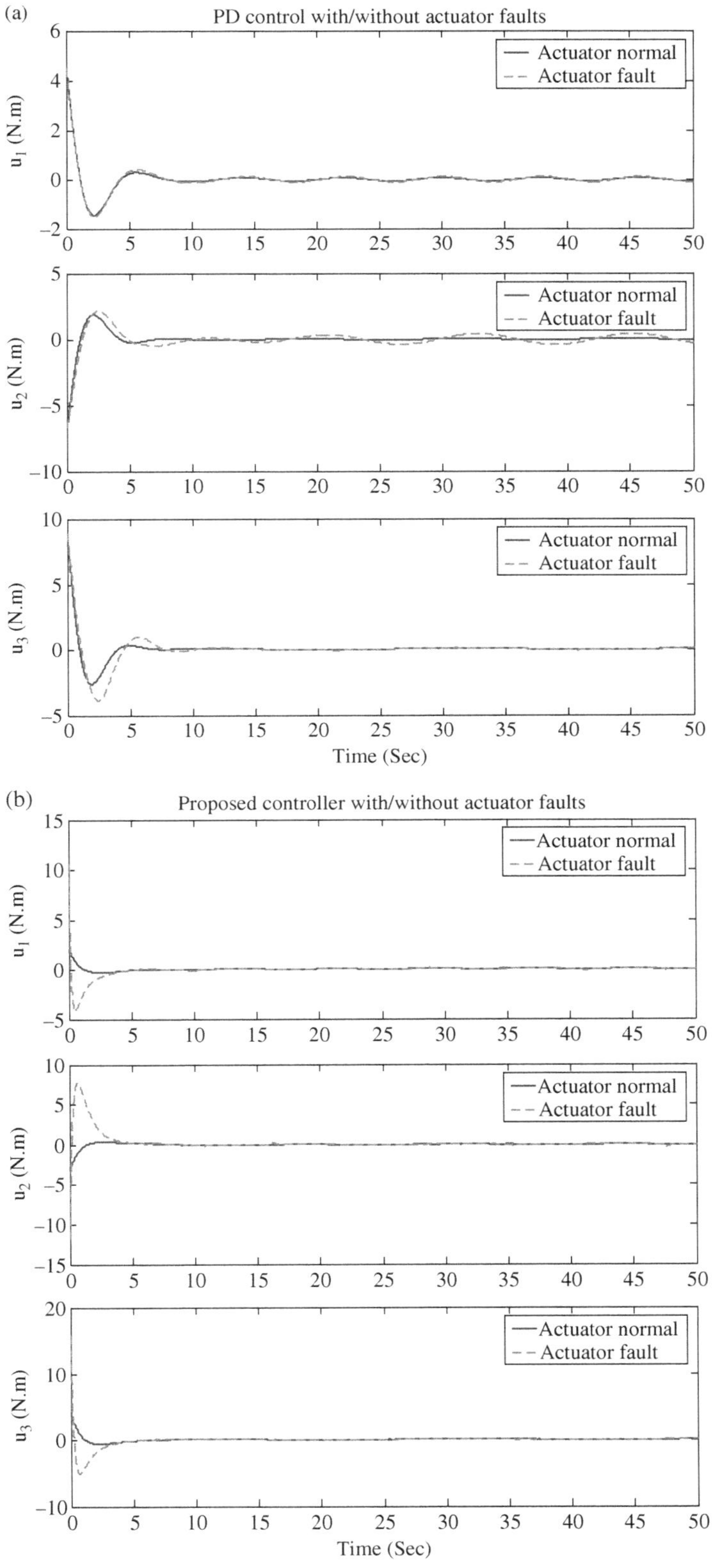

Figure 19.5 Time histories of control input.

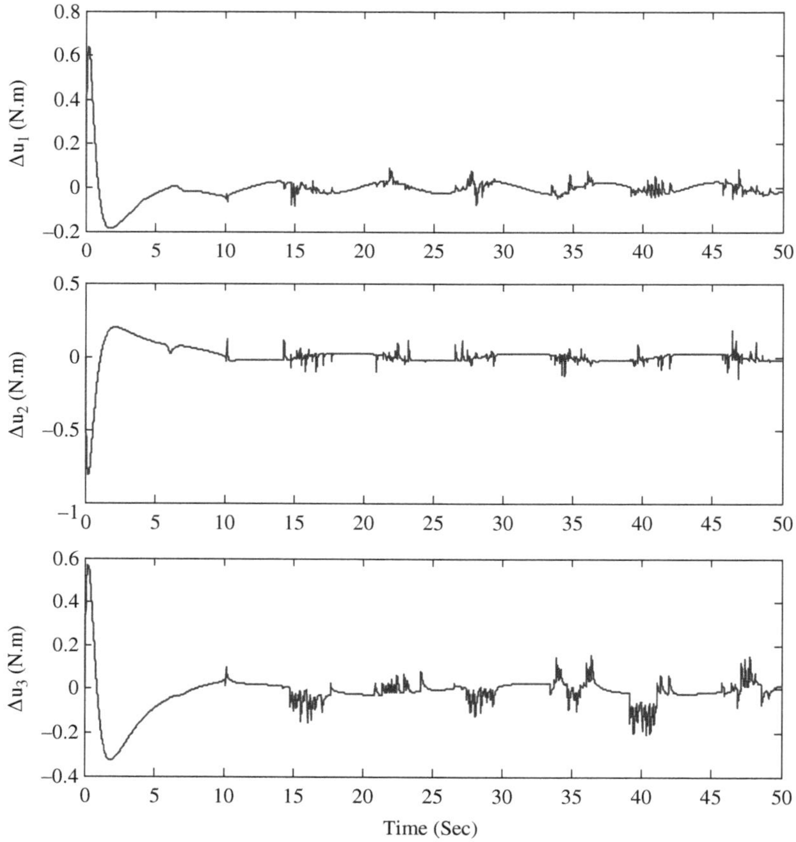

Figure 19.6 Time histories of tracking error Δu.

here merely illustrate formulations for attitude stabilization, further testing would be required to reach any conclusions about the efficacy of the control and adaptation laws for tracking arbitrary manoeuvres. In addition, this fault-tolerant control scheme places no restrictions on the magnitude of the desired control, and designs where actuator limits are specifically considered are also being investigated.

References

1 Bai X.L. and Junkins J.L., 'New results for time-optimal three-axis reorientation of a rigid spacecraft,' *Journal of Guidance Control and Dynamics*, **32** (4), 2009, pp. 1071–1076.
2 Sharma R. and Tewari A., 'Optimal nonlinear tracking of spacecraft attitude maneuvers,' *IEEE Transactions on Control Systems Technology*, **12** (5), 2004, pp. 677–682.
3 Hu Q.L., 'Robust adaptive sliding mode attitude maneuvering and vibration damping of three-axis-stabilized flexible spacecraft with actuator saturation limits,' *Nonlinear Dynamics*, **55** (3), 2009, pp. 301–321.

4 Zheng Q. and Wu F., 'Nonlinear H-infinity control designs with asymmetric spacecraft control,' *Journal of Guidance Control and Dynamics*, **32** (3), 2009, pp. 850–859.
5 Li Z.X. and Wang B.L., 'Robust attitude tracking control of spacecraft in the presence of disturbances,' *Journal of Guidance Control and Dynamics*, **30** (4), 2007, pp. 1156–1159.
6 Zhang Y.M. and Jiang J., 'Bibliographical review on reconfigurable fault-tolerant control systems,' *Annual Reviews in Control*, **32** (2), 2008, pp. 229–252.
7 Benosman M., 'A survey of some recent results on nonlinear fault tolerant control,' *Mathematical Problems in Engineering*, **2010**, pp. 1–25.
8 Zhang X.D., Parisini T., and Polycarpou M.M., 'Adaptive fault-tolerant control of nonlinear uncertain systems: an information-based diagnostic approach,' *IEEE Transactions on Automatic Control*, **49** (8), 2004, pp. 1259–1274.
9 Zhang Y.W. and Qin S.J., 'Adaptive actuator fault compensation for linear systems with matching and unmatching uncertainties,' *Journal of Process Control*, **19** (6), 2009, pp. 985–990.
10 Wang M. and Zhou D.H., 'Fault tolerant control of feedback linearizable systems with stuck actuators,' *Asian Journal of Control*, **10** (1), 2008, pp. 74–87.
11 Boskovic J.D., Jackson J.A., Mehra R.K., and Nguyen N.T., 'Multiple-model adaptive fault-tolerant control of a planetary lander,' *Journal of Guidance Control and Dynamics*, **32** (6), 2009, pp. 1812–1826.
12 Jung B., Kim Y. and Ha C., 'Fault tolerant flight control system design using a multiple model adaptive controller,' *Proceedings of the Institution of Mechanical Engineers Part G – Journal of Aerospace Engineering*, **223** (1), 2009, pp. 39–50.
13 Lombaerts T.J.J., Smaili M.H., Stroosma O., et al., 'Piloted simulator evaluation results of new fault-tolerant flight control algorithm,' *Journal of Guidance Control and Dynamics*, **32** (6), 2009, pp. 1747–1765.
14 Ye S.J., Zhang Y.M., Wang X.M. and Rabbath C.A., 'Robust fault-tolerant control using on-line control reallocation with application to aircraft,' *Proceedings of American Control Conference*, New York, 2009, pp. 5534–5539.
15 Cai W.C., Liao X.H., and Song Y.D., 'Indirect robust adaptive fault-tolerant control for attitude tracking of spacecraft,' *Journal of Guidance Control and Dynamics*, **31** (5), 2008, pp. 1456–1463.
16 Ji J., Ko S., and Ryoo C.K., 'Fault tolerant control for satellites with four reaction wheels,' *Control Engineering Practice*, **16** (10), 2008, pp. 1250–1258.
17 Jiang Y., Hu Q.L. and Ma G.F., 'Adaptive backstepping fault-tolerant control for flexible spacecraft with unknown bounded disturbances and actuator failures,' *ISA Transactions*, **49** (1), 2010, pp. 57–69.
18 Godard, K.D. Kumar, and B. Tan, 'Fault-tolerant stabilization of a tethered satellite system using offset control,' *Journal of Spacecraft and Rockets*, **45** (5), 2008, pp. 1070–1084.
19 Zhang T.P. and Ge S.Z. S., 'Adaptive neural control of MIMO Nonlinear state time-varying delay systems with unknown dead-zones and gain signs,' *Automatica*, **43** (6), 2007, pp. 1021–1033.
20 Marino R. and Tomei P., 'Adaptive control of linear time-varying systems,' *Automatica*, **39** (4), 2003, pp. 651–659.
21 Zhang Y.P., Fidan B., and Ioannou P.A., 'Backstepping control of linear time-varying systems with known and unknown parameters,' *IEEE Transactions on Automatic Control*, **48** (6), 2003, pp. 1908–1925.

22 Zhou J., Wen C., and Zhang Y., 'Adaptive output control of a class of time-varying uncertain nonlinear systems,' *Journal of Nonlinear Dynamic and System Theory*, **5** (3), 2005, pp. 285–298.
23 Sidi M.J., *Spacecraft Dynamics and Control*, Cambridge University Press, 1997, pp. 25–40.
24 Ge S.Z. S., Hong F., and Lee T.H., 'Adaptive neural control of nonlinear time-delay systems with unknown virtual control coefficients,' *IEEE Transactions on Systems Man and Cybernetics Part B-Cybernetics*, **34** (1), 2004, pp. 499–516.
25 M. Krstic, I. Kanellakopoulos, and P.V. Kokotovic, *Nonlinear and Adaptive Control Design*, Wiley, 1995, pp. 40–65.
26 Ryan E.P., 'A universal adaptive stabilizer for a class of nonlinear systems,' *Systems and Control Letters*, **16** (3), 1991, pp. 209–218.

2.2 Quad-rotor Aircraft

20

Novel Concepts in Multi-rotor VTOL UAV Dynamics and Stability

Emaid A. Abdul Retha

Unmanned Vehicle University, Phoenix, AZ, USA

20.1 Introduction

The multi-rotor is an unmanned aerial vehicle (UAV) lifted and propelled by two or more motors, usually electric, with propellers. The multi-copter (as it is also called) is classified as a rotary UAV, which in contrast to fixed-wing UAVs uses multiple rotating profiles or propellers for flight. This vehicle is characterized by a simple design with control of the craft delivered by varying the rotation of its rotors. In recent years, the perception of the uses of the multi-rotor has increased due to its flexibility. It can to meet many of the requirements of vertical take-off and landing (VTOL) UAVs for applications in reconnaissance and surveillance.

The different sites that need reconnaissance, surveillance and monitoring demand changes in the shape and configuration of VTOL UAVs, so as to guarantee operations in each set of special circumstances; for example, in narrow areas between buildings, in dense forests, in high winds and in areas of critical infrastructure with difficult and dangerous access. Effective multi-rotor electromechanical design modifications can increase the under-actuation of conventional multi-rotors, leading to an increase in manoeuvrability and the ability to resist wind and fly in tight spaces.

The unmanned aerial system (UAS) industry is highly technological, and defines the development level of any state. The existence of a UAS industry requires development of the country's infrastructure, including professional aviation educational institutions, high technology industries and highly skilled personnel in the industrial sphere. UASs are a promising and dynamic industry. UASs are in demand and are essential to cover a wide range of applications (military, security and civilian). UAS markets are very promising, and businesses involved can reach breakeven in reasonable periods.

UASs' small size, remote or automatic control, low energy consumption, high safety in emergency situations, as well as the ability to use electric and alternative power sources, are characteristic.

Advanced UAV Aerodynamics, Flight Stability and Control: Novel Concepts, Theory and Applications,
First Edition. Edited by Pascual Marqués and Andrea Da Ronch.

Companion website: http://www.wiley.com/go/marques/advanced_UAV_aerodynamics

The multi-rotor VTOL UAV is a vehicle with many lifting rotors. Multi-rotors are often equipped with fixed-pitch propellers; control is achieved by varying the revolutions per minute (RPM) of each rotor. A quad-rotor is a multi-rotor with four rotors. A typical quad-rotor is electronically controlled; there are also novel types that use additional means of control and these are referred to as electromechanically controlled quads. These will be the focus of this chapter, in which we will analyse novel concepts in multi-rotor dynamics and stability, such as tilting rotors, variable pitch control and flap vector thrust, and consider how these concepts can be used to improve the quad-rotor's performance.

20.1.1 Motivation

Multi-rotors in general, and the quad-rotor in particular, have problems with their under-actuation and autonomy. The problem of under-actuation is related to the fact that the quad-rotor has six degrees of freedom, but only four actuations (in other words, lift from the four variable rotors due to their variable RPM). This shortage is dealt with by increasing the actuations by adding electro-mechanical controls. The energy required is stored in batteries, which are heavy and therefore limit the autonomy of the quad-rotor. To solve this problem, a hybrid propulsion capability is required and the use of electro-mechanical design allows this to happen, enhancing the vehicle's autonomy.

Adding electro-mechanical concepts to a multi-rotor has many advantages:

- The quad-rotor becomes a fully-controlled system which can track any arbitrary trajectory; hovering with controlled pitch and roll angle, and motion with desired orientation.
- Navigation around obstacles in a cluttered environment is possible.
- Manoeuvring in only the horizontal plane is possible, useful for obstacle avoidance [1]; see Figure 20.1.
- Multi-rotor input coupling can be achieved.
- The advantages of a 45° control method due to simultaneous cross-coupling control of longitudinal, lateral in horizontal to all directions quad-rotor's movements can be demonstrated.

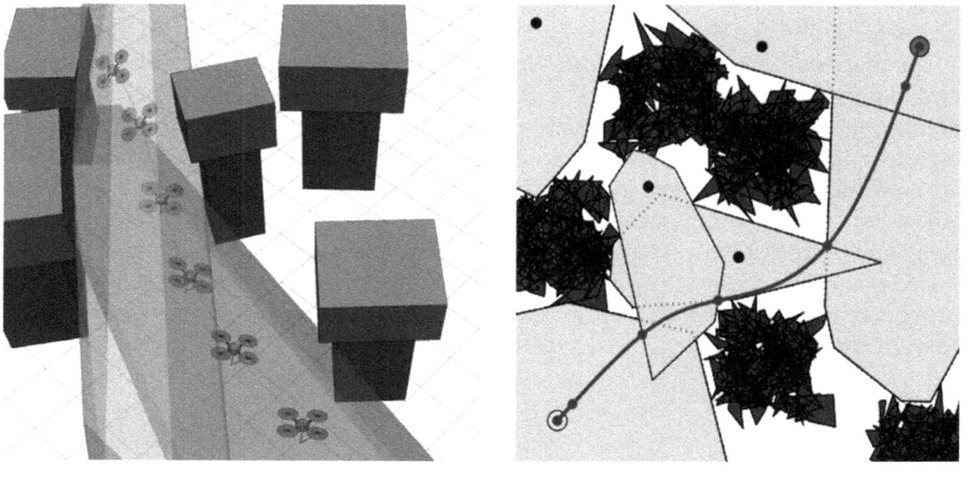

Figure 20.1 Quad-rotor obstacle avoidance.

For a quad-rotor to hover in the wind or move opposite to the wind's direction, it is necessary to equal or overcome the drag forces: this happens by tilting the quad-rotor's body into the downwind direction. Logically, the higher the wind speed the higher the quad-rotor tilting required; the tilting will keep increasing until the saturation of two motors opposite the wind direction is reached (they reach their maximum RPM). Therefore, in this chapter we will discuss this case and the use of electromechanical control: rotor tilting with either flap thrust-vectoring control or with RPM control.

20.1.2 Review of Quad-rotor Developments

In the last few decades, small-scale UAVs have become common in many applications. Since the 1990s, with the development of microelectromechanical system (MEMS) technology, the multi-rotor vehicle, especially the quad-rotor vehicle, have been widely studied. The need for a UAV with greater manoeuvrability and hovering ability has led to the current rise in quad-rotor research. The four-rotor design allows quad-rotors to be relatively simple in design, yet highly reliable and manoeuvrable. Cutting-edge research is continuing to increase the viability of quad-rotors by making advances in multi-craft communication, environment exploration and manoeuvrability. If all of these developments can be combined, quad-rotors will become capable of advanced autonomous missions that are currently not possible.

20.1.3 Contributions

The objective of this chapter is to highlight novel concepts in multi-rotor dynamics and stability. It looks at a particular well-developed quad-rotor, which has been modified with specific electromechanical rotor controls to enhance its flying performance. The chapter will give:

- a full technical description of the modified quad-rotor to overview the basic concepts
- a total mathematical model of the quad-rotor, deriving kinetics and dynamics with regards to the addition of the electromechanical systems
- the basics of tilting quad-rotor control, the control loops for each part of the control system and a state-of-the-art mixing mechanism for tilting rotors (what is called a thrust-vectoring quad-rotor).

20.1.4 Chapter Overview

After this introduction, Section 20.2 is about general concepts of multi-rotors, with a particular focus on the quad-rotor, its mathematical modelling and its applications. Section 20.3 describes novel quad-rotor concepts, types and designs, and the modelling and applications of electromechanical multi-rotors. Section 20.4 provides some conclusions.

20.2 Multi-rotors

The interest in UAVs for military and civil applications is growing fast; cheaper and more capable machines are in demand. Among the multi-rotor layouts, the quad-rotor has been widely chosen by many researchers as a promising vehicle for indoor and outdoor navigation. Multidisciplinary concepts are necessary because this type of rotorcraft attempts to achieve stable hovering and precise flight by balancing the forces produced by the four rotors. Nowadays, the design of multi-copters with more than four rotors – the hexa-copter and octo-copter – is developing thanks to the possibility they bring of managing one or more engine failures and of increasing the total payload.

20.2.1 General Concept

The electronically controlled multi-rotor, such as the quad-rotor, has the components shown in Figure 20.2.

Electronically controlled multi-rotors have several configurations (Figure 20.3 and Table 20.1).

Figure 20.2 Multi-rotor and its components: (a) propellers are usually fixed-pitch; (b) airframe; (c) brushless motors are mounted rigidly to the airframe structure; (d) flight controller, stabilization board (containing an IMU, mainly mounted in a core hub section, communications, other assisting equipment, and parts for multi rotor control, stabilization and navigation); (e) landing gear; (f) payload; (g) radio control; (h) power (lithium polymer batteries); (i) ground station; (j) communications; (k) accessories.

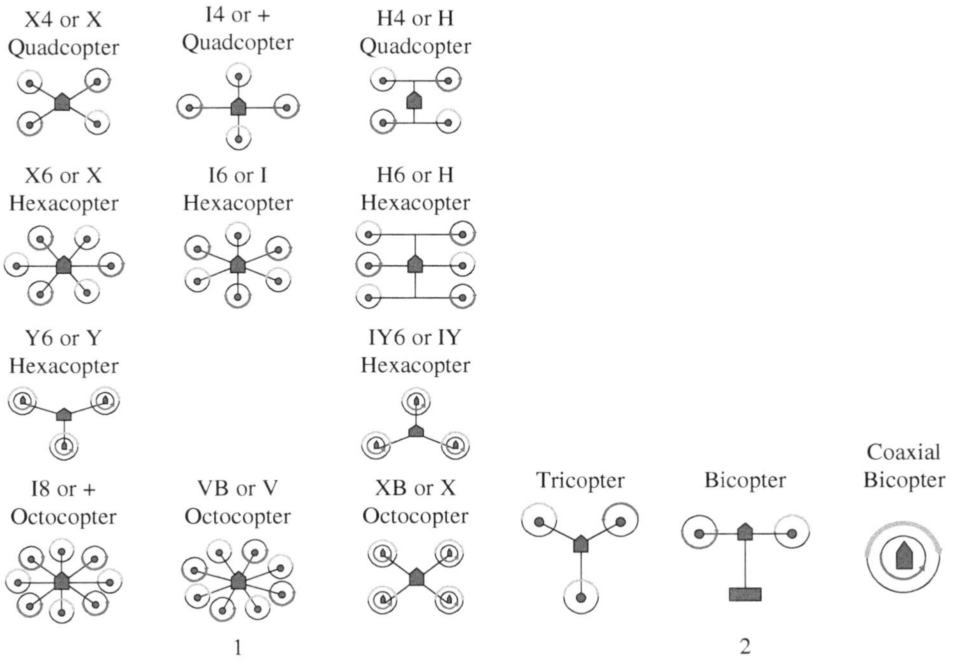

Figure 20.3 Multi rotor configurations: (1) conventional; (2) unconventional.

Table 20.1 Multi-rotor configurations: advantages and disadvantages.

Type	Motors	Coaxial	Advantages	Disadvantages
X4 or X	4	No	Simple and cheap	No redundancy
I4 or +	4	No	Simple and cheap	no redundancy
H4 or H	4	No	Simple and cheap	no redundancy
X6 or X	6	No	Limited redundancy and larger payload	Larger footprint and pricier
I6 or I	6	No	Limited redundancy and larger payload	Larger footprint and pricier
H6 or H	6	No	Limited redundancy and larger payload	Larger footprint and pricier
Y6 or Y	6	Yes	Small, high stability and wind resistance	Poor efficiency and complex
IY6 or IY	6	Yes	Small, high stability and wind resistance	Poor efficiency and complex
I8 or +	8	No	True redundancy and horsepower	large and expensive
V8 or V	8	No	True redundancy and horsepower	large and expensive
X8 or X	8	Yes	Very high lift capacity and wind resistance	Inefficient

20.2.2 Quad-rotor

Multi-rotors, and in particular quad-rotor configurations, have achieved their increasing popularity because of their mechanical simplicity, low cost and the availability of reliable components off the shelf. This makes them ideal for many mission scenarios. A quad-rotor is regarded as a base design for multi-rotor VTOL UAVs. Unlike most helicopters, quad-rotors use two sets of identical fixed-pitched propellers; two clockwise and two counter-clockwise. Propellers use variation of RPM to control lift and torques, which are generated aerodynamically due to the propeller blades' rotation. Control of vehicle motion is achieved by altering the rotation rate of two or more rotors, thereby changing the torque load and thrust/lift characteristics. More recently, quad-rotors have become popular in UAV research. These vehicles use an electronic control system and electronic sensors to stabilize the UAV. With their small size and agile manoeuvrability, these quad-rotors can fly indoors as well as outdoors. Flight control in electronically controlled quad-rotors is achieved using a minimum of four control channels. One channel is usually the throttle, and increases or decreases power to all motors equally. This causes the aircraft to ascend or descend. The other three channels (as in radio control, called the aileron, elevator and rudder) control the roll, pitch, and yaw axes, respectively (Figure 20.4). These three control inputs work by causing a change in the quad-rotor's attitude (tilt or direction).

The main mechanical components needed for construction are the airframe, propellers (either fixed-pitch or variable-pitch) and the electric motors. For the best performance and simplest control algorithms, the motors and propellers should be placed equidistant. Recently, carbon fibre composites have become popular due to their light weight and structural stiffness. The electrical components needed to construct a working quad-rotor are similar to those needed for a modern radio-controlled (RC) electric helicopter. These are the electronic speed control module, on-board computer or controller board and the batteries. Typically, an RC transmitter is also used to allow for human operator inputs.

Quad- and other multi-rotors can often fly autonomously. Many modern flight controllers use software that allows the use of GPS to mark 'way-points' on a map, to which the quad-rotor will fly and perform tasks, such as landing or gaining altitude. The

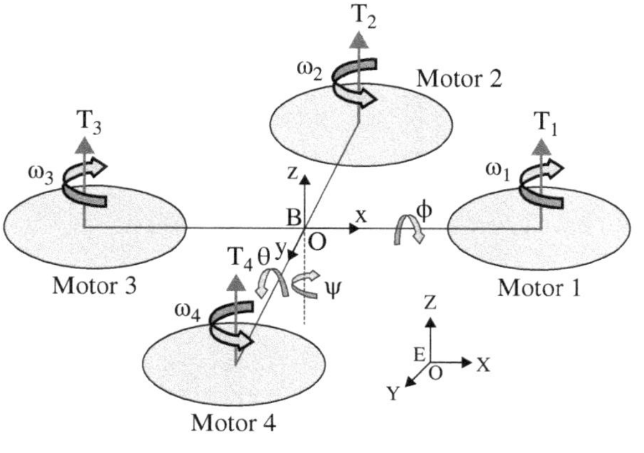

Figure 20.4 Quad-rotor frames, forces, torques and control.

mathematical model of such an aerial vehicle represents a dynamical system with a 12-dimensional state and 4-dimensional inputs control.

20.2.3 Quad-rotor Mathematical Modelling

For the mathematical modelling of novel concepts it is essential to start with a conventional quad-rotor and then analyse electromechanical quad-rotors as a development of them. By varying the motors' speed and direction of rotation, the quad-rotor can change its position and orientation. There are 12 (four groups × three parameters) states that describe the quad-rotor's dynamic behaviour:

- space position: $P = [x\ y\ z]$
- linear velocity: $V = [u_x\ v_y\ w_z]$
- rotational angles: $\Omega = [\varphi\ \theta\ \psi]$
- angular velocities: $\omega = [p_\varphi\ q_\theta\ r_\psi]$

These values can be considered as the quad-rotor's outputs while the inputs are the applied forces and torques generated from the four motors' rotation. The control model is shown in Figure 20.5. This shows the inputs and outputs for the quad-rotor plant as a whole mechanical system. Generally, Newton's second law of motion is working to represent the motion of the quad-rotor [2].

Therefore, the net force and the net moment acting on the quad-rotor are written as:

$$Fnet = d/dt\left[mV\right] + \omega\, x\left[mV\right] \tag{20.1}$$

$$Mnet = d/dt\left[I\,\omega\right] + \omega\, x\left[I\,\omega\right] \tag{20.2}$$

where m is the quad-rotor mass and I is the inertia matrix.

For a quad-rotor, the lift force is distributed in four shares, one to each rotor and these forces are transmitted to the propellers (F_1, F_2, F_3 and F_4) according to the motor numbering (Figure 20.4). To find the values of these forces, the blade element momentum theory is used and they are a function of the rotation speeds of the propellers (denoted ω_1, ω_2, ω_3 and ω_4, to match the motors). The relationship used is $F_i = k\,\omega_i^2$, $i = 1, 2, 3, 4$), where k is a proportionality coefficient). The thrust values depend on the propeller shape and air density, and so on. Other values affect the quad-rotor, such as its mass (m) and weight (mg). It is natural that moments accompany propeller rotation (denoted M_1, M_2, M_3 and M_4), and their value is calculated as $M_i = L \times F_i$, $i = 1, 2, 3, 4$, where L is the arm distance from the rotor axis to quad-rotor centre of gravity (CG). It is very important to know the forces and moments to define the quad-rotor modelling specifications, and to find

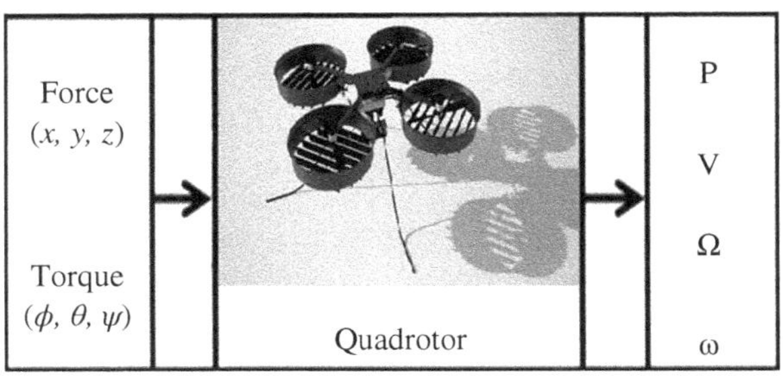

Figure 20.5 Quad-rotor control.

the dynamic changes of these forces and moments is the foundation of quad-rotor dynamic mathematical modelling, which in turn is the basis of quad-rotor design and control [3].

The most important features of quad-rotor design are:

1) The reference systems (Figure 20.4), of which there are two:
   - The Earth reference systems, or the Earth frame (x_E, y_E, z_E).
   - The quad-rotor reference system, or the body frame (x_B, y_B, z_B).
2) The Euler angles(φ, θ, ψ), which define the transformation between the two systems:
   - Roll φ: angle of rotation along axis x_B to x_E.
   - Pitch θ: angle of rotation along axis y_B to y_E.
   - Yaw ψ: angle of rotation along axis z_B to z_E.
3) The angular speeds: the derivatives of (φ, θ, ψ) with respect to time are the angular rotation speeds of the system:
   - $\dot{\varphi}$, Roll rate
   - $\dot{\theta}$, Pitch rate
   - $\dot{\psi}$, Yaw rate.

Analysis of quad-rotor flight requires an elaborate mathematical model. The model needs to reflect the quad-rotor's behaviour in different flight phases (climb, descent, forward flight, manoeuvres, and so on). The mathematical model incorporates body motion dynamics and propulsion system aerodynamics. The propulsion system aerodynamics modelling tasks include: momentum theory, blade element theory, ground effect, vortex ring state and windmill break state. We need to know the quad-rotor forces and torques, reference systems, Euler angles and angular speeds. We can then start assimilation of the process: the analysis, design and writing of the quad-rotor mathematical model (the equations of motion). Analysis will be carried out on the various phases of the quad-rotor's flight.

Hovering over a point Four conditions must be fulfilled:

- equilibrium of thrust: the sum of the rotor thrusts must equal the quad-rotor's weight
- no directional motion: the rotor thrusts are identical in each other's direction as well as in the direction of gravity.
- balance of moments: the sum of moments equals zero.
- the sum of rotor rotation speeds equals zero:

$$\left(\omega 1+\omega 3\right)-\left(\omega 2+\omega 4\right)=0 \tag{20.3}$$

Thus, the following conditions must be achieved: $\dot{\varphi} = 0$, $\dot{\theta} = 0$, $\dot{\psi} = 0$; $\varphi = 0$, $\theta = 0$, $\psi = 0$.

Manoeuvres in the vertical direction There must be no equilibrium of thrust. The sum of the rotor thrusts does not equal the quad-rotor weight. Euler angles and rates must also remain equal to zero.

Yaw rotation The sum of rotors rotation speeds is not equal to zero:

$$\left(\omega 1+\omega 2\right)-\left(\omega 3+\omega 4\right)\neq 0 \tag{20.4}$$

As a consequence: $\psi = \int \dot{\psi}\, dt$.

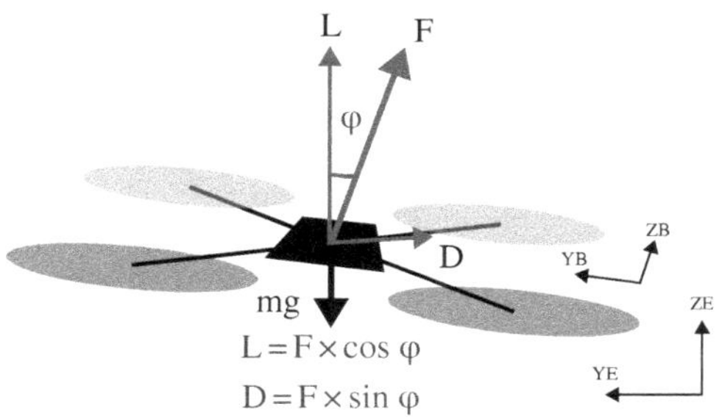

Figure 20.6 Quad-rotor roll rotation and translated flight.

Roll rotation To achieve quad-rotor rolling we have to disturb the moment's equilibrium, which is achieved by unbalancing propeller speeds:

$$\left(\omega 1+\omega 4\right)-\left(\omega 2+\omega 3\right)\neq 0 \tag{20.5}$$

So $\varphi = \int \dot{\varphi}\, dt$. This also means that the direction of all rotor lift forces does not act parallel to the direction of gravity g (no equilibrium of directions), so the quad-rotor undergoes a roll through some angle. The roll rotation and translated flight means that the total thrust emanating from the quad-rotor CG can be decomposed into a lift force = thrust × cos φ and drag force = thrust × sin φ. It is necessary to note here that in a roll manoeuver with further transition the quad-rotor will slip down due to the lack of lifting thrust (the deficiency equal to $F \times \sin \varphi$). To avoid diving it is necessary that lift force = thrust × cos φ = $-mg$; so in translational flight we need more power to achieve hover or yawing (Figure 20.6). At this nominal hover state, the force produced from each propeller must satisfy:

$$F_i = \tfrac{1}{4}\left(mg\right) \tag{20.6}$$

where $F_i = k\,\omega_i^2$; motor speeds are given by:

$$\omega_i = \omega_{\text{hover}} = \sqrt{\left(mg/\,4k\right)} \tag{20.7}$$

Pitch rotation To make the quad-rotor pitch, we have to disturb the moment equilibrium, which is achieved by unbalancing the propeller speeds:

$$\left(\omega 1+\omega 2\right)-\left(\omega 3+\omega 4\right)\neq 0 \tag{20.8}$$

So for pitching, ($\theta = \int \dot{\theta} dt$). Also, by analogy to rolling, the total thrust is decomposed, so we need more power for quad-rotor pitching.

Euler angle transformations are defined by ψ, θ and φ and a combined transformation matrix from body coordinates to the earth coordinates is obtained by three successive rotations (quad-rotor kinetics).

$$R_{B\text{ to }E} = \begin{bmatrix} c\psi c\theta & c\psi s\theta s\varphi & -s\psi c\varphi & c\psi s\theta c\varphi & +s\psi s\varphi \\ s\psi c\theta & s\psi s\theta s\varphi & +c\psi c\varphi & s\psi s\theta c\varphi & -c\psi s\varphi \\ -s\theta & & c\theta s\varphi & & c\theta c\varphi \end{bmatrix} \tag{20.9}$$

where 'c' and 's' denote 'cos' and 'sin' respectively. The Newtonian method is the most popular was of modelling rigid bodies with six degrees of freedom and has been used extensively for the modelling of many types of multi-rotors. The derived model has six dynamic (three linear and three rotational) system equations (20.10) and (20.11):

$$
\begin{aligned}
m\ddot{x} &= \sum F_i\left(s\psi s\varphi + c\psi s\theta c\varphi\right) - C_{D1}\,\dot{x} = \left(s\psi s\varphi + c\psi s\theta c\varphi\right)\left(k\left(\omega_1^2+\omega_2^2+\omega_3^2+\omega_4^2\right)\right) - C_{D1}\dot{x}\\
m\ddot{y} &= \sum F_i\left(s\psi s\theta c\varphi - c\psi s\varphi\right) - C_{D2}\dot{y} = \left(s\psi s\theta c\varphi - c\psi s\varphi\right)\left(k\left(\omega_1^2+\omega_2^2+\omega_3^2+\omega_4^2\right)\right) - C_{D2}\dot{y}\\
m\ddot{z} &= \sum F_i\left(c\theta c\varphi\right) - mg - C_{D3}\dot{z} = \left(c\theta c\varphi\right)\left(k\left(\omega_1^2+\omega_2^2+\omega_3^2+\omega_4^2\right)\right) - g - C_{D3}\dot{z}
\end{aligned}
\tag{20.10}
$$

where, C_{D1}, C_{D2} and C_{D3} are drag coefficients, and $(k\,(\omega_1^2+\omega_2^2+\omega_3^2+\omega_4^2)) = u_1$ are the quad-rotor control inputs. Euler equations are related to angular accelerations as follows:

$$
\begin{aligned}
I_x \times \ddot{\varphi} &= 1\left(F_3 - F_1 - C_{1DR}\dot{\varphi}\right) = l\left(\dot{\theta}\dot{\psi}\left(I_y - I_Z\right)\right) + k\left(\omega_4^2 - \omega_2^2\right) - C_{1DR}\dot{\varphi})\\
I_y\ddot{\theta} &= 1\left(F_4 - F_2 - C_{2DR}\dot{\theta}\right) = l\left(\dot{\varphi}\dot{\psi}\left(I_z - I_x\right)\right) + k\left(\omega_3^2 - \omega_1^2\right) - C_{2DR}\dot{\theta})\\
I_z\,\ddot{\psi} &= M_1 - M_2 + M_3 - M_4 - C_{3DR}\dot{\psi} = \dot{\varphi}\,\dot{\theta}\left(I_x - I_y\right) + K_m\left(\omega_1^2 - \omega_2^2 + \omega_3^2 - \omega_4^2\right) - C_{3DR}\dot{\psi})
\end{aligned}
\tag{20.11}
$$

where l is the distance of each rotor from the vehicle's CG. I_x, I_y and I_z are moments of inertia along each of the axes and C_{1DR}, C_{2DR} and C_{3DR} are rotational drag coefficients. M_i, $(i = 1,2,3,4)$ are rotor moments produced by the angular velocity of the rotors and given by: $M_i = K_m\,\omega_i^2$; K_m is a constant. $u_2 = k\,(\omega_4^2 - \omega_2^2)$, $u_3 = k\,(\omega_3^2 - \omega_1^2)$ and $u_4 = K_m\,(\omega_1^2 - \omega_2^2 + \omega_3^2 - \omega_4^2)$, where u_2, u_3 and u_4 are the angular quad-rotor control inputs.

20.2.4 Quad-rotor Applications

Despite their many uses, quad-rotors can be categorized in a simple way as follows:

Military and law enforcement Quad-rotors are used for surveillance, reconnaissance, different intelligences, target acquisition, marches and demonstrations, and some sorties of dropping and delivery by military and law enforcement units, disaster and earthquake search and rescue missions in urban environments, geography and mapping.

Civil The most notable of civilian applications are electric power line observation, geological or buildings survey, oil slick monitoring, communications, forest observation and meteorology services.

Research and education The quad-rotor is one of the most VTOL UAV requested and desirable at universities and research institutions involved in research related to control, stability, guidance and navigation. Modelling, simulations and new systems identification are also typical investigations.

Commercial In the fields of sports, photography, cinema, hobbies and economically feasible projects such as quad-rotor production for play and fun, which represents a wide public area.

20.3 Novel Quad-rotor Concepts

Here we describe a patented quad-rotor with tilting rotors, which built as a proof of concept of one type of electromechanically controlled quad-rotor. This tilting quad-rotor design is based on an electronically controlled conventional quad-rotor. Therefore, we will first explain traditional quad-rotors in a comprehensive manner, which will simplify the explanation of electromechanical control.

Electronical control of quad-rotors entails the use of only the motors to control the attitude and transitions. This is the basic control for quad-rotors but it is limited in performance because the vehicle is under-actuated (there are four actuations to treat six degrees of freedom). To achieve innovative multi-rotor VTOL UAV dynamics and stability, performance enhancements are required. This is achieved by increasing the actuations and adding electromechanical mechanisms to ensure improvements in multi-rotor performance. The types, design, modelling and applications of different types of electromechanical multi-rotors are discussed in the next subsections.

20.3.1 Electro-mechanical Quad-rotor

Multi-rotor electromechanical designs have a variety of mechanisms of action. They are currently mostly trials or at the proof of concept stage: with the exception of the tri-copter, they have not yet attained the popularity of pure electronic designs. The motivation for the use of electromechanical quad-rotors is to address the shortcomings of conventional quad-rotors for applications such as urban flights, obstacle avoidance, side slip imagery and windy-flight capability.

20.3.1.1 Multi-rotor with Variable-pitch Propeller

These multi rotors utilize the same type of variable-pitch rotor and swash plate as a conventional helicopter. This allows both very agile control and the potential to replace individual electric motors with belt-driven rotors and propellers, which are hooked to a central electric motor or hollow-shaft electric motor installed on each rotor arm's end, with a control rod inside the hollow shaft to change the blade's angle of attack. Internal combustion engines with belt drive can also be used; this is very important to enhance electric multi-rotor performance parameters such as range and endurance. Variable pitch is not unusual, appearing in several multi-rotor prototypes (Figure 20.7).

20.3.1.2 Multi rotor with Servo Thrust-vectoring

The tilting of a rotor-propeller vehicle, such as the bi-copter, tri-copter, and some multi rotor VTOL UAVs (Figure 20.8), utilizes both differential thrust and rotor tilting, with a servo mechanism to change its orientation. The bi-copter and the tri-copter are tilting-rotor UAVs that are very popular alternative to electronically controlled multi-rotors, which operate on pure lift by throttle control.

20.3.1.3 Flap Thrust-vectoring

In fully actuated rotorcraft UAVs, whenever it is possible to rotate a motor-prop combination or tilt rotors, it is also possible to redirect the flow using control vanes in the propeller downwash; in other words, using aerodynamically generated side force to drive the multi-rotor to the desired direction. Flap thrust-vectoring is not a common

Figure 20.7 Different types of variable-pitch propellers.

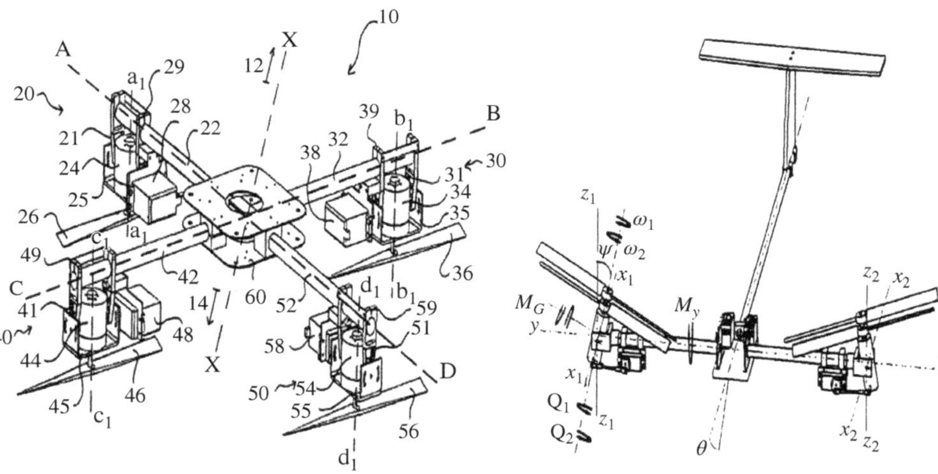

Figure 20.8 Different types of multi-rotors with servo thrust-vectoring. Source: author's patents (1) US 20130105635 A1; (2) US 20130105620 A1.

solution on commercial UAVs, but it is present in several custom-built VTOL UAVs, and there are many promising research and production projects (Figure 20.9).

20.3.2 Variable-pitch Quad-rotor

In traditional fixed-pitch quad-rotors, stability and flight control are achieved by changing the speed of each of the four motors. While quad-rotor differential RPM control is sufficient for most flight regimes and can result in agile and aggressive flight

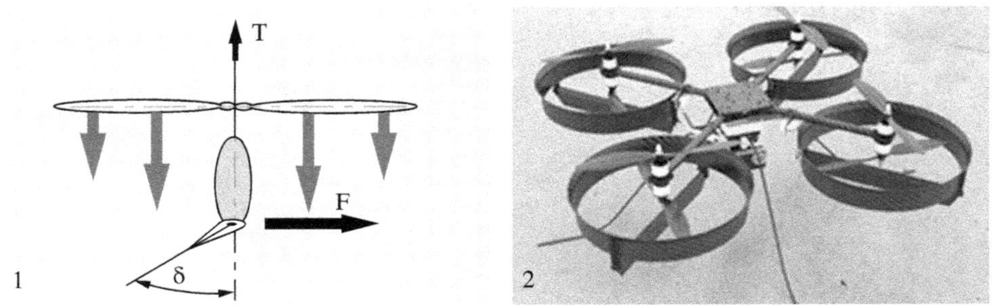

Figure 20.9 Flap thrust-vectoring: left, principle; right, a design.

Figure 20.10 Variable pitch quad-rotor.

patterns, it places fundamental constraints on the vehicle as the flight envelope expands. In particular, control bandwidth is limited by the rotational inertia of the motors and fixed-pitch quad-rotors are unable to efficiently achieve reverse thrust. These constraints limit the aerobatic manoeuvers a quad-rotor can perform, therefore limiting the future applicability of quad-rotors in intensive agile missions. These limitations in fixed-pitch quad-rotors are overcome with the addition of variable-pitch propellers. While variable-pitch propellers add complexity to an otherwise simple and relatively robust quad-rotor, the advantages of increased controller bandwidth and reverse-thrust capabilities justify such a design when aggressive and agile flight is required (Figure 20.10).

Controller bandwidth can be a significant problem for quad-rotors, which is an issue for their stability as their size increases. Larger quad-rotors require larger motors which, in turn, have larger inertias and cannot be controlled as quickly as smaller motors. Eventually, as the size increases sufficiently, the quad-rotor can no longer be stabilized through RPM control alone because the torque required to change the rotational velocity of the motor quickly exceeds the capacity of the motor. Thus variable-pitch blades may be necessary for larger quad-rotors merely for stabilization purposes. The design of the variable-pitch quad-rotor in Figure 20.11 uses four digital high-speed servos to control the pitch angle of the

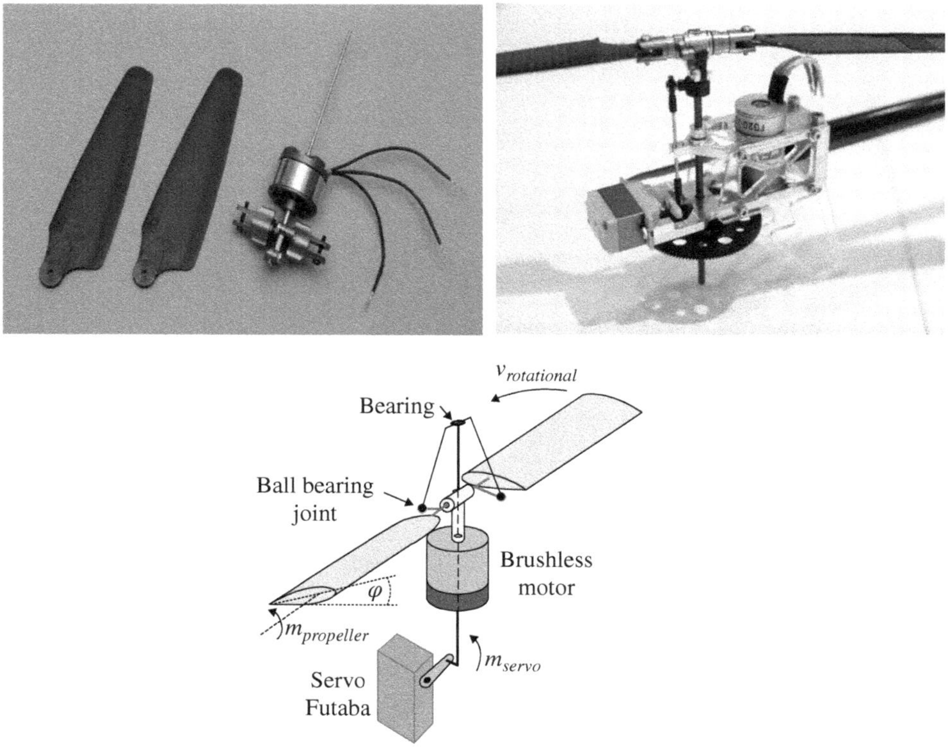

Figure 20.11 Variable pitch rotors: top left, motor with hollow shaft; top right, motor with gear; bottom, variable-pitch propeller design. Key: 1, servo-machine; 2, ball bearing joint; 3, bearing; 4, hollow shaft; 5, brushless motor; 6, link bar.

blade via a control rod that runs through a hollow motor shaft. Brushless motors are driven by electronic speed controllers. Inertial measurement, filtering, and high-rate attitude stabilization is achieved using an autopilot, with combined optimum control achieved through simultaneous changes to propeller pitch angle and RPM of the motors.

Thrust actuation with fixed-pitch propellers based on thrust produced by the propellers is almost constant with constant motor RPM (assuming the quad-rotor is near hover). The way to change propeller's thrust is by varying the voltage supplied to the motors, which varies propeller's RPM. Adding variable-pitch propellers to the quad-rotor platform results in an additional degree of freedom in its actuation by changing the pitch angle of the propeller's blade, thus varying the thrust produced by each motor-propeller combination. With variable-pitch propellers, thrust can be changed by either changing the blade pitch or by changing the rotational rate of the motors. These two actuations, to a large extent, can overlap. For instance, with variable-pitch propellers the quad-rotor can hover using high RPM and low blade pitch, or using low RPM and high blade pitch, or any combination in between. The output of the quad-rotor attitude

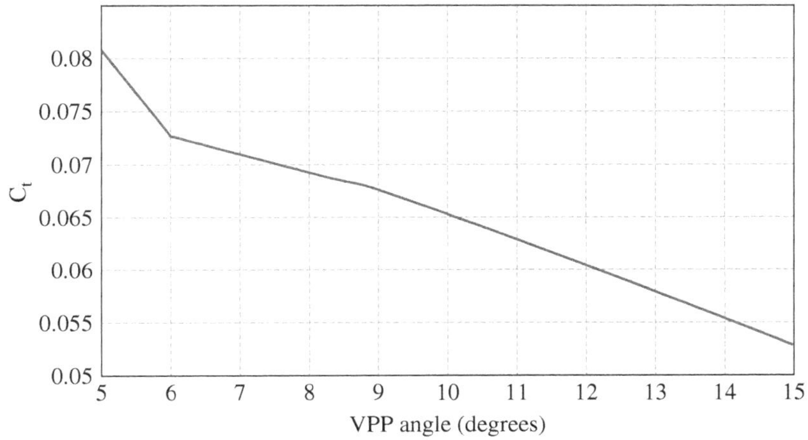

Figure 20.12 C_t at different (VPP) angles.

control loop is assumed to be a desired thrust. The lift force of each variable-pitch rotor can be considered the rotor's thrust and can be calculated using:

$$T = C_t \rho n^2 D^4 \tag{20.12}$$

where, C_t is the thrust coefficient, ρ is the density of the air, n is the RPM of the motor, and D is the diameter of the propellers. The thrust coefficient is a function of the pitch angle of the propeller ϑ. The thrust coefficient in a linear (C_t = f (ϑ)) region can be calculated as:

$$C_t = C_t^{\vartheta} \vartheta \tag{20.13}$$

where, is C_t^{ϑ} a derivative, which represents the thrust slope with respect to the variable pitch propeller (VPP) angle. This derivative is estimated using specialized software, which uses the number of blades, RPM, diameter of the propellers, and the velocity and the power of the motor. The value of C_t for an operational (VPP) angle range of $5° \leq \vartheta \leq 15°$ is shown in Figure 20.12.

Quad-rotor VPP mathematical modelling uses the same proceedure as for conventional quad-rotors except that the constant thrust of a fixed-pitch propelleris replaced with the variable thrust as a function of the changing blade pitch and RPM, with the condition that VPPs have no tilt.

One recent attractive practical development of VPP quad-rotors is the incorporation of internal combustion engines (ICEs) with hydrocarbon fuels. This will lead to marked increases in flight duartion (a weakness of electric quad-rotors). With ICEs, different quad-rotor or multi-rotor configurations can be organised (Figure 20.13):

- single ICE (with generator), mechanical drive (belts or gears with shafts), multiple VPPs.
- multiple ICEs (with generators), direct drive, multiple or single VPP.

Figure 20.13 VPP quad-rotor with ICEs.

20.3.3 Quad-rotor Thrust-vectoring

This section concerns a small tilt-rotor VTOL UAV quad-rotor; we will also use the term *quad-tilt-rotor*. The work is based on a patent by author 'Quad tilt rotor vertical take-off and landing (VTOL) unmanned aerial vehicle (UAV) with 45 degree rotors'. The patent is for a quad-tilt-rotor for research and proof of concept (Figures 20.8 and 20.14). Here we first discuss the quad-tilt-rotor design and its structure, and secondly we reflect on the quad-tilt-rotor's modelling and control with regards to the specific impact of the tilting on the vehicle's forces and moments as they directly affect the quad-rotor dynamics and stability.

20.3.3.1 Quad-tilt-rotor Design

Quad-rotor servo thrust-vectoring or rotor-tilting is a novel concept of multi-rotor UAV control. Typical quad-rotors are limited in their mobility and flight abilities because of their basic under-actuation. The design motivation is to have a fully functional quad-rotor without inclination of the body frame for those flight conditions that require a horizontal body frame and no stressed motor operation, such as when the quad-rotor is confronting the wind horizontally; when minimum drag is the objective. The design should also ensure the correct position of the quad-rotor's CG, so that the trend line of the horizontal component passes through the CG; this avoids the generation of longitudinal and lateral moments because of the distance between the trend line and the CG is not equal to zero. Therefore, in some quad-rotor designs motors with propellers are directed downwards; that is, the rotor is at a lower position. Also, the control algorithm is designed so that all the motors are able to tilt and/or rotate independently from each other.

The quad-rotor prototype has the ability to control the alignment of its rotors, thus making it possible to overcome the under actuation and behave as a fully-actuated flying vehicle. Driven by these limitations, a novel concept of a quad-rotor UAV with tilting rotor-propellers (propellers which can be actively rotated along the arm axes) is proposed. This will give full controllability of the quad-rotor pose in 3D space. Developing this novel design, the goal is to transfer quad-rotors from pure flying

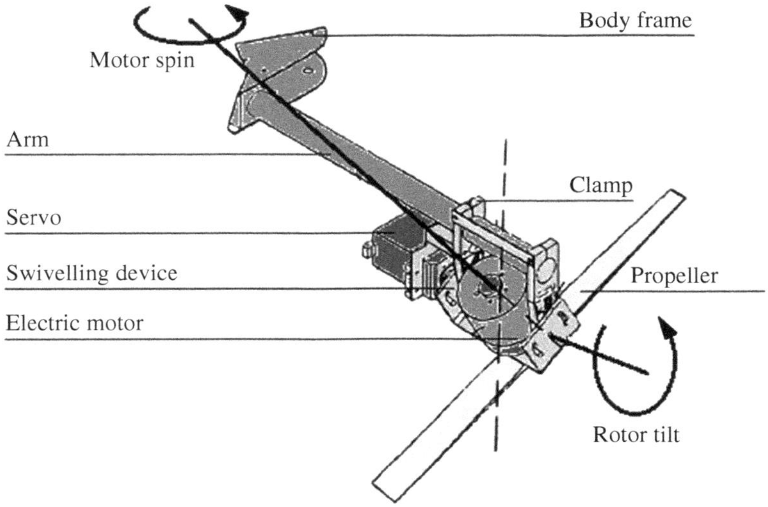

Figure 20.14 Quad-tilt-rotor component arrangement.

vehicles to fully flying robots with independent control over all six degrees of freedom of the main body and the ability to apply forces in arbitrary directions [4, 5].

The reason to choose a quad-rotor with tilting rotors that is similar to a conventional quad-rotor is that the equations of motion of the conventional quad-rotor are simple and can help to reveal the differences needed to manage the required performance of the tilting quad-rotor. Also, their mathematical model helps clarify the goals and benefits when rotor tilting is added to the design. In common with conventional quad-rotors, the body frame consists of three main parts: central hub, carbon tubes (arms installation is at 45° relative to the x, y horizontal axes) and propulsion unit. In our case the fundamental difference is the propulsion unit. The electric motor is replaced with an electric-motor tilting mechanism. The tilting mechanism contains the motor with the propeller, the clamp with swivelling device and the servo motor allowing the rotor to tilt in the plane perpendicular to the quad-rotor arm at a certain angle (usually ±20°; Figure 20.14). This design gives convenient tuning, maintenance and replacement of parts.

The control of the quad-rotor with such a tilting design is executed by two control loops: the first controls the rotation of the motors in a manner identical to conventional quad-rotor; stability and motion are controlled by the quadrotor's inclination as the motors are controlled by their RPM. The second control loop is to control the quad-rotor transition in the horizontal plane (without inclination) in all directions. The transition is executed by tilting the rotors at different angles relative to the quad-rotor arms (Figure 20.15); for example, to move the quad-rotor accurately forward (along the x-axis) we have to tilt the front rotors (rotors 1 and 2) inside towards the x-axis and the backwards rotors (rotors 3 and 4) away from the x-axis. If we need to move accurately right (along the y-axis) we need to tilt the right rotors inside toward the y-axis and the left motors away from the y-axis. Logically if we need to move backwards or left the tilts are inversed. If we want to move to some in-between direction, we must tilt the motors proportionally inside or outside; for example, if we need to move at 45° (forward and

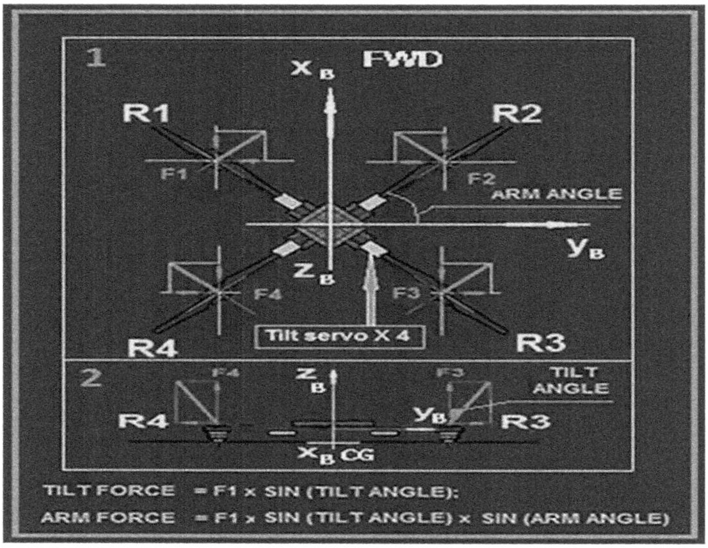

Figure 20.15 Quad-tilt-rotor principle with 45° tilting; (1) top view (2) side view.

right) we tilt only rotors 1 and 3 inside the arm to the right; rotors 2 and 4 will be not tilted (in a flight test the quad-tilt-rotor combination showed very good matching). Note that quad-rotor's horizontal motion direction is the same as with tilted lift resultant inclination. To move quad-rotor forward in the horizontal plane (X_B–Y_B), the resultants with same (X_B) direction will be added together for the four rotors and the resultants of each pair of motors in the Y_B direction will cancel (for four rotors their sum will be equal zero). This will result in the quad-rotor moving forward.

When we needed to get full horizontal-plane movement, the quad-rotor's CG vertical location must be at same level as the horizontal components of the lift force; otherwise, we will obtain inclined quad-rotor motion (itself also a very interesting phenomenon for future analysis) due to an imbalance between the tilting force resultants' moments (vertical and horizontal resultants). $M_{vert}= F_{vert} \times L$, where L is the distance from the rotor to the CG, $M_{hor}= F_{hor} \times H$, where H is the horizontal distance between the rotor and the CG. The first loop control will balance the propeller's lift difference, which keeps the quad-rotor in level flight. In other words, the distance is equal to zero, and there is no moment, so no disturbance for the first loop controller; meaning that the quad-rotor moves level on the horizontal.

The basic concept of the novel quad-rotor is to achieve over-actuation (four spinning velocities plus four tilting velocities for a total of eight commands (shown in Figure 20.14 for one arm).

20.3.3.2 Quad-tilt-Rotor Modelling and Control

The first step in designing a control system for a given rotor vehicle is to derive a model that captures the key dynamics of this vehicle in the operational range of interest. Dynamic modelling is affected by the forces and torques that act on the flying vehicle. The conventional quad-rotor, as an under-actuated vehicle, has input from only the four lifting forces originating from the rotating propellers. Through adding tilting capability to all four rotors, the quad-tilt-rotor is over-actuated, with eight actuators and six degrees of freedom.

From physics, we know that a tilting spinning body will generate different forces and torques. Therefore, it is necessary to present them:

- *Force vector*: The force generated by the quad-rotor, which is the resultant force of the thrusts generated by the four propellers.
- *Gyroscopic moments:* Tilting of spinning propellers (rotation movements around two axes) creates gyroscopic moments that are perpendicular to these axes.
- *Propeller torques:* As the blades rotate, drag forces will be generated, which produce torques around the propeller's aerodynamic centre. These torques act in the opposite direction to their direction of rotation.
- *Thrust-vectoring torques:* These are the torques exerted by the force vector.
- *Adverse reactionary torque:* This torque acts when the rotor is suddenly tilted around certain axes. It depends especially on the propeller inertia and on tilting rate. The torque direction is opposite to the direction of tilting.

The dynamic model of the quad-tilt-rotor is derived by considering the vehicle as a rigid body that moves in a 3D space, through the main thrust and the pitch, roll and yaw torques. The quad-rotor's generalized coordinates are expressed as:

$$Q=\left[x\ y\ z\ \psi\ \vartheta\ \varphi\right]^T \in R^6 \tag{20.14}$$

where the vector $[x\ y\ z]^{\mathrm{T}} \in R^3$ (R: 3D space) corresponds to the position of the centre of mass of the quad-rotor with respect to a fixed inertial frame. The $[\psi\ \theta\ \varphi]^{\mathrm{T}} \in R^3$ vector represents the Euler angles for the yaw, pitch and roll, around the x-, y- and z-axes respectively, which correspond to the quad-rotor's orientation in 3D space. The control of the quad-tilt-rotor is executed by two loops: first combine the quad-rotor attitude control with motor RPM, quad-rotor transition with quad-rotor yaw control by differential motor RPM (that is, the first loop acts as a conventional quad-rotor control). By the second loop, the quad-rotor will transit strictly horizontally (x–y plane) without torques relative to the horizontal component of the lifting component direction. The arrangement of forces and torques is shown in Figure 20.16.

Under the assumption that the quad-rotor is a rigid body, the CG is located in a plane of horizontal components (according to the requirements of translation in horizontal flight so that the CG location and quad-rotor's controller do not generate any torques). Gyroscopic moments are neglected to simplify the analysis of tilting effects. In Figure 20.15, we have two views of the quad-rotor with the body and rotor systems of coordinates. Four rotors R_i (i = 1,2,3,4) tilt around the arms in the perpendicular plane at independent angles, with electric motors and propellers generating lift forces F_i (i = 1,2,3,4). Each rotor can tilt from vertical to an angle of 20° to either side. To explain how the tilting mechanism acts, we set the arm angle in this quad-rotor design case to 45°; the tilt angle (α) and the motor have a lower position. In Figure 20.17, another of the author's prototype designs shows the upper rotor arrangement.

Now we analyse whether forward transition is required (direction X_{B}). First, assume a quad-rotor in hover, so we have four equal lift forces F_i that in sum equal the quad-rotor's weight. Second, we tilt rotor 1 to the left and rotor 2 to the right; in other words,

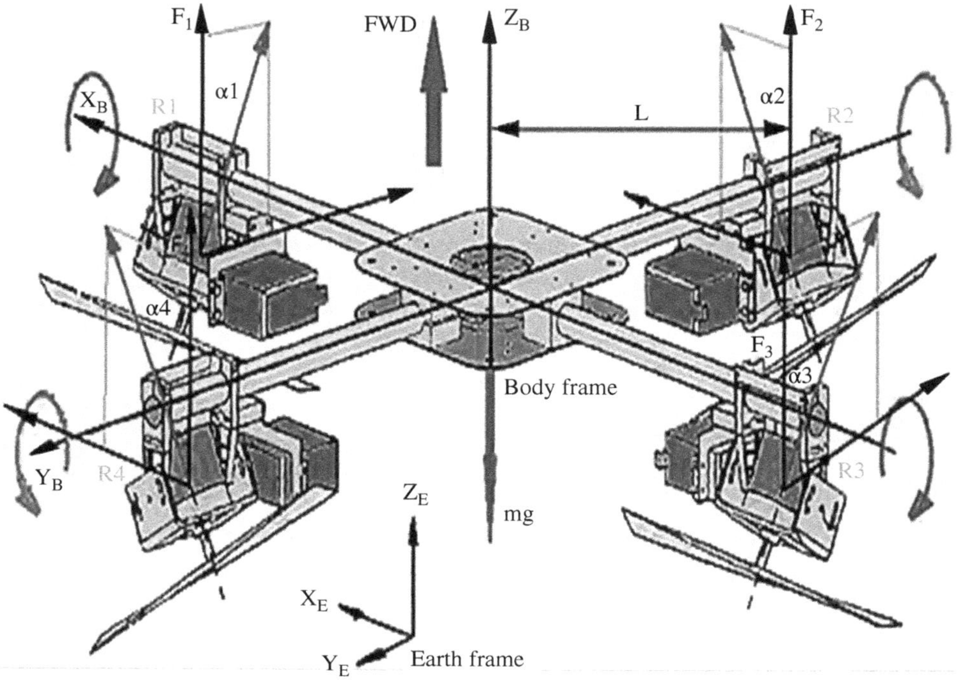

Figure 20.16 Quad-tilt-rotor: arrangement of forces and torques.

Figure 20.17 Upper position of rotors.

inside the forward arms at tilt angle = $\alpha = 0° \pm 20°$. Rotor 3 tilts to the right and rotor 4 to the left, outside the backward arms at tilt angle = $\alpha = 0° \pm 20°$). We have four tilted components; the vertical one equals $F_i \cos \alpha$, four horizontal components each equal $F_i \sin \alpha$, each of which is in turn divided into two components, the first parallel to axis X_B (quad-rotor's arm force = $F_i \sin(\pm\alpha)\sin(\pm 45°)$), the second parallel to axis Y_B (quad-rotor arm force = $F_i \sin(\pm\alpha)\sin(\pm 45°)$. Components parallel to axis X_B will be added and push the quad-rotor horizontally forward; components parallel to axis Y_B will oppose each other and their sum will equal zero, which means that no side motion will take place (in the direction of Y_B). In the same manner, the quad-rotor can transit to an arbitrary heading using different combinations of tilts (also, all moments for yaw are equal, so the quad-tilt-rotor experiences no yawing). Knowing the characteristics of the quad-tilt-rotor forces, we can now analyse the vehicle's mathematical model.

Generally Newton's second law of motion (the Newton–Euler approach will be used) represents the motion of the quad-rotor. Therefore, the net force and the net moment acting on the quad-rotor are as in Eqs. (20.1) and (20.2). These dynamic equations could be obviously divided into translational and rotational equations of motion. For a quad-tilt-rotor, four more variables are included, which represent rotor combinations of tilting with angles α_i. Regulation of these angles results in transition in all directions, which improves vehicle manoeuvrability, capability of hovering at a tilted angle and effectiveness in resisting the influence of the wind. It can be seen that rotor combinations are free to tilt around axes that are coincidental with the arm axes. The tilting rotors are slanted by angles α_i relative to the vertical; α_i (i = 1,2,3,4) is the tilt angle of the rotor. It is noted that the forces generated by the propellers are perpendicular to these particular planes of rotation, with regard to the previously mentioned rotational matrix and forces F_i (i = 1,2,3,4), and we can write translational equations of motion in the Earth frame as follows:

$$
\begin{aligned}
m(x'') = {} & (F1\, s\alpha 1\, c\psi\, c\theta - F3\, s\alpha 3\, c\psi\, c\theta - F4\, s\alpha 4\, c\psi\, s\theta\, s\varphi \\
& + F4\, s\alpha 4\, s\psi\, c\varphi + F2\, s\alpha 2\, c\psi\, s\theta\, s\varphi - F2\, s\alpha 2\, s\psi\, c\varphi \\
& + F1\, c\alpha 1\, c\psi\, s\theta\, c\varphi + F2\, c\alpha 2\, c\psi\, s\theta\, c\varphi \\
& + F3\, c\alpha 3\, c\psi\, s\theta\, c\varphi + F4\, c\alpha 4\, c\psi\, s\theta\, c\varphi + F1\, c\alpha 1\, s\psi\, s\varphi \\
& + F2\, c\alpha 2\, s\psi\, s\varphi + F3\, c\alpha 3\, s\psi\, s\varphi + F4\, c4\, s\psi\, s\varphi)\, s450 - \left(CD1\, x^{\cdot}\right) s45°; \\[1ex]
m(y'') = {} & (F1\, s\alpha 1\, c\psi\, c\theta - F3\, s\alpha 3\, s\psi\, c\theta - F4\, s\alpha 4\, s\psi\, s\theta\, s\varphi \\
& + F2\, s\alpha 2\, s\psi\, s\alpha 2\, s\varphi - F4\, s\alpha 4\, c\psi\, c\varphi + F2\, s\alpha 2\, c\psi\, c\varphi \\
& + F1\, c\alpha 1\, s\psi\, s\theta\, c\varphi + F2\, c\alpha 2\, s\psi\, s\theta\, c\varphi \\
& + F3\, c\alpha 3\, s\psi\, s\theta\, c\varphi + F4\, c4\, s\psi\, s\theta\, c\varphi - F1\, c\alpha 1\, c\psi\, s\varphi \\
& - F2\, c\alpha 2\, c\psi\, s\varphi - F3\, c\alpha 3\, c\psi\, c\varphi - F4\, c\alpha 4\, c\psi\, s\varphi)\, s450 - \left(CD2\, y^{\cdot}\right) s45^{\circ}; \\[1ex]
m(z'') = {} & (-F1\, s\alpha 1\, s\theta + F3\, s\alpha 3\, s\theta - F4\, s\alpha 4\, c\theta\, s\varphi \\
& + F2\, s\alpha 2\, c\theta\, s\varphi + F1\, c\,\alpha 1\, c\theta\, c\varphi + F2\, c\alpha 2\, c\theta\, c\varphi \\
& + F3\, c\alpha 3\, c\theta\, c\varphi + F4\, c\alpha 4\, c\theta\, c\varphi)\, s450 - mg - CD3(z');
\end{aligned}
\tag{20.15}
$$

The rotational equations of motion are derived based on Euler equations:

$$
\begin{aligned}
I_x\ (\varphi'') &= \left(L\left(F_3 c\alpha_3 - F_1 c\alpha_1 - C_{1DR}\right) + \left(Q_1 s\alpha_1 - Q_3 s\alpha_3\right) + \left(M_{2T} + M_{4T}\right)\right) s45^{\circ}; \\
I_y\ (\theta'') &= (L\left(F_4 c\alpha_4 - F_2 c\alpha_2 - C_{2DR}\right) + \left(Q_4 s\alpha_4 - Q_2 s\alpha_2\right) + \left(M_{1T} + M_{3T}\right) s45^{\circ}; \\
I_z\ (\psi'') &= \begin{pmatrix} L\left(F_1 s\alpha_1 + F_2 s\alpha_2 + F_3 s\alpha_3 + F_4 s\alpha_4 - C_{3DR}\right) \\ +\left(Q_1 c\alpha_1 - Q_2 c\alpha_2 + Q_3 c\alpha_3 - Q_4 c\alpha_4\right) \end{pmatrix} s45^{\circ};
\end{aligned}
\tag{20.16}
$$

where m is the total mass of the quad-rotor; g is the gravitational acceleration; x, y and z give the quad-rotor's position in Earth frame coordinates; C_{D1}, C_{D2} and C_{D3} are drag coefficients; F_i (i = 1,2,3,4) is the rotor's lift force; the 45° angle refers to the cross quad-rotor design, with its arms at 45° relative to the body axes; L is distance of each rotor from the vehicle's CG; I_X, I_Y and I_Z are moment of inertia around the x-, y- and z-axes, respectively; C_{1DR}, C_{2DR}, C_{3DR} are rotational drag coefficients; Q_i (i = 1,2,3,4) is the rotor moments due to their rotation; and M_{iT}, (i = 1,2,3,4) is the tilt angle of reaction moments.

On the subject of quad-tilt-rotor control, it is necessary to address mathematical modelling and simulation, due to their importance in design, creation and development of any prototype. Modelling and simulation must be carried out in a certain sequence as this is important in the design process and the quality of the final product. Appropriate quad-rotor mathematical modelling consists of two parts. First, we must design the quad-rotor controller and, second, we must create a simulation model. Development of the controller is carried out by running the model through software such as MATLAB or SIMULINK with data from real flight tests on an actual flyable prototype (as in our case). The control performs a key role in the quad-rotor's stability, making it possible to control precisely the attitude and altitude states. Its main goal is to make the quad-rotor move to a new desired position (called the 'reference') and also to react to external disturbances quickly and in a controlled way. Attitude control is the key element to maintain stability during flight. The quad-rotor in space is represented through 12 state vectors, which are obtained by altering the mathematical model's differential equations to the form:

$$\dot{X} = \left[\ddot{\varphi}\,\dot{\varphi}\,\ddot{\theta}\dot{\theta}\,\ddot{\psi}\,\dot{\psi}\,\ddot{z}\,\dot{z}\,\ddot{x}\,\dot{x}\,\ddot{y}\,\dot{y}\right] = f\left(X,U\right) \tag{20.17}$$

where $X = [\varphi\,\dot{\varphi}\;\theta\,\dot{\theta}\;\psi\,\dot{\psi}\;z\,\dot{z}\;x\,\dot{x}\;y\,\dot{y}]^T$; X is a state vector, U is the inputs vector; and T is the state space.

Now if assume the main quad-rotor dynamic system $\dot{X} = f(X, U)$ is composed from two subsystems – the angular rotational and linear transitional Figure 20.18, top – we can see that three inputs (u_2, u_3, u_4) affect six angular subsystems (six quad-rotor attitude outputs: angles and angular rates). The other three inputs (u_1, u_5, u_6) affect six linear transition subsystems (six quad-rotor position outputs: coordinates and their rates or velocities). The u_1 input acts on the electric motor's thrust at vertical axes to change and maintain altitude. The u_5, u_6 inputs control the tilt of the rotor angles for steering and position changes in the horizontal plane. The transition subsystem also has attitude-angle inputs to enhance dynamic response precisely; as the transition subsystem has output contain coordinates and their rates the quad-rotor controller can operate dynamically. The lower part of Figure 20.18 shows the strategic control of the tilting quad-rotor; it is the relationship between the input vectors (u_2, u_3, u_4), input vector u_1 and both linear transition vectors (u_5, u_6) with the quad-rotor's linear and rotational output parameters.

The aim of this strategy is not only to control the position of the vehicle over a trajectory in three dimensions, as for a conventional quad-rotor, but also to control its position in horizontal level hover and during path tracking. The controller inputs are the four independent speeds of the propellers and their tilt about the axes parallel to the quad-rotor arms. Referring to Figure 20.19, we can see how four tilt servos mix for inclined control about the axes of tilting in the horizontal transition plane. In this case, we

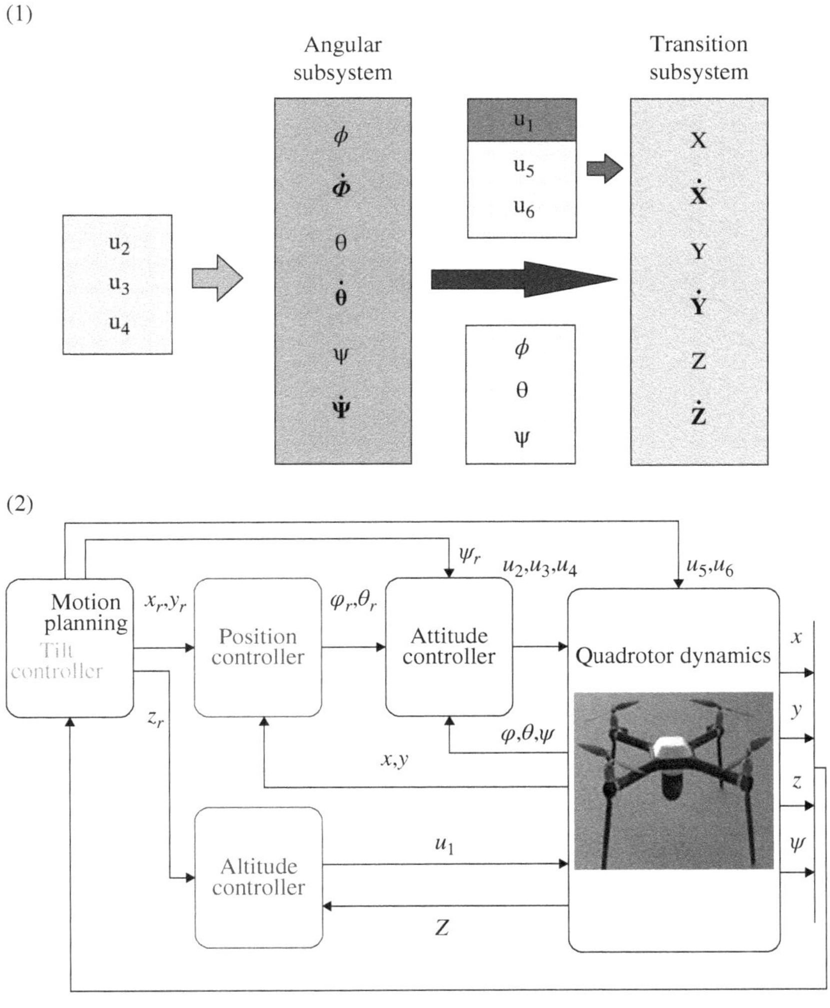

Figure 20.18 (1) Quad-tilt-rotor state vectors with input vector relationships (angular, transition subsystems); (2) Quad-tilt- rotor control strategy.

have three commands each one in two directions (normal and its opposite). The command could be received from a manual controller or from an autopilot; for conceptual understanding, suppose that commands can be given in pulse width modulation (PWM) format from minimum to maximum values or staying in the middle (neutral). The PWM commands, whatever their values, pass three mixers simultaneously (these mixers work using the same principle as the well-known V-tail RC mixer, namely two inputs and two outputs). The connections between commands, mixers and servos are organized in such way that the rotors will tilt in a direction that ensures the quad-tilt-rotor responds in such a way as to provide a horizontal plane level transition in any direction of motion or to face the wind in the horizontal plane by rotor tilting towards the wind.

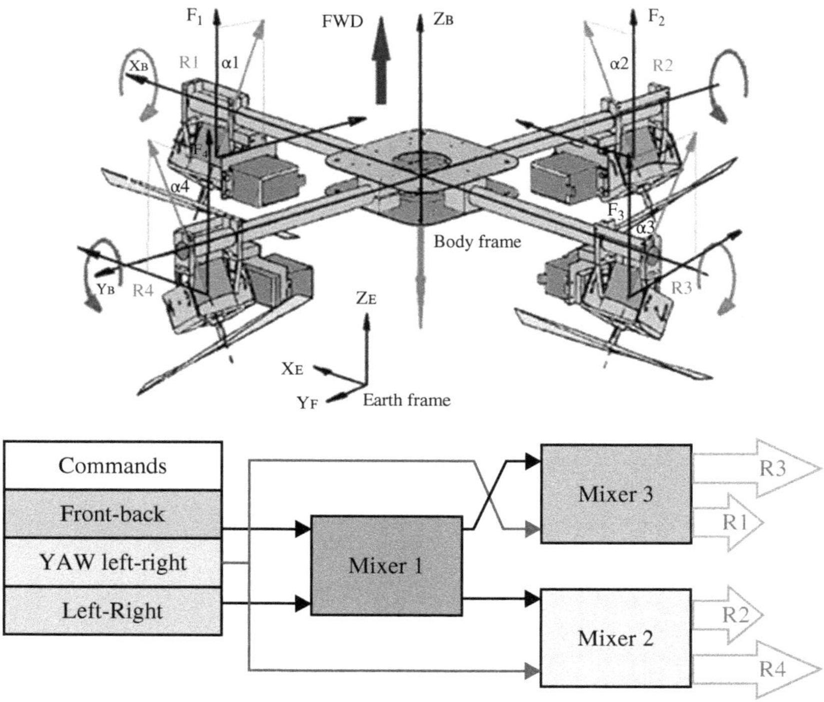

Figure 20.19 Quad-tilt-rotor control with control mixing.

20.3.4 Flap Thrust-vectoring

It is important for multi-rotor VTOL UAVs to be able to change position without changing attitude and to maintain attitude without changing position, which is needed when flying in windy conditions. That cannot be achieved until the multi-rotor has full actuation control. Above, we mentioned a configuration that allows an increase of actuation, namely rotor servo thrust-vectoring or rotor-tilting. However, there is another unique method for increasing multi-rotor actuation: flap thrust-vectoring. This method uses the flow stream from an upper positioned propeller to generate aerodynamic forces on the flap (or the vane) and to have a component directed horizontally, parallel to the propeller's plane of rotation. These forces are used to control or steer the quad-rotor in the same manner as the horizontal component of a tilt rotor [6].

From Figure 20.20, we note that the airfoil with a flap is blown by the propeller downwash to create aerodynamic force, while the flap is deflected (according to wing theory). These aerodynamic forces increase the actuation (more forces or servos make more degrees of freedoms). From wing theory, many factors affect the value of the aerodynamic force, such as airfoil shape, downwash velocity, flap deflection angle and flap surface area. As per the variable-pitch propeller case, the aerodynamic forces can be calculated as follows:

$$F = \tfrac{1}{2}\rho V^2 S\, C_L \tag{20.18}$$

where F is the generated aerodynamic force, ρ is the air density, V is the stream speed below the propeller; S is the airfoil total surface area; C_L is air foil aerodynamic

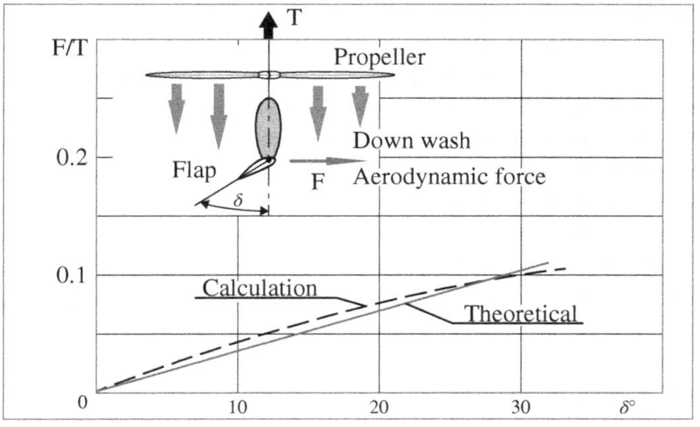

Figure 20.20 Flap thrust-vectoring principle and parameters.

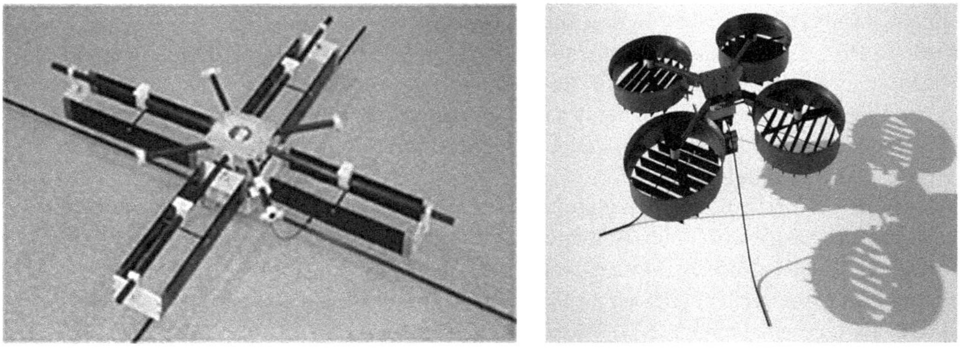

Figure 20.21 Quad-rotor with flap thrust-vectoring: left, with props not shown; right, complete prototype.

coefficient when the flap is deflected: $C_L = C_L\delta\ \delta$ and $C_L\delta$ is a derivative that represents the airfoil aerodynamic coefficient slope with respect to flap deflection angle δ.

The dynamical modelling is implemented as for servo thrust-vectoring configurations, by reflecting in the mathematical model the relationship between flap deflection angles and the resulting forces and torques. Figure 20.21 shows flap thrust-vectoring in a quad-rotor.

20.4 Conclusions

UAVs of different shapes and sizes use electric propulsion systems to generate the required thrust. An efficient design of the propulsion system enhances performance, maximizes endurance, increases payload capabilities and prolongs the flight times of missions.

A conventional quad-rotor mathematical model was presented as a basis for the model of a novel multi-rotor (quad-rotor) concept.

Different types of novel electromechanical quad-rotors (variable-pitch propellers, servo thrust-vectoring; flap thrust-vectoring) are described, focusing on servo thrust-vectoring because of their common features.

State-of-the-art electromechanical quad-rotor mathematical modelling and their tilt-mixing methodology was presented.

Based on our issued US patent [7] and research, and many successful flight tests of different quad-tilt-rotor prototypes, there is a good match of these quad-tilt-rotor concepts with the results of test flights, vehicle identification, and quad-tilt-rotor flight mixers for future autopilot developments.

We present the prospects for future research and developments projects, such as flap thrust-vectoring.

References

1 R. Deits and R. Tedrake. Efficient mixed-integer planning for UAVs in cluttered environments. *IEEE International Conference on Robotics and Automation (ICRA)*, 2015.
2 P. Castillo Garcia, R. Lozano, and A. Dzul, *Modelling and Control of Miniflying Machines*, Springer, 2005.
3 S. Bouabdallah, *Design and Control of Quad Rotors with Application to Autonomous Flying*. PhD thesis, Ecole Polythechnique Federale De Lausanne, 2007.
4 M.-D. Hua, T. Hamel, and C. Samson, 'Control of VTOL vehicles with thrust-direction tilting,' in *Proceedings of the 19th IFAC World Congress*, 2014.
5 A. Gibiansky. Quadcopter dynamics and simulation. Blog post, Andrew.Gibiansky.com, 23 November 2012. URL: http://andrew.gibiansky.com/blog/physics/quadcopter-dynamics/.
6 J. van den Berg, D. Wilkie, S.J. Guy, M. Niethammer, and D. Manocha LQG-obstacles: Feedback control with collision avoidance for mobile robots under uncertainty. Web page 2011. URL: http://gamma.cs.unc.edu/CA/LQGObs/.
7 Quad tilt rotor vertical take-off and landing (VTOL) unmanned aerial vehicle (UAV) with 45 degree rotors. US 2A1 (Patent).

Further reading

1 Sided performance coaxial vertical take-off and landing (VTOL) UAV and pitch stability technique using oblique active tilting (OAT). US 2A1 (Patent).
2 E.A. Abdul-Retha. *UAV Propulsion Methods and Selection*. Amazon, 2013.
3 E. Altug, J.P. Ostrowski and R. Mahony. 'Control of a quadrotor helicopter using visual feedback,' *Proceedings of the 2002 IEEE International Conference on Robotics and Automation*, pp. 72–77, 2002.
4 F. Archer, A. Shutko, T. Coleman, A. Haldin, E. Novichikhin, and I. Sidorov. 'Introduction, overview, and status of the microwave autonomous copter system MACS,' *Proceedings of the 2004 IEEE International Geoscience and Remote Sensing Symposium*, 2004.
5 R.W. Beard. 'Quadrotor dynamics and control,' Lecture notes, Brigham Young University, 19 February 2008. http://www.et.byu.edu/groups/ece490quad/control/quadrotor.pdf.
6 E.B. Nice. *'Design of a Four Rotor Hovering Vehicle*. MSc Thesis, Cornell University, 2004.
7 P. Harrop. Electric unmanned aerial vehicles (UAV) 2015–2025. Technical report, IdTechEx.

21

System Identification and Flight Control of an Unmanned Quadrotor

Wei Wei[1], *Mark B. Tischler*[2], *Nicholas Schwartz*[3] *and Kelly Cohen*[3]

[1] *Department of Electrical Engineering, University of Cincinnati, Cincinnati, OH, USA*
[2] *US Army Research, Development and Engineering Command, Moffett Field, CA, USA*
[3] *Department of Aerospace Engineering and Engineering Mechanics, University of Cincinnati, Cincinnati, OH, USA*

21.1 Introduction

In recent years, there has been great interest in using unmanned aerial vehicles (UAVs) for many different mission profiles, including package delivery, fire-fighting, and many others. With modern advances in electronics, the cost and effectiveness of UAV configurations makes them very feasible to develop. The popularity of quadrotor UAVs has recently skyrocketed. The simple and rugged control via propeller RPM, the ability to scale down the configuration, and its low cost make it an optimal choice for many different types of mission. However, the quadrotor configuration is naturally unstable in flight, so a feedback control system is necessary so that it can be easily piloted.

In the aerospace industry, there is a major focus on high-fidelity aircraft and rotorcraft simulation. Accurate simulations greatly reduce the costs of pilot training and control system development. Having access to accurate dynamic models of aircraft or rotorcraft configurations allows flight control engineers to efficiently develop control systems for their particular configuration. To utilize this capability, however, it is crucial to obtain and develop an accurate dynamic model of the system being analyzed.

For this project, the dynamic models of a bare-airframe quadrotor in hover flight conditions and its controller systems were to be obtained using system identification from flight-testing data. The models were used for simulation and control system development purposes. This model would then be validated and simulated, along with the controller, in Simulink. The PID controller is also optimized in the CONDUIT environment and flight testing was conducted to verify the improved dynamic response.

Advanced UAV Aerodynamics, Flight Stability and Control: Novel Concepts, Theory and Applications,
First Edition. Edited by Pascual Marqués and Andrea Da Ronch.

Companion website: http://www.wiley.com/go/marques/advanced_UAV_aerodynamics

21.2 Quadrotor System and Experimental Infrastructure Overview

21.2.1 Airframe

For this project, an AeroQuad Cyclone quadrotor was selected; this is shown in Figure 21.1. The AeroQuad frame is constructed with aluminum plates in the center, which houses most of the electrical components. Four hollow square aluminum tubes make up the motor arms, each having dimensions of 5/8 × 5/8 × 13 inches. Attached at the end of each motor arm is one A2217 950-kV motor powered by a 30-A electric speed controller. These four motors directly drive APC 12 × 3.8 inch propellers, which control the AeroQuad.

21.2.2 Instrumentation

The sensors on board the AeroQuad include an ITG-3200 gyroscope, ADXL345 accelerometer, HMC5883L magnetometer, BMP085 barometer, and MaxSonars EZ0 ultrasonic sensor. The sensor configuration is shown in Figure 21.2 below. This sensor suite provides adequate data-collection capability and records the parameters required for system identification at a rate of 40 Hz. Thanks to the Arduino controller platform onboard, the data provided from these sensors can be recorded in real time. For this project, it was desired to be able to record the quadrotor's response wirelessly.

Due to its configuration, there is a great deal of vibration on the AeroQuad due to the rotating propellers. To suppress vibration noise and not compromise onboard sensor performance, each of the four motor mounts and the electronic bay was isolated from the main frame using rubber vibration insulators.

With the instrumentation onboard, the pilot inputs, motor command, Euler angles (ϕ, θ, ψ), angular rates (p, q, r), linear accelerometers (a_x, a_y, a_z), altitude, battery level, and elapsed time were directly measured during flight testing. These

Figure 21.1 Quadrotor in flight.

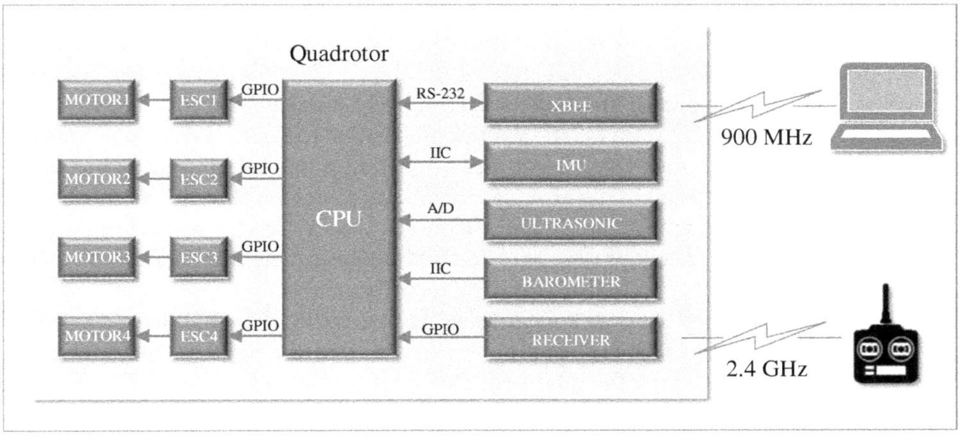

Figure 21.2 System layout of quadrotor instrumentation.

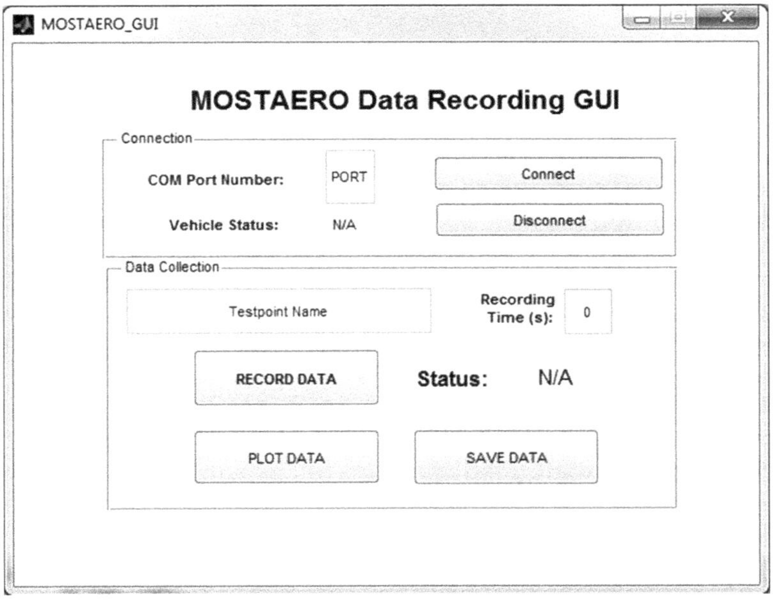

Figure 21.3 MOSTAERO data-recording GUI.

parameters were satisfactory to extract bare-airframe models using system identification and for validating the flight controller models.

For wireless flight data transmission, a pair of Digi XBee-PRO 900 MHz modules were utilized: one onboard the quadrotor and the other on a ground station computer. The XBee protocol allows accurate and fast data transmission over a large range, ideal for our setup. On the ground, the data output from the quadrotor was collected and recorded using a MATLAB graphical user interface (GUI) created specifically for this project; this is shown in Figure 21.3.

The GUI can automatically connect and disconnect from the quadrotor system. The status icons update the current status of both the AeroQuad connection and data collection processes, so that the current state of the data collection process can be easily identified. After the quadrotor is connected, a test point name is input by the user, along with the desired recording time for the data point. One of the main features of the GUI is the ability to plot the measured input and output data immediately after recording each flight testing data point. This provides the ability to ensure that the data collected is acceptable for system identification and that no data dropouts have occurred, before saving the data. If the collected data was deemed unacceptable, this feature gives the pilot the chance to correct any errors in flying technique, and the engineers the chance to resolve any issues with the data collection system. Overall, this GUI proved to be an efficient tool that minimized the amount of time needed for flight testing and post-processing, while maximizing the quality of data being collected.

Due to the instrumentation setup, it was not possible to record real-time propeller RPM during flight testing. As such, the axis commands are used as inputs for the bare-airframe model identification. Therefore, the motor dynamics are included in the extracted dynamic model as effective time delays.

21.2.3 Controller

The onboard controller consists of an ATmega2560 Arduino-based platform. The Arduino platform was suitable for our project, as it is open-source. This feature made the data collection process simpler than if a closed-source software had been used. The Arduino board works hand-in-hand with an onboard PID controller, which used the sensors described above to help stabilize the quadrotor in flight. A Futaba T12MZ 2.4 GHz radio transmitter coupled with a R6008HS receiver was used to control the quadrotor during all phases of flight testing. The R6008HS receiver was connected to the onboard PID controller, the gains of which were set for good controllability during the project's initial phase. After the PID gains were set, they were not altered throughout the entire system identification process, so that no discrepancies would be present in the identified closed-loop model.

Due to the open-source nature of the Arduino board, many different parameters could be recorded during flight testing, including direct motor commands from the controller, pilot commands, and quadrotor responses. This unique feature allowed us to not only carry out a closed-loop system identification process, but also an open-loop/bare-airframe model extraction as well, as shown in Figure 21.4. For the closed-loop model, the pilot commands are the dynamic model input, while the quadrotor responses are the outputs. However, since the post-controller motor commands were recorded as well, the bare-airframe model could also be extracted, using the axis commands as inputs and the quadrotor responses as the output. Since all parameters were recorded during all flight testing points, the same flight testing data was used to extract both the closed-loop and bare-airframe models, which greatly reduced flying time and resulted in a more complete quadrotor dynamic model.

21.2.4 Quadrotor Modeling

In this chapter, system identification is used to extract the dynamic model of the quadrotor. System identification uses cost-function optimization to extract a linear model

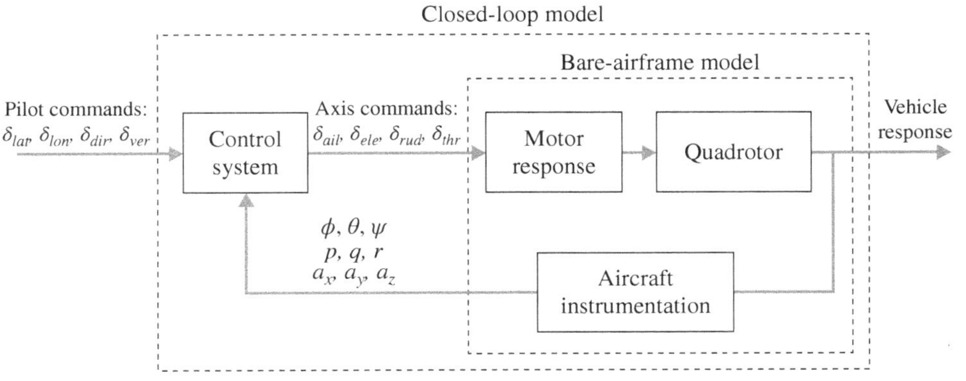

Figure 21.4 Quadrotor control structure.

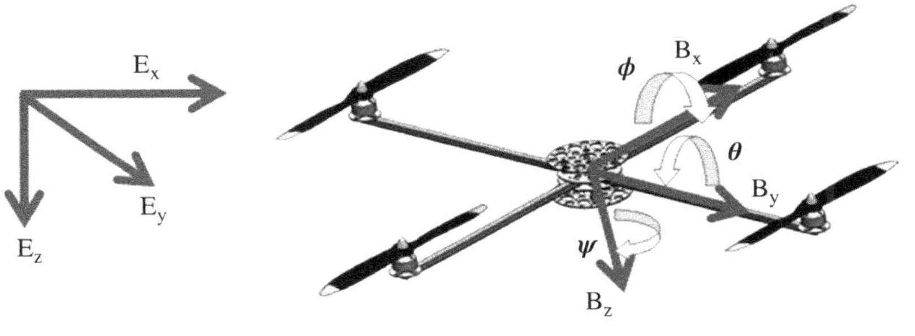

Figure 21.5 Quadrotor body-axis definition.

based on input and output data collected from the system. Nevertheless, the quadrotor system is treated as a 'grey box,' instead of black box in this process. In other words, the model structure of the quadrotor system is initially based on first principles (in other words, Newton's second law), and model parameter values are acquired through system identification. Model structure refinement is then conducted as part of the identification process. This chapter elaborates the kinematic and dynamic analysis and the model structure determination process.

21.2.5 Kinematics and Dynamics

In this section, we will discuss quadrotor control, leading up to the linearized state-space model. For our project, we wished to describe the quadrotor motion in the body axes, as shown in Figure 21.5.

For simplifying purposes, the following assumptions are made prior to the dynamic analysis:

- The quadrotor frame is assumed to be rigid and symmetrical.
- The propellers are assumed to be rigid, due to their relatively small diameter (12 inches).

- Wind and ground-effect disturbances are neglected, as the quadrotor will fly in a calm environment at a sufficient altitude.
- The center of gravity (CG) of the quadrotor is assumed to be at the center of the assembly, as all four motor arms are symmetrical and all other electrical components are mounted in the center.
- The gyroscopic effect is neglected, as the mass of the propellers (18.4 g) has a negligible contribution toward the total quadrotor weight of 1850 g.

The quadrotor body-axis equations of motion can be derived as (McRuer et al. 1973):

$$\boldsymbol{F} = \begin{bmatrix} X \\ Y \\ Z \end{bmatrix} = m\begin{bmatrix} \dot{u} + qw - rv \\ \dot{v} + ru - pw \\ \dot{w} + pv - qu \end{bmatrix} \tag{21.1}$$

$$\boldsymbol{M} = \begin{bmatrix} L \\ M \\ N \end{bmatrix} = \begin{bmatrix} I_{xx}\dot{p} - (I_{yy} - I_{zz})qr \\ I_{yy}\dot{q} - (I_{zz} - I_{xx})pr \\ I_{zz}\dot{r} - (I_{xx} - I_{yy})pq \end{bmatrix} \tag{21.2}$$

As stated earlier, the quadrotor uses a fixed-pitch propeller on each motor. Therefore, the overall control is achieved by directly altering the rotation speed of the individual motors via the flight control commands, which in turn provides lift force and rotational torque differences that control the quadrotor.

To quantify the amount of force and torque produced from each propeller, Padfield (2008) gives equations that predict the rotor thrust, T, and torque, Q, from a single propeller by integrating the lift and drag along a given blade.

$$T = C_T \rho \pi R^4 \Omega^2 = k_T \Omega^2 \tag{21.3}$$

$$Q = C_Q \rho \pi R^5 \Omega^2 = k_Q \Omega^2 \tag{21.4}$$

Here, C_T and C_Q are the rotor thrust and torque coefficients, ρ is air density, R is the propeller radius, Ω is the angular rate of the rotor, and k_T and k_Q are the resulting thrust and torque coefficients. Using the motor numbering scheme for the quadrotor shown in Figure 21.6, the input matrix for the quadrotor can be derived as:

$$\begin{bmatrix} \delta_{thr} \\ \delta_{ail} \\ \delta_{ele} \\ \delta_{rud} \end{bmatrix} = \begin{bmatrix} k_T\left(\Omega_1^2 + \Omega_2^2 + \Omega_3^2 + \Omega_4^2\right) \\ k_T l\left(-\Omega_2^2 + \Omega_4^2\right) \\ k_T l\left(\Omega_1^2 - \Omega_3^2\right) \\ k_Q\left(-\Omega_1^2 + \Omega_2^2 - \Omega_3^2 + \Omega_4^2\right) \end{bmatrix} \tag{21.5}$$

where δ_{thr}, δ_{ail}, δ_{ele} and δ_{rud} represent the thrust, lateral, longitudinal, and directional inputs respectively.

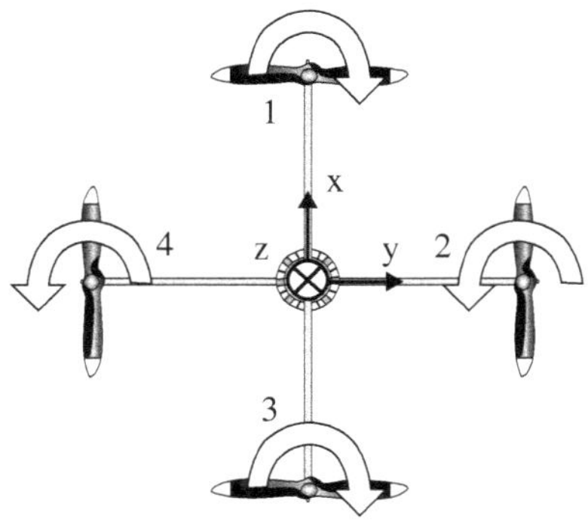

Figure 21.6 Quadrotor motor numbering.

The force arm l is defined as the distance from the propeller center to the quadrotor center of gravity. Since the quadrotor configuration is symmetrical and the center of gravity is at the center, the force arm length is the same for all four motors.

Incorporating the gravitational force in the quadrotor body-axis equations of motion, a complete set of equations that describe the quadrotor dynamics is obtained.

$$\begin{cases} \dot{u} = rv - qw - g\sin\theta + a_x \\ \dot{v} = pw - ru + g\sin\phi\cos\theta + a_y \\ \dot{w} = qu - pv + g\cos\phi\cos\theta + \dfrac{\delta_{thr}}{m} \\ \dot{p} = \dfrac{(I_{yy} - I_{zz})qr + \delta_{ail}}{I_{xx}} \\ \dot{q} = \dfrac{(I_{zz} - I_{xx})pr + \delta_{ele}}{I_{yy}} \\ \dot{r} = \dfrac{(I_{xx} - I_{yy})pq + \delta_{rud}}{I_{zz}} \end{cases} \tag{21.6}$$

21.2.6 Linearized State-space Model Structure

The quadrotor equations of motion in Eq. (21.6) are non-linear, which is a concern when discussing control system analyses, which are often oriented toward linear systems. Using small perturbation approximation techniques, the equations of motion can be linearized.

The lateral linear velocity equation below is used as an example for the small perturbation approximation, where the velocity, u, is disturbed from the trim condition, u_0, and has an increment of Δu. The higher-order terms are neglected, as they have non-dominant effects on the system.

$$u(t) = u_0 + \Delta u \tag{21.7}$$

Applying the trim conditions to all the quadrotor state variables, which are zero in hover, the linearized equations of motion are derived.

$$\begin{cases} m\Delta\dot{u} = \Delta X - mg\Delta\theta \\ m\Delta\dot{v} = \Delta Y + mg\Delta\phi \\ m\Delta\dot{w} = \Delta Z \\ I_{xx}\Delta\dot{p} = \Delta L \\ I_{yy}\Delta\dot{q} = \Delta M \\ I_{zz}\Delta\dot{r} = \Delta N \\ \Delta p = \Delta\dot{\phi} \\ \Delta q = \Delta\dot{\theta} \\ \Delta r = \Delta\dot{\psi} \end{cases} \tag{21.8}$$

These equations of motion are described using only aerodynamic forces and moments, as the perturbations are the only terms that disturb the quadrotor during flight. Eq. (21.8) can be represented in terms of the unknown mass or moments of inertia of a given system, shown as an example below.

$$X_u = \frac{1}{m}\frac{\partial X}{\partial u} \tag{21.9}$$

$$L_p = \frac{1}{I_{xx}}\frac{\partial L}{\partial p} \tag{21.10}$$

Then, the final linearized perturbations are represented as:

$$\begin{cases} \Delta X = X_u u + X_q q + X_{\delta_{ele}}\delta_{ele} \\ \Delta Y = Y_v v + Y_p p + Y_{\delta_{ail}}\delta_{ail} \\ \Delta Z = Z_w w + Z_{\delta_{thr}}\delta_{thr} \\ \Delta L = L_v v + L_p p + L_{\delta_{ail}}\delta_{ail} \\ \Delta M = M_u u + M_q q + M_{\delta_{ele}}\delta_{ele} \\ \Delta N = N_r r + N_{\delta_{rud}}\delta_{rud} \\ \Delta p = \Delta\dot{\phi} \\ \Delta q = \Delta\dot{\theta} \\ \Delta r = \Delta\dot{\psi} \end{cases} \tag{21.11}$$

The linear state-space representation of the bare-airframe quadrotor dynamics in hover flight can then be derived.

$$\begin{bmatrix} \dot{u} \\ \dot{v} \\ \dot{w} \\ \dot{p} \\ \dot{q} \\ \dot{r} \\ \dot{\phi} \\ \dot{\theta} \\ \dot{\psi} \end{bmatrix} = \begin{bmatrix} X_u & 0 & 0 & 0 & X_q & 0 & 0 & -g & 0 \\ 0 & Y_v & 0 & Y_p & 0 & 0 & g & 0 & 0 \\ 0 & 0 & Z_w & 0 & 0 & 0 & 0 & 0 & 0 \\ 0 & L_v & 0 & L_p & 0 & 0 & 0 & 0 & 0 \\ M_u & 0 & 0 & 0 & M_q & 0 & 0 & 0 & 0 \\ 0 & 0 & 0 & 0 & 0 & N_r & 0 & 0 & 0 \\ 0 & 0 & 0 & 1 & 0 & 0 & 0 & 0 & 0 \\ 0 & 0 & 0 & 0 & 1 & 0 & 0 & 0 & 0 \\ 0 & 0 & 0 & 0 & 0 & 1 & 0 & 0 & 0 \end{bmatrix} \begin{bmatrix} u \\ v \\ w \\ p \\ q \\ r \\ \phi \\ \theta \\ \psi \end{bmatrix} + \begin{bmatrix} 0 & 0 & X_{\delta_{ele}} & 0 \\ 0 & Y_{\delta_{ail}} & 0 & 0 \\ Z_{\delta_{thr}} & 0 & 0 & 0 \\ 0 & L_{\delta_{ail}} & 0 & 0 \\ 0 & 0 & M_{\delta_{ele}} & 0 \\ 0 & 0 & 0 & N_{\delta_{rud}} \\ 0 & 0 & 0 & 0 \\ 0 & 0 & 0 & 0 \\ 0 & 0 & 0 & 0 \end{bmatrix} \begin{bmatrix} \delta_{thr} \\ \delta_{ail} \\ \delta_{ele} \\ \delta_{rud} \end{bmatrix}. \tag{21.12}$$

21.3 System Identification Using CIFER

21.3.1 System Identification Introduction

There are two main methods for obtaining dynamic models of systems: traditional physics-based simulation methods and system identification. Physics-based simulation methods rely on the predicted aerodynamic, inertial, and structural models of the system. These models are then analyzed through multiple simulations, with the dynamic responses compared to flight testing data. Typically, traditional dynamic models are modified in an iterative process as the design matures to match the dynamic characteristics displayed in flight. This process can be one that is time consuming and in some cases results in a low-fidelity model.

On the other hand, system identification is a process that can give the dynamic model in a fraction of the time required for traditional methods. Rather than estimating and fine-tuning a dynamic model before and during flight testing, system identification utilizes the flight testing data to develop the dynamic model. In this process, the system's input and output time histories are measured during flight and used to develop the dynamic model (Tischler and Remple 2012). A general comparison of system identification and traditional simulation methods is shown in Figure 21.7. The system identification process is an inverse to simulation and is very useful when an aircraft can be flight tested. This approach to model determination is both quicker and produces a more accurate model than simulation methods.

Rather than starting with assumptions, system identification begins with the actual measured aircraft data and builds from the data itself. In addition, a major convenience of using system identification is that the structural and inertial characteristics of the system do not need to be precisely estimated. By measuring the inputs and outputs of the aircraft configuration, the transfer function of the aircraft system can be computed, which inherently includes all of the inertial characteristics of the system. This saves a lot of time in the preliminary model generation by eliminating the need to provide accurate inertial characteristics, which can be tough to accurately predict with a complex system.

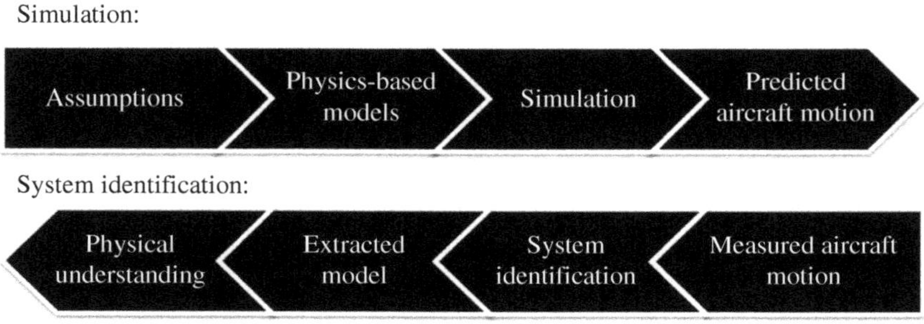

Figure 21.7 Dynamic modeling: top, simulation method; bottom, system identification.

21.3.2 CIFER

In regards to system identification, the process can take place in the time- or frequency-domain. Frequency-domain analyses minimize errors associated with bias effects and processing noise compared to time-domain analyses. In addition, the bare-airframe dynamics of the rotorcraft are unstable (very unstable for the quadrotor), and so the time-domain integration of the equations of motion will diverge. These problems are avoided when using frequency-domain system identification methods. As a result, the dynamic model of the quadrotor was extracted in this study using the frequency-domain system identification process.

For this project, the Comprehensive Identification from FrEquency Response (CIFER) program was used to extract both the closed-loop and bare-airframe quadrotor dynamic models. Since its development in the early 1990s (Tischler and Remple 2012) by the US Army Aviation Development Directorate, CIFER has been used to extract dynamic models of many commercial and military aircraft and rotorcraft configurations, such as the XV-15, Bell-214ST, BO-105, AH-64, UH-60, V-22, AV-8 Harrier, and OH-58D. In addition, dynamic models of scaled-down miniature aircraft and single-rotor model helicopter configurations have also been successfully extracted using CIFER.

The frequency-sweep input is the ideal test signal for the frequency-response identification method, and was easily accomplished with the closed-loop quadrotor rotorcraft configuration. Then, using the total control inputs (pilot plus controller), the unstable bare-airframe dynamics were determined with excellent accuracy. The frequency-sweep procedure is further described in Section 21.4.1.

A very important aspect of using flight data for system identification is ensuring the system is well excited over the frequency range of interest (0.3–20 rad/sec for flight dynamics and control). A key aspect to ensure good data quality is a good signal-to-noise ratio for the primary sweep channel, without highly correlated off-axis inputs (Tischler and Remple 2012). This is achieved with adequate excitation, pilot practice to avoid large correlated off-axis inputs, and testing in conditions of low atmospheric turbulence. Also very important to the system identification results are adequate instrumentation characteristics: low noise, good precision, and minimum signal drift.

An important measure of the validity of the data in the CIFER frequency-response identification method is the *coherence functions*, which give an indication of the signal-to-noise ratio and the linearity of the input-to-output response. The coherence is defined as:

$$\Upsilon_{uy}^{2}(\omega)=\frac{\left|\Phi_{uy}(\omega)\right|^{2}}{\left|\Phi_{uu}(\omega)\right|\left|\Phi_{yy}(\omega)\right|} \tag{21.13}$$

where $\Phi_{uu}(\omega)$, $\Phi_{yy}(\omega)$, and $\Phi_{uy}(\omega)$ represent the two auto-spectral densities (input and output) and the input/output cross-spectral density, respectively.

Coherence values are within a range of zero to one. A perfect linear relationship between the input and output signals (and no measurement noise) would result in a coherence value of $\Upsilon_{uy}^{2}(\omega)=1.0$, while an acceptable coherence value is $\Upsilon_{uy}^{2}(\omega)\geq 0.6$. After flight testing data is collected and input to CIFER, the coherence can be validated for the frequency ranges of interest. If poor coherence is displayed in the data, the data should be discarded and the system instrumentation or flight testing procedures must be re-evaluated. In the system identification process, the transfer function of each axis will be identified first, followed by a state-space representation, and finally a complete system analysis. The single-input, single-output (SISO) system identification cost function J_{TF} is written as,

$$J_{TF}=\frac{20}{n_{\omega}}\sum_{\omega_1}^{\omega_{n_\omega}}W_{\gamma}\left[W_{g}\left(\left|\hat{T}_{c}\right|-\left|T\right|\right)^{2}+W_{p}\left(\angle\hat{T}_{c}-\angle T\right)^{2}\right] \tag{21.14}$$

In Eq. (21.14), n_{ω} represents the number of frequency points, ω_1 and ω_{n_ω} are the start and end frequencies of fit, $||$ and $\angle$ are the magnitude (dB) and phase (°) at each frequency, and W_{γ}, W_{g}, and W_{p} are the weights of overall, magnitude, and phase, respectively.

The SISO cost function in Eq. (21.14) can be used to define the multi-input, multi-output (MIMO) cost function, where n_{TF} represents the number of transfer functions to be optimized.

$$J_{ss}=\sum_{l=1}^{n_{TF}}\left\{\frac{20}{n_{\omega}}\sum_{\omega_1}^{\omega_{n_\omega}}W_{\gamma}\left[W_{g}\left(\left|\hat{T}_{c}\right|-\left|T\right|\right)^{2}+W_{p}\left(\angle\hat{T}_{c}-\angle T\right)^{2}\right]\right\} \tag{21.15}$$

21.4 Flight Testing

21.4.1 Flight Testing Procedures

Since CIFER was chosen to be the software used in the system identification process, a flight testing procedure was devised based on previously determined frequency-domain system identification guidelines (Tischler and Remple 2012). For frequency-domain system identification, each individual axis was excited using frequency-sweep maneuvers. To extract a complete dynamic model, frequency sweeps in the pitch, roll, yaw, and thrust axes were carried out during flight testing. One doublet maneuver was also carried out in each axis for time-domain verification purposes. Verifying the model using data that was gathered during flight testing but was not used in the system identification process ensures that an accurate model has been extracted.

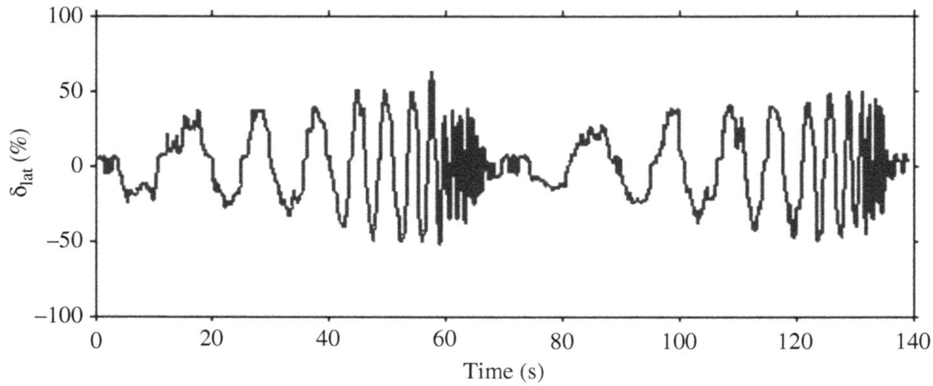

Figure 21.8 Concatenated lateral axis frequency-sweep input.

During flight testing, each test point began with the quadrotor in a steady hover, remaining there for approximately 5 s, after which the axis of interest was excited in a periodic motion, starting at a low frequency, and slowly increased until it reached a defined maximum, driven by the instrumentation or vehicle limits. When the sweep was complete, the quadrotor was returned to a hover until data collection ended. Figure 21.8 displays an example of two frequency sweep inputs in the lateral axis. For each axis of interest, the four sets of gathered sweep data were evaluated. The best two sets of data for each axis were kept and concatenated for system identification use in CIFER.

21.4.2 Flight Data Processing

Because linear velocity (u, v, w) data is not usually measured directly with sufficiently accuracy and/or adequate frequency content, they are best reconstructed from the other measurements:

$$\begin{cases} \dot{u} = rv - qw - g\sin\theta + a_x \\ \dot{v} = pw - ru + g\cos\theta\sin\phi + a_y \\ \dot{w} = qu - pv + g\cos\theta\cos\phi + a_z \end{cases} \tag{21.16}$$

The reconstructed linear velocity data suffers from drift due to sensor error and integration calculation, so the data is reconstructed to remove the drift influence. One linear velocity reconstruction example is shown in Figure 21.9.

Analysis of the reconstructed velocity in the time-domain showed problems with data consistency. A further investigation confirmed that this problem was due to poor low-frequency accuracy of the attitude data. Figure 21.10 shows frequency response of $(p/\phi)s^{-1}$, which has a 5.13-dB magnitude and −27° phase shift at the frequency of 0.3765 rad/s; they should be 0 dB and 0°, respectively. This kinematic inconsistency between the attitude and rate response is likely due to the low-frequency attitude measurement. A linear state-space model that correctly simulates the kinematics is to be identified, so the data must be consistent.

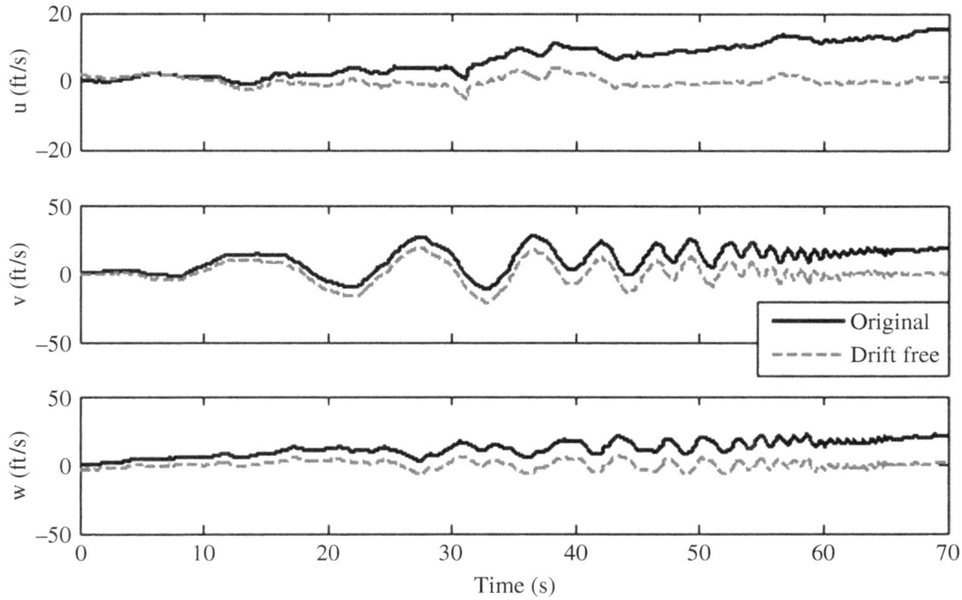

Figure 21.9 Time-domain linear velocity reconstruction.

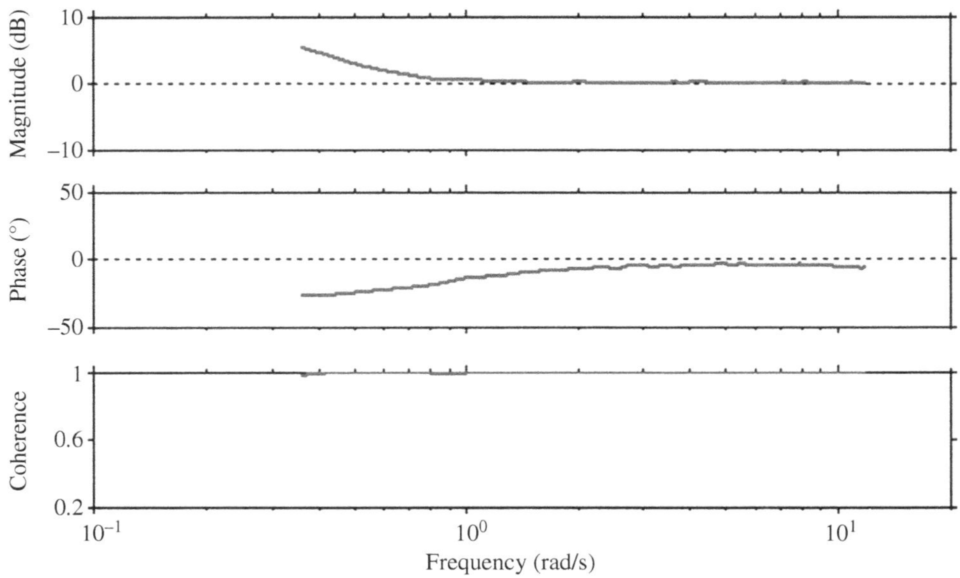

Figure 21.10 Frequency response of $(p/\phi)s^{-1}$.

By reconstructing the frequency response of the linear velocity rate (body-axis acceleration) in the frequency- rather than the time-domain, kinematic error in the attitude measurement is averted. The lateral body-axis acceleration is then determined in the frequency domain, from the linear form of $\frac{\dot{v}}{\delta_{ail}}$, Eq. (21.17):

$$\frac{\dot{v}}{\delta_{ail}} = \frac{a_y}{\delta_{ail}} + g\left(\frac{p}{\delta_{ail}}\right)s^{-1} \tag{21.17}$$

where $\frac{\dot{v}}{\delta_{ail}}$, $\frac{a_y}{\delta_{ail}}$, and $\frac{p}{\delta_{ail}}$ are the frequency responses of each channel with respect to the input *ail*. The same method is used to determine the body-axis acceleration in the other axes. Then the velocity responses are simply determined by correcting with (1/s) in the frequency domain.

A comparison of body-axis acceleration responses $\frac{\dot{v}}{\delta_{ail}}$ based on the time- and frequency-domain reconstructions is shown in Figure 21.11. The solid line shows the frequency-domain reconstruction result and the dashed line is from the time-domain reconstruction result. As can be seen, the two methods are in excellent agreement, except at low frequency, as expected. The frequency-domain reconstructed signal is thus used to ensure that the database is kinematically consistent.

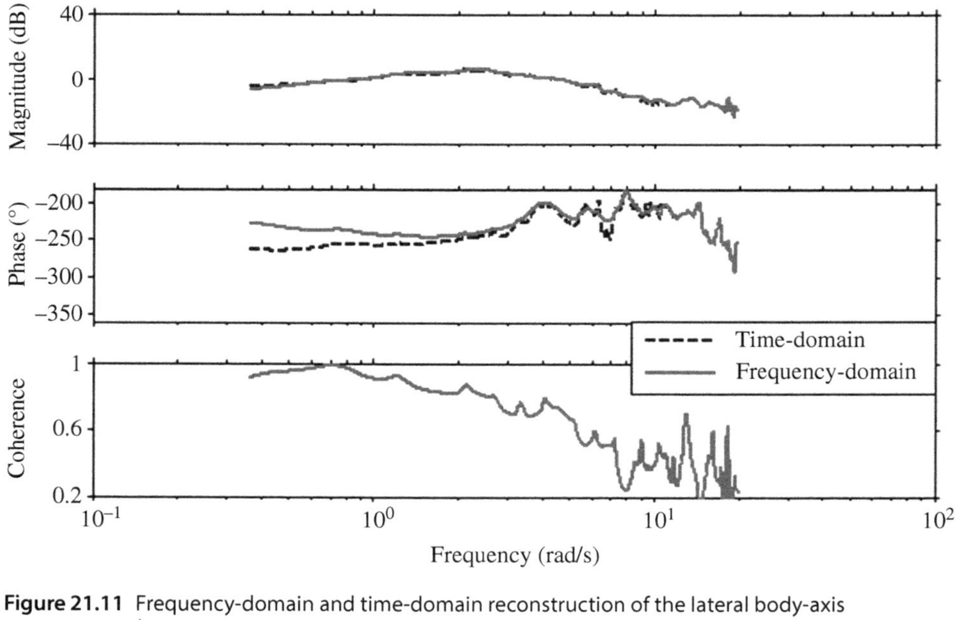

Figure 21.11 Frequency-domain and time-domain reconstruction of the lateral body-axis acceleration $\frac{\dot{v}}{\delta_{ail}}$.

21.5 System Identification Results and Discussion

Both closed-loop and bare-airframe dynamics are identified. The closed-loop results can be reviewed in Wei et al. (2014). In the identification process, the inputs are reconfigured to be a ratio of the total travel of the specific axis in percentage terms. In the closed-loop identification, the pilot's stick inputs are converted to a percentage of total stick travel. In the bare-airframe identification, the axis commands are converted to a percentage of maximum value.

The bare-airframe system identification uses commands in each axis and measured system outputs to build the dynamics model. The identified model represents dynamics of the AeroQuad airframe combined with motor responses. Due to the naturally unstable dynamics of the quadrotor airframe, an unstable model is expected.

21.5.1 Bare-airframe System Identification Results

The lateral and longitudinal dynamic of the quadrotor can be taken from the linearized state-space model and written in the following form:

$$\begin{bmatrix} \dot{v} \\ \dot{p} \\ \dot{\phi} \end{bmatrix} = \begin{bmatrix} Y_v & 0 & 32.17 \\ L_v & 0 & 0 \\ 0 & 1 & 0 \end{bmatrix} \begin{bmatrix} v \\ p \\ \phi \end{bmatrix} + \begin{bmatrix} Y_{\delta_{ail}} \\ L_{\delta_{ail}} \\ 0 \end{bmatrix} \delta_{ail}\left(t - \tau_{\delta ail}\right) \tag{21.18}$$

$$\begin{bmatrix} \dot{u} \\ \dot{q} \\ \dot{\theta} \end{bmatrix} = \begin{bmatrix} X_u & 0 & -32.17 \\ M_u & 0 & 0 \\ 0 & 1 & 0 \end{bmatrix} \begin{bmatrix} u \\ q \\ \theta \end{bmatrix} + \begin{bmatrix} X_{\delta_{ele}} \\ M_{\delta_{ele}} \\ 0 \end{bmatrix} \delta_{ele}\left(t - \tau_{\delta ele}\right) \tag{21.19}$$

Unlike a conventional flapping-rotor helicopter, the quadrotor has negligible angular rate damping (L_p, $M_q \approx 0$). The low Cramer–Rao bounds indicated in Table 21.1 show that the pitch and roll parameters are identified with high confidence (Tischler and Remple 2012).

The time delays, of 91 ms and 75 ms respectively, in roll and pitch (Table 21.1), reflect the total effective lag associated with the measurement and motor dynamics. The six eigenvalues of the bare-airframe angular responses are shown in Table 21.2. These constitute the unstable hovering cubics for the lateral and longitudinal angular dynamics that are common to all hovering vehicles. These unstable responses are also seen in the phase rise of all of the of frequency responses (Figure 21.12 and Figure 21.13) at the modal natural frequency of about 2.5 rad/sec. In the case of the small quadrotor, the associated average time-to-double amplitude in pitch and roll $\tau_{double} = \dfrac{0.693}{-\zeta\omega} \cong 0.55\,sec$ renders the bare airframe essentially unflyable without a feedback control system. There is also a high degree of symmetry between the pitch and roll derivatives and associated eigenvalues, as expected for this symmetrical quadrotor configuration.

Table 21.1 Identified parameters for lateral and longitudinal bare-airframe models.

	Value	Cramer–Rao %	Insens. %
Y_v	−0.0429	53.75	6.009
L_v	−0.4376	5.969	1.797
$Y_{\delta_{ail}}$	−0.2016	18.34	2.52
$L_{\delta_{ail}}$	0.7066	4.64	1.38
X_u	−0.0429	53.75	6.009
M_u	0.5241	6.02	1.819
$X_{\delta_{ele}}$	0.2269	13.21	2.341
$M_{\delta_{ele}}$	0.6662	4.751	1.421
$\tau_{\delta ail}$	0.09077	5.487	2.722
$\tau_{\delta ele}$	0.07460	6.713	3.328

Table 21.2 Lateral and longitudinal model eigenvalues.

λ#	Eigenvalue location
1–2	1.1931 ± 2.0911i (ζ = −0.4956; ω = 2.4075)
3	−2.4291
4–5	1.2679 ± 2.2206i (ζ = −0.4958; ω = 2.5571)
6	−2.5786

The identified bare-airframe directional model is a simple damped first-order system, as shown in Eq. (21.20).

$$\begin{bmatrix} \dot{r} \\ \dot{\psi} \end{bmatrix} = \begin{bmatrix} N_r & 0 \\ 1 & 0 \end{bmatrix} \begin{bmatrix} r \\ \psi \end{bmatrix} + \begin{bmatrix} N_{\delta_{rud}} \\ 0 \end{bmatrix} \delta_{rud}\left(t - \tau_{\delta_{rud}}\right) \tag{21.20}$$

The yaw parameters are shown in Table 21.3 and the good frequency response agreement of this simple model is seen in Figure 21.14.

The identified heave model is also a simple damped first-order system:

$$\dot{w} = Z_{\delta_{thr}} \delta_{thr}\left(t - \tau_{\delta_{thr}}\right) \tag{21.21}$$

The parameters are shown in Table 21.4. The frequency response is depicted in Figure 21.15. The first-order dynamics for yaw and heave complete the eight eigenvalues of the quadrotor's dynamics (with six degrees of freedom).

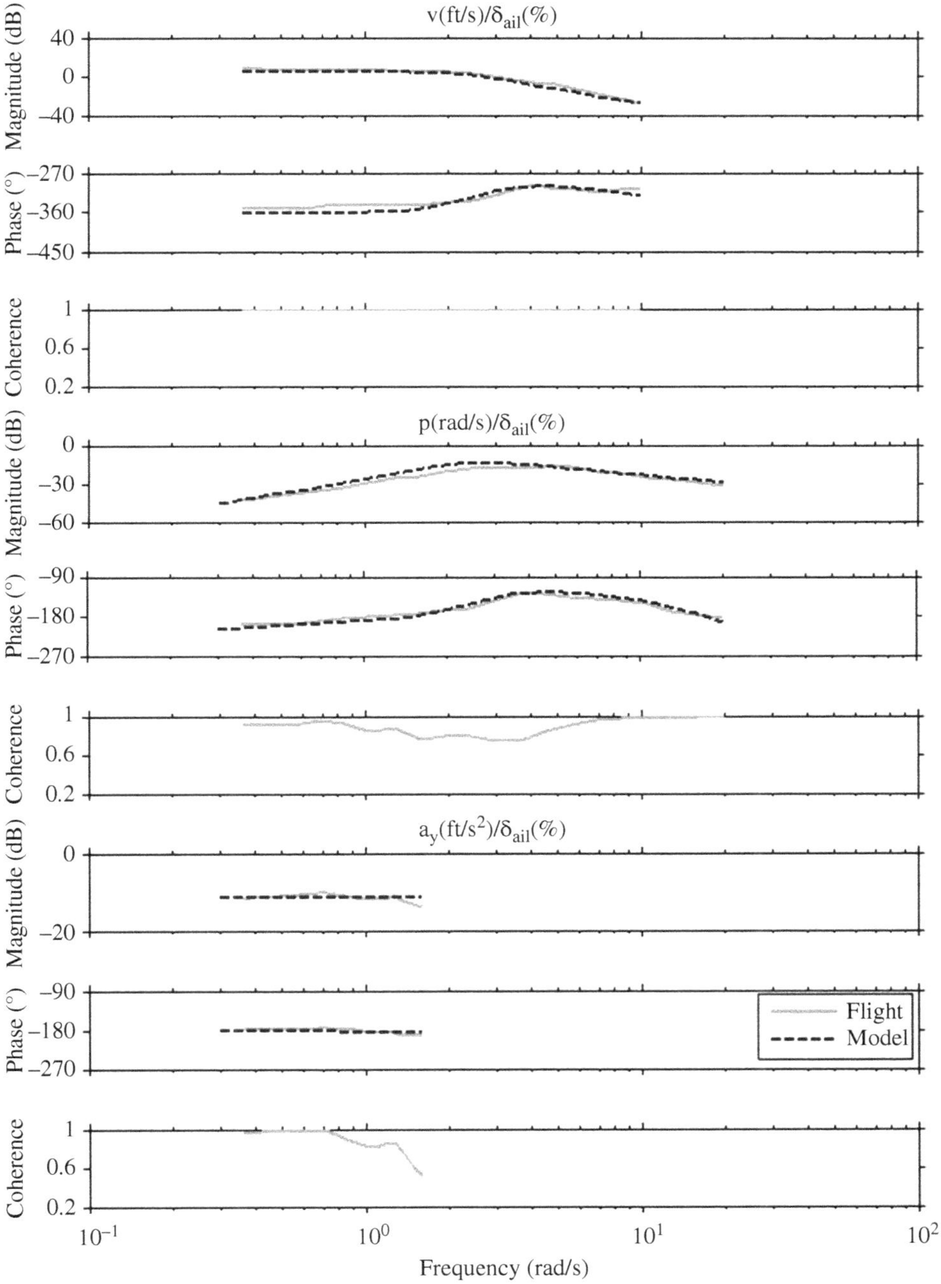

Figure 21.12 Bare-airframe lateral frequency response.

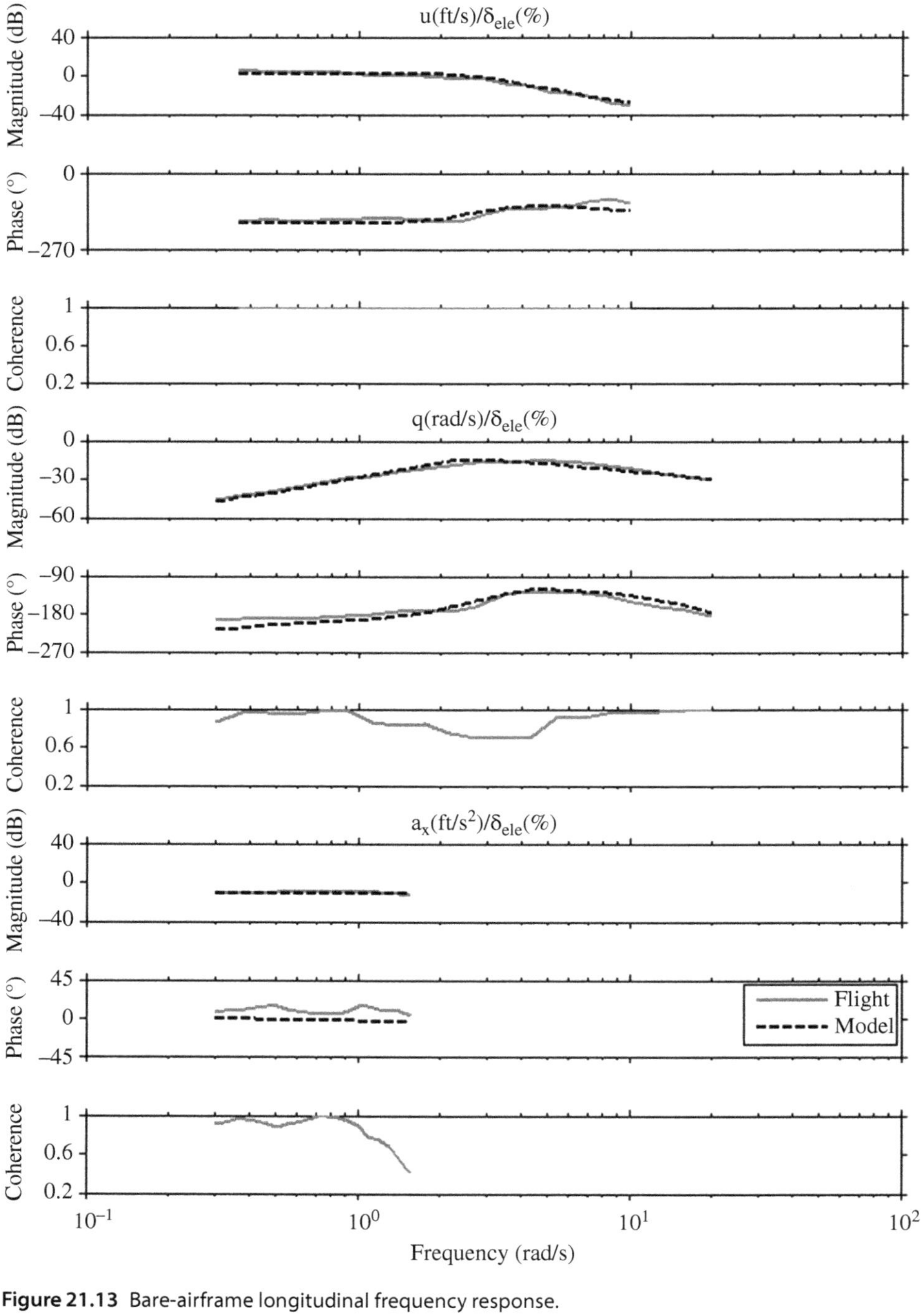

Figure 21.13 Bare-airframe longitudinal frequency response.

Table 21.3 Identified parameters for directional bare-airframe model.

	Value	Cramer–Rao %	Insens. %
N_r	−0.5231	14.42	6.750
$N_{\delta_{rud}}$	0.1306	3.954	1.864
$\tau_{\delta_{rud}}$	0.02295	14.86	7.372

The resulting state-space model is presented in Eq. (21.22):

$$\begin{bmatrix} \dot{u} \\ \dot{v} \\ \dot{w} \\ \dot{p} \\ \dot{q} \\ \dot{r} \\ \dot{\phi} \\ \dot{\theta} \\ \dot{\psi} \end{bmatrix} = \begin{bmatrix} -0.0429 & 0 & 0 & 0 & 0 & 0 & 0 & -g & 0 \\ 0 & -0.0429 & 0 & 0 & 0 & 0 & g & 0 & 0 \\ 0 & 0 & 0 & 0 & 0 & 0 & 0 & 0 & 0 \\ 0 & -0.4376 & 0 & 0 & 0 & 0 & 0 & 0 & 0 \\ 0.5241 & 0 & 0 & 0 & 0 & 0 & 0 & 0 & 0 \\ 0 & 0 & 0 & 0 & 0 & -0.5231 & 0 & 0 & 0 \\ 0 & 0 & 0 & 1 & 0 & 0 & 0 & 0 & 0 \\ 0 & 0 & 0 & 0 & 1 & 0 & 0 & 0 & 0 \\ 0 & 0 & 0 & 0 & 0 & 1 & 0 & 0 & 0 \end{bmatrix} \begin{bmatrix} u \\ v \\ w \\ p \\ q \\ r \\ \phi \\ \theta \\ \psi \end{bmatrix} + \begin{bmatrix} 0 & 0 & 0.2269 & 0 \\ 0 & -0.2016 & 0 & 0 \\ -0.7414 & 0 & 0 & 0 \\ 0 & 0.7066 & 0 & 0 \\ 0 & 0 & 0.6662 & 0 \\ 0 & 0 & 0 & 0.1306 \\ 0 & 0 & 0 & 0 \\ 0 & 0 & 0 & 0 \\ 0 & 0 & 0 & 0 \end{bmatrix} \begin{bmatrix} \delta_{thr} \\ \delta_{ail} \\ \delta_{ele} \\ \delta_{rud} \end{bmatrix} \quad (21.22)$$

Note again that the heave and angular rate damping of the vehicle are zero (L_p, M_q, and $Z_w \approx 0$) as mentioned earlier, due to the negligible effect of these derivatives for the quadrotor.

The frequency-domain cost for the identification is summarized in Table 21.5. They are all within or near the cost guidelines for identification ($J_{ave} \leq 100$), indicating that the model identification accuracy is adequate. Improved model accuracy would require a higher-fidelity instrumentation system and/or increased model complexity, which is not deemed warranted.

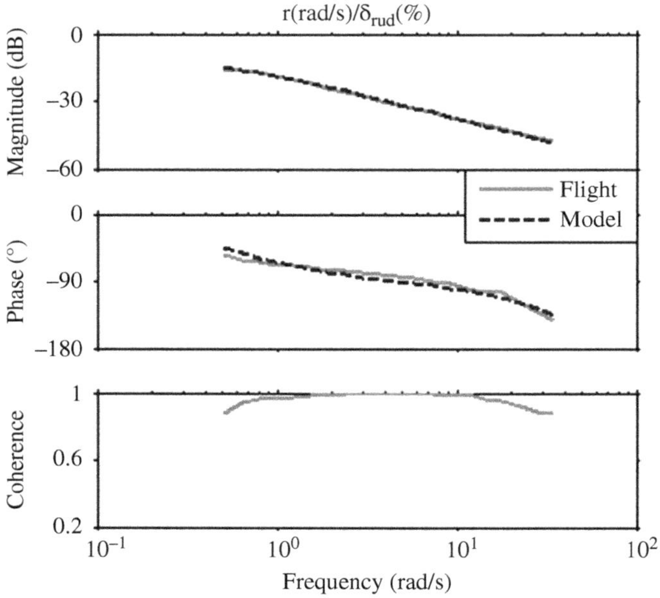

Figure 21.14 Bare-airframe directional frequency response.

Table 21.4 Identified parameters for heave bare-airframe model.

	Value	Cramer–Rao %	Insens. %
$Z_{\delta_{thr}}$	−0.7414	3.961	1.980
$\tau_{\delta_{thr}}$	0.2884	3.393	1.697

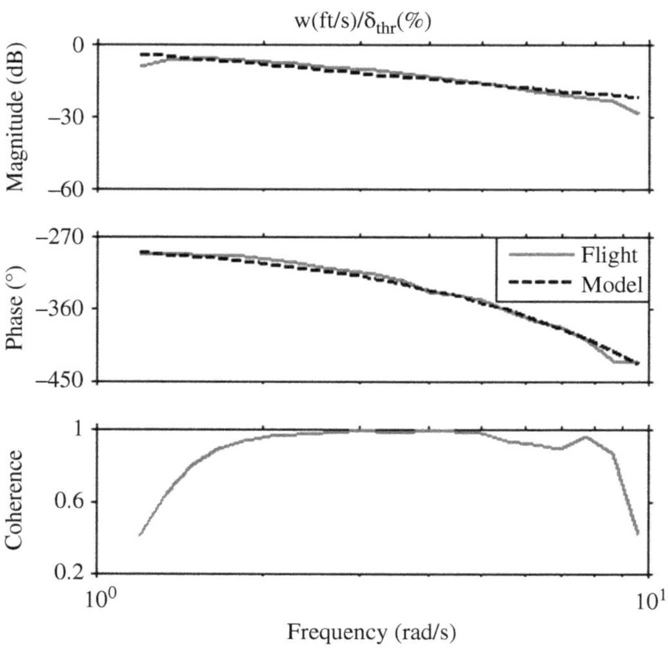

Figure 21.15 Bare-airframe heave frequency response.

Table 21.5 Cost summary.

	Lateral	Longitudinal	Directional	Vertical
Identification	120.8	120.8	15.8	54.41
Verification	0.437	0.378	1.165	0.157
TIC	0.166	0.164	0.054	0.127

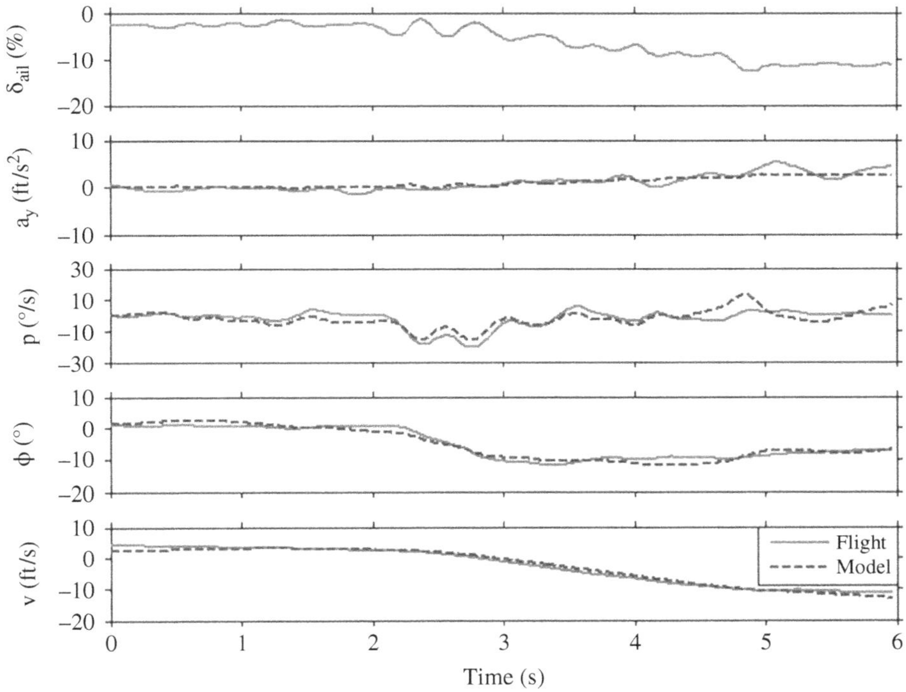

Figure 21.16 Time-domain verification of lateral bare-airframe model.

21.5.2 Bare-airframe Model Verification

In the previous section, the dynamics of the three axes of interest were identified. In order to validate these models, time-domain verification was conducted using doublet maneuvers collected during flight testing. These were used for model verification only. From Figures 21.16–21.19, the model and flight data agree very well in all axes. The time-domain verification cost functions are based on the root-mean squared function J_{rms}, proposed by Tischler and Remple (2012), but they are scaled by 1/5 due to large relative responses of the small-sized quadrotor. The Theil inequality constant (TIC) is a normalized metric, and is not scaled. The resulting verification cost values shown in Table 21.5 indicate good predictive accuracy for the identified models, based on the guidelines values of $J_{rms} \leq 2$ and TIC ≤ 0.25.

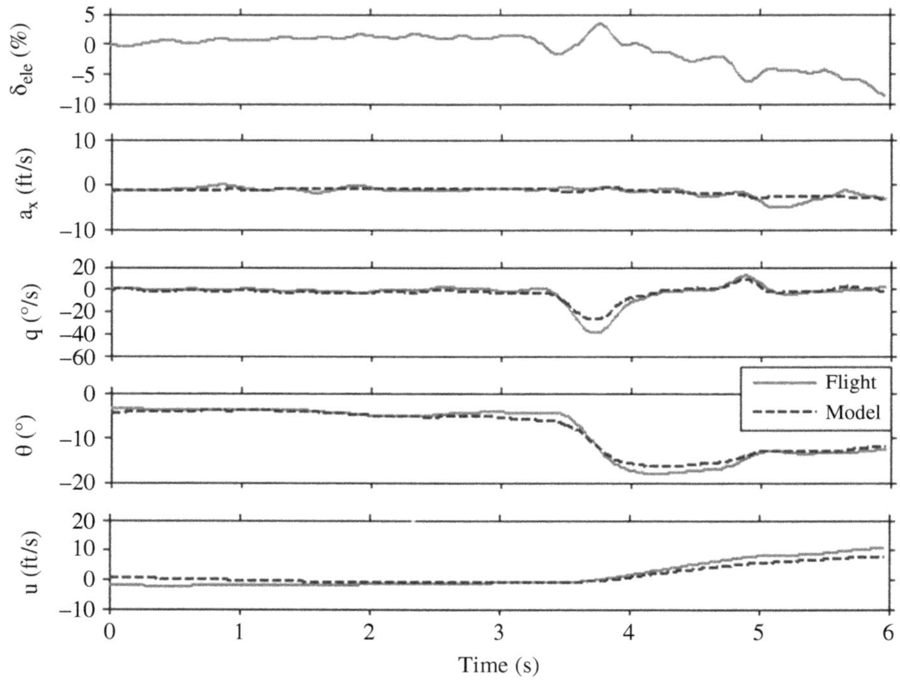

Figure 21.17 Time-domain verification of longitudinal bare-airframe model.

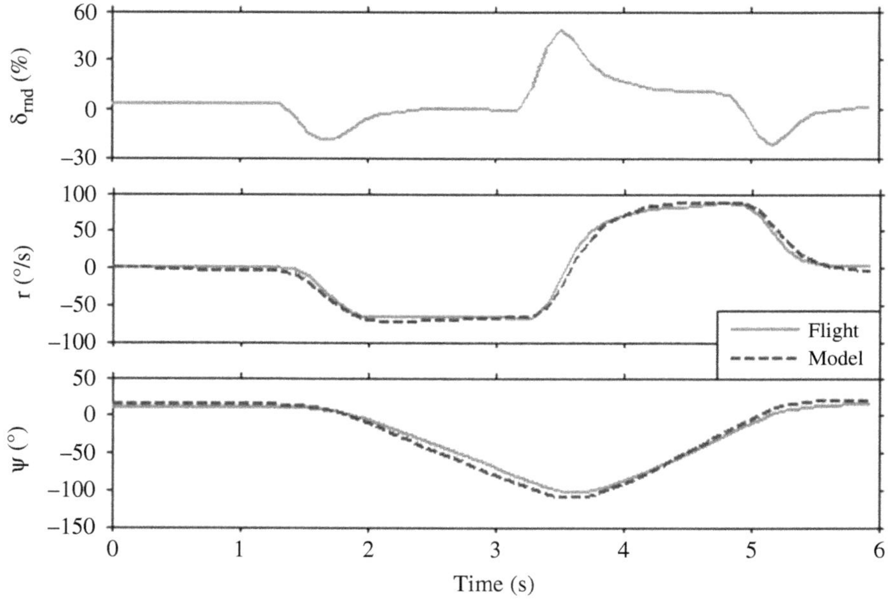

Figure 21.18 Time-domain verification of directional bare-airframe model.

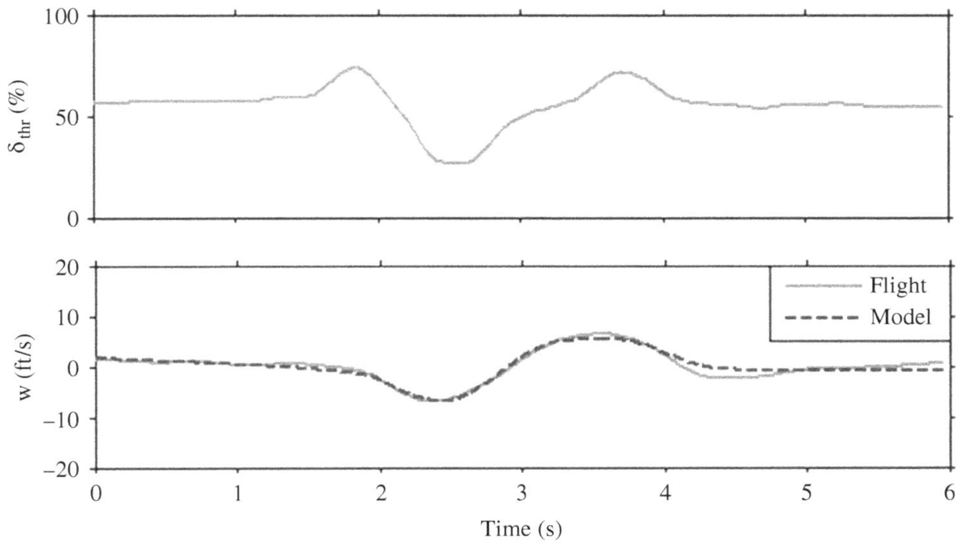

Figure 21.19 Time-domain verification of vertical bare-airframe model.

21.6 Controller Modeling and Validation

The dynamic model of the quadrotor derived from system identification using CIFER sets an excellent foundation for quadrotor controller development. But before this happens, the current controller needs to be accurately modeled in Simulink to create an "anchor point", proving that the physical system as a whole, including the controller and the airframe, can be accurately represented in the simulation environment.

21.6.1 Quadrotor Controller Modeling

The quadrotor control system applies a dual PID control system, illustrated in Figure 21.20. Taking the lateral axis as an example, the attitude PID controller takes the difference of required attitude (δ_{lat}) and current roll angle (ϕ) as inputs, and outputs a roll-rate command (att_out). The rate PID's proportional and integral block takes the difference of the rate command and the current roll rate (p), and generates the roll-axis command. The derivative block of the rate PID controller takes p instead, so that the spike due to changes of roll-rate command, also known as a "derivative spike", can be avoided.

The gain parameters are shown in Figure 21.20. The nominal (also referred to here as "legacy") values, shown in Table 21.7, provide acceptable (but not optimal) handling qualities for flight testing. All flight data for the system identification was collected using this nominal controller configuration.

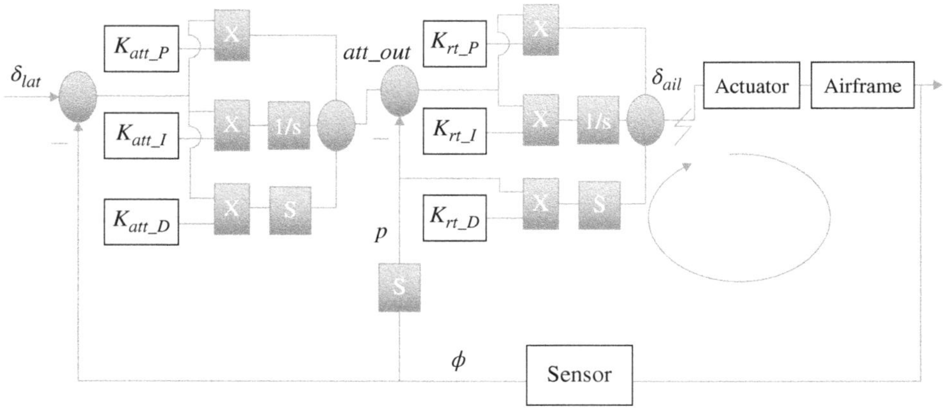

Figure 21.20 Quadrotor dual PID controller.

21.6.2 Controller Validation

The controller optimization, stability analysis, and flight evaluation will be based on the controller model in Simulink. With both the airframe and controller modeled, it is necessary to check the characteristics of the controller model as well as the overall model against flight data, making sure that the physical system is correctly represented in Simulink. For consistency, the validation needs to be conducted in both frequency- and time-domains. This process establishes the anchor point for control system optimization and stability analysis.

For the frequency-domain verification, both broken-loop and closed-loop frequency responses are checked. The broken-loop response cuts the loop open before the actuator block in Figure 21.20. Sweep signals from flight data are used to excite the actuator block, and the rate PID commands are collected as output. The results are compared to broken-loop responses from flight data to determine the fidelity of the model. Closed-loop responses are also evaluated.

Both the broken-loop and closed-loop frequency response model verification results are shown in Figure 21.21. As can be seen, the match between the model and flight data is very good. This indicates the stability margins of the physical system can be accurately represented through model analysis, so the optimization of the controller for the quadrotor can be completed using the identified model.

The time-domain verification starts with the controller output. Lateral input commands (δ_{lat}) from pilot, and p, ϕ, values from flight data are fed into the controller model, and both att_pid and rate_pid model outputs are collected. As shown in Figure 21.22, the controller model is in good agreement with the flight data. Next, the loop is closed by feeding the quadrotor bare-airframe model outputs (p, ϕ) into the controller model, and the same δ_{lat} command is used to excite the model. The model outputs (p, ϕ) are compared to flight data in Figure 21.23. Note that aggressive maneuvers were included in the model verification to demonstrate fidelity for piloted inputs. The results show that the model achieves excellent time-domain predictions, consistent with the good frequency-response fits.

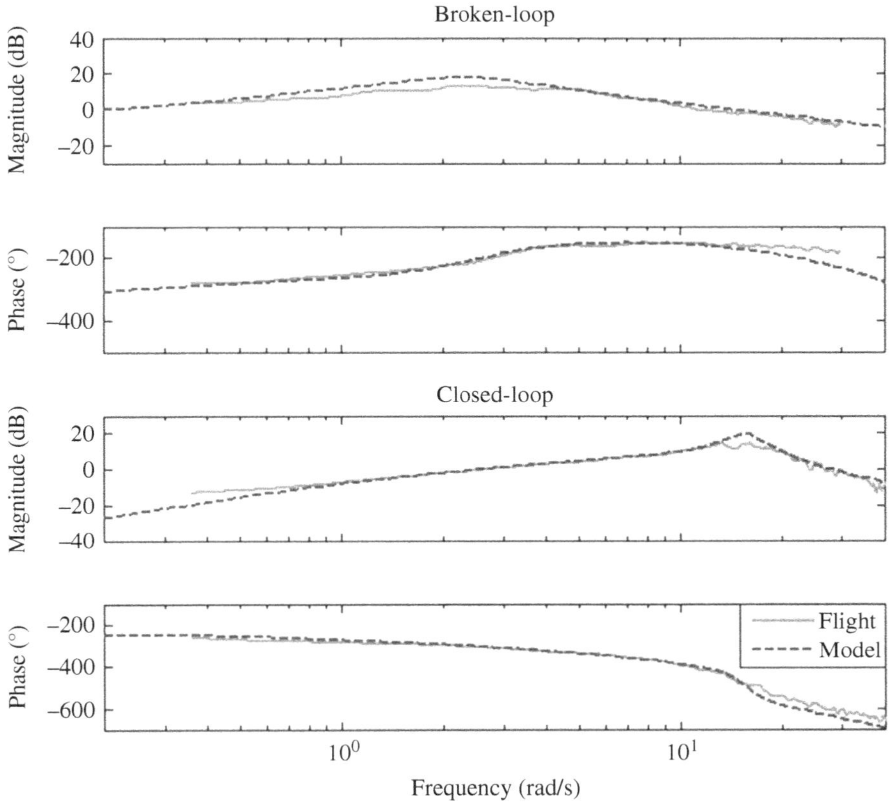

Figure 21.21 Frequency-domain verification of lateral axis model.

Now that the model is verified in both frequency- and time-domains, a critical anchor point is completed and the flight controller optimization and evaluation can be performed using the model.

21.7 Controller Optimization in CONDUIT

21.7.1 CONDUIT Overview

The software package COntrol Designer's Unified Interface (CONDUIT) is another tool developed by US Army AFDD for flight controller design, analysis, and optimization. This software has been successfully used in many flight control design applications, such as the AH-64D (Harding et al. 2006), UH-60A (Mansur et al. 2009, 2013), and the RASCAL helicopter (Ivler et al. 2012).

One of the many challenges for modern flight control design is that with many coupled states and tunable parameters in the control diagrams, tuning one of the parameters always affects many specifications, which makes the fine-tuning of design parameters time consuming. What further complicates this process is that many

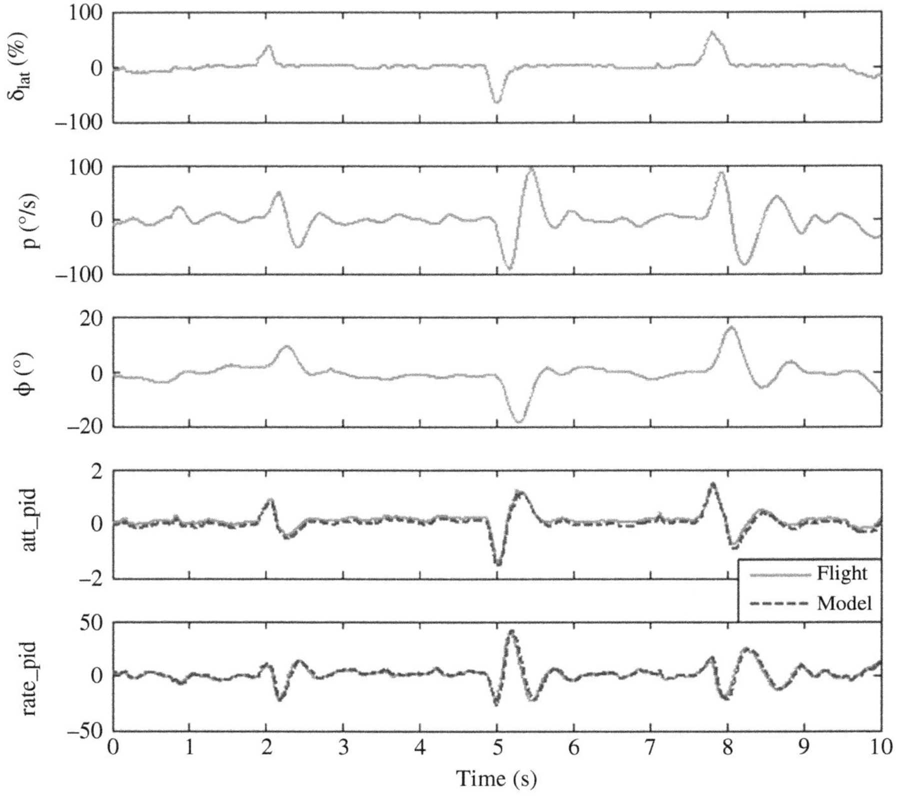

Figure 21.22 Controller model verification of lateral axis.

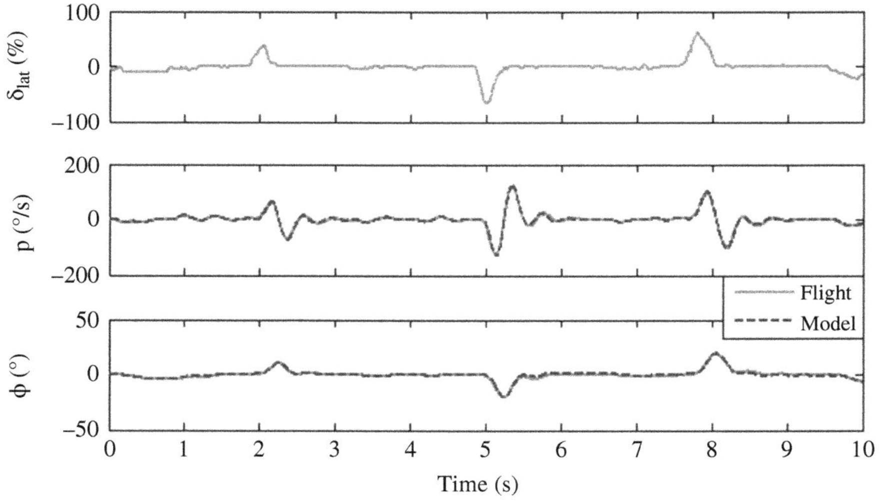

Figure 21.23 Overall model verification of lateral axis.

design specifications (user requirements) are in competition with one another. So a comprehensive understanding of trade-offs of the design parameters and specifications is required, and the process can be so computationally intensive that manual tuning for an optimization solution can be frustrating or even impractical. In CONDUIT, the specifications are defined first and the program determines the design parameter values that achieve the specifications with minimum over-design.

CONDUIT reaches the optimum solution using a multi-phase multi-objective feasible sequential quadratic programming algorithm (Tischler et al. 2008). CONDUIT also has built-in libraries from which users can select applications for fixed-wing aircraft (Anon 1997) or rotorcraft (Anon 2000). The libraries cover all aspects of flight control system evaluation, including handling qualities for piloted inputs, disturbance rejection, stability, and control usage. A numerical score is assigned to indicate how the system complies with each of the specifications, and a plot displays this correlation.

Other features of CONDUIT include (Anon 2005):

- customizable specifications
- quick access to detailed information
- integration with CIFER database
- optional batch mode.

21.7.2 Design Parameters

In CONDUIT, the parameters to be tuned in the control block diagrams are referred to as "design parameters". In a PID controller, the design parameters are the proportional (P), integral (I), and derivative (D) gains. The system identification results show similar dynamics between the lateral and longitudinal axes, which is expected for a quadrotor configuration. Therefore, for the attitude-stabilized controller, we focused on the lateral axis, and uses the same controller configuration for the longitudinal axis.

As can be seen in Figure 21.20, the controller has a dual PID configuration. Because the K_{att_D} has the same effect as K_{rt_P}, and K_{rt_I} as K_{att_P}, so both

$$K_{att\_D} \text{ and } K_{rt\_I} \text{ are set to } 0.$$

The integral gain of the attitude controller K_{att_I}, was tied to proportional gain K_{att_P}, such that the ratio of K_{att_I} to K_{att_P} is equal to 1/10 of the crossover frequency, as suggested by Mansur et al. (2013). This ensures that the integral effect does not degrade the phase margin, and also that the integral effect is not too small to quickly recover a steady-state error. So in this design, there are three design parameters: K_{att_P}, K_{rt_P}, and K_{rt_D}.

21.7.3 Specifications

As stated in the previous section, in CONDUIT, flight controller design specifications are established in the beginning and checked throughout the design process. This serves as the foundation of the design problem setup.

Specifications selected for this design are shown in Table 21.6. They are divided into three categories:

- hard constraints (H)
- soft constraints (S)
- summed objectives (J).

Table 21.6 Design specifications.

Name	Description	Type
EigLcG1	Eigenvalues	H
NicMgG1	Nichols margin	H
StbMgG1	Stability margin	H
BnwRoH1	Bandwidth	S
DstBwG1	Disturbance rejection bandwidth	S
DstPkG1	Disturbance rejection peak	S
EigDpG1	Generic damping ratio	S
CrsMnG2	Minimum crossover frequency	S
OlpOpG1	Open-loop operating point rate limit saturation	S
CrsLnG1	Crossover frequency	J
RmsAcG1	Actuator RMS	J

Hard constraints are specifications that relate to system stability and stability margins, soft constraints are specifications that relate to handling qualities, and the summed objectives constraints prevent over-design and minimize the cost of feedback.

The specifications in CONDUIT applies the Cooper–Harper scale (Cooper and Harper 1969) to rate the compliance as:

- satisfactory without improvement (Level 1)
- deficiencies that warrant improvement (Level 2)
- requiring improvement (Level 3).

The goal of the flight control design is to achieve Level 1 in all design specifications with minimal control usage.

21.7.4 Optimization Strategy

The design specifications selected in CONDUIT libraries are for full-scale rotorcraft rather than small (2 kg) rotorcrafts. To address this, the criteria are based on a baseline evaluation of the legacy controller setup, as shown in Figure 21.24. As can be seen, the initial design has very limited stability margins and a low closed-loop damping ratio. Another specification in Level 3 (red zone) is the open-loop onset point specification, which indicates a potential for actuator saturation, which could in turn lead to pilot-induced oscillation. Based on this nominal controller performance, the Level 1/Level 2 stability margin boundary was set to 3 dB gain margin and 15° phase margin, and the damping ratio Level 1/Level 2 boundary was set to ζ =Z0.2.

The optimization process is completed in three phases. Because stability related (hard) constraints take priority over all other constraints, in Phase 1 the design parameters are optimized to achieve all hard constraints. Then in Phase 2, all handling-quality (soft) constraints are satisfied, while the hard constraints are maintained. At this point, the design meets all stability and handling quality requirements, but might not be an optimal design in terms of the use of the actuator authority. Therefore, in Phase 3, the

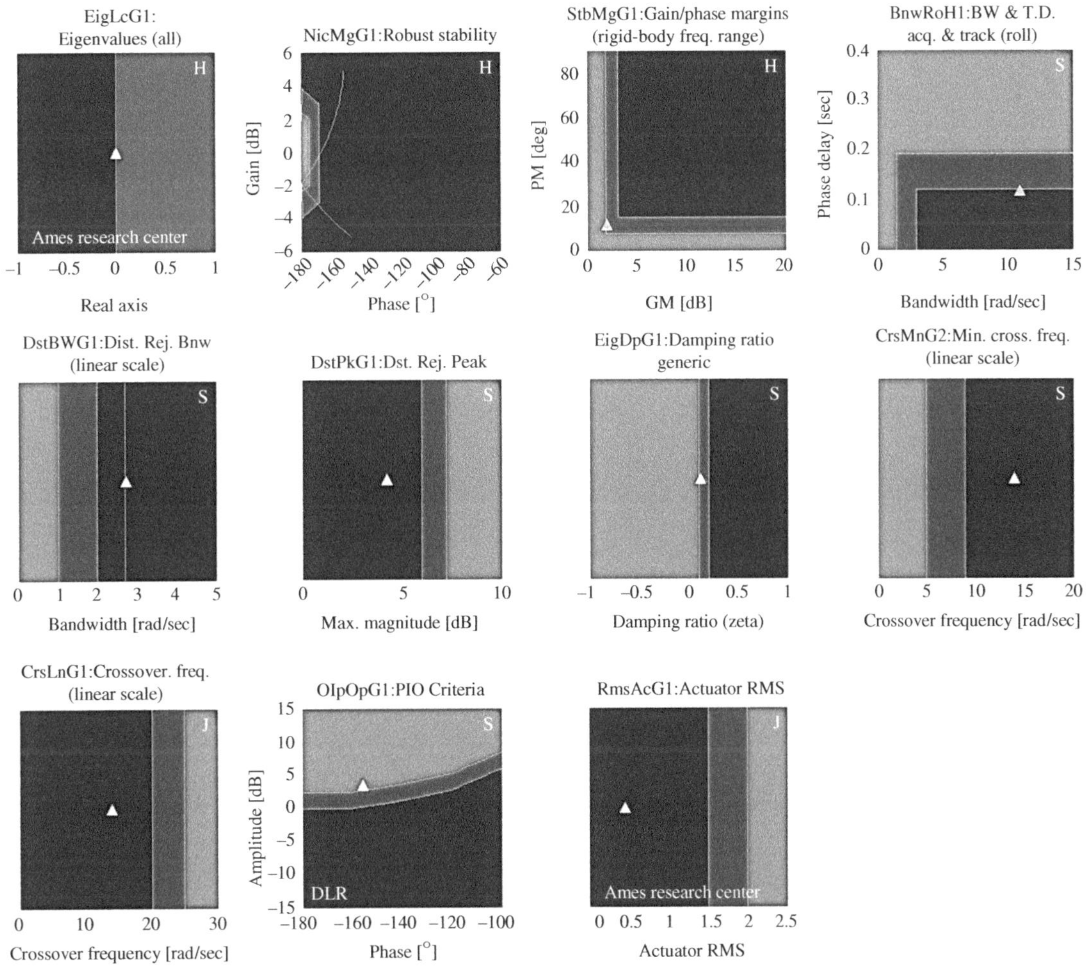

Figure 21.24 CONDUIT handling quality window for initial design. (*See insert for color representation of this figure.*)

summed objectives (crossover frequency and actuator RMS) are minimized, while all of the hard and software constraints are maintained. The final optimized CONDUIT solution meets the requirements with the most economical use of the actuators – this is the "Pareto optimum" solution (Tischler et al. 2008).

In CONDUIT there is a design margin optimization (DMO) function that allows user to push the Level 1/Level 2 boundary forward to a certain level into the Level 1 region in a batch mode for system robustness or better performance. In this design, DMO was applied to the disturbance rejection bandwidth (DstBwG1) and minimal crossover frequency (CrsMnG2) specifications. The DMO level was increased until an optimal solution could not be reached.

21.7.5 Results

A comparison of the initial and final design parameters and performance is shown in Table 21.7. It can be seen that the K_{rt_P} value was reduced from 100 to 64.95, while the optimization algorithm did not alter the other design parameters. This solution provides more stability margin and more damping, at the price of slightly reduced bandwidth.

The handling-quality window for the final design is shown in Figure 21.25. Note that after the optimization, every specification is at Level 1 (blue).

Figure 21.26 compares the closed-loop frequency response of the legacy (nominal) controller and the final controller design as optimized by CONDUIT. The final design has a reduced closed-loop response frequency of about 10 rad/sec, and an associated reduced magnitude peak and more gradual phase roll-off associated with the increased closed-loop damping.

21.7.6 Flight Testing Validation

Flight testing was conducted with the final design values uploaded to the flight controller. Flight data was recorded, and input data were used to excite the legacy controller for comparison. As shown in Figure 21.27, the flight data and simulation of the final design are in excellent agreement, and have much less oscillation than the legacy (nominal) controller, with only a slight reduction in response speed. This also validates the frequency-domain analysis results.

Table 21.7 Lateral controller design summary.

Design parameters and flight characteristics	Initial design	Final design	Change
K_{att_P}	3.5	3.53	0.86%
K_{rt_D}	3	3	0%
K_{rt_P}	100	64.95	−35.05%
Gain margin (dB)	1.93	5.80	200.52%
Phase margin (°)	11.93	24.41	104.61%
Bandwidth (rad/s)	10.97	7.69	−29.9%
Damping ratio	0.18	0.45	150%
DRB (rad/s)	2.72	2.76	1.47%

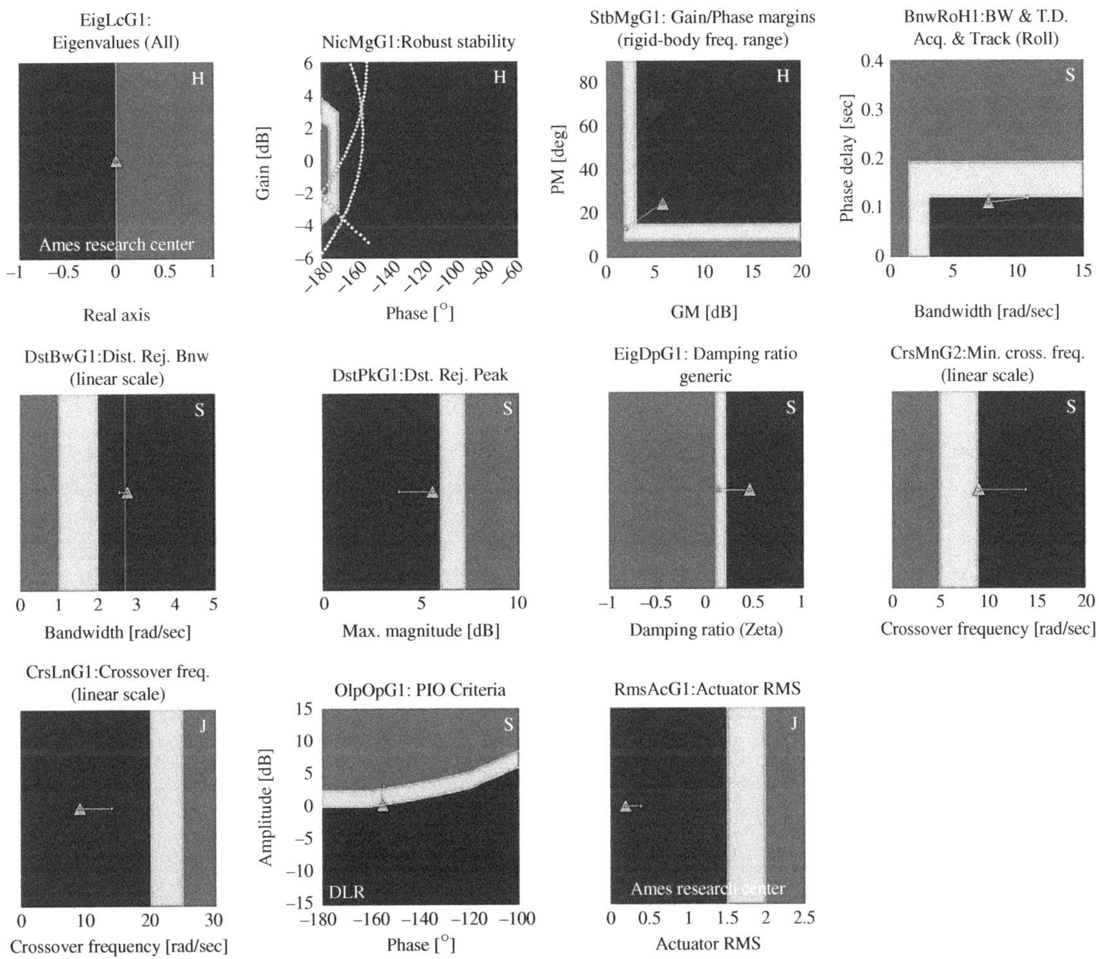

Figure 21.25 CONDUIT handling quality window for final design. (*See insert for color representation of this figure.*)

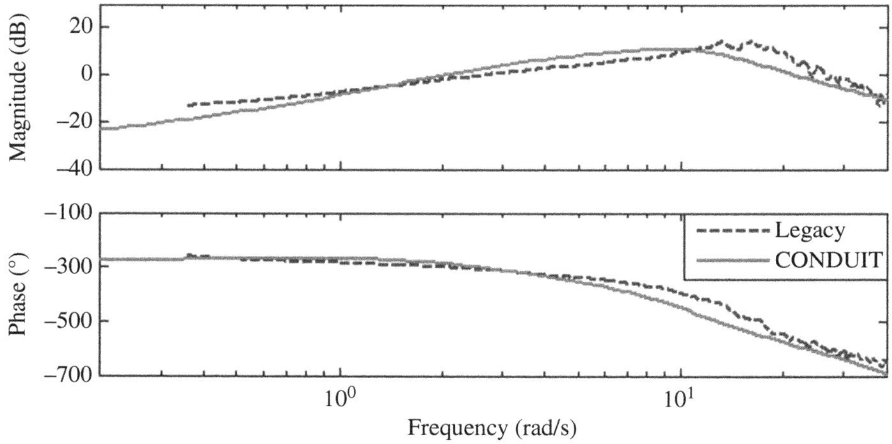

Figure 21.26 Lateral axis closed-loop frequency response of initial and final designs.

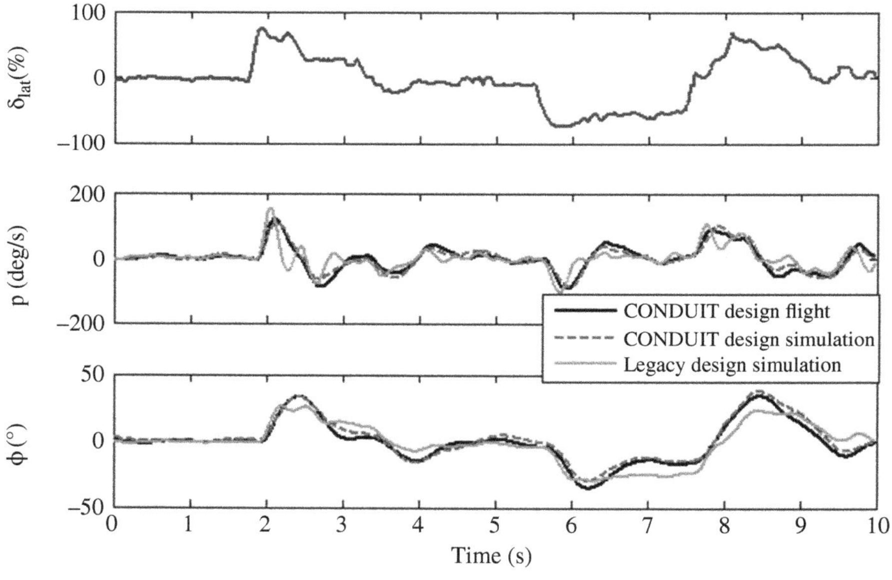

Figure 21.27 Lateral axis design evaluation.

21.8 Conclusion

This chapter describes the system identification and flight control optimization of an unmanned quadrotor. The process includes hardware setup, flight testing, flight data collection and processing, system identification, model validation, controller optimization and validation. The main conclusions are:

1) Quadrotor system dynamics are accurately determined using frequency-domain system identification methods and very simple state-space models.

2) Bare-airframe dynamics exhibit highly unstable and classical hovering cubic characteristics that are essentially perfectly symmetrical between pitch and roll, as expected. The responses are dominated by the control and speed stability derivatives. Bare-airframe yaw and heave dynamics are stable and accurately characterized by simple first-order dynamic systems.
3) Accurate time-domain verification of the identified models with input data not used in the identification shows that the models are very well suited to flight control design.
4) The optimized controller demonstrates significantly improved closed-loop dynamic responses in the simulation and flight testing of the quadrotor.

References

Anon. (1997). Flying qualities of piloted aircraft, Technical report USAF MIL-HDBK-1797, US Dept. of Defense.

Anon. (2000). Aeronautical design standard, handling qualities requirements for military rotorcraft, Technical report USAAMCOM ADS-33E-PRF, US Army Aviation and Missile Command, Huntsville, Alabama.

Anon. (2005). CONDUIT User's Guide, Version 5.0. University Affiliated Research Center, Moffett Field, CA.

Cooper GE, and Harper RP. (1969). The use of pilot rating in the evaluation of aircraft handing qualities. Technical report TN D-5153, NASA.

Harding JW, Moody SJ, Jeram GJ, Mansur HM, and Tischler MB. (2006). Development of modern control laws for the AH-64D in hover/low speed flight. *American Helicopter Society 61st Annual Forum.* Grapevine, TX.

Ivler CM, Powell JD, Tischler MB, Fletcher JW, and Ott C. (2012). Design and flight testing of a cable angle/rate feedback flight control system for the RASCAL JUH-60 helicopter. *American Helicopter Society 68th Annual Forum.* Fort Worth, TX.

Mansur MH, Lusardi JA, Tischler MB, and Berger T. (2009). Achieving the best compromise between stability margins and disturbance rejection performance. *American Helicopter Society 65th Annual Forum.* Fort Worth, TX.

Mansur MH and Tischler MB. (2013). Flight test comparison of alternate strategies for multi-loop control law optimization. *American Helicopter Society 69th Annual Forum.* Phoenix, AZ.

McRuer DT, Ashkenas IL, and Graham D. (1973), *Aircraft Dynamics and Automatic Control.* Princeton University Press, Princeton, NJ.

Padfield GD. (2008). *Helicopter Flight Dynamics.* John Wiley & Sons.

Tischler MB, and Remple RK. (2012). *Aircraft and Rotorcraft System Identification,* 2nd edn. American Institute of Aeronautics and Astronautics.

Tischler MB, Ivler, CM, Mansur MH, Cheung K, Berger T, and Berrios M. (2008). Handling-qualities optimization and trade-offs in rotorcraft flight control design. *AHS Specialist's Meeting on Rotorcraft Handling-Qualities.* Liverpool, UK.

Wei W, Tischler MB, Schwartz N, and Cohen K. (2014). Frequency-domain system identification and simulation of a quadrotor controller. *AIAA Modeling and Simulation Technologies Conference.* National Harbor, MD.

Index

Note: Page references in '*italics*' refer to figures, those in '**bold**' refer to tables.

Advanced UAV Aerodynamics, Flight Stability and Control: Novel Concepts, Theory and Applications, First Edition. Edited by Pascual Marqués and Andrea Da Ronch.

Companion website: http://www.wiley.com/go/marques/advanced_UAV_aerodynamics

b

d

e

i

m

t

u

www.ingramcontent.com/pod-product-compliance
Ingram Content Group UK Ltd.
Pitfield, Milton Keynes, MK11 3LW, UK
UKHW050842101026
15624UKWH00018B/428

9 781118 928684